Anti-Abrasive Nanocoatings

Related Titles

Coatings for biomedical applications
(ISBN 978-1-84569-568-2)

Metals for biomedical devices
(ISBN 978-1-84569-434-0)

Advances in marine antifouling coatings and technologies
(ISBN 978-1-84569-386-2)

Woodhead Publishing in Materials

Anti-Abrasive Nanocoatings

Current and future applications

Edited by

Mahmood Aliofkhazraei

ELSEVIER

AMSTERDAM • BOSTON • HEIDELBERG • LONDON
NEWYORK • OXFORD • PARIS • SANDIEGO
SAN FRANCISCO • SINGAPORE • SYDNEY • TOKYO
Woodhead Publishing is an imprint of Elsevier

WP WOODHEAD PUBLISHING

Woodhead Publishing Limited is an imprint of Elsevier
80 High Street, Sawston, Cambridge, CB22 3HJ, UK
225 Wyman Street, Waltham, MA 02451, USA
Langford Lane, Kidlington, OX5 1GB, UK

ISBN: 978-0-85709-211-3 (print)
ISBN: 978-0-85709-217-5 (online)

British Library Cataloguing-in-Publication Data
A catalogue record for this book is available from the British Library

Library of Congress Cataloging-in-Publication Data
A catalog record for this book is available from the Library of Congress

Library of Congress Number: 2014955911

For Information on all Woodhead Publishing publications
visit our website at http://store.elsevier.com/

Typeset by MPS Limited, Chennai, India
www.adi-mps.com

Printed and bound in the US

Contents

List of figures

List of tables

About the editor

Dr. Mahmood Aliofkhazraei works about corrosion, coatings and surface engineering in Tarbiat Modares University. His research interests include nanotechnology and its use in surface and corrosion science. One of his main interests is plasma electrolysis and he published more than 40 papers and a book in this area. He has given some invited talks including several keynotes in several countries. Aliofkhazarei has received several awards including the Khwarizmi award, IMES medal, INIC award, best thesis award (multiple times), best book award (multiple times) and the best young nanotechnologist award of Iran (2 times). He is on the advisory editorial board of several materials science and nanotechnology journals. He is a member of the national association of surface sciences, Iranian corrosion association, and national elite foundation of Iran.

About the contributors

Sunil Mohan is Professor of Metallurgy in IIT (BHU), Varanasi, India. He completed his B. E., M.E. and PhD from University of Roorkee/IITR. He is teaching at IIT (BHU) for over 22 years His area of teaching is Composite Materials, Surface Engineering, Transport Phenomena, Modelling and Simulation. He is working in the area of alloy development, synthesis of composites by different routes and tribology of materials for more than 26 years. He has published more than sixty research papers in National and International refereed Journals. He has produced several M. Techs. and PhDs. He is life member of Tribological Society and Electron Microscopy Society of India.

Anita Mohan is Asst. Professor in IIT (BHU). She did her M.Phil. from IITR and PhD from IIT(BHU) in Physics. Her main research area is Solar Physics and Space Physics. She is life member of International Astronomical Union and Astronomical Society of India. She is teaching Electromagnetics, Semiconductor Physics, Instrumentation, Atmospheric Physics, etc. for over ten years. She is also pursuing research in the area of Composite materials and Tribology. She guided fifteen M.Tech. students and one student is pursuing PhD under her guidance. She has published forty eight research papers in National and International refereed Journals, two book chapters and four popular science articles.

Associate Professor Dr. **Xuesong Han** received his M.E. degree from Tianjin University, China in 2002, and also received his PhD from Tianjin University, China in 2005. He was a Postdoctoral Fellow at Tsinghua University (2005–2007), China. He is currently the Associate Professor at the Department of mechanical engineering, Tianjin university of China. Professor Han is American Nano Society Full Member and American Ceramic Society Member. Being a visiting Scholar, he study at The University of Toledo and Northwestern University (USA) for two years. He has conducted a number of research projects related to Micro/Nano Machining. He authored more than 40 technical publications including 30 international journal papers indexed by Science Citation Index (SCI) and six books. Many of his publications cited by researchers form Ohio State Univwesity (USA), Tsinghua University (China), The University of New South Wales (AUS) and so on. He is the recipient of the Excellent Postdoctor Research Fellow (2007) of Tsinghua University. His research interests include Micro/Nano Machining, Micro/Nano Tribology, Surface Engineering, High Performance Computing Techniques, Numerical Analysis Method (Molecular Dynamics, Finite Element, Mesh Free Method, Multiscale method etc.).

Shengzhao Wang was awarded a Ph.D. in Materials Physics and Chemistry from Tongji University in 2010, after which he worked at Nanyang institute of Technology as associate professor. His research interests centre on nano materials and preparation of antireflective film and multi-layer thin film by sol-gel technology. He was very concerned about investigating stress of nano-optical thin films by sol-gel method.

Dr. **Surekha Krishnan** is currently working as Project Manager in the South African Institute of Welding (SAIW), Johannesburg, South Africa. She is currently involved in the certification of companies and audits according to EN ISO 3834, failure analysis and preparation of various course notes. She was awarded a PhD in Metallurgical Engineering from Indian Institute of Technology Madras, India in 2009. She is basically a Metallurgical Engineer from Madras University (2000) and had her masters degree in Materials Science (2002) from National Institute of Technology (NIT) Tiruchirapalli, India. She worked as a post-doctoral fellow in Nelson Mandela Metropolitian University (2009–2011) and University of Johannesburg (2012). She has wide teaching, research and industrial experience in India and South Africa. Her research interests includes nano materials, solid state welding, arc welding, lasers, surface engineering, materials characterization and friction surfacing. She is an active reviewer of various scientific journals. She has 20 publications to her credit including book chapters, peer reviewed journals, international conferences and books.

K. R. C. Soma Raju has a master's degree in Mechanical Engineering from Osmania University, Hyderabad, India. He worked in cement bonded wood particle board manufacturing and polyester yarn production for three years prior to joining "International advanced research center for powder metallurgy and new materials (ARCI)", Hyderabad an autonomous R & D unit under Dept. of Science and Technology, Govt. of India where he has put up a service of 20 years in various fields of surface engineering. His research interests include thermal spray coatings such as detonation spray coating (DSC), high velocity oxy-fuel flame spray (HVOF), and others including electrospark coating (ESC), micro arc oxidation (MAO) and sol-gel coatings. He has been involved in design and fabrication of various models of DSC, ESC coating systems for introducing these technologies into the Indian market. He has authored 5 book chapters and published 22 journal articles on the above subjects.

Raghavan Subasri obtained her Masters degree in Chemistry from IIT, Madras, India in 1993 and Ph.D in Chemistry in 1999 from the University of Madras, Tamil Nadu, India with the research work carried out at Indira Gandhi Centre for Atomic Research, Kalpakkam, Tamil Nadu, India. After spending couple of years as a post doctoral fellow at the Functional Ceramics Group at Max Planck Institut für Metallforschung, Stuttgart, Germany and two years at National Institute for Materials Science, Tsukuba, Japan, she returned to India and joined Advanced

Research Centre International (ARCI) Hyderabad in March 2005 as a senior scientist. She has been leading the Centre for Sol-Gel Coatings at ARCI, Hyderabad since April 2006. Her team has set up a unique and state-of-the-art comprehensive facility for demonstration of organic-inorganic hybrid sol-gel based coating technology for commercialization in the Indian/Global market. She has 65 publications in peer reviewed international journals, two book chapter contributions, one Indian patent (granted) and seven Indian patent applications to her credit. She is a member of professional bodies like International Sol-Gel Society, The Electrochemical Society Inc., and Indian Institute of Metals. She is also a Max-Planck-India Fellow. Her research interests include sol-gel coatings, microwave sintering of ceramics and solid state electrochemistry.

Dr. **Kal Renganathan Sharma,** P.E. received his BTech degree in chemical engineering from Indian Institute of Technology, Chennai, India, in 1985 and MS and PhD degrees in chemical engineering from West Virginia University, Morgantown, WV, in 1987 and 1990, respectively. His post-doctoral research training was under former chair and Prof. R. Shankar Subramanian, Clarkson University, Potsdam, NY, Chemical Engineering. He is the author of 14 books, 550 conference papers, 7 Book-Chapters, 46 journal articles, 3 review articles, and 113 other presentations. He has instructed 2,953 students in deemed universities in India and at HBCUs and technical colleges in this country, United States in 109 courses. He is a Fellow of Indian Chemical Society and listed in Who's Who in America. He has reviewed over 51 journal articles. He serves on editorial boards of 6 journals that are accessible via internet such as *Journal of Laser Optics and Photonics, ChemXpress, Caspian Journal of Applied Science and Research, Journal of Scientific Research and Reports, Journal of Pharmaceutical and Biomedical Sciences* and *International Journal of Advanced and Applied Science.* He has served at high-level positions in academia in India and this country, USA. Recently served titled are *Adjunct Professor*, Department of Physics at Texas Southern University, Houston, TX, *Adjunct Professor*, Departments of Engineering Technology, Machining and Manufacturing, Petroleum Instrumentation at Lone Star College, North Harris, Houston, TX and *Adjunct Professor*, with the departments of Chemical Engineering, Civil & Environmental Engineering and Mechanical Engineering at Prairie View A & M University, Prairie View, TX. He has received cash awards from Monsanto Plastics Technology, Indian Orchard, MA; SASTRA University, Thanjavur, India; and Prairie View A & M University, Prairie View, TX. He has received 14 press citations. His works have been cited more than 300 times in refereed journals are are searchable by google Scholar Citations. One of his papers on "Reactions in Circle Representation for Systems that Obey Michaelis and Menten kinetics and Krebs Cycle Kinetics" has been downloaded more than 1050 times. His books have been cataloged in more than 3,500 libraries world over. He has introduced four dimensionless groups and one statistical distribution into the literature and has served as a cochair at nine conferences.

A paper presented in an American Chemical Society conference in 1997 at Las Vegas, NV, on small rubber particle morphology in ABS thermoplastics has been cited in Encyclopedia of Chemical Technology, Encyclopedia of Polymer Sci. & Technology , 2003/2004. Bioinformatics: Sequence Alignment and Markov Models, McGraw Hill Professional, New York, NY, 2009, has been cataloged in 530 libraries according to the Worldcat.org. This book has been cited in creation Wikipedia under bioinformatics in English and Wikipedia in Portuguese under Proteina and MAFFT. Nanostructuring Operations in Nanoscale Science and Engineering has been cataloged in 520 libraries. This book has been listed as required text in the syllabus of Electronics Engineering at the University of Mumbai, India.

Véronique Vitry studied at the Engineering Faculty of Mons (Blegium) were she was awarded an Engineering degree in 2003. After a brief stint working in the field of optical thin films, she became a FRIA (Fonds pour la Recherche dans l'Industrie et l'Agriculture) Research Fellow and obtained her PhD in Materials Science in 2010 at the University of Mons (UMONS). She is, since 2012, Assistant Professor in the Metallurgy Laboratory, at the Engineering Faculty of the University of Mons, Belgium, were she teaches extractive metallurgy and nanotechnologies. She also teaches part time at the University of Valenciennes Hainaut-Cambresis (UVHC) in France.

Fabienne Delaunois obtained her Engineering degree in metallurgy and her PhD from the The Engineering Faculty of Mons (FPMs) in 1993 and 2002 respectively. She's been working of electroless nickel-boron for nearly 20 years and is the authors of about 20 publications in that field. She is nowadays Associate Professor and Department Head in the Metallurgy Laboratory, at the Engineering Faculty of the University of Mons, Belgium, where she teaches Physical Metallurgy. She also teaches physical metallurgy at the University of Valenciennes Hainaut-Cambresis in France. She chairs the Chemistry and Materials Science from the Engineering Faculty of UMONS since 2010 and is a member of the board of Materia Nova and BCRC-INISMA.

Bernard Riedl obtained a B.Sc. (1974) from Université de Montréal, à M.Sc. From Université du Québec à Montréal (1980) and a Ph.D. from Université de Sherbrooke (1984), all in chemistry. After post –doctoral studies at Université Laval (Québec, Canada), he was admitted as assistant professor in 1987. Since then, he has published about 200 peer-reviewed papers and graduated about 50 graduate students. He has established a laboratory dedicated to adhesives and surface adhesion in relation to wood surfaces. Current interests are: general studies on adhesion, adhesives and coatings based on bio-based materials, inclusion on cellulose nanocrystals in these and the effect of plasma treatments of wood on adhesion. Professor Riedl is a member on the American Chemical Society, le groupe français d'études des polymers, the Adhesion Society and the Forest Products Society.

M. J. Jackson is Director of Research and Chief Scientist at Micro Machinists and at Bonded Abrasives Group. He was Professor of Mechanical Engineering at the Center for Advanced Manufacturing at Purdue University and is heavily involved in developing micro and nanotechnologies associated with manufacturing such as nanocoatings for use in micro cutting tools. Dr. Jackson gained his engineering degrees at Liverpool and has served on the faculty at the Universities of Liverpool and Cambridge. Dr. Jackson was a visiting research fellow at the University of Oxford, MIT, and at Worcester Polytechnic Institute in Massachusetts.

R. G. Handy is Professor of Engineering Technology at the University of North Carolina at Charlotte. Dr. Handy was previously at the Center for Advanced Manufacturing at Purdue University where he developed a program of teaching and research in the area of environmental health and safety and micro and nanotechnologies.

M. D. Whitfield is a senior manufacturing engineer at Senior Aerospace Katema in San Diego, California. Michael was manufacturing engineer at Doncaster's and at Solar Turbine focusing on the application of nanostructured coatings for machining aerospace alloys. He was a graduate assistant at Purdue University where he studied the machining mechanics of nanostructured coatings on tungsten carbide tooling.

J. W. Burgess is a manufacturing engineer at the US Army's Picatinny Arsenal in New Jersey. He is responsible for Howitzer production and was previously at the US Army's Rock Island Arsenal in Illinois. John was previously at Purdue University where he studied the machining mechanics of nanostructured coatings machining tungsten-tantalum alloys used in armor penetrating rounds.

G. M. Robinson is a concurrent engineer at Medtronic Spine, Carolina Beach, North Carolina. Grant was previously at the Center for Advanced Manufacturing at Purdue University and was involved in developing micro and nanotechnologies associated with manufacturing such as nanocoatings for use in micro cutting tools. Grant continued his research in a consultancy role with the Purdue Technical Assistance program where he applied his knowledge of nanostructured coatings to solving industrial problems associated with machining.

T. Novakov is co-owner of DentaMax in Santiago, Chile. Tamara was previously at TP Orthodontics in Indiana and at Center for Advanced Manufacturing at Purdue University where she was involved in developing micro and nanotechnologies associated with manufacturing such as nanocoatings for use in micro cutting tools and understanding the computational mechanics of micromachining. She holds the degree of Ph. D. from Purdue University and Bachelor of Science and Master of Science degrees in Mechanical Engineering from the University of Novi Sad in Serbia.

W. Ahmed, A.M. Elhissi, S.J. Crean and D. A. Phoenix, Institute of Nanotechnology and Bioengineering, School of Medicine and Dentistry, University of Central Lancashire, Preston PR1 2HE, UK.

Waqar Ahmed is Head of the Institute of Nanotechnology and Bioengineering, School of Medicine and Dentistry, University of Central Lancashire, Preston, UK and Chair: Nanotechnology & Advanced Manufacturing at the University of Central Lancashire (UK). Previously, Prof. Ahmed was Chair, Nanotechnology and Advanced Materials in the Northern Ireland Bioengineering Centre Research (NIBEC), University of Ulster (UK), and he has contributed to the wider industrial adoption of surface coating solutions through fundamental research and modelling of gas phase processes in CVD and studies of tribological behaviour. Prof. Ahmed's research interests have expanded to include medical and dental materials. He has edited/authored fifteen books and wrote more than a three hundred scientific papers.

Abdelbary Elhissi is a Senior Lecturer in Pharmaceutics and the Head of Drug Delivery Division and Deputy Head of the Institute of Nanotechnology and Bioengineering. Abdelbary completed his PhD in 2005 and a Postdoctoral Fellowship in 2006 at the UCL School of Pharmacy. This was followed by working at Xenetic Biosciences in London as a Senior Scientist for one year. Abdelbary's research is mainly focused on nanomedicines for respiratory drug delivery and the medical applications of nano and biomaterials. He is particularly interested in designing proliposome formulations for treatment of asthma and cancer. Abdelbary is also interested in Clinical Pharmaceutics and researching the contemporary issues of pharmacy education.

St. John Crean BDS MBBS FDSRCS FFGDP(UK) FRCS FRCS(OMFS) PHD. His specialist training is in oral and maxillofacial surgery and has held posts of senior lecturer in oral and maxillofacial surgery at University of Cardiff School of Dentistry before being appointed as Professor of Oral and Maxillofacial Surgery at Queen Marys, Barts and the London School of Medicine and Dentistry in 2003. In 2005 he was appointed as Consultant in Oral Maxillofacial Surgery to North Glamorgan NHS Trust with NHS duties at Royal Glamorgan and Prince Charles Hospitals whilst holding an Honorary Professorship at University of Cardiff School of Dentistry.Alongside his role at UCLan, StJohn is currently the Robert Bradlaw advisor in the Faculty of Dental Surgery at the Royal College of Surgeons of England. He is also the current President of the British Society of Oral Implantology.

Professor **David Andrew Phoenix**, OBE, AcSS, DSc studied Biochemistry at degree and doctoral level at Liverpool University which in 2009 awarded him a Doctor of Science for his impact on the field. In 2000 he was appointed Professor of Biochemistry, at the University of Central Lancashire (UCLan) and has held visiting chairs in Canada, China and Russia. He has over 200 publications as well as a number of edited collections and monographs focused on the structure function relationships of bioactive amphiphilic molecules. He is a Fellow of the Royal Society of Chemistry, The Society of Biology, The Institute of Mathematics and Its Applications and the Royal Society of Medicine. He was elected to Fellowship of the Royal College of Physicians (Edinburgh) for his contribution to medical research and education in 2013 and in 2013 was also appointed Vice Chancellor of London South Bank University. He was made an Officer of the Most Excellent Order of the British Empire in 2010 for services to Science and Higher Education and recognized as an Academician by the Academy of Social Sciences in 2012.

Álisson Rocha Machado, Marcio Bacci da Silva, School of Mechanical Engineering, Federal University of Uberlandia, Av. João Naves deAvila, 2121, CEP: 38400–902, Uberlandia, MG, Brazil.

Álisson Rocha Machado is Professor of Mechanical Engineering at the School of Mechanical Engineering, Federal University of Uberlandia, Brazil. Alisson is conducting research and teaching in the area of machining and received his Ph. D. in machining from the University of Warwick in the United Kingdom under the supervision of Dr. Wallbank. Alisson is also working with international groups from Purdue and London universities in the area of machining.

Marcio Bacci da Silva is Professor of Mechanical Engineering at the School of Mechanical Engineering, Federal University of Uberlandia, Brazil. Marcio is conducting research and teaching in the area of micromachining and received his Ph. D. in machining from the University of Warwick in the United Kingdom under the supervision of Dr. Wallbank. Marcio is active in the field of nanotechnology as applied to machining and cutting processes.

Katalin Halasz received her M.Sc. (polymer science) and her Ph.D. (material science and technology) from the University of West Hungary. She is currently a scientific assistant at the Institute of Wood Based Materials and Technologies at the University of West Hungary. Her research interests are in the area of biodegradable polymers, LbL nanocoating with green processed nanoparticles, material science and cellulose modification.

George Grozdits received his M.Sc. from the University of British Columbia, Vancouver, Canada, and his Ph.D. from Virginia Tech, USA. He is currently a professor emeritus at Louisiana Tech University, USA. He is one of the inventors of nanocoating process on lignocellulosic pulp fibers. He has several patents and great experience on pulp and paper production and wood processing technologies.

Levente Csoka received his two M.Sc.-s (pulp, paper technology and wood sciences) and his Ph.D. from the University of West Hungary. He is currently an associate professor at the Institute of Wood Based Materials and Technologies at the University of West Hungary. His research interests are in the area of finite element modelling, numerical transformations, developing multifunctional pulp and paper, cellulose fiber products, nanotechnology and cavitation technology.

Velaphi Msomi was born Highflats, KwaZulu Natal in South Africa. He did his Undergraduate degree from 2002 to 2005 and Postgraduate degrees from 2006 to 2008 at University of Zululand. He then joined Cape Peninsula University of Technology (CPUT) in 2009 for Doctoral Degree under the Department of Mechanical Engineering. He then started to write journal articles since 2010. In 2012, he was appointed as editorial board member for Journal of Engineering and Science and Global Advanced Research Journal of Engineering, Technology and Innovation. He was employed permanently by CPUT in 2012 as Mathematics lecturer. He is currently active in research and submitted number of articles for publication in well-known journals. He is focusing on mathematical modelling of smart materials and investigation of microstructural behaviour during phase transformation.

Vladimir N. Malyshev graduated in 1976 from the Ufa Petroleum Institute (UPI), majoring in "Machinery and Equipment for Oil and Gas Fields". In 1984 he post-graduated at the Gubkin Moscow Petroleum Institute and defended his thesis "Enhancement of the wear resistance of friction knots by microarc oxidation" and in 1999, his doctoral thesis on "Hardening of friction surfaces by microarc oxidation", at the Gubkin Russian State University of Oil and Gas.

He has worked as an engineer, as a Junior Scientific Collaborator in UPI (1976–1980), Senior Engineer of Lab for Special Materials in the Kharkiv branch of VNIPImorneftegaz (1984–1986) and Assistant Professor of Brest Polytechnic Institute (1986–1994).

Since 2000, he has been Associate Professor, then from 2004 Professor of Chair for "Tribology and Repair Technologies of Oil and Gas Equipment" at the Gubkin Russian State University of Oil and Gas. His main research areas are materials science, tribology, wear resistant and protective ceramic coatings on metals, and synergetics.

Dr. **Malyshev** is the author of more than 200 scientific papers, including 3 textbooks, a monograph, 18 copyright certificates and 6 patents of the Russian Federation, more than 150 articles and abstracts and 8 learning methodical developments. He was awarded with Sign "Inventor of the USSR" in 1982 and Sign of Laureate of All-Russian Contest "Engineer of Year" in 2012.

Dr. **Prashant K Sarswat** finished his PhD from University of Utah in 2012. He has made major strides in development of CZTS solar cells, an alternative technology for solar cell production that is not only more economical but also uses earth abundant materials rather than the ordinarily used rare materials in short supply.

Currently, he holds Research Assistant Professor Position in the Department of Metallurgical Engineering, University of Utah. His primary research is in the area of cost effective synthesis of chalcogenide solar cell materials, sensors and optical materials. He is investigating modified kesterite nanocrystals, which can be used for applications such as light emitting diode, quantum dot solid, and biosensor fabrication. Prior to this, he spent couple of years developing famatinite /oxide based semiconducting materials. He also spent few years developing nanostructures using focused ion beam assisted deposition and milling, these nanostructures can be utilized for fabrication of ultra sensitive sensors such as femtogram, squid devices, and nano-cantilever. Dr. Sarswat is editorial board member of several materials science and engineering journals as well as member of Alpha Sigma Mu scientific society.

Kelvii Wei Guo is Professor of Materials Processing and Engineering. Currently in the Department of Mechanical and Biomedical Engineering, City University of Hong Kong. He has been engaged in different fields including: academia, industry and military service. His research interests include: *Micro/Nanofabrication/ Machining*; *Surface Engineering*; *MEMS/NEMS*; *Lab on Chip* (*μTAS*); *Biomaterials*; *Biosensory*; *Analytical and Bioanalytical Chemistry*; *Green Synthesis Techs*; *Green Energy/Environment*; *Wastewater Treatment*; *Femtosecond/ Picosecond Laser*; *Advanced Manufacturing Processes*; *Intelligent Control and Learning, Intelligent Modeling*.

Yuri Fedorovich Migal is a head of chair at the Rostov State Transport University since 2002, and a head of department at the Southern Scientific Centre of the Russian Academy of Sciences since 2004. His works are devoted to the quantum theory of polyatomic systems and application of quantum chemical methods in tribology. Now he is engaged in research of atomic interactions on a surface of metals and the compatibility of chemical elements with iron on grain boundaries of steel.

J.M. Antunes is Professor in the Mechanical Engineering Department of the Polytechnic Institute of Tomar. Also, is member of the Mechanical Engineering Center from University of Coimbra. Is co-author of more than forty publications about the mechanical characterization of the materials.

Nataliya Sakharova is a Research Assistant in Mechanical Engineering Research Center of University of Coimbra, Portugal. She obtained her PhD degree in Physical and Mathematical Sciences (Solid State Physics) by the V.N. Karazin Kharkiv National University in Kharkiv, Ukraine in 1999. In 2004 she took an opportunity to continue her research activity in CEMUC (University of Coimbra's Mechanical Engineering Research Centre), Portugal. Her areas of scientific activity are mechanical properties of bulk materials and thin films: modelling and numerical simulation, inverse analysis and experimental aspects; experimental mechanical characterization of materials: mechanical tests and structural studies by different techniques with high experience in Transmission Electron Microscopy; mechanical properties of nanomaterials: modelling and experimental aspects; metal forming: fundamental aspects, modelling, constitutive laws, inverse analysis, numerical simulation and applications. She is author of around 80 publications (chapters of books, papers in international and national journals, conferences).

JV Fernandes is full Professor at the Mechanical Engineering Department of the Faculty of Sciences and Technology of the University of Coimbra. Coordinator of the Centre for Mechanical Engineering of the University of Coimbra (CEMUC), Unity nº 285 – Portuguese Foundation for Science and Technology – FCT). Main research interests: Mechanical properties of metals and thin films – Fundamental and experimental aspects, Modelling, Inverse analysis and Applications to metal forming."

Cinzia Caliendo, was born in Rome, Italy, in 1960 and graduated in Physics from the University of Rome *"La Sapienza"*. She is senior researcher at the Institute of Acoustics and Sensors *Corbino*, IDASC, of the National Research Council of Italy. Her research interests include physical acoustics, and acousto-optic devices, surface and bulk acoustic waves propagation in layered media, thin piezoelectric films and surface and bulk acoustic wave-based sensors. She has participated to many national and international research projects in the sensors field and served as a referee for many leading international journals. Main topics of scientific activity: Chemical and physical acoustic wave sensors; Thin film technology for acoustic applications; Theoretical models applied to thin film characterization for acoustic applications; Growth and characterization of piezoelectric thin films.

Fabio Lo Castro was born in Rome, Italy, in 1971. He received the Laurea degree in Electronic Engineering and the Ph.D. degree in *Sensorial and Learning Systems Engineering* at the University of Rome Tor Vergata. His research activities are concerned with the development of sensor interfaces, applications and data processing. Since 2006 he worked at the CNR – Institute of Acoustics, focusing his research activities on the environmental acoustics and underwater acoustics, and collaborated with the Sensors and Microsystems Group of University of Roma Tor Vergata in the development of physical, chemical and biological sensors. He is author or co-author of more than 30 papers in journal and conference proceedings and more than 90 technical reports.

Rodrigo Sávio Pessoa has undergraduate degree in Physics from "Universidade Estadual Paulista Julio de Mesquita Filho, UNESP" in 2003, Master of Science degree in the field of Plasma Physics from Technological Institute of Aeronautics in 2005 and Doctorate in Sciences in the field of Plasma Physics from Technological Institute of Aeronautics in 2009. He is currently research professor at the University of Paraiba Valley and collaborator at the Technological Institute of Aeronautics. He has experience in physics, with emphasis on Condensed Matter

Physics and Plasma Physics, Materials Science and Microelectronics acting on the following topics: thermal and non-thermal plasma technology, plasma diagnostic techniques, deposition processes and etching of materials by plasma, surface treatment by plasma, techniques for materials characterization, simulation/modeling of cold plasmas, and plasma gasification.

Mariana Amorim Fraga is Professor at the Faculty of Technology of São Paulo, Brazil. She holds a Ph.D. in Mechanical Engineering from the Technological Institute of Aeronautics and M.Sc. degree in Electrical Engineering and Microelectronics from the University of São Paulo. Her research interests are semiconductor materials, thin films, plasma processes, silicon carbide and related materials, piezoresistive sensors for harsh environments applications and MEMS devices.

Lúcia Vieira Santos has undergraduate degree in Chemistry from "Universidade de Mogi das Cruzes" in 1993, Master of Science degree in the field of Materials from Technological Institute of Aeronautics in 2000 and Doctorate in Sciences in the field of Materials from Technological Institute of Aeronautics in 2004. She is currently research professor at the University of Paraiba Valley and collaborator at the Technological Institute of Aeronautics. Has experience in Materials Science, focusing on Plasma Deposition of Diamond-like Carbon films and Materials diagnostic techniques, acting on the following subjects: plasma, magnetron sputtering, DLC films.

Nierlly Karinni de Almeida Maribondo Galvão has undergraduate degree in Materials from " Centro Federal de Educação Tecnológica do Rio Grande do Norte" in 2004, Master in Materials Engineering from "Universidade Federal do Rio Grande do Norte" in 2007 and Doctorate in Materials Engineering from "Universidade Federal do Rio Grande do Norte" in 2011. She has experience in the area of Materials and Metallurgical Engineering, with emphasis on materials processing by plasma, acting on the following topics: cold plasmas, plasma technology, plasma deposition processes, surface treatment by plasma techniques and materials characterization.

Marcos Massi is Visiting Professor at the Universidade Federal de São Paulo. He has Ph.D. in Electrical Engineering and Microelectronics from the University of São Paulo.

Homero Santiago Maciel has bachelor's at Engenharia Eletrônica from Instituto Tecnológico de Aeronáutica in 1976, master's at Physic from Instituto Tecnológico de Aeronáutica in 1980 and doctorate at Electrical Discharges and Plasmas from University of Oxford in 1986. He is currently research professor at the University of Paraiba Valley and collaborator at the Technological Institute of Aeronautics. Has experience in Physics, focusing on Plasma Physics and Electrical Discharges, acting on the following subjects: plasma, magnetron sputtering, dlc films, reactive ion etching and plasma etching.

George Kim received his Bachelor degree in Metallurgical Engineering and Doctorate degree in Chemical Engineering from McGill University. He has over 20 years of surface engineering experience, including thermal spray technology. Since May 2000, he has focused most of his efforts on consulting in materials and surface engineering, with an emphasis on nanostructure technology for military and industrial applications. He is a co-founder of a company which is currently being funded by the United States Navy and Army on the economical production of nanostructured powder for structural and coatings applications. Dr. Kim currently has two patents and one patent-pending on nanostructured coatings, including the first thermal sprayed nanostructured coating to be applied industrially, and used in 10 mines around the world.

G. Di Girolamo is an Italian researcher working at Italian National Agency for New Technologies, Energy and Sustainable Economic Development. He has a PhD in mechanical engineering from University of Calabria (Italy). His research activity is focused on the development of advanced materials for high-temperature applications as well as materials for hydrogen storage, In the last years he has published in highly respected journals several articles on the fabrication and characterisation (microstructural, mechanical and thermal) of advanced materials and components. In his area of expertise he has also worked to several research projects.

Yiqing Chen was a Research Fellow at the School of Mechanical and Manufacturing Engineering, the University of New South Wales (UNSW), Australia. She has been conducting research on polishing of diamond materials since her PhD study in 2003 at The University of Sydney, and has 25 publications in this field, including a book titled "*Polishing of Diamond Materials: Mechanisms, Modelling and Implementation*", in Engineering Materials and Processes series by Springer-Verlag London Limited. Her research focuses on the characterization of advanced materials and development of advanced methods for processing of difficult-to-machine materials. She has received 7 academic and research awards, including a best paper and a best presentation award in two prestigious international conferences, and has a worldwide patent in "Method and apparatus for polishing diamond and diamond composites".

Dr. **Mei Liu** is a research associate in the laboratory of PNPT (Precision & Nano Processing Technologies) in school of Mechanical Engineering, the University of New South Wales. Her main research focus is characterizing and modeling of the residual stresses and micro-structural defects in the thin film system. She is also dedicated to the research areas such as surface enhance and defect-free precision manufacturing. Her expertise is the x-ray diffraction residual stress analysis and the electron microscopy microstructure analysis.

Weidong Liu is Research fellow at the Laboratory for Precision & Nano Processing Technologies, UNSW Australia. Prior to UNSW, He received his Ph. D in solid mechanics in 2011 at Peking University in Beijing, China. Dr Liu's research emphasizes on the mechanics of amorphous materials, constitutive modelling, multi-scale numerical simulation, and impact dynamics. His work on metallic glass has led to significant advances in understanding their complex deformation mechanisms and in establishing novel processing techniques. The research outcomes have been disseminated in a series of publications, including 17 refereed articles in top international journals, 1 book chapter, 1 patented technology, and 10 conference papers.

Liangchi Zhang is Scientia Professor, Professor of Mechanical Engineering, and Head of Laboratory for Precision and Nano Processing Technologies, the University of New South Wales (UNSW), Australia. He was awarded a higher doctorate degree by the University of Sydney in 2005 and was elected the Fellow of the Australian Academy of Technological Sciences and Engineering in 2006. Prior to UNSW, he has worked in the University of Cambridge, National Mechanical Engineering Laboratory Japan and University of Sydney; and has been the Director of Graduate School of Engineering and Associate Dean of Engineering at the University of Sydney. He has received many academic awards, prizes and honours. He is currently carrying out research in the interdisciplinary area of precision and nano manufacturing, solid mechanics and characterisation of advanced materials.

Kausala Mylvaganam is a Research Fellow at the School of Mechanical and Manufacturing Engineering, The University of New South Wales (UNSW), Australia. She received her PhD from the University of Cambridge, UK for her work on the *ab-initio* calculation of molecular properties. Prior to UNSW, she has worked in the University of Cambridge, UK on a Commonwealth Fellowship and in the University of Sydney. Currently she is doing research in Nanotechnology with a particular focus on the multiscale modelling of ultra-precision machining and characterization of advanced materials.

Preface

The first thin film is probably created using the electrolysis method in 1838. Researchers fabricated thin metallic layers using the chemical reaction and glow discharge sputtering. Moreover, Faraday (1857) developed a thin metallic layer using the thermal evaporation of a wire with a high electric current passing through it. The useful properties of the thin films and the interest paid to behavioral studies of two-dimensional solids have led to the scientific and technological importance of the thin films. Today, modern and complicated optic-electrical, electronic and industrial equipments are fabricated by nanostructured thin films.

Indeed, nanocoatings consist of thin films with nanometric dimensions or a matrix, in which the fine nanometric particles are distributed and give specific properties to the coatings. Coating process, followed by fabrication of nanostructured materials, is among the present widely and effectively used nanotechnologies in many industries. The investigations conducted on nanocoatings indicate that properties of these materials are considerably improved as compared to the ordinary coatings. In comparison to the micrometric coatings, nanocoatings indicated superior thermal expansion coefficient, hardness, ductility with appropriate corrosion and wear resistance.

By now, the majority of works conducted on nanocoatings are for hard and super hard coatings. Super hard coatings are those with hardness usually greater than 40 GPa. Nanocoatings usually are divided into four groups, nanocrystalline coatings, super lattice and multilayer coatings, thin films, and nanocomposite coatings. Different methods used for fabrication of nanocrystalline coatings. For instance, alumina nanocoating with purity of 99.5% and average crystallite size of 36 nm can be used in scratch resistant, semiconductor, and UV-proof coatings. Moreover, alumina can be produced in different colors without any change in their chemical composition. Usage of nanocoating results in a lighter weight and longer life of the equipment.

Alumina and titania nanoparticles are also used in coating of the military equipments, as their use in marine facilities considerably reduces the costs and damages induced by corrosion. Furthermore, zinc oxide, titanium oxide, and copper oxide nanoparticles can be used for fabrication of UV-proof layers. Among the advantages of these coatings, one can name their high resistance to cracking and abrasion, as well as their proper transparency. Using the nickel electrodes in multilayer capacitors, it is possible to obtain high capacities by low volumes. However, these electrodes suffer from oxidation at high temperatures. Through coating the nickel electrodes by stable oxides such as $BaTiO_3$, these oxides can be protected against oxidation. Nevertheless, it must be noted that these coatings must have higher thickness, unless they affect the properties of the layer. To reach the optimum thickness, $BaTiO_3$ coating is deposited using the sol–gel method on nickel electrodes.

Multilayer coatings consist of several layers, each with thickness of 1–5 nm. Each layer has specific crystalline structure and is composed of elements such as Ni, Ti, V, and Al. These coatings are highly compacted and possess high density and are known as multilayer dense coatings. Implantations of these coatings on industrial pieces improve their physical and chemical properties. AlN/ZrN, CrN/AlN, and TiN/CrN are among the most important lattice coatings.

Thin films are successive layers composed of different phases. These layers involve high hardness and modulus of elasticity and desired wear properties. The main reason for the increased hardness of multilayer thin films is the stack of extremely thin layers with different dislocation lengths on each other and, as a result, reaching their strength to theoretical values. Due to the different dislocation length of the lines in each layer, dislocations cannot move from one layer to another. Another cause for this behavior is the extremely small thickness of this layer which prevents independent activity of dislocation sources. Thus, the hardness of these coatings indicated a remarkable growth. The common deposition methods of thin films are physical vapor deposition (PVD), chemical vapor deposition (CVD), and electrochemical deposition (ECD). Another technique for coating of thin layers is "layer-by-layer" method which is established on basis of formation of a multilayer on a substrate material, where both layers have opposite charges. Thin films have been recently receiving high attention as it can be applied for fabrication of aerospace parts and artificial bone.

Among the four abovementioned groups of nanostructured coatings, nanocomposite coatings have higher rate of application since their fabrication allows developing unique chemical and physical properties on surface of pieces. These coatings consist of two matrix and reinforcing phases, where the nano-phase (reinforcing phase) is loaded within the matrix phase. Here, the amorphous phase can be composed of diamond-like carbon, carbonitride, or some other compounds with proper hardness and modulus of elasticity. In addition, materials such as AlN, TiN, and Si_3N_4 can be used as the nanostructured reinforcing phase. For instance, by loading TiN nanoparticles with grain size of 8–11 nm in DLC substrate, hardness can be enhanced to 50–70 GPa.

The size and distribution of reinforcing phase in the matrix is a key role in properties of this group of nanocoatings. The smaller size of reinforcing material results in the more delayed formation of dislocations and less severe plastic deformation. The excessively short spacing of these nanoparticles also results in the reaction between atomic plates of nanocrystalline grains. Therefore, size, volumetric percentage, and distribution of these nanoparticles are among the important factors in design of this type of nanocomposite coatings, as their changes would affect ductility and hardness of the coatings.

Techniques including magnetron sputtering, thermal spray, and CVD are among the most important techniques applied for fabrication of nanocomposite coatings. The high deposition rate and high uniformity of the coating developed by CVD are among the advantages of this method. To create these coatings, typically thermal expansion and magnetron sputtering methods are utilized, as these techniques can be easily preformed at low temperatures. Besides, the texture and size of nanoparticles can be controlled using these techniques. Using the thermal expansion method, it is possible to deposit single-phase ceramic coatings and composite coatings with ceramic matrix on industrial pieces.

This just elucidates an aspect of nanocoatings: products improving. Probably, the key importance of nanocoatings is producing new materials or new usage of conventional materials, which are just resulted in advanced methods. This book collects chapters about those nanocoatings and their fabrication methods for using as wear- and abrasion-resistant materials. I like to appreciate all of contributors to this book and thank them for their high quality manuscripts. I wish this collection help all researchers to benefit and develop from this important type of nanocoatings.

Mahmood Aliofkhazraei
Tarbiat Modares University
November 2014

Wear, friction and prevention of tribo-surfaces by coatings/nanocoatings

1

Sunil Mohan and Anita Mohan
Indian Institute of Technology (BHU), Varanasi, India

1.1 Introduction

Tribology is at the forefront of modern research of practical importance. It deals with the science and technology of wear, friction and lubrication, which has applications in our day-to-day life in different ways. In prehistoric times, men created fire using frictional heating of stones. With time our knowledge has increased immensely. There are numerous examples of tribology in our day-to-day life. Since the scope of this chapter is limited, emphasis will be given to the tribology of mechanical systems exposed to wear and friction during solid–solid contact (Lubrication Engineering Working Group, 1966).

Leonardo da Vinci was the first to postulate the concept of friction in 1519, and this was further advanced by French physicist Guillaume Amontons by the end of the seventeenth century. It was, however, elaborated by C.A. Coulomb, in 1785, who distinguished between static and kinetic friction (Szeri, 1998) (cf. Table 1.1).

In general, wear refers to the damage to the surface or loss of material from one or both solid bodies in contact. Such situations arise when surfaces are in sliding or rolling contact. Adhesive, abrasive, fatigue and corrosion are the most common types of wear in tribological applications. Wear is a perennial phenomenon of surface damage in contacting bodies, which may be solid–solid, solid–liquid or solid–air. Whatever the system may be we cannot get away from its complexity; as stated by (Rigney et al., 1984, p. 195): 'wear is just like a jigsaw puzzle in which pieces fit together here and there but the overall pattern is far from clear'. Wear process in solid–solid contact is very complex. It should be noted here that with the available data, equations generalization of the process is still elusive. Wear in solid–solid contacts takes place because of adhesion and abrasion. The rate at which the contacting surfaces wear off mainly depends on various characteristics of each surface, applied load, sliding velocity, distance slid, temperature and environmental factors.

Wear and friction lead to tremendous losses in the form of energy and material. If we combine both, they amount to heavy financial loss. For example an expert estimate shows that in 1978 in US only the energy loss was about 4.22×10^6 T Joule, which amounted to $20 billion (Szeri, 1998). Even after more than three decades, it is not possible to eliminate losses completely because of wear and

Anti-Abrasive Nanocoatings.

Table 1.1 Progression in friction concept

Progression in concept	Date
Leonardo da Vinci was the first to introduce the concept of friction	1519
Guillaume Amontons further clarified the concept	Late seventeenth century
C.A. Coulomb distinguished conceptualized static and kinetic friction	1785
Bowden and Tabor did extensive work on the mechanism of friction and provided major evidence of adhesion theory, including plastic deformation of asperities	Mid-twentieth century
Further additions in the coefficient of friction were made by N. P. Suh and Larsen-Basse	End of the twentieth century

friction. However, this can be minimized to a large extent by proper materials choice, altering material properties and providing suitable surface treatment or coatings/nanocoatings.

1.2 Friction of materials

The basic mechanism of friction depends on various factors such as physical, chemical and materials properties. Over the years, several mechanisms have been proposed with modifications. However, the basic mechanism remains the same. In actuality, it is situation-based or system-based property. The mechanisms involved at microscopic level are adhesive forces between mating surfaces, mechanical interactions of asperities, the ploughing of one surface due to the harder asperities of other surface, deformation of surfaces or fracture in oxide film and/or introduction of a third body as a result of debris or environmental effects. Various attempts have been made to explain friction (cf. Bowden and Tabor, 1964; Dowson, 1979; Ludema, 1987).

Bowden and Tabor (1964) did extensive research into friction. Although adhesion is the main cause of friction, adhesion and deformation both contribute to the energy dissipation in friction. Bowden and Tabor (1964) provided major evidence supporting adhesion theory including plastic deformation of asperities. They proved experimentally that the mechanical properties of contacting surfaces play an important role. As a result of their seminal contribution in this field, adhesion theory is attributed to Bowden and Tabor.

Bowden and Tabor (1964) assumed that interacting asperities undergo deformation up to plastic flow regime and reach pressure equal to the indentation hardness of the material. In such case, the real area of contact (A_r) is determined as:

$$A_r = \frac{N}{H} \tag{1.1}$$

It is further assumed that friction is due to shearing of bonds during sliding. This means that the coefficient of friction (μ) is given by:

$$\mu = \frac{F}{N} = \frac{A_r\tau}{A_r H} = \frac{\tau}{H} \tag{1.2}$$

where N is normal load in newton, H is flow hardness in N/m^2 and τ is shear stress.

This expression satisfies both of the Amontons laws in which both load and contact area are eliminated. Hardness is almost three times the flow stress (σ_y) of the material and shear stress (τ) is almost equal to 0.5 to $0.6\sigma_y$. Therefore, the universal value of coefficient of friction should be 0.17 to 0.2 (Larsen-Basse, 1992). As a matter of fact, it is true for clean pure metals but values are much higher in a vacuum when there is no protective layer. In a further study, Bowden and Tabor (1950) took three pairs of sliders of a material against three different flat surfaces (cf. Figure 1.1) to see the effect of a contacting surface with a different shearing force on the coefficient. The results are summarized in Table 1.2.

Al–Pb composites also exhibit the same nature (Figure 1.2) when the lead content increases, the friction decreases. This is because shearing takes place in thin film of lead and load is supported by aluminium matrix. However, as the lead reaches an optimum value film thickness of lead increases at the interface. Apart

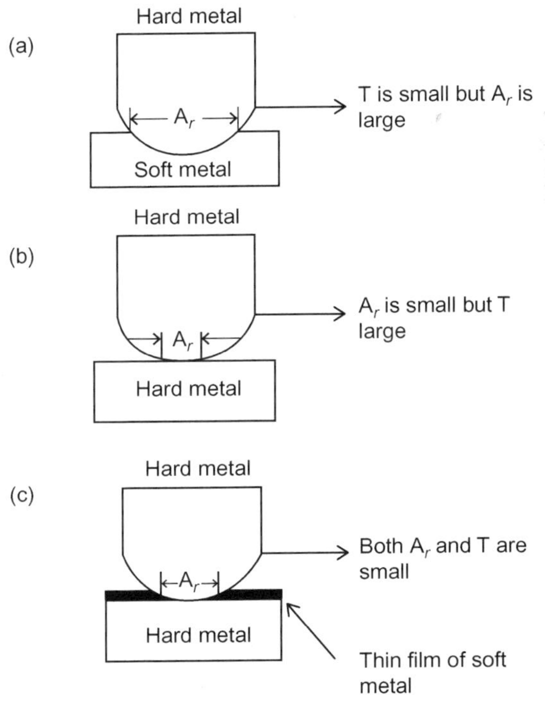

Figure 1.1 Schematic of friction force between two metal surfaces in contact.

Table 1.2 **Coefficient of friction with different contacting surfaces**

Type of surfaces in contact	Observation	Coefficient of friction
Hard slider on soft surface (steel ball on flat indium surface)	τ of indium is low but A_r is large	0.6 to 1.2
Hard slider on hard surface (steel ball on steel flat)	τ of steel is large while A_r is small	0.6 to 1.2
Hard slider on hard surface with thin film of soft metal (steel ball on steel flat with thin indium film)	Both τ and A_r are small because shearing takes place in indium film while load is supported by steel	0.6

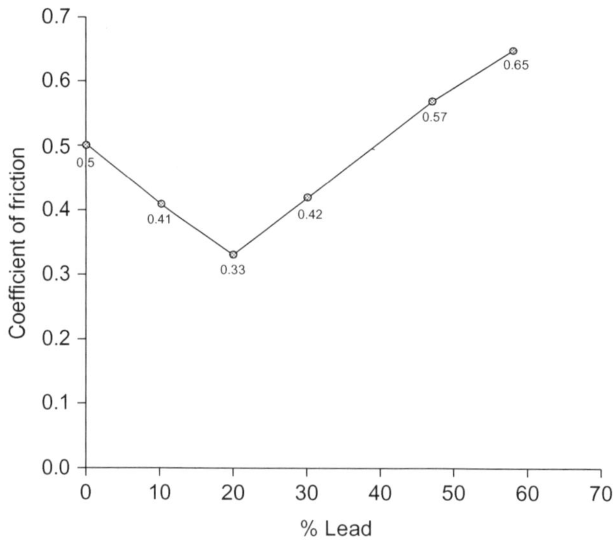

Figure 1.2 Variation of coefficient of friction with percentage of lead content (Mohan et al., 1990).

from shearing, load is also supported by lead film enhancing the friction (Mohan et al., 1992).

The coefficient of friction acts differently for different combinations of materials, surface conditions and operating conditions. It is also affected by inputs resulting from internal effects, external existing conditions or imposed conditions. Examples of internal effects are a different morphology or surface deformation as a result of a rise in temperature through frictional heating. Examples of external factors are environmental effects, use of lubricants or external members at the interface in the form of debris. To be specific, the coefficient of friction is affected by both mechanical action and chemical effects. Hence, the value of coefficient of

Table 1.3 Typical values of kinetic, μ_k coefficient of friction of unlubricated metals and alloys at room temperature in air

Material combination	Coefficient of friction	
	On self	On mild steel
Pure metals		
Precious metals		
Au, Pt	1–1.5	0.4–0.5
Ag	0.8–1.2	0.3–0.5
Soft metals		
In, Pb, Sn	0.8–2	0.5–0.8
Metals		
Al	0.8–1.2	0.5–0.6
Co	0.5–0.6	0.4–0.5
Cr	0.5–0.6	0.4–0.5
Cu	0.8–1.2	0.6–0.7
Fe	0.8–1.5	0.8–1.5
Mg	0.5–0.6	0.4–0.6
Mo	0.5–0.6	0.4–0.6
Ni	0.7–0.9	0.6–0.9
Sn	0.8–1	0.6–0.8
Ti	0.5–0.6	0.4–0.6
W	0.7–0.9	–
Zn	0.5–0.6	–
Zr	0.7–0.8	–
Alloys		
Lead base white metal or Babbitt (Pb, Sn. Sb)	–	0.25–0.6
Leaded brass (Cu, Zn, Pb)	–	0.2–0.4
Leaded bronze (Cu, Sn, Pb)	–	0.2–0.4
Gray cast iron	0.8–1	0.3–0.5
Mild steel	0.7–0.9	–
Intermetallic alloys		
Co-based alloy (stellite, Tribaloys)	0.3–0.5	–
Ni-based alloys (Haynes alloy 40)	0.6–0.9	–

Courtesy of Bhushan (1999).

friction is relative, not absolute. A number of reference works are available providing such data (Avallone and Baumeister, 1987; Bhushan, 1999; Bhushan and Gupta, 1997; Blau, 1992, 1996; Neale, 1973). Table 1.3 presents the typical values of coefficient of friction for different materials.

1.2.1 Friction of metals, alloys and composites

Metals without any contamination exhibit a high coefficient of friction because of high adhesive forces. The asperities may cold weld together to form atomic bonds across the interface even when there is no external load imposed. But the presence of contaminants may form film at the contacting surface restricting metal to metal contact and coefficient of friction is drastically reduced (Bowden and Tabor, 1950, 1964; Buckley, 1981).

In clean surfaces strong adhesive forces act especially in a high degree of vacuum. Adhesion problems have been studied extensively (Buckley, 1981). Work is primarily focused on space applications where adhesion and seizure are very important. Clean metal surfaces undergo seizure or cold welding when they are in sliding contact in a vacuum.

Asperities on contact may undergo elastic and plastic deformation to release stresses. It is proposed that the coefficient of friction has two components: adhesion and ploughing. Ploughing is due to energy dissipation in a plastic deformation regime. Harder asperities plough softer surfaces creating deep grooves, whereas the adhesion component is controversial except for with clean surfaces in a high degree of vacuum (Bowden and Tabor, 1950). An additional component of friction was further added by Suh, 1986, and finally μ is given as:

$$\mu = \mu_{asp} + \mu_{plou} + \mu_{part} \tag{1.3}$$

where μ_{asp} is due to asperity deformation, μ_{plou} is due to ploughing and μ_{part} is due to the presence of wear particles. This model has been supported by studies on copper–copper systems (Suh, 1986), where the friction coefficient rises from low value as sliding distance increases. In this case, if particles detached are removed well before agglomeration then the friction remains low. An additional factor of coefficient of friction (μ_e) has also been observed, hence, the friction coefficient may be given as (Larsen-Basse, 1992):

$$\mu = \mu_a + \mu_p + \mu_e + \mu_{part} \tag{1.4}$$

where μ_a is due to adhesion, which is important and dominating in clean surfaces in a vacuum, though under normal conditions its role is limited, μ_p is plastic deformation and ploughing due to hard asperities creating grooves, μ_e is the contribution of elastic deformation, which is important in cold worked conditions of the surfaces and μ_{part} is the contribution due to third body particles trapped between the mating surfaces.

Though there may be agreements or disagreements on different theories on the coefficient of friction, it is evidently a compound property of a system and its prediction with just first principle is not possible.

Generally, alloys have a lower coefficient of friction compared to pure metals. The 10% addition of chromium in cobalt improves its oxidation and corrosion resistance, whereas the addition of tungsten and molybdenum improve the strength, friction and wear properties. The addition of silicon, iron, zinc, lead, tin indium and so on, improves the friction of aluminium alloys, which may be due either to the formation of hard phases increasing the overall hardness of the alloy or the presence of soft metal forming a layer at the contacting surface restricting direct contact and increasing shearing strength. Likewise, lead-based and tin-based babbitt, brass, bronze and gray cast iron also exhibit low coefficient of friction due to the presence of elements with low shearing strength. Aluminium base composites also show low coefficient of friction compared to base alloys (Bhushan, 1999; Mohan et al., 1992).

1.2.1.1 Effect of operating parameters

In addition to the inherent properties of the metals, alloys and composites; the friction is also affected by operating parameters such as sliding velocity, pressure conditions and contact point temperature and environmental effects such as relative humidity or through any imposed gaseous environment (Blau, 1996; Bowden and Tabor, 1964; Buckley, 1981; Hutching, 1992; Peterson et al., 1960; Rabinowicz, 1995; and references cited therein).

When surfaces are exposed to air, oxidation of the surface takes place, reducing the friction in comparison to clean surfaces. With the increase in the normal load, oxide films may crack and friction may increase, as in the case of copper on copper. But under the same load range steel on aluminium shows an almost constant coefficient of friction (Whitehead, 1950). In case of AISI 440C stainless steel on Ni_3Al alloy, the coefficient of friction drops continuously (Blau, 1992). Oxidation kinetics of different metals and alloys are different at the same temperature and with a change in temperature the oxidation kinetics also act differently. Hence, the nature of the coefficient of friction is also different for different pairs of materials with load. The coefficient of friction also changes with sliding velocity. It is governed by the material pair in question. Frictional heating increases with an increase in sliding velocity, which may lead to localized melting of low melting components forming a layer at the interface and a reduction in friction. The formation of an oxide layer, hardening of surface and sub-surface layer also reduces friction. But very high velocities may crack the oxide layer and lead to three-body wear increasing the friction. Working temperature also affects the coefficient of friction. A rise in temperature leads to a softening of the surface; phase transformation may also take place which may cause surface hardening or softening, depending on the history of the material, which may either improve or cause a deterioration in the mechanical properties. At high temperatures, if low melting point metals such as lead and indium are present, they may melt and form a layer of low shear strength, reducing the friction. Surface hardening may also lower the friction (Mohan et al., 1992).

With the exception of some noble metals, other metals generally react with the oxygen present in the environment and form a thin oxide layer. This oxide film restricts the adhesion between mating surfaces and the coefficient is reasonably reduced. However, during sliding, cracks may develop in oxide films due to high applied loads or sliding velocities and the coefficient of friction may increase.

1.3 Wear in metals, alloys and composites

Wear in material is generally expressed in terms of wear rate or bulk wear. At the beginning of a process, wear rate is generally high, known as *running-in regime*, and once it is stabilized, it is known as *steady state wear*. The initial period depends on the inherent characteristics of the material as well as the outward conditions of the contacting material pairs. Wear is also dependent on material properties and initial surface condition (which is likely to change with time), and definitely on operating and environmental conditions. Wear coefficient was introduced to predict the wear data but it provides data in relative magnitude and not in absolute value. Several reference works provide wear rates of different pairs of materials in the form of dimensionless number wear coefficient, k as given in Table 1.4 (Bhushan, 1999; Bhushan and Gupta, 1997; Peterson and Winer, 1980).

Similar materials in contact show high wear rates compared to dissimilar materials because of high adhesive forces. Generally, metals, alloys and composites with clean surfaces in contact have a high degree of adhesion and show high wear rate, which further increases under a vacuum because the chances of any kind of contamination forming any film at the interface is minimized (Buckley, 1981; Rigney, 1988; Bhushan, 1996).

1.3.1 Effect of operating parameters

Operating parameters such as sliding distance, sliding velocity, applied load, working temperature, relative humidity and other environmental factors affect the wear process. With sliding distance, initially high wear rate is observed in most of the

Table 1.4 **Typical range of wear coefficient**

Material combination	Wear coefficient range	Specific comment
Self-mated metal pairs	$>5 \times 10^{-3}$ (high)	Undesirable in dry contacts
Dissimilar metal pairs	10^{-4} to 10^{-3} (moderate)	Easy to fabricate, low cost
Alloy–alloy pairs	10^{-6} to 10^{-3} (moderate)	For high-temperature applications, for lowest wear requirements, self-mated ceramic pairs desirable, good for unlubricated conditions
Ceramic–metal pairs	10^{-7} to 10^{-4} (very low)	
Ceramic–ceramic pairs	10^{-7} to 10^{-5} (very low)	
Polymer–hard surface pairs	10^{-6} to 10^{-3} (low)	For corrosive environment and low-friction applications, good for low loads, good for unlubricated conditions
Solid lubricant–hard surface pairs	10^{-4} to 10^{-3} (low)	For lowest friction requirement

Courtesy of Bhushan (1999).

materials. However, their working load and sliding velocity range may differ. An extensive study conducted to understand the operative mode and mechanism shows that even with constant applied load and sliding velocity, wear pattern changes with distance slid. Debris of composite at a lesser distance displays a mainly oxidative nature and wear track is smooth with a thin oxide layer. But at a larger distance, debris comprises different oxides with metallic particles, wear track is observed with a thick oxide layer with a deeper track and wear rate is higher. However, it is evident that at low velocity, the effect of load dominates and at higher velocity the process is dominated by sliding velocity (Mohan and Srivastava, 2006).

In aluminium base materials, wear rate is observed to increase continuously with applied load because of surface delamination, three-body wear, development of grooves and so on. While studying debris and wear tracks at different loads for a particular velocity, debris changes from oxidative to oxidative-metallic to metallic as load applied is increased, and wear track observes broken oxide film, deep grooves and delamination of surface takes place at low sliding velocity, and a more severe effect with larger debris particles is observed at higher sliding velocity. But in steels, at times, in the presence of alloying elements even at higher loads, wear rate may decrease due to martensitic transformation (Mohan et al., 2002).

In aluminium base materials, wear rate decreases initially with sliding velocity as a result of oxide film formation restricting surface-to-surface interactions. With an increase in velocity, however, cracks may be initiated in the oxide film, leading to gross material transfer in the form of metallic wear, the formation of deep grooves and an increase in wear rate (Figure 1.3). This process enhances the wear rate drastically (Mohan et al., 1991, 1992, 2002; Mohan and Srivastava, 2006). When two bodies are in sliding contact, there is continuous removal of material

Figure 1.3 Scanning electron micrograph (SEM) of wear track of Al−1.67 wt% Fe composite showing broken oxide particles with deep grooves at high load of 5 kg and high sliding velocity of 4 m/s (Mohan and Mohan, 2008).

from the surface, but the extent may vary depending on the materials in contact, load applied, sliding velocity, temperature and so on. In general, under the same working conditions, materials in contact with the same micro-constituent phases show more wear. In the present work, sliding surfaces are different and material observes deformation up to sub-surface level starting from the asperities present.

1.4 Materials and their selection for wear and friction applications

There are numerous materials which have been used or are in use, such as cast irons, carbon and alloy steels, stainless steels, bearing steels, tool steels, hard-facing alloys, co-base alloys, intermetallic alloys, aluminium alloys (Al–Si, Al–Sn, Al–Pb, Al–G), cemented carbides, ceramics, thermoplastics and composites/nanocomposites.

1.4.1 Cast irons

Cast iron is a dense material and has long history of use. A variety of cast irons are available such as gray, chilled, white, malleable and nodular. These materials are inexpensive and provide excellent castability, good damping capacity, machinability and wear resistance. Gray (flake) iron is used in piston rings and cylinder bores for compressors and internal combustion engines, machine bed tools and sideways, brake blocks and low-stress gears. Nodular cast iron is used in crankshafts, gears and highly stressed piston rings. Chilled iron is used in cams and tappets where hertzian stress is high and lubrication is border line. Cast and heat treated white cast iron is used in grinding balls, chute and mill plates and tiles, cone, feeders, deflectors, and breaker liner plates and grizzlies. Hard-facing white iron is used in dragline buckets, shrouds, pick holders, side plate protection, hard-facing of steel diggers teeth for showels and so on (Eyre, 1992).

1.4.2 Steels

Steels have diversified use in wear applications. With modification in composition and suitable heat treatments, AISI 52100 steels are used in ball and roller bearings. Hadfield manganese steel is used in the mining, earth-moving and railroad industries requiring a high level of toughness, impact resistance and wear resistance. These steels are used in load bearing parts such as shafts, gears, housings, cables and so on by coating them with wear-resistant materials. Carbon steels with small additions of manganese are used in needle rollers, while with a small addition of manganese, silicon, chromium, nickel and molybdenum, steels can be used for large roller bearings, and for the applications requiring high shock and high fatigue resistance and so on (Glaeser, 1992).

Stainless steels are broadly classified with chromium content above 10.5% and are used primarily in corrosive media from mild to severe conditions, but their wear and galling resistance is poor. These steels are used in coal-handling equipment where the abrasivity of coal combines with the corrosive action of slurry. These are also used in the food and drug industries. All other industries where lubricants cannot be used, such as the oil and gas exploration, these steels are used in drill collars, valve parts and so on. Programmed valves redirect exhaust gases away from the carburettor in automotive industries. Stainless steel is used as a stem material to work at high temperatures. These steels are also used in structural application such as fasteners, pin and hanger assemblies and also in nuclear reactors. Bearing and tool steels are generally used in bearings and tools applications (Magee, 1992; Maloney, 1992; Moyer, 1992).

1.4.3 Other bearing alloys

Other bearing alloys such as tin-based and lead-based alloys are used as bimetal surface layers, generally cast on steel backs. Copper-based alloys are used as bushings and washers and also as bimetal surface layers and trimetal intermediate layers.

Components prone to wear are hard-faced by weld overlay or thermal spray coatings. Weld overlay materials are classified as build-up alloys (carbon steels, manganese steels); metal-to-metal alloys (easy-purpose, general-purpose and high-temperature steels) and metal-to-earth abrasion alloys (steels with high chromium and small amounts of molybdenum, manganese and silicon). In addition to this some non-ferrous alloys are also used. Build-up alloys are generally used in tractor rails, railroad rail ends, steel mill table rolls, large slow-speed gears and so on. Metal-to-metal wear alloys are used in the under-carriage components of tractors and power shovels, steel mill work rolls, crane wheels and so on. Metal-to-earth alloys are used for shovel teeth, rock crusher parts, ploughshares and auger flights, and non-ferrous hard-facing alloys are used in agriculture, mining, the construction industries, especially in valve seating surfaces, pump parts, extrusion screw flights, rock bit bearings, marine bearings, glass moulding hardware and so on (Kingsbury, 1992).

Among the aluminium alloys, Al–Si is the most commercially used for tribological applications. These alloys are widely used in applications related to the automotive industries – for example in the pistons of internal combustion engines, engine blocks, cylinder liners, roller type valve rocker arms and valve lifters. These alloys also find some applications in the aerospace industries (Pathak and Mohan, 1998; Shabel, 1998).

1.4.4 Metal-matrix composites and nanocomposites

The ever increasing demand for property optimization in view of the stringent requirements for specific applications and the global energy crises has led to the development of lightweight and energy-efficient wear materials. The contacting surfaces of metallic components may lead to surface damage under severe loads or excessive high sliding velocities (Figure 1.3) and thus to high wear rate (Figure 1.4). Additionally, external factors may also contribute to the oxidation of the surface due to a rise in temperature at the contact (Stott, 2002). Wear rate can

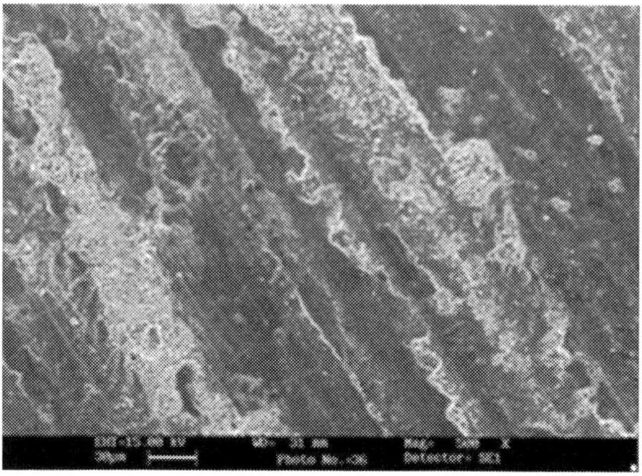

Figure 1.4 Variation of wear rate with sliding velocity at 1 kg load for Al–Fe composites (Mohan and Srivastava, 2006).

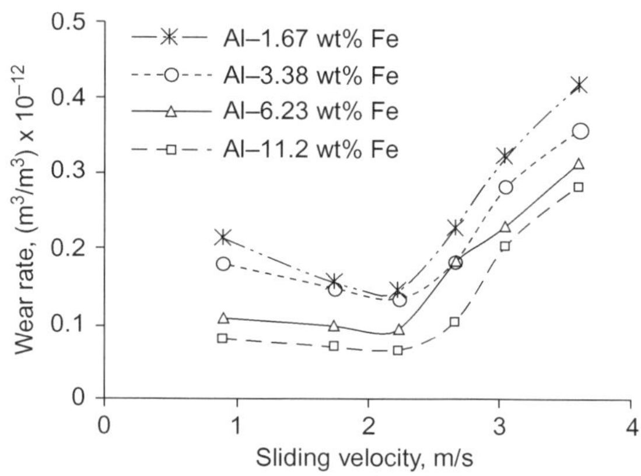

Figure 1.5 SEM of wear track of Al–1.67 wt% Fe composite showing oxide layer at low load of 1.0 kg and low sliding velocity of 1 m/s (Mohan and Mohan, 2008).

be very well reduced by the oxide layer (Figure 1.5) formation restricting metal-to-metal contact and the wear mode may be changed from severe to mild even in dry sliding conditions at higher temperatures (Archard and Hirst, 1956; Jiang et al., 1998; Stott et al., 1973, 1985).

Reinforcement of ceramic whiskers, fibres and particles in aluminium/aluminium alloys has extensively been explored in the context of tribo-materials. These could be used successfully in typical applications of pistons and cylinder liners of automobile engines. Comprehensive studies (Alpas and Zhang, 1992; Mohan and Srivastava, 2006; Agarwal et al., 2014; Surappa et al., 1982) on dry sliding behaviour of metal-matrix composites (MMCs) and nano-MMCs depict that the principal tribological parameters affecting their performance can be categorized as mechanical and physical factors or material factors. The intrusion of hard ceramic particles support the load, largely reduce the wear rate, and also extend the mild wear region compared to aluminium alloys, resulting in enhanced working range.

Aluminium matrix composites (AMC) find diverse applications in the structural and automotive sectors because of their superior strength-to-weight ratio, good thermal and electrical conductivity, good corrosion resistance and good tribological properties (Eliasson and Sandstrom, 1995; Lloyd, 1994). The most widely used reinforcing materials for AMCs include ceramics such as α-Al_2O_3 and SiC because of their high hardness, high melting point, low density and good corrosion resistance. However, ceramics usually have low fracture toughness, poor wettability with a metal matrix and very poor ductility (Han et al., 1993; Kacar et al., 1991). Improvement in properties is observed when composites are prepared using mechanical alloying (Gautam et al., 2013). The problem of ductility with enhanced strength can be minimized by dispersing intermetallics to a large extent. Intermetallic Al_3Fe is a thermodynamically stable equilibrium phase with high elastic modulus. Usually, strength and ductility follow reverse trends. If one is enhanced, the other is automatically reduced. This paradox is taken care of by reducing the intermetallic grain size to nanoscale, where strength further increases and as a result of higher grain boundary diffusivity (larger surface-to-volume ratio of nanoparticles), ductility is also increased (Kozlov et al., 2008; Agarwal et al., 2014).

1.4.5 Selection

While selecting a material for such application, hardness (Archard's equation) is the most important property. But other properties of the materials which may play key role in selection of material cannot be ignored; these include density, which becomes very important (minimizes fuel consumption) in applications such as pistons, cylinder liners and so on. Likewise, thermal conductivity is very important in high-temperature applications. There are many other physical properties, such as thermal expansion coefficient, heat capacity, specific strength and bending strength, which may have an important role in specific applications. Further, the condition in which a material is procured for end usage is equally important. This may include surface finish, porosity level, residual stresses, microstructure and crystallography of the material. Last but not the least, surface treatment such as coatings, nanocoatings and surface hardening techniques are also used to combat wear and friction. Hence, proposing or changing a material for any application without proper evaluation of a system may lead to excessive damage in the form of wear. Therefore, we need to consider all properties that are important for a specific application, and not just wear properties.

1.5 Coatings/nanocoatings and surface treatments

Nanostructured or nanocomposite coating is an emerging area because of its exceptional physical and mechanical properties to combat wear and friction. These coatings have two components, one of which is nanocrystalline and the other can be amorphous or nanocrystalline. Mechanical and tribological properties of nanocomposite coatings depend on interactions of the composite constituents (Gusev, 1998). Nitride coatings have been used to improve wear resistance (Matthes et al., 1990; Sue and Troue, 1990; Takadoum and Bennani, 1997). These nanocoatings can be hard, superhard and ultra-hard in nature depending on their hardness level. Hardness of these coatings may vary from 36 Gpa for CrN–Ni (Karvánková et al., 2001) to 103 Gpa for TiN/Si_3N_4 (Veprek, 1999). These coatings have wide applications in cutting and polishing tools, moulds, dies and so on, as they enhance the wear resistance and lower the friction coefficient (Wang and Zhang, 2010). Coatings are generally exposed to different environments and working conditions, hence they may require a set of properties such as high hardness, high toughness, high corrosion resistance and oxidation resistance. The transition metal nitride based multilayer coatings are suitable for various applications because of their enhanced hardness, toughness and so on (Kustas et al., 1997; Prengel et al., 1998).

In addition to the properties mentioned above, adhesion of the surface plays a key role in determining the tribological properties of coatings. Poor adhesion between substrate and coating may cause spalling and delamination of the coating, which can be largely controlled by relieving residual stresses from the coatings (Barshilia et al., 2010).

There are various ways to control wear and friction (Table 1.5). It can be done by varying material properties through composition change, applying a suitable treatment, by external treatments to the contacting surfaces such as nanocoatings or coating through thermal spray, electroplating, ion implantation, or vapour deposition processes like physical vapour deposition (PVD) and chemical vapour deposition (CVD), or by suitable surface treatments such as carburizing, nitriding, nitro-carburizing and laser surface processing.

1.5.1 Thermal spray coatings

Thermal spray coatings are used to improve friction and wear and, as a result, the life span of machine parts is enhanced. It is a versatile process that can use a large range of materials starting from pure metals to alloys, carbides, ceramics and even composites. It is economically viable and has wide scope of applications. There are several methods of thermal spray coatings such as oxy fuel wire (OFW), electric arc wire (EAW), oxy fuel powder (OFP), plasma arc powder (PAP) and highvelocity oxy fuel (HVOF). The material can be coated in a molten or plastic state through propulsion of particles or through the impact onto a workpiece. This technique is used to minimize adhesive, abrasive, fretting and erosive type of wear (Kushner and Nowinski, 1992; Novinski, 1981).

1.5.2 Electroplated coatings

This kind of coating can be tailored to meet the demand of specific applications related to tribology. Electroplating is basically of two types: hard and soft. Hard

Table 1.5 **Various techniques for wear and friction prevention**

Technique	Overview
Thermal spray coatings	Coating material can be coated in molten or plastic state through propulsion of particles or by the impact onto a work piece. This technique is used to minimize adhesive, abrasive, fretting and erosive type of wear.
Electroplated coatings	Hard coatings are generally used for adhesive, abrasive or erosive conditions, whereas soft coatings are desirable in bearing applications requiring low shear strength. Materials such as chromium, nickel, gold, palladium, cobalt, tin, silver, lead metals and alloys such as Co–P, Co–Ni–P can be plated using this technique.
Ion implantation	This technique involves creation of charged particles in an ion source, accelerating ions to high velocity and directing the beam of these ions onto substrate. These ions interact with the surface and create changes to the surface, improving wear and friction.
PVD and CVD	PVD has higher deposition rates, whereas CVD has higher throwing power. Soft metals are used to lower fiction coefficient and nitrides, carbides and oxides of metals are used to minimize wear.
Carburizing	Carburizing is used in power train gears for helicopters, earth moving equipment, heavy trucks and passenger cars. Most of the roller bearings are made of carburized steels.
Nitriding and nitro-carburizing	It provides good resistance to wear and corrosion. Powders, salt baths, gaseous mixtures or plasma-assisted processes are used as a source of nitrogen. A layer of compounds such as Fe_4N, $Fe_{2\text{-}3}(N,C)$ is formed on the surface of steels. Other complex nitrides of aluminium, vanadium, molybdenum, chromium and titanium can also be formed for special applications.
Laser surface processing	Localized heating brings about the transformation in microstructure and surface properties are altered.

coatings are generally used for adhesive, abrasive or erosive conditions with some degree of toughness to avoid crack generation in the coating, whereas soft coatings are desirable in bearing applications requiring low shear strength. Materials such as chromium, nickel, gold, palladium, cobalt, tin, silver, lead metals and alloys like Co–P and Co–Ni–P can be plated using this technique (Ide and Vanhumbeeck, 1987; Luce and Foulke, 1974; Weil and Sheppard, 1992).

1.5.3 Ion implantation

This is a viable and economical surface modification process for improving tribological properties. The technique involves the creation of charged particles in an ion source, accelerating ions to high velocity and directing the beam of these ions onto the substrate (work piece). These ions interact with the surface and create changes to the tribological surface, improving wear and friction (Fenske, 1992).

1.5.4 PVD and CVD

PVD requires producing vapours of material to be plated by means of sputtering, evaporation or laser ablation and subsequent condensation of the produced vapour onto a substrate (the material that has to be coated). CVD deposits a solid material from the vapour phase onto a heated substrate and undergoes a chemical reaction. The PVD technique uses conventional methods such as deposition of metals by evaporation, sputter deposition or ion plating, while plasma assisted deposition (PAPVD) uses compounds by activated reactive evaporation (ARE), reactive sputtering (RS) or reactive ion plating (RIP). Generally, PVD processes have higher deposition rates and CVD processes have higher throwing power. To counter tribological problems generally soft metals are used to lower fiction coefficient, and nitrides, carbides and oxides of metals are used to counter wear (Bhushan and Gupta, 1997; Bunshah, 1992).

1.5.5 Carburizing

This is one of the oldest techniques for hardening the surface of steels. Gas carburizing is the most widely used method in modern times. Other methods are also available, such as packing in charcoal and heating in a closed container, vacuum and plasma carburizing using hydro-carbon gas and salt bath carburizing. Carburizing can be used in power train gears for helicopters, earth-moving equipment, heavy trucks and passenger cars; most of the roller bearings are made of carburized steels (Stickels, 1992).

1.5.6 Nitriding and nitro-carburizing

Nitriding and nitro-carburizing are thermochemical processes used for surface hardening which provide good resistance to wear and corrosion. Other advantages are good fatigue strength and less distortion. Originally, gaseous ammonia was used but now powders, salt baths, gaseous mixtures or plasma assisted processes are used as the source of nitrogen. A layer of compounds such as Fe_4N, $Fe_{2\text{-}3}(N,C)$ is formed on the surface of steels. Other complex nitrides of aluminium, vanadium, molybdenum, chromium and titanium can also be formed for special applications (Hoffmann and Mayr, 1992).

1.5.7 Laser surface processing

Laser transformation hardening is already in commercial use and laser cladding is in a development stage which uses high power lasers. It is a chemically clean light source providing a controlled amount of energy to a desired region. Localized heating brings about a transformation in microstructure, and surface properties are altered. There are different modification methods for surfaces such as laser transformation hardening, laser melting, laser alloying, laser cladding and laser particle injection (Cooper, 1992).

1.6 Conclusion

We have presented an overview of the up-to-date concepts, mechanisms and operating parameters involved during sliding contact of different materials such as metal, alloys, composites and nanocomposites for wear and friction. Prevention of tribo-surfaces through different techniques of surface modification, coatings and nanocoatings such as thermal spray, electroplating, ion implantation, PVD, CVD, carburizing, nitriding, nitro-carburizing and laser techniques have also been elaborated for different types of wear and friction. Illustrations have been used wherever possible.

Acknowledgements

The authors sincerely thank Prof. B.N. Dwivedi, Department of Physics, Indian Institute of Technology (BHU), for critical reading and valuable suggestions.

References

Agarwal, R., Mohan, A., Mohan, S., Gautam, R.K., 2014. Synthesis and characterization of Al/Al_3Fe nanocomposite for tribological applications. ASME J. Tribol. 136, 012001-1–012001-9.

Alpas, A.T., Zhang, J., 1992. Wear rate transitions in cast aluminium – silicon alloys reinforced with SiC particles. Scr. Met. Matr. 26, 505–509.

Archard, J.F., Hirst, W., 1956. The wear of metals under unlubricated conditions. Proc. R. Soc. Lond. A. 236, 397–400.

Avallone, E.A., Baumeister, T., 1987. Standard Handbook for Mechanical Engineers, ninth ed. McGraw Hill, New York.

Barshilia, H.C., Deepthi, B., Rajam, K.S., 2010. Transition metal nitride-based nanolayered multilayer coatings and nanocomposite coatings as novel superhard materials. In: Zhang, S. (Ed.), Nanostructured Thin Films and Coatings Mechanical Properties. CRC Press, New York, pp. 427–480.

Bhushan, B., 1996. Tribology and Mechanics of Magnetic Storage Devices. second ed. Springer-Verlag, New York.

Bhushan, B., 1999. Principles and Applications of Tribology. John Wiley & Sons, New York.

Bhushan, B., Gupta, B.K., 1997. Handbook of Tribology-Materials, Coating and Surface Treatments. McGraw Hill, New York, Reprinted with corrections Malabar, FL: Krieger.

Blau, P.J., 1992. Friction and wear of ordered intermetallic alloys of NI_3Al. In: Henry, S.D. (Ed.), ASM Handbook: Friction Lubrication and Wear Technology, vol. 18. ASM International, Metals Park, OH, p. 27.

Blau, P.J., 1996. Friction Science and Technology. Marcel Dekker, New York.

Bowden, F.P., Tabor, D., 1950. Friction and Lubrication of Solids, Part I. Clarendon Press, Oxford.

Bowden, F.P., Tabor, D., 1964. Friction and Lubrication of Solids, Part II. Clarendon Press, Oxford.

Buckley, D.H., 1981. Surface Effects in Adhesion, Friction, Wear and Lubrication. Elsevier, Amsterdam.

Bunshah, R.F., 1992. PVD and CVD coatings. In: Henry, S.D. (Ed.), ASM Handbook: Friction Lubrication and Wear Technology, vol. 18. ASM International, Metals Park, OH.

Cooper, K.P., 1992. Laser surface processing. In: Henry, S.D. (Ed.), ASM Handbook: Friction Lubrication and Wear Technology, vol. 18. ASM International, Metals Park, OH, p. 861.

Dowson, D., 1979. History of Tribology. Longman Group, London.

Eliasson, J., Sandstrom, R., 1995. Applications of aluminium matrix composites. Key Eng. Mater. 104–107, 3–36.

Eyre, T.S., 1992. Wear and friction of cast irons. In: Henry, S.D. (Ed.), ASM Handbook: Friction Lubrication and Wear Technology, vol. 18. ASM International, Metals Park, OH, p. 695.

Fenske, G.R., 1992. Ion implantation. In: Henry, S.D. (Ed.), ASM Handbook: Friction Lubrication and Wear Technology, vol. 18. ASM International, Metals Park, OH, p. 850.

Gautam, R.K., Mohan, S., Mohan, A., 2013. Fabrication and characterization of Al-Al_2O_3 composites by mechanical alloying. Mater. Sci. Forum. 736, 81–97.

Glaeser, W.A., 1992. Friction and wear of carbon and alloy steels. In: Henry, S.D. (Ed.), ASM Handbook: Friction Lubrication and Wear Technology, vol. 18. ASM International, Metals Park, OH, p. 702.

Gusev, A.I., 1998. The effects of nanocrystalline state in solids. Uspekhi Fizicheskikh Nauk. 168, 55–83.

Han, D.S., Jones, H., Atkinson, H.V., 1993. The wettability of silicon carbide by liquid aluminium: the effect of free silicon in the carbide and of magnesium, silicon and copper alloy additions to the aluminium. J. Mater. Sci. 28, 2654–2658.

Hoffmann, F.T., Mayr, P., 1992. Nitriding and nitrocarburising. In: Henry, S.D. (Ed.), ASM Handbook: Friction Lubrication and Wear Technology, vol. 18. ASM International, Metals Park, OH, p. 878.

Hutching, I.M., 1992. Tribology: Friction and Wear of Engineering Materials. CRC Press, Boca Raton, FL.

Ide, G.L., Vanhumbeeck, J.B., 1987. Palladium and palladium alloy electroplating for contact applications. In: Romankiw, L.T., Turner, D.R. (Eds.), Proceedings of the Symposium on Electrodeposition Technology, Theory and Practice. The Electrochemical Society, Pennington, NJ, pp. 179–190.

Jiang, J., Stott, F.H., Stack, M.M., 1998. The role of tribo-particulates in dry sliding wear. Tribol. Int. 31, 245–256.

Kacar, A.S., Rana, F., Stefanescu, D.M., 1991. Kinetics of gas-to-liquid transfer of particles in metal matrix composites. Mater. Sci. Eng. A. 135, 95–100.

Karvánková, P., Manning, H.-D., Eggs, C., Veprek, S., 2001. Thermak stability of CCZrN–Ni and CrN–Ni super-hard nanocomposites coating. Surf. Coat. Technol. 146–147, 280–285.
Kingsbury, G.R., 1992. Friction and wear of sliding bearing materials. In: Henry, S.D. (Ed.), ASM Handbook: Friction Lubrication and Wear Technology, vol. 18. ASM International, Metals Park, OH, p. 741.
Kozlov, E.V., Zhdanov, A.N., Koneva, N.A., 2008. Deformation mechanisms and mechanical properties of nanocrystalline materials. Phys. Mesomech. 11, 42–50.
Kushner, B.A., Nowinski, E.R., 1992. Thermal spray coatings. In: Henry, S.D. (Ed.), ASM Handbook: Friction Lubrication and Wear Technology, vol. 18. ASM International, Metals Park, OH, p. 829.
Kustas, F.M., Fehrehnbacher, L.L., Komanduri, R., 1997. Nanocoatings on cutting tools for dry machining. CIRP Ann. Manuf. Tech. 46, 39–42.
Larsen-Basse, J., 1992. Basic theory of solid friction. In: Henry, S.D. (Ed.), ASM Handbook: Friction Lubrication and Wear Technology, vol. 18. ASM International, Metals Park, OH, p. 27.
Lloyd, D.J., 1994. Particle reinforced aluminium and magnesium matrix composites. Int. Mater. Rev. 39, 1–23.
Lubrication Engineering Working Group, 1966. Dept. of Ed. and Sci., Lubrication (Tribology). HMSO, London.
Luce, B.M., Foulke, D.G., 1974. Silver. In: Lowenheim, F.A. (Ed.), Modern Electroplating, third ed. John Wiley & Sons, New York, pp. 358–376.
Ludema, K.C., 1987. Friction – a study in the prevention of seizure. ASTM Standard News. May, 54–58.
Magee, J.H., 1992. Wear of stainless steels. In: Henry, S.D. (Ed.), ASM Handbook: Friction Lubrication and Wear Technology. ASM Handbook, vol. 18. ASM International, Metals Park, OH, p. 710.
Maloney III, J.L., 1992. Friction and wear of tool steels. In: Henry, S.D. (Ed.), ASM Handbook: Friction Lubrication and Wear Technology, vol. 18. ASM International Metals Park, Metals Park, OH, p. 734.
Matthes, B., Broszeit, E., Kloos, K.H., 1990. Fundamental and tribological properties of sputtered TiN coatings in plastic manufacturing model wear tests. Surf. Coat. Tech. 43–44, 688–698.
Mohan, A., Mohan, S., 2008. Unpublished work.
Mohan, S., Srivastava, S., 2006. Surface behaviour of as-Cast Al–Fe intermetallic composite. Tribol. Lett. 22, 45–51.
Mohan, S., Agarwala, V., Ray, S., 1990. The effect of lead content on the wear characteristics of a stir-cast aluminium–lead alloy. Wear. 140, 83–92.
Mohan, S., Agarwala, V., Ray, S., 1991. Liquid–liquid dispersion for fabrication of Al–Pb metal–metal composites. Mater. Sci. Eng. A. 144, 215–219.
Mohan, S., Agarwala, V., Ray, S., 1992. Friction characteristics of stir-cast Al–Pb alloys. Wear. 157, 9–17.
Mohan, S., Prakash, V., Pathak, J.P., 2002. Wear characteristics of HSLA steel. Wear 252, 16–25.
Moyer, C.A., 1992. Friction and wear of bearing steels. In: Henry, S.D. (Ed.), ASM Handbook: Friction Lubrication and Wear Technology, vol. 18. ASM International Metals Park, Metals Park, OH, p. 725.
Neale, M.J., 1973. Tribology Handbook. Newnes-Butterworths, London.

Novinski, E.R., 1981. Application of thermal spray processes to conserve critical materials, Workshop on conservation and substitution technology for critical materials, conference proceedings, Vanderbilt University, USA.
Pathak, J.P., Mohan, S., 1998. Abrasive wear characteristics of as-cast aluminium silicon alloys under dry sliding against corundum disc. Indian Foundry J. 44, TP-5-10.
Peterson, M.B., Winer, W.O. (Eds.), 1980. Wear Control Handbook. ASME, New York.
Peterson, M.B., Florek, J.J., Lee, R.E., 1960. Sliding characteristics of metals at high temperatures. Trans. ASLE. 3, 101–115.
Prengel, H.G., Pfouts, W.R., Santhanam, A.T., 1998. State of art in hard coatings for carbide cutting tools. Surf. Coat. Tech. 102, 183–190.
Rabinowicz, E., 1995. Friction and Wear of Materials. second ed. Wiley, New York.
Rigney, D.A., 1988. Sliding wear of metals. Ann. Rev. Mat. Sci. 18, 141–163.
Rigney, D.A., Chen, L.H., Naylor, M.G.S., Rosenfield, A.R., 1984. Wear processes in sliding wear systems. Wear 100, 195–219.
Shabel, B.S., Granger, D.A., Truckner, W.G., 1998. Friction and wear of aluminium–silicon alloys. In: Henry, S.D. (Ed.), ASM Handbook: Friction, Lubrication and Wear Technology, vol. 18. ASM International, Materials Park, OH.
Stickels, C.A., 1992. Carburising. In: Henry, S.D. (Ed.), ASM Handbook: Friction Lubrication and Wear Technology, vol. 18. ASM International Metals Park, Metals Park, OH, p. 873.
Stott, F.H., 2002. High-temperature sliding wear of metals. Tribol. Int. 35, 489–495.
Stott, F.H., Lin, D.S., Wood, G.C., 1973. The structure and mechanism of formation of the 'glaze' oxide layers produced on nickel-based alloys during wear at high temperatures. Corros. Sci. 13, 449–471.
Stott, F.H., Glascott, J., Wood, G.C., 1985. Models for the generation of oxides during sliding wear. Proc. R. Soc. Lond. A. 402, 167–186.
Sue, J.A., Troue, H.H., 1990. Friction and wear properties of TiN coating in sliding contact with AISI 01 steel. Surf. Coat. Tech. 43–44, 709–720.
Suh, N.P., 1986. Tribophysics. Prentice-Hall, Upper Saddle River, NJ.
Surappa, M.K., Prasad, S.C., Rohatgi, P.K., 1982. Wear and abrasion of cast Al-alumina particle composites. Wear 77, 295–312.
Szeri, A.Z., 1998. Fluid Film Lubrication – Theory and Design. Cambridge University Press, Cambridge.
Takadoum, J., Bennani, H.H., 1997. Influence of substrate roughness and coating thickness on adhesion, friction and wear of TiN films. Surf. Coat. Tech. 96, 272–282.
Veprek, S., 1999. The search for novel, superhard materials. J. Vac. Sci. Technol. A. 17, 2401–2420.
Wang, H., Zhang, S., 2010. Toughness and toughening of hard nanocomposite coatings. In: Zhang, S. (Ed.), Nanostructured Thin Films and Coatings Mechanical Properties. CRC Press, New York, pp. 99–145.
Weil, R., Sheppard, K., 1992. Electroplated coatings. In: Henry, S.D. (Ed.), ASM Handbook: Friction Lubrication and Wear Technology, vol. 18. ASM International Metals Park, Metals Park, OH, p. 834.
Whitehead, J.R., 1950. Surface deformation and friction of metals at light loads. Proc. R. Soc. Lond. A. 201, 109–124.

An investigation into the tribological property of coatings on micro- and nanoscale

2

Xuesong Han
School of Mechanical Engineering, Tianjin University, China

2.1 Drivers of studying the origin of tribology behavior

Many objects in our daily lives, including the house in which we live and the materials we use (e.g., toothbrushes, pots and pans, refrigerators, televisions, computers, cars, furniture) all come under the "umbrella" of coated materials. Likewise, fields such as military applications – for example, vehicles, artilleries, and invisible radars – and aerospace products – for example, aircraft, satellites, and solar panels – all involve the widespread use of coated materials. Recently, industry made an increasing use of surface treatments and coatings on friction and wear control (Holmberg and Matthews, 1998; Halling, 1986; Achanta et al., 2005; Ahmed et al., 2003; Baek and Khonsari, 2005; Bandorf et al., 2003; Bhushan and Gupta, 1991; Caron et al., 2006; Cavaleiro and Hosson, 2006; Batista et al., 2002). This has led to, and to some extent been fuelled by, the growth of a new discipline called surface engineering. The surface has a dominant influence on lifetime cost and performance, including machinery maintainability. The surface may also have other functionally important attributes, not confined to mechanical or chemical properties, such as thermal, electronic, magnetic and optical characteristics that influence the choice of surface material. The retention of these physical surface properties is clearly essential throughout the life of the product. It is a further reason why surface durability enhancement by appropriate coating selection is critical to any product's effectiveness, and therefore its saleability. Like all products, mechanical components and tools are today facing higher performance requirements (Table 2.1). Developing a fundamental understanding of the behavior of surfaces in contact will provide a theoretical basis for the prediction of the desirable attributes of surfaces.

Tribology behavior such as friction, wear and lubrication has already been considered as one of the most important issues in the science and technology. In reality, the study of tribology has already transformed from qualitative research to quantitative research, which means tribology has already become a mature discipline. While many researchers have written excellent books on tribology, most of the efforts have focused on analyzing the macroscopic aspects, with only slight attention paid to the rich interplay going on between the atoms and molecules at the contacting surfaces, as these have historically been poorly understood. Friction is one of the oldest

Anti-Abrasive Nanocoatings.

Table 2.1 New challenges

New coating and treat methods with novel characteristics and tribochemical property
The recognition of the importance of lifetime cost and performance
Higher performance requirements
Different requirement on surface and bulk materials
Mass magnetic storage system
Microelectromechanical system
Micro- and nanomachining or micro- and nanomanipulation

subjects of research and the definition of friction is the resistance to the relative motion or the tendency of relative motion between two solid bodies where a loading force acts perpendicular to the contacting surface. The resistive force which is parallel to the direction of motion is called friction force. When two solid bodies slide against each other, the friction force opposing to this motion is generated at the interface. The tangential force necessary for initiate sliding is called static friction force while the tangential force required to keep sliding is called kinetic friction force. In reality, two basic laws (*Da Vinci–Amonton* law) are obeyed in many situations, which are listed as follows:

1. The first law of friction: The first law which is also referred to as *Amonton* laws states that the friction force is proportional to the normal load.
2. The second law of friction: The second law states that the friction is independent of the apparent area of the contact. The first law of friction can be expressed as the following:

$$F = \mu \times N \quad (2.1)$$

here μ is the friction coefficient between the two materials that make up the friction pair. This expression also implies the second friction law, that friction is independent of the apparent area of contact. However, experiments under the conditions of a small contact area have shown that these empirical laws of friction do not always hold. Reasons may be related to the large surface-to-volume ratio and the greater importance of adhesion, surface structure and surface chemistry at the molecular/atomic scale. The origin of tribological phenomena has an intense relationship with many cutting-edge industrial applications and this has also focused more attention on this area.

Magnetic recording is an important technology in IT and is expected to grow by a factor of five or more in the next decade. It is accomplished by relative motion between the magnetic head and magnetic medium. For high-performance magnetic storage devices, physical contact between the head and the tape during start–stop operations in tape drives can result in high friction, and the tape motion may not occur when the tape-drive motor is started. It is necessary to optimize geometry and materials for various tape-path components in order to minimize the friction in the tape path. High friction will affect the acceleration during starts and stops or cause instability in the head slider, which will lead to head damage and slider buckling deformation. High friction generally results in excessive local heating, high wear of the interface materials and excessive power requirements. High static friction at the head–disk interface has already been a serious problem for the computer disk drive industry for several years. Investigation into the origin of friction at atomic scale can improve the fundamental understanding of tribology and provide theoretical guidance for minimizing and controlling friction at the head–disk interface.

Fabrication of micro/nanosized complex structures and devices has been made possible thanks to the advancement in micro- and nanofabrication technology such as lithographic techniques. Micro/nanoelectromechanical systems (MEMS/NEMS) form an integral part of the microelectronic industry. MEMS/NEMS devices form a group of products ranging in size from a micrometer to a centimeter that combine mechanical and electrical structures. On miniaturizing any device or system it is critical to have a good understanding of the scaling properties of the overall design, materials, and the fabrication processes involved. The scaling properties related to any one of these components could present a formidable barrier to achieving adequate performance or economic feasibility. Tribological issues such as stiction, friction, and wear are of prime concern around normal function realization of microscale components. A major challenge for MEMS designers is to overcome the effects of stiction, which is where microscopic structures tend to adhere to each other when they come into contact. With the increasing complexity of devices, an in-depth investigation into this phenomenon is essential to counteract it. Stiction is also a leading cause of failure in many MEMS/NEMS applications, including accelerometers used in air-bag devices in automobiles. The friction force is still the limiting factor to a successful operation and can cause the parts in relative motion to each other in MEMS to be unreliable. As the devices range in micrometers, the coatings or the protective layers on silicon should be only a few monolayers or nanometers thick. Fundamental investigation of tribological mechanisms in such materials will play an important role in ensuring a normal useful-life and function realization of MEMS/NEMS.

While macroscopic investigations into tribology could have been carried out for centuries, investigations targeting nanoscale frictional behavior have only be possible over the last 20–30 years as a result of the introduction of new experimental tools such as scanning probe microscopy (SPM). Although the initial objective of SPM is to improve people's ability to observe micro- and nanoscale phenomenon, it has already been extended from a pure observation tool to a micro- and nanoscale manipulation platform which is used for atom/molecular movement, micro- and nanoscale structure assembly, micro- and nanoscale machining, and so on. Classical physical principles such as *Newtonian mechanics* are not suitable for the phenomenon in this area, while quantum mechanics and surface force will dominate this process. Therefore, tribology behavior related to the above mechanical dynamic process exhibits special features different from that on the macroscopic scale. A lack of a basic understanding of the tribology process makes it difficult to accurately control operation parameters and the adversely affects the repeatability of experimental results. Different control mechanisms dominate the tribology behavior related to SPM as the working conditions fall into the micrometer or nanometer scale.

An understanding of the origin of novel tribology phenomenon happening at this scale is of great value to science and its application. Experiments with small contacts have shown that these empirical laws of friction do not always hold. This may be related to the large surface-to-volume ratio and the greater importance of adhesion, surface structure and surface chemistry. Therefore, it is necessary to gain a better understanding of the origin of friction to learn how to quantify and eventually control it. In fact, the study of the origin of tribology comes from two requirement. The first is the character of the fundamental scientific research – exploring the ultimate principle of the natural phenomenon; the second is that the present theory cannot give a reasonable explanation of tribology-related phenomena in microscopic devices/manipulation/machining. On a macroscopic scale, classical laws of tribology can give a reasonable explanation and the details of surface morphology can be evaluated using statistical average. Variations in microscopic-scale atomic interaction force can often be omitted comparing with the magnitude of the force leading to the elastic/plastic deformation or stress in the bulk materials. However, the variation of force cannot be omitted as the size of microscopic details is approximately equal to the size of the peak (the total contact area at microscopic scale). Therefore, the control mechanism may be quite different between different scales. Furthermore, the number of defects in the local small area are very few, which makes the strength of surface/subsurface higher than that of the general state and thus influences the tribology behavior at the interface.

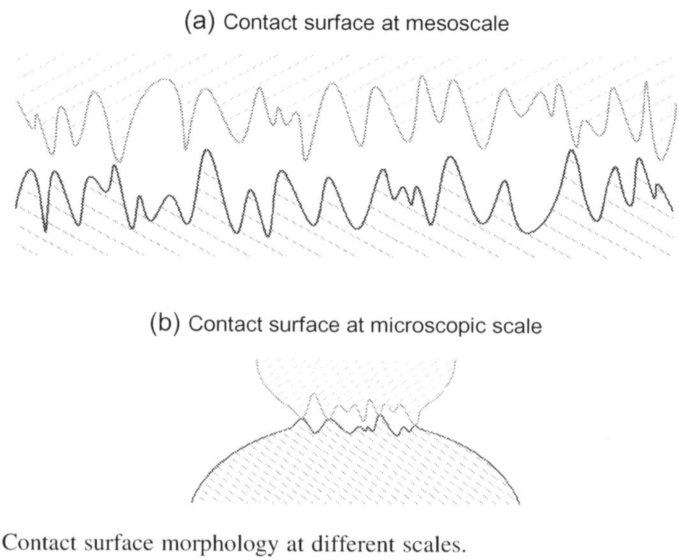

Figure 2.1 Contact surface morphology at different scales.

At present, hard coatings such as titanium nitride, titanium carbide and aluminium oxide are commonly used on cutting tools in the manufacturing industry. Chromium nitride and molybdenum disulphide coatings are used on metal forming tools.

Very hard but also low-friction diamond-like carbon coatings are deposited for wear protection on magnetic storage devices produced for computers. Optical lenses are produced with hard erosion-resistive thin transparent coatings. Various carbon-based coatings are used on components in the automotive industry to reduce energy consumption. The coatings in some applications are deposited as multicomponent coatings, multilayers, gradient layers, superlattice structures and duplex-treated surfaces with various material combinations. While coatings for the applications cited are in commercial use, there are many others that are still at a developmental stage. Coatings are widely used to control friction and wear in all kinds of contacts. Many excellent results in industrial sliding applications (Suh, 1986; Arnell, 1990) have been achieved. In the most successful solutions the coefficient of friction has been reduced by more than one order of magnitude, e.g. by coating a steel surface, which slides against a steel counterface, with a thin molybdenum disulphide or diamond-like carbon coating; also the wear has been reduced by several orders of magnitude. The results have mainly been achieved by trial and error concerning both the choice of material and the tribological coating parameters, such as coating thickness, hardness and surface roughness. This is due to the lack of a general theoretical concept that can satisfactorily describe the prevailing friction and wear mechanisms in tribological processes with coated surfaces involved.

In summary, the origin of tribology behavior is of great value in developing a basic understanding of the interfacial phenomena at a small scale and also helpful in understanding and optimizing working conditions in nanocoating-related technology such as magnetic storage systems, MEMS, micro- and nanomachining or micro- and nanomanipulation, and so on. The components used in micro- and nanoscale applications are very light and operate under very small loads. As a result, tribology behaviors on nanometer scales with lightly loaded components are highly dependent on the surface interactions. In this chapter, the author will present some example studies of contact, friction, adhesion, scratching, wear and indentation and after that put forward some fundamentals of tribology.

2.2 Contact at nanometer scale

Theoretically speaking, general contact happens when two surfaces are placed together while restricted contact means two surfaces placed together with definite external pressure exerted along their common normal direction. As for tribology-related problems, the restricted contact is the focus, which is also the key issue in the origin of tribology behavior. Therefore, all of the contact problems within tribology are induced under an external load, which can also be called indentation contact. When two nominally flat surfaces are placed in contact, the real contact occurs in only a small part of the surface area; the sum of the areas of all the contact spots makes up the true area of contact. Investigation into the variation of contact area and elastic/plastic deformation occurring at discrete contact spots will play an important role in determining the origin of tribology behavior and evaluate the micromechanical property of nanomaterials. Theoretically speaking, the force exerted upon a small contact area can be assumed as a case of stress concentration which can affect many aspects of friction. A very high contact stress is often generated at the contact spot because real contact is occurring in only a small part of the surface area. High contact stress and sliding velocity together will cause high flash temperature and result in the modification of the physical properties of the surface layer and formation of cold-welding. The subsurface layer will also be subjected to high strains due to mechanical and thermal action which cause cracks and growth and finally result in a surface or subsurface fracture.

Indentation tests are perhaps the most commonly applied means of testing the mechanical properties of materials and the contact property at the nanometer scale. In such a test, a hard tip, typically a diamond, is pressed into the sample with a known load. After some time, the load is removed. The area of the residual indentation in the sample is measured and the hardness, H, is defined as the maximum load, F_{max}, divided by the residual indentation area, A_r, or $H = F_{max}/A_r$. Thin film is one of the fast growing sciences while one of its dimensions lies in the nanometer scale. Thin films often support very high stresses during service, which can lead to the distortion of the devices containing the film. Although many materials are selected for their electrical, magnetic or optical properties, it is often the the mechanical properties that limit the fabrication and reliability of potential advanced devices. Therefore, knowledge of the mechanical properties of thin films is critical for the design, fabrication and application of advanced devices. The principal goal of such mechanical testing is to obtain values for elastic modulus and hardness of the specimen material from experimental readings of indenter load and depth of penetration. The forces involved are usually in the millinewton range and are measured with a resolution of a few nanonewtons. The depths of penetration are in the order of nanometers, hence the term "nanoindentation." Investigation of the nanoindentation can provide the knowledge of surface contact nanomechanics required in the nanoscale wear processes, such as those occurring in materials manufacturing processes, nanofabrication technologies based on atomic force microscope (AFM), and nanoimprint lithography and MEMS. Nanoindentation has proven to be the simplest and most direct way to determine the mechanical properties of thin films. The idea of nanoindentation arose from the realization that an indentation test is an excellent way to measure the properties of very small volumes of materials. In principle, if a very sharp tip is used, the contact area between the sample and the tip, and thus the volume of material that is tested, can be made arbitrarily small. The only problem is determining the indentation area. It is easy to make an indentation that is so small but it is difficult to see it without a powerful microscope.

Experimental procedure is appropriate for investigating the mechanical properties of bulk materials in the larger length scale. However, microhardness testing instruments available cannot apply low enough forces to give penetration depths less than the required 10% of the film thickness so as to avoid influence on the hardness measurement from the presence of the substrate. Even if they could, the resulting size of the residual impression cannot be determined with sufficient accuracy to be useful. For example, the uncertainty in a measurement of a 5 μm diagonal of a residual impression made by a Vickers indenter is in the order of 20 percent when using an optical method and increases with decreasing size of indentation, and can be as high as 100 percent for a 1 μm impression. Therefore, determining the intrinsic properties of the films can be difficult because of the influence of the substrate on measured properties.

Since the spatial dimensions of the contact area are not conveniently measured, modern nanoindentation techniques typically use the measured depth of penetration of the indenter and the known geometry of the indenter to determine the contact area. For such a measurement to be made, the depth measurement system needs to be referenced to the specimen surface, and this is usually done by bringing the indenter into contact with the surface with a very small "initial contact force," which, in turn, results in an inevitable initial penetration of the surface by the indenter that must be accounted for in the analysis. Additional corrections are required to account for irregularities in the shape of the indenter, deflection of the loading frame, and piling-up of material around the indenter. These effects contribute to errors in the recorded depths and, subsequently, the hardness and modulus determinations. On the other hand, we need to recognize that material properties measured by nanoindentation are generally evaluated according to contact mechanics of isotropic elastic bodies. For instance, if the material under investigation is indented by a spherical indenter, *Hertzian* elastic solution (Hertz, 1992) is always applied to evaluate its elastic properties. This approach is widely used and yields reasonable results in certain cases. However, the assumption of isotropy may lead to critical errors in the case of metals and materials with strongly anisotropic structures. Furthermore, the scale of deformation in a nanoindentation test becomes comparable to the size of material defects such as dislocations and grain sizes, and the continuum approximation used in the analysis can become less valid. There is no doubt that as the scale of mechanisms becomes smaller, the nature of surface forces and adhesion will continue to increase. Therefore, a powerful theoretical toolkit which can describe the microscopic world effectively is needed to investigate the physical property of thin film at small scales.

Molecular dynamics (MD) simulation, by virtue of its high temporal and spatial resolution, can offer an ideal approach for gaining insights into atomic-scale processes and understand their mechanisms. A remarkable enhancement in computational capability (computer hardware) and high performance computation techniques (parallel computation) have enabled us to employ a large-scale classical MD method to investigate the nanoindentation process and gain insights into this atomistic behavior. This section aims to verify the effects of size scales on the physical property of nanoscale materials (copper coatings) and the control mechanism of contact using a high-precision numerical simulation method. The observed numerical experimental trends exhibit a strong size dependence, which cannot be rationalized on the basis of continuum concepts. Furthermore, the mechanism of elastic-plastic behavior and anisotropy properties is investigated accordingly. By correlating observations from computations, some physical mechanisms responsible for these are discussed.

2.2.1 Methodology

The atomic configuration of the studied system is illustrated in Figure 2.2. The diamond indenter is assumed as an ideal rigid body without any deformation. Copper atoms of the workpiece are initially arranged in fcc structure with a constant lattice parameter of 3.61 Å. The dimension of copper workpiece is $100 \times 80 \times 100$ Å^3 along x, y, and z directions. The moving speed is 30 m/s, the environment temperature is 293 K. The indenter radius is 15 Å, the indent depth is 10 Å.

The embedded-atom method (EAM) potential (Daw and Baskes, 1984) is used to depict the interaction between the copper atoms as follows:

$$E_i = F_\alpha\left(\sum_{j\neq i}\rho_\alpha(r_{ij})\right) + \frac{1}{2}\sum_{j\neq i}\phi_{\alpha\beta}(r_{ij}) \tag{2.2}$$

where F is the embedding energy which is a function of the atomic electron density, ϕ is a pair potential interaction, α and β are the element types of atoms i and j. With Eq. (2.2), the interaction force between copper atoms can be obtained by calculating the negative gradient of E_i. The interaction between the copper atom and the diamond atom is modeled by the Morse Potential (Morse, 1929) as follows:

$$\phi_{ij} = D[e^{-2\alpha(r_{ij}-r_0)} - 2e^{-\alpha(r_{ij}-r_0)}] \tag{2.3}$$

The interaction force between copper atom and diamond atom is calculated by the negative gradient of ϕ.

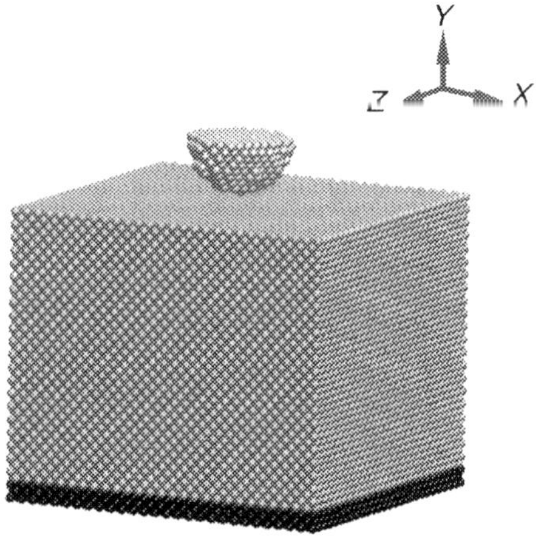

Figure 2.2 MD simulation.

The Velocity-Verlet method is a symplectic algorithm which can prevent energy dissipation and has high computational efficiency. This research adopted this method as follows:

$$r_i^{n+1} = r_i^n + hv^n + \frac{h^2}{2m}F_i^n \tag{2.4}$$

$$v_i^{n+1} = v_i^n + \frac{h(F_i^n + F_i^{n+1})}{2m} \tag{2.5}$$

where r_i^{n+1}, v_i^{n+1}, and F_i^{n+1} are position, velocity and force at the $n+1$ step of the ith atom while r_i^n, v_i^n, and F_i^n are position, velocity and force at the n step of the ith atom, h is timestep, and m is mass of atom.

2.2.2 Results and discussion

Figures 2.3–2.7 give the numerical simulation results of the indentation of single crystal copper film. Figure 2.3 indicates the variation tendency of substrate potential energy over the complete nanoindentation cycle. In this nanoscale, material behavior is mainly controlled by surface-driving effects rather than by bulk effects. Surface contact of micromechanical structures can contribute to stiction (adhesive contact), damage, pitting, and surface hardening over the local contact area. The presence of stiction is due to high compliance and large surface-to-volume ratio of micromechanical structures which do not have enough restoring force to overcome the surface interaction after mechanical contact. Stiction can be classified into two

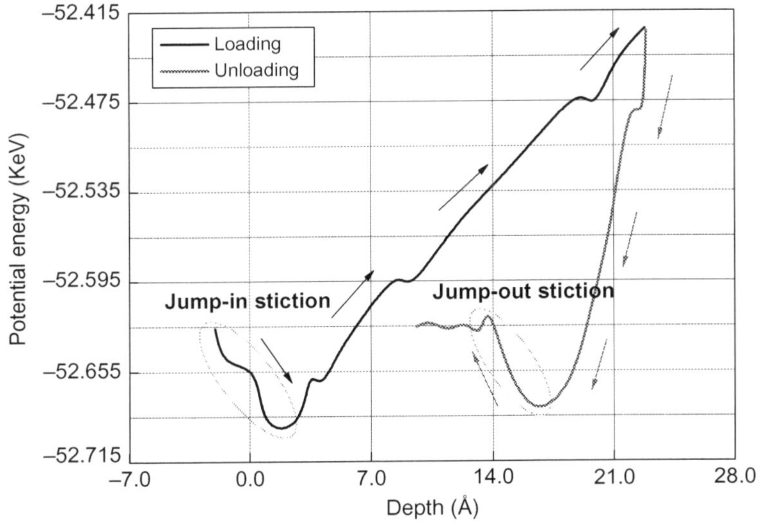

Figure 2.3 Variation of substrate.

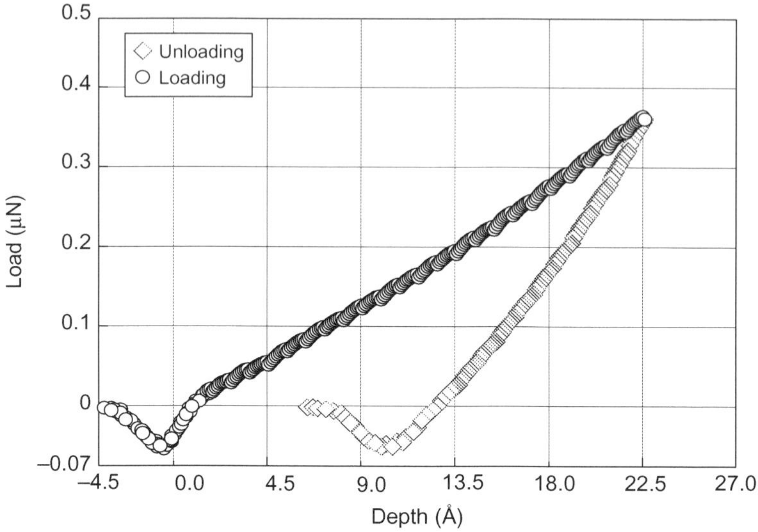

Figure 2.4 The relationship between load and displacement.

Figure 2.5 Layer penetration.

categories: jump-in stiction and jump-out stiction. Under each of the two stiction conditions the system potential energy tends to evolve in opposite ways. The potential energy gradually decreases in the jump-in stage as the indenter is having a negative effect on the copper film while the it is having a positive effect on the copper film in the jump-out stage, which results in an increase in system potential energy. Figure 2.4 indicates the relationship between the load, that is, the force experienced by the indenter, and the indentation depth over the complete nanoindentation cycle. As the rigid indenter approaches substrate, surface atoms

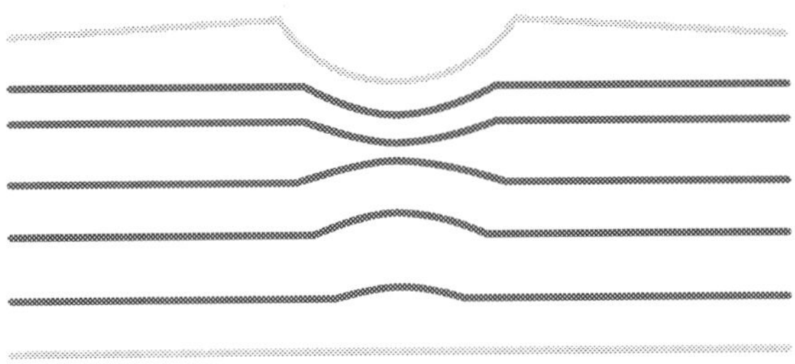

Figure 2.6 Layer deformation.

Figure 2.7 Number of atoms with specified nearest number of neighbors.

first jump into contact with the tip and the interfacial force is attractive. The adhesion force reaches its maximal value after the indenter moves downward 4.157 Å (approximately 0.0518 μN). The length that the chemical bond is kept constant indicates that the adhesion-contact status is induced at the interface. The contact load increases quickly after the indenter penetrates the surface of copper film. The force response during loading exhibits some fluctuations as a result of the inducing of complex dislocation in the materials. Another notable feature is the decaying of the load with time at maximum displacement, which causes the slope of the load-displacement curve to diverge there. This phenomenon justifies the elastic-plastic deformation property of copper film in the nanometer scale and there must be plastic deformation energy stored in the copper film. The surface atoms gradually lump upward as the system temperature increases from 293 K to 427 K. After that, the temperature of the contact area gradually drops to 300 K as the heat dissipates away by means of random colliding. In the nanometer or subnanometer scale, it is the surface force that dominates the material's physical properties, which makes their micromechanical property different from the corresponding macromechanical property. During the unloading process, there are still attractive forces after the indenter is retracted from the deformed copper thin film. The transformation of control mechanisms in this small length scale (such as surface force and thermal fluctuating) may exert great effect upon materials' microhardness.

The hardness of the single crystal copper along a (010) plane calculated from MD simulation is approximately 12.3 GPa while the corresponding hardness calculated from finite element simulation is approximately 1.483 GPa. This result indicates that a size effect exists in the material's mechanical properties, namely, the nanohardness of copper film is greater than the corresponding macrohardness.

This is also the advantage of using nanocoatings as an anti-wear protective layer. Furthermore, the nanohardness of copper film gradually increases as the indentation load increases. This phenomenon illustrates that there exists indentation hardening behavior in the thin film materials. There may be several factors which can account for this phenomenon. First, the probability of encountering a stress-reducing defect decreases as the size of the material decreases. Second, the elastic recovery after the removal of the indenter will lead to the decrease in the residual indentation size, which will lead to the increase of the computation material's hardness. Third, smaller samples are expected to have less free volume than the bulk materials, therefore if the volumes are small enough, shear band propagation or dislocation movement, which is a primary mechanism of plasticity in the metal, might not even be possible. Plastic deformation confined within a very small volume underneath the indenter may result in the creation of non-uniform stresses and strains within the sample, thereby setting up strong gradients of strain, which are also thought to be one of the factors responsible for the larger material's nanohardness.

The deformation condition inside the film is shown in Figures 2.5 and 2.6. In Figure 2.5, every other layer of the substrate materials are revealed for clarity. First, the top layer is elongated and gradually results in plastic fracture. After that, part of the atoms belonging to this layer are deposited on the adjacent layer. The workpiece surface gradually sinks when the indenter keeps feeding and results in the downward protuberance of the atomic layer. This trend terminates at somewhere beneath the material surface (shown in Figure 2.6). The protuberant behavior in the opposite direction beneath the surface may be the effect of stiction between the adjacent layers. This abnormal phenomenon should have a close relationship with the length scale. The different deformation behavior of crystal layers inside the films may be the potential key factor of its breaking off from the substrate. This phenomenon should be some kind of collective behavior, namely, all of the atoms in the same layer, but not any discrete atom behavior. It is claimed by many researchers (Izabela et al., 2005; Elmustafa et al., 2004; Wang et al., 2005) that the ductile material surface will lump upward when the indenter approaches the workpiece and thus influences the depth of indentation. The animation movie of the MD simulation shows that there still exists stiction effect beneath the material surface in the nanoindentation process. Compared with the macro-indentation process, this phenomenon should be directly affected by the size effect and have a great effect on the material's micromechanical property.

The atom structure at the local indentation region is shown in Figure 2.7. In this figure, only the local deformation substrate atoms are revealed. The perfectly coordinated atoms ($N = 12$) far away from the indentation region have been removed from the snapshots for clarity, such that only surface and imperfectly coordinated atoms are visible. We notice that in the case of the indentation, huge pressure is being exerted on the local contact area and results in the increasing of the atom coordinate numbers (13 and 14). Most of the atoms in the normal region hold 12 neighbors while this value gradually increases to 14 at the symmetrical center of the workpiece (intense deformation area). This result shows that the material structure in this region has undergone solid-state amorphization and the film hardening just originates from the rearrangement of the crystal structure at the atomic level.

There are basically two factors which can induce the anisotropy phenomenon: the indenter and the substrate. The spherical indenter is adopted to eliminate the influence of the indenter geometry. Therefore, this anisotropy phenomenon should be entirely induced by the physical property of the substrate and belongs to planar mechanical behavior, as the thickness of the film can be ignored compared with the other two dimensions of the film. As for substrate, there are two kinds of anisotropy: structure anisotropy and texture anisotropy. The anisotropy deformation in the thin film belongs to structure anisotropy, as the atomic arrangement in the crystal materials has its definite order. As shown in the simulations, material from the damaged region is essentially noncontiguous, irregularly shaped lumps at several direction illustrated in Figure 2.8. Initially, the contact between two bodies belongs to the elastic contact and the deformation is identical in every direction. Gradually, four sets of separated parts ("handles") are generated in the in-plane (101) directions. This result shows that strong anisotropy deformation exists in the copper film. The anisotropy deformation has nothing to do with the material's inhomogeneities, as the structure and the property are identical at any point of crystal interior. The anisotropy also has strict preferred orientation distribution. First, this anisotropy deformation of surface is symmetrically distributed about the (100) and (001) plane. Second, there is more intense plastic deformation generated in the in-plane [101] directions than in any other direction. Finally, the anisotropy deformation gradually minimizes alongside the increase in plastic deformation.

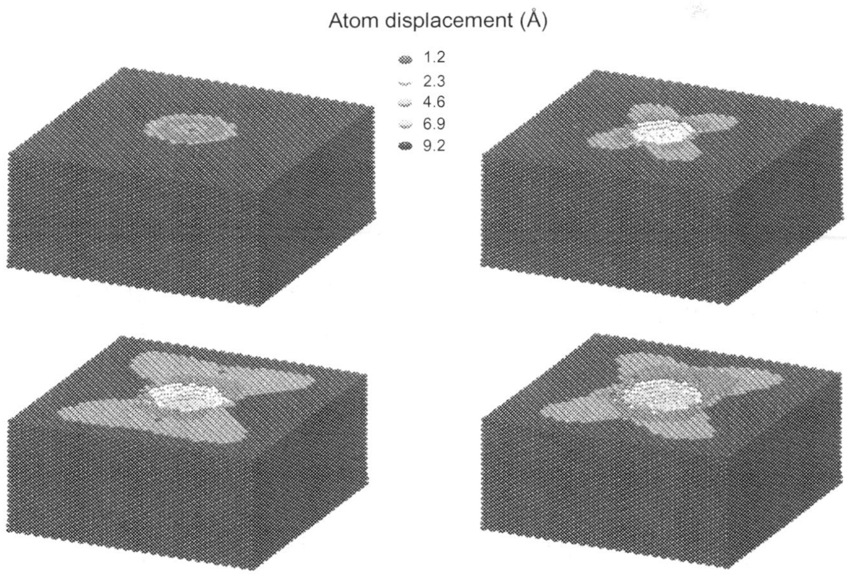

Figure 2.8 The displacement of surface atoms.

As the copper is a typical elastic-plastic material, there is only local surface lumping instead of the materials pileup that occurs with the indentation of a rigid-plastic workpiece. The elasticity of single crystal copper plays an important part in the plastic nanoindentation process. When the yield point is first exceeded the plastic zone is small and fully contained by material remains elastic so that the plastic strains are almost of the same order of magnitude as the surrounding elastic strains. In these circumstances the material displaced by the indenter is accommodated by an elastic expansion of the surrounding solid. As the indentation becomes more severe, either by increasing the load on a curved indenter or by using a more acute-angled wedge or cone, an increasing pressure is required beneath the indenter to produce the necessary expansion. Eventually the plastic zone breaks out to the free surface and the displaced material is free to escape by plastic flow to the sides of the indenter and results in the materials lump (Figure 2.8).

2.2.3 Conclusion

Nanoindentation has proven to be the simplest and most direct way to determine the mechanical properties and the corresponding contact property at the nanometer scale. Numerical simulation of nanoindentation has been carried out to investigate the nature of contact in the thin materials at the atomic scale. MD simulation results in this work shows that there still exists a similar stiction phenomenon in bulk materials in the elastic-plastic contact process. The protuberant behavior in the opposite direction beneath the surface may be the effect of stiction between the adjacent layers. The different deformation behavior of crystal layers inside the film may be the potential key factor to its breaking off from the substrate. The potential energy gradually decreases in the jump-in stage as the indenter has a negative effect on the copper film while the indenter has a positive effect on the copper film in the jump-out stage and results in an increasing in system potential energy. There is a little anisotropy phenomenon in the elastic contact stage while there is obvious anisotropy phenomenon induced in the plastic contact process. The results of this work show that classical Hertzian contact theory (Hertz, 1992) is not fit for the investigation of plastic deformation at the nanometer scale where the isotropy assumption is not available.

Currently, the study of the elastic properties of nano-objects is very challenging both from an industrial and a theoretical point of view. From the industrial side, investigating the mechanical properties of nano-objects is critical because of the lack of reliable methods to quantitatively measure and evaluate the anti-wear property of the coatings at the nanoscale. The problems are related to spatial and force resolution, instrument calibration as well as not well defined surface shape and chemistry (and at this scale each atom makes a difference). From the theoretical side, developing a theory of contact at the nanoscale is an intriguing challenge, which lies at the cross-over between the atomic level and the continuum.

2.3 Atomic friction with zero separation

Strictly speaking, phenomena involved in the friction process are far from well understood. When two materials are pressed against each other, initial solid–solid contact occurs at the surface high points. This area where the asperities touch is usually an extremely small fraction of the total area covered by the surfaces, but the forces generated between the contacting atoms in this small area are responsible for most tribological phenomena-friction, for example, wear and adhesion. Consequently, understanding how forces acting on asperities distort the material around the points of contact provides an important basis for understanding tribology. Until now, the most classical laws of sliding friction were put forward by Amonton who states that the friction force is proportional to the normal load at the macroscopic level. This enables the coefficient of friction μ to be constant and depend upon the materials of the friction pair. While macroscopic friction involves interactions between numerous asperities of the two contacting surfaces, the friction force has an intense relationship with the surface roughness of the friction pair and is dominated by many mechanisms such as interlocking friction, plough friction, and so on. As the parts become microsized, the molecular adhesion forces begin to dominate over gravity and inertia forces. So, while loading forces and bulk hardness may be the major factors determining friction and wear of macrosized objects running in dry contact, they are less important for minute objects where molecular adhesion forces are comparable to loading forces. The laws of nanometer scale friction differ drastically from those of macroscopic scale friction. The unlubricated sliding is important in many mechanical devices covering a wide range of sliding velocities. However, the effect of sliding velocity on the tribological behavior of unlubricated materials has not been widely studied. In particular, it has been known from the time of Coulomb that the force of friction between two macroscopic bodies in contact is independent of their relative velocity. In contrast, friction force on the nanoscale exhibits a nontrivial velocity dependence which will be the subject of the present contribution.

While numerous researchers have pointed out that adhesion occurs between two materials when they are brought into contact and that these interatomic forces should contribute to sliding friction, it was Bowden and Tabor who in the 1930s and 1940s developed a model to describe the connection between adhesive forces and sliding friction forces and performed experiments to verify these concepts (Bowden and Tabor, 1950). The attractive and repulsive interatomic forces between the atoms are generated when the surface atoms are in intimate contact. As tangential force is applied to slide one object over the other, shear stresses develop over the junction interfaces to resist this force. At low shear stresses, the interatomic forces between the atoms are strong enough to prevent the atoms from sliding over each other, and the material around the contact junctions deforms elastically to counter the applied tangential force. At some critical shear stress, the applied force acting on the individual atoms exceeds the sum of all the interatomic forces that are trying to keep the atoms in their positions, and they start to slide over each other. However, researchers cannot provide a convincing explanation about the adhesion behavior generation and under what conditions the junction can fracture between the molecularly smooth surfaces. Furthermore, the relationship between adhesion and friction has not been investigated because the correlation between both properties is usually delicate. However, better predict of friction behavior and associated phenomena like wear or lubrication at the fundamental stage is full of challenge. Traditional continuum methods cannot describe this atomistic discrete process. Although many excellent works have focused on tribology, most have been limited to analyzing the macroscopic aspects, with only slight attention paid to the rich interplay going on between the atoms and molecules at the contacting surfaces, as these have historically been poorly understood. New powerful theoretical tools are necessary to investigate this tribological phenomena occurring at small scales – the atomic to the micrometer scale. MD simulation, by virtue of its high temporal and spatial resolution, can offer an ideal approach for gaining insights into the atomic scale processes and understanding their mechanisms. A remarkable enhancement in computational capability (computer hardware) and high-performance computation techniques (parallel computation) has enabled us to employ the large-scale classical MD method to investigate the nanometric tribology process and gain insights into this atomistic tribology behavior. Investigating the atomic-scale tribology mechanism is beneficial to uncovering the physical essence of the macroscopic tribological phenomenon.

2.3.1 Methodology

The copper atoms of a friction pair are initially arranged in fcc structure with a constant lattice parameter of 3.61 Å. The dimension of the substrate (lower part) is 50 Å × 20 Å × 10 Å and the dimension of the upper part is 12 Å × 10 Å × 10 Å along x, y, and z direction separately. The separation between the two parts is zero and the last few layers of the lower block are fixed. The moving speed is set from 50 m/s to 100 m/s, the environment temperature is 293 K, the computation timestep is 2.5×10^{-15} s.

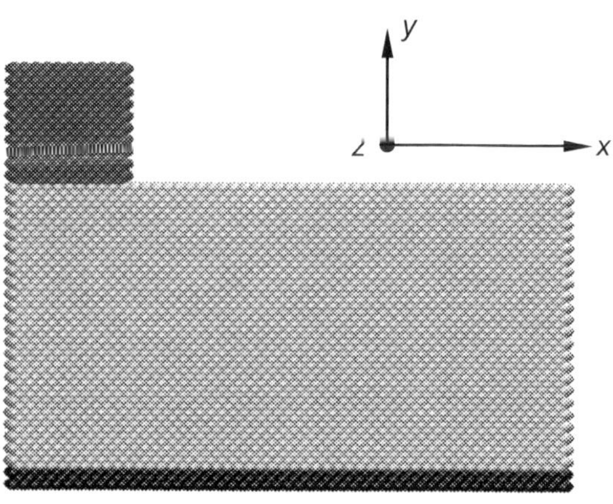

Figure 2.9 MD simulation model.

The EAM potential is used to depict the interaction between the copper atoms, which is illustrated in Eq. (2.2). This investigation is carried out along a path with constant energy (E), constant particle number (N) and constant volume (V) in the phase space, namely, the NVE ensemble. In reality, the energy of the system does not remain unchanged in the total simulation process. To adjust the system to a given energy, the total system energy is either drained or added, which is accomplished by scaling the velocities. The scaling process is performed by integrating the equations of motion for a certain number of MD steps which is done with the following coefficients:

$$\beta = \sqrt{\frac{E^*}{48k_B \sum_i v_i^2}} \tag{2.6}$$

The procedure is repeated until the desired energy is reached.

The Velocity-Verlet method is a symplectic algorithm which can prevent energy dissipation and has high computational efficiency, which is used in this research. The initial velocity is drawn from a boltzmann distribution with definite temperature. The velocity of atoms of the upper part is affected by two factors: the velocity comes from the macroscopic translation (sliding speed) and from particle interaction at a microscopic scale while the velocity of the lower part is only affected by particle interaction. In reality, the friction can only be induced with relative motion at the interface. To mimic the real working conditions, the upper part moves ahead at each timestep (sliding); after that the relative position between atoms of the friction pair is updated, which induces the interaction force, and thus the new atom velocity and the atom position can be acquired.

2.3.2 Results and discussion

MD simulation of sliding friction is shown in Figures 2.10–2.13. Obvious elastic deformation normal to the sliding direction at the local area and edge dislocation are observed in the bulk materials. Furthermore, the longitudinal shear deformation

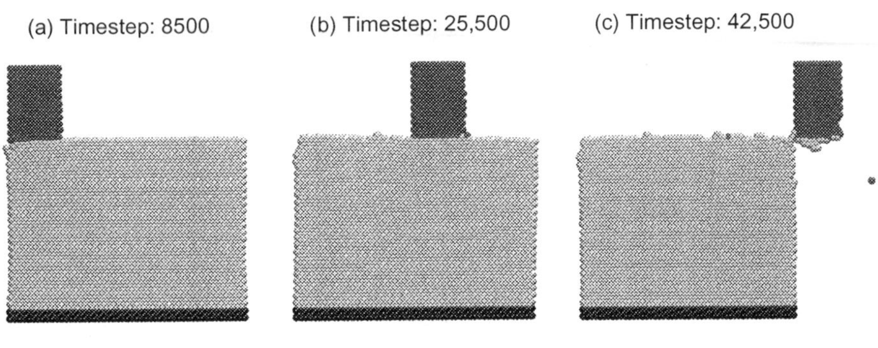

Figure 2.10 MD simulation results (velocity = 50 m/s).

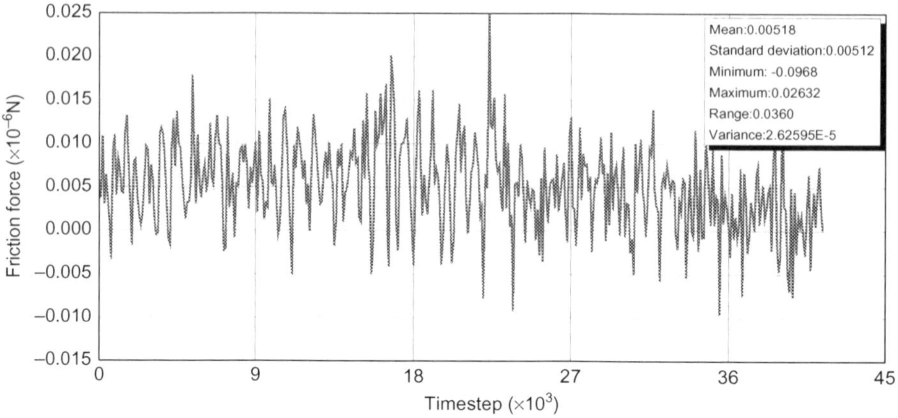

Figure 2.11 Variation of friction force.

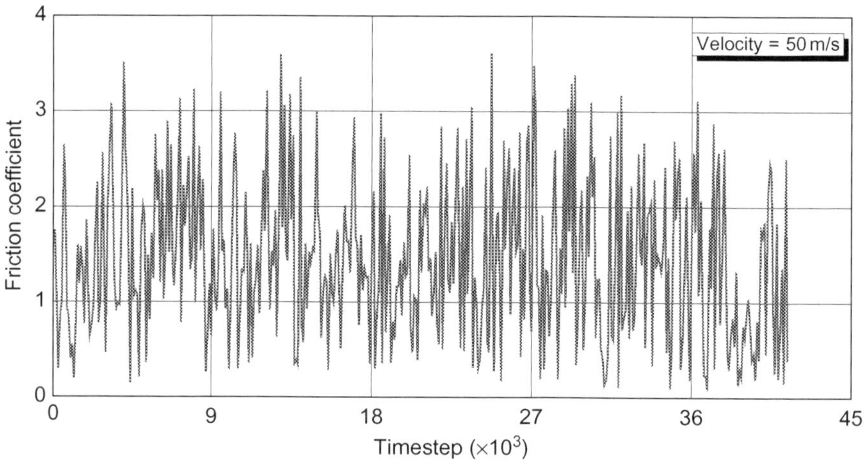

Figure 2.12 Variation of friction coefficient.

in the lower part is observed while this deformation is recovered at the end of sliding friction which justifies its elastic deformation essence. Also the permanent lateral plastic deformation is induced at the left side of the lower part as a result of movement of edge dislocation. Adhesion wear occurs when the attractive force among the atoms of a friction pair is larger than the interaction forces within themselves. This phenomenon also means no wearless-friction existed in the dry friction between solid materials. However, there exists different wear control mechanism corresponding to different surface morphology. Materials with molecularly smooth surfaces have high surface energy and thus result in very strong interaction in the friction pair which results in the adhesion effect between them. The continuous generation and fracture of a molecular junction between a friction pair ought to be the

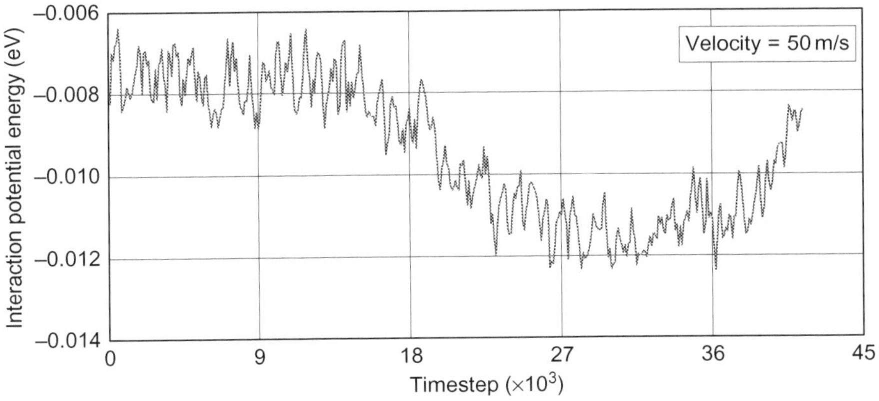

Figure 2.13 Variation of interaction energy at interface.

key reason for variations in friction force. The friction force in Figure 2.11 can be achieved as follows:

$$F = \sum_{j=1}^{m} \sum_{i=1}^{n} f_{ij} \tag{2.7}$$

where f_{ij} is the interaction force between the atom of the upper part ($j = 1,\ldots m$) and the atom of the lower part ($i = 1,\ldots n$). The friction force is the component of f_{ij} along the direction of sliding.

Friction would be induced when two surfaces come into contact and are sliding relative to each other. So, the friction pair with zero separation is of great value when analyzing some critical phenomenon. According to Coulomb's laws of dry friction, the friction force is given by material parameter (friction coefficient) times the normal force. In the classical friction process, the force necessary to start sliding is always equal to or larger than the force necessary to keep sliding at a constant velocity. The friction force in the atomic-scale stick-slip process is not a typical feature of sated friction force in the macroscopic scale, which indicates that some special mechanism dominated at this zero separation contact status. There are no obvious boundaries between static friction and dynamic sliding and this can be understood as follows: the contact area will gradually increase with time due to the integral effect of adhesion and external force. On the other hand, the contact area in dynamic sliding friction is less than that of static friction, which leads to little friction force compared to static friction. There are no external forces loaded on this single peak, so the contact area remains unchanged and no obvious difference is observed in the friction between static sliding friction and dynamic sliding friction. Some nanoscale junctions are periodically generated and fractured in the subsequent dynamic sliding friction because lateral shear deformation cannot be observed if no temporary stick (nanoscale junction) is formed. The integral surface of the friction pair in this investigation has molecular/atomic morphology and the local disturbed atomic structure induced by sliding friction results in the force being composed of many high frequency peaks at the shorter scale and low frequency features at the longer scale, as shown in Figure 2.11. Low-frequency features should have intense relationships with relatively stable surface interaction between friction pairs under conditions of whole smooth surface, while the high-frequency features should be the result of irregular local surface interaction with unstable active atomic arrangement. Also the bifurcation is observed on top of some single peaks in the graph, which may be the result of some kind of wear at the atomic scale and means that the sliding friction and wear are induced together in one feeding motion. Furthermore, both attractive and repulsive forces are observed in the dynamic sliding force variation graph, which also justifies the origin of this atomic-scale tribology behavior – the intermolecular interaction force. The friction coefficient in this case lies in the range $0 \sim 4$, as both the normal force and the tangential force come from the intermolecular interaction, which is different from the friction at macroscopic scale. The friction coefficient in Figure 2.12 can be achieved as follows:

$$\mu = \frac{f_p}{f_n} \tag{2.8}$$

where μ is the friction coefficient, f_p is the component of f_{ij} parallel to the direction of sliding and f_n is the component of f_{ij} normal to the direction of sliding.

Theoretically speaking, friction includes two basic processes: mass transfer and heat transfer. Mass transfer in friction mainly means the wear of friction, while heat transfer is related to energy generation and dissipation. Currently, researchers know little about the energy dissipation process at the atomic scale and cannot give a convincing explanation for it because they cannot see what is taking place at the interface during sliding. Figure 2.13 gives the dynamic interaction potential energy ($E = \sum_{j=1}^{m}\sum_{i=1}^{n} u_{ij}$, u_{ij} is the interaction potential between the atom of the upper part ($j = 1,\ldots m$) and the atom of the lower part ($i = 1,\ldots n$)). First of all, the negative interaction energy between the two surfaces remains constant, which means no energy dissipation is induced in the initial stage. Then, the interaction energy quickly decreases and after that increases again at the end of the sliding process, which also results in a variation in the system energy. Compared with Figure 2.10 it can be found that materials wear is generated in the second stage of the sliding friction, which changes the surface atom structure and the interaction between the friction pair. The decrease in interaction potential means that the two surfaces gradually come into fusion but this process is gradually weakened with the continually relative sliding between friction pair and finally leads to the fracture of the nanoscale junction and recovery of the potential interaction energy. This mechanical nonlinear dynamic behavior is one of the key factors which influence the energy dissipation of the system. As the above process does not occur under equilibrium conditions, this sliding process is irreversible and the energy lost cannot be reused to assist the sliding process. In more physical terms, the strain energy put into

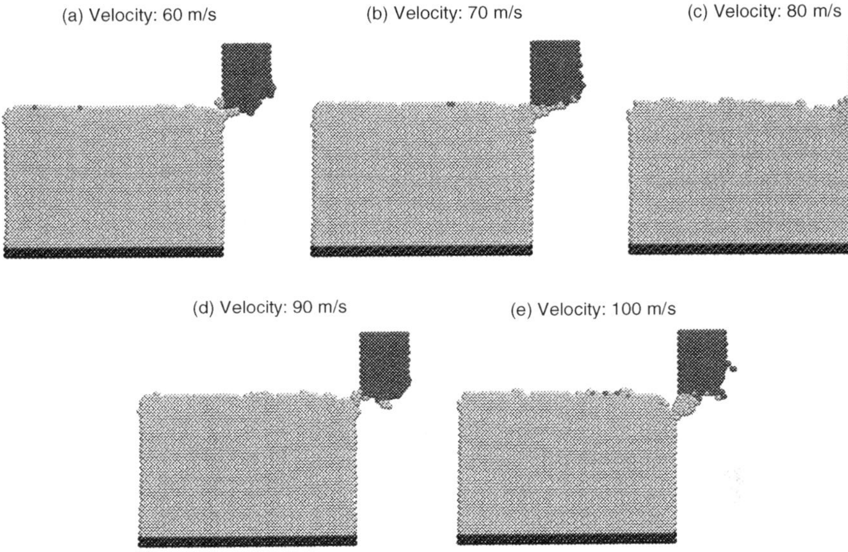

Figure 2.14 Simulation results under different velocity.

stretching materials surface atoms is not recovered totally; instead, it should be converted into molecular vibrational motion which dissipates into the substrate/environment as heat energy.

Figures 2.14–2.18 show a MD simulation of sliding friction with different velocity, with end failure observed in all of the cases. However, the dimension of the end failure does not increase linearly with the increasing of velocity, that is to say, the dimension of the end failure decreases if the velocity a larger than some critical value. This should have some relationship to the lateral plastic flow process in the lower substrate. The simulation results in Figure 2.14 justify this conclusion because permanent plastic deformation at the left side of the lower substrate is observed with all sliding friction velocity. The intense relationship between end failure and the sliding velocity lies in the fact that the interaction between the friction surfaces is weakened with the increasing of the relative sliding velocity. This is reasonable as the contact area would increase accordingly with time, which would give rise to a steady increase of the friction force in static contact. The dynamic variation of end failure under different sliding velocity should be the integral result of the thermal activation effect, which improves the materials flow and the weakening of interaction between the friction pair with the increase in sliding velocity. Figure 2.18 is the schematic of classic materials transfer/wear model put up by Archard which is appropriate to explain the rough surface contact where contact is generated at discrete surface peaks. The location of the two areas will form full contact at some point in time. A short time later the contact area is reduced to zero but, in addition, it is assumed that, at this point in time, a new similar contact area will be fully established somewhere on the surface. In each asperity contact, chemical inter-bonding and diffusion can lead

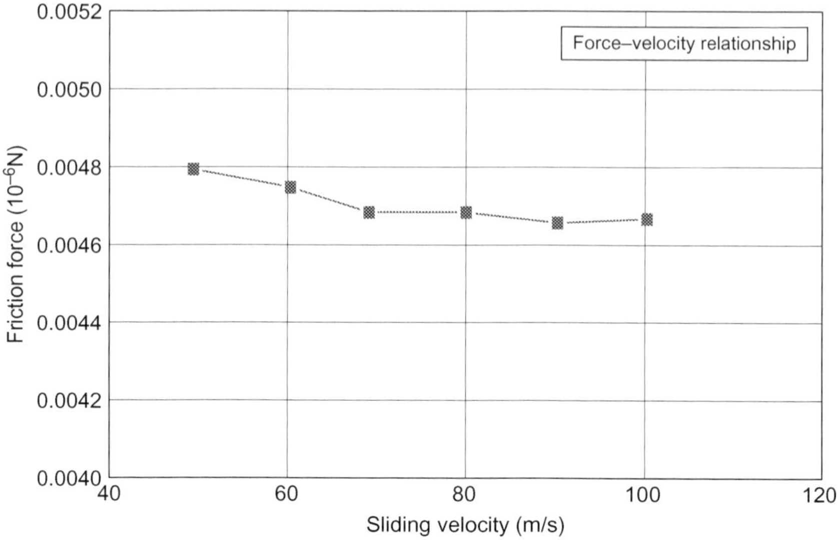

Figure 2.15 Friction force versus velocity.

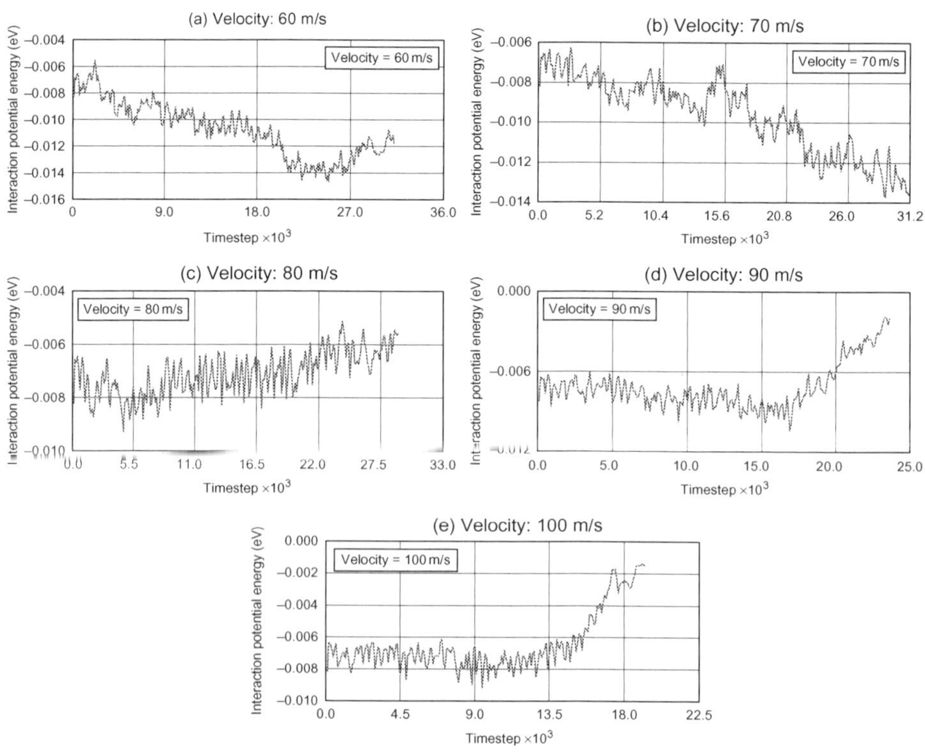

Figure 2.16 Potential energy versus velocity.

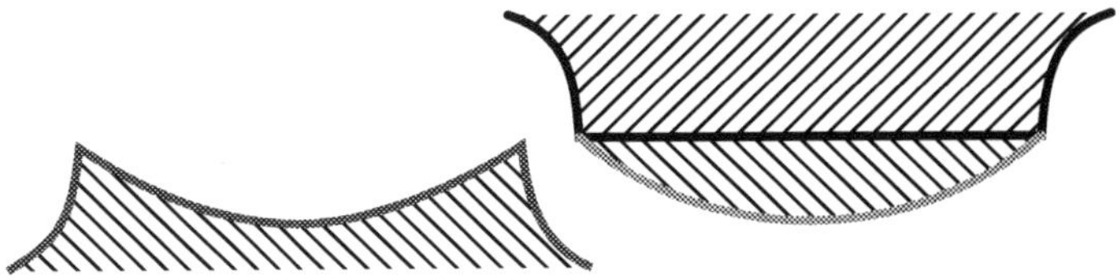

Figure 2.17 Archard materials transfer model.

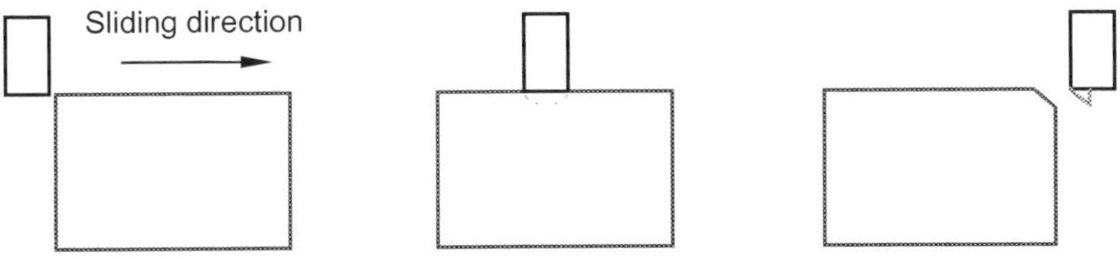

Figure 2.18 Materials transfer model in this study.

to local sticking. The continuous relative motion can, therefore, result in a separation of the asperities at a surface different from the original one, leading to surface damage and wear by material transfer. The Archard model (Archard, 1953) considers that the volume of material transferred from one asperity to another per unit of sliding length is simply proportional to the contact area when the contact is fully established. The important feature of this model is that the friction pair holds equivalent contact areas and the materials are transferred from the softer part to the harder part. In this work, a friction pair with a different cross section is constructed to mimic working conditions of scanning probe microscopy. The different materials transfer mode (Figure 2.18) is observed in all of the friction process with different sliding velocity. According to the Archard model, different mass transfer modes should have an intense relationship with the different cross sections at the interfaces.

In this nondirect contact friction, the adjacent bodies are separated by a potential barrier thick enough to prevent electrons or other particles from tunneling across it, but allowing interaction via the long-range electromagnetic field, which is always present in the gap between bodies and can have different origins. The presence of an inhomogeneous tip – for example, a sample electric field – is difficult to avoid, even under the best experimental conditions and produces friction, which is termed electrostatic friction. The irregular variation tendency of the friction force is the result of a dynamic changeable atomic structure induced by a complex potential interaction field.

The force generated in this atomic-scale sliding friction differs from classical tribology theory, which claims that the friction force will be same with a normal load while the materials are the same for the friction pair. The reason for this is that the fluctuation of friction force induced at atomic scale is negligible in the case of macroscopic friction. The decrease in the average friction force with the increase in sliding velocity should be the result of changeable contact conditions induced by different sliding velocity, which is shown in Figure 2.15. This should be the result of a gradual increase in difficulty on generation of a stable weld joint, as there is not enough time for diffusion at the interface, and the reduction in heat conduction with the increase in sliding velocity.

The tendency toward variation in interaction potential at the interface in the MD simulation process is shown in Figure 2.16. Some obvious variations in the potential interface interaction under different sliding velocities are observed. First, the strength of the potential interaction energy gradually decreases with successive feeding when the velocity is less than 70 m/s. This phenomenon shows that fusion tendency between friction surfaces is strengthened with the increasing sliding velocity, which leads to a decrease in interaction energy. After that, with a further increase in sliding velocity, the surface atom activity is increased to such degree that the tendency to shear (as a result of successive changes in contact conditions) overwhelms the tendency toward surface fusion and the material's transfer is strengthened, which leads to a fresh atomic surface structure and a gradual increase in interaction energy at the interface. The increase in interaction potential at the interface is observed in the subsequent cases when the sliding velocity is greater than 70 m/s, which also means that the adhesive effect is gradually decreased.

2.3.3 Conclusion

Conventional tribology techniques used for large objects can be ineffective in the micro- and nanoworld, where different criteria and control mechanisms are needed. The difficulties in realizing an efficient control of friction are related to the lack of a fundamental understanding of many of the underlying physical processes which take place within the microscopic separation between two materials. In order to understand the behavior of two real surfaces in relative motion while still in contact, one needs to look first into what goes on at the "single asperity" level. These challenges have motivated various theoretical efforts, both numerical and analytical, but many phenomena remain unexplained or still controversial. The author investigated the physical essence of the sliding friction process between molecularly smooth surfaces using the MD simulation method, after that finding some novel mechanisms. Both elastic deformation normal for the sliding direction at the local area and edge dislocation are observed in bulk materials. Furthermore, the longitudinal shear deformation of the lower part is observed on one side of the friction pair and this deformation is recovered at the end of sliding friction which proves its elastic deformation property. Materials with a molecularly smooth surface have high surface energy and very strong interactions are quickly formed in the case of contact which results in adhesion behavior between a sliding friction pair. The successive generation and fracture of molecular interaction between a friction pair ought to be the key reason for variations in friction force. End failure is observed under both larger and smaller sliding velocity. But the dimension of end failure is not always increased linearly with the increase in velocity, that is to say, the dimension of end failure will decrease if the velocity is larger than some critical value. The dynamic variation of end failure under different sliding velocities should be the integral result of both the thermal activation effect, which improves the material's plastic flow, and the weakening of interaction between the friction pair with the increasing of sliding velocity. The friction force of sliding friction at this atomic scale is different from classical tribology theory, which shows some statistical properties and must be the result of other contributions originated at atomic scale such as elastic deformation of substrate and the entropy of the microscopic degrees of freedom.

Investigation of durable and/or low friction surfaces has become an important factor in the miniaturization of moving components in many technological devices, as the simple old empirical laws of friction do not always hold. This breakdown stems from the high surface-to-volume ratio and from neglecting the important role of surface chemistry, adhesion and surface structure or roughness. This work uncovers some novel features of friction at small scale, but further study with larger models still needs to be carried out to prove the size effect at longer scales applicable to some real conditions.

2.4 Scratching wear at atomic scale

2.4.1 Methodology

Theoretically speaking, wear is the loss of material from one or both of the contacting surfaces when subjected to relative motion. Wear is a complex process, making it one of the most difficult aspects of tribology to study. The generic "wear of materials" has been classified in different modes, in accordance with its type of occurrence and with its phenomenology (abrasion, adhesion, erosion, fretting, etc.); the damage mechanisms of rubbing surfaces have been identified (cutting, plowing, cracking, delamination, fatigue, oxidation, etc.) and mathematical models for wear, with increasingly wide application ranges, have been developed in many high-technology industries.

New developments on integrated-circuit-related technologies have led to rapid progress in the development of microdynamic systems. These systems are based on the science, technology and design of moving micromechanical devices, and are a subclass of what is known as MEMS. Mechanical materials for some of the microdynamic systems that have thus far been demonstrated consist of deposited thin films of silicon, silicon nitride, aluminum, polyimide, and tungsten among other materials. Continued research on the mechanical properties of the electrical materials forming microdynamic structures (which previously had exclusively electrical uses), on the scaling of mechanical design, on tribological effects, on coatings, and on the effective uses of computer aids is now under way. Surface micromachining is an important MEMS microfabrication technique which can be used to create movable microstructures on top of a silicon substrate. The main advantage of surface micromachining is that extremely small sizes can be obtained. In addition, it is relatively easy to integrate the micromachined structures with on-chip electronics for increased functionality. In contrast to substrate micromachining in which mechanical parts are sculpted from the wafer itself, surface micromachining makes mechanical elements out of materials deposited on the wafer surface.

Thermal spray coatings are considered as one promising solution for anti-wear applications, specifically in situations where sliding wear and abrasive particles are present. Examples of these types of applications are the pump seals and bearing surfaces which can be coated by thermal-sprayed coatings to increase their resistance. Meanwhile, there are several obstacles in the way of using coatings. Finding a solution for these problems can have a huge impact on the application of coatings. These problems are usually rooted in the tribology properties of coatings. In order to use thermal spray nanocoatings in industry, it is important to know about their capabilities. Several mechanical properties of nanostructured coatings need to be investigated and analyzed.

In reality, wear is the basis of the advanced manufacturing processes which can be dated back to when humans fashioned their first tools by cutting, grinding, and chipping various sticks and stones they found lying around into more useful shapes. Study the origin of tribology is necessary for developing a fundamental understanding of interfacial phenomena on a small scale and studying interfacial phenomena in surface microstructure machining. Friction and wear of lightly loaded AFM tips or machining tools are highly dependent on surface interactions (few atomic layers). Investigation into the origins of tribology is also valuable in constructing a bridge between science and engineering. Currently, an AFM tip is always used to study single-asperity contact-based wear/friction with a solid or lubricated surface. AFMs become the basic tools used for fabricating the various three-dimensional microstructures. In situ surface characterization of local deformation of materials and thin coatings can also be carried out using a tensile stage inside an AFM. Furthermore, mechanical properties such as hardness, Young's modulus of elasticity and creep/relaxation behavior can be determined on the micro- to picoscales using a depth-sensing indentation system in an AFM. Microcrack/pit is always generated in typical hard-brittle machining when the depth of cut and the structure dimension lies in the range of a micrometer or greater; materials removal (wear) and corresponding plastic deformation in the nanometer scale must be different from the mechanism at macroscopic scale. This small scale deformation and the tribology behavior are far from providing a convincing explanation. Traditional continuum theory cannot uncover the physical essence underlining this nano-fabrication process. A new powerful toolkit is needed to investigate this atomic-scale phenomenon. MD simulation, by virtue of its high temporal and spatial resolution, can offer an ideal approach for gaining insights into atomic-scale process and understanding their mechanisms. A remarkable enhancement in computational capability (computer hardware) and high-performance computation techniques (parallel computation) have enabled us

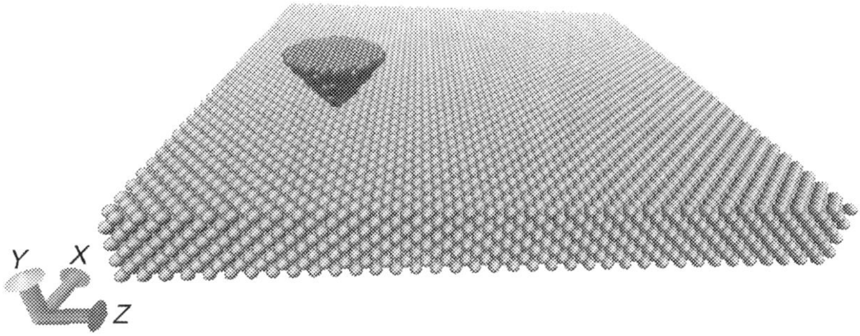

Figure 2.19 Initial model.

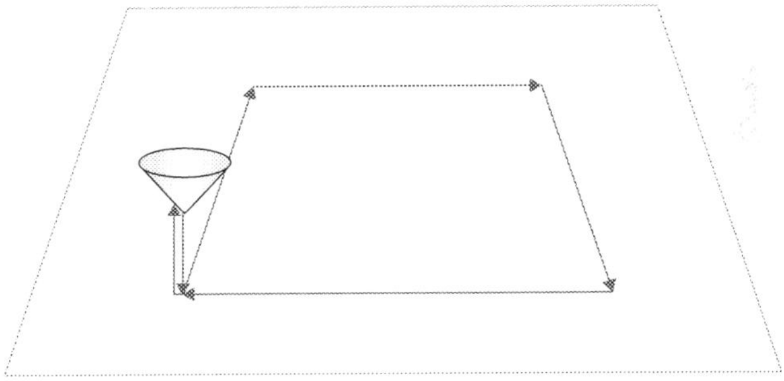

Figure 2.20 The corresponding process route.

to employ a large-scale classical MD method to gain insights into this atomistic behavior.

Figure 2.19 shows the MD simulation model of silicon nanoscratching (wear) using AFM. The whole process is composed of six stages: downward movement (along the negative *Y* axis), two pairs of parallel linear movement along the *X* and *Z* axes, and the upward movement (along the positive *Y* axis) which is shown in Figure 2.20. These three moving directions represent the three series representative crystal orientation, namely, [100], [010], and [001].

2.4.2 Results and discussion

There is no brittle crack or dislocation band generated in the bulk material or in the surface of the shallow groove. The scratch processes described above involves plastic deformation of material in small localized regions where the opposing surfaces or hard particles make contact. It turns out that the onset of plasticity for such small regions can be dramatically different from the plastic yield stress determined by macroscopic measurements. In reality, macroscale plastic deformation within crystalline materials mainly occurs through the motion of flaws and defects in the crystal structure, called dislocations. When the size of the region undergoing plastic flow becomes much smaller than the typical distance between dislocations, however, the contribution of dislocations to the yield stress becomes insignificant; instead, the yield stress is governed by the force needed to slide one plane of atoms over another.

Figure 2.22 (a cross-sectional view of the substrate surface after scratching) also shows that no brittle crack or dislocation is observed in the materials. The volume V_L of materials displaced when the asperity slides a distance L is given by:

$$V_L = \frac{1}{2} L h^2 \tan\beta \tag{2.9}$$

If V_a is the volume of silicon atom, then the number of removed atoms (N) after scratching is:

$$N = V_L / V_a \tag{2.10}$$

In fact, the value of the removed atoms (213) computed by Eqs (2.9) and (2.10) is far beyond the real value (51), which means that the structure rearrangement/transformation induced by local high pressure in this atomic-scale abrasive wear dominate process. There is not obvious materials removal in the whole process except at the start/end point where the AFM indent or lift from the substrate and the adhesive wear dominate this stage. The residual depth of the groove is minor compared with the original scratch depth, which shows that AFM-based scratching-wear would be some kind of little-wear friction at the small scratch depth (less than 10 nanometers) and belongs to the elastic-plastic deformation regime. At the same time the lateral deformation is larger, which makes the final mark similar to a welding seam, as shown in Figure 2.23.

The little-wear tribology behavior could result in local disorder of the crystal structure or amorphous layer, which can change the material's electrical/magnetic property. Larger plastic deformation or higher compression pressure is induced because little defect is included in the small volume which greatly improves the fracture strength comparing with the corresponding value at macroscopic scale. Larger fracture strength makes it difficult for materials removal and thus a high external load is needed for scratching to proceed. The extremely high pressure at the local area ahead of AFM tip can lead to structure transformation and finally result in the ductile plastic deformation of single crystal silicon. Furthermore, the final microstructure acquired by an AFM-based technique only has a similar shape to the concept design because of the self-organization property of nanomaterials. Also, the effect of this AFM-based scratching-wear (nanolithography) does not extend far beyond the surface which keeps the original material's atomic structure a long way from the working area.

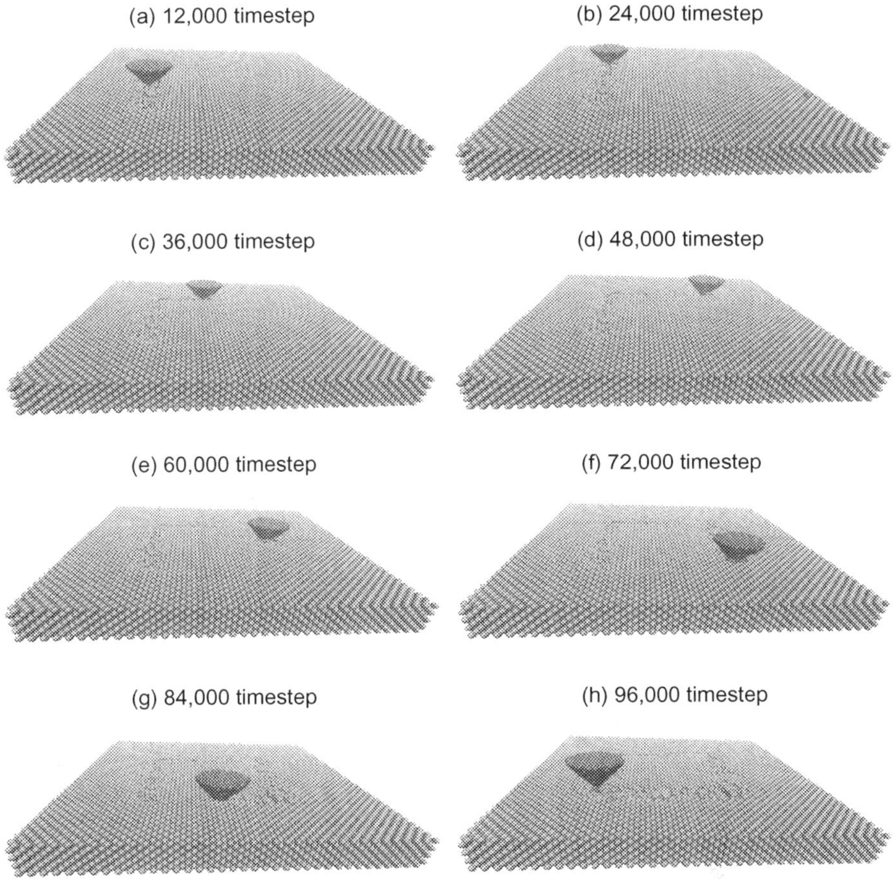

Figure 2.21 Simulation results.

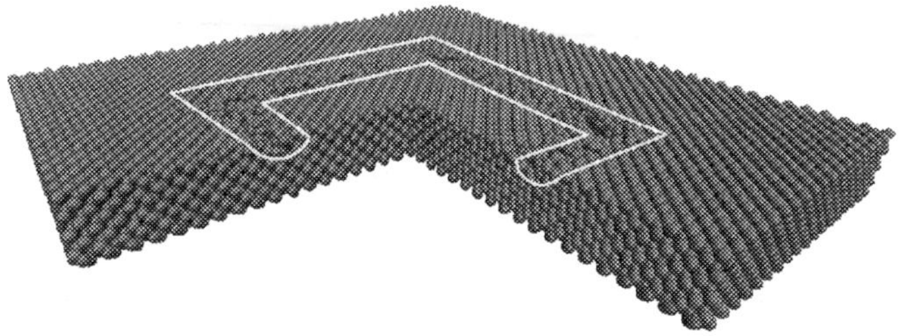

Figure 2.22 Corresponding process route.

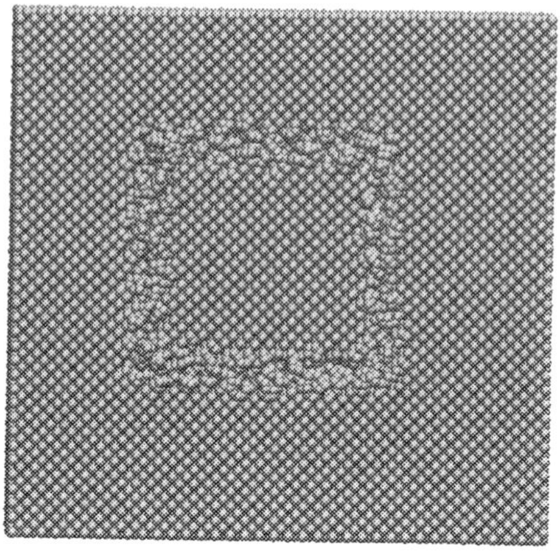

Figure 2.23 Vertical view of the scratching track.

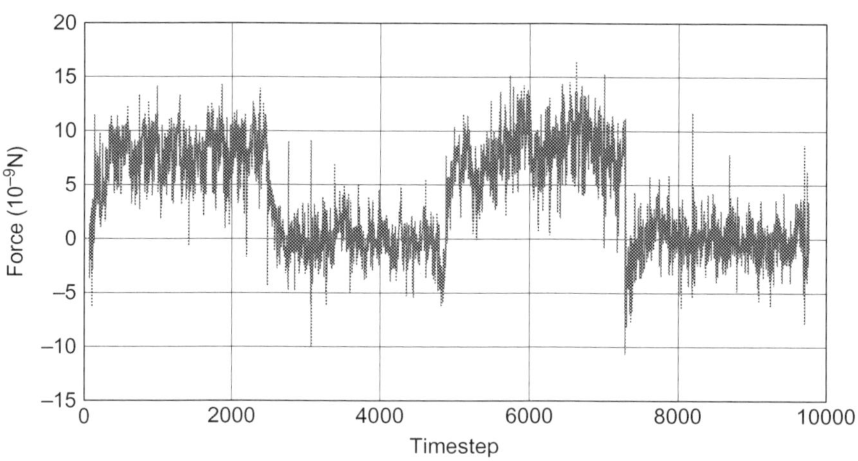

Figure 2.24 Variation of scratching force.

During scratching, the AFM tip moves discontinuously over the substrate surface and jumps with discrete steps from one potential minimum (well) to the other. This leads to a saw-tooth-like pattern for the lateral motion (force) with a periodicity of the lattice constant (Figure 2.24). This motion which is called stick-slip movement of the tip accompanying all of the atomic-scale wear process. As just mentioned, stick-slip on the atomic scale is the result of the energy barrier that must be overcome to jump over the atomic corrugations on the sample surface. This corresponds to the energy required for the jump of the tip from a stable equilibrium position on the surface into a neighboring position. The perfect atomic regularity of the surface guarantees the periodicity of the lateral force signal, independent of actual atomic structure of tip apex. Few atoms (based on the magnitude of the friction force) on a tip sliding over an array of atoms on the sample are expected to go through stick-slip. As the sample surface slides against the AFM tip, the tip remains stuck initially until it can overcome the (potential) energy barrier, which is illustrated by some definite kinds of interaction potential experienced by the tip. After some motion, there is enough energy stored in the spring to lead to slip into the neighboring stable equilibrium position. During the slip and before attaining stable equilibrium, stored energy is converted into vibrational energy of the surface atoms in the range of 10^{13} Hz (phonon generation) and decays within the range of 10^{11} s into heat. The stick-slip phenomenon, resulting from irreversible atomic jumps, can be further theoretically investigated with discrete mechanical models.

The mechanisms and dynamics of the interactions of between contacting solids during relative motion, ranging from atomic to microscale, need to be understood in order to develop fundamental understanding of AFM-based nanolithography processes and the origin of tribology such as the anti-wear mechanism of nanocoatings, which may be helpful about optimization/controlling the technique and improving the useful-life of parts.

2.4.3 Concluding remarks

Currently, many apparatus are appropriate for the study of phenomena at the nanometer scale such as the AFM, which has already been extended to construct tools used for digging out microstructures in the substrate surface or manipulation tools employed to extract/move atoms. The complexity of such phenomena and the precision with which AFM can attempt to explore and understand them makes the field of AFM processing techniques very challenging and intriguing. It has become clear that the atomic process cannot be neglected when interpreting AFM-based processing techniques. Furthermore, there is always wear debris or microscale hard dust embedded in the small separation between coupled surfaces which accelerate the failure of coatings. Different from the macroscopic experience, this particle holds the properties of both self-organization and condensed matter. Particle-surface interaction is full of challenges and critical for normal useful-life of nanocoatings. Traditional continuum mechanics cannot account for the atomic-scale process, which is closely related to understand the silicon patterning process. Macroscopic quantities such as shear modulus or pressure fail to describe the mechanical behavior of this atomic wear process. Microscopic modeling using MD simulation is carried out to study the material's deformation and tribology behavior at the nanometer scale and after that some novel phenomena are observed. In the case of the atomic wear process, wear and friction together, friction plays an important role, which is reflected in force variation. The anisotropy of friction (wear) behavior is observed in the nano-scratching process. In this case, the deformation along ssthe [100] direction is more severe than that along the [001] direction. It turns out that the onset of plasticity for small regions in the nano-scratching can be dramatically different from the plastic yield stress determined by macroscopic measurements. The contribution of dislocations to the yield stress becomes insignificant; instead, the yield stress is governed by the force needed to slide one plane of atoms over another. The residual depth of the groove is minor compared with the original scratch depth which shows that particle-based scratching would be little-wear friction at the small scratch depth (less than 10 nanometers) and includes both elastic and plastic deformation.

2.5 Conclusion

The coatings-related tribology process plays an important role in industry, although there are many other kinds of applications related to surface coating. As the length scale goes into the micro- and nanoscale, there are many novel mechanisms dominating the tribology behavior. Tribology behavior occurs in a small region and contains only a few layers of molecules, so it can consequently be atomistic, or discrete in nature, rather than continuous, as is assumed in conventional continuum mechanics. In studies of such atomistic processes, which are difficult to investigate experimentally, computer simulation by molecular techniques is useful. MD simulation, like other simulation techniques can play a significant role in addressing a number of surface problems at the atomic scale. It may be noted that atomic simulations are providing new data and exciting insights into various phenomenon in micro- and nanotribology processes that cannot be obtained readily by any other theory or experiment.

We shall take as our definition of tribological coatings those which are sufficiently thin for the substrate material to play a role in the friction and wear performance. Thus, we exclude coatings which are so thick that there is little or no substrate influence on the tribological behavior – the coating in effect acts as a bulk material. With this definition of tribological coatings we focus on solid surface films that are typically in a thickness range of 0.1–10 mm. Coatings tribology is at the forefront of developments in most mission-critical components, regardless of the sector in which they have to operate. When the technological demands are coupled with the economic imperative to reduce production costs and reliability, it is clear that the challenges facing tribologists and coating process developers are enormous. Nevertheless, the payoff from success in meeting these goals is huge, with great societal impact, and it is my hope that this chapter will assist those who seek to achieve new solutions and breakthroughs.

References

Achanta, S., Drees, D., Celis, J.P., 2005. Friction and nanowear of hard coatings in reciprocating sliding at milli-Newton loads. Wear. 259, 719–729.

Ahmed, S.I.U., Bregliozzi, G., Haefke, H., 2003. Microfrictional properties of diamond-like carbon films sliding against silicon, sapphire and steel. Wear. 254, 1076–1083.

Archard, J.F., 1953. Contact and rubbing of flat surfaces. J. Appl. Phys. 24 (8), 981–988.

Arnell, R.D., 1990. The mechanisms of tribology of thin film systems. Surf. Coat. Technol. 43, 674–687.

Baek, D.K., Khonsari, M.M., 2005. Friction and wear of rubber coating in fretting. Wear. 258, 898–90.

Bandorf, R., Luthje, H., Wortmann, T., Staedler, T., 2003. Influence of substrate material and topography on the tribological behaviour of submicron coatings. Surf. Coat. Technol. 174–175, 461–464.

Batista, J.C.A., Joseph, M.C., Godoy, C., Matthews, A., 2002. Micro-abrasion wear testing of PVD TiN coatings on untreated and plasma nitrided AISI H13 steel. Wear. 249, 971–979.

Bhushan, B., Gupta, B.K., 1991. Handbook of Tribology – Materials, Coatings and Surface Treatments. McGraw-Hill, New York.

Bowden, F.P., Tabor, D., 1950. The Friction and Lubrication of Solids. Clarendon Press, Oxford.

Caron, I., Gras, R., Ganne, T., Monicault, J.M., 2006. Coatings on Ti6Al4V as palliatives for fretting fatigue in cryogenic environment. Tribol. Int. 39, 1045–1051.

Cavaleiro, A., Hosson, J.T.M., 2006. Nanostructured Coatings. Springer, New York.

Daw, M.S., Baskes, M.I., 1984. Embedded-atom method: derivation and application to impurities, surfaces, and other defects in metals. Phys. Rev. B. 29 (12), 6443–6453.

Elmustafa, A.A., Anda, A.A., Elmahboub, W.M., 2004. Bilinear behavior in nano and microindentation tests of fcc polycrystalline materials. J. Eng. Mater. Technol. 126, 353–359.

Halling, J., 1986. The tribology of surface coatings, particularly ceramics. J. Mech. Eng. Sci. 200 (1), 31–40.

Hertz, H., 1992. On the contact of elastic solids. Journal für die reine und angewandte Mathematik. 92, 156–171.

Holmberg, K., Matthews, A., 1998. Coatings tribology – contact mechanisms and surface design. 1997 World Tribol. Congr. 31 (1–3), 107–120.

Izabela, S., Rajiv, K.K., Aiichiro, N., Priya, V., 2005. Atomistic mechanisms of amorphization during nanoindentation of SiC: a molecular dynamics study. Phys. Rev. B. 71, 174113.1–174113.11.

Morse, P.M., 1929. Diatomic molecules according to the wave mechanics II vibrational levels. Phys. Rev. 34, 57–65.

Suh, N.P., 1986. Tribophysics. Prentice Hall, Englewood Cliffs, NJ.

Wang, Y.M., Hodge, A.M., Biener, J., Hamza, A.V., Barnes, D.E., et al., 2005. Deformation twinning during nanoindentation of nanocrystalline Ta. Appl. Phys. Lett. 86, 101915.1–101915.3.

Stress on anti-abrasive performance of sol-gel derived nanocoatings

3

Shengzhao Wang
Nanyang institute of Technology, China

Thin film nanocoatings enhance the performance and quality of a variety of substrates giving more functionality and aesthetic value. Improved light transmission can reduce energy costs and increase solar cell efficiencies (Wang and Shen, 2010; Jun et al., 2008). Protection can also be provided against corrosion or to improve anti-abrasive performance (Li and Shen, 2011). These functional coatings on different substrates such as glass will be reviewed together with their added value as they become essential to the performance of engineered constructions such as buildings, automobiles, containers, solar cells, and multi-touch panels.

Nanocoatings can be prepared by many methods. Higher-density hard coatings are obtained at high production rates in the Chemical Vapor Deposition method; however, yields and spectral performance are lower and chemical by-products can be more harmful to the environment. Sputtering Physical Vapor Deposition (PVD) magnetron processes deposit one or more nanocoatings of metal oxide under a vacuum to finished glass. PVD coatings are better reflectors of ultraviolet (UV) and infrared (IR) wavelengths of light. However, these coatings are not as hard and dense. The very important factors of these large area coatings are inherently limited in the design space of anti-abrasive nanocoatings as well as being sensitive to intrinsic stress due to the nature of deposition (Belleville et al., 2000; Shen et al., 2009; Liang et al., 2008).

The sol-gel derived nanocoating process is one of the best processes and used extensively. Compared with other conventional coating methods, the sol-gel method has many advantages: (1) Different combinations of ingredients can be easily achieved; the temperature of the sol-gel method process is low. (2) The nanocoating is homogeneous and of high purity. (3) Its reaction process is controlled easily; the sol film is easy to control. (4) In particular, its unique low-temperature processing characteristic provides unique opportunities to make pure and well-controlled compositions of organic/inorganic hybrid materials through the incorporation of low molecular weight and oligomeric/polymeric organic molecules with appropriate inorganic moieties at temperatures under which the organics can survive. (5) Sol-gel derived inorganic networks lead to a high anti-abrasive performance and to the possibility of quick thermal curing by three-dimensional polycondensation-based cross-linking by heat (Uhmann et al., 1997; Jianye and Wilkes, 1996; Schmidt et al., 1988; Iwamoto and Mackenzie, 1995; Wu et al., 2002; Yang et al., 2000; Shen et al., 2000).

Anti-Abrasive Nanocoatings.

In fact, one of the applications of sol-gel derived nanocaoting is to protect samples which have not enough hardness or abrasion resistance and easily damaged surfaces. Without doubt, the resistance to accidental scratch damage depends on the mechanical properties and other non-material factors, including the size and shape of the scratching object, the coating thickness, and the test environment and speed. While it is important to understand the effect of all these factors on abrasive failure, the current work will focus mainly on the mechanical properties and the thickness of the coatings. The mechanical properties for consideration include Young's modulus, hardness, fracture toughness, adhesion toughness and thin film stress to the substrate (Friedrich and Schlarb, 2008).

The following sections give a brief overview of the mechanics, including stress theory and some scratch tests.

3.1 Classical curvature stress for thin films on plate substrates

Abrasive action exerts a relationship with the coating stress. For the convenience of understanding the mechanical analysis, we begin with a summary of classical curvature stress for thin films on plate substrates. The applicability in various circumstances of the classical equation-related substrate curvature to film stress for the case of a thin film on a plate substrate. Theoretical treatments, based on elementary plate-bending theory and general elasticity theory, are given on substrate curvature. Thin solid films of many kinds, deposited or grown onto substrates, are used in today's microelectronic devices and for other purposes. Mechanical stresses in these films arise from various causes, including thermal expansion differences between film and substrate materials, the physical nature of the as-deposited material, and chemical or structural changes occurring in the films upon processing after deposition. These stresses can lead to cracking or delamination of the films and can induce defect formation in the substrate (Oliver et al., 2012). In order to experimentally determine mechanical stresses in thin films on substrates, it is common practice to use the formula:

$$\sigma = \left(\frac{E_s}{1-\nu_s}\right)\frac{t_s^2}{6Rt_f} \qquad (3.1)$$

where σ is the film stress, R is the radius of curvature of the surface, E_s is Young's modulus of the substrate, ν_s is Poisson's ratio for the substrate, and t_f and t_s are the thickness of the film and substrate, respectively. This describes the impact of the stress on the radius of curvature of the optic surface, leading to changes in the flatness and corresponding optical performance of the component. While mechanically stiff substrate materials of sufficient thickness t_s will exhibit minimal bending from film stress, tensile stresses remain a problem if they lead to cracking, or crazing, of the coating (Stoney, 1909).

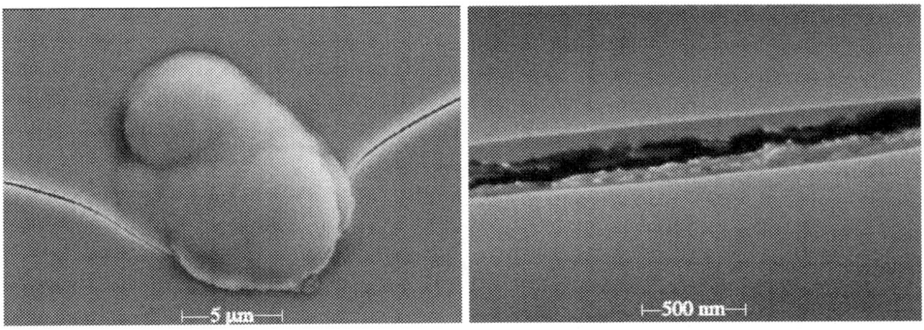

Figure 3.1 Scanning electron microscope (SEM) imaging of the initiation site and crack formed as a result of the high tensile stress in the film. A defect site in the coating provides an initiation site for tensile stress failure, while tearing of the film is evident within the crack that forms.

Fracture of the surface of a coating results from sufficiently high tensile stresses such that the fracture toughness of the film is exceeded. Fracture will initiate at a defect in the coating, whether initiated by a scratch at the edge from coating tooling or optic mounting, or at a defect within the film such as shown in Figure 3.1. Fracture also requires a sufficiently thick film, such that the stress in the film can be relieved through fracture at the surface, given the strain at that point. This relationship is given by Hutchison and Wu (1992); Hutchinson and Suo (1992):

$$h_c = \frac{\Gamma \overline{E}_f}{Z\sigma^2} \tag{3.2}$$

Eqs. (3.1) and (3.2) is given. It is important to reduce the film stress such that optical performance of the component is preserved while fracture of the coating is avoided. Thicker substrates may aid in maintaining flatter optical surfaces, and thinner coatings help to prevent cracking, but both of these approaches result in a cost in substrate size and/or achievable coating performance (Hutchison and Suo, 1992; Tikhonravov et al., 2004). To provide optimal performance of the optical coating, the magnitude of the tensile film stress must be kept low or ideally moved to a compressive state. Compressive stress also must be kept low in order to maintain surface flatness of the optical component in accordance with Eq. (3.1).

Stresses in a multilayer coating are a function of the film thickness of each constituent material in the coating, as well as the stress from all sources in each material. Eq. (3.1) may be modified to account for the individual stress in each layer by (Ohring, 2002):

$$\sigma_1 t_1 + \sigma_2 t_2 + \ldots + \sigma_n t_n = \left(\frac{E_s}{1-\nu_s}\right)\frac{t_s^2}{6R} \tag{3.3}$$

where σ_i and t_i are now the stress and thickness of each layer of the coating, respectively. Likewise, the total stress in the multilayer will be given by

$$\sigma_{\text{total}} = \frac{\sigma_1 t_1 + \sigma_2 t_2 + \ldots + \sigma_n t_n}{\sum_{\text{i}} t_i} \tag{3.4}$$

since the individual stress contributions are simply weighted by the relative layer thicknesses of each. Modification of selected layers provides a means of adjusting the overall stress, with the use of three materials providing the ability to calculate the stress according to

$$\sigma_{\text{total}} = \frac{\sigma_H t_H + \sigma_L t_L + \sigma_A t_A}{T_H + T_L + T_A} \tag{3.5}$$

where T_i is now the total thickness of a given material in the entire multilayer, since the stress is assumed to be constant for all layers of the same material deposited in the same manner. Subscripts H, L, and A can denote hafnia (high refractive index), silica (low refractive index), and alumina, respectively. Given individual material stresses, coating designs may then be modified to yield the desired overall stress.

Eq. (3.5) assumes that the stresses in individual layers are identical for a given material, and that the stress is homogeneous throughout the thickness of each layer. Given that electron-beam deposited films exhibit inhomogeneity, roughness, and columnar-growth structure, the influence of anisotropic properties in the composite film stress is expected to be present but of unknown magnitude. The influence of other layers, through roughness or capping effects, is also unknown, as are interfacial strains. The complexity in film structure for porous, evaporated coatings is anticipated to influence the film stress, suggesting additional complexity in quantitative understanding beyond that in Eq. (3.5).

Of course, from the curvature of the substrate K, the residual stresses in the deposited thin films can be evaluated by the formula proposed by Atkinson, that is (Klein, 2000; Shao et al., 2003)

$$\sigma = \frac{E'_s t_s^3}{6 t_f^2 (1 + t_s/t_f)} K \tag{3.6}$$

if $t_f < t_s$, and the radii of curvature before and after deposition are R_1 and R_2, respectively, this formula can be written as

$$\sigma = \frac{E_s t_s^2}{6(1 - \upsilon_s) t_f} \left(\frac{1}{R} - \frac{1}{R_0} \right) \tag{3.7}$$

where E_s is Young's modulus of the substrate, and t_s and t_f are the thickness of the substrate and the thin film, respectively. R_1 and R_2 were measured by different methods.

3.2 Thermal stress of thin films

The total stress of a thin film is the sum of all the stress components that result from intrinsic stress, thermal stress, and extrinsic stress, caused by crystalline or volume transformations within the coating or from an external load and a physical–chemical component that describes stress changes caused by adsorption phenomena such as water into a porous columnar structure or grain boundary oxidation (Wang and Zeng, 2008).

The intrinsic stress is basically related to the chemical composition of the film, to the morphological and crystalline structure developed in the coating, and to coating–substrate bonding. This stress property can be strongly influenced by deposition parameters such as substrate temperature, growth rate, pressure, and residual gas atmosphere.

The thermal stress is caused by the difference between the thermal expansion coefficients of the coating α_{film} and the substrate α_{sub} materials and by the temperature difference between deposition temperature T_d and ambient environmental temperature T, as follows:

$$\sigma_{th} = \frac{E}{(1-\nu)_{\text{film}}}(\alpha_{\text{sub}} - \alpha_{\text{film}})(T - T_d) \tag{3.8}$$

where E and ν are the Young's modulus and the Poisson's ratio, respectively, of the film and $(E/(1-\nu)_{\text{film}})$ is the biaxial modulus of the film. Because of the morphological and crystalline microstructure, both the biaxial modulus and the thermal expansion coefficient of a thin-film material will be different from those values of the mother bulk material. For calculation of thermal stress it is therefore essential to determine these two film properties separately. The biaxial modulus as well as the thermal expansion coefficient of a thin film can be estimated from the slopes $(d\sigma/dt)$ of the measured stress–temperature curves of identical films deposited onto substrates with different thermal expansion coefficients in the case of purely elastic response behavior of the coatings during heating and cooling. The total stress of a coating will cause deformation of the substrate. Therefore the change in substrate curvature before and after deposition can be used as a measure of the film-induced stress, such as Eq. (3.8).

3.3 Why do drying films crack?

Evaporating solvent from a suspension creates many apparently solid surfaces. As solvent leaves dispersion, particles consolidate and in many cases the films will display cracks. The stresses play a significant role in causing cracking and have been studied using a cantilever technique. This method provides the transverse stress in a film averaged laterally. For a drying dispersion, the meniscus of the air–water interface between particles gives a capillary pressure in the fluid that is below atmospheric. The magnitude of this pressure can be enormous, scaling inversely

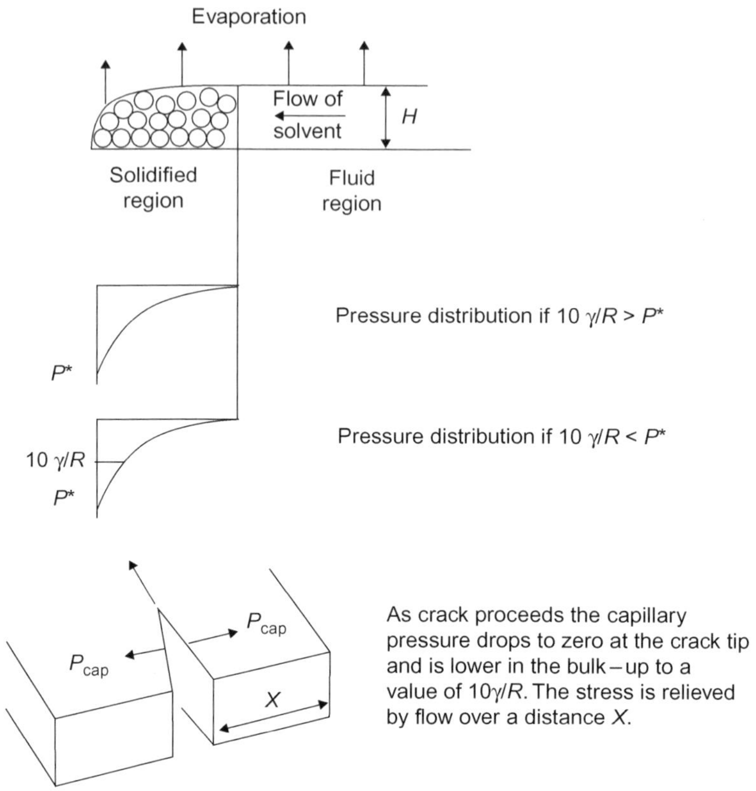

Figure 3.2 Drying front, pressure distribution, and fracture mechanism.

with the particle radius (Lee and Routh, 2004). Atmospheric pressure, pushing on the film surface, puts the entire film into compression. If the film compresses, or yields, the exudation of solvent results in a reduction, or even elimination, of capillary pressure until evaporation increases it again. Recent work by Dufresne et al. (2003) showed cracking films to be wet, except at the position of the cracks, demonstrating that the capillary pressure is responsible for the film failure.

A large body of work has been performed where dispersions are constrained between two glass plates with evaporation from one end (Allain and Limat, 1995; Komatsu and Sasa, 1997). This results in a progression of cracks from the free surface into the bulk. Here we examine a slightly different scenario, with fracture in thin films subject to evaporation at their free surface. This conceptually simple problem is complicated by horizontal drying fronts. At the edge of drying films, the reduced thickness causes particles to consolidate into a close-packed configuration. Continued evaporation from this solid region leads to a large transverse flow, bringing particles toward the packed region and, hence, propagating the close-packed front. In addition the flow of solvent through the packed region leads to a distribution of pressure. This process is sketched schematically in Figure 3.2 and described

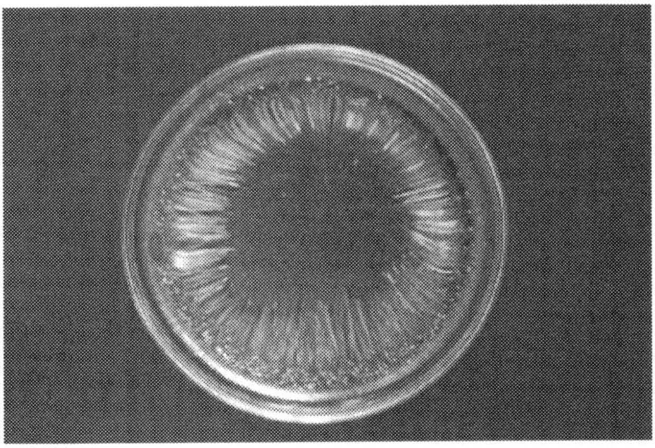

Figure 3.3 Picture of a dried film in a Petri dish. Here, 40-nm silica particles have been dried from a volume fraction of 20%. The radial cracks can be seen at the edge of the Petri dish, and the spacing is easily measured. Cracks actually propagate throughout the film, although debonding from the base allows the cracks at the edge to be observed.

fully in a previous publication. The characteristic horizontal length scale for drying fronts is found from a balance between surface tension driven flow and evaporation and is often called the capillary length scale. The speed of crack propagation can have a crucial influence over the cracking process. For stresses induced by differences in temperature this crack speed can be easily controlled. For the case of evaporation, Dufresne et al. examined the dynamics of cracking and showed cracks to follow the particle compaction front.

To minimize the effect of horizontal drying the experiments described here were performed in a Petri dish. A front of close-packed particles propagates from the center outward with cracks following the particle front. This simple drying procedure has been used previously to examine drying precipitates. A picture of a dried film is shown in Figure 3.3. The proposed cracking mechanism, presented here, is shown schematically in Figure 3.2. A crack is pulled apart because the capillary pressure in the solvent is below atmospheric. Once the material yields, the reduction in available volume eliminates the capillary pressure in the vicinity of the crack. This causes a flow of solvent toward areas of lower pressure in the bulk. The distance of flow sets the spacing between cracks. The capillary pressure, pulling cracks apart, is the same as that proposed by Scherer in sol-gel processing (Lee and Routh, 2004; Scherer, 1992).

3.4 Cracks by stress come from constraint of shrinkage by the substrate

An adherent film cannot contract in the plane of the substrate, so all of the shrinkage caused by drying and firing must be accommodated by decreasing the thickness of the film. This leads to development of tensile stresses in the film parallel to the interface with the substrate (Tirumkudulu and Russell, 2004). During drying, the total stress in the plane of the film is roughly equal to the capillary pressure (tension) in the liquid (Scherer, 1989; Brinker and Scherer, 1990), but there is relatively little residual drying stress, because most of the shrinkage is accommodated by viscous or plastic deformation. In contrast, the shrinkage caused by removal of OR groups (hydroxy or alkoxy) (von Preissig, 1989) occurs while the gel is rigid, so it causes high stresses. The magnitude of the stresses can be determined by depositing a film on a flexible substrate, and measuring the warping of the substrate as the film contracts (von Preissig, 1989). Using this technique, Garino and Bowen (1990) found that the stress in a piezoelectric ceramic transducer (PZT) film reached a maximum of about 200 MPa on heating to 350°C, as the organics were burned out; further heating caused the stress to decrease, because the amorphous network crystallized to pyrochlore, which has a higher thermal expansion coefficient than the silicon substrate. For a silica film, which remained amorphous, Parrill (1994) found that the stress rose to 200 MPa at 200°C and reached a maximum of 550 MPa at 500°C where most of the organics and hydroxyls were eliminated; if the film was cooled from 500°C, the stress relaxed as the newly formed siloxane bonds rehydrolyzed. Fardad et al. (1995) also found that most of the stress in a silica film resulted from pyrolysis of the organics; the higher the water content in the sol, the lower the residual organic content in the gel film, and the lower the stresses produced during pyrolysis.

The stresses are so high that it is remarkable that the films do not always crack. In fact it is found that there is a critical thickness for a film above which it will crack, and below which it cannot crack because there is not enough strain energy stored in the film to propagate the flaw (Hu and Evans, 1989). The accuracy of this picture has been demonstrated in experimental studies of drying byAtkinson and Guppy (1991) a critical thickness has also been observed for pyrolysis by Innocenzi et al. (1994).

If the film does not crack during drying or pyrolysis, it is not likely to crack during sintering or crystallization, because those processes require high atomic mobility; therefore, the stresses can be accommodated by viscous flow or diffusive creep. For an amorphous film, theory predicts that the sintering rate will be reduced (Scherer and Garino, 1985), but densification will still go to completion; this prediction has been verified for amorphous films. For crystalline films the situation is more complicated, because the substrate inhibits sintering, but not grain growth; therefore, the densification can be severely retarded, and even arrested by coarsening (Garino and Bowen, 1990). Unfortunately, there is no satisfactory theory for the sintering kinetics of a crystalline film, and no existing theory takes proper account of the anisotropy of the films.

Most sol-gel films are amorphous as-deposited, but many crystallize during sintering. Since crystallization requires shrinkage, the constraint of the substrate creates tensile stresses that affect the kinetics of the phase change. If the volume of a material changes from V_1 to V_2, the corresponding linear freestrain (ε_f) is

$$\varepsilon_f = \ln(V_2/V_1)^{1/3} \approx (V_2 - V_1)/(3V_1) \tag{3.9}$$

For a film on a rigid substrate undergoing such a strain, the stress in the plane of the film is

$$\sigma_x = -\frac{E\varepsilon_f}{1-\nu} \approx \frac{E}{1-\nu}\frac{V_1 - V_2}{3V_1} \tag{3.10}$$

where E and υ are the Young's modulus and Poisson's ratio for the porous film (not for the solid phase itself). The hydrostatic pressure in the film is $p = -(\sigma_x + \sigma_y + \sigma_z)/3$, and since $\sigma_x = \sigma_y$ and $\sigma_z = 0$,

$$p \approx \frac{2}{9}\frac{E}{1-\nu}\frac{V_2 - V_1}{V_1} \tag{3.11}$$

According to the Clausius–Clapeyron equation (Swalin, 1962), the temperature of the phase change would be shifted by

$$\Delta T \approx \frac{T(V_2 - V_1)p}{\Delta H} \tag{3.12}$$

where ΔH is the enthalpy for the phase change. If the phase change involves a contraction of a few percent (typical for crystallization of a glass) and the modulus of the porous film is on the order of about 10 GPa, the pressure caused by the phase change is on the order of about 100 MPa. Tension of that magnitude could significantly favor the phase with lower density (which will generally inhibit crystallization). Of course, at the temperature of crystallization, diffusive mechanisms can relax some of the stress, so the elastic analysis presented here will overestimate the actual stress.

After the thermal treatment is done, very large stresses can be generated on cooling, because of the thermal expansion mismatch between the film and substrate. Garino and Bowen (1990) and Swalin (1962) showed that tensile stresses of 350 MPa developed on cooling a PZT film from 550°C, where the pyrochlore phase appeared, because of the high thermal expansion coefficient of that phase. In contrast, if the film was fired to 650°C, the stress decreased on cooling, because the perovskite phase formed at that temperature, and it has a lower thermal expansion coefficient than the silicon substrate. Syms and Holmes (1994) deposited silica–titania films on silicon, and found that the residual stresses were tensile in films heated to $T \leq 975$°C (by which point OR groups were removed and some densification occurred), but the stresses were compressive in films heated to $T \geq 1075$°C. In this case, viscous flow relieved the tension created by decomposition and densification, so the room-temperature stresses were dominated by the thermal expansion difference between the film and substrate.

3.5 Rapid sol-gel fabrication to confront tensile trailing cracks

A versatile and rapid sol-gel technique for the fabrication of high-quality one-dimensional photonic bandgap materials was developed. Silica/titania multilayer materials are fabricated by a sol-gel chemistry route combined with dip-coating onto planar or curved substrates. A shock-cooling step immediately following the thin-film heat-treatment process is introduced. This step was found crucial in the prevention of film crack formation, especially in silica/titania alternating stack materials with a high number of layers (Barhoum et al., 2011).

The dominating challenge in fabricating high-quality thin film single and multi-layer structures is the suppression of crack formation. Crack formation is the result of stresses occurring during and after the deposition process that exceed the thin film strength (Bockmeyer and Löbmann, 2006, 2007). For inorganic thin films derived by sol-gel chemistry cracks can form at multiple processing stages. We designed a sol-gel dip-coating method optimized to address these challenges at the three most important processing stages: the initial film drying, the preannealing step, and the annealing (heat-treatment) process. The major steps of our process for fabricating highly reproducible, large-area, crack-free, single and multilayer structures of silica and titania, as well as stacks of alternating layers of these two compounds, are schematically shown in Figure 3.4.

During the drainage process, which starts immediately after the film withdrawal in the dip-coating process, inhomogeneous solvent evaporation within the film creates a pressure gradient (Woignier et al., 1988). The gradient causes a differential shrinkage of the thin-film network, whereby the exterior of the network shrinks faster than the interior. This leads to the build-up of tensile stresses that can fracture the thin film upon ongoing solvent evaporation. It was found that this problem could be overcome by designing the dip-coating process such that the freshly deposited film is withdrawn into an environment of high solution vapor pressure. This is achieved by simply using a tall vial that is only half-filled with the deposition solution (Figure 3.4a). The deposited film is subsequently subjected to a drying step in a flowing hot-air stream generated by a heat gun (Figure 3.4b), resulting in a rapid and homogeneous solvent evaporation. We found the combination of these two steps counteracts the differential shrinkage within the deposited layer and results in the formation of crack-free films. In addition, the heat-gun treatment greatly reduces the occurrence of edge effects. Edge effects in deposited thin films are a known phenomenon especially associated with dip-coating processes, and are caused by substrate vibrations and inhomogeneous solvent evaporation in the center and at the edges of deposited films during the withdrawal process. Heat-treatment of the deposited films is the final stage of the fabrication process. During heat-treatment residual solvent and other organic components of the precursor solution are pyrolytically removed and the loosely formed oxide network is further densified. In the case of titania films, heat-treatment also leads to partial crystallization of the initially amorphous network. However, these reactions during heat-treatment

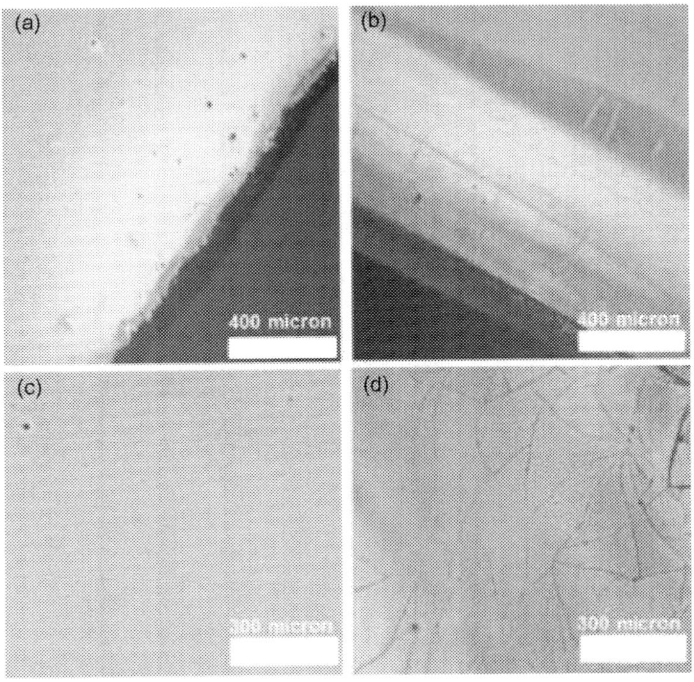

Figure 3.4 Optical micrographs of 12 alternating silica/titania thin-film layers deposited onto a silicon (100) wafer. Top: Samples fabricated (a) including and (b) omitting the dry-blowing step in the deposition cycle. Bottom: Samples fabricated (c) including and (d) omitting the shock-cooling step in the deposition cycle.

also lead to further film shrinkage, accompanied by an increase in tensile stresses, potentially causing crack formation. On the other hand, it is known that rapid heating rates increase the viscoelasticity of the initially forming amorphous oxide network (Bockmeyer and Löbmann, 2006, 2007). This network relaxation counteracts the tensile stress formation and suppresses crack formation.

It was found that placing the deposited and hot-air treated films into a preheated oven at 350–500°C for 1 minute gave the best results (Figure 3.4c). Similar to the heating process the cool-down stage must be optimized to create high-quality films. While this stage of the heat-treatment is often neglected, we found it to be of particular importance for fabricating crack-free thin-film stacks composed of alternating layers of silica and titania. Several factors are responsible for the formation of cracks during the cool-down process. One source is the difference in expansion coefficients between the alternating deposited layers as well as between the substrate and the deposited film. Another source of cracking is reorganization of the thin film metal oxide network. This reorganization is caused by phase transitions of the amorphous or amorphous/polycrystalline network upon cooling, leading to a build-up of stresses that can fracture the film network. We studied the cool-down process in detail and found that similar to the heating stage rapid cooling rates are beneficial in the prevention of crack formation. Although perhaps counterintuitive, we found the best results were obtained by shock-cooling the film samples to room temperature immediately following the 1-minute heating period. Shock-cooling was achieved by placing the hot samples onto a Pyrex glass substrate at room temperature or immersing them into a water/isopropanol mixture (Figure 3.4d). While the exact mechanism is still under investigation, we propose rapid cooling minimizes the occurrence of interfacial stresses (because of the difference in thermal expansion coefficients of alternating layers). In addition, temperature quenching freezes high-temperature amorphous/polycrystalline phases (Woignier et al., 1988), maintaining high viscoelasticity of the oxide network at room temperature. To conclude, the versatility of sol-gel chemistry combined with our inexpensive and rapid thin-film processing route (a 20-layer silica/titania stack can be fabricated in less than 2 hours) makes our method an attractive alternative to conventional physical/chemical deposition and sputtering techniques. The fabricated multilayer photonic bandgap structures possess excellent structural and optical properties and should be interesting candidates for use in optical and electronic applications.

3.6 Anti-abrasive SiO_2 film in application: self-assembling covalently bonded nanocoating

Room temperature hydrolysis and condensation reactions utilize nanotechnology to deliver site-specific performance attributes. Self-assembling molecules bond covalently and form nanoscale films that can enhance chemical and mechanical performance. Silane chemistry is typically utilized and customer-specific applications require different functional groups, activation promoters and suitable solvent/carriers. A typical hydrolysis and condensation reaction is illustrated in Figure 3.5.

Molecules settle on the glass substrate surface and covalently bond while self-assembling into a continuous thin film less than 500 nm. Durable film-forming compositions can be formed from amphiphilic molecules consisting essentially of RmSiXn where R is a polar or fluorinated alkyl group of about 10–30 carbon atoms, m plus n equals 4, and X is selected from the group consisting of halogens, hydroxyl, and alkoxy groups. Invisible and optically clear films with visible light transmission of >98% can be achieved. The amphiphilic chemistry of the R molecule lowers surface energy and

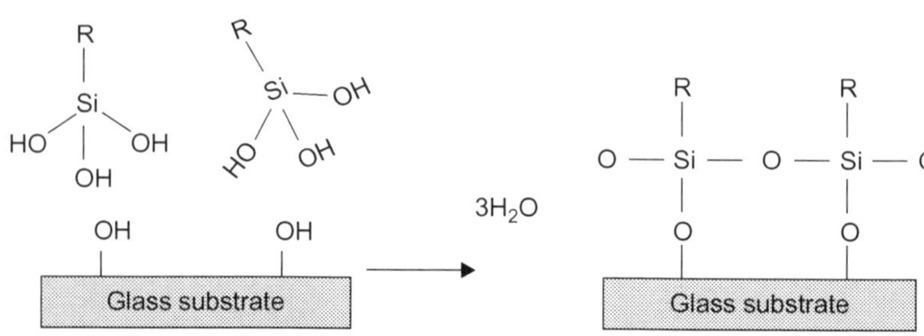

Figure 3.5 Self-assembling hydrolysis condensation reaction.

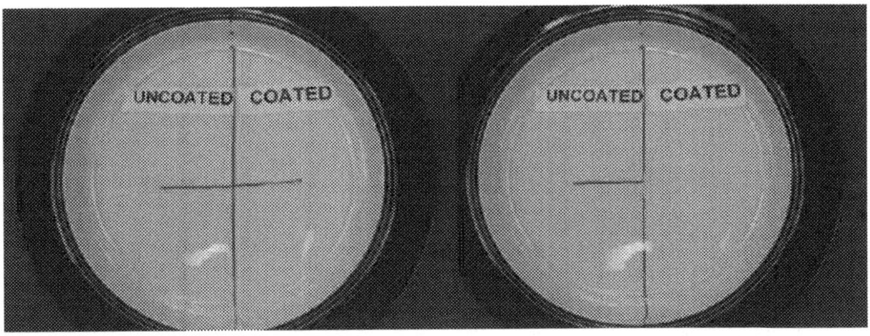

Figure 3.6 Permanent marker pen is removed by Scotch Tape on coated Glass.

Figure 3.7 Treated and un-treated automotive glass in a rainstorm.

can repel water and oils creating hydrophobic and oleophobic functionality making glass surfaces harder to soil, easier to clean, stain resistant, and chemical resistant (Figure 3.6). Mechanical properties include mar resistance and enhanced micro-scratch resistance. Continuous long exposures to elevated temperatures, UV, chemicals, and mechanical action will eventually erode these molecular layers. However these coatings are easy to apply with rapid room temperature curing and can be easily refurbished as needed in the field.

Applications include automotive windshields and side lights. Visibility in a rainstorm is improved because the rain is repelled (Figure 3.7). This increases the driver's vision 34% on a rainy night, which can add about 1 second to a driver's response time.

3.7 Abrasive test

The scratch test has been used for many years to provide a measure of coating/substrate adhesion (Perry, 1981; Steinmann and Hintermann, 1985; Valli, 1986; Hintermann, 1984; Perry, 1983; Bull and Rickerby, 1990). In the normal configuration of the test a diamond stylus is drawn across the coated surface under an increasing load until some well-defined failure occurs at a load which is often termed the critical load, L_c. Many different failures are observed which include coating detachment, through-thickness cracking, and plastic deformation or cracking in the coating or substrate (Benjamin and Weaver, 1960; Bull 1995). In fact, it is usual that several different failure modes occur at the same time and this can make the results of the test difficult to interpret.

The failure modes observed in the scratch test depend on many factors and are most easily characterized in terms of the hardness of both substrate and coating (Figure 3.8). In the case of a typical Rockwell 'C' diamond indenter (1208 cone with 200 mm hemispherical tip), for soft coatings and soft substrates the test is dominated by plastic deformation and groove formation and little or no cracking is observed except at very high loads. For hard coatings on soft substrates deformation of the substrate is predominantly plastic while the coating may plastically deform or fracture as it is bent into the track created by plastic deformation of the substrate. Soft coatings on a harder substrate tend to deform by plastic deformation and some extrusion of the coating from between the stylus and the substrate may occur. Considerable thinning of the coating by plastic deformation will occur before plastic deformation and fracture of the substrate becomes significant. For hard coatings on hard substrates plastic deformation is minimal and fracture dominates the scratch response.

The scratch test is usually only regarded as semi-quantitative as there are a number of intrinsic and extrinsic parameters which are known to affect the measured critical load. Many of these intrinsic factors are instrument-specific and require a careful calibration approach if results are to be compared between instruments. However, the extrinsic factors such as coating thickness and substrate hardness must also be known if the results of the test are to be understood. These parameters, together with the residual stress in the coating and its Young's modulus, are an important requirement for the models of the failure mode used to generate

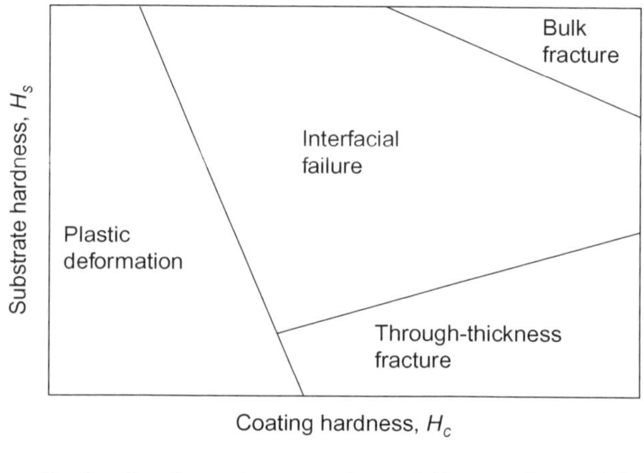

Figure 3.8 Schematic showing the various scratch test failure modes, which dominate as a function of coating and substrate hardness (H_c and H_s, respectively).

interfacial fracture toughness. There are thus four requirements for a quantitative scratch adhesion test:

1. an adhesion-related failure mode;
2. a well-defined failure mechanism;
3. a method to identify that adhesion failure has occurred, where it is located and the size of the failure;
4. a method for determining the stresses which cause failure.

Meanwhile, the determination of adhesion and the scratch resistance is essential for the mechanical characterization of thin and hard films both for research and industry. To evaluate the performance of coatings the scratch test with a Rockwell-C diamond stylus as indenter is a commonly used test for the characterization of hard coatings on predominant metallic substrates because of its comparatively robust experimental setup, easy use, and representative results (Figure 3.9). However, the versatility as compared to other mechanical test methods (Sander et al., 2011) is limited for several applications. The scratch test was modified by using bearing balls instead of the Rockwell-shaped indenter. This allows to easily change the contact radius and the material of the indenter and therefore to influence the stress field forming in the composite of coating and substrate (Figures 3.10 and 3.11).

First measurements were performed and demonstrated on hard layers of CrN and TiN deposited on a soft polymeric substrate of ASA. The comparison between the scratches with the Rockwell-C indenter and the bearing balls in different sizes during the tests of the hard coating on soft substrates shows clearly that the obtained failure modes can vary significantly between the two test methods. The comparatively sharp Rockwell indenter could not generate a failure mode at a reasonable load range in this case. This could lead to the wrong assumption, that the adhesion and scratch resistance of the coated surfaces are very good and suitable for most applications. Enlarging the indenter radius by using bearing balls as indenters, the stress field in

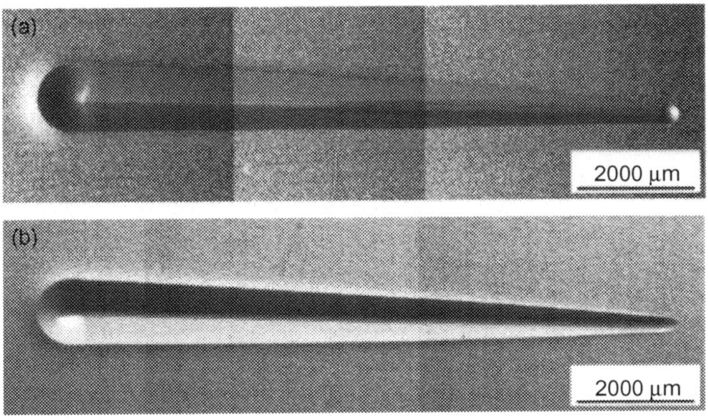

Figure 3.9 Scratch test with common Rockwell-C indenter, linearly increasing load 0–100 N on soft ASA-substrate and thin hard layers (a) CrN and (b) TiN.

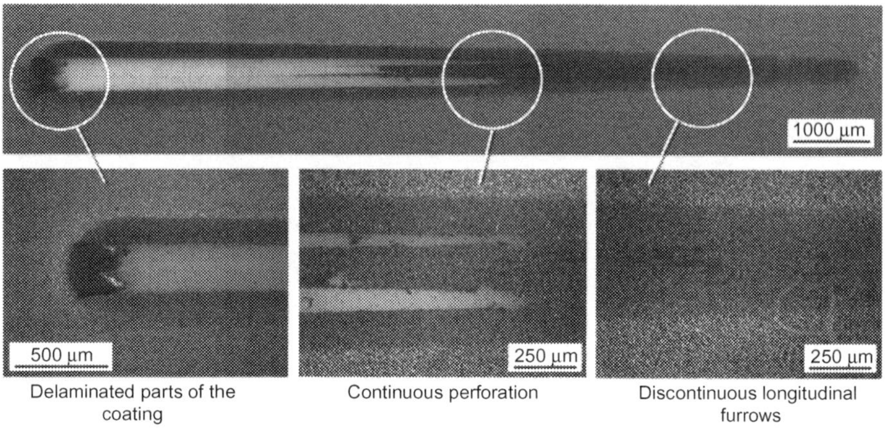

Figure 3.10 Modified scratch test of CrN on ASA substrate with tip radius of $R = 1.9845$ mm and the occurring failure modes.

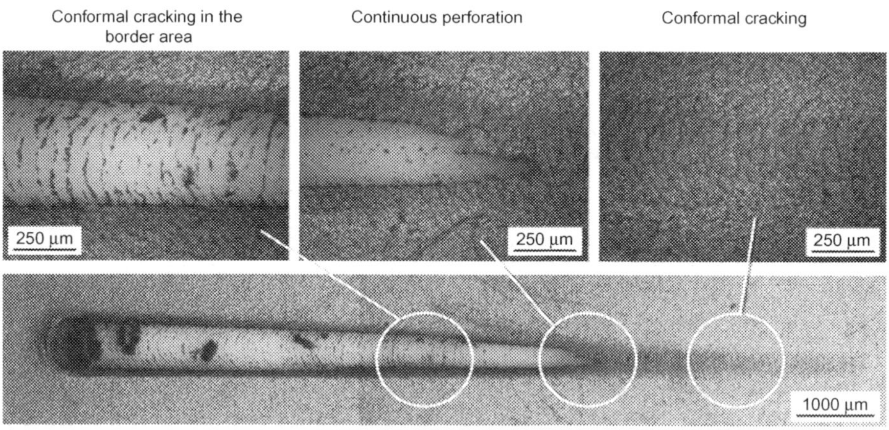

Figure 3.11 Modified scratch test of TiN on ASA substrate with tip radius $R = 1.389$ mm.

the contact area is shifted from bending stress more to compressive stress. The critical forces of the resulting failure modes can be quantitatively measured in order to evaluate the performance of the coating-substrate-composite (Sander et al., 2011).

3.8 Anti-abrasive performance of sol-gel nanocoatings

3.8.1 The process of obtaining sol-gel nanocoatings

A flow chart of the double-layer coating synthesis is shown in Figure 3.12. Tetraethylorthosilicate (TEOS) was mixed with ethanol (EtOH), distilled water (H_2O)

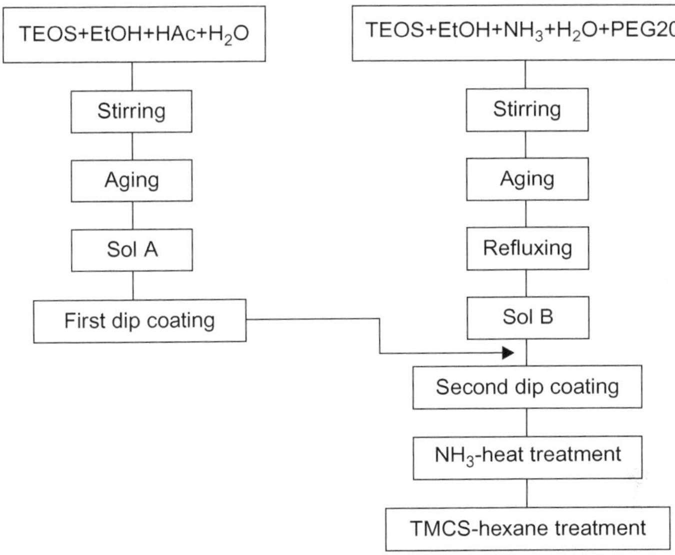

Figure 3.12 Coating synthesis flow chart.

and acetic acid (HAc) [pH = 1]. The molar ratio was TEOS:EtOH:H_2O:HAc = 1:38:2:0.5. This clear, acidic solution was considered ready-for-use after seven days' ripening at room temperature. Dense and high-index silica film could be obtained from this acidic sol via dip coating. TEOS, ethanol, distilled water, polyethylene glycol (PEG200), and ammonia were mixed together for the base-catalyzed sol. The molar ratio was TEOS:EtOH:H_2O:NH_3:PEG200 = 1:38:2:0.54:0.15. These solutions were aged at room temperature for 3–5 days and refluxed at 80°C for 6–20 hours. Porous and low-index silica film could be obtained from this basic sol via dip coating. The glasses (BK7) were coated using these two kinds of sols, and the desired thickness could be adjusted by the withdrawal rate. The low-index layer was deposited after just a 2-minute wait for the high-index layer drying in the air. Coated glasses were then put into the atmosphere of NH_3 and H_2O, with surrounding temperature elevated slowly from room temperature to 200°C and kept for more than 5 hours. After that, samples were soaked in TMCS solution for 6 hours, and then in ethanol solution for half an hour. At last, samples were dried by blowing with N_2.

3.8.2 Cross-sectional morphology

The SEM image of the cross section of this double-layer coating (silicon wafer as substrate) is shown in Figure 3.13. The interface between the two layers is clear, which means no obvious infiltration happened during the preparation. The low-index layer appears loose and of a brighter color than the other layer and

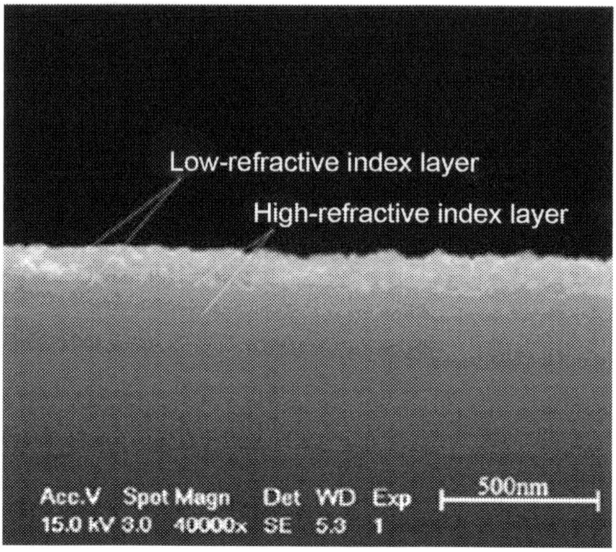

Figure 3.13 The SEM image of the cross section of the double-layer coating.

the substrate, which corresponds to its porous structure. As a comparison, both the high-index layer and the substrate are dense and compact. Usually, double-layer anti-reflection AR coating derived from the sol-gel method needs a heat-treatment process after each deposition. However, in our experiment it is not necessary to heat the coating after the first deposition, just 1–2 minutes of waiting is enough for the second step deposition. This is because the acid-catalyzed silica coating is dense enough to prevent the infiltration coming from the outer layer. Omitting one heat treatment during the preparation process may save much time and cost when the coating is industrialized.

3.8.3 Surface morphology

The two-dimensional atomic force microscope (AFM) image of the double-layer coated glass surface is shown in Figure 3.14. The obvious fluctuation on the surface indicates the existence of particle and pore structures, which is in accord with the base-catalyzed silica film. Usually, the acid-catalyzed silica film features a very flat surface with the RMS (Root Mean Square) roughness less than 1 nm (Wang and Shen, 2010). As a comparison, the RMS roughness of this double-layer coating is about 4.32 nm, similar to that of the single layer base-catalyzed silica film (Jun et al., 2008). This means that the surface morphology only depends on the porous film. Although the fluctuation is violent compared with that of the dense film, it is still very small relative to the wavelength we are concerned with, so the scattering waste can be neglected.

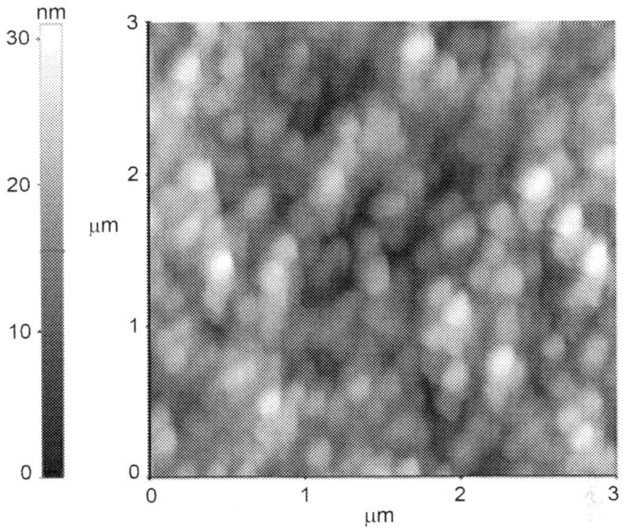

Figure 3.14 Two-dimensional AFM image of the coating's surface.

3.8.4 Mechanical performance

It is widely known that the dense acid-catalyzed silica film presents a good adhesion property, while the as-deposited porous colloidal silica film is not scratch-resistant (Li and Shen, 2011). Usually, NH_3 atmosphere treatment at room temperature can be employed to enhance the mechanical property, but not in a large degree. In our experiment, we elevated the surrounding temperature when the atmosphere treatment was underway, and found that the scratch-resistance of the coating was greatly improved after NH_3 heat-treatment with ambient temperature of 200°C. This should be attributed to the large amount of firm Si—O—Si bond generated between the two films as well as the silica particles. The chemical reaction can be described as follows:

$$\mathrm{RS}i - \mathrm{OH} + \mathrm{OH}^- \xrightarrow[200^\circ\mathrm{C}]{\mathrm{NH_3 + H_2O}} \mathrm{RS}i - \mathrm{O}^- + \mathrm{H_2O} \tag{3.13}$$

$$\mathrm{RS}i - \mathrm{OH} + \mathrm{RS}i - \mathrm{O}^- \xrightarrow[200^\circ\mathrm{C}]{\mathrm{NH_3 + H_2O}} \mathrm{RS}i - \mathrm{O} - \mathrm{S}i\mathrm{R} + \mathrm{OH} - \tag{3.14}$$

The first step is deprotonation of the peripheral silanols of the base-catalyzed silica film. In the second, particle-bonding occurs leading to surface condensation.

The temperature rise of 200°C, which accelerates the water release, enhances the formation of the Si—O—Si bond. The abrasion-resistance test was carried out using the method suggested by Floch et al. – namely, rubbing the coating 100 times with cotton balls immersed with dust and ethanol, then comparing the transmission spectra before and after rubbing.

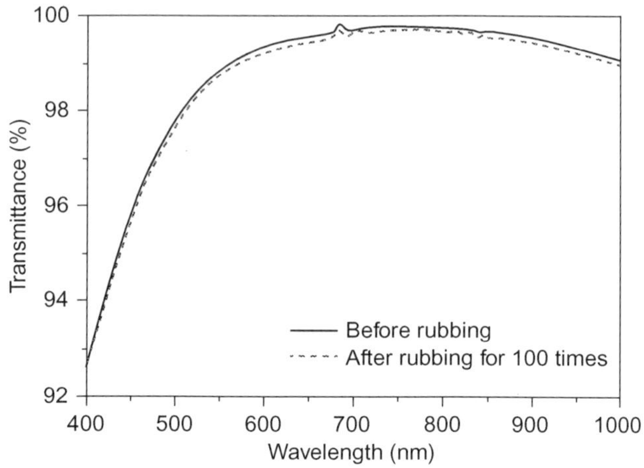

Figure 3.15 Transmittance of the coated glass before and after rubbing.

This test method has been employed by many researchers (Belleville et al., 2000) and is considered very suitable for comparison purposes. In our experiment, no obvious damage could be found on the coating by visual inspection after rubbing 100 times, and the transmission spectra before and after rubbing are very close to each other, as shown in Figure 3.15.

The average transmittance after scratching decreased just by 0.12% (400–1000 nm). In comparison, the coating without treatment can be totally scratched off within rubbing three times using the same method, leaving naked glass with about 4% reflectance of one side. The 'hunchback' at about 675 nm in Figure 3.15 may come from some fortuitous noise of the spectrophotometer sensor.

3.9 Conclusion

The sol-gel process can be carried out at low temperatures. In this process, an organic functional group or organic component can easily be introduced in the inorganic or organic backbone network. The nanocoatings of traditional materials can acquire new specific features through the sol-gel method. One of its applications is to obtain a nanocoating with sufficient hardness and abrasion resistance for protecting a sample. However, stress is an important factor to be considered. Thin film stress can affect the performance of scratch-resistant films. We can improve the impact of film stress with theory and a variety of ways such as heating, ammonia treatment, and other surface modification methods. We can even prepare nanocoatings with different composite materials to diversify the film's surface. In short, we are still improving film stress to enhance the anti-abrasiveness of sol-gel derived nanocoatings as much as possible.

Acknowledgments

Many works have been provided by Professor Shen Jun, Dr Xiaodong Wang, and Xiaoguang Li. Zhiyong Xie is gratefully acknowledged. The authors also express their appreciation for the help of Zhong Chen, Linda Y.L., Jianda Shao, Shuying Shao, J.B. Oliver, H.G. Floch, F.J. von Preissig, Dong Wu, Wai Peng Lee, Alexander F. Routh, Claude A. Klein, Yuhan Sun, Liping Liang, and G. Sakoske. The encouragement and patience of Prof. Z.A. Niknam, one of the editors of this book, is very much appreciated. Shengzhao Wang particularly acknowledges the guidance and help he has received over the years from his PhD supervisor, Professor Jun Shen, a great scholar.

References

Allain, C., Limat, L., 1995. Regular patterns of cracks formed by directional drying of a colloidal suspension. Phys. Rev. Lett. 74 (15), 2981–2984.

Atkinson, A., Guppy, R.M., 1991. Mechanical stability of sol-gel films. J. Mater. Sci. 26 (14), 3869–3873.

Barhoum, M., Morrill, J.M., Riassetto, D., Bartl, M.H., 2011. Rapid sol-gel fabrication of high-quality thin-film stacks on planar and curved substrates. Chem. Mater. 23 (23), 5177–5184.

Belleville, P.F., Bonnin, C., Priotton, J.J., 2000. Room-temperature mirror preparation using sol-gel chemistry and laminar-flow coating technique. J. Sol-Gel Sci. Technol. 19 (1–3), 223–226, http://dx.doi.org/10.1023/A:1008788322168.

Benjamin, P., Weaver, C., 1960. Measurement of adhesion of thin films. Proc. Roy. Soc. Lond. Ser. A. 254, 163–176.

Bockmeyer, M., Löbmann, P., 2006. Densification and microstructural evolution of TiO_2 films prepared by sol-gel processing. Chem. Mater. 18, 4478.

Bockmeyer, M., Löbmann, P., 2007. Crack formation in TiO2 films prepared by sol-gel processing: quantification and characterization. Thin Solid Films. 515, 5212.

Brinker, C.J., Scherer, G.W., 1990. Sintering of sol-gel films. J. Sol-Gel Sci. Technol. 8 (1–3), 353–363.

Bull, S.J., 1995. Spallation failure maps from scratch testing. Mater. High Temp. 13, 169–174.

Bull, S.J., Rickerby, D.S., 1990. New developments in the modelling of the hardness and scratch adhesion of thin films. Surf. Coat. Technol. 42, 149–164.

Dufresne, E.R., Corwin, E.I., Greenblatt, N.A., Ashmore, J., Wang, D.Y., et al., 2003. Flow and fracture in drying nanoparticle suspensions. Phys. Rev. Lett. 81 (22), 4501.

Fardad, M.A., Yeatman, E.M., Dawnay, E.J.C., Green, M., Horowitz, F., 1995. Effects of H2O on structure of acid-catalysed SiO2 sol-gel films. J. Non-Cryst. Solids. 183, 260–267.

Friedrich, K., Schlarb, A.K., 2008. In: Briscoe, B.J. (Ed.), Tribology of Polymeric Nanocomposites: Friction and Wear of Bulk Materials and Coatings, Chapter 14, Scratch Resistance of Protective Sol-Gel Coatings on Polymeric Substrates (Chen, Z., and Wu, L.Y.L.), First ed. Elsevier, Amsterdam, The Netherlands, p. 326.

Garino, T.J., Bowen, H.K., 1990. Kinetics of constrained-film sintering. J. Am. Ceram. Soc. 73 (2), 251–257.

Hintermann, H.E., 1984. Adhesion, friction and wear of thin hard coatings. Wear. 100, 381–397.

Hu, M.S., Evans, A.G., 1989. The cracking and decohesion of thin films on ductile substrates. Acta Metall. 37 (3), 917–925.

Hutchison, J.W., Suo, Z., 1992. Mixed mode cracking in layered materials. In: Hutchison, J.W., Wu, T.Y. (Eds.), Advances in Applied Mechanics, vol. 29. Academic Press, Waltham, MA, pp. 63–191.

Hutchison, J.W., Wu, T.Y. (Eds.), 1992. Advances in Applied Mechanics, vol. 29. Academic Press, Waltham, MA, p. 29.

Innocenzi, P., Abdirashid, M.O., Guglielmi, M., 1994. Structure and properties of sol-gel coatings from methyltriethoxysilane and tetraethoxysilane. J. Sol-Gel Sci. Technol. 3, 47–55.

Iwamoto, T., Mackenzie, J.D., 1995. Ormosil coatings of high hardness. J. Mater. Sci. 30, 2566–2570.

Jianye, W., Wilkes, G.L., 1996. Organic/inorganic hybrid network materials by the sol-gel approach. Chem. Mater. 8, 1667.

Jun, S., Zhiyong, X., Xiaoxian, W., Ling, O., Yiqun, X., 2008. Sol-gel derived SiO2 antireflective (AR) coating used in solar cells. Rare Met. Mater. Eng. 37 (Suppl. 2), 47–50.

Klein, C.A., 2000. How accurate are stoney's equation and recent modification. J. Appl. Phys. 88, 5487.

Komatsu, T.S., Sasa, S., 1997. Pattern selection of cracks in directionally drying fracture. Jpn. J. Appl. Phys., Part 1. 36, 391–395.

Lee, W.P., Routh, A.F., 2004. Why do drying films crack? Langmuir. 20 (23), 9885–9888.

Li, X., Shen, J., 2011. A scratch-resistant and hydrophobic broadband antireflective coating by sol-gel method. Thin Solid Films. 519, 6236–6240.

Liang, L., Xu, Y., Zhang, L., Wu, D., Sun, Y, 2008. Polyvinylpyrrolidone/ZrO2-based sol-gel films applied in highly reflective mirrors for inertial confinement fusion. J. Sol-Gel Sci. Technol. 47 (2), 173–181, http://dx.doi.org/10.1007/s10971-008-1784-2.

Ohring, M., 2002. Materials Science of Thin Films: Deposition and Structure. second ed. Academic Press, Waltham, MA, pp. 723–730.

Oliver, J.B., Kupinski, P., Rigatti, A.L., Schmid, A.W., Lambropoulos, J.C., et al., 2012. Stress compensation in hafnia/silica optical coatings by inclusion of alumina layers. Opt. Express. 20 (15), 16596–16610, http://dx.doi.org/10.1364/OE.20.016596.

Parrill, T.M., 1994. Heat treatment of spin-on acid-catalyzed sol-gel silica films. J. Mater. Res. 9 (3), 723–730.

Perry, A.J., 1981. The adhesion of chemically vapour deposited hard coatings to steel – the scratch test. Thin Solid Films. 78, 77–93.

Perry, A.J., 1983. Scratch adhesion testing of hard coatings. Thin Solid Films. 107, 167–180.

Sander, T., Tremmel, S., Wartzack, S., 2011. A modified scratch test for the mechanical characterization of scratch resistance and adhesion of thin hard coatings on soft substrates. Surf. Coat. Technol. 206, 1873–1878.

Scherer, G.W., 1989. Drying gels: VIII. Revision and review. J. Non-Cryst. Solids. 109, 171–182.

Scherer, G.W., 1992. Drying of ceramics made by sol-gel processing. In: Mujumdar, A.S. (Ed.), Drying '92. Elsevier Science, New York, pp. 92–113.

Scherer, G.W., Garino, T., 1985. Viscous sintering on a rigid substrate. J. Am. Ceram. Soc. 68 (4), 216–220.

Schmidt, H., Seiferling, B., Philipp, G., et al., 1988. Development of organic–inorganic hard coatings by the sol-gel process. In: Mackenzie, J.D., Ulrich, D.R. (Eds.), Ultrastructure of Process of Advanced Ceramic. John Wiley & Sons, New York, p. 651.

Shao, S., Fan, Z., Shao, J., He, H., 2003. Evolutions of residual stress and microstructure in ZrO2 thin films deposited at different temperatures and rates. Thin Solid Films. 445 (1), 59–62.

Shen, J., Zhang, Q., Wang, J., Yang, T., Deng, Z., et al., 2000. Sol-gel processing of Zirconia coating for HR mirrors with high laser damage threshold. J. Sol-Gel Sci. Technol. 19, 271–274, http://dx.doi.org/10.1023/A:1008761010781.

Shen, J., Wang, S.-Z., Wang, X.-D., Liu, Y., Li, X.-G., 2009. Refractive index controllable coatings for optical detectors. Proc. SPIE. 7381, 7381C1–7381C8, http://dx.doi.org/10.1117/12.833909.

Steinmann, P.A., Hintermann, H.E., 1985. Adhesion of TiC and TiCN coatings on steel. J. Vac. Sci. Technol. A3, 2394–2400.

Stoney, G.G., 1909. The tension of metallic films deposited by electrolysis. Proc. R. Soc. Lond., A Contain. Pap. Math. Phys. Character. 82 (553), 172–175.

Swalin, R.A., 1962. Thermodynamics of Solids. Wiley, New York, pp. 74–75.

Syms, R.R.A., Holmes, A.S., 1994. Deposition of thick silica-titania sol-gel films on Si substrates. J. Non-Cryst. Solids. 170, 223–233.

Tikhonravov, A., Trubetskov, M., Amotchkina, T., Kokarev, M., 2004. Key role of the coating total optical thickness in solving design problems. In: Amra, C., Kaiser, N., Macleod, H.A. (Eds.), Advances in Optical Thin Films Bellingham, 5250. SPIE, WA, pp. 312–321.

Tirumkudulu, M.S., Russell, W.B., 2004. Role of capillary stresses in film formation. Langmuir. 20 (7), 2947–2961.

Uhmann, D.R., Teowee, G., Boulton, J., 1997. The future of sol-gel science and technology. J. Sol-Gel Sci. Technol. 8, 1083–1091.

Valli, J., 1986. A review of adhesion test methods for thin hard coatings. J. Vac. Sci. Technol. A4, 3001–3014.

von Preissig, F.J., 1989. Applicability of the classical curvature-stress relation for thin films on plate substrates. J. Appl. Phys. 66 (9), 4262–4268.

Wang, J., Zeng, S., 2008. Thermal stresses analysis in adhesive/solder bonded bimaterial assemblies. J. Appl. Phys. 104 (11), 113508-1–113508-8.

Wang, X., Shen, J., 2010. Sol-gel deriveddurable antireflective coating for solar glass. J. Sol-Gel Sci. Technol. 53 (2), 322–327.

Woignier, T., Phalippou, J., Hdach, H., Scherer, G.W., 1988. In: Zelinski, B.J.J., Brinker, C.J., Clark, D.E., Ulrich, D.R. (Eds.), Better Ceramics Through Chemistry III. Mateial Research Society, Pittsburgh, PA, pp. 1087–1099.

Wu, G., Shen, J., Zhou, B., Ni, X., Wang, J., 2002. Sol-gel derived scratch-resistant nanoporous silica films and their anti-reflective properties. Paper presented at Shanghai international Nanotechnology Cooperation Symposium, 30 July–1 August.

Yang, X.F., Tallman, D.E., Gelling, V.J., Bierwagen, G.P., Kasten, L.S., 2000. Use of a sol-gel conversion coating for aluminum corrosion protection. Surf. Coat. Technol. 140, 44–50.

Self-cleaning glass

4

Krishnan Surekha and Sureshbabu Sundararajan†*
*Department of Mechanical Engineering Science, University of Johannesburg, South Africa; †Ndizani Civil Works, Durban, South Africa

4.1 Introduction

Nanotechnology has set foot in various fields, such as medicine, electronics, textiles, cosmetics, food, automobile, energy and building and construction. Drug delivery for cancer cells, nanosensors and nanoelectro mechanical systems (NEMS), water- or stainproof cloth, sunscreens and face cream with nanoparticles and self-cleaning windows for automobile and construction are a few applications of nanotechnology in the above fields. Table 4.1 lists some of the applications of nanotechnology in diferent fields.

Glass is widely used in building and construction, the transport industry, solar energy glass products, various appliances including microscopes and tablet computers and furniture. The advantages of using glass in building and construction are (1) it controls light by letting in the good rays and keeping out the bad ones; (2) it saves energy costs; (3) it provides natural light during the day; (4) it harmonizes a structure with its environment (The Glass Guide, 2002). These advantages have made glass a viable material for construction. Hence, the modern construction industry uses glass in areas that are very hard to reach for cleaning. As ‘necessity is the mother of invention’, scientists and researchers have invented a new type of glass called *self-cleaning glass*, which needs minimal maintenance. Those wearing spectacles will welcome the glare-free glass that does not mist up when drinking hot beverages in winter or while cooking. Besides this, with anti-fogging glass, tablet computers can be used near swimming pools and anti-reflective glass will enable the user to use a cell-phone handset or a laptop in broad daylight. Glass windows and doors for offices and homes needs minimal cleaning with these types of self-cleaning glass (SCG).

SCG has a very thin layer of titanium oxide (TiO_2) or silicon dioxide (SiO_2) in the order of 10–25 nm coated on the its surface to achieve the special cleaning properties. Besides self-cleaning properties, SCG is expected to possess multifunctional properties such as anti-reflective, anti-fogging and anti-abrasive properties. The history, types, methods of fabrication, mechanics and advantages and disadvantages of SCG are discussed in this chapter.

Anti-Abrasive Nanocoatings.

Table 4.1 **Applications of nanotechnology**

Field	Category	Research carried out
Medicine	Reproductive medicine	Nanomaterials used as experimental tools for the detection and treatment of malignant and benign reproductive conditions (Barkalina et al., 2014).
	Dentistry	Nanotechnology has been in dentistry for tooth sealants and fillers that use nanosized particles to improve their strength and lustre, and resist wear. In future nanotechnology will be able to repair and rebuild damaged enamel (Subramani and Ahmed, 2012).
	Gastrointestinal disorders	The main objectives for integrating nanoparticle and nanodevice applications for medical diagnosis and disease treatment in the context of intestinal diseases are reviewed (Laroui et al., 2013).
	Drug delivery	Problems encountered in oral chemotherapy of anti-cancer drugs and the pharmaceutical nanotechnology solutions such as prodrugs, nano-emulsions, dendrimers, micelles, liposomes, solid lipid nanoparticles and nanoparticles of biodegradable polymers are listed (Mei et al., 2013).
	Bio-sensors	Nanotechnology-based biosensing and imaging techniques that are improving screening and diagnosis of cancers are reviewed (Zeineldin, 2013).
	Neurology	Early detection of Alzheimer's disease is possible by nanoparticulate entitites and nanoparticles (Brambilla et al., 2011).
	Urology	Wide range of research and clinical applications of nanomedicine in the field of urology is discussed (Gommersall et al., 2007).
	Osteology	An overview of tissue engineering material systems used in bone healing is given. Nanotechnology is used to produce scaffolds with an interconnected porous network of micro- and nanoscale to allow extensive body fluid transport through the pores (Stylios et al., 2007).
Electronics	Data storage	About 31.6 Gb/in^2 surface data-writing density was realized successfully on thin polymethyl methacrylate (PMMA) film by using a silicon nanomechanical electro-thermal probe (Yang et al., 2006).
Textiles	UV ray protection	Swimming fabrics which will not absorb UV rays and have high hydrophobicity are produced by impregnating 3–5% ZnO in polyester (Broasca et al., 2013).
	Antibacterial cloth	Gamma irradiation with $AgNO_3$ chitosan solution and immobilization 12 nm diameter silver nanoparticles into cotton fabric yielded anti-bacterial cloth (Hanh et al., 2014).

(Continued)

Table 4.1 (Continued)

Field	Category	Research carried out
	Flame retardant	Flame retardant clothing is developed by depositing nanocoatings on the fabric surface through nanoparticle adsorption, layer-by-layer assembly, sol-gel and dual-cure processes or plasma deposition (Alongi et al., 2014).
	Smart fabrics	Ten bio-inspired clothes, such as stain-resistant, tear-resistant, touch-sensitive and breathable, are reviewed (Singh et al., 2012).
Cosmetics	Preservative	Ag nanoparticles when added as preservative to cosmetics increase the shelf life without any sedimentation (Kokura et al., 2010).
	UV ray protection	Addition of vegetable oil nano-carriers of 100 and 120 nm to cosmetics improved the UV ray protection when compared to conventional cream (Lacatusu et al., 2014).
Food	Packaging	Possibility of developing antimicrobial glass surfaces coated with nano-encapsulated cinnamaldehyde for active packaging of liquid foods was studied (Makwana et al., 2014).
		Silver based nano-engineered materials are currently the most commonly used in commodities because of their antimicrobial capacity. Copper, zinc and titanium nanostructures are also showing promise in food safety and technology (Llorens et al., 2012).
	Preservation	Coating jujubes with chitosan + nano-silicon dioxide will be a promising alternative if nano-silicon dioxide is allowed for food use (Yu et al., 2012).

4.2 History of glass

Glass dates back to around 3500 BC and has its origin in Egypt and Eastern Mesopotamia, and the original glass objects were non-transparent. The discovery of glass was serendipitous. When calciferous sand was put into an overheated kiln, it combined with soda to form a coloured glaze on ceramics. Glass making was born as a craft and then was then spread across the world by sailors. It was in the late nineteenth century that the glass-making which was considered as a craft was industrialized. German scientist Otto Schott (1851–1935) is considered the forefather of modern glass research. By 1950, glass reached the status of being used as large curtain walls in buildings. From then, scientists and researchers focused on developing solar control glass, which can control harmful UV rays while at the same time allowing natural light to flow in. The other special types of glass that came into existence were fire protection glass and self-cleaning glass.

4.3 Self-cleaning glass

Glass is used in various sectors including building and construction, the automotive industry and solar energy cells, hence new demands on glass-making are high. Glass is used in locations which are quite hard to reach for daily maintenance in buildings and hence nanotechnology gave birth to SCG, a new type of glass. In SCG, a thin layer of about 10–25 nm of titania or silica is coated on the glass surface through a top-down or bottom-up approach. Self-cleaning ability is achieved by controlling the wettability of the surfaces. The first limit is complete drying, where liquid drops remain spherical on the solid surface, that is, a hydrophobic surface. This is usually accomplished by controlling surface roughness or by applying low surface energy components to the surface. Applying a thin coating of SiO_2 makes the surface hydrophobic. The second limit is complete wetting, in which a liquid on a solid surface makes a film, that is, a hydrophilic surface. For example, photocatalytic TiO_2 coatings produce hydrophilic surfaces which utilize sunlight and rinsing water to achieve self-cleaning. By coating TiO_2, the surface becomes hydrophilic.

A surface coating on a solid substrate creates new interfaces between dissimilar materials. Wettability, spreading, interface evolution and adhesion properties between the interfaces should be taken into account. The wettability of a solid by a liquid is defined in terms of the angle of contact that the liquid makes on the solid (Murr, 1974; Neumann and Good, 1979). The contact angle, θ, is obtained from a balance of interfacial tensions as shown in Figure 4.1 and is defined from Young's equation:

$$\sigma_{sv} = \sigma_{sl} + \sigma_{lv} \cos\theta \quad (4.1)$$

where σ_{sv}, σ_{sl} and σ_{lv} are the corresponding interfacial energies between the solid–vapour, solid–liquid and liquid–vapour phases, respectively. θ denotes the contact angle defined by the tangent on a liquid droplet with respect to the solid surface. The condition $\theta < 90°$ indicates that the solid is wet by the liquid, and $\theta > 90°$ indicates non-wetting, with the limits $\theta = 0°$ and $\theta = 180°$ defining complete wetting and complete non-wetting, respectively (Asthana and Sobzack, 2000). Depending upon the contact angle, coatings are of two types: hydrophilic and hydrophobic.

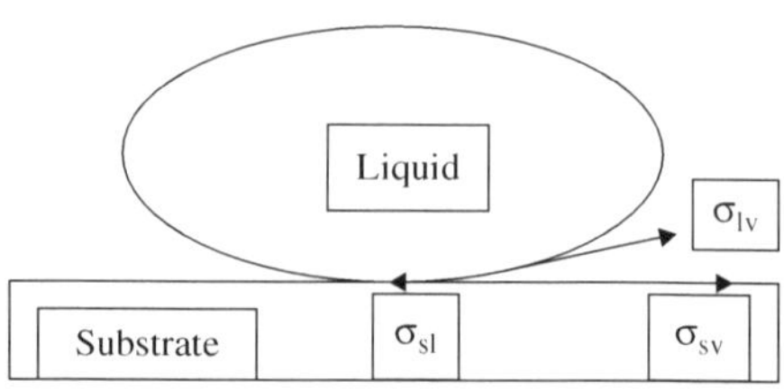

Figure 4.1 Interfacial tensions on a solid substrate.
Source: Asthana and Sobzack (2000). © 2000 The Minerals, Metals & Materials Society. Reprinted with permission.

However, Young's equation is valid only for smooth surfaces. For rough surfaces, Wenzel (1936) introduced the following relationship:

$$\cos \theta_w = r \cos \theta_\gamma \quad (4.2)$$

where r is the roughness parameter of the surface; r is defined as r = rough area/smooth area. Wenzel's relation shown in equation (4.2) indicates that if the intrinsic contact angle for a liquid on a solid surface is below 90° (hydrophilic), then an additional 'roughening' of this surface will reduce the effective contact angle. In the same way, the contact angle of a hydrophobic surface will increase even more as the surface roughness increases (Nadargi et al., 2010).

4.4 Hydrophilic coating

A surface whose water contact angle (CA) is less than 90° is called a hydrophilic surface and if the CA is less than 50°, it is called superhydrophilic surface. The water spreads out and forms a thin layer on a hydrophilic surface. TiO_2, WO_3, ZrO_2, ZnO, SnO_2, SiO_2 and CdS are the various self-cleaning materials available (Gao et al., 2004: 198–216). Of these, TiO_2 is widely used for self-cleaning applications because of its advantages over its counterparts.

Fujishima and Honda's (1972) discovery of the photo-electrochemical splitting of water to hydrogen and oxygen by TiO_2 under UV radiation led to the wide exploration of diverse applications based on TiO_2 photocatalysis such as photovoltaics, photocatalysis, self-cleaning coatings, advanced oxidation and photoelectrocatalytic degradation of organic compounds (Qiu et al., 2011).

TiO_2 is a transparent, semiconductor material with a high band gap (3.2–3.41 eV), high refractive index and good mechanical performance.

Refractive index n is a dimensionless number that is defined as:

$$n = c/v \quad (4.3)$$

where c is the speed of light in vacuum and v is the speed of light in the substance.

TiO_2 is chemically stable within the wavelength range of 0.35 to 12 mm. Titania exists in three crystalline forms – anatase, rutile and brookite – with rutile having the highest refractive index of the three (2.61–2.90). Rutile is therefore the most interesting for many optical applications. It is also the most thermodynamically stable form at all temperatures. However, it is the anatase form that grows preferentially at the lower temperatures that are often required for film formation on thermally sensitive substrates. Hence, it is either amorphous or crystalline anatase at temperatures below 400°C that are produced for self cleaning glasses. The transition temperature for anatase to rutile range from 700 to 1100°C (Nakata and Fujishima, 2012).

4.5 Anti-reflective coating

Fujishima in Japan (Fujishima et al., 1996) and Guiselin et al. (1999) were the pioneers that patented methods for depositing transparent, abrasion-resistant and photocatalytically efficient TiO_2 thin films onto glass surfaces. Currently, several SCG are available commercially. Bioclean from Saint Gobain, Hydrotech™ from TOTO, Activ™ from Pilkington Glass and Thermotech™ from Viridian are just a few. Besides these, there are user-finished self-glazing products that are available either as liquid or as wipes. These self-cleaning products can be applied on a normal glass and the glass will then act as a SCG. BalcoNano™ from Balcony Systems Solutions, Rain Racer™ from Rain Racer Developments, ClearShield™ from Ritec International are the examples of user-finished products available commercially (Midtdal, 2012). The SCG can be used in the facades, buildings and offices. A high refractive index is needed for enhanced photocatalytic activity and anatase coatings are used because of low temperature processing.

However, in many applications, such as solar cells, smart phones and spectacles, besides self-cleaning properties, anti-reflection, anti-fogging and anti-abrasive properties are also required for the glass. Anti-reflective (AR) coatings on glass substrates were first fabricated by Fraunhofer in 1817. The AR phenomenon is defined as the destructive interference between light reflected from the coating substrate and the air-coating interfaces by Fresnel and Poisson (Fowles, 1989). Constructing a single-layer coating with a low refractive index is one method of producing AR coatings. Materials with such a low refractive index are rare and expensive to obtain in nature (Hattori, 2001).

The volume-averaged refractive indices of materials can be effectively decreased by constructing porous nanostructures. By varying the degree of porosity in the coatings, anti-reflective coatings can be achieved. Increasing surface roughness enhances both the hydrophilic and hydrophobic property of coating. However, diffuse scattering in rough surfaces decreases the transmittance, thereby increasing the reflection.

The transmittance of a sample is the ratio of the intensity of the light that has passed through the sample to the intensity of the light when it entered the sample ($T = I_{out}/I_{in}$). Hence, when the transmittance increases, photocatalytic activity is decreased as the intensity of the light decreases. However, anti-reflective surfaces are desirable in SCG. Hence, optimum surface roughness should be chosen to maintain both anti-reflectivity and self-cleaning properties.

The static contact angle and the dynamic sliding angle are used to judge the wettability of a solid surface. The critical factor is surface roughness and chemical functionalization.

4.6 Porous materials

Porous materials increase the specific surface area and thus increase the photocatalytic activity. Hence, porous self-cleaning coatings are fabricated. It has been reported that compared to the titania nano particle, the band gap becomes narrower for titania porous materials (Dittrich, 2000; Reddy et al., 2002). The band gap of titania nano particle is 3.2 eV and that of the nanoporous material is 2.95 eV (Reddy et al., 2002). Hence, lower energy is sufficient for photocatalysis of porous titania. Similar behaviour on an increase in surface energy for bulk nanoparticle and a decrease in energy for porous material was also observed for Al_2O_3-MgO ceramics (Wunderlich and Takahashi, 2002). The amount of crystalline material, appropriate pore structure, thickness of the coating and UV light source must be controlled to obtain an enhanced photocatalytic activity.

The difference between the nanoparticles and nano-porous materials on the the physical properties such as band gap and surface energy can be explained with Figure 4.2.

It can be seen from Figure 4.2 that for any physical property (e.g. the bandgap or the surface energy) there is a change in the property depending on whether the structure consists of nanoparticles or nanopores.

In terms of the band gap energy of titania nanoparticle and nano pores, it can be seen that when the diameter d of the particles or pores drops below a critical

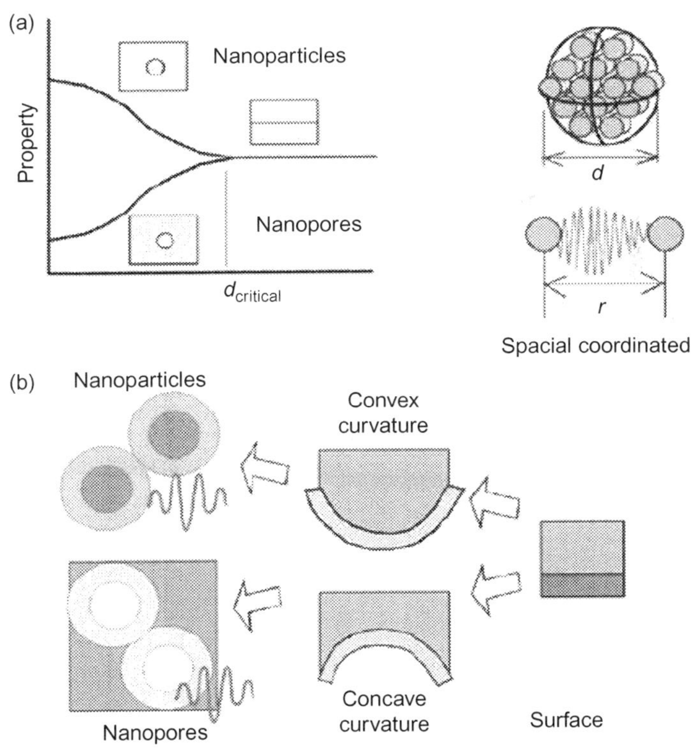

Figure 4.2 Schematic illustration of the relationship between nanoparticles and nanopores. *Source:* Wunderlich et al. (2006). Reprinted with permission.

limit, the band gap energy increases for nanoparticles and decreases for nanopores. This is attributed to the fact that the structural diameter d of the pore or the particle becomes similar in size to the interaction distance between the atoms r (Figure 4.2(a)). Nanopores are differentiated from the nanoparticles by the bending of the surface (Figure 4.2(b)). When the surface bends into a convex shape, it becomes a particle and when it bends into a concave shape, a pore is formed (Wunderlich et al., 2006). Because of the concave shape, the specific surface area increases and hence an increased photocatalytic activity is observed in porous materials compared to the nanoparticles.

4.7 Photocatalytic activity of TiO_2

When UV light falls on the TiO_2, absorption of ultraviolet (UV) light results in formation of photo-generated charge carriers (hole and electron) corresponding to the band gap as shown in Figure 4.3. The photo-generated holes in the valence band diffuse to the TiO_2 surface and react with adsorbed water molecules, forming hydroxyl radicals (OH^-) as shown in Figure 4.4. The photo-generated holes and the hydroxyl radicals oxidize nearby organic molecules on the TiO_2 surface. Meanwhile, electrons in the conduction band typically participate in reduction processes, which typically react with molecular oxygen in the air to produce superoxide radical anions (O_2^-). This photocatalysis is helpful in self cleaning ability of the glasses.

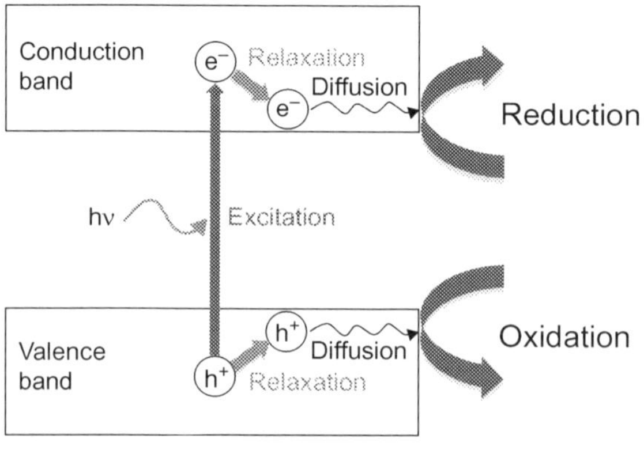

Figure 4.3 Schematic illustration of the formation of photo generated charge carriers (hole and electron) upon absorption of ultraviolet (UV) light.
Source: Reprinted from Nakata and Fujishima (2012). © 2012 Elsevier. Reprinted with permission.

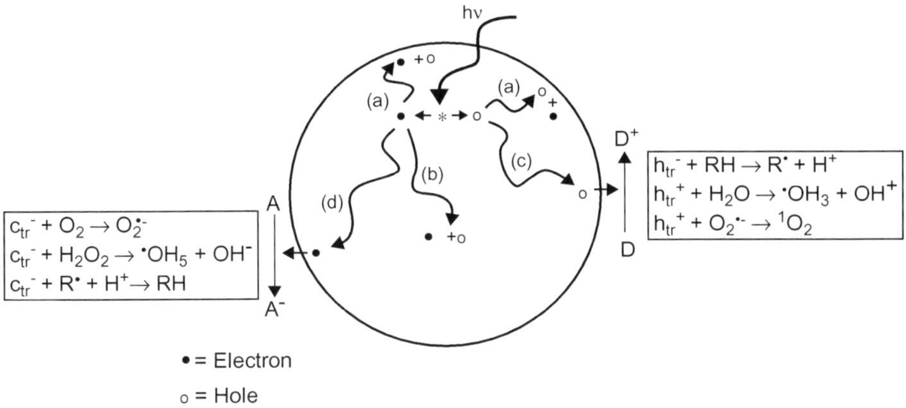

Figure 4.4 Reactions occurring on bare TiO_2 after UV excitation.
Source: Reprinted from Nakata and Fujishima (2012). © 2012 Elsevier. Reprinted with permission.

4.8 Hydrophobic coatings

A surface with water contact angle larger than 150° and sliding angle lower than 10° is generally known as a superhydrophobic surface (Fujishima and Honda, 1972). Lotus leaf (*Nelumbo nucifera*), taro (*Colocasia esculenta*) and butterfly wings are naturally occurring hydrophobic materials. Wettability plays a critical role in deciding the self-cleaning property of a solid surface which is influenced by the chemical composition and the morphology of the surface. Static water contact angle is a criterion for evaluation of the hydrophobicity and water sliding angle is used to estimate the self-cleaning property. A surface with a high static water contact angle does not always show a low sliding angle. The sliding angle behaviour of water droplets should be evaluated separately from the contact angle for self-cleaning surfaces. Sliding angle is defined as the critical angle where a water droplet begins to slide down the inclined surface.

The hydrophobic coatings on external surfaces of biological structures such as animals, insects and plants possess excellent abilities and comply with various conflicting demands. These natural structures are the inspiration to design functional artificial structures. For example, the surface of lotus leaves is covered by microscale papillae with branch-like nanostructures. These structures yield self-cleaning superhydrophobic surfaces (contact angle for water higher than 150°C) and many researchers have focused on the fabrication of micro and nanoscale hierarchical structures that is analogous to the lotus leaves. The microscopical examination of lotus (*Nelumbo nucifera*) leaves shows double-scaled roughness; about 10 μm protruding nubs with a gap of 10–30 μm covered with tiny wax crystals of about 0.2–2 μm (Barthlott and Neinhuis, 1997). Hydrophobic materials should possess high roughness and low surface energy. The high roughness decreases the mechanical properties and other properties such as transparency.

The interaction of micro- and nanostructures with water is a critically important factor in influencing the function of biological surfaces. This interaction can be described by one of the following two states:

1. The Wenzel state, in which a liquid droplet completely wets the asperities of a rough surface (Wenzel, 1936);
2. The Cassie–Baxter state, in which a liquid droplet is suspended on the top of a rough surface, leaving air pockets inside the texture (Cassie and Baxter, 1944).

Droplets can transit from the higher energy Cassie–Baxter state to the lower energy Wenzel state if they overcome the energy barrier between the two states by filling liquid into the asperities and forming more stable homogeneous interface (Patankar, 2004). The Cassie–Baxter state generally corresponds to superhydrophobicity, which is strongly related to both the chemical composition and the geometry of the surface and which is useful for applications requiring surfaces that are self-cleaning, waterproof and/or have low drag in water. Stability of the composite solid–air–liquid interface is very important to maintain superhydrophobicity.

Superhydrophobic surfaces can be produced in two ways:

1. By creating a rough structure on a hydrophobic surface (CA $> 90°C$) through mechanical stretching, laser/plasma/chemical etching and lithography;
2. By modifying a rough surface by compounds with low surface free energy, such as fluorinated or silicon compounds, organic materials (polyethylene, polystyrene, etc.) and inorganic materials (ZnO and TiO_2).

The water–solid contact angle varies with the surface energy and roughness of the solid surface. The surface energy of a solid is determined by the surface chemistry, which in turn depends on the chemical composition and atomic arrangements at or near the surface. Structure relaxation, surface restructuring and composition segregation can all reduce the surface energy and thus result in an increase in contact angle. Increased roughness will result in an increase in true surface area and thus lead to an increased nominal surface energy. Accordingly, the contact angle varies as dictated by Young's equation (Eq. 4.2). It has been demonstrated that the contact angle will increase with increased roughness of a hydrophobic surface.

4.9 Fabrication of self-cleaning glass

A thin layer of TiO_2 or SiO_2 is applied on the surface glass substrate to make it hydrophilic or hydrophobic.

The fabrication method of nanomaterials is generally classified into two sub-divisions: (1) top-down and (2) bottom-up.

In the top-down approach, progressive removal of material from a massive structure occurs until the desired nanomaterial is obtained. An example of such a process is lithography. This can be compared to a carpenter making a doll from a block of wood.

In the bottom-up approach, gradual building of atoms or molecules occurs until the desired nanomaterial or nanocoating is achieved. This is compared to building a house using lego bricks. The bottom-up approach is again sub-divided into two methods: (1) gas phase; and (2) liquid phase.

Plasma arc evaporation and chemical vapour deposition (CVD) belong to the gas phase method, while sol-gel and molecular self-assembly belong to the liquid phase method.

4.9.1 Top-down approach

4.9.1.1 Lithography

This is a process which involves series of fabrication techniques in which an image is transferred from a mask to a substrate. The process involves three stages:

1. Coating a substrate (glass in this instance) with a resist.
2. Exposing the resist to light, electron or ion beams.
3. Developing the resist with proper developer material.

Polymers are widely used as resist materials. However, the maximum resolution obtained using polymer resist is 100 nm. Hence, inorganic resists such as glass (silica glass, chalcogenide glass and amorphous peroxopolytungstic acid) and polycrystalline (polycrystalline fluorides and $Al-Al_2O_3$ cermets) are considered as resist materials.

Chalcogenide glass is photo-sensitive and hence they are widely used. Chalcogenide glass resists are suitable for selective etching of SiO_2, TiO_2, Si, Cr and so on. Changing the composition of the developer, certain chalcogenide glass can be used as a positive or a negative photo resist.

4.9.2 Bottom-up approach

4.9.2.1 Chemical vapour deposition

CVD is a process in which thin films or coatings are produced by the dissociation and/or chemical reactions of gaseous reactants in an activated (heat, light, plasma) environment. The deposition involves homogeneous gas phase reactions, which occur in the gas phase, and/or heterogeneous chemical reactions which occur on/near the vicinity of a heated surface leading to the formation of powders or films, respectively. No post-treatment is required and the process is capable of depositing high-quality crystalline thin films. The film structure can be easily controlled by deposition conditions and porous film can also be obtained.

CVD is widely used for making TiO_2 coatings on glass substrate. Lee et al. (2011) deposited TiO_2 nanoparticles on glass substrate by CVD. It was found that coating time was less and the amount of TiO_2 deposited was controlled via the coating time. A uniform shape of the coating was also obtained. Kim et al. (2004) used CVD to obtain a thin coating of TiO_2 on Al_2O_3 substrate. An ultrasonic atomizer was used to spray TiO_2 mist on the substrate. It was observed that the film thickness was independent on the substrate temperature but dependent on the deposition time. The surface roughness gradually increased with increasing the substrate temperature and the deposition time.

4.9.2.2 Plasma arc evaporation

This is one of the physical vapour deposition processes in which high-temperature vacuum evaporation with subsequent condensation or plasma sputter bombardment occurs to deposit thin films. The process is purely physical and no chemical reaction occurs. The source material used is in a solid state and it is evaporated to deposit on the substrate.

Miyata et al. (2006) used vacuum plasma arc evaporation (VAPE) to prepare anatase TiO_2 thin film on glass substrate. The schematic representation of the experimental set up used is shown in Figure 4.4. TiO_2 pellets are used as source material. The target material, sintered TiO_2 powder is bombarded by plasma beam as shown in Figure 4.5 and this process vapourizes the target. The vapourized atoms move towards the substrate and are deposited. The effects of substrate temperature varying from room temperature to 450°C and the effect of oxygen flow rate on electrical properties were studied. It was found that the electrical resistivity increased as the O_2 gas flow rate increased. It was further found that anatase TiO_2 thin films prepared at a substrate temperature of 300°C exhibited high photocatalytic activity and photo-induced hydrophilicity.

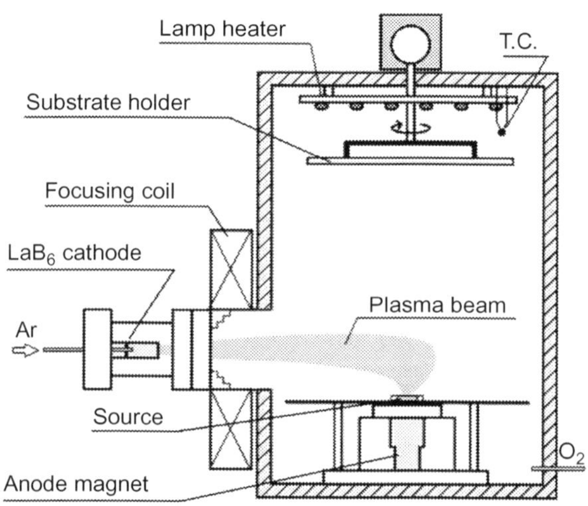

Figure 4.5 Schematic diagram of the VAPE apparatus.
Source: Miyata et al. (2006). © 2006 Elsevier. Reprinted with permission.

4.9.2.3 *Sol-gel process*

Sol-gel is a widely used process to fabricate both hydrophobic and hydrophilic coatings. In this process a precursor material is hydrolysed/polymerized to obtain a colloidal sol. The sol is deposited on the substrate and a solid film is obtained by heating the substrate. The sol-gel process is used to get ultrafine or spherical shaped powders, thin film coatings, fibrous, porous or dense materials and extremely porous aerogel materials.

The synthesis of thin films by sol-gel process involves the following steps:

1. Pre-treating the substrate for good quality film.
2. Preparing of the sol.
3. Applying the sol to the substrate.
4. Annealing the as-deposited film.

The general formula of metal alkoxides is M(OR)*n*, where M = metal, R = alkyl group and *n* is the valence of the metal atom. Titanium isopropoxide (TIP) Ti[OCH $(CH_3)_2]_4$ and tetramethoxysilane (TMOS) are widely used alkoxides for TiO_2 and SiO_2 thin film preparation, respectively. The sequence of reactions occurring in the sol-gel process is shown in Figure 4.6.

Hydrolysis

$$M(OR)n + 4H_2O \xrightarrow[\text{Acid catalyst}]{} M(OH)n + nROH$$

Water condensation

$$nM(OH)n \xrightarrow[\text{Base catalyst}]{} nMO_2 + H_2O$$

MO_2 gel network/particles or both

The prepared sol can be applied on the substrate by spray, dip or spin coating methods.

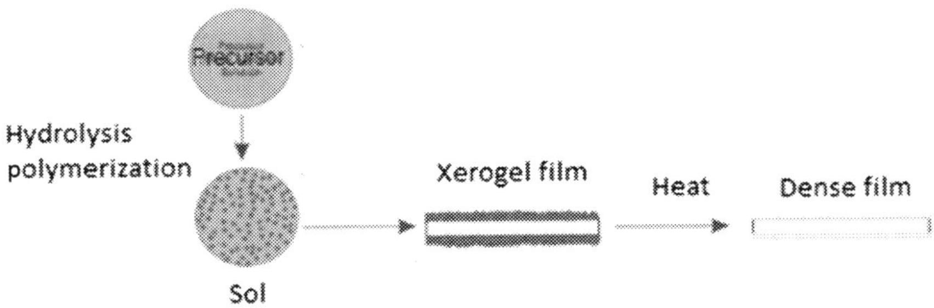

Figure 4.6 Schematic representation of the sol-gel process.
Source: Adapted from chematvision.com. Copyright © 2010 Chemat Technology. Reprinted with permission.

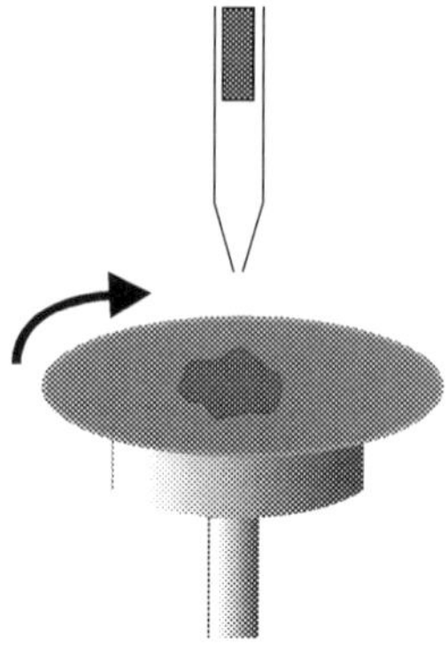

Figure 4.7 Schematic representation of spin coating process.
Source: Brewer Science, Inc. (n.d.). Reprinted with permission.

4.9.2.3.1 Spin coating

Spin coating is widely used for creating thin films. Figure 4.7 shows a schematic representation of spin coating. A spin coating process has three steps:

1. Dispense, in which the resin is deposited on the substrate.
2. High speed spin, in which thinning of the resin occurs.
3. Drying, to remove the excess solvent from the film.

The dispense step has two sub-divisions: static and dynamic. In the former, a small puddle of resin is deposited on or near the centre of the substrate; in the latter the substrate is rotated at a lower speed of 500 rpm. The amount of resin deposited is dependent on the viscosity of the resin and the substrate to be coated.

In the second step, the substrate is rotated at a speed depending upon the final thickness of the coat. The substrate speed can be varied up to 6000 rpm. The spin speed and spin time determines the final coating thickness. Higher spin speeds and longer spin times create thinner films.

The final step is drying step in which the film is allowed to dry to remove the excess solvents. It is especially applicable to thicker coatings. Without the drying step, the coating may pour off the substrate when it is removed from the spin.

The physical properties of the resin such as viscosity, drying rate, percentage of solids and surface tension also influence the final film thickness.

4.9.2.3.2 Dip coating

In dip coating, the substrate is dipped in the tank containing the colloidal sol and the final coating thickness is determined mainly by the deposition speed and solution viscosity. Besides these two major factors, the functionalization of the initial substrate surface, immersion time, withdrawal speed, number of cycles dipped, solution composition, concentration and temperature, number of solutions in each dipping sequence and environment humidity have an influence on the final coating. Dip coating can be used to coat on complex structures.

Aqueous sol-gel process involves the precursor being hydrolysed with water. However, titanium alkoxides have a greater reactivity towards water and they form unstable sols. Hence, a non-aqueous sol-gel process is used in which the alkoxide is hydrolysed by the moisture from the air. TiO_2 sols with lower reactivity and long-term stability are obtained by non-aqueous sol-gel (Buscema et al., 2002). Haimi et al. (2011) fabricated anatase (TiO_2) thin films grown by non-aqueous sol-gel dip-coating using titanium (IV) *n*-butoxide as precursor and 1-butanol as solvent. Defect-free film of 100 average film thickness was obtained at a high withdrawal speed of 4.7 mm/s after subsequent heat treatments.

Shokuhfar et al. (2012) could coat automotive wind shields with SiO_2-TiO_2 nano film by sol-gel dip coating. The effect of annealing temperature was studied. It was found that increasing heat treatment temperature from 500 to 700°C, decreased the photocatalytic activity and increased the contact angle of the films. This is attributed to the conversion of anatase to rutile at 700°C. It was further reported that too low temperatures would not eliminate the organic residues and 500°C was the optimum annealing temperature.

4.9.2.3.3 Spray coating

In spray coating, the colloidal sol is sprayed on the substrate with a spray gun. This process can be used to coat any complex shapes. Spray coating process is faster and less wastage of sol occurs. Conventional air spraying, high volume low pressure (HVLP) and electrostatic spraying are different kinds of spray coating.

Flow coating, capillary coating and roll coating are other types of coating techniques, but they are not as widely used as the above three processes.

Fabrication of TiO_2 thin films for self-cleaning applications by dip-coating, spray-coating or spin-coating requires the post-annealing process to crystallize into an anatase structure. It has been noticed that the post-annealing process diffuses sodium ions from the window glass into TiO2 film thereby degrading the photocatalytic efficiency. Diffusion of sodium ions is dominant when the temperature is over 450°C.

Sodium ion diffusion effect in TiO_2 is dominant when the temperature is over 450°C and the typical post-annealing temperature is above 600°C. Use of a SiO_2 diffusion barrier was presented to prevent the sodium ion diffusion during the post-annealing process.

Nadargi et al. (2010) prepared hydrophobic SiO_2 through a sol-gel and dip coating process. Tetramethyl orthosilicate (TMOS) was used as the precursor and trimethyl chlorosilane (TMCS) as a sylilation agent. It was found that uniform, strongly adherent and highly transparent coatings were obtained for a TMOS: MeOH:TMCS:H_2O molar ratio of 1:14.8:1.4:3.3 with 4M ammonia solution serving as the base catalyst. Macroporous SiO_2 was obtained.

Segota et al. (2011) synthesized TiO_2 thin film on glass substrate by the sol-gel dip coating method. The effect of surface roughness was studied with and without the addition of polyethylene glycol (PEG). It was found that roughness was higher with the addition of PEG compared to films without PEG.

4.9.2.4 Self-assembly

Self-assembly is a reversible process in which disordered components are arranged into definite patterns. It can be classified as either static or dynamic. Static self-assembly is when the ordered state occurs when the system is in equilibrium and does not dissipate energy. Dynamic self-assembly is when the ordered state requires dissipation of energy. Examples of self-assembling systems include weather patterns, solar systems and self-assembled monolayers.

Molecular self-assembly is the assembly of molecules without guidance or management from an outside source. There are two types of self-assembly: intramolecular and intermolecular.

Intramolecular self-assembling molecules are often complex polymers with the ability to assemble from the random coil conformation into a well-defined stable structure (secondary and tertiary structure). An example of intramolecular self-assembly is protein folding. Intermolecular self-assembly is the ability of molecules to form supramolecular assemblies (quarternary structure).

Layer-by-layer (LbL) assembly is also widely used for fabrication of thin films. This is a process in which positively and negatively charged materials are deposited in layers. Li et al. (2013) fabricated anti-reflective, water-repellent coatings via LbL, arranging silica nanoparticles and polyelectrolytes in alternate layers. A porous, 20 nm thick coating was obtained. The thickness of the coating depended on the number of deposition cycles. AR coatings had enhanced the light transmittance and decreased the light reflection.

Liu et al. (2008) fabricated superhydrophilic and superhydrophobic micropatterns on the TiO_2 nanotube structured film by electrochemical self-assembly and photocatalytic lithography. A patterned photomask can control the site-selective decomposition, thereby converting superhydrophobic surface to superhydrophilic because of the roughness on the aligned nanotube structure. This micropatterned wettability is based on a self-assembled monolayer (SAM) process on the special nanostructured TiO_2 film to achieve superhydrophobicity and the region-optional photocatalytic decomposition of perfluoro octyl-triethoxysilane (PTES) SAM to obtain superhydrophilicity. This type of micropattern can be used in functional materials and micro-nano devices.

4.9.2.5 SiO_2-TiO_2 coating

Besides a self-cleaning property, multifunctional properties such as anti-reflectivity and photocatalysis are essential in many applications like solar cells and smart phones. Glop et al. (2008) and Liu et al. (2008) reported a photoactive anti-reflection coating prepared by sol-gel process and pulse magnetron sputtering, respectively.

The outer self-cleaning coating of TiO_2 increases the reflectance of the glass or plastic substrate because of its relatively high refractive index (c. 2.5 for the anatase phase). Thus anti-reflectivity and self-cleaning ability are potentially incompatible unless the composition and the structure are conveniently modulated. Prado et al. (2010) fabricated multifunctional coatings consisting of two layer stacks with an inner mesoporous

SiO_2 AR layer and a dense/mesoporous TiO_2 outer layer. A marked enhancement in self-cleaning capacity of the multifunctional coatings is reported and it was 25–30% greater than the photo-degradation degree achieved with the commonly used porous or compact TiO_2 coatings layer.

Glass is widely used in solar power plants and solar collectors. The intensity of the incident sunlight determines the power of conversion of solar light to electricity or water heating. However, the glass material used results in reflection losses though convection and radiation losses are minimized.

Refractive index (n) is an important property in self-cleaning glasses. To make the surfaces anti-reflective for solar cover glasses, a low refractive index, such as SiO_2 whose $n \leq 1.4$, is needed. To enhance the photcatalytic activity of hydrophilic self-cleaning glasses, high refractive index materials such as titania ($n > 2.0$) is essential. Helsch and Deubener (2012) investigated the feasibility of combining these two functionalities on a single glass by sol-gel dip coating. A high transmittance is also desirable. Combinations of anti-reflective with photocatalytic properties have been achieved by preparing SiO_2/TiO_2 double layers. On silica glass porous coatings $x TiO_2 \cdot (100 - x) SiO_2$ with titania concentrations up to 50 wt% were prepared by sol-gel dip coating. In the compositional range from $x = 7.5–20$ compatibility of anti-reflective and photocatalytic properties was achieved. An increase in 2.3% solar transmittance was observed in this porous coating compared to uncoated silica glass. Compared to nanoporous film of pure SiO_2, these bi-functional coatings have 20 times greater degradation rate of the air borne contaminants.

Highly porous and low density materials have combined the advantage of high surface area and high pore volume as well as larger pore size with the accessible diffusion pathways associated with nanoporous structures.

Porous silica layers are widely used as anti-reflection coatings. Helsch et al. (2006) found that the light transmission of borosilicate glass at 550 nm increased from 92 to 97% with film thickness of 110 nm and a porosity of about 35%.

Transparent self-cleaning TiO_2 films on glass substrates have a high potential for practical applications such as mirrors, window glasses and windshields of automobiles.

Anti-reflective porous coatings will be exposed to harsh environmental conditions during service, such as sandstorms, hail and salt atmospheres, dust particles and airborne volatile organic compounds. These above factors decrease the solar transmittance if they contaminate or damage the AR coating. Cathro et al. (1981) reported that adsorption of airborne contaminants increased the refractive index of porous thin films. Oil vapour contamination also increases the refractive index of porous antireflective coatings (Pareek et al., 2008).

The principle of AR is interference of the reflected light at the air–coating and coating–substrate interfaces. In order to achieve the lowest reflectance of a coating on a cover glass, the refractive index of the coating should be close to 1.21–1.25 based on the rule:

$$n_c = (n_o n_s)^{1/2} \tag{4.4}$$

where n_c, n_o and n_s are the refractive indices of the coating, incident medium and substrate, respectively (Miao et al., 2013). Porous SiO_2 films with homogenous pore distributions are suitable candidates for use as anti-reflective coatings. Precise control of the pore size and the pore distribution of the SiO_2 coating are crucial for achieving a low refractive index.

In a substrate which is covered by a thin film of only TiO_2, the contact angle of water almost becomes zero during UV irradiation. The contact angle rises immediately in a dark place and is restored to its original value. However, it is desirable that the contact angle increases slowly and stays low for a longer duration for the surface to be hydrophilic on a cloudy or rainy day when there is no sunlight. It was found that adding SiO_2 decreases the contact angle, and increases the hydrophilic and photocatalytic activity of TiO_2 (Guan et al., 2003; Yu et al., 2001).

Guan (2005) found that the amount of SiO_2 in a TiO_2/SiO_2 film surface determines whether a coating can have more photocatalytic activity and less superhydrophilic activity, or vice versa:

1. Addition of 10–20 mol% SiO_2 to the film gives optimum photocatalytic character. 30–40 mol% SiO_2 addition was the optimum for hydrophilicity.
2. Up to 30–40 mol% SiO_2 addition, the amount of adsorbed organic substances decreases with the increase in SiO_2 in TiO_2 films.
3. Photocatalytic activity is closely related to hydrophilicity; both activities reinforce each other and their mutual action maintains the self-cleaning effect.
4. Addition of SiO_2 to TiO_2 increases the acidity on the surface. During the process of competitive adsorption on the surface between soiling compounds and water, the water (OH groups) is preferentially adsorbed and the contamination is limited on the surface (Patankar, 2004).

4.9.2.6 *Visible light*

The photocatalytic activity of TiO_2 is feasible only in the presence of UV radiation. When a light of wavelength ≤ 390 nm falls on TiO_2, electron/hole pairs are generated. It was found that the photcatalytic activity of TiO_2 can be extended to a visible light region by doping non-metals such as carbon, nitrogen, sulphur, fluorine and boron (Asahi et al., 2001; Gorska et al., 2008; Li and Shang, 2010; Herrera and Pulgarin, 2010; Subagio et al., 2010; Zaleska et al., 2008) to TiO_2 by narrowing the band gap from 3.2 eV. Yang et al. (2008) synthesized nitrogen and carbon-doped TiO_2 films by heating a TiO_2 gel in an ionized N_2 gas. The as-synthesized $TiO_{2-x-y}N_xC_y$ films showed an improvement over titanium dioxide in optical absorption and photocatalytic activity such as photodegradation of methyl orange under visible light. Oxygen vacancy induced by the formation of Ti^{3+} species and nitrogen and carbon doped into substitution sites of TiO_2 have been proven to be indispensable for the enhancement of photocatalytic activity in the visible light region.

Wang et al. (2011) synthesized a novel carbon-sensitized and nitrogen-doped TiO_2 (C/N-TiO_2) by sol-gel method. Titanium butoxide was used as both titanium precursor and carbon source and nitric acid as nitrogen source. The film obtained could be illuminated in visible light region.

4.10 Application of self-cleaning glasses

Self-cleaning coatings have useful applications for car windshields, satellite antennas, anti-snow coatings, hulls of ships, glass used in buildings and tubes or pipes. Greenhouse roofs, solar cells, tablet computers, spectacles and the construction industry (e.g. windows, doors, shower screens, bath tubs) also employ SCG.

4.10.1 Advantages

SCG has several advantages compared to ordinary glass:

1. Minimal maintenance in hard-to-reach areas is sufficient to keep the surface clean.
2. SCG helps to protect the environment, as less water and no detergents are used for cleaning.

4.10.2 Disadvantages

As every coin has two sides, SCG is not without some disadvantages:

1. SCG is not effective in cleaning oily contaminants.
2. TiO_2 needs the presence of UV sunlight to make it active.
3. The self-cleaning ability is lost when oily contaminants are adhered to the surface.

Acknowledgements

The authors acknowledge the help rendered by Pramita Nayak in the preparation of this chapter. The authors are indebted to Elseiver for allowing us to use their published diagrams.

References

Alongi, J., Carosio, F., Malucelli, G., 2014. Current emerging techniques to impart flame retardancy to fabrics: an overview. Polym. Degrad. Stab. 106, 138–149.

Asahi, R., Morikawa, T., Ohwaki, T., 2001. Visible-light photocatalysis in nitrogen-doped titanium oxides. Science. 293 (5528), 269–271.

Asthana, R., Sobzack, N., 2000. Wettability, spreading and interfacial phenomena in high-temperature coatings. JOM-e. 52 (1). Available from: <http://www.tms.org/pubs/journals/JOM/0001/Asthana/Asthana-0001.html> (accessed 14.06.13.).

Barkalina, N., Charalambous, C., Jones, C., Coward, K., 2014. Nanotechnology in reproductive medicine: emerging applications of nanomaterials. Nanomedicine. 10 (5), 921–938.

Barthlott, W., Neinhuis, C., 1997. Characterization and distribution of water repellent, self-cleaning plant surfaces. Ann. Bot. 79, 667–677.

Brambilla, D., Le Droumaguet, B., Nicolas, J., Hashemi, S.H., Wu, L.P., et al., 2011. Nanotechnologies for Alzheimer's disease: diagnosis, therapy, and safety issues. Nanomedicine. 7 (5), 521–540.

Brewer Science, Inc., n.d. Spin-coating theory, <http://www.brewerscience.com/research/processing-theory/spin-coating-theory> (accessed 10.09.14.).

Broasca, G., Borcia, G., Dumitrascu, N., Vrinceanu, N., 2013. Characterization of ZnO coated polyester fabrics for UV protection. Appl. Surf. Sci. 279, 272–278.

Buscema, C.L., Malibert, C., Bach, S., 2002. Elaboration and characterization of thin films of TiO2 prepared by sol-gel process. Thin Solid Films. 418 (2), 79–84.

Cassie, A.B.D., Baxter, S., 1944. Wettability of porous surfaces. Trans. Faraday Soc. 40, 546–551.

Cathro, K.J., Constable, D.C., Solaga, T., 1981. Durability of porous silica antireflection coatings for solar collector cover plates. Sol. Energy. 27, 491–496.

Dittrich, T., 2000. Porous TiO2: electron transport and application to dye sensitized injection solar cells. Phys. Status Solidi A. 182, 447–455.

Fowles, G., 1989. Introduction to Modern Optics. Dover Publication, New York.

Fujishima, A., Honda, K., 1972. Electrochemical photolysis of water at a semiconductor electrode. Nature. 238, 37–38.

Fujishima, A., Hashimoto, K., Iyoda, T., Fukayama, S., (Kanagawa Academy of Science and Technology), 1996. Titanium oxide photocatalyst structure and method of manufacturing the same. European patent application 0737513A1. 16 October 1996.

Gao, L., Zheng, S., Zhang, Q.H., 2004. Nanophotocatalyst. Wunan Press, Taiwan.

Glop, D., Frach, P., Gottfried, C., Klinkenberg, S., Liebig, J., et al., 2008. Multifunctional high-reflective and anti-reflective layer systems with easy-to- lean properties. Thin Solid Films. 516, 4487–4489.

Gommersall, L., Shergill, S.I., Ahmed, U.H., Hayne, D., Arya, M., et al., 2007. Nanotechnology and its relevance to the urologist. Eur. Urol. 52 (2), 368–375.

Gorska, P., Zaleska, A., Kowalska, E., 2008. TiO_2 photoactivity in vis and UV light: the influence of calcination temperature and surface properties. Appl. Catal. B. 84 (3–4), 440–447.

Guan, K.S., 2005. Relationship between photocatalytic activity, hydrophilicity and self-cleaning effect of TiO_2/SiO_2 films. Surf. Coat. Technol. 191, 155–160.

Guan, K.S., Lu, B.J., Yin, Y.S., 2003. Enhanced effect and mechanism of SiO_2 addition in super-hydrophilic property of TiO_2 films. Surf. Coat. Technol. 173, 219–223.

Guiselin, O., Macquart, P., Rondeau, V., (Saint-Gobain Vitrage), 1999. Transparent substrates coated with a stack of thin layers having reflection properties in the infrared and/or in the solar radiation range. US Patent Application US 5965246 A Oct 12, 1999.

Haimi, E., Lipsonen, H., Larismaa, J., Kapulainen, M., Krzak-Ros, J.C., et al., 2011. Optical and structural properties of nano crystalline anatase (TiO_2) thin films prepared by non-aqueous sol-gel dip-coating. Thin Solid Films. 519, 5882–5886.

Hanh, T.T., Phu, D.Y., Thu, N.T., Quoc, L.A., Duyen, D.N.B, et al., 2014. Gamma irradiation of cotton fabrics in AgNO3 solution for preparation of antibacterial fabrics. Carbohydr. Polym. 101, 1243–1248.

Hattori, H., 2001. Anti-reflection surface with particle coating deposited by electrostatic attraction. Adv. Mater. 13, 51–54.

Helsch, G., Deubener, J., 2012. Compatibility of antireflective coatings on glass for solar applications with photo catalytic properties. Sol. Energy. 86, 831–836.

Helsch, G., Krzyzak, M., Heide, G., Frischat, G.H., 2006. Adherent anti reflection coatings on borosilicate glass for solar collectors. Glass Technol. Eur. J. Glass Sci. Technol. A. 47, 153–156.

Herrera, R.J.A., Pulgarin, C., 2010. Photocatalytic activity of N, S co-doped and N-doped commercial anatase TiO_2 powders towards phenol oxidation and E. coli inactivation under simulated solar light irradiation. Sol. Energy. 84 (1), 37–43.

Kim, B.H., Lee, J.Y., Choab, Y.H., Higuchi, M., Mizutani, N., 2004. Preparation of TiO2 thin film by liquid sprayed mist CVD method. Mater. Sci. Eng. B. 107, 289–294.

Kokura, S., Handa, O., Takagi, T., Ishikawa, T., Naito, Y., Yoshikawa, T., 2010. Silver nanoparticles as a safe preservative for use in cosmetics. Nanomed. Nanotechnol. Biol. Med. 6 (4), 570–574.

Lacatusu, I., Niculae, G., Badea, N., Stan, R., Popa, O., et al., 2014. Design of soft lipid nanocarriers based on bioactive vegetable oils with multiple health beneficies. Chem. Eng. J. 246, 311–321.

Laroui, H., Rakhya, P., Xiao, B., Viennois, E., Merlin, D., 2013. Nanotechnology in diagnostics and therapeutics for gastrointestinal disorders. Dig. Liver Dis. 45 (12), 995–1002.

Lee, H., Song, M.Y., Jurng, J., Park, Y.K., 2011. The synthesis and coating process of TiO2 nanoparticles using CVD process. Powder Technol. 214, 64–68.

Li, Q., Shang, J.K., 2010. Composite photocatalyst of nitrogen and fluorine codoped titanium oxide nanotube arrays with dispersed palladium oxide nanoparticles for enhanced visible light photocatalytic performance. Environ. Sci. Technol. 44 (9), 3493–3499.

Li, X., He, J., Liu, W., 2013. Broadband anti-reflective and water-repellent coatings on glass substrates for self-cleaning photovoltaic cells. Mater. Res. Bull. 48, 2522–2528.

Liu, Z., Zhang, X., Murakami, T., Fujishima, A., 2008. Sol-gel SiO_2/TiO_2 bilayer films with self-cleaning and anti-reflection properties. Sol. Energy Mater. Sol. Cells. 92, 1434–1438.

Llorens, A., Lloret, E., Picouet, A.P., Trbojevich, R., Fernandez, A., 2012. Metallic-based micro and nano-composites in food contact materials and active food packaging. Trends Food Sci. Technol. 24 (1), 19–29.

Makwana, S., Choudhary, R., Dogra, N., Kohli, P., Haddock, J., 2014. Nanoencapsulation and immobilization of cinnamaldehyde for developing antimicrobial food packaging material. LWT–Food Sci. Technol. 57 (2), 470–476.

Mei, L., Zhang, Z., Zhao, L., Huang, L., Yang, X.L., Tang, J., et al., 2013. Pharmaceutical nanotechnology for oral delivery of anticancer drugs. Adv. Drug Deliv. Rev. 65 (6), 880–890.

Miao, L., Su, L.F., Tanemura, C.S., Fisher, A.J., Zhao, L.L., et al., 2013. Cost effective nanoporous SiO2–TiO2 coatings on glass substrates with antireflective and self-cleaning properties. Appl. Energy. 112, 1198–1205.

Midtdal, K., 2012. Self-Cleaning Glazing Products: A State-of-the-Art Review and Future Research Pathways. Norwegian University of Science and Technology, Trondheim, Norway.

Miyata, T., Tsukada, S., Minami, T., 2006. Preparation of anatase TiO2 thin films by vacuum arc plasma evaporation. Thin Solid Films. 496, 136–140.

Murr, L.E., 1974. Interfacial Phenomena in Metals and Alloys. Addison Wesley, New York.

Nadargi, D.Y., Gurav, J.L., El Hawi, N., Rao, A.V., Koebel, M., 2010. Synthesis and characterization of transparent hydrophobic silica thin films by single step sol–gel process and dip coating. J. Alloys Compd. 496, 436–441.

Nakata, K., Fujishima, A., 2012. TiO2 photocatalysis: design and applications. J. Photochem. Photobiol., C. 13, 169–189.

Neumann, A.W., Good, R.J., 1979. Methods of measuring contact angles. In: Good, R.J., Stromberg, R.R. (Eds.), Surface and Colloid Science. Plenum Press, New York, pp. 31–91.

Pareek, R., Kumbhare, M.N., Mukherjee, C., Joshi, A.S., Gupta, P.D., 2008. Effect of oil vapour contamination on the performance of porous silica sol–gel anti reflection-coated optics in vacuum spatial filters of high-power neodymium glass laser. Opt. Eng. 47, 023801-1–023801-5.

Patankar, N.A., 2004. Transition between superhydrophobic states on rough surfaces. Langmuir. 20, 7097–7102.

Prado, R., Beobide, G., Marcaide, A., Goikoetxea, J., Aranzabe, A., 2010. Development of multifunctional sol–gel coatings: anti-reflection coatings with enhanced self-cleaning capacity. Sol. Energy Mater. Sol. Cells. 94, 1081–1088.

Qiu, J., Zhang, S., Zhao, H., 2011. Recent applications of TiO2 nanomaterials in chemical sensing in aqueous media. Sensor Actuat. B. 160, 875–890.

Reddy, K.M., Manorama, S.V., Reddy, A.R., 2002. Bandgap studies on anatase titanium dioxide nanoparticles. Mater. Chem. Phys. 78, 239–245.

Segota, S., Curkovic, L., Ljubas, D., Svetlic, V., Houra, I.F., et al., 2011. Synthesis, characterization and photocatalytic properties of sol–gel TiO_2 films. Ceram. Int. 37, 1153–1160.

Shokuhfar, A., Alzamani, M., Eghdam, E., Karimi, M., Mastali, S., 2012. SiO2-TiO2 nanostructure films on windshields prepared by sol-gel dip-coating technique for self-cleaning and photocatalytic applications. Nanosci. Nanotechnol. 2 (1), 16–21.

Singh, A.V., Rahman, A., Kumar, N.V.G.S., Aditi, A.S., Galluzzi, M., et al., 2012. Bio-inspired approaches to design smart fabrics. Mater. Des. 36, 829–839.

Stylios, G., Wan, T., Giannoudis, P., 2007. Present status and future potential of enhancing bone healing using nanotechnology. Injury. 38 (1), S63–S74.

Subagio, D.P., Srinivasan, M., Lim, M., 2010. Photocatalytic degradation of bisphenol-A by nitrogen-doped TiO2 hollow sphere in a vis-LED photoreactor. Appl. Catal. B. 95 (3–4), 414–442.

Subramani, K., Ahmed, W., (Eds.) 2012. Nanotechnology and the future of dentistry. In: Emerging Nanotechnologies in Dentistry. William Andrew Publishing, Boston, MA, pp. 1–14.

The Glass Guide, 2002. Why glass is red hot green building material, The Glass Magazine. Available from: <http://www.glasswebsite.com/techcenter/articles/glass%20guide%200602.pdf> (accessed 14.06.13.).

Wang, P., Zhou, T., Wang, R., Lim, T.T., 2011. Carbon-sensitized and nitrogen-doped TiO2 for photocatalytic degradation of sulfanilamide under visible-light irradiation. Water Res. 45, 5015–5026.

Wenzel, R.N., 1936. Resistance of solid surfaces to wetting by water. Ind. Eng. Chem. 28, 988–994.

Wunderlich, W., Takahashi, M., 2002. Characterization of nano-particles during Mg-Al-spinel-formation calculated by MD-Simulations. Ceram. Trans. 133, 189–194.

Wunderlich, W., Hue, N.T., Tanemura, S., 2006. Fabrication of nano-structured titania – thin-films on carbon-coated nickel sheet. AZo J. Mater. Online. http://dx.doi.org/10.2240/azojomo0208.

Yang, J., Bai, H., Jiang, Q., Lian, J., 2008. Visible-light photocatalysis in nitrogen–carbon-doped TiO2 films obtained by heating TiO2 gel–film in an ionized N2 gas. Thin Solid Films. 516, 1736–1742.

Yang, Z., Yu, Y., Li, X., Bao, H., 2006. Nano-mechanical electro-thermal probe array used for high-density storage based on NEMS technology. Microelectron. Reliab. 46 (5–6), 805–810.

Yu, J.G., Yu, J.C., Zhong, G.R., Han, J.J., Zhao, Q.N., 2001. The grain size and surface hydroxyl content of super-hydrophilic TiO2/SiO2composite nanometer thin films. J. Mater. Sci. Lett. 20, 1745–1748.

Yu, Y., Zhang, S., Ren, Y., Li, H., Zhang, X., et al., 2012. Preservation using chitosan film with nano-silicon dioxide. J. Food Eng. 113 (3), 408–414.

Zaleska, A., Sobczak, J.W., Grabowska, E., Hupka, J., 2008. Preparation and photocatalytic activity of boron-modified TiO2 under UV and visible light. Appl. Catal. B. 78, 92–100.

Zeineldin, R., 2013. Nanotechnology for cancer screening and diagnosis. In: Park, K. (Ed.), Biomaterials for Cancer Therapeutics: Diagnosis, Prevention and Therapy. Woodhead Publishing, Cambridge, pp. 137–164.

Sol-gel nanocomposite hard coatings

5

Ramachandra S.R. Kalidindi and Raghavan Subasri
Centre for Sol-Gel Coatings, International Advanced Research Centre for Powder Metallurgy and New Materials (ARCI), Andhra Pradesh, India

5.1 Introduction

The unique properties of nanostructured materials through a sol-gel wet chemical method are now well acknowledged. Sol is a colloidal suspension prepared as a result of hydrolysis and condensation of metal/silicon alkoxide precursors in the presence of water, where suitable catalysts control the reaction rates. When one or more hydrolyzable precursors are hydrolyzed in the presence of water, nanocrystalline or amorphous nanoscale solid particles are formed as a result of the synthesis and a solid in liquid phase is obtained (sol); subsequent condensation results in gelation where a complete solid network forms having pores filled with liquid (gel) (Brinker and Scherer, 1990). This whole process can be controlled in such a way that the sol can either be immediately coated on a work piece and the subsequent drying results in a three-dimensional metal/silicon oxide network or can be stored for a longer duration before subsequent use. The major advantage of the process is that tailor-made solutions are possible for different engineering applications. When the sol-gel process was first invented, completely hydrolyzable precursors were employed and, as a result, pure inorganic coatings were developed that had to be cured necessarily at higher temperatures close to 600°C–1000°C. Moreover, these coatings were quite thin with thicknesses ranging from few hundreds of nanometer to a maximum of 1 micrometer. Residual tensile stresses were observed to be the reason for not being able to build-up higher coating thicknesses. Coatings were hard and able to provide good wear and corrosion resistance, in addition to other specific properties depending on the precursors employed. However, coatings could be deposited only on high melting metal or alloys, glass, and ceramics (Aegerter and Mennig, 2004; Aegerter, 2005). The coatings were brittle, limiting the application segment of the novel process.

Subsequently, sol-gel science evolved and organic–inorganic hybrid coatings were successfully developed with curing temperatures as low as room temperatures to a maximum of 500°C, which is the upper limit of the temperature stability of the organic groups (Kickelbick, 2007; Romero and Sanchez, 2004). Now coatings can even be densified using ultraviolet (UV) or near-infrared (NIR) radiation with addition of a suitable photoinitiator for UV curing or an NIR absorber in the sol. This has increased the application segment of hybrid coatings enormously starting

Anti-Abrasive Nanocoatings.

from low temperature melting plastics such as polymethylmethacrylate (PMMA) and polycarbonate (PC) to any metal. It has been reported that the structural properties of the coating are strongly influenced by the synthesis and processing parameters such as precursors used, precursor type, precursor-to-water molar ratio, catalyst, solvent, modifying agent, pH, viscosity, chamber temperature during coating, humidity, curing technique, curing temperature, and heating rate (Barbe et al., 2005; Estella et al., 2007; Wang and Bierwagen, 2009). Low densification temperature, the possibility of introducing trace elements and employing radiation curing are some of the salient features of nanocomposite hybrid coatings (Schmidt et al., 1998).

5.2 Sol-gel nanocomposite hard coatings

Organic–inorganic hybrid coatings are further subdivided into two classes (Plinio et al., 2005; Schottner, 2001). Class I has organic molecules or polymers embedded in an inorganic matrix. While the inorganic network is generated as a result of the hydrolysis and condensation of the alkoxy groups, the organic compound that is added occupies the pores in the inorganic network. Organic–inorganic hybrid materials where the interaction between the organic and inorganic moieties is regulated by hydrogen bonding and van der Waals forces come under class I hybrid materials. A porous oxide doped with organic dye not grafted to the pore surface is an example of this category, whereas in the class II hybrid materials, the organic and inorganic components are strongly linked through iono-covalent bonds. Hybrid coatings can easily be generated in-situ through sol-gel process by the use of appropriate organically modified silanes or organo(alkoxy)silanes. Here, the organic groups in the organically modified silanes could be non polymerizable, like the alkyl/aryl group or they could be polymerizable like the epoxy, acryl or vinyl groups. As mentioned above, polymerization of these groups can be initiated by thermal or radiation techniques.

As a single material/component cannot meet all the functional requirements in many practical applications, composites were introduced which provide the best properties of each of the component. Nanostructured materials are materials whose structural elements such as crystallites, molecules or clusters have dimensions in the range of 1 to 100 nm. Hybrid nanocomposite coatings can be generated by external addition of nanoparticles or by in-situ generation of nanoparticles in the hybrid matrix through use of appropriate transition metal alkoxides that are completely hydrolyzable. In all of the nanocomposite hard coatings, the main skeleton network is drawn from the metal alkoxide precursors, that is selected based on the end property requirement such as optical properties; hard, scratch, abrasion, and corrosion resistance; and so on. Organic muscle over this skeleton can be drawn from a large variety of available organo(alkoxy)silanes, organic monomers, and polymers. While the former precursor dictates the necessary hardness, the latter precursor provides the flexibility by reducing the brittleness and enhances the final coating thickness. External addition of metal oxide nanoparticles further enhances the hardness and other mechanical properties of the composite coatings. The end-user, however, needs to optimize the quantity of externally added inorganic particles with respect to the target property requirements. The selection of metal alkoxide precursors and organo(alkoxy)silanes, however, is also based on the energy requirement for curing the coatings, as these conditions ultimately decide on the suitability of sol for a given substrate metal or a plastic. A sol prepared from a combination of more than one metal alkoxide precursors results in nanocomposite hard coatings with in-situ prepared oxide networks such as SiO_2, TiO_2, Al_2O_3, and ZrO_2. A much simpler alternative is a sol prepared from hydrolysis and condensation of organo(alkoxy)silane that results in silica as basic network, and the additional nanosized component may be chosen from SiO_2, TiO_2, ZrO_2, CeO_2, SnO_2, Al_2O_3, In_2O_3, La_2O_3, Fe_2O_3, Ta_2O_3, Nb_2O_5, V_2O_5, Cu_2O, MoO_3, WO_3, ZnO, CdO, and so on, and added externally based on functional requirements.

The commonly used precursors to generate the hybrid nanocomposite coatings are drawn from metal salts, oxides, hydroxides, complexes, alkoxides, and amines, out of which alkoxides are more common in usage. Since metal alkoxides (e.g. titanium tetraisopropoxide, zirconium-*n*-propoxide) are more reactive toward water than alkoxysilanes, they require sensitive synthesizing procedures (Kasemann and Schmidt, 1994; Schmidt et al., 1998) when used. 3-glycidoxypropyltrimethoxysilane (GPTMS) with an epoxy functional group, 3-methacryloxypropyltrimethoxysilane (MPTMS) containing methacrylate functional group and vinyltrimethoxysilane (VTMS) containing vinyl functional groups are some of the most extensively studied and commercially successful organically modified alkoxides with polymerizable organic functional groups (Plinio et al., 2005). Other alkoxysilanes that are routinely being used include tetraethyl orthosilicate (TEOS), tetramethyl orthosilicate (TMOS), methyl triethoxysilane (MTES), methyl trimethoxysilane (MTMS), 3-aminopropyl trimethoxysilane (APTMS), and γ-methacryloxypropyl trimethoxysilane (γ-MAPTS) (Jianguo et al., 2006; Wang and Bierwagen, 2009).

However, the coating quality strongly depends on the substrate preparation such as solvent cleaning, sand blasting, or surface activation through plasma treatment, and so on. These techniques not only help thorough cleaning, for example removal of dust and organic material such as oil stains, but also improve adhesion through the addition of functional groups to the substrate, such as —OH groups, during surface activation in the case of plasma treatment and increase the effective surface area and induce residual compressive stresses in the case of shot blasting. The former enhances the wettability and hence uniformity of coating thickness, while the latter enhances the critical coating thickness. Critical coating thickness is the thickness beyond which the coatings are not able to retain the tensile stresses formed as a result of coating shrinkage during the drying process and cracking. This can also be enhanced as a result of the induced compressive stresses prior to coating application. The shot blasting could result in certain surface roughness, which may reduce the gloss of the substrate surface, subsequently reducing the aesthetic appeal after coating deposition. Hence, there should be a trade-off between the coating thickness and the aesthetic appeal.

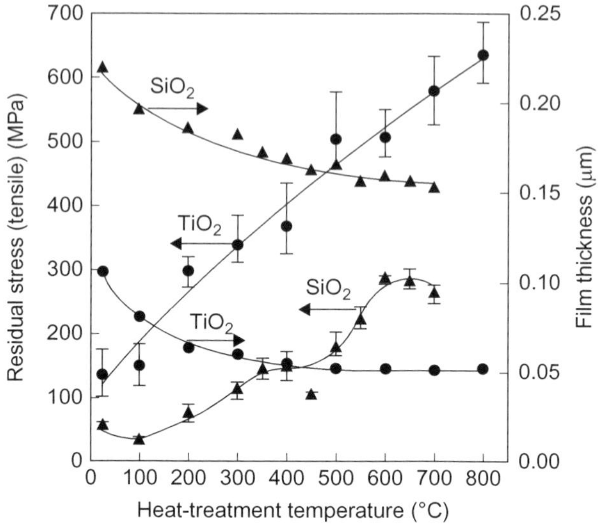

Figure 5.1 Influence of heat treatment temperature on residual stress and coating thickness of silica and titania coatings produced by spin coating on silica glass substrate.
Source: Reproduced from Kozuka et al. (2003).

It has been reported that the stress is induced in the wet film right from the stage of coating deposition and increases as the coatings are consolidated in the subsequent curing or heat treatment condition (Aegerter, 2005). The total stress in the coating is the sum of the stress developed due to shrinkage of coating as a result of evaporation of the solvents during drying (Marzolin et al., 1998) and subsequent heat treatment. This in turn depends on the changes occurring in microstructure, thickness and any impurities, and the stress arising during heat treatment due to the difference in coefficient of thermal expansion of coating material and the substrate. It was observed that these stress levels decrease over a period of time due to internal material reorientation, diffusion, and the absorption of water from the surrounding atmosphere (Ottermann and Bange, 1993). It was also noticed that the film stress levels are strongly dependent on density of the coating material. As can be seen from Figure 5.1, the prominent parameter that affects the stress build-up and coating thickness is heat treatment temperature. It was found to be tensile in nature and increases with the heat treatment temperature. Water-to-precursor ratio could be another important factor that affects the stress level, which increases with an increase in the ratio. Other ingredients that are added during sol preparation such as chelating agents too aid in the change of stress levels (see Figure 5.2): for example, residual stress σ of titania coating for a heat treatment temperature of $T_s = 700°C$ changes from $\sigma = 340 \pm 70$ MPa without any chelating agent to $\sigma = 490 \pm 60$ MPa with acetic acid and is further enhanced to $\sigma = 1090 \pm 70$ MPa for diethanolamine (Kozuka et al., 2003). Other important parameters are the coating thickness and heating rate. Stress develops and increases with increased coating thickness and

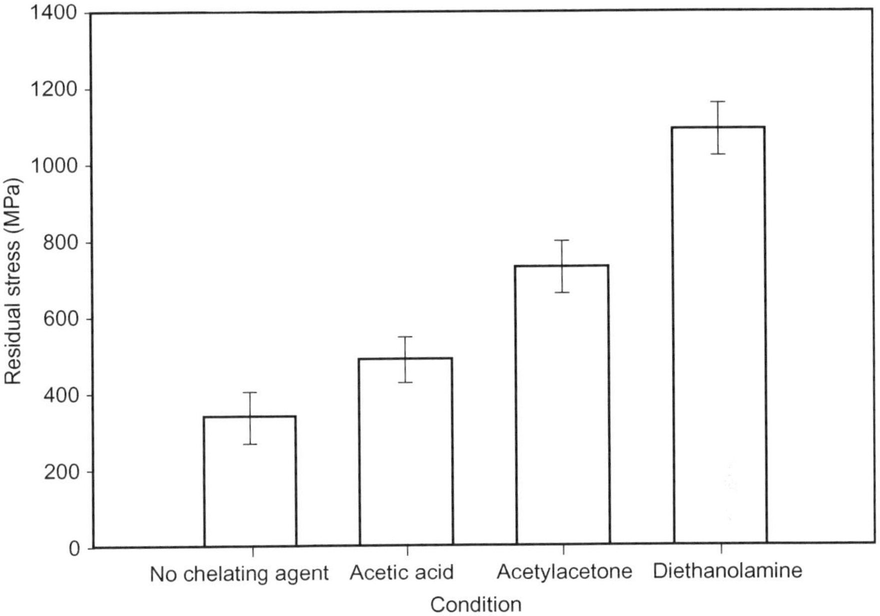

Figure 5.2 Effect of addition of chelating agent on residual stress of titania coatings applied on silica glass substrate.
Source: Redrawn from Kozuka et al. (2003).

increased heating rates even at lower temperatures. Such cracking was found to occur in situ during heating itself. As the residual stress is an important parameter that affects all the mechanical properties, it has to be kept as low as possible by either inducing compressive stress prior to the coating deposition using surface activation techniques, or the sol composition should be managed in such a way that the stress levels are suppressed due to the presence of organic polymers in the composition (Schmidt et al., 1998).

Sol-gel coatings particularly ORMOCER®s and ORMOSIL®s are found to be the most promising coatings for many commercial applications. ORMOSIL is an organically modified silicate, a noncrystalline solid between organic and inorganic material, $CH_3{-}Si{-}CH_3$ (Wilkes et al., 1985). Whereas ORMOCER is an organically modified ceramic, where other oxides such as ZrO_2 and TiO_2 are bonded to organic groups (Schmidt, 1985). This set of coatings is elastic to withstand the strain caused due to the difference in coefficient of thermal expansion that results during curing of the coatings and hence does not result in cracking until coating thicknesses are built over multiples of a few micrometers. Plastic substrates cannot be elevated to the temperatures that are normally used for metallic or ceramic substrates for curing/densification of coatings. As a result, the coatings when applied to plastics are deemed to provide lower hardness and adhesion when compared to the coatings applied to metals or ceramics. But use of organic–inorganic hybrid coatings that can be densified at low temperatures can be applied to plastics and their mechanical properties can be improved either by addition of nanoparticles into a basic hybrid sol or through an in-situ formed nanostructured hybrid coating. The following is a brief compilation of the commonly evaluated mechanical properties of sol-gel derived nanocomposite hard coatings.

5.3 Mechanical property studies of sol-gel hard coatings on various substrates

5.3.1 Adhesion

Coating adhesion to the substrate could result from a purely mechanical bond held together due to anchoring as a result of increased effective surface area or a chemical bond, or the combination of both. It is always assessed by the amount of force required to separate the coating from the substrate. Coating could fail at the interface of the coating and substrate if the adhesion to the substrate is poor, or in the coating if the cohesion is poor.

Adhesion can be assessed by various techniques, which are described below.

5.3.1.1 Indentation technique

Fracture toughness can be measured from the interfacial cracking when the film cracks during indentation. As per the ASTM definition of adhesion (ASTM D907-12a, 2012), two surfaces are held together due to interfacial forces that may consist of valence forces or interlocking forces or both. This bonding force may be van der Waals forces, electrostatic forces, or chemical bonding forces that are effective across the interface. The soft and flexible organic constituent of the sol-gel coatings enhance the adhesion of the coating to the substrate.

Quantification of the adhesion strength can be accomplished by multiple indentations on the coating and measurement of average crack length as a function of load. Resistance offered by the coating to the propagation of crack will be the measure of adhesion. Fracture toughness K_{li} can be derived from the linear portion of the plot drawn from the above data and using the following equation (Jindal et al., 1987):

$$K_{\mathrm{li}} = \left(\frac{G_{\mathrm{li}} E_c}{1 - V_c^2}\right)^{1/2} \tag{5.1}$$

where G_{li} is the slope of the linear portion of the plot, E_c Young's modulus, and V_c Poisson's ratio. The advantage of the proposed model is that the adhesion parameters are insensitive to the substrate hardness. Of the available methods, it was considered that the indentation test method is one of the best methods available for assessing the coating adhesion in relation to the performance of real engineering components in service (Rickerby, 1988).

Another study has employed a diamond Rockwell C indenter for indenting a sol-gel nanocomposite hard coating applied on a SS 316 substrate until it penetrated the substrate through the coating. Plastic deformation of the substrate leads to generation of annular cracks and the radius of the annular crack was measured for calculating the interface toughness. Moreover, it was also observed that sand-blasted substrate preparation and higher heat treatment temperatures were found to enhance the adhesion of the coating to the substrate (Xie and Hawthorne, 2003).

Yet another study used a rigid indenter to penetrate through the coating resulting in cracking of coating and subsequently its propagation and delamination. A compressive stress is induced due to indentation leading to plastic deformation of the coating and even the substrate as well. Cracking of the coating and its delamination from the substrate is said to depend on the stored elastic energy U (Latella et al., 2012).

$$U = \frac{1 - V_f}{E_f} \sigma_f^2 t \tag{5.2}$$

where V_f is the Poisson's ratio, E_f is the Young's modulus, σ_f is the normal stress, and t is the coating thickness.

5.3.1.2 Scratch testing

Scratch testing is a simple and rapid method of characterizing coatings, but results obtained are influenced by various factors such as coating thickness, substrate mechanical properties, interfacial bond strength, and test conditions such as scratch speed, load, and indenter tip radius.

Here, the work piece is held stationary and, commonly, a Rockwell diamond indenter (Figure 5.3) with a tip radius of 200 μm is used to scratch the work piece for a predefined length with a predefined normal load. The load can be fixed or increasing progressively, depending on the requirement. The output is measured in terms of

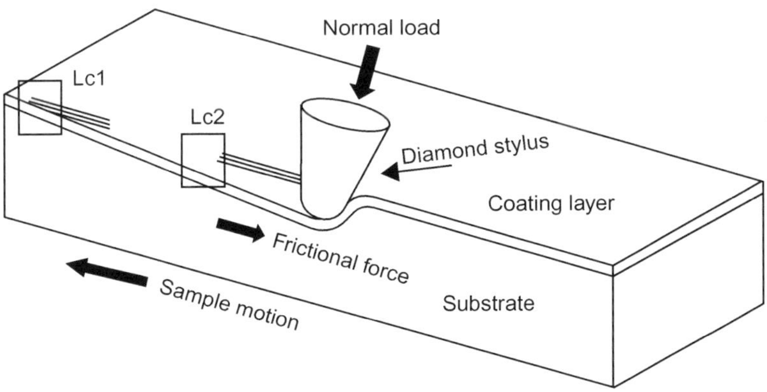

Figure 5.3 Schematic showing an indenter scratching the work piece. Lc1: Critical load for the first crack and Lc2: critical load for severe cracking.

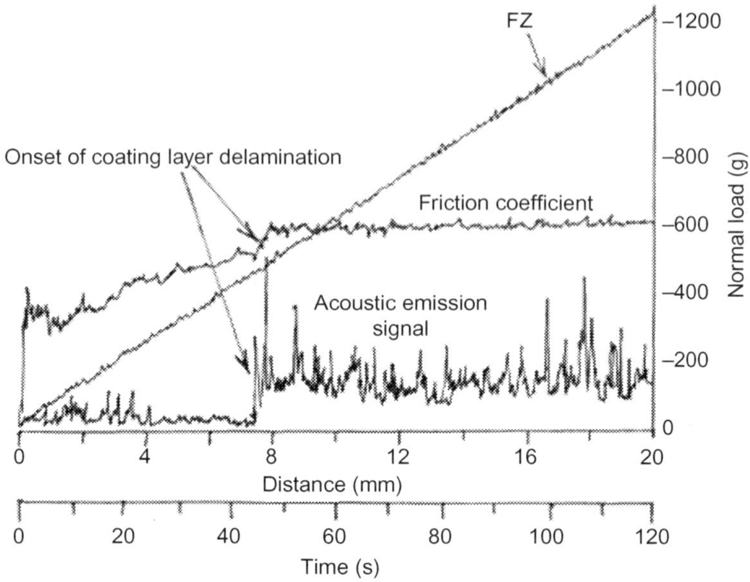

Figure 5.4 Typical scratch test data plots.

acoustic emission, penetration depth, and tangential frictional force. Transducers are used for measuring the frictional force acting at the tip of the indenter. The onset load for coating cracking is commonly referred to as the critical load and, as can be seen in Figure 5.4 is associated with the rise in value of coefficient of friction and acoustic emission used to assess the coating-substrate adhesion. The morphology of the scratch track and transverse cracks provide information on failure mechanisms such as cohesive, interfacial, conformal, and brittle chipping. The load at which the first crack appears is called a "lower critical load" and is a measure of the cohesion strength of the coating. The load at which total peeling-off of the coating from the substrate surface occurs is called as "upper critical load" and is a measure of adhesion strength of the coating to the substrate (Soutar et al., 2008).

The adhesion strength measured by scratch testing of sol-gel fluorinated hydroxyapatite (HA) coating applied in Ti_6Al_4V titanium alloy has shown marked improvement with an increase in firing temperature (Zhang et al., 2006). While the average adhesion strength of a nonfluorinated HA coating cured at 500°C was found to be 0.37 N which rose to 0.47 N when cured at 700°C, the adhesion of a fluorinated HA coating cured at 500°C increased from 0.38 N to 0.57 N when cured at 700°C. Figure 5.5 shows a micrograph of the scratch tested nanocomposite hard coating synthesized from MPTMS or M, zironium-*n*-propoxide (Z), and tetraethoxy orthosilicate (T) sol, deposited on PC and Figure 5.6 presents how it was able to withstand an average scratching load of 4.3 N (Sowntharya et al., 2013). An in-situ formed —Si—O—Si— network in a hybrid sol (MZ + T) was found to be better than an ex-situ addition of acrylic

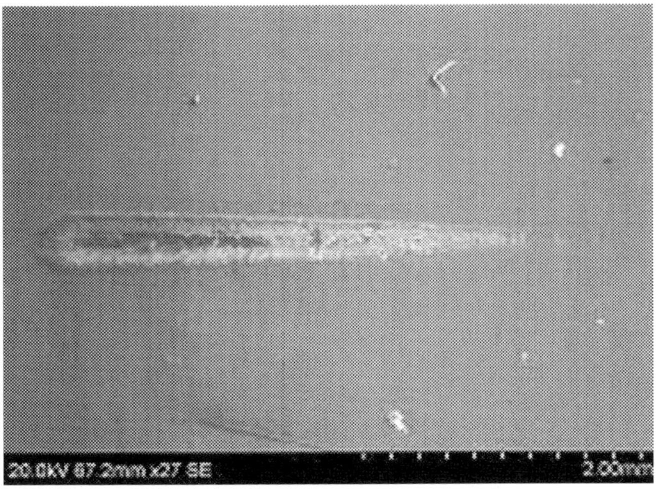

Figure 5.5 Scanning electron micrograph of the scratch on nanocomposite hard coating without any addition of silica nanoparticles addition and applied on PC (Sowntharya et al., 2013).

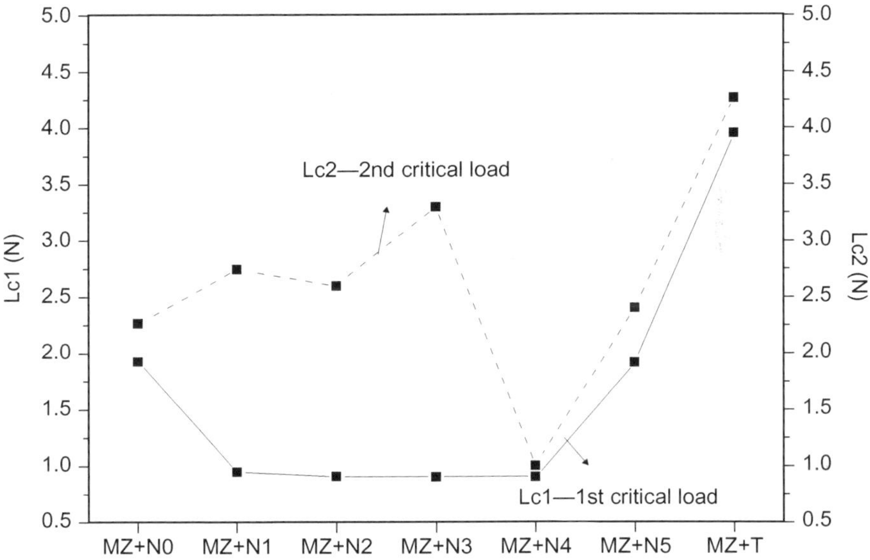

Figure 5.6 First and second critical loads of the scratch test data performed on the coated PC (Sowntharya et al., 2013).

surface-modified silica nanoparticulate dispersion (N) with a scratching load of 1 N that remained unchanged even with an increase in amount from 0.01 to 0.25 weight ratio to that of the basic sol synthesized from MPTMS and Zr-*n*-propoxide (MZ + N0-N5).

Extensive work has been carried out on assessing energy release rate G and stress of the coating on substrate due to the contact load of the indenter and friction on scratch track σ (Chen et al., 2008; Malzbender and With, 2001):

$$G = A\frac{\sigma^2 t}{E} \tag{5.3}$$

$$\sigma = \sigma_b + \sigma_t \tag{5.4}$$

$$\sigma_b = B\frac{P}{t^2} \tag{5.5}$$

$$\sigma_t = C\frac{\mu P}{wt} \tag{5.6}$$

$$P = \sqrt{E\Gamma t}\frac{(wt)}{(A(Bw + C\mu t))} \tag{5.7}$$

where G is the energy release rate; σ is stress, t is film thickness; E is elastic modulus; P is vertical load transferred to the coating from the indenter tip; w is scratch width; μ is friction coefficient; σ_b is bending stress; σ_t is tensile stress due to friction in the scratch track; Γ is fracture toughness of the coating while A, B, and C are the coating material-related factors. When energy release rate is equated to the coating fracture toughness and substituting Eqs. (5.5) and (5.6) in (5.4) and subsequently in (5.3) we will get Eq. (5.7) from which the critical scratch load can be calculated. The above equations hold good for both mechanized scratch testing and pencil scratch testing and it can be understood that higher modulus of elasticity and higher coating thickness enhances scratch resistance of the coating.

5.3.1.3 Cross-hatch cut and scotch tape test

The ASTM D3359-02 (2000) standard refers to a test method for measuring adhesion by tape test. A cutter, as shown in Figure 5.7, with sharp cutting edges spaced at 1 mm from each other is pressed against the coating and scratched to make deep grooves. Sufficient load is applied for the cutter to penetrate through the coating to the interface and a similar cut is made normal to the previous cut so as to get a square cut grid. A piece of sticky tape is pressed on to the grid surface and pulled quickly with a sudden jerk at a normal angle. After the tape removal, the surface is inspected under the microscope in white light to examine how much material has been removed from the square grid. Based on the amount of material removed, adhesion of the coatings is ranked from 5B (excellent adhesion) to 0B (very poor adhesion). 5B refers to no material removed, 4B refers to <5% material removal, 3B refers to 5–15% removal, 2B refers to 15–35% removal, 1B refers to 35–65%, and 0B greater than 65%.

Figure 5.7 Cross-hatch cutter used for making the grids on the coating.

In one of the recent investigations, nanocomposite hybrid coatings showed marked improvement in scratch hardness when nanoscaled inorganic particles were added to the sol. A new concept of heat impregnation treatment was introduced for PC substrates. A precursor solution was prepared in a molar ratio of methyl trimethoxysilane:3-aminopropyl triethoxysilane:ethanol:ethylmethylketone:water:acetic acid of 1.0:1.0:14.0:4.0:6.0:0.4. The sample was loaded into a heated and evacuated chamber at 70°C in an inclined position and the precursor solution was then poured onto the substrate surface. The coated substrate was then pressurized in an inert atmosphere to 3 bar and held for a prescribed time period. The sample was then cleaned and further cured at 80°C for an hour before application of a hybrid sol-gel coating. PC undergoes aminolysis in the above procedure and is chemically bonded to the substrate, resulting in 5B rank adhesion (Wu et al., 2009).

In a similar test method, it was observed that the adhesion strength of the PMMA–silica composite in terms of percentage of coated area removed on SS 316L substrate has shown enhanced performance when compared to a pure PMMA coating. It was also noticed the adhesion strength was further enhanced with an increase in the TEOS component (Caselis et al., 2012).

Simple silica-based nanocomposite hybrid coatings deposited on AA5083 grade aluminum substrate resulted in an excellent adhesion property of 5B depending on low organic–inorganic molar ratio (GPTMS/TEOS), higher hydrolysis water content, 60°C drying temperature, and 130°C curing temperature (Najafabadi et al., 2013). In another study it was shown that the composite sols prepared from an acid-catalyzed MTEOS and aqueous colloidal silica dispersion with or without fluoropolymer have an excellent adhesion property of 5B (Lakshmi et al., 2012). It was also observed that the addition of plasticizer such as tri(butoxyethyl) phosphate (TBOEP) or glycerol enhances the adhesion of nanostructured silica–zirconia coatings applied to stainless steel substrates to 5B even after subjecting them to severe thermal cycling by dipping in liquid nitrogen and bringing them back to room temperature (Audebert et al., 1997). Phase separation behavior of sols synthesized from each of phenyltriethoxysilane (PhTES), methyltriethoxysilane (MTES), and 1,4-bis(triethoxysilyl)benzene (BTEM) with TEOS was found to affect the adhesion of coatings on PET, PC, PE, PP, and PVC substrates (Mizuta et al., 2013).

5.3.1.4 Pull-off adhesion test

Pull-off adhesion strength is assessed as per ASTM D4541-09 (2009) standard where an adhesive of known adhesion strength is applied to the face of a dolly and pressed against the coated flat surface. Excess adhesive squeezed out needs to be wiped and the sample with the dolly is allowed to dry for a prescribed time until the adhesive is cured. Drying in an oven is an alternative to cure the adhesive quickly and save time. Once the adhesive is cured, the dolly is held in a loading fixture, aligned to apply tension normal to the test surface, and pulled with a gradually increasing normal load. A hydraulic system is employed to exert a normal load and the greatest perpendicular force that the surface withstands before the dolly detaches from the coating can be recorded. Dollies are usually available in different diameters and the most commonly used one is of 20 mm in diameter. This test method exerts tensile stress while all other methods apply shear stress and hence results of samples from the same test method only are to be used for comparative studies. The coated surface around the dolly is cut with the help of a cutter so that the perpendicular load exerted is applied only to the coated surface that is exactly equal to the contact area of the dolly. Various commercial products are available to exert the normal load with the help of a hydraulic system either manually or with the help of a hydraulic pump system. One of the commercially available products is presented in Figure 5.8. Figure 5.9 shows the close-up view of the test procedure.

As per one of the variants of the above test procedure, adhesion strength was measured with the help of an epoxy-coated stud adhered to the coating and cured at 150°C for 1 hour and subsequently pulled similar to the dolly test (Kim et al., 2005). While the stud dimension in this investigation was 2.69 mm, the dolly of ASTM standard has a diameter of 20 mm. Adhesion strength was measured from the maximum load applied prior to failure divided by the contact area (Kim et al., 2005). As can be seen from Figure 5.10, HA, HA composite with titania, and titania sol-gel coatings on titanium substrate exhibited increasing adhesion strength. The composite coating adhesion strength remained between 37 MPa for HA to 70 MPa for titania.

In a similar pull-off adhesion testing method using an 8 mm diameter stud, the adhesion strength of magnesium containing fluorinated HA coating on Ti_6Al_4V titanium alloy substrate increased significantly to 18 MPa from nonmagnesium containing HA coating with an adhesion strength of 12 MPa (Cai et al., 2011).

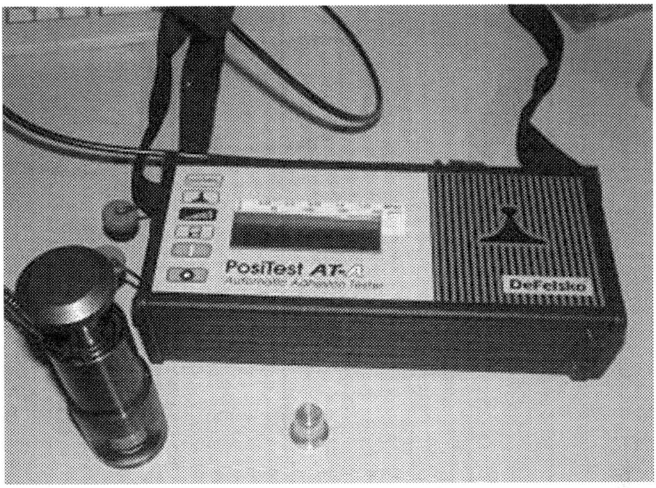

Figure 5.8 Commercially available pull-off adhesion tester.

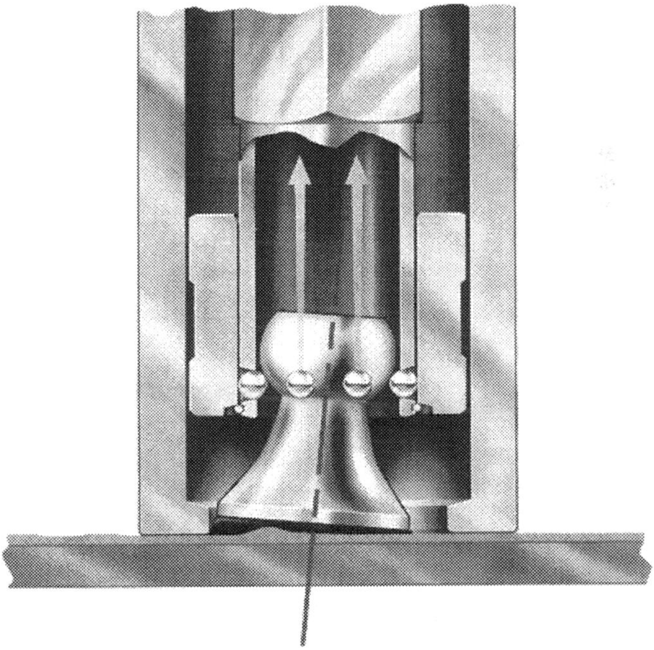

Figure 5.9 Close-up of the arrangement of the dolly over the coated specimen. *Source:* Reprinted with permission from ASTM D4541-09 (2009). Copyright ASTM International, West Conshohocken, PA.

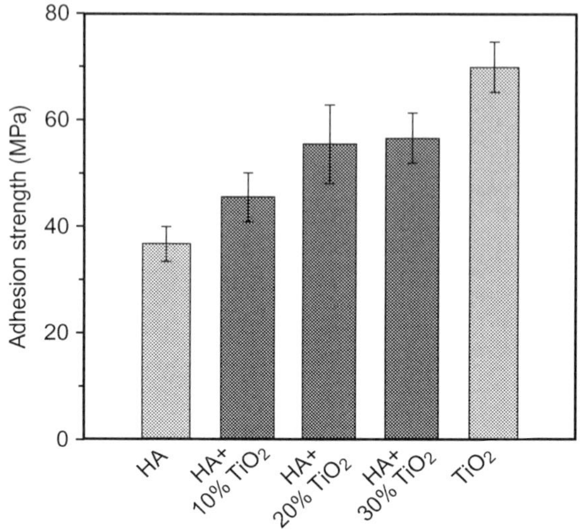

Figure 5.10 Average adhesion strength of the composite coatings applied on titanium substrate (Kim et al., 2005).

Adhesion enhancement of sol-gel coatings on a polymer surface can be possible in two ways (Chou and Cao, 2003):

1. Formation of chemical bonds through surface condensation reactions: Functional groups such as –COOH, –OH, –(C–O)–, –(C = O)–, –(C = O)–O–, and so on, can be introduced onto a polymer surface with the help of air or oxygen plasma surface activation (Gururaj et al., 2011). These groups readily form hydrogen and covalent bonds with the sol and exhibit excellent adhesion property. Wet chemical routes can oxidize polymer surfaces. Etching polyethylene surfaces with sulfuric acid solution is one of such possibility.
2. Development of interlocked ceramic and polymeric networks through diffusion of alkoxide precursors: A polyester substrate when immersed in a 1:1 tetramethylorthosilicate: ethanol solution at 90°C for 5 hours results in diffusion of silicon alkoxide. This results in the formation of silicon alkoxide within polyester substrates, which then undergoes hydrolysis and condensation process with the help of water available from ambient conditions inside polymer matrix forming silica network. This extends to the surface and forms –Si–O–Si– bonds with the sol-gel derived top coating. Adhesion assessment is done by thermal cycling between 150°C in an oven and −20°C in a freezer.

Adhesion enhancement in metallic substrates is possible by carrying out a shot-blasting surface pre-treatment or by a surface activation through plasma pre-treatment (Kumar et al., 2012a,b; Kiruthika et al., 2010). In the case of aluminum–aluminum alloys, a fresh and thick oxide layer is formed on the aluminum surface after air–oxygen plasma pre-treatment (Kim et al., 2003) that improves the adhesion of sol-gel coatings deposited after plasma surface pre-treatment. In the case of mild steel, in addition to removing any surface contamination due to a hydrocarbon layer, the plasma treatment promotes increase in surface concentration of Fe^{3+} and a decrease in the density of the reduced species, that is, the electron donor species. Due to this modification, the reduction of atmospheric oxygen on the steel surface that is responsible for corrosion is expected to slow down (Fracassi et al., 2003).

5.3.2 Pencil hardness

Pencil scratch hardness measurement is a quick check tool to measure hardness of sol-gel and other organic coatings on any rigid substrate. In this method, according to ASTM D3363-05 (2000), a set of pencils with increasing order of hardness 9B–8B–7B–6B–5B–4B–3B–2B–B–HB–F–H–2H–3H–4H–5H–6H–7H–8H–9H are used to make a scratch on the coated substrate by holding the pencil firmly in a rolling holder at 45° to the ground and pushed for a scratch length of 6.5 mm. The test usually starts with the hardest pencil and is then carried out with a series of softer pencils. The hardness number is the same as the number of the pencil with one step lower hardness than the pencil that scratches the coating – in other words, the number of the hardest pencil that will not rupture or scratch the coating is the scratch hardness. Another important term used is gouge hardness, which refers to the number of the hardest pencil that will leave the coating without a cut for a minimum stroke length of 3 mm out of 6.5 mm stroke length.

Figure 5.11 depicts a commercial pencil scratch hardness testing device holding a pencil. The holder has to be placed on coated surface and wheels are held between fingers and the body is rested over the sample. This test is being used extensively for hybrid nanocomposite coatings as a quick evaluation tool to assess whether the film has sufficient scratch hardness on various substrate materials. It was reported that the pencil scratch hardness of the coatings is thickness dependent (Sowntharya et al., 2012; Wu et al., 2008) and normally it attains maximum when it reaches the critical coating thickness (Soma Raju et al., 2012). Photocurable cyanate ester containing

Figure 5.11 Pencil scratch testing device.

hybrid coatings generated by an anhydrous sol-gel technique on aluminum substrate have resulted in a scratch hardness of 5H (Bayramoglu et al., 2012). Precursor solution impregnation has helped in enhancing thickness-dependent pencil scratch hardness to a remarkable 8H at 20-μm coating thickness (Wu et al., 2009). Radiation curing techniques such as using UV light, NIR light, and microwave (MW) are commercially attractive due to reduced cycle times or adaptability to online processing, especially for the first two processes. One of the recent studies shows pencil scratch hardness of PC was enhanced from 2B to a maximum of 3H for coatings generated using epoxy-modified silane–titania (Sowntharya et al., 2012). Another study by the same group showed that alternative and rapid curing techniques such as UV light and MW radiation can be used to cure coatings to obtain scratch hardness comparable to the thermally cured silica–titania nanocomposite coatings, thereby saving a lot of processing time (Sowntharya and Subasri, 2013).

In a study where hybrid sol-gel coatings containing about 15% surface-modified silica nanoparticles embedded in poly(2-hydroxyethyl methacrylate (PHEMA) enhanced glass transition temperature by 36°C and the pencil hardness up to 5H (Chang et al., 2012), coating thickness, elastic modulus, fracture toughness, and intrinsic hardness of coating materials together influence the pencil scratch hardness. Figure 5.12 shows that coating thickness has significant influence on enhanced pencil hardness while other parameters were optimized (Wu et al., 2008). The hardest reported sol-gel coatings on plastics is 8H and on metals is >9H. Even though surface-modified nanoparticles are utilized in nanocomposite hard coatings, coatings often result in poor scratch resistance if nanoparticles are not homogenously distributed in the hybrid matrix. In-situ prepared

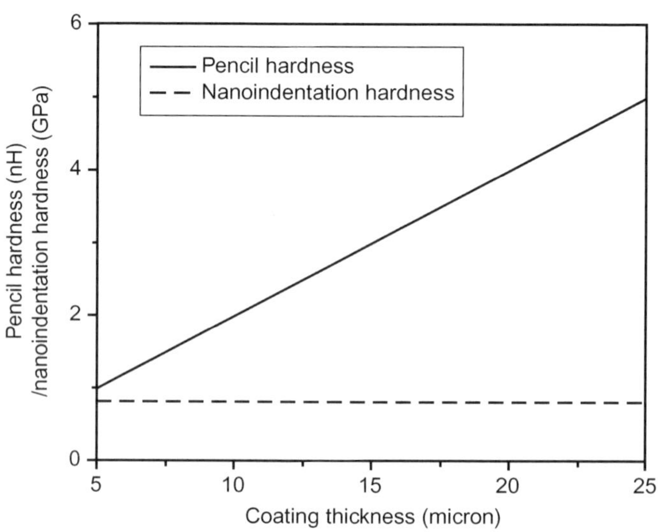

Figure 5.12 Effect of coating thickness on pencil hardness and nanoindentation hardness of nanocomposite hard silica coatings with 30% colloidal silica (Wu et al., 2008).

nanoparticles perform better than externally added nanoparticles because of the networking nature of former.

5.3.3 Nanoindentation hardness

Surface hardness is an important mechanical property of coatings and is a measure of the resistance of the coating to a scratch tip. Nanoindentation is normally used to characterize the coating-substrate system with respect to hardness. The basic principle behind the measurements is to indent the sample and to record the force as a function of the penetration depth. Conventionally, hardness is measured by pressing an indenter of a certain shape into the test surface and imaging the indent after releasing the load and retracting the indenter. However, in the case of coatings, especially those with a thickness of only few microns, this method is not particularly convenient. Hence, an instrumented indentation technique, also known as nanoindentation is used, where a load is applied to a sharp indenter with a Berkovich tip and the displacement into the test specimen is recorded to probe into the mechanical properties of surfaces by accurately measuring displacements of only a few nanometers using sensitive transducers. It is generally accepted that when the depth of indentation is $1/10^{th}$ or less of the coating thickness, the substrate effect can be avoided. Analysis of the load (*P*) Vs displacement (*h*) curves yields the main mechanical property parameters, such as the hardness and Young's modulus without the need to visualize the indent imprint.

Figure 5.13 presents the schematic representation of the nanoindentation test and the related load displacement curve. The elastic modulus (*E*), hardness (*H*), elastic recovery (R_e), and plastic index (*PI*) or inverse of brittleness index (*BI*) can be determined from the following equations:

$$H = P_{max}/A \tag{5.8}$$

$$R_e = (h_{max} - h_c)/h_{max} * 100 \tag{5.9}$$

$$E = (1 - \upsilon^2)/(1/E_r - (1 - \upsilon_i^2/E_i)) \tag{5.10}$$

$$PI = E_c/H \tag{5.11}$$

where E_r, E_i, υ and υ_i are the reduced elastic modulus, tip elastic modulus, Poisson's ratio of the coating, and Poisson's ratio of the tip, respectively.

Several groups have reported on the nanoindentation studies on sol-gel derived hybrid nanocomposite coatings on different substrates, namely transparent plastics such as PC (Sowntharya et al., 2012, 2013; Yahyaei and Mohseni, 2013), PMMA (Blanc et al., 2006; Chantarachindawong et al., 2012), silicon (Uhlmann et al., 2013; Yaacoub et al., 2012), glass (Almaral-Sanchez et al., 2006; Etienne et al., 2004), and metals/stainless steels (Ballare et al., 2009; Cambon et al., 2012; Ramezanzadeh and Mohseni, 2012).

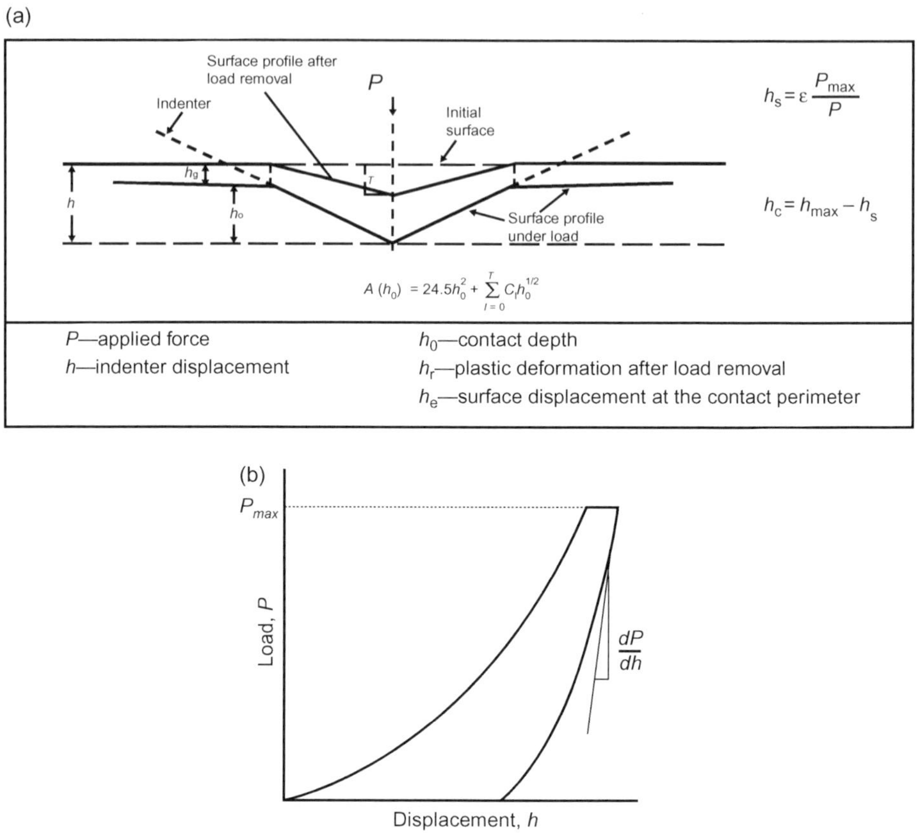

Figure 5.13 Schematic representation of a nanoindentation test (a) events happening during and after the test and (b) typical load–displacement curve obtained after the test (Ramezanzadeh and Mohseni, 2012).

The mechanical properties of the hybrid sol-gel coatings have been shown to be a function of a variety of parameters such as the ratio and nature of the organic/inorganic precursors, the reaction conditions such as pH, catalyst, solvent, and the coating deposition and curing parameters (Blanc et al., 2006). In addition, the mechanical properties of the coatings are found to be different when the nanocomposite coating is generated from a physical dispersion of the nanoparticles in the sol-gel matrix when compared to an in-situ generated metal oxide network. In case of the former, a strong covalent bond formation between the –OH surface groups of the nanoparticles with the sol by condensation shown in Figure 5.14 can be envisaged. Addition of oxide/nonoxide nanoparticle filler to the hybrid sol-gel matrix is expected to improve the mechanical properties of the coatings.

However, the successful application of nanoparticles depends upon both the synthesis methodology and the nature of the surface of these nanoparticles. The major

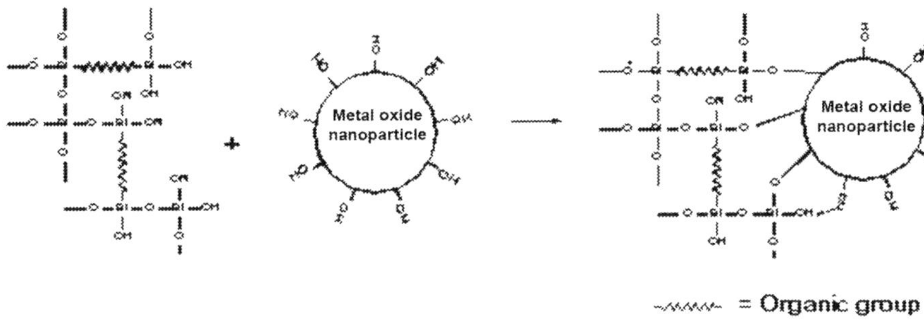

Figure 5.14 Schematic representation of incorporation of metal oxide nanoparticles into an organic–inorganic hybrid sol.

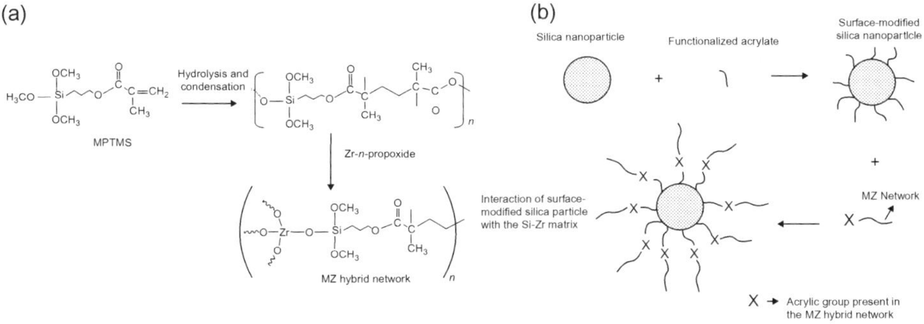

Figure 5.15 Schematic representation of (a) a hybrid network formed between methacryloxypropyltrimethoxy silane (M) and zirconium-*n*-propoxide (Z), and (b) hybrid MZ matrix with surface-modified silica nanoparticles dispersed (Sowntharya et al., 2013).

problem in dispersing the nanoparticles into an organic–inorganic hybrid matrix is the aggregation, which results in poor mechanical properties of the coatings. A macroscopic separation between the organic and inorganic moieties should be prevented. Also, increase in concentration of the nanoparticles rapidly increases the sol viscosity, which affects the critical thickness of the coating (Amerio et al., 2008). In order to overcome these problems, the surface of the nanoparticles has to be modified so as to improve their interaction with the matrix. Usually, surface modification is carried out using the same organic group that is present as the polymerizable group in the organic–inorganic matrix, so that there is compatibility between the organic groups from the nanoparticles and the hybrid sol-gel matrix. A typical example has been provided in Figure 5.15, where surface-modified (acrylic-modified) silica nanoparticles have greater compatibility with the hybrid sol-gel matrix (silica–zirconia matrix labeled as MZ) and lead to lesser phase separation when compared to a direct addition of nanoparticles without any surface modification into the matrix (Sowntharya et al., 2013).

Presence of chemical bonds of high strength contributes to greater hardness of the coating. Accordingly, a multimetal oxide network, such as a silica–zirconia network, in the coating is reported to exhibit greater hardness. Typical hardness and reduced modulus values reported for MPTMS/ZrO_2 coatings on PMMA are 337–440 MPa and 3.8–6 GPa, as compared to 170–178 MPa and 3.6–4 GPa, respectively, for the bare PMMA substrates (Blanc et al., 2006).

Incorporation of an organic constituent in an inorganic network alters mechanical properties in a significant manner. As can be seen from Figure 5.16, the addition

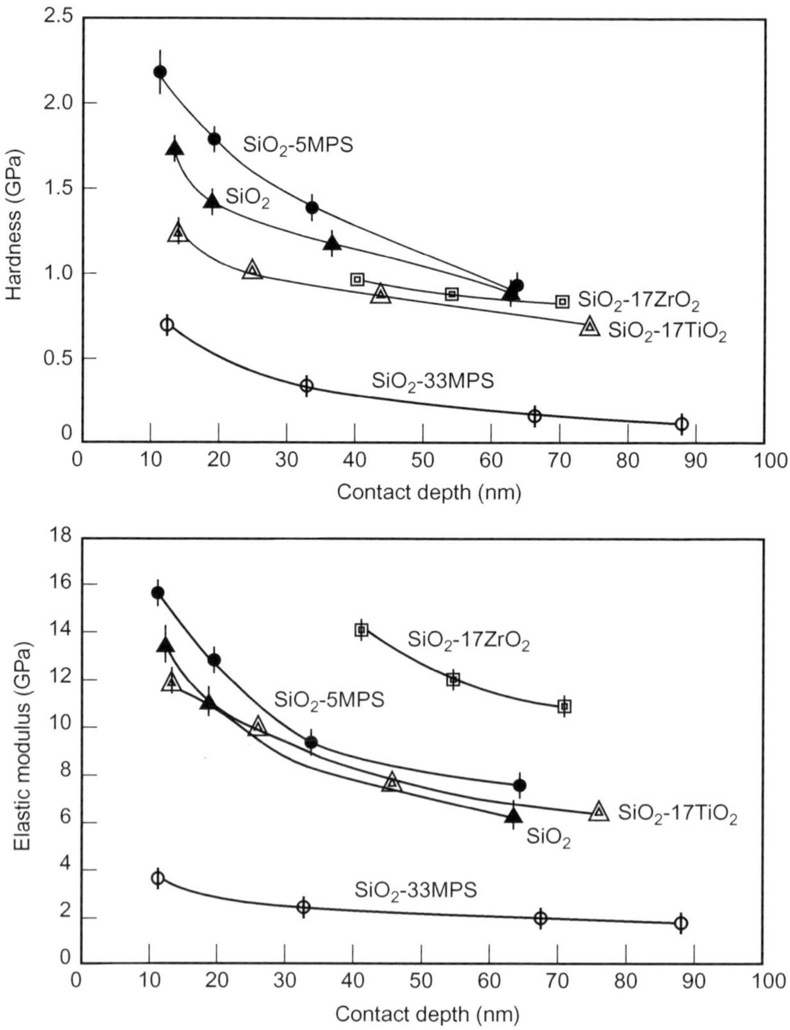

Figure 5.16 Hardness and elastic modulus of ambient dried various nanocomposite coatings deposited on polyester (Chan et al., 2000).

of a small quantity enhanced the hardness and elastic modulus because of a denser microstructure. But further addition resulted in softer but flexible coatings. Normally the addition of harder materials such as titania and zirconia would increase the hardness of the coating, but often the result is reduced hardness because of the enhanced porosity that has successfully been demonstrated through nitrogen sorption analysis on xerogels. Enhanced porosity is found to be due to the catalytic effect of titanium or zirconium alkoxides, which considerably reduce the gelation time. Further, it was also shown that the substrate hardness has significant influence on the coating properties. The nanoindentation results presented in Figure 5.17 show that both hardness and elastic modulus increased with

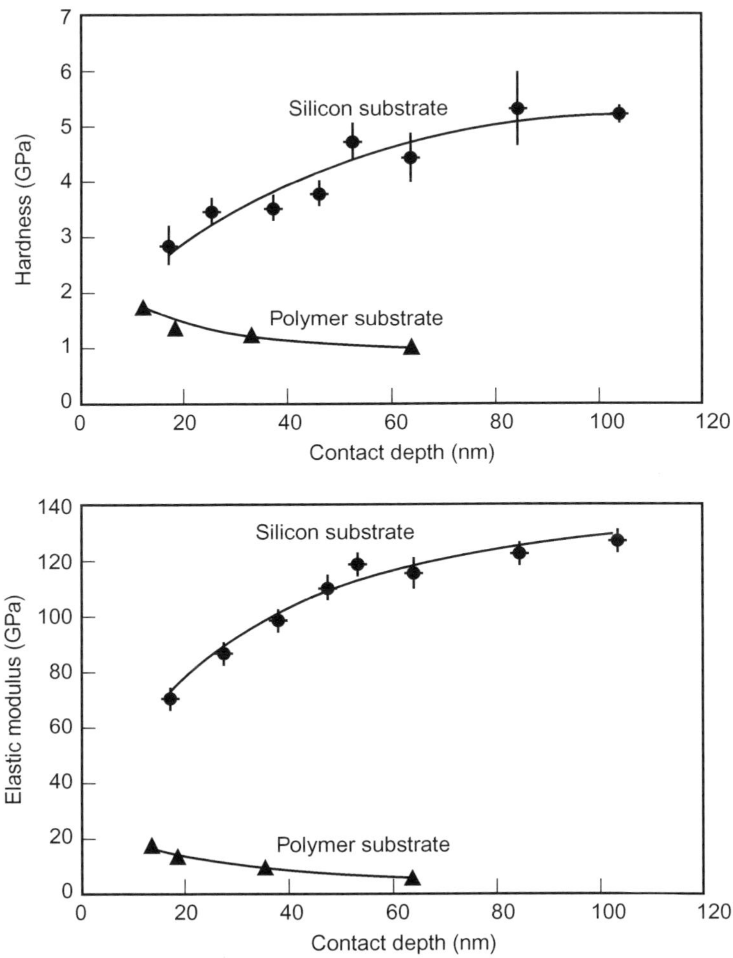

Figure 5.17 Hardness and elastic modulus of silica coatings applied on polyester and silicon substrates (Chan et al., 2000).

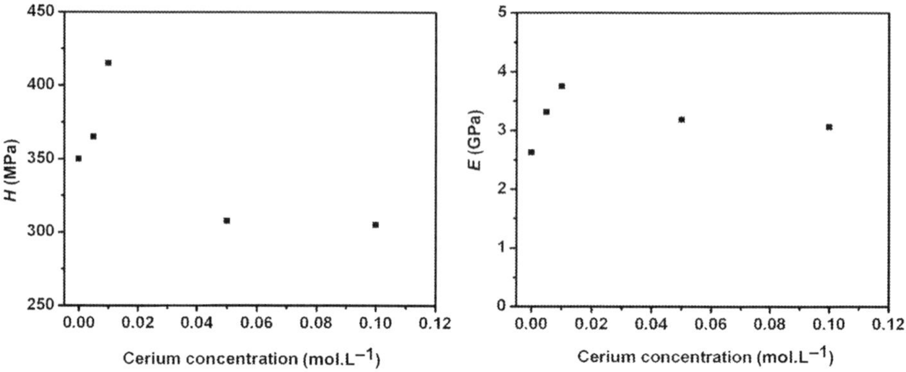

Figure 5.18 Evolution of (a) hardness and (b) elastic modulus as a function of Ce^{3+} added to the GPTMS–alumina hybrid sol (Cambon et al., 2012).

increased depth on silicon wafer (12.2 ± 0.04 GPa) while both of them reduced with increased depth on polymer (0.2960.04 GPa) substrate (Chan et al., 2000).

Nanoindentation investigations have also recently been used to corroborate results obtained from corrosion studies (Cambon et al., 2012). The optimum concentration of Ce^{3+}, added as an inhibitor to a hybrid sol-gel matrix, was determined as 0.01 M using nanoindentation experiments and verified by electrochemical characterization of the coatings. As can be seen from Figure 5.18, there was an excellent correlation shown between the evolution of the barrier effect and mechanical properties because of the addition of Ce^{3+}.

Results of investigations from the different research groups show that the elastic modulus and hardness are important parameters in judging the mechanical properties, mainly the scratch resistance of a coating. Greater hardness and elastic modulus increase the resistance of coatings to plastic deformation, leading to resistance to the entry of the scratch tip. However, high values of elastic modulus and nanoindentation hardness reduce energy loss leading to stress concentration and thus to fracture scratch formation. Hence, increased hardness and elastic modulus can play an influential role only up to a certain degree without a fracture pattern (Yahyaei and Mohseni, 2013). Hence, medium content of organic and inorganic species is needed to obtain the maximum mechanical properties.

5.3.4 Abrasion resistance

While there may be many methods for assessing the abrasion resistance of coatings, crockmeter, and taber testers are most common options for evaluating abrasion resistance of the sol-gel nanocomposite coatings.

5.3.4.1 Crockmeter method

Originally this test method (ASTM F1319-94, 2011) was intended for measuring the smudge resistance of images obtained from business copy machines and for

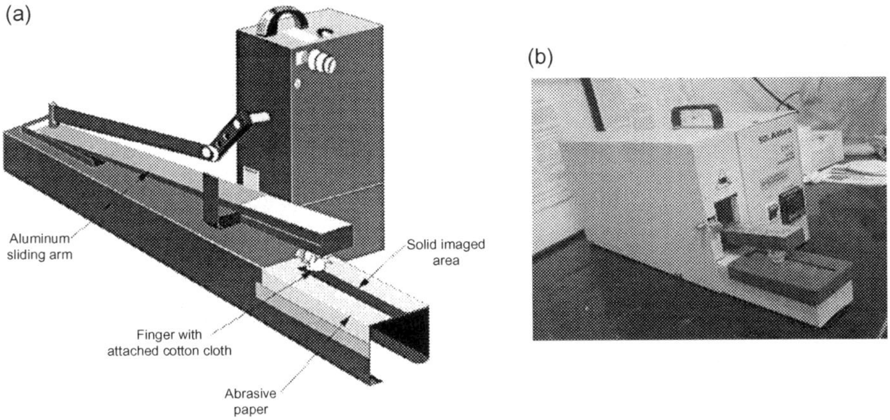

Figure 5.19 (a) Schematic and (b) photograph of a crockmeter.
Source: Reprinted with permission from ASTM F1319-94 (2011). Copyright ASTM International, West Conshohocken, PA.

measuring dye-fastness of colored fabrics, but subsequently employed to qualitatively assess the abrasion resistance of hybrid nanocomposite sol-gel coatings. This method employs a 10 mm diameter finger rubbing on a flat work piece in a reciprocating motion as shown in Figure 5.19(a) with a normal load of 4.5 N and for a minimum stroke length of 20 mm. A half coated work piece of 3.5″ × 5.5″ dimensions is fixed under the finger so as to be abraded by the finger. A 4-ply cotton cloth is used to cover the face of the finger coming in contact with the sample and the ability of the image/coating to withstand the frictional forces attempting to remove the material from the sample is measured by the change in optical quality. Commercial units are provided with a counter to fix the number of cycles required for the test. Steel wool or an abrasive cleaning pad can also be used in place of cloth as the abrading medium in the case of hard coatings. In case of coatings on transparent plastics, variation in haze or transmittance of the coatings after abrading shall be measured with respect to the number of cycles. In the case of coatings on opaque substrates such as stainless steel/metals, the substrates after crockmeter test are observed visually and compared for scratches on the uncoated and coated side, as shown in Figure 5.20. Relatively softer coatings are reported to be tested with a hard rubber finger having a shore hardness of A.

5.3.4.2 Taber test method

Taber testing is a commonly used test method (ASTM D1044-08, 2012) for assessing abrasion resistance of organic coatings and especially hybrid nanocomposite sol-gel coatings. The advantage of this method over the crockmeter method is quantification of the worn surface by measuring the weight of the sample before and after the test. Transparent coatings are preferably assessed with the change in haze

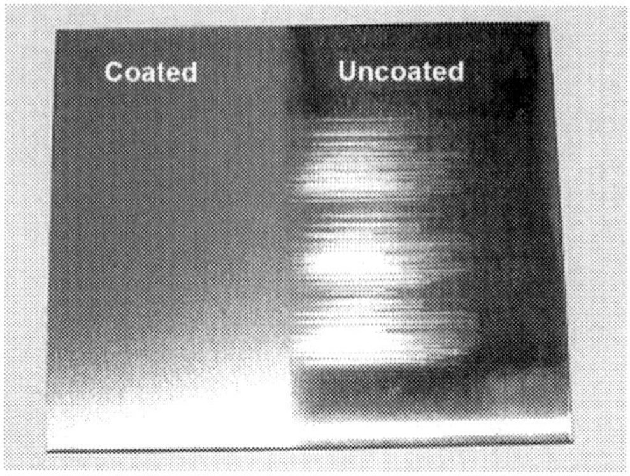

Figure 5.20 Sample after 1000 cycles of aggressive crockmeter test using scotch brite, revealing the coated area to be completely devoid of scratches.

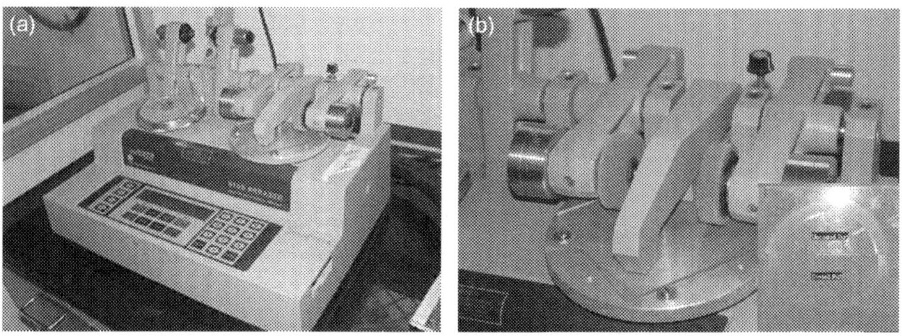

Figure 5.21 Taber tester (a) apparatus and (b) close-up view showing sample with wear track as inset.

(ASTM D1003, 2013), that is, measurement of light scattered by the tested coating surface. The commercially available machine consists of a horizontal turntable over which a round sample with a maximum diameter of 100 mm or a square sample can be fixed. As shown in Figure 5.21, two loading arms are provided for applying normal load and holding abrasive wheels of 52 mm diameter and a width of 12.7 mm on the sample.

There are different grades of abrasive wheels, such as CS17 for hard coatings, and CS10 and CS10F for relatively soft coatings but hard for plastics such as PMMA and PC. While the sample rotates, the abrasive wheels also rotate but in an anticlockwise direction because of friction. The test is conducted for a specified number of rotations and the change in weight (in the case of coatings on opaque substrates) or change in haze (in case of coatings on transparent plastics or glass) is measured after testing. The values are compared between coated and uncoated samples to gauge the abrasion resistance of the coatings.

It was noticed that the abrasion resistance in terms of percentage change in haze for PC could be reduced from 40% to <6% with the epoxy-modified silane–titania coating when tested for 500 cycles using CS10F abrading wheels under a 2 × 500 g load (Sowntharya et al., 2012). A nanocomposite hard coating synthesized from MPTMS, zironium-*n*-propoxide, and tetraethoxy orthosilicate (TEOS) sol and deposited on PC exhibited enhanced abrasion resistance as compared to the coatings obtained from an ex-situ addition of acrylic surface-modified silica nanoparticulate dispersion (Sowntharya et al., 2013). One recent study concluded that the surface activation improves adhesion but does not show any appreciable improvement in abrasion resistance because it is a surface phenomenon unlike scratch and nanoindentation, where the indenter penetration is likely to extend to the substrate (Soma Raju et al., 2012).

5.4 Possible applications of hard coatings

A thorough survey conducted to discover companies commercializing sol-gel coating technology reveals that sol-gel coating products are found to be quite successful wherever large industrial houses are involved (Abraham, 2006). The main advantages are as follows:

1. The sol-gel process makes it possible to obtain compositions that are not obtainable by other methods such as integrated multilayer fabrication.
2. The technology is available at low cost relative to vacuum coating.
3. It has a low quantity raw material requirement.
4. It involves a low temperature process.

Another added advantage is that multifunctional coatings can be generated in a single layer. The applications that have been cited include architectural coatings, abrasion/scratch-resistant coatings, high-porosity coatings, hybrid coatings, water-repellent coatings, protective coatings, anti-reflective coatings, colored coatings for glass, cookware, abrasion/scratch-resistant coatings for plastics, coatings for optical applications, and anti-bacterial systems (Sakka, 2006; Uhlmann and Teowee, 1998; Yamane, 2006).

Another review opined that hybrid polymers or nanocomposites derived from the huge range of available precursors such as organo(alkoxy)silanes, main group or transition metal alkoxides, and various other constituents, are versatile and find applications such as protection of transparent plastics, metals, and glasses (Schottner, 2001). Hard poly(methylsiloxane)s and related colloidal silica containing coating formulations developed several decades ago still have industrial applications in improving the wear and abrasion resistance of engineering plastics such as on headlamp covers in cars or side windows in buses/aircraft made from PMMA. The biocompatibility of the hybrids has also been demonstrated and two products are already commercially available (Degussa-Dental Co., Frankfurt, Germany; Admira, Voco Co., Cuxhaven, Germany). In yet another review, the sol-gel technology sustainability assessment shows that the sol-gel process is more competitive than the powder coating process for decorative coatings on car wheels (Yang et al., 2009).

Sol-gel nanocomposite hard coatings have also been successfully demonstrated for:

- high O_2 permeability and scratch resistance on contact lenses;
- scratch resistance on ophthalmic lenses;
- scratch and abrasion resistance on pavement markers, and automotive/aircraft/helicopter windshields;
- easy-to-clean and tarnish resistance on automotive exhausts;
- improved aesthetic appeal and scratch, tarnish and corrosion resistance on modular kitchen panels;
- tarnish resistance on stove tops;
- antifogging effect on mirrors;
- colored or transparent scratch-resistant coatings on plastic and glassware for automobile and household articles;
- self-cleaning properties in glass and ceramics;
- anti-microbial activity together with scratch resistance in surgical and personal aid devices; and
- low-friction, anti-fingerprinting on iron shoes and steel switches, and so on.

A summary of the previous work carried out by various researchers on different topics in the field of sol-gel based nanocomposite coatings is presented in Table 5.1 for ease of reference.

Table 5.1 Summary of the previous work carried out in the field of sol-gel based nanocomposite coatings

Sl. No	Author name and year of publication	Section no.	Topic
1	Brinker and Scherer, 1990	5.1	Basics of sol-gel processing
2	Aegerter, 2005; Aegerter and Mennig, 2004	5.1	Inorganic coatings
3	Kickelbick, 2007; Romero and Sanchez, 2004	5.1	Hybrid coatings
4	Barbe et al., 2005; Estella et al., 2007; Wang and Bierwagen, 2009	5.1	Influence of process parameters
5	Schmidt et al., 1998	5.1	Radiation curing and introduction of trace elements
6	Plinio et al., 2005; Schottner, 2001	5.2	Classification of hybrid nanocomposite coatings
7	Schmidt et al., 1998	5.2	Introduction on metal alkoxides
8	Plinio et al., 2005	5.2	Use of organically modified alkoxides
9	Jianguo et al., 2006; Wang and Bierwagen, 2009	5.2	Use of alkoxysilanes

(*Continued*)

Table 5.1 **(Continued)**

Sl. No	Author name and year of publication	Section no.	Topic
10	Aegerter, 2005; Marzolin et al., 1998; Ottermann and Bange, 1993; Kozuka et al., 2003; Schmidt et al., 1998	5.2	Residual stress
11	Wilkes et al., 1985; Schmidt, 1985	5.2	"ORMOCER®" and "ORMOSIL®"
12	Jindal et al., 1987; Rickerby, 1988; Xie and Hawthorne, 2003; Latella et al., 2012; Zhang et al., 2006; Sowntharya et al., 2013; Malzbender and With, 2001; Chen et al., 2008; Najafabadi et al., 2013; Lakshmi et al., 2012; Audebert et al., 1997; Mizuta et al., 2013; Kim et al., 2005; Cai et al., 2011; Chou and Cao, 2003; Gururaj et al., 2011; Kumar et al., 2012a,b; Kiruthika et al., 2010; Kim et al., 2003; Fracassi et al., 2003	5.3.1.1	Adhesion strength assessment through various techniques
13	Wu et al., 2008, 2009; Caselis et al., 2012; Sowntharya et al., 2012, 2013; Soma Raju et al., 2012; Bayramoglu et al., 2012; Chang et al., 2012	5.3.1.2 to 5.3.1.4	Scratch hardness
14	Yahyaei and Mohseni, 2013; Sowntharya et al., 2012, 2013; Blanc et al., 2006; Chantarachindawong et al., 2012; Yaacoub et al., 2012; Uhlmann et al., 2013; Almaral-Sanchez et al., 2006; Etienne et al., 2004; Ramezanzadeh and Mosheni, 2012; Ballare et al., 2009; Cambon et al., 2012; Chan et al., 2000; Amerio et al., 2008	5.3.2 to 5.3.3	Nanoindentation study
15	Sowntharya et al., 2012, 2013; Soma Raju et al., 2012	5.3.4	Abrasion resistance
16	Abraham, 2006; Uhlmann and Teowee, 1998; Schottner, 2001; Yang et al., 2009	5.4	Coating applications

5.5 Summary

A summary of various commonly used characterization techniques for evaluating adhesion, scratch, abrasion resistance, hardness, and nanoindentation have been presented. Basic knowledge of the following three aspects is essential for the successful product development in sol-gel nanocomposite hard coatings.

1. Reactivities of precursors toward hydrolysis and condensation under given conditions.
2. Identifying the resulting structures.
3. Correlation of the structure with material properties.

Two routes were identified for preparing sol-gel nanocomposite hard coatings. The first involves in-situ preparation of the nanocomposite, involving co-condensation of more than one precursor during sol preparation. The second involves external addition of nanoparticles in a hybrid sol-gel matrix. Both have their merits and demerits in terms of their properties. However, it was ascertained that whichever route is adopted for preparing nanocomposite hard coatings, their mechanical properties determined during performance evaluation are impressive, whether it be improving scratch or abrasion resistance in plastic substrates or enhancing abrasion and tarnish resistance in metals.

Acknowledgments

The authors would like to thank Dr G. Sundararajan and Dr G. Padmanabham for their constant support and encouragement throughout the writing of this chapter and Mr. C. Karunakar for his help in redrawing some of the figures. The authors are thankful to Elsevier, Cambridge University Press, Springer Science and Business Media LLC, John Wiley and Sons, and ASTM International for granting the necessary permissions to reproduce some of the figures.

References

Abraham, T., 2006. Sol-gel Processing of Ceramics and Glass. Market research report AVM016E, BCC Research.

Aegerter, M.A., 2005. Characterization of the mechanical properties of sol-gel coatings. In: Almeida, R.M., Sakka, S. (Eds.), Handbook of Sol-Gel Science and Technology, Vol II: Characterisaton and Properties of Sol-Gel Materials and Products. Kluwer Academic Publishers, USA, pp. 297–313.

Aegerter, M.A., Mennig, M., 2004. Sol-Gel Technologies for Glass Producers and Users. Kluwer Academic Publishers, USA.

Almaral-Sanchez, J.L., Lopez-Gomez, M., Ramirez-Bon, R., Munoz-Saldana, J., 2006. Mechanical behaviour of hybrid SiO_2-PMMA coatings measured by nanoindentation. Adv. Technol. Mater. Mater. Process. 8, 81–86.

Amerio, E., Fabbri, P., Malucelli, G., Messori, M., Sangermano, et al., 2008. Scratch resistance of nano-silica reinforced acrylic coatings. Prog. Org. Coat. 62, 129–133.

ASTM D1003, 1003. Standard Test Method for Haze and Luminous Transmittance of Transparent Plastics. ASTM International, West Conshohocken, PA.

ASTM D1044-08, 1044. Standard Test Method for Resistance of Transparent Plastics to Surface Abrasion. ASTM International, West Conshohocken, PA.

ASTM D3359-02, 3359. Standard Test Methods for Measuring Adhesion by Tape Test. ASTM International, West Conshohocken, PA.

ASTM D3363-05, 3363. Standard Test Method for Film Hardness by Pencil Test. ASTM International, West Conshohocken, PA.

ASTM D4541-09, 4541. Standard Test Method for Pull-Off Strength of Coatings Using Portable Adhesion Testers. ASTM International, West Conshohocken, PA.

ASTM D907-12a, 2012. Standard Terminology of Adhesives. ASTM International, West Conshohocken, PA.

ASTM F1319-94, 1319. Standard Test Method for Determination of Abrasion and Smudge Resistance of Images Produced from Business Copy Products. ASTM International, West Conshohocken, PA.

Audebert, P., Bertrand, S., Bresson, F., Tribillon, G., 1997. Stable and adhesive hybrid fluorescent sol-gel coatings on pre-treated stainless steel: application to the realization of optical fluorescence sensors. Appl. Surf. Sci. 119, 207–214.

Ballare, J., Jimenez-Pique, E., Anglada, M., Sergio, A.P., Cavalieri, A.L., 2009. Mechanical characterization of nano-reinforced silica based sol-gel hybrid coatings on AISI 316L stainless steel using nanoindentation techniques. Surf. Coat. Technol. 203, 3325–3331.

Barbe, C.J., Cassidy, D.J., Triani, G., Latella, B.A., Mitchell, D.R.G., et al., 2005. Sol-gel bonding of silicon wafers, Part I: influence of the processing temperature on final bond morphology and interfacial energy. Thin Solid Films. 488, 153–159.

Bayramoglu, G., Ahbab, I., Kahraman, M.V., Gungor, A., 2012. Photocurable cyanate ester containing hybrid coatings by an anhydrous sol-gel technique. Prog. Org. Coat. 74, 511–519.

Blanc, D., Last, A., Franc, J., Pavan, S., Loubet, J.L., 2006. Hard UV-curable organo-mineral coatings for optical applications. Thin Solid Films. 515, 942–946.

Brinker, C.J., Scherer, G.W., 1990. Sol-Gel Science: The Physics and Chemistry of Sol-Gel Processing. Academic, San Diego, CA.

Cai, Y., Zhang, S., Zeng, X., Qian, M., Sun, D., et al., 2011. Interfacial study of magnesium containing fluoridated hydroxyapatite coatings. Thin Solid Films. 519, 4629–4633.

Cambon, J.B., Esteban, J., Ansart, F., Bonino, J.P., Turq, V., et al., 2012. Effect of cerium on structure modifications of a hybrid sol-gel coating, its mechanical properties and anti-corrosion behaviour. Mater. Res. Bull. 47, 3170–3176.

Caselis, J.L.V., Rosas, E.R., Meneses, V.M.C., 2012. Hybrid PMMA-silica anticorrosive coatings for stainless steel 316L. Corr. Sci. 47, 131–137.

Chan, C.M., Cao, G.Z., Fong, H., Sarikaya, M., 2000. Nanoindentation and adhesion of sol-gel-derived hard coatings on polyester. J. Mater. Res. 15 (1), 148–154.

Chang, C.C., Oyang, T.Y., Hwang, F.H., Chen, C.C., Cheng, L.P., 2012. Preparation of polymer/silica hybrid hard coatings with enhanced hydrophobicity on plastic substrate. J. Non-Cryst. Solids. 358, 72–76.

Chantarachindawong, R., Luangtip, W., Chindaudom, P., Ostachan, T., Srikhirin, T., 2012. Development of the scratch resistance on acrylic sheet with basic colloidal silica (SiO_2)-methyltrimethoxysilane (MTMS) nanocomposite films by sol-gel technique. Can. J. Chem. Eng. 90, 888–896.

Chen, Z., Wu, L.Y.L., Chwa, E., Tham, O., 2008. Scratch resistance of brittle thin films on compliant substrates. Mater. Sci. Eng. A. 493, 292–298.

Chou, T.P., Cao, G., 2003. Adhesion of sol-gel-derived organic–inorganic hybrid coatings on polyester. J. Sol-Gel Sci. Technol. 27, 31–41.

Estella, J., Echeverria, J.C., Laguna, M., Garrido, J.J., 2007. Effects of aging and drying conditions on the structural and textural properties of silica gels. Microporous Mesoporous Mater. 102, 274–282.

Etienne, S., Duri, A., Etienne, P., 2004. Fracture study of organic–inorganic coatings using nanoindentation technique. J. Non-Cryst. Solids. 344, 60–65.

Fracassi, R.D., Agostino, F., Palumbo, E., Angelini, S., Grassini, F., et al., 2003. Application of plasma deposited organosilicon thin films for the corrosion protecton of metals. Surf. Coat. Technol. 107, 174–175.

Gururaj, T., Subasri, R., Raju, K.R.C.S., Padmanabham, G., 2011. Effect of plasma pretreatment on adhesion and mechanical properties of UV-curable coatings on plastics. Appl. Surf. Sci. 257, 4360–4364.

Jianguo, L., Gaoping, G., Chuanwei, Y., 2006. Enhancement of the erosion–corrosion resistance of dacromet with hybrid SiO_2 sol-gel. Surf. Coat. Technol. 200, 4967–4975.

Jindal, P.C., Quinto, D.T., Wolfe, G.J., 1987. Adhesion measurements of chemically vapor deposited and physically vapor deposited hard coatings on WC-Co substrates. Thin Solid Films. 154, 361–375.

Kasemann, R., Schmidt, H., 1994. Coatings for mechanical and chemical protection based on organic–inorganic sol-gel nanocomposites. New J. Chem. 18, 1117–1123.

Kickelbick, G., 2007. Hybrid Materials: Synthesis, Characterization and Applications. Wiley-VCH, Germany.

Kim, H.W., Kim, H.E., Salih, V., Knowles, J.C., 2005. Hydroxyapatite and titania sol-gel composite coatings on titanium for hard tissue implants; mechanical and in vitro biological performance. J. Biomed. Mater. Res. B Appl. Biomater. 72B, 1–8.

Kim, M.C., Yang, S.H., Boo, J.H., Han, J.G., 2003. Surface treatment of metals using an atmospheric pressure plasma jet and their surface characteristics. Surf. Coat. Technol. 174–175, 839–844.

Kiruthika, P., Subasri, R., Jyothirmayi, A., Sarvani, K., Hebalkar, N.Y., 2010. Effect of plasma surface treatment on mechanical and corrosion protection properties of UV-curable sol-gel based GPTS-ZrO_2 coatings on mild steel. Surf. Coat. Technol. 204, 1270–1276.

Kozuka, H., Takenaka, S., Tokita, H., Hirano, T., Higashi, Y., et al., 2003. Stress and cracks in gel derived ceramic coatings and thick film formation. J. Sol-Gel Sci. Technol. 26, 681–686.

Kumar, N., Jyothirmayi, A., Subasri, R., 2012a. Effect of plasma surface pretreatment on Ce^{3+}-doped GPTMS-ZrO_2 self-healing coatings on aluminum alloy. ISRN Corr. 2012, Article ID: 506560.

Kumar, N., Jyothirmayi, A., Soma Raju, K.R.C., Subasri, R., 2012. Effect of functional groups (methyl, phenyl) on organic–inorganic hybrid sol-gel silica coatings on surface modified SS316. Ceram. Int. 38, 6565–6572.

Lakshmi, R.V., Bharathidasan, T., Bera, P., Basu, B.J., 2012. Fabrication of superhydrophobic and oleophobic sol-gel nanocomposite coating. Surf. Coat. Technol. 206, 3888–3894.

Latella, B.A., Swain, M.V., Ignat, M., 2012. Indentation and fracture of hybrid sol-gel silica films. Intech 134–160. Available from: <http://dx.doi.org/10.5772/48140> (accessed 01.08.14.).

Malzbender, J., With, G.D., 2001. Scratch testing of hybrid coatings on float glass. Surf. Coat. Technol. 135, 202–207.

Marzolin, C., Terfort, A., Tien, J., Whitesides, G.M., 1998. Patterning of a polysiloxane precursor to silicate glasses by microcontact printing. Thin Solid Films. 315, 9–12.

Mizuta, Y., Daiko, Y., Mineshige, A., Yazawa, T., 2013. Effect of plastics substrate on phase separation behaviour and adhesion for R-Si(OC_2H_5)$_3$-Si(OC_2H_5)$_4$ coatings prepared by sol-gel process. Ceram. Int. 39, 925–930.

Najafabadi, A.H., Mozaffarinia, R., Rahimi, H., Razavi, R.S., Paimozd, E., 2013. Sol-gel processing of hybrid nanocomposite protective coatings using experimental design. Prog. Org. Coat. 76, 293–301.

Ottermann, C.R., Bange, K., 1993. Correlation between the density of TiO_2 films and their properties. Thin Solid Films. 286, 32–34.

Plinio, I., Giovanna, B., Massimo, G., Florence, B., 2005. Structural characterization of hybrid organic–inorganic materials. In: Almeida, R.M., Sakka, S. (Eds.), Handbook of Sol-gel Science and Technology, Vol II: Characterisaton and Properties of Sol-Gel Materials and Products. Kluwer Academic Publishers, USA, pp. 139–157.

Ramezanzadeh, B., Mohseni, M., 2012. Preparation of sol-gel based nano-structured hybrid coatings: effects of combined precursor's mixtures on coating's morphological and mechanical properties. J. Sol-Gel Sci. Technol. 64, 232–244.

Rickerby, D.S., 1988. A review of the methods for the measurement of coating-substrate adhesion. Surf. Coat. Technol. 36, 541–557.

Romero, P.G., Sanchez, C., 2004. Functional Hybrid Materials. Wiley-VCH, Germany.

Sakka, S., 2006. Current sol-gel activities in Japan. J. Sol-Gel Sci. Technol. 37, 135–140.

Schmidt, H., 1985. New type of non-crystalline solids between inorganic and organic materials. J. Non-Cryst. Solids. 73, 681–691.

Schmidt, H.K., Geiter, E., Mennig, M., Krug, H., Becker, C., et al., 1998. The sol-gel process for nano-technologies: new nanocomposites with interesting optical and mechanical properties. J. Sol-Gel Sci. Technol. 13, 397–404.

Schottner, G., 2001. Hybrid sol-gel-derived polymers: applications of multifunctional materials. Chem. Mater. 13, 3422–3435.

Soma Raju, K.R.C., Sowntharya, L., Lavanya, S., Subasri, R., 2012. Effect of plasma pre-treatment on adhesion and mechanical properties of sol-gel nanocomposite coatings on polycarbonate. Compos. Interfaces. 19, 259–270.

Soutar, A.M., Chen, Q., Raja Khalif, R.E., 2008. Evaluation of commercially available transparent hard coating for polycarbonate, SIMTech technical report (STR_V9_N3_09_STG) 9 (3): 161–165.

Sowntharya, L., Subasri, R., 2013. A comparative study of different curing techniques for SiO_2-TiO_2 hybrid coatings on polycarbonate. Ceram. Int. 39, 4689–4693.

Sowntharya, L., Lavanya, S., Ravi Chandra, G., Hebalkar, N.Y., Subasri, R., 2012. Investigations on the mechanical properties of hybrid nanocomposite hard coatings on polycarbonate. Ceram. Int. 38, 4221–4228.

Sowntharya, L., Ravi, C.G., Soma Raju, K.R.C., Subasri, R., 2013. Effect of addition of surface modified nanosilica into silica-zirconia hybrid sol-gel matrix. Ceram. Int. 39, 4245–4252.

Uhlmann, D.R., Teowee, G., 1998. Sol-gel science and technology: current state and future prospects. J. Sol-Gel Sci. Technol. 13, 153–162.

Uhlmann, I., Hawelka, D., Hilderbrandt, E., Pradella, J., Rödel, J., 2013. Structure and mechanical properties of silica doped zirconia thin films. Thin Solid Films. 527, 200–204.

Wang, D., Bierwagen, G.P., 2009. Sol-gel coatings on metals for corrosion protection. Prog. Org. Coat. 64, 327–338.

Wilkes, G.L., Orler, B., Huang, H., 1985. Ceramers: hybrid materials incorporating polymeric/oligomeric species into inorganic glasses utilizing a sol-gel approach. Polym. Prepr. 26 (2), 300–302.

Wu, L.Y.L., Chwa, E., Chen, Z., Zeng, X.T., 2008. A study towards improving mechanical properties of sol-gel coatings for polycarbonate. Thin Solid Films. 516, 1056–1062.

Wu, L.Y.L., Boon, L., Chen, Z., Zeng, X.T., 2009. Adhesion enhancement of sol-gel coating on polycarbonate by heated impregnation treatment. Thin Solid Films. 517, 4850–4856.

Xie, Y., Hawthorne, H.M., 2003. Measuring the adhesion of sol-gel derived coatings to a ductile substrate by an indentation-based method. Surf. Coat. Technol. 172, 42–50.

Yaacoub, S., Calas, S., Jabbour, J., Tauk, R., Zaouk, D., et al., 2012. Study of the mechanical properties of two organic–inorganic hybrid systems: GPTMS/colloidal silica and GPTMS/TEOS. J. Non-Cryst. Solids. 358, 3036–3041.

Yahyaei, H., Mohseni, M., 2013. Use of nanoindentation and nanoscratch experiments to reveal the mechanical behaviour of sol-gel prepared nanocomposite films on polycarbonate. Tribol. Int. 57, 147–155.

Yamane, M., 2006. Unique commercial applications of the sol-gel process in Japan. J. Sol-Gel Sci. Technol. 40, 273–279.

Yang, Q.Z., Chua, B.H., Nemeth, S., Tan, S.H., 2009. Technical, economic and environmental sustainability assessment of sol-gel coating process, SIMTech technical reports (STR_V10_N2_10_STA) 10 (2): 118–125.

Zhang, S., Xianting, Z., Yongsheng, W., Kui, C., Wenjian, W., 2006. Adhesion strength of sol-gel derived fluoridated hydroxyapatite coatings. Surf. Coat. Technol. 200, 6350–6354.

Process considerations for nanostructured coatings

6

Kal Renganathan Sharma
Texas Southern University, Houston, TX & San Jacinto College District North Campus, Houston, TX

6.1 Overview

Nanos is a Latin word that means dwarf. *Technolgia* means a systematic treatment of an art or craft. 1 nanometer (nm) is 1 billionth of a meter (10^{-9} m). The covalent radius of helium is 28 pm. 1000 pico meter makes 1 nm. Graphene is the thinnest material created and can be made in single layers in copper foil (Geim, 2010, 2009). References are summarized in Table 6.1 below.

Nanocoatings with thickness less than 100 nm can offer superior performance properties than conventional coatings. One of the first commercial application of nanocoating was Nanotuff. Triton systems developed this product on a five-year contract from the US Navy. Sharma (2010) The application was advanced abrasion resistant and chemical resistant coating for transparent substrates.

Nanosized particles suspended in an epoxy matrix were used in the coating mix. This coating was found to be four times more durable than conventional coatings. Surfaces treated with Nanotuff were found to be free from crazing, cracking, and shattering upon impact. Some flame retardancy was incorporated into the product. Nanoceramic coatings with higher toughness than conventional coatings were developed in the 1990s. The US Department of Defense (DoD) encouraged partnerships among academia, industry, and the Pentagon to evaluate nanocoatings for use in the seas. A total of 100 research and development (R&D) awards were given out. The DoD estimated a reduction of $20 million in maintenance costs by using nanocoatings in air valves. Wear-resistant coatings can be applied to personal cars and heavy industrial machinery.

Alternates for cadmium (Cd) plating, have been solicited by the DoD with the sole purpose of protecting aircraft components from corrosion. Side-by-side evaluations of whether the coatings are absolutely essential and whether the same set of properties can be met using a steel of a different alloy composition are needed. Reports from the operating experience of aircraft have been used to show that, often, the real issue for maintenance is service failure attributable to stress corrosion cracking. Questek came up with a steel grade made up of nanophase precipitated hardened material equal to 300 M that can be used to address the concerns of service failure. Concerns about Cd corrosion were also allayed. Nanocoatings can offer an alternate to this critical technology. Reduction in cost and time to market of an order of magnitude can be realized.

Anti-Abrasive Nanocoatings.

Table 6.1 **References at a glance**

Ref.	Authors	Title	Journal/Source	Year	Other information
[1]	A. K. Geim	Random Walk to Graphene	Nobel Lecture	2010	http://nobel.se
[2]	A.K. Geim	Graphene: Status and Prospects	Science	2009	Vol. 324: 1530
[3]	K. R. Sharma	Nanostucturing Operations in Nanoscale Science and Engineering	New York: McGraw Hill Professional	2010	n/a
[4]	J. C. Gunalan		*ACS Macro Letters*		DOI:10.1021/mz400105e.
[5]	H. Legg	Overview of Nanocoating Technology: Nanocoatings IP Landscape	Rowan Technology Group	2003	Chicago, IL
[6]	A.K. Sood et al.	Development of Nanostructured AR Coatings for EO/IR Sensor and Solar Cell Applications	*Materials Sciences and Applications*	2012	Vol. 3: 633–9
[7]	D. J. Poxson et al.	Quantification of Porosity and Deposition Rate of Nanoporous Films Grown by Oblique Angle Deposition	*Applied Physics Letters*	2008	Vol. 93: 101914–23
[8]	J. Q. Xi et al.	Optical Thin-Film Materials with Low Refractive Index for Broadband Elimination of Fresnel Reflection	*Nature Photonics*	2007	Vol. 1: 176–9
[9]	S. Bertaux et al.	Pearlescent Pigments	US Patent 6,719,837	2004	Merck, Darmstadt, DE

[10]	T. N. Lo et al.	Using E-Beam and X-ray Lithography to Fabricate Zone Plates for Hard X-ray	*AIP Conference Proceedings*	2007	Vol. 879: 1466–9
[11]	B. K. Jha et al.	High Performance Alumina Nano-Alumina Grafted Waterborne Polyurethane Coating	*International Journal of Nanoscience*	2011	Vol. 10(4,5): 1119–24
[12]	Y. Wang et al.	Abrasive Wear Characteristics of Plasma Sprayed Nanostructured Alumina/Titania Coatings	Wear	2000	Vol. 237: 176–85
[13]	U. Steiner	Structure Formation in Polymer Thin Films: From μm to Sub 100 nm Length Scales, in W. T. S. Huck (ed.) Nanoscale Assembly – Chemical Techniques	Nanostructure Science and Technology, Springer	2005	New York
[14]	K. R. Sharma	Polymer Thermodynamics: Blends, Copolymers and Reversible Polymerization	CRC Press	2012	Boca Raton, FL
[15]	Y. Lu and D. Wang	Process for the Preparation of Metal-Containing Nanostructured Films	US Patent 7,001,669	2006	New Orleans, LA
[16]	K. R. Sharma	Graphene Nanomaterials	Momentum Press	2014	Happuage, NY

Nanocoating with 30 nm thickness was developed to be used in fighting furniture fires. This coating is better for the environment than the flame retardants currently used to protect cushions used in furniture. The coating is made of a sulfur-based polymer and a carbohydrate polymer found in crustacean shells. The polyurethane used to make foams used as cushions are flammable. Fire safety guidelines require the use of flame retardant chemicals. These chemicals have been shown to act as endocrine disrupters or lead to neurological problems. The European Union, Canada, and the United States have begun to scrutinize the use of these chemicals.

Researchers at Texas A&M University have developed formulations that do not affect the stiffness of the foam and have reduced toxicity. The coating is made of polyvinyl sulfonic (PVS) acid and chitosan. Chitosan is a long carbohydrate molecule derived from the shells of shrimp and other crustaceans. The coating is prepared by dipping foam in aqueous forms of the two polymers. Positive charges are possessed by chitosan and negative charges are possessed by PVS. The two polymers become entangled by the attraction of unlike charges and the coating is created (Laufer et al., 2013). The weight increase of the foam is 5.5% without an alteration in the stiffness of the foam. Tests confirmed that it would take three times as much conventional flame retardant material to achieve the same effect. As PVS burns, sulfur dioxide (SO_2) vapors are generated that create a gas blanket on the foam surface, thereby reducing oxygen and thus starving the fire. Another research group at National Institute of Standards and Technology (NIST) developed clay-based nanocoatings that can be used to make materials fire proof.

Some of the critical issues in the preparation of nanocoatings are as follows (Legg, 2003):

1. the process for making optical flakes that are precursors to inks that are used in currencies;
2. functionalization of nanoparticles in the polymer coating;
3. particulation for hard, lubricious razor blade coatings;
4. materials and methods for barrier, clear, and sacrificial coats;
5. meeting required standards, such as for aerospace vehicles.

6.2 Anti-reflection coatings

An increase in transmittance of more than 10% can be observed upon use of nanostructured coating on both sides of a glass slide (Sood et al., 2012) compared with uncoated glass. Nanostructured silica is used as anti-reflection (AR) coating. For angle of incidences of less than 70° the transmittance of nanostructured coated glass was found to be about 99%, up from 92% for broadband transmittance between 350 and 1800 nm. The effect of less reflection through use of nanostructured coating is shown in Figure 6.1. The incident optical beam when going from a rarer to a denser medium refracts with an angle of refraction less than the angle of incidence. This would mean less reflection from the surface and more transimittance. It has been shown that when coatings with lower refractive indices are used, such as shown in

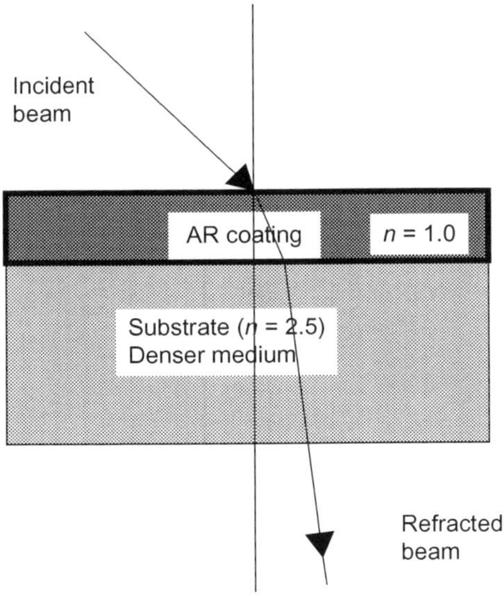

Figure 6.1 Less reflection using AR coating.

Figure 6.1, Fresnel reflection can be virtually eliminated. Oblique angle SiO_2 nanowires and nanorods can be used to prepare advanced nanostructured AR coatings. These can be used in next-generation sensors and optical windows. In this approach the reflection losses are minimized. AR coatings are critical to the development of electro-optical/infra-red (EO/IR) sensors and photovoltaic (PV) sources. They can be designed using a genetic algorithm (GA) and fabricated using oblique angle deposition (Sood et al., 2012). They have developed an analytical model that can be used to predict the porosity and deposition rate of nanoporous film grown by oblique-angle deposition. One fitting parameter is used to account for geometrical factors and surface diffusion phenomena. The porosity and deposition rate were obtained from the measured refractive index and thicknesses of SiO_2 and indium tin oxide (ITO) nanoporous film as a function of various incident angles of deposition. The model can be used for predictive control of refractive index, porosity, and deposition rate for a wide range of deposition angles and materials.

The refractive index of the coating is an important parameter in the design of solar cells and other optical devices where the coating is applied. Materials with refractive index, n, in the range of $1-1.4$ are limited. Aerogels may fit the bill, but control of their thickness is difficult. Professor F. Schubert's research group (Poxson et al., 2008) deposited silica nanorods by evoporation. They obtained good control of the refractive index by varying the angle of incidence. They obtain a material with refractive index of 1.1 for an incident angle of 80°, material with $n = 1.07$ at an incident angle of 85°. A world record for material with low $n = 1.05$ was obtained at an incident angle of 87°. They conducted research into the development of graded-index AR coatings for a new class of optical materials. Standard AR coating is 0.25λ thick and zero reflectivity is seen at one wavelength. This can

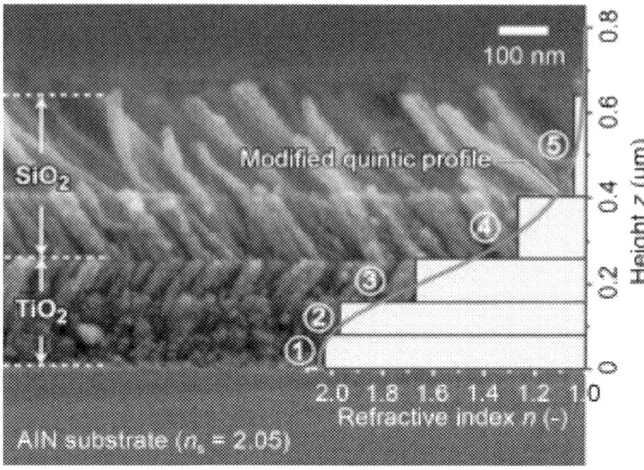

Figure 6.2 Cross-sectional scanning electron microscopy of TiO_2 and SiO_2 step-graded index nanowire/nanoroad coating that approximates a modified quintic profile.

be used in some applications. The electromagnetic spectrum from the sun is broader in scope, ranging from ultraviolet (UV) and IR.

The principle of total internal reflection is used to obtain zero transmission losses. When an optical ray travels from a denser to rare medium the angle of refraction will be greater than the angle of incidence. At some higher angle of incidence the angle of refraction would be greater than 90°. This would mean the ray that was to be refracted has been totally internally reflected. They deposited five layers of decreasing refractive indices. The difference in refractive index between two adjacent layers is small. The output efficiency of light emitting diodes (LEDs) can be improved by using of AR coating. Another application is in the area of super luminescent diodes. High-reflectance mirrors and Bragg reflectors can use AR coating.

Fresnel reflection can be eliminated from an interface of aluminum nitride (AlN)/air, aluminum nitride.

Xi et al. (2007) showed the feasibility of obtaining high-performance AR coating on AlN substrate. Figure 6.2 shows the cross-sectional SEM of TiO_2 and SiO_2 step-graded index nanowire/nanorod coating that approximates a modified quintic profile. This has been demonstrated for UV LED applications. Three layers of SiO_2 and two layers of TiO_2 are used. The refractive indices of SiO_2 layers are 1.05, 1.27, and 1.67, respectively, and that of TiO_2 layers are 1.97 and 2.03, respectively. As the light ray from the air travels through from a rarer to a denser medium the angle of refraction will be successively less than the angle of incidence. This would decrease the probability of reflection. Studies have confirmed the elimination of Fresnel reflection in this manner.

E-beam evaporation was used (Sood et al., 2012) to grow porous nanostructured materials by using the oblique angle deposition technique. Directional vapor flux is used to create a variety of optical structures. Surface diffusion, self-shadowing phenomena may have a significant effect in the deposition process. A shadow region is created from random growth fluctuations on the substrate where the incident vapor flux did not arrive. Shine regions are where the incident light is found to deposit in a directional manner. An oriented structure such as a nanorod is formed. High porosity is a feature of this method. Self-organization is evident and only some materials possess this capacity. A GA is used to optimize the AR coating performance. Multilayer coatings with co-sputtered and porous nanostructures can be designed. A porosity of 90% at $n = 1.05$ was achieved.

6.3 Fluidized bed method

Nanocoatings with nitride and oxynitride layers with thicknesses in the 5–500 nm range are used in pearlescent pigment formulation (Bertaux et al., 2004). Optimal interferences are seen at 50–350 nm. A liquid–solid suspension in water is made from 100 gm of mica muscovite with 10–50 μm diameter for 2 liters of water. With the suspension maintained at 75°C a tantalum pentachloride, $TaCl_5$ solution in hydrochloric acid (HCl) is slowly added to the reactor. The pH is maintained at 9 in the kettle. The preparation is filtered off, water washed, and dried at 110°C for 12 hours. For every 100 gm of mica, 5 gm of $TaCl_5$ is precipitated onto the solid.

The precursor material is calcined in ammonia at 1000°C for 48 hours in a fluidized bed reactor. A red tri-tantallumpenta nitride (Ta_3N_5)-coated mica pigment is produced. These pigments are semitransparent with an angle-dependent optical effect. Because the calcination is performed in a fluidized bed, a product with less agglomerates is formed at reduced reaction times. The process can be performed either in batch mode or in continuous mode. Ammonia is an interesting choice for nitridation. The conversion from oxides to nitrides/oxynitrides depends on a number of parameters, such as gas flow rates, reaction times, and temperature profiles. A higher nitride-to-oxide ratio is obtained at higher reaction times. The structure of the product obtained depends on the reaction time. The color of the compound and the color strength of the compound is found to be dependent on the structure. Bubbles, slugs, and other instabilities are reduced by use of vibrators/agitators. The substrate particle size range is between 1 and 500 μm.

After a pigment of the desired thinness is obtained a surface treatment step can be used to increase the weather stability and chemical stability of the pigment. It can be used to facilitate the application of pigment in different media. Pearlescent pigments can be used for coloring plastics, glass, ceramic products, agricultural foils, cosmetic formulations, automotive coatings, ink, and printing ink. Printing ink for intaglio printing, bronzing, and flexographic printing can be made out of nanocoated pigments. They can also be used for laser-making paper, plastic, and in national security applications such as banknotes, ID cards, concert tickets, and plastic film.

6.4 Electroplating

Large fabrication precision and high aspect ratio are the technical challenges in nanomanufacturing. State-of-the-art X-ray lithography and electroplating techniques were used (Lo et al., 2007) to create a zone plate with a thickness of 500 nm and with zone dimensions less than 50 nm. Photoresist selection and the capability to electroplate a thick Au layer into small trenches are critical parameters in the performance of the device. The substrate used in fabrication of the X-ray mask is prepared by a deposition process of Si_3N_4 film on an Si wafer. The desired zone plate pattern is created using E-beam lithography with the capability of writing less than 10 nm on photoresists. This step is followed by electrodeposition of the gold layer onto the nanopatterened photoresist structure of sufficient thickness. The resist is then removed. Potassium hydroxide, KOH, is used in order to etch the Si wafer down to the Si_3N_4 film. The desired Z-ray hard mask is produced. The zone plate pattern with the X-ray lithography method is defined by a hard mask. Precisions of up to 30 nm outermost zone in the mask can be obtained (Figure 6.3).

High penetration into the material is one reason for the choice of X-rays with wavelengths of less than 1 nm. The resolution of the device cannot be any better than the smallest zone. For an image resolution of less than 50 nm, the smallest zone has to be less than 50 nm. E-beam lithpgraphy is the one feasible approach for structures less than 500 nm. X-ray lithography can be used for structures thicker than 50 nm. A typical zone plate has concentric circles with alternate empty and filled zones. The width of the zone decreases when the diameter of the outermost zone increased.

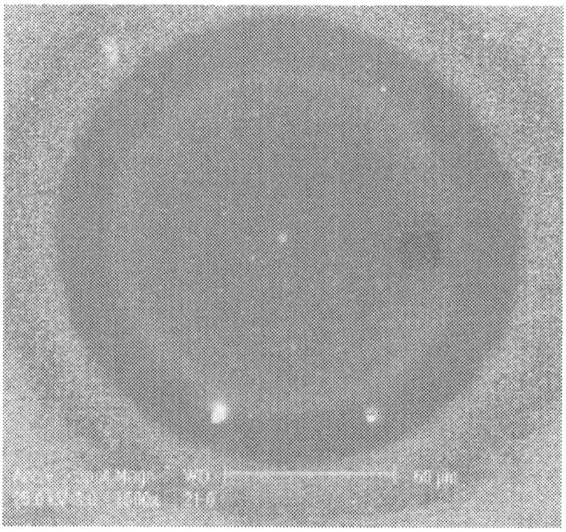

Figure 6.3 Zone plate pattern after X-ray lithography as seen using SEM.

6.5 Nanografting

The properties of polyurethane (PU) coating such as mechanical and corrosion resistance can be significantly improved by grafting techniques. The interfacial adhesion between nanoalumina and an organic substrate is increased by grafting. Nanostructured alumina is incorporated in the PU conventional coating and applied on mild steel. Bifunctional polymers (Jha et al., 2011) such as methyl-trimethoxysilane (MTMS) and 3-glyxidoxt-propyltriethoxysilane (GPTMS) were grafted onto the surface of the nanostructured alumina. This prevents agglomeration of the dispersed particles. Dispersion of particles in coating was achieved by ultra-sonication. Atomic force microscopy (AFM) analysis was used to confirm that grafting nanoalumina particles results in reduction of agglomeration of the dispersed phase. The root mean square (RMS) surface roughness parameter was also found to decrease. The coating formulation comprises of four steps:

1. Aqueous uranedio and hexamethoxymethyl-melamine are mixed in molar ratio of 3:1 and ultrasonicated for 10 minutes at room temperature.
2. 10% each of the mixture, alumina and MTMS and GPTMS are dispersed separately.
3. 0.4 wt% of blockes acid catalyst *para*-toluenesulfonic acid (*p*-TSA) are added and the resulting solution ultrasonicated for 20 minutes at room temperature.
4. The resulting solution is applied to a cleaned mild steel surface with a brush at room temperature and placed in a furnace at 150–160°C for 30 minutes for final curing.

The average dry coating thickness was found in the range of 15–20 μm for brush coating.

6.6 Plasma spray coating

Nanostructured alumina/titania powders can be used to improve the abrasive wear resistance of ceramic coatings. A 400% improvement in the performance of the coatings was reported (Wang et al., 2000). Alumina is an effective coating used in the tool industry. Alumina-coated carbide tools have a speed capability 250% higher than uncoated carbide tools. Nanoparticulated powders can be reformed into spherical granules. This is achieved by dispersion of the nanopowders into a colloid. A binder is subsequently added and granules are formed by spray drying. Sprayable granules can be from metals, ceramics, and composite powders. A Metco 9 MB plasma gun was used (Wang et al., 2000) in the thermal spraying process. Coating thickness achieved was 250–600 μm. Mild steel coupons were used as substrate and were blasted in order to remove dust. A GM-Fanuc six axis thermal spray robot was used and controlled using a computer. The plasma deposition parameters were: (1) primary argon gas pressure (690 kPA); (2) secondary hydrogen gas pressure (380 kPa); (3) argon gas flow rate (120); (4) powder carrier gas flow rate (40–70); (5) powder feed rate (400–900 g/h); (6) current (600 amps); (7) voltage (65 V); (8) moving speed of gun (500 mm/s); (9) spray distance (100 mm); (10) thermal spraying 10 passes after one cycle.

6.7 Nanostructuring in thin films

In some applications in industrial electronics polymer structures are required on length scales down to individual molecules. A "bottom-up" approach is preferred to a "top-down" approach in order to achieve this. A lateral resolution less than 100 nm can be created by allowing for surface instabilities and pattern formation in polymer films. Steiner (2005) discussed the methodology for the demixing of a polymer blend and pattern formation. Nanostructure formation is by capillary instabilities.

A compatible blend is one when some improvement of property of belnds is shown in the application when two polymers are mixed. Different from this is the molecular mixing achieved in miscible blends. Nanostructures can result from control of the morphology of phase-separated polymer blends. Polymer blends can also be found to be partially miscible (Sharma, 2012). Miscible polymer blends a single glass transition temperature much like a homopolymer made out of a single monomer. Immiscible polymer blends will exhibit the same two glass transition temperatures as the homopolymers that went into the make of the blend. The partially miscible blends can be expected to exhibit glass transition temperatures distinctly different from the homopolymer glass transition temperatures.

When polymers undergo phase separation in thin films the kinetic and thermodynamic effects are expected to be pronounced. The phase-separation processes can be controlled to effect desired morphologies. Under suitable conditions a film deposition process can lead to pattern replication. Demixing of polymer blends can lead to structure formation. The phase-separation process can be characterized by the binodal and spinodal curves. UCST is the upper critical solution temperature, which is the temperature above which the blend constituents are completely miscible in each other in all proportions. Not found that often in other systems other than among polymers is the LCST behavior. The LCST is the lower critical solution temperature. This is the temperature above which the polymers that were miscible below this temperature now exhibit immiscibility. The binodal and spinodal curves can be calculated by the stability criteria of Gibbs free energy:

$$\Delta G = \Delta H - T\Delta S$$

$$\Delta G \leq 0$$

$$\frac{\partial^2 G}{\partial \phi^2} > 0$$

where G is the Gibbs free energy, H is the enthalpy, S is the entropy, and ϕ is the phase composition. When the free energy change is negative the polymer blend can be expected to be miscible. The immiscible and miscible regions are separated by the binodal curve. The spinodal curve can be found within the binodal where the curvature of the free energy curve becomes negative. When the material falls on the spinodal region of the phase diagram spinodal decomposition can be expected to occur.

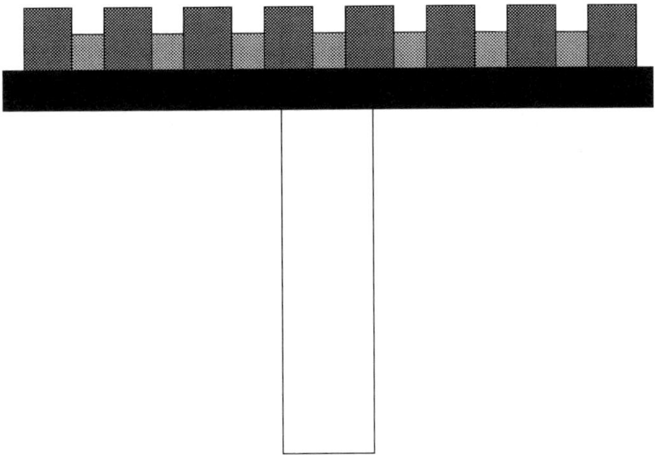

Figure 6.4 Polymer thin film formation from spin coating solvent.

Phase *morphology* with a *single characteristic length scale* can be synthesized by quenching a partially miscible polymer blend below the critical temperature of demixing. A well-defined spinodal pattern is formed and becomes larger with the passage of time. Polymer films can be made by the spin coating solvent casting method (Figure 6.4). The polymers and solvent form a homogeneous mixture at first. Solvent evaporation during spin coating causes an increase in polymer phase volume resulting in traversal to the spinodal region of the phase diagram. This can be expected to lead to polymer–polymer demixing. Characteristic phase morphology can be found in the polymer film. The polystyrene and PV pyridine are mixed in the tetrahydrafuran (THF) solvent.

Different phase-separated morphologies can be found in different polymer-solvent systems. The pattern formation consists of several stages. In the initial stage phase-separation results in a layered morphology of the two solvent-swollen phases. As more solvent evaporates this double layer is destabilized in two ways:

1. capillary instability of the interface; and
2. surface instability.

Each of the mechanisms results in different morphological length scales. Core shell spherical domains in phase-separated ternary systems have also been found. The shell thickness can be a few nanometers.

6.8 Electrochemical deposition

Electrodeposition can be used to prepare metal chalcogenide nanostructured film. Thin film of metals and semiconductors is needed in advanced technologies. Precise control of the nanostructuring can lead to improvement in performance of thin film-based devices such as nanosensors, magnetic storage media, and nanoscale optical devices. Wet methods for nanostruturing include chemical bath deposition (CBD) and electrochemical deposition (ED). Template-based synthesis is a more efficient method of preparation of nanostructured thin film nanowire arrays. A variety of nanostructures have been electrochemically deposited within porous alumina and polycarbonate templates providing arrays of nanoscale wires. Nanowires possess dimensions of about 20–250 nm and a length of 1–10 mm. Materials used in the preparation of nanowires include metals such as gold, silver, cobalt, copper, nickel, palladium, and platinum; compound semiconductors such as cadmium sulfide (CdS), cadmium sellenide (CsSe), indiumphospide (InP), gallium arsenide (GaAs); and conductive polymers such as polyacetylene.

An electrodeposition method was patented by the University of Tulane (Lu and Wang, 2006) to make nanostructured films within the pores of mesoporous silica by formation of a metal silica nanocomposite. Nanostructured film needs to possess high structural integrity and mechanical strength. The nanocomposite is obtained by annealment in order to strengthen the deposited metallic composition. The deposition temperature is at a fraction of the melting temperature of deposited metal. Removal of silica from the nanocomposite is by dissolution in a suitable etching solvent such as hydrofluoric acid (HF).

The dimensions of nanoscale wires of metal chalcogenide thin films assumed are determined by the dimensions of the pores in the template used to prepare the film. The pore structure is mesoporous. Templates may be hexagonal with arrays of straight pores. The pore structure is interconnected, continuous and forms a nanomesh. Segments with nanoscale cross-sectional dimension are present. Cubic mesoporous silica templates having a network of three-dimensional interconnected pores can be used.

A mesoporous silica film is prepared by deposition of a surfactant micelle-templated silica sol onto an electrically conductive substrate. The next step is calcination of the deposit in order to provide a film that is surfactant-free. The pore structure of the film is a function of the choice of the surfactant and the concentration of surfactant. The film is deposited on a conductive metal surface such as Cu/Au/silver in order to provide an electrode for electrodeposition of metal chalcogenide within the pores of the film. The film with its conductive substrate is immersed in a solution of metal ion complex and elemental chalcogen such as sulfur, selenium, or tellurium and a voltage potential is applied between the conductive substrate in contact with the mesoporous silica template and a counter electrode in the solution. The nanowires grow outward in a single direction through the pores of the silica forming a nanocomposite of silica and metal chalcogenide. The nanocomposite is later annealed and the structural integrity of the film is imparted in this step. The silica is removed by solvent-etching techniques.

These nanostructured films can be used in solar cells' catalytic membranes and information storage media. A self-supporting metal chalcogenide nanomesh film can be used as a photoelectronic component in PV solar cells. The solar cell comprises cadmium sellenide (CdSe) nanomesh film and a layer of CdSe nanomesh in contact with an optically transparent indium-oxide (ITO) substrate. The pores of the nanomesh are filled with a hole transport agent such as polythiopene, a conductive polymer. A reflective metallic layer is coated over the nanomesh and hole transport agent. Semiconductive CdSe nanomesh is activated by the light entering through the ITO substrate. An electron is generated that is injected into the hole transport agent and migrates to the conductive metal layer generating an electric current. This is expected to increase the PV efficiency of solar cells from 14% to the order of 40–50% through use of nanostructured components.

6.9 Anti-corrosion coating

Graphene has been used to make the thinnest coating known the world over and can be used for protecting metals against corrosion (Sharma, 2014). The potential use of graphene as anti-corrosion coating was discussed in the journal *ACS Nano* by D. Prasai and his colleagues. Rusting and other corrosion of metals is an important global problem. Contact of metal with air, water, or other substances can cause corrosion. Graphene is a single layer of carbon atoms. It is evaluated for use as anti-corrosion coating. An ounce of graphene arranged in a single layer and comprised of rows of benzene rings can fill the size of 28 football fields. Graphene, whether made directly on copper or nickel or transferred onto another metal, can be used to prevent corrosion. Copper coated with graphene by chemical vapor deposition (CVD) was found to corrode seven times faster than uncoated copper. Nickel coated with multiple layers of graphene was found to corrode at a rate of 20 times slower than uncoated Nickel. The same amount of corrosion protection can be obtained using a single layer of graphene that was obtained using five layers of organic coatings. Graphene coatings may be ideal for applications in industrial microelectronics. They can be used in aircraft, implants and as interconnects.

6.10 Infrared transparent electromagnetic shielding

Graphene is an interesting material of choice in naval applications to demonstrate an infrared-transparent, electromagnetic shielding coating that can be applied to electro-optic sensor windows and domes. Transmission greater than 90% in the 3–5 micron wavelength region can be achieved. Sheet resistance can be made less than 10 Ω per square. The coating can be found to be chemically stable in the atmosphere and in daylight and exhibit good resistance to rain and solid particles. The coating use temperature can be as high as 600°C.

Electromagnetic shielding of electro-optic sensor electronics in military systems is currently provided by metal grids that are electrically conductive applied to the dome or sensor window. However, the optical system is compromised via geometric blockage, diffraction, and unwanted reflection of light. Grids provide electrical shielding but are difficult to deposit on curved shapes. Damage is caused by erosion from rain and particle impact on the external surface of the dome or window.

Graphene thin film coating with high electrical conductivity and optical transparency can provide adequate electromagnetic shielding and superior optical performance. With its hardness, erosion resistance can be expected to be high. The graphene layer can be part of a stack of layers in order to minimize reflection losses and provide high erosion resistance. The coating can be deposited on an infrared-transparent substrates such as sapphire or silicon. This allows for infrared optical properties to be measured. Silicon can be used only if it is chemically inert under the film deposition conditions.

6.11 Underlying science – self-assembly

The science underlying nanocoatings is being studied. One interesting area to track is self-assembly. The Priestley Medal was awarded to Professor P. J. Stang, Distinguished Professor at University of Utah, by the American Chemical Society (ACS) in 2013. This is the highest honor bestowed by the ACS. Among other things he was recognized for his research contributions in molecular self-assembly to make supersized lego-type molecules. Self-assembly, according to Professor Stang, is the spontaneous reaction and organization of simple building units into well-defined more complex ensembles. This is based upon the recognition elements embedded in the building units. A number of noncovalent interactions can be found in natural systems. These interactions include hydrophobic/hydrophilic, van der Waals, Pi orbital stacking, dipole interactions, electrostatic, London forces, Keesome forces, and hydrogen bonding. Molecular self-assembly can be found in living organisms. Protein folding, secondary structure of proteins, nucleic acid structure, ribosomes, chromosomes, and phospholipid membranes are examples where self-assembly is critical. The protein coats of viruses, encasing their genetic material, comprise self-assembled capsids resembling polyhedrons such as dodecahedrons and icosahedrons.

Attempts to mimic natural systems in the lab took some time. Use of coordination chemistry and dative metal ligand bonds obviate the need for directionality of weak interactions and precise positioning of these interactions in order to obtain functional assemblies. High directionality is provided by use of metal ligand interactions due to d orbital geometry. Metal–ligand interactions have bond energies of about 15–25 kcal. These values lie between that of covalent bonds (60–120 kcal.mol^{-1}) and weak interactions of 0.5–12 kcal.mol^{-1}. Kinetics of coordination bond formation may be controlled. Reversible, self-repairing superstructures can be formed by thermodynamic control. Several hydrogen bonds can be replaced with one dative metal–ligand bond in the self-assembly process. Over two decades Professor Stang has developed by exploiting coordination chemistry and metal–ligand interactions, wide variety of large, nanoscale, complex two- and three-dimensional assemblies with well-defined shapes and sizes. Some of these shapes are trigonalbipyramid, tetrahedron, cube, anamantanoid, double square, dodecahedron, truncated tetrahedron, and cuboctahedron.

6.12 Conclusions

Nanocoatings with less than 100 nm thickness can be developed with improved performance compared with conventional coatings. Nanotuff was the first commercial application and was developed by Triton systems with improved chemical and abrasion resistance. The DoD estimated a reduction of $20 million in costs upon development of nanocoatings for valves. Nanocoatings with 30 nm thickness have been development to reduce furniture fires. Critical issues in the development of nanocoatings are the process for precursors, functionalization of nanoparticles in coatings, particulation, materials and methods for sacrificial and other coats, and meeting specifications.

Nanostructured silica is used as AR coating. For angles of incidence less than 70° the transmittance of nanostructured coated glass was found to be about 99%, up from 92% for broadband transmittance between 350 nm and 1800 nm. Five layers of nanowire/nanorod coatings of TiO_2 and SiO_2 were used to obtain AR coatings that can be used to eliminate Fresnel reflection. Oblique angle SiO_2 nanowires and nanorods can be used to prepare advanced nanostructured AR coatings. These can be used in next generation sensors and optical windows. Such multilayered coatings can be designed using GA and made using oblique angle deposition.

A fluidized bed can be used to obtain the nanocoating. The precursor material is calcined in ammonia at 1000°C for 48 hours in a fluidized bed reactor. A red tritantallumpenta nitride (Ta_3N_5)-coated mica pigment is produced. Bubbles and slugs and other instabilities formed during the operation of the fluidized bed are reduced by using vibrators/agitators. The substrate particle size range is between 1 and 500 μm. After a pigment of the desired thinness is obtained a surface treatment step can be used to increase the weather stability and chemical stability of the pigment.

Large fabrication precision and high aspect ratio are technical challenges in nanomanufacturing. State-of-the-art X-ray lithography and electroplating techniques were used to create a zone plate with thickness of 500 nm and with zone dimensions less than 50 nm.

The interfacial adhesion between nanoalumina and an organic substrate is increased by grafting. Nanostructured alumina is incorporated in the PU conventional coating and applied to mild steel. Mechanical and corrosion resistance obtained from PU coating can be increased by nanografting.

Nanostructured alumina/titania powders can be used to improve the abrasive wear resistance of ceramic coatings by 400%. A Metco 9 MB plasma gun, GM-Fanuc, 6 axis robot was used and the spray coating parameters were controlled using a computer.

Polymer nanocoatings can be developed using principles of phase behavior and capillary instability. Desired morphologies can be obtained by controlling the phase-separation process. The free energy of mixing of polymer blends, and enthalpic and entropic considerations can be used to design nanocoatings for the electronic industry. Spin coating can be used to cast the polymer film.

A variety of nanostructures have been electrochemically deposited within porous alumina and polycarbonate templates providing arrays of nanoscale wires.

Nanowires possess dimensions of about 20–250 nm and a length of 1–10 mm. Materials used in preparation of nanowires include metals such as gold, silver, cobalt, copper, nickel, palladium, and platinum; compound semiconductors such as CdS, CsSe, InP, and GaAs; and conductive polymers such as polyacetylene.

Graphene has been used to prepare the thinnest coating in the world. Copper coated with graphene by CVD was found to corrode seven times faster than uncoated copper. Nickel coated with multiple layers of graphene was found to corrode at a rate of 20 times slower than uncoated nickel.

Graphene is an interesting material of choice in naval applications to demonstrate an infrared-transparent, electromagnetic shielding coating that can be applied to electro-optic sensor windows and domes. Transmission greater than 90% in the 3–5 micron wavelength region can be achieved.

Self-assembly principles can be used to explain the science behind nanocoating technology.

References

Bertaux, S., Reynders, P., Wichmann, J.U., Schweda, E., 2004. Pearlescent Pigments. US Patent 6,719,837, Merck Patent Gesellschaftmitbeschrankter Haftung, Darmstadt, DE.

Geim, A.K., 2009. Graphene: status and prospects. Science. 324, 1530.

Geim, A.K., 2010. Random Walk to Graphene. Nobel Lecture, December. Available from: <http://nobel.se> (accessed summer 2013.).

Jha, B.K., Bhadu, M.K., Khanna, A.S., 2011. High-performance nano-alumina-grafted waterborne polyurethane coating. Int. J. Nanosci. 10 (4, 5), 1119–1124.

Laufer, G., Kirkland, C., Morgan, A.B., Grunalan, J.C., 2013. Exceptionally flame retardant sulphur-based multilayer nanocoating for polyurethane prepared from aqueous polyelectrolyte solutions. ACS Macro Lett. 2 (5), 361–365.

Legg, K., 2003. Overview of Nanocoating Technology: Nanocoatings IP Landscape. RowanTechnology Group, Chicago, IL. Available from: <http://www.rowantechnology.com> (accessed summer 2013.).

Lo, T.N., Chen, Y.T., Liu, C.J., Chang, W.D., Lai, T.Y., et al., 2007. Using e-beam and x-ray lithography techniques to fabricate zone plates for hard x-ray. AIP Conf. Proc. 879, 1466–1469.

Lu, Y., Wang, D., 2006. Process for the Preparation of Metal-Containing Nanostructured Films. US Patent 7,001,669, 2006, Tulane Educational Fund, New Orleans, LA.

Poxson, D.J., Mont, F.W., Schubert, M.F., Kim, J.K., Schubert, E.F., 2008. Quantification of porosity and deposition rate of nanoporous films grown by oblique-angle deposition. Appl. Phys. Lett. 93, 101914–101923.

Sharma, K.R., 2010. Nanostructuring Operations in Nanoscale Science and Engineering. McGraw Hill Professional, New York.

Sharma, K.R., 2012. Polymer Thermodynamics: Blends, Copolymers and Reversible Polymerization. CRC Press, Boca Raton, FL.

Sharma, K.R., 2014. Graphene Nanomaterials. Momentum Press, Happuage, NY.

Sood, A.K., Sood, A.W., Welser, R.E., Pethuraja, G.G., Puri, Y.R., et al., 2012. Development of nanostructured antireflection coatings for EO/IR sensor and solar cell applications. Mater. Sci. Appl. 3, 633–639.

Steiner, U., 2005. Structure formation in polymer films: from μm to Sub 100 nm length scales. In: Huck, W.T.S. (Ed.), Nanoscale Assembly – Chemical Techniques. Nanostructure Science and Technology. Springer, New York.

Wang, Y., Jiang, S., Wang, M., Wang, S., Xiao, T.D., et al., 2000. Abrasive wear characteristics of plasma sprayed nanostructured alumina/titania coatings. Wear. 237, 176–185.

Xi, J.Q., Schubert, M.F., Kim, J.K., Schubert, E.F., Chen, M., et al., 2007. Optical thin-film materials with low refractive index for broadband elimination of Fresnel reflection. Nat. Photonics. 1, 176–179.

Part Two

Nanostructured electroless nickel-boron coatings for wear resistance

7

Véronique Vitry and Fabienne Delaunois
University of Mons, Belgium

7.1 Introduction

Electroless nickel-boron coatings are obtained by a variation of electroless nickel where boron is incorporated into the coating during deposition. The process for electroless nickel deposition has been used since 1946 (Brenner and Riddel, 1956) and electroless nickel-boron was developed a few years later (Warf and Schaltegger, 1955) on the basis of the work of Schlessinger and Brown (1949). It is renowned for its ability to coat conducting and insulating substrates, forming a continuous deposit of homogeneous thickness (Agarwala and Agarwala, 2003; Riedel, 1991; Sahoo and Das, 2011).

Electroless nickel-boron coatings are good candidates for the protection of metals against abrasive wear for the following reasons:

1. Their continuity and thickness homogeneity allow deposition on machined-to-size parts, with very good control of the size tolerance. Moreover, if stripping of the worn part is used, parts can be refurbished for wear by re-plating without the need for further machining (Riedel, 1991).
2. Their typical cauliflower-like surface is favourable to load bearing (Delaunois et al., 2000) and can retain lubricants.
3. Their hardness is high and their wear resistance is good (Agarwala and Agarwala, 2003; Riedel, 1991; Sahoo and Das, 2011).
4. Their adhesion to metallic substrate is good to very good (Delaunois, 2002).

As a result, they find applications in various industries such as the chemical and food industries, aeronautic and automobile parts, moulds, cutting tools (e.g. saws) and firearms (McComas, 2004).

7.2 Synthesis of electroless nickel-boron coatings

Most of the properties of electroless nickel-boron deposits are the result of their synthesis method, which influences the chemistry, morphology and structure of the material and allows the plating of various kinds of conductive and insulating substrates, with good thickness homogeneity.

Anti-Abrasive Nanocoatings.

For this reason, it is important to have an overall view of the deposition method of these materials and of the way the plating parameters can influence the deposit properties.

7.2.1 General principle of electroless deposition

Electroless nickel-boron coatings are obtained by the electroless plating method, which is one of the three electrochemical methods for the formation of metals from salts. Contrary to electrolysis, where the electrons used to reduce the ions to their metallic form are given by an external power source, electroless plating uses only chemical energy (and heat). While the general process of electroless plating can be compared to displacement plating (where the deposition of the metal is induced by the oxidization of another metal, generally the substrate), they differ in that electroless plating is not limited by complete coverage of the substrate.

The electroless plating bath contains metallic salts that are the source of the deposited metal. Those salts are reduced by a chemical agent, the reducing agent that is also present in the bath. The principle of the method – reduction of metallic salts by a chemical agent present in the solution – is very simple. However, its application is a lot more complicated because several conditions must be reunited in order to make the process reliable and usable in industrial contexts.

7.2.2 Thermodynamic and kinetic conditions for electroless deposition of metals

Electroless deposition is only possible if the thermodynamic and kinetic conditions are respected (Ohno, 1991; Ohno et al., 1985; Paunovic, 1983; Vitry, 2010).

The thermodynamic condition is easily deduced from the deposition reaction. This reaction is written in Eq. 7.1 for a general case:

$$\mathrm{Me}^{n+} + \text{reducing agent} \rightarrow \mathrm{Me} + \text{oxidized form of the reducing agent} \qquad (7.1)$$

Eqs. 7.2 and 7.3 are the partial reactions involved in the deposition process, reduction of metallic salts and oxidation of the reducing agent respectively.

$$\mathrm{Me}^{n+} + ne^- \rightarrow \mathrm{Me} \qquad (7.2)$$

$$\text{Reducing agent} \rightarrow \text{oxidized form of the reducing agent } (^{z+}) + ze^- \qquad (7.3)$$

Redox potential of the metal/metal ions and of the reducing agent couple can be obtained using Nernst's Law (Eqs. 7.4 and 7.5).

$$E_{\mathrm{Me/Me}^{n+}} = E^{\circ}_{\mathrm{Me/Me}^{n+}} + (2.3^{*}\mathrm{RT})/(n^{*}\mathrm{F})\log(\mathrm{Me}^{n+}) \qquad (7.4)$$

$$E_{\mathrm{red/ox}} = E^{\circ}_{\mathrm{red/ox}} + (2.3^{*}\mathrm{RT})/(n^{*}\mathrm{F})\log((\mathrm{ox})/(\mathrm{red})) \qquad (7.5)$$

The thermodynamic condition for electroless condition is the following: The reaction is spontaneous and must thus have a negative ΔG. Knowing the expression of ΔG of electrochemical reactions, $\Delta G = -nF\Delta U_{eq}$, where N is the number of exchanged electrons, F the Faraday constant and ΔU_{eq} is the potential difference between the couples (Eq. 7.6). The condition is thus respected if $E_{Ni/Ni^{2+}} > E_{red/ox}$, that is if the metal is more noble than the reducing agent couple. The reaction becomes easier when the potential difference increases (due to an increase of affinity). However, maximizing this value is not always beneficial, as a too important potential difference could lead to bath destabilization due to spontaneous oxidation of the reducing agent.

$$\Delta U = E_{Me/Me^{n+}} - E_{red/ox} \tag{7.6}$$

The kinetic condition comes from the need for preferential reaction on the substrate of the sample: the thermodynamic condition only ensures that the spontaneous reaction is possible. However, it does not ensure selectivity of the reaction. Without precautions, a bath could be made where reaction occurs in the bulk of the solution. This would lead to the formation of small metallic particles in suspension in the bath and/or to anarchic deposition on all exposed surfaces. This case is typically encountered when the potential difference between the metal and the reducing agent is too big.

The kinetic condition, which can be considered as a catalytic condition, guarantees that deposition only occurs on selected surfaces (on the substrate to be plated). The phenomenon used to establish this condition is overpotential and can be explained as follows.

When an electrochemical reaction takes place, the anodic and cathodic polarization curves for each of the couples can be drawn, as shown on Figure 7.1. In the case of a spontaneous reaction, the system can only react at a potential (called mixed potential U_M) where the anodic and cathodic current are equals. This enables the electrical neutrality of the system.

In electroless deposition, the cathodic polarization curve (related to the reduction of metallic salts) is not influenced by overpotential. However, the anodic polarization of the reducing agent is influenced by this in the following way. If the surface where the electron exchange takes place (ideally the substrate) is catalytic, the overpotential is low and the mixed potential is established in conditions were the deposition current is high (U_{M1} and I_{D1}). However, if the surface is not catalytic, the overpotential for the anodic polarization will be high and the mixed potential will be displaced to more positive values of potential, closer to the equilibrium potential of the Me/Me^{n+} couple, where the deposition current is much smaller (U_{M2} and I_{D2}).

The kinetic condition can thus be formulated as such: overpotential must be as low as possible on the substrate to allow plating and must be as high as possible everywhere else to avoid anarchic deposition. In other words, the substrate must be catalytic for the oxidation of the reducing agent. This must be completed by the following considerations: deposition must continue after the substrate is fully covered

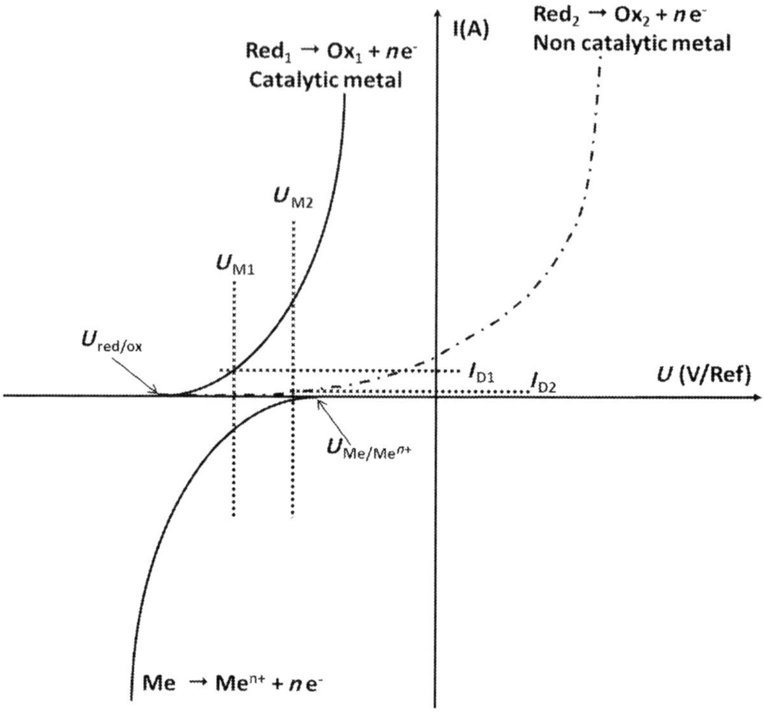

Figure 7.1 Mixed potential determination and effect of the catalytic activity of the metal on the deposition current.

Table 7.1 Summary of thermodynamic and kinetic conditions for electroless deposition

Thermodynamic condition	At the beginning of the plating process	$E_{Ni/Ni^{2+}} > E_{red/ox}$ Substrate must be catalytic for the oxidation of the reducing agent or must have been catalysed
Kinetic condition	During steady state plating	Deposited metal must be catalytic for the oxidation of the reducing agent

to obtain sufficient thickness. This means that the deposited material must adhere to a similar specification to the substrate material.

Table 7.1 summarizes the thermodynamic and kinetic conditions for electroless nickel-boron.

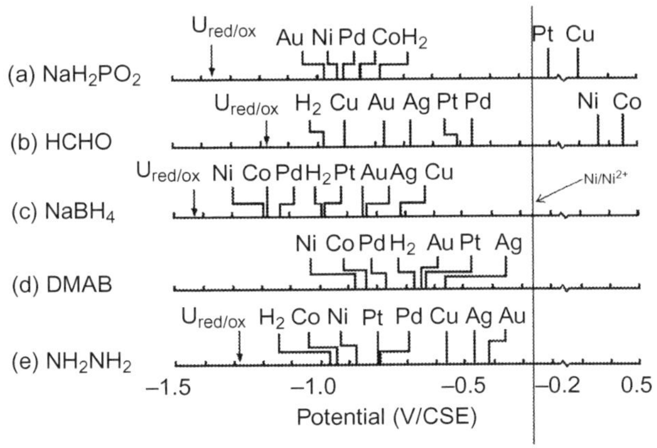

Figure 7.2 Catalytic activity of various metals at 25°C, under an anodic current of 10^{-4} A.cm^2, for diverse reducing agents.
Source: Adapted from Ohno (1991), with permission.

7.2.3 Possible reducing agents for electroless nickel deposition

The choice of the metal-reducing agent couple is very important for electroless deposition, as attested by the kinetic condition: catalytic activity of the substrate, then of the plated metal is required for the process. Of course, activity of a metal varies according to the reducing agent that is used, as was shown by Ohno (1991). He studied the oxidation of several reducing agents on various metals by galvanostatic tests carried out in a solution of the reducing agent under an anodic current of 10^{-4} A.cm^2, as shown on Figure 7.2.

The catalytic activity of a metal against a given reducing agent is characterized by the potential this metal takes in a solution of the reducing agent. If this potential is lower than the equilibrium potential of the couple of the metal that has to be reduced, plating is possible. Otherwise, the establishment of the mixed potential is not possible and plating does not occur. Five reducing agents, coupled with various metals are shown on Figure 7.2 and the potential of the Ni/Ni^{2+} couple has been added to the figure. Electroless plating is only possible if the potential of Nickel in the solution of the reducing agent is lower than the equilibrium potential of Ni/Ni^{2+}. This is the case for most of the reducing agents, except formaldehyde (HCHO).

The choice of the reducing agent not only influences the feasibility of the plating: some of those agents contain metalloids that are included in the deposit during the plating process. This allows the synthesis of pure nickel (from hydrazine, NH_2NH_2), nickel-phosphorous alloys (from sodium hypophosphite, NaH_2PO_2) and nickel-boron alloys (from either dimethylamineborane (DMAB) or sodium borohydride ($NaBH_4$)).

7.2.4 Components of an electroless nickel-boron bath and their role

There are more components in an electroless nickel-boron plating bath than just a source of nickel ions and a reducing agent. A typical electroless deposition bath is thus composed of at least five components (Anik et al., 2008; Bielinsky et al., 1990; Cheong et al., 2004; Hari Krishnan et al., 2006; Huang and Cui, 2007; Riedel, 1991): a source of nickel ions, a reducing agent, a complexing agent, a stabilizer and a pH regulator.

The source of nickel ions can be any nickel salt, but nickel chloride hexahydrate ($NiCl_2.6H_2O$) and nickel sulphate are the most popular (Riedel, 1991).

The reducing agent has to be chosen carefully to ensure a good plating process. There are two reducing agents suitable for synthesizing nickel-boron deposits: DMAB and sodium (or potassium) borohydride ($NaBH_4$). Those reducing agents are used in very different baths and the resulting samples vary greatly in boron content. Coatings synthesized with $NaBH_4$ as a reducing agent usually contain a significant amount of boron (from 1 to 8 wt%) while those synthesized with DMAB contain 0.5 to 1.5 wt% boron (Riedel, 1991).

The borohydride ion is the most powerful reducing agent available for electroless nickel deposition: the oxidation state of boron in this ion is B^{5-}, and the boron end product is BO_2^- in which boron is in the B^{3+} state. This means that borohydride can provide up to eight electrons for the reduction of nickel, while sodium hypophosphite can only provide two. Moreover, the molecular weight of $NaBH_4$ is much lower than that of sodium hypophosphite, which means that, for a given weight of reducing agent, the difference in efficiency between hypophosphite and borohydride will be even greater (Gao and Bai, 1987). The amount of borohydride needed to reduce one kilogram of nickel is close to 600 g. However, borohydride suffers from one major inconvenience: the ion is not stable in acid and neutral conditions. This ion becomes stable enough for practical use in industrial processes only for a pH higher than 11 (with a half-life of 10.2 hours) and most users prefer working at the highest possible pH to ensure maximal stability of borohydride (Gervasio et al., 2005).

A DMAB-based bath can operate in either acid or alkaline conditions. It is thus more stable and flexible as a reducing agent than borohydride. However, its reduction efficiency is lower (Riedel, 1991). The amount of reducing agent in a nickel-boron plating bath is usually in the range 0.4–1.2 g/l but can be as high as 2.2 g/l.

The complexing agent plays multiple roles in the electroless nickel-boron bath: first, it is used to increase the solubility of the metallic ions in the plating solution and regulate the free ion concentration. Second, by decreasing the concentration of free ions, the complexing agent displaces the equilibrium potential of the nickel/nickel ions couple towards more negative potentials and thus decreases the gap between the metal and reducing agent couples. This means that the stability of the solution is increased and the bath is easier to control. Third, the complexing agent has a buffering effect on the bath. Last but not least, it allows the avoidance of precipitation of insoluble nickel hydroxides in alkaline media. A strong complexing agent is thus indispensable to the operation of borohydride reduced electroless nickel-boron plating baths (Vitry, 2010). Ethylene diamine is a popular complexing agent for electroless nickel-boron baths.

The stabilizing agent is added to the plating bath to limit the risk of uncontrolled reactions. As the whole plating method is based on a spontaneous catalysed reaction, it is important to keep the reaction localized on the desired sites (substrate) and avoid it in the bulk of the solution, on the surface of the cell and on precipitates and impurities. To control the kinetic of the reaction, parts of the catalytically active sites are blocked by adsorption of the stabilizing agent. The most popular agents are organic molecules (like thiourea) and heavy metals salts. Thallium, in its nitrate or acetate form is one of the most efficient heavy salts stabilizers and is thus popular, but its toxicity is very high (Delaunois et al., 2000; Henry, 2006; Riedel, 1991; Sankara Narayana and Seshadri, 2004). Recent baths operate with less toxic alternatives such as $PbWO_4$ (Delaunois et al., 2000; Delaunois and Lienard, 2002; Vitry et al., 2012a) or $PbNO_3$ (Dervos et al., 2004; Ziyuan et al., 2004). However, as all stabilizing agents have negative side effects and are not environment friendly, the stabilizers are used in very small quantities (Riedel, 1991; Delaunois et al., 2000; Vitry et al., 2012a).

The last indispensable component of the electroless nickel-boron bath is the pH regulator (and buffer). It is used to maintain the bath in the optimal pH range. In the case of alkaline electroless baths, the pH adjuster can be sodium hydroxide or ammonia.

Finally, there are optional components, such as accelerating agents that counteract the decrease of plating rate brought by the complexing and stabilizing agents, and shine agent that are used in industrial bath and make it possible to achieve smoother, shinier surfaces.

The nature of the agents used in electroless nickel-boron plating baths as well as their concentration in several baths' chemistries are presented in Tables 7.2 and 7.3.

As can be seen, baths based on amineborane compounds (Table 7.3) operate up to 70°C, in a relatively large range of pH. However, a baths based on borohydride (Table 7.2) operate at high pH, in a relatively wide temperature range, with a higher concentration of borohydride in the low-temperature baths.

7.2.5 Reactions in nickel-boron deposition baths, control parameters

The plating baths based on sodium (or potassium) borohydride are by far more popular than those based on amineborane compounds, as attested by Tables 7.2 and 7.3. In consequence, we will examine in greater detail the phenomena happening in the former.

The formation of nickel-boron deposits by reduction of salts by the borohydride ion is based on two main reactions: the reduction of metallic salts (nickel, but also the heavy metal salts used as stabilizers) by the borohydride ion (Eq. 7.7) and the formation (and co-deposition) of metallic boron (Eq. 7.8) (Riedel, 1991).

Table 7.2 **Chemistry of electroless nickel-boron plating baths reduced by sodium borohydride**

Bath	Nickel source (g/l)		Reducing agent	Complexing agent (g/l)	Stabilizer(g/l)						pH adjustor (g/l)		Others (g/l)			pH	Temperature (°C)	Plating rate (μm/h)	B content (wt%)
	Nickel chloride	Nickel methane sulphonate	Sodium borohydride	Ethylene diamine	Thallium sulfate	Thallium nitrate	Thallium acetate	Lead nitrate	Lead tungstate	Ethylene thiourea	Sodium hydroxide	Ammonia (28% NH_3)	Disodium tartrate	Polyamide	Sodium citrate				
Henry, 2006	20		0.4	90	0.4						90					14	95	15–20	
Lacourcelle, 1979	24		0.4									120				10 to 12	60	1.3	
Dervos et al., 2004	20		0.48	59				0.0145			39								
Delaunois et al., 2000	24		0.48	59					0.02		39					14	95	15	6 (1wt.% Pb)
Lacourcelle, 1979	30		0.6	50	0.07						40					12 to 14	90	12.5	
Lacourcelle, 1979	20		0.67	44							40					12 to 14	97	8.8	
Sankara Narayana and Seshadri, 2004	30		0.8	90			0.016				90			0.4		14	95		

Krishnaveni et al., 2005	30		0.8	90			0.016				90					14	95		
Zhang et al., 2008	30		0.8	90						0.0010	90					13	85		
Anik et al., 2008	30		1	90			0.018				90					14	95		
Delaunois et al., 2000	20		1.05	90		0.11					90					14	95	15	4 (6%Tl)
Henry, 2006	31		1.2	52		0.022					42					>13	93–95		
Srinivasan et al., 2010		6	1.2	60						0.0002	90					13	95	20.2	
Bulbul et al., 2012	40		1.2	90				0.0145			90					13.5	80		
Lacourcelle, 1979	36		2.2									132				10 to 12	40	0.7	
Correa, 2012	20		8	35						0.0001	110					>12	80		
Sudagar et al., 2013	30		12	60		0.07					40					14	90	20–25	
Baskaran et al., 2006	30		0.2–1	15			0.016				40		40			13	45	1–10	0.6–3.2
Kumar and Nair, 1994	30		0.3 to 0.7	70							40				50	13	87		2.34–4.32
Ziyuan et al., 2004	15–40		0.5–2	55–70				0.01–0.05			40–60						95		

Table 7.3 **Chemistry of electroless nickel-boron plating baths reduced by amineborane compounds**

	Reference	Henry, 2006	Sudagar et al., 2013	Sudagar et al., 2013	Henry, 2006	Henry, 2006	Henry, 2006	Saito et al., 1998	Saito et al., 1998	Saito et al., 1998
Nickel source (g/l)	Nickel chloride			30		24–48	30			
	$NiSO_4.6H_2O$	50	30		25			26	26	26
Reducing agent (g/l)	DMAB	3	3	4	4	3–4.8		3	3	1.5–6
	DEAB			3			3			
Others (g/l)	Isopropanol						50			
	Ammonium citrate		15							
	Sodium succinate			20			20			
	Potassium acetate				15	18–37				
	Methanol			40						
	Sodium acetate			20						
	Ammonium chloride		15							
	Sodium pyrophosphate	100								
	Sodium citrate			10			10	60	30	30
	Sodium malonate								16	16
	Lead acetate				0.002					
	2-Mercaptobezothiazole		0.0002							
pH		10	7.5	5–6	5.9	5.5	5–7	6	4–7.5	6
Temperature (°C)		25	25–35	50–60	26	70	65	60	60	60
Plating rate (μm/h)			7–12	15–20		7–12	7–12			

$$2Ni^{2+} + NaBH_4 + 4OH^- \rightarrow 2Ni + NaBO_2 + 2H_2O + 2H_2\uparrow \quad (7.7)$$

$$2BH_4^- + 2H_2O \rightarrow 2B + 2OH^- + 5H_2\uparrow \quad (7.8)$$

The reduction of metallic salts by $NaBH_4$ has one major by-product, $NaBO_2$, that is accumulated in the bath and needs to be removed for continued use of the bath (i.e. in industrial operations). Otherwise, the build-up of sodium borate causes problems with bath operations and deposit quality.

In some cases, the spontaneous formation of nickel boride Ni_2B is possible, according to Eq. 7.9.

$$8BH_4^- + 4Ni^{2+} + 18H_2O \rightarrow 2Ni_2B + 6H_3BO_3^- + 25H_2\uparrow \quad (7.9)$$

Some other, unwanted, reactions are possible in the electroless bath, for example the hydrolysis of $NaBH_4$, which is very detrimental to electroless deposition because the hydrolysed $NaBH_4$ is not available any longer for the reduction of nickel. This hydrolysis happens when the pH of the bath decreases under a value of 12, which can be a consequence of the deposition reaction that decreases the pH of the bath. For this reason, when $NaBH_4$ is used as a reducing agent, a strong alkaline compound is used to ensure that the pH value stays in the adequate range and the pH is checked regularly.

7.2.6 Importance of substrate preparation

Substrate preparation is very important in electroless nickel plating because of the catalytic nature of the process. The ideal surface state for electroless deposition is a perfectly clean and catalytic surface with an adequate roughness. The main ways to get to this kind of surface are described below.

The cleaning of the substrate is clearly influenced by its nature, as different substrates will react differently to different cleaning methods. However, there are general principles that can be applied to all kinds of substrates (Collin et al., 2010; Hajdu, 1990; Riedel, 1991).

First, all traces of grease have to be removed from the part. This can be done through the use of several degreasing agents such as acetone or isopropanol. After this, the sample is cleaned in an alkaline cleaning medium and/or with electrochemical cleaning. These operations can be repeated to get the cleanest possible surface. The sample is carefully rinsed with distilled water after each cleaning operation.

In the case of inadequate cleaning several problems can arise, the most noticeable being blisters and peelers. In the most stringent cases, the problems are detected directly after plating and the parts can be rebutted. However, inadequate surface preparation can lead to more subtle problems such as premature loss of corrosion protection and adhesive failure in use (Riedel, 1991; Collin et al., 2010).

The creation of a catalytically active surface differs according to the nature of the substrate material, as shown on Figure 7.3:

- Some metals are intrinsically catalytically active for electroless deposition, such as nickel and iron (steel). In this case, the last step of the surface preparation process is an

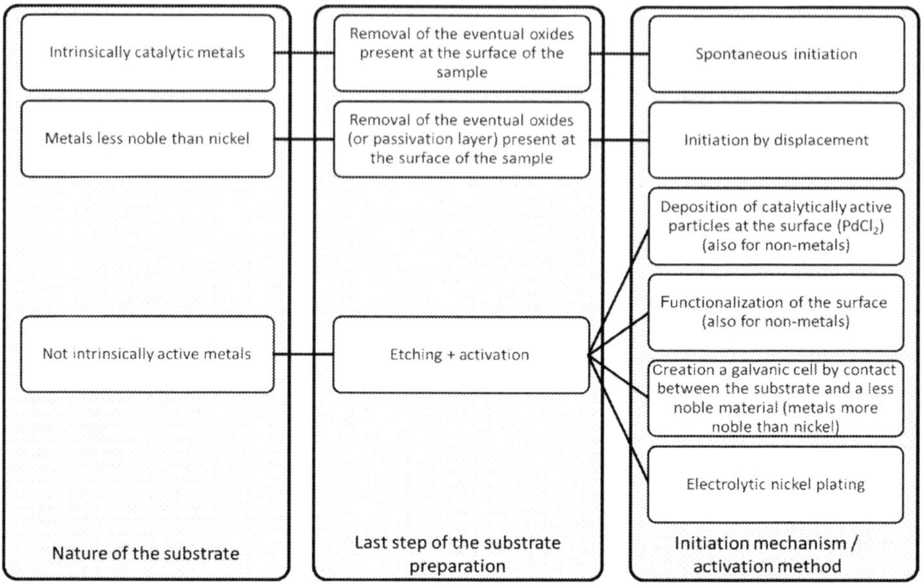

Figure 7.3 Chart of the types of surface preparation needed as a function of the activity of the substrate.

activation that allows removal of the eventual oxides present at the surface of the sample. This is usually carried out by immersion in acid (Collin et al., 2010; Hajdu, 1990; Riedel, 1991).

- Some metals are less noble than nickel. In this case, initiation of the plating process takes place due to a displacement reaction that causes the formation of a thin layer of nickel (that is intrinsically catalytic for the process) on the substrate. For those metals, to ensure that the displacement reaction takes place in an efficient manner, a complete removal of the passivation layer (if one is present) is needed. This can be carried out by immersion in various acids (according to the nature of the substrate) (Hajdu, 1990; Riedel, 1991).
- The remaining metals are not intrinsically active and must be activated. There are three ways of doing this:
  - Depositing catalytically active particles at the surface of the sample by immersion in an adequate solution containing, for example, $PdCl_2$ (Charbonnier et al., 2006; Khoperia, 2003; Matsubara et al., 2002, 2006; Riedel, 1991) or by functionalization of the surface (Charbonnier et al., 2006). This applies also to non-metallic materials, but other steps, such as etching before the activation, are needed (Riedel, 1991; Sudagar et al., 2013).
  - In the case of a metal more noble than nickel, creating a galvanic cell by establishing a contact between the substrate and a less noble material. This will induce the deposition of a thin layer of catalytically active nickel (Riedel, 1991; Vitry, 2010).
  - Applying a thin coating of electrolytic nickel (often named 'nickel strike'). This can be done using any electrolytic nickel formulation, such as Wood's nickel or nickel sulfamate (Hajdu, 1990; Riedel, 1991).

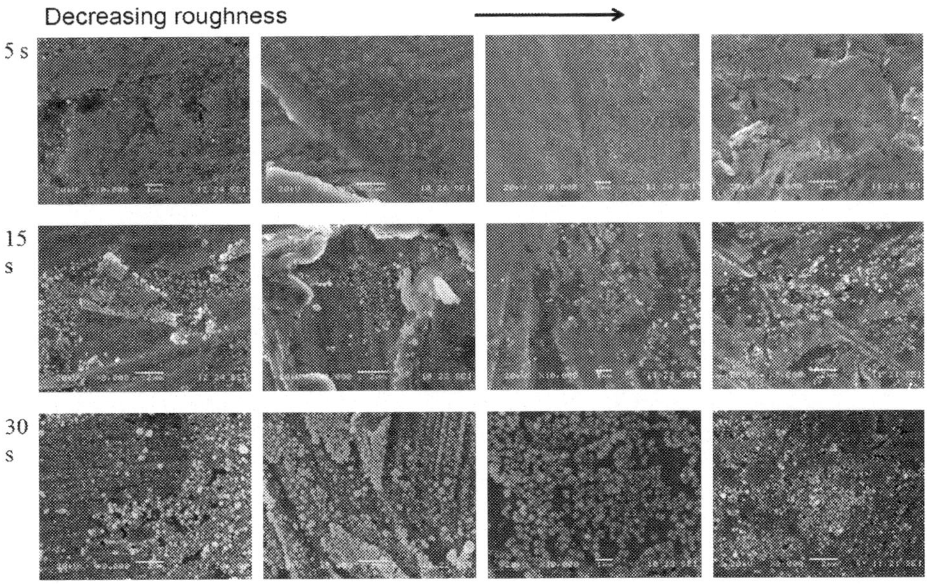

Figure 7.4 Effect of roughness on initial deposition of electroless nickel-boron deposits from a borohydride-reduced bath on steel substrate (Vitry et al., 2010).

The roughness of the substrate is also an important parameter that has an influence on the plating process and on coating properties (Lee, 2008; Liu and Gao, 2006), such as appearance, which is important in terms of loss of revenue (Williams and Feist, 1994), but it also influences the final roughness of the plated part (Vitry, 2010; Sahoo and Das, 2011). However, the control of substrate roughness in industrial applications is not easy because parts are supplied in various surface states and may have complex geometries that impede mechanical preparation of the substrate. In any case, knowledge and modification, if needed, of the roughness of the part is essential. Vitry et al. (2010) have shown, in a paper investigating the initiation of electroless nickel-boron plating of samples with varying roughness that initiation occurs quicker on rougher substrates (as shown on Figure 7.4) due to an increase of the number of active sites (Sahoo and Das, 2011), but the subsequent growth of the deposit is then slower.

7.2.7 Preparation methods for the most common alloys

The surface preparation methods described in this section are appropriate for alkaline electroless nickel-boron baths. For methods appropriate for acid nickel-boron baths (using DMAB) and for electroless nickel-phosphorous, please see Riedel (1991).

Mild steel is easily prepared for electroless deposition, as are most low alloy steels and irons. The preparation process includes eventual mechanical preparation, degreasing, alkaline cleaning and activation, for example in concentrated hydrogen chloride (HCl) for a short time. After the activation step, the substrate is rinsed one last time and directly immersed in the plating bath (Collin et al., 2010; Vitry, 2010).

Aluminium alloys are a very singular case as far as the formation of electroless nickel-boron deposition is concerned: aluminium is considered as catalytically active for plating. However, the spontaneous formation of a passivation layer on its surface (that is, not catalytically active) impedes plating. For this reason, aluminium is submitted to the double zincate process, during which a very thin layer of zinc is deposited on the sample (Alexis, 2000). This layer is later dissolved when the sample is put in the electroless plating bath, allowing initiation of the plating process. It is interesting to note that, in the case of aluminium, direct electroless nickel-boron plating with borohydride-reduced baths is not possible because of the high pH of the bath. A nickel strike or a protective layer of acid electroless nickel (usually nickel-phosphorous) must be applied before alkaline electroless nickel-boron plating (Delaunois et al., 2000; Delaunois and Lienard, 2002; Sudagar et al., 2013; Vitry, 2010). A flow sheet for the surface preparation of aluminium is shown on Figure 7.5.

The preparation of stainless steel for electroless plating is more difficult than that of other steels. In most cases, after the substrate degreasing and cleaning, a nickel strike is needed (Hajdu, 1990). However, it is possible, on the lab scale, to realize direct electroless plating on austenitic and duplex stainless steel after activation in concentrated HCl (Vitry, 2010).

Creating an active surface, free of passivation layer is as difficult on magnesium alloys as on aluminium because they oxidize even more quickly (Gray and Luan, 2002; Sudagar et al., 2013). However, contrary to aluminium, it has been possible to proceed to direct electroless nickel-boron plating on magnesium alloys, without a protective layer (even if a protective layer is often used).

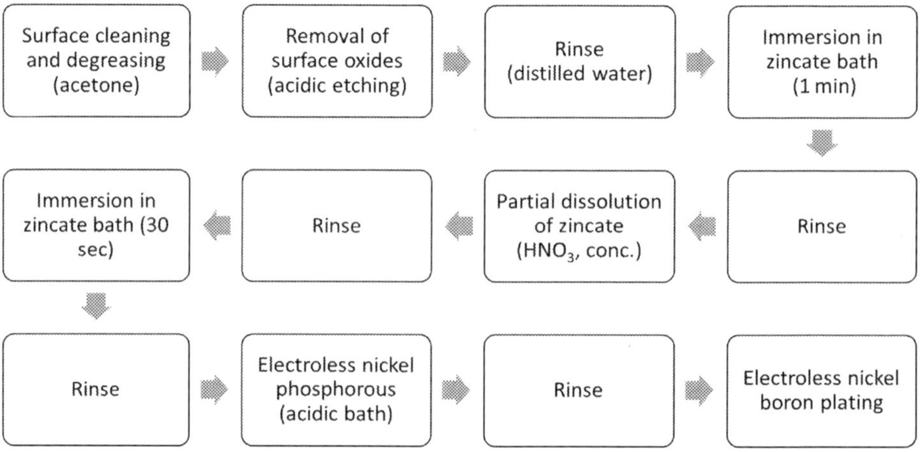

Figure 7.5 Flow sheet for the preparation of aluminium alloys before electroless nickel-boron plating.

Usually, surface activation of magnesium alloys for electroless plating includes chromate or fluorides: after an eventual chromate treatment (Fan et al., 2010), a protective conversion film, constituted of magnesium fluorides, is formed on the surface of the sample (Wang, 2012; Peixoto Barbosa and Knörnschild, 2009; Gu et al., 2005). The use of a protective Ni–P coating is also reported (Fan et al., 2010), as well as variations of the zincate process (Chang et al., 2011; Wu et al., 2010).

Correa (2012) proposed this procedure, based on fluorides, for pure Mg and AZ91D alloy: mechanical preparation, degreasing, drying, alkaline cleaning and activation in 200 g/l NH_4HF_2 for 10 minutes, followed by direct immersion in an electroless Ni–B bath containing NH_4HF_2. Wang et al. (2012) proposed the following procedure that avoids the use of fluoride and chromates: mechanical grinding, alkaline cleaning, acid pickling in a bath containing acetic acid and sodium nitrate, which allows removal of the superficial oxides, and finally direct electroless plating in a bath containing nickel acetate.

7.3 Morphology and structure of electroless nickel-boron coatings

Electroless nickel-boron does resemble nickel-phosphorous in several, but not all, aspects. The morphology and structure of both types of coatings are amongst the differences.

7.3.1 General aspect and plating rate

Electroless nickel-boron deposits have a metallic appearance, with a bright aspect, as shown in Figure 7.6. Due to the catalytic nature of the plating process and the absence of an external current source, the process allows the formation of coatings with a very homogeneous thickness, even on complex shapes and blind holes, contrary to electroplating, as shown on Figure 7.7.

As can be seen in Table 7.2, the plating rate of electroless nickel-boron baths using $NaBH_4$ as a reducer varies between 9 and 20 mm/h for high temperature baths (90°C and over) and is much lower for low temperature baths.

During electroless plating, the rate of coating formation decreases with time, as the reactive is progressively used by the plating reaction, as can be seen in Figure 7.8. To keep a high plating rate throughout the process industrial plating facilities use continuous or batch replenishment of the solution and add reactive to counteract their depletion.

7.3.2 Morphology, microstructure and chemistry

The surface texture of electroless nickel-boron coatings presents a typical cauliflower-like surface texture (that some authors also describe as blackberry-like), as can be seen in Figures 7.9 and 7.10 (Contreras et al., 2006; Vitry et al., 2010).

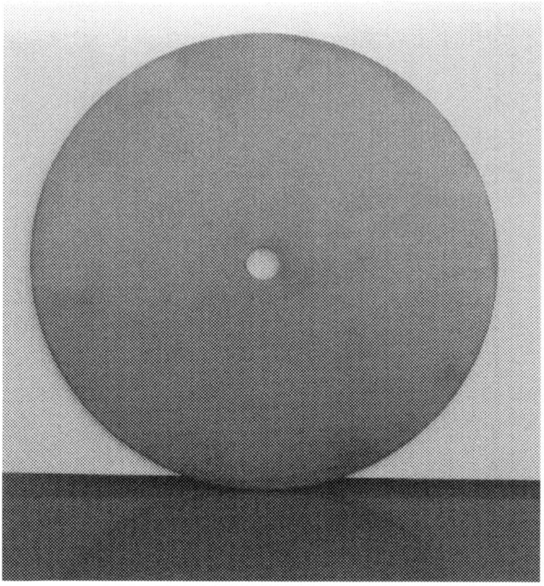

Figure 7.6 Aspect of an nanocrystalline nickel-boron plated part.

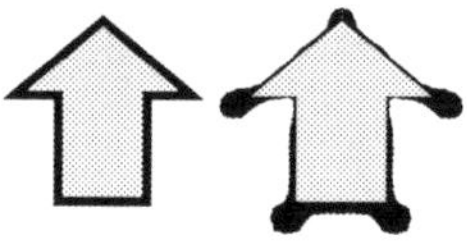

Figure 7.7 Schematic comparison of thickness obtained with, left, electroless plating, right, electroplating.

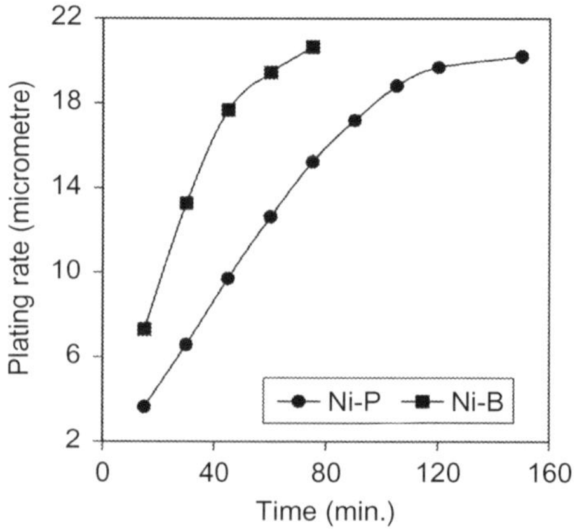

Figure 7.8 Evolution of plating rate as a function of plating time.
Source: SankaraNarayanan et al. (2003), reproduced with permission.

Figure 7.9 Typical cauliflower-like surface texture of an electroless nickel-boron coating.

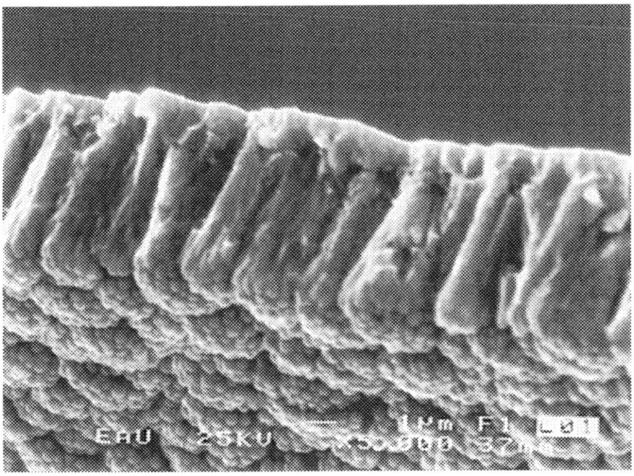

Figure 7.10 Surface and cross-sectional view of an electroless nickel-boron deposit.
Source: Bulbul et al. (2012), reproduced with permission.

This singular surface texture is reported as being favourable to wear applications, first because of the associated roughness type (that will be discussed later) and second because it can easily retain lubricants (Delaunois, 2002).

The surface aspect is clearly linked to the growth mode of electroless nickel-boron coatings: the deposition begins by the formation of nodules that then grow in a columnar morphology, as can be seen in Figure 7.11. The tops of the columns form the typical cauliflower-like texture.

Rao et al. (2005) showed that the typical columnar morphology is caused by the effect of the thickness of the diffusion layer, combined with lateral confinement

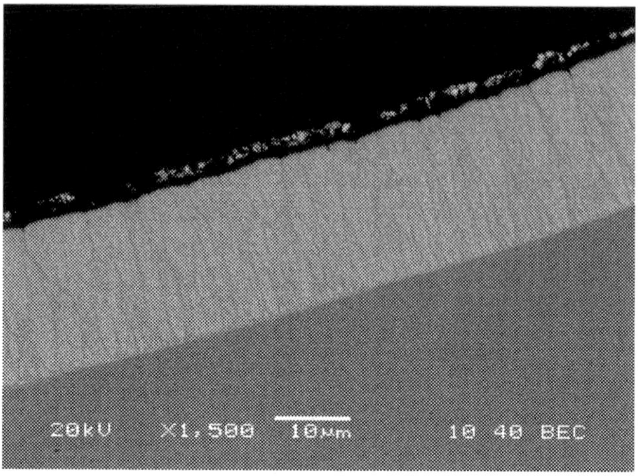

Figure 7.11 Columnar morphology of an electroless nickel-boron deposit.

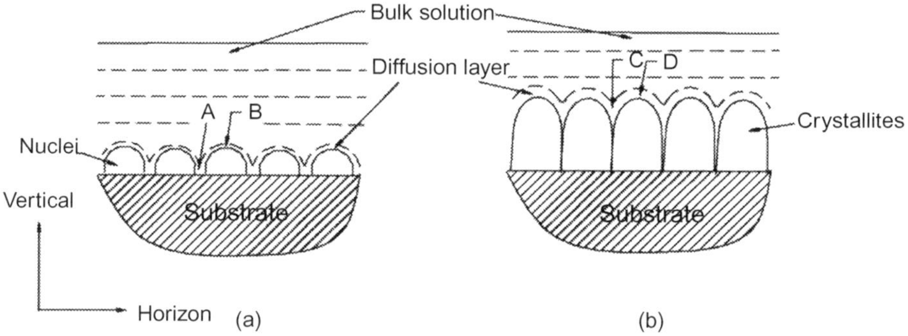

Figure 7.12 Diffusion layer on top of nickel-boron deposit during deposition and effect on the deposit formation.
Source: Rao et al. (2005), reproduced with permission.

due to the presence of other growing columns: the diffusion layer is thicker at the edge of columns, where it is less likely to be disrupted by convection, than on their tops, as shown on Figure 7.12. The growth of electroless nickel is thus quicker at the top of columns, which leads to the typical cauliflower-like surface.

The chemistry of electroless nickel deposits is not homogeneous. This is not surprising given their growth method: the coating is formed by a catalysed chemical reaction in a bath that can be considered (except for the injection of heat) as a closed system and is thus subject to reactive depletion (Vitry et al., 2012a) causing variations of plating with time (Vitry et al., 2012b). Moreover, the catalytic nature of the reaction and the means employed to avoid destabilization (the addition of a

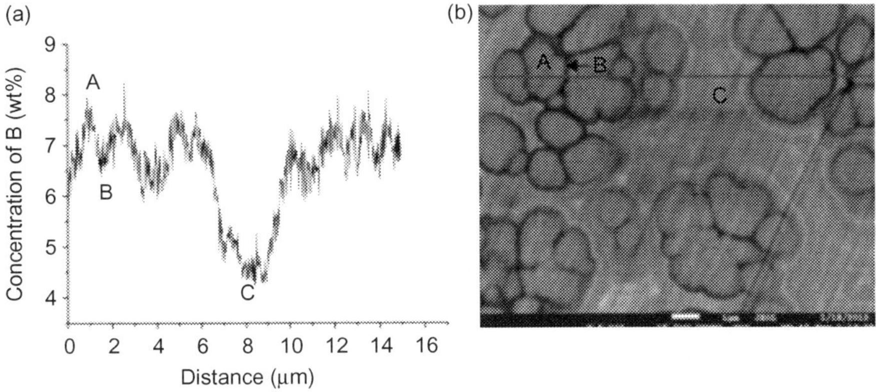

Figure 7.13 Local chemical variations inside and electroless nickel-boron coating.
Source: Pal et al. (2011), reproduced with permission.

stabilizer that acts as a catalytic poison and adsorbs on the substrate to block parts of the active sites) suggest variations of plating with position.

The chemical heterogeneity of electroless nickel-boron coatings was shown by Pal et al. (2011) who performed local analysis on the sub-surface of electroless nickel deposits, as shown in Figure 7.13. He was able to observe that the concentration of boron was higher in the middle of the columns than at their edges, with a very low concentration in the intercolumnar spaces. If related to the timeline of electroless coating formation, it appears that the parts formed earlier (i.e. the centre of columns) contain more boron than the parts formed later. This confirms the results of Vitry et al. (2012c) who observed variations of boron content measured by XPS on samples with various plating time.

7.3.3 Structure

The structure of electroless nickel-boron is obviously linked to their chemistry and thermal history, as is the case of all materials. Due to the important modifications brought by heat treatment, the structure of the electroless deposit before and after heat treatment will be examined separately.

7.3.3.1 Structure of as-deposited electroless nickel-boron

The structure of electroless nickel-boron in the as-deposited state is the result of the method of synthesis: the energy available during electroless deposition is not sufficient to allow reaching the equilibrium state. As a result, as-deposited electroless nickel-boron deposits are supersaturated solid solutions of boron in nickel (Vitry, 2010; Delaunois, 2002). The grain size of electroless nickel-boron decreases with increasing boron content and it has been accepted that it is

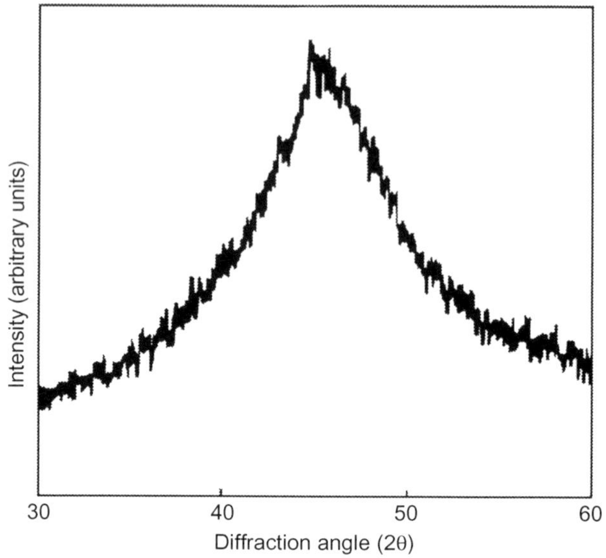

Figure 7.14 Typical X-ray diffraction pattern for an as-deposited electroless nickel-boron coating.
Source: Krishnaveni et al. (2005), reproduced with permission.

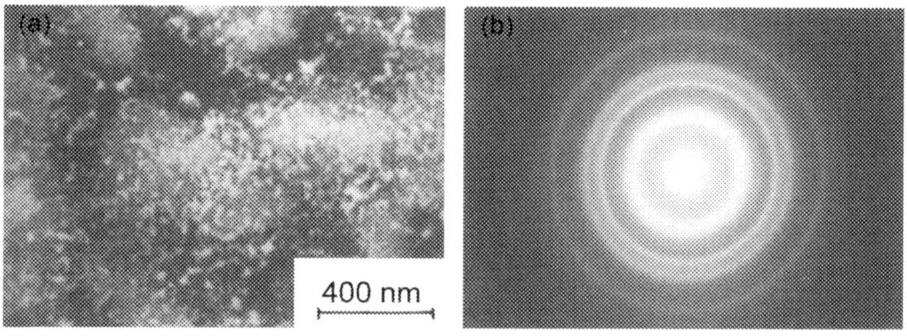

Figure 7.15 (a) TEM micrograph from electroless nickel-boron deposit with 4.32 wt% B, as deposited, parallel to the surface; (b) SAD pattern corresponding to micrograph presented in (a).
Source: Kumar and Nair (1994), reproduced with permission.

amorphous for boron content higher than 5 wt% (Krishnaveni et al., 2005; Riedel, 1991; Watanabe and Tanabe, 1983), as suggested by X-ray diffraction data (see Figure 7.14). However, several researchers have shown the existence of structural order in as-deposited electroless nickel-boron with high boron content by transmission electron microscopy (TEM) carried out parallel to the coating surface and on cross sections (Kumar and Nair, 1994; Vitry et al., 2012d), as shown respectively in Figures 7.15 and 7.16. A grain size close to 1 nm has been

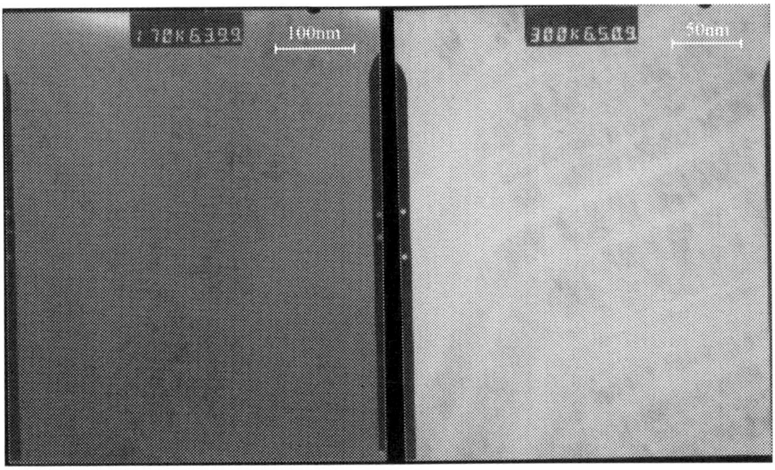

Figure 7.16 TEM micrographs from as-plated electroless nickel-boron deposit with 6 wt% B, in cross section.
Source: Vitry et al. (2012d), reproduced with permission.

determined for samples containing roughly 5 wt% boron (Vitry et al., 2012d). In conclusion, the grain size of nanocrystalline electroless nickel-boron deposits decreases with boron content and the grain size obtained for high boron content is so small that some would consider the coating amorphous, but it keeps low distance order.

7.3.3.2 Structure of heat treated electroless nickel-boron

Heat treatment of nanocrystalline electroless nickel-boron deposits can bring enough energy for the coating to revert to equilibrium state, which is, according to the nickel-boron phase diagram, either a mix of Ni and Ni_3B or Ni_3B and Ni_2B, depending on the coating chemistry (Delaunois and Lienard, 2002; Vitry, 2010; Hamid et al, 2010). However, as can be seen on Figure 7.17, nanocrystalline Ni, Ni_3B and Ni_2B can be found simultaneously in heat treated coatings (Delaunois and Lienard, 2002; Das and Sahoo, 2012), which is not predicted by the phase diagram. Pal et al. (2011) attributed this phenomenon to the heterogeneities in coating chemistry that are not supressed by the heat treatment due to its relatively low temperature.

Usually, heat treatment for electroless nickel-boron coating is carried out at 400°C for 1 hour, under either air or a neutral atmosphere (Delaunois and Lienard, 2002; Ziyuan et al., 2004). In these conditions, grain growth occurs, but in a limited manner and the coating stays nanocrystalline, as can be seen on Figure 7.18, with a grain size close to 40 nm (Vitry et al., 2012d).

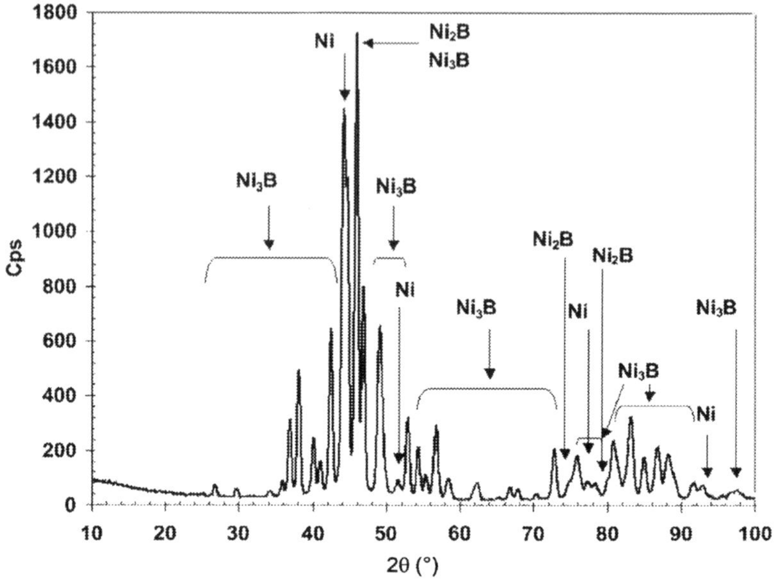

Figure 7.17 X-ray diffraction pattern for an heat treated (400°C, 1 hour) electroless nickel-boron coating.
Source: Delaunois and Lienard (2002), reproduced with permission.

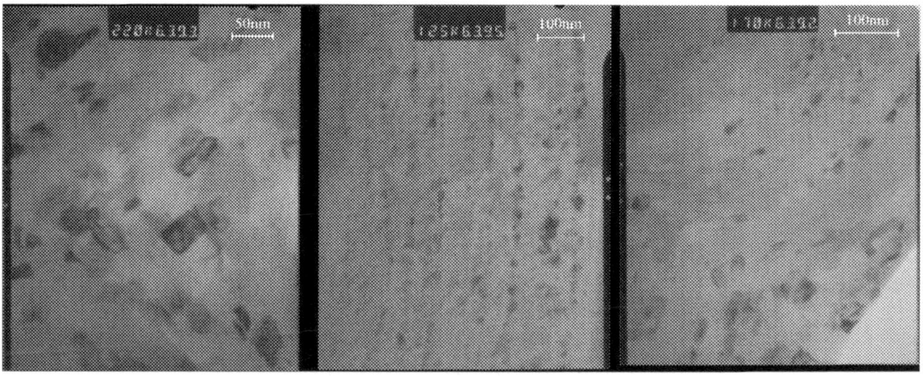

Figure 7.18 TEM micrographs from heat treated (400°C, 1 h) electroless nickel-boron deposit with 6 wt% B, in cross section.
Source: Vitry et al. (2012d), reproduced with permission.

7.4 Mechanical and wear properties of nanocrystalline electroless nickel-boron coatings

The mechanical and tribological properties of nanocrystalline electroless nickel-boron deposits are one of their main points of interest because they are usually destined for mechanical or wear protection applications. Below, their hardness, roughness, scratch test behaviour (which provides information about adhesion) and wear resistance will be discussed.

7.4.1 Hardness

It is well-known that nanocrystalline electroless nickel-boron coatings are harder than their phosphorous containing counterparts. In the as-deposited state, values of 700 hv_{100} and more are reported (Anik et al., 2008; Delaunois and Lienard, 2002; Dervos et al., 2004; Oraon et al., 2008; Pal et al., 2011; Vitry et al., 2008, 2011). Optimal heat treatment (usually at 400°C for one hour (Sahoo and Das, 2011)) brings an important increase in the hardness of the coating and values in the order of 1200 hv_{100} have been reported (Anik et al., 2008; Delaunois and Lienard, 2002; Dervos et al., 2004; Oraon et al., 2008; Pal et al., 2011; Vitry et al., 2008, 2011; Ziyuan et al., 2004).

However, the actual situation of knowledge about the hardness of electroless nickel-boron deposits is not so easily described.

Table 7.4 presents hardness values obtained by various authors, in the as-deposited and heat treated state, as well as the nature of the substrate and the composition of the coating (when available).

In the as-deposited state, values ranging from 500 to 1170 hv_{100} have been reported. It is, however, difficult to compare those values as there are variations in chemistry of the coating (and thus of grain size), substrate material and hardness measurement method. The best reported results do not mention any information about substrate material or coating chemistry, which makes them difficult to compare to others. To get a better idea of the influence of coating chemistry, several authors, including Anik et al. (2008), have studied the effect of boron content on hardness of electroless nickel-boron deposits. As shown in Figure 7.19, hardness of nanocrystalline electroless nickel-boron coatings increases with increasing boron content at least up to 8 wt% boron.

It is possible, using the nanoindentation method, to measure hardness profiles across the nickel-boron coating. As can be seen on Figure 7.20, the hardness is stable across the deposit in the as-deposited and heat treated state. This means that eventual local variations of chemistry, structure and microstructure in the coating do not have a significant influence on the measure of hardness.

Hardness values after heat treatment vary from 790 to 1760 hv_{100}. It is once again difficult to compare those results because they were obtained using various substrates, chemistries and measurement methods, as well as various heat treatments (which induce differences in grain size inside the nanocrystalline coating).

The lowest value (Contreras et al., 2006) was obtained for a coating that contains unusually high boron content, and that was treated at a rather low temperature. The highest values were obtained by an unusual heat treatment (Dervos et al., 2004), an unknown measurement method (Pal et al., 2011), or on coatings for which the boron content is unknown (Oraon et al., 2008; Ziyuan et al., 2004), which make them difficult to compare to others. It is, however, possible to affirm that values

Table 7.4 **Hardness, in as-plated and heat treated conditions, of various electroless nickel-boron coatings**

Reference	As-deposited hardness	Heat treatment	Heat treated hardness	Coating composition	Substrate
Vitry et al., 2008	854 hv_{100}	400°C, 1 h, Ar (5%H_2)	1302 hv_{100}	6 wt% B	Aluminium with Ni–P protective layer
Vitry et al., 2011	891 hk_{25}	400°C, 1 h, Ar (5%H_2)	1229 hk_{25}	6 wt% B–1 wt% Pb	Mild steel
Dervos et al., 2004	1000 hv_{100}	5 min, 850°C, high vacuum	1760 hv_{100}	5 wt% B	Carbon steel
Ziyuan et al., 2004		350°C, 2 h	1400 hv_{30}		Copper (>99%)
Oraon et al., 2008	1171 hv_{100}	300°C, 1 h	1402 hv_{100}		
Delaunois and Lienard, 2002	793 hk_{25}	400°C, 1 h	1133 hk_{25}	4 wt% B–6.9 wt% Tl	Aluminium alloys (1050 and AS7G06)
Delaunois and Leinard, 2002	726 hk_{25}	400°C, 1 h	1152 hk_{25}	5.75 wt% B–1.25 wt% Pb	Aluminium alloys (1050 and AS7G06)
Sankara Narayanan et al., 2003	570 hv_{100}	450°C, 1 h	908 hv_{100}	6.5 wt% B–0.3 wt% Tl	Mild steel
Contreras et al., 2006	520–535 hv_{100}	300°C, 10 h	790 hv_{100}	8.35 wt% B	Mild steel
Anik et al., 2008	764 hv_{100}	540°C, 1 h	1122 hv_{100}		Low carbon steel and stainless steel (AISI 304)
Pal et al., 2011	7.5 Gpa	300°C, 5 h	15 Gpa	6 wt% B–1 wt% Pb	Mild steel
Hamid et al., 2010	500 hv_{50}	400°C, 2 h	1000 hv_{50}		Low carbon steel and pure copper
Correa et al., 2012	616 hv_{50}				AZ91D Magnesium alloy
Srinivasan et al., 2010	682 hv_{50}	400°C, 1 h	1102 hv_{50}		Copper and mild steel

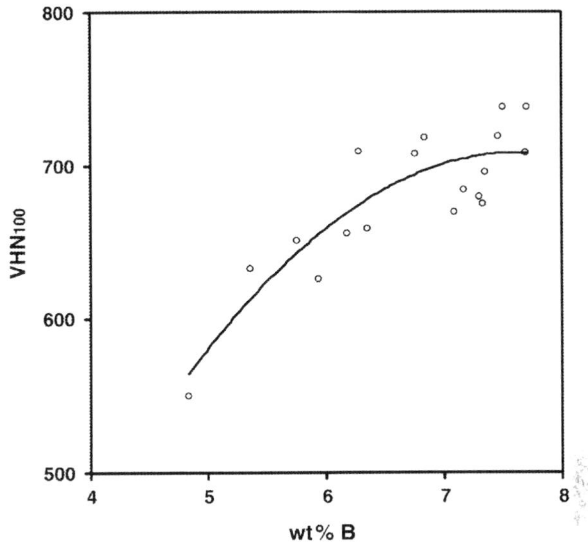

Figure 7.19 Effect of boron content on hardness of electroles nickel-boron coatings.
Source: Anik et al. (2008), reproduced with permission.

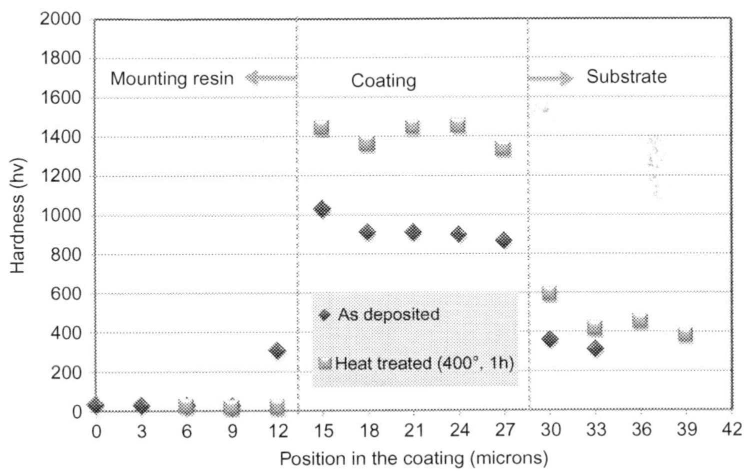

Figure 7.20 Nanohardness profile on as-plated and heat treated electroless nickel-boron coatings.
Source: Vitry et al. (2011), reproduced with permission.

between 1100 hv_{100} and 1300 hv_{100} are usual for heat treated nanostructured nickel-boron coatings.

To get real information about the influence of heat treatment on hardening of nanocrystalline electroless nickel-boron coatings, it is easier to use results obtained by a single group of researchers, usually working with a single nickel-boron chemistry. Some of the more recent and complete results about the influence of heat treatment on hardness of electroless nickel-boron were obtained by Pal et al. (2011) and are presented on Figure 7.21. From this it appears that, for their electroless nickel-boron chemistry (6 wt% B, 1 wt% Pb), the best temperature (for 1 hour heat treatment) is close to 350°C, while the optimal heat treatment time at 300°C is 5 hours. Experiments carried out on a similar chemistry by Delaunois and Lienard (2002) are in accordance with this.

The reasons for the hardening of electroless nickel-boron due to heat treatment are not completely elucidated. It is obvious that the important structural reorganization

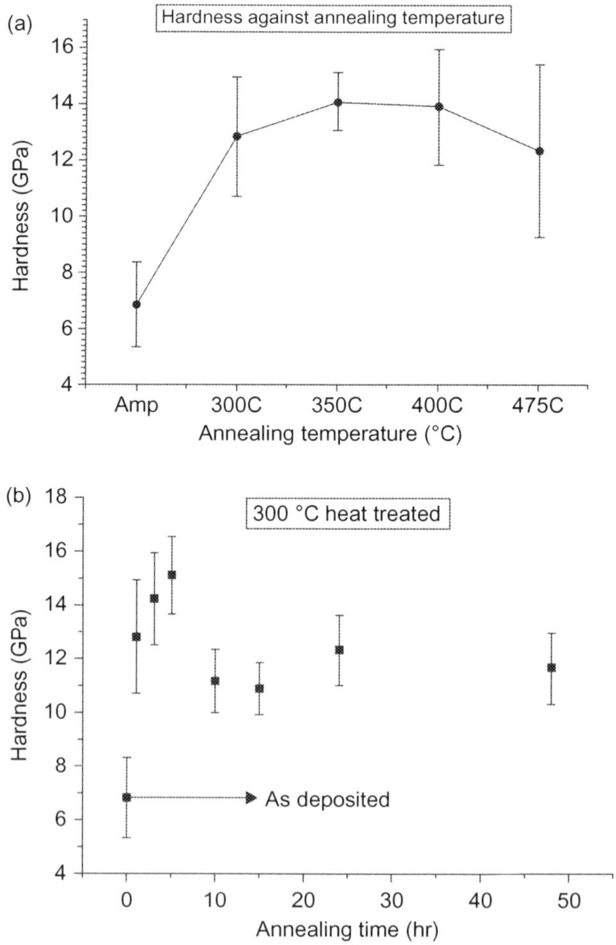

Figure 7.21 Influence of heat treatment time and temperature on hardness of electroless nickel-boron coatings.
Source: Pal et al. (2011), reproduced with permission.

induced by heat treatment plays a prominent role in hardening. However, several sources (Anik et al., 2008; Evans and Schlesinger, 1994; Oraon et al., 2008; Riedel, 1991; Sankara Narayanan and Seshadri, 2004; Ziyuan et al., 2004) attribute hardening to the creation of grain boundaries in an originally amorphous material, and thus to the Hall–Petch effect. This hypothesis is not the best for the following reason: electroless nickel-boron coatings are not fully amorphous in the as-deposited state and even if their initial grain sizes are small enough to allow inverse Hall–Petch effect (Meyers et al., 2006; Suryanarayana and Koch, 2000; Tjong and Chen, 2004), it would not explain such an important hardening effect. However, the Hall–Petch effect can be used as an explanation for the softening observed in coatings that are treated at high temperature for relatively long times, leading to an important increase in grain size (sometimes outside of the nano range).

One potential explanation of hardening that is rarely advocated is the phase transformation occurring during heat treatment: as-deposited electroless nickel-boron is a nanostructured supersaturated solid solution of boron in face-centred cubic nickel, which is a rather soft phase. After heat treatment, boride phases such as Ni_3B and Ni_2B are present. The structure of Ni_3B is orthorhombic and is very similar to that of cementite (Fe_3C iron carbide). It can thus be easily extrapolated that, in the same way that carbon allows important hardening of iron by the transformation of (part of) the soft cubic metallic phase in hard iron carbide, the addition of boron in nickel would harden the latter by transformation of part of the soft face-centred cubic nickel into then hard orthorhombic nickel-boride phase (Vitry, 2010).

7.4.2 Roughness

Roughness is an essential feature for wear applications of nanocrystalline electroless nickel-boron coatings. Those coatings are reputed to be relatively smooth, which is confirmed by results from different groups. Shaffer and Rogers (2007) reported Ra values of 0.62 μm for a NanoCEM™ electroless nickel-boron coating with a thickness of 14–16 μm.

Dervos et al. (2005) obtained values of 0.21 μm for Ra, 1.9 μm for Rt and 1.2 μm for Rz on a substrate with initial values of 0.58, 4.7 and 3.8 μm, respectively.

Das and Sahoo (2009) studied the influence of plating parameters on roughness of the coating with the aim of reducing roughness. They chose the following parameters: Ra, Rq, Rsk and Rku, Rsm, and were able to show that concentration of reducing agent and bath temperature were the most important parameters for the optimization of roughness. However, their study was carried out at the lab scale and did not take into account important parameters such as substrate roughness and ageing of the bath.

Bath ageing is usually accompanied by important modifications to coating properties, including an increase of the roughness (Riedel, 1991).

While it is well-known that nanocrystalline electroless nickel-boron coatings are usually smooth, it is less well-known that their roughness is very much influenced by the roughness of the substrate, due to the growth mechanism of the coating (Clerc, 1986): the columns that grow on the substrate are locally perpendicular to the surface. Consequently, they'll only be parallel if the substrate is already smooth.

The influence of substrate roughness on coating roughness has been confirmed by Vitry (2010): as shown on Figure 7.22, the magnitude of substrate average roughness Ra is conserved after plating. This means that electroless nickel-boron will be able to maintain the shape and state the plated parts. However, a study of only average roughness is not sufficient to fully characterize the roughness of nanocrystalline electroless nickel-boron coatings. For this reason, several other parameters are shown on Figure 7.22 as well: the quadratic average (Rq), maximum peak to valley distance (Rz), peak height (Rp), valley depth (Rv) and total roughness (Rt). All those parameters follow a similar trend, which is to decrease with the initial average roughness.

Other roughness parameters, more specific to wear applications were also followed as a function of substrate roughness: skewness and kurtosis. Skewness is the third central moment of profile amplitude probability density function over the assessment length. This parameter allows measuring the symmetry of the roughness profile and is sensible to occasional important variations such as high peaks and deep valleys. A profile with a similar amount of peaks and valleys will have a null skewness, while a profile with deep grooves will have a negative value and one with high peaks or filled valleys will present a positive skewness. A negative skewness is supposed to be more favourable to wear applications. Kurtosis is the fourth

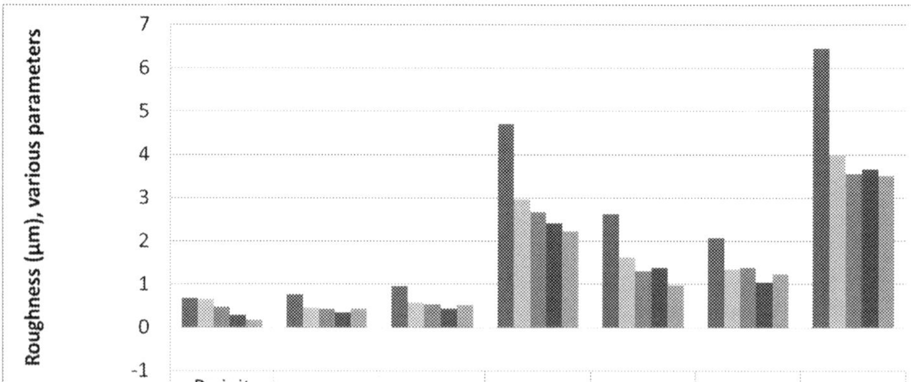

	Ra init (μm)	Ra (μm)	Rq (μm)	Rz (μm)	Rp (μm)	Rv (μm)	Rt (μm)
Ra init = 0.67	0.67	0.76	0.96	4.70	2.62	2.08	6.46
Ra init = 0.65	0.65	0.45	0.57	2.96	1.62	1.34	3.98
Ra init = 0.48	0.48	0.42	0.54	2.67	1.30	1.38	3.55
Ra init = 0.29	0.29	0.34	0.44	2.41	1.38	1.03	3.66
Ra init = 0.18	0.18	0.43	0.52	2.22	0.97	1.25	3.51

Figure 7.22 Influence of initial substrate roughness on electroless nickel-boron deposit roughness parameters.
Source: Adapted with permission from Vitry (2010).

central moment of profile amplitude density function over the assessment length. It is used to describe sharpness of the profile. Kurtosis values under 3 are related to profiles with few high peaks and low valleys (platykurtoic) and values over 3 are related to profiles with many high peaks and valleys (leptokurtoic).

As can be seen in Figure 7.23, skewness and kurtosis values vary with initial average roughness in a similar manner. The smoother substrate provides a better surface state to the coating due to the negative skewness (more valleys than peaks) that is expected for the cauliflower-like surface texture, and to the low kurtosis.

On initially rougher substrates, the skewness and kurtosis values are less favourable but it is probably due to the presence of a few sharp peaks on an otherwise cauliflower-like coating.

Roughness of nanocrystalline electroless nickel-boron coatings is also influenced by heat treatment. This does not come as a surprise as those treatments are accompanied by important structural modifications in the coating. As can be seen on Figure 7.24, heat treatment of electroless nickel-boron deposits appears to be accompanied by a decrease in coating roughness, which is more important for heat treatments at higher temperatures. This is in accordance with the results of Dervos et al. (2005) who observed a reduction from 0.21 to 0.20 μm for Ra and 1.9 to 1.7 for Rt after a treatment at 850°C for 5 minutes under vacuum.

7.4.3 Adhesion and scratch testing behaviour

To be able to provide good protection against wear and other aggressions, coatings need to have a good adhesion to the substrate (Sahoo and Das, 2011). The measure

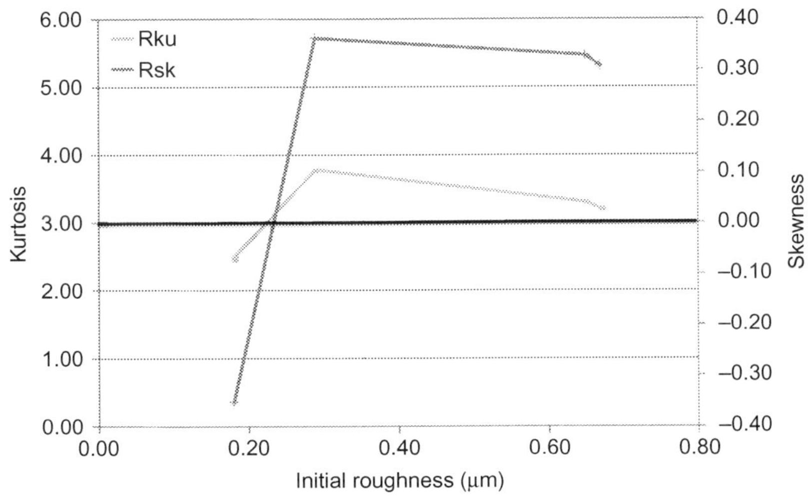

Figure 7.23 Influence of initial substrate roughness on skewness and kurtosis of electroless nickel-boron deposit.
Source: Adapted with permission from Vitry (2010).

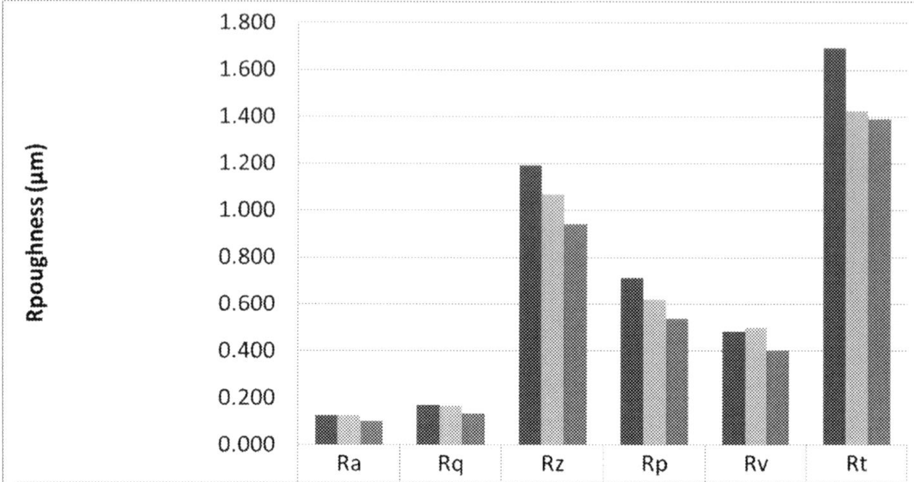

	Ra	Rq	Rz	Rp	Rv	Rt
15 µm as deposited	0.125	0.170	1.193	0.712	0.481	1.693
15 µm treated 180°C, 4 h	0.127	0.167	1.069	0.619	0.499	1.425
15 µm treated 400°c, 1 h	0.101	0.135	0.941	0.537	0.405	1.391

Figure 7.24 Influence of heat treatment on roughness parameters of electroless nickel-boron coatings.
Source: Adapted with permission from Vitry (2010).

of this adhesion for electroless nickel, and particularly for electroless nickel-boron, has for many years been, and is still to a certain extent, an unsolved problem. The most popular adhesion methods, such as pull-off test, ring-shear test, peel test and three-point bend test provide only semi-quantitative or qualitative results (Sahoo and Das, 2011). Moreover, the strong metal-to-metal bonding formed during electroless deposition makes those coatings more adherent than electrodeposited coatings, and their adherence is usually outside the range of the testing methods. For instance, Shaffer and Rogers (2007) measured adhesion of electroless nickel-boron coating with Test Method B of ASTM D3359−02, usually called the 'tape test' method and did not observe any failure. Delaunois (2002) carried out an adhesion test by gluing 2 Nickel-boron plated samples together with Araldite™ and performing a three-point bend test on the system. She observed failure in the glue and was thus unable to quantify adhesion of the coating to the substrate.

Adhesion of electroless nickel-boron must thus be observed by indirect methods, such as scratch testing. The principle of this method is the following: an indenter similar to that of a Rockwell hardness tester is used to generate a scratch on the coated system with a load that is either fixed or continuously increasing. When the continuous load increase is used, it is possible to observe and identify damage in the coated system and to link it to a critical load at which this damage occurs by measuring the distance between the beginning of the scratch and the observed damage (Delaunois, 2002; Vitry et al., 2011).

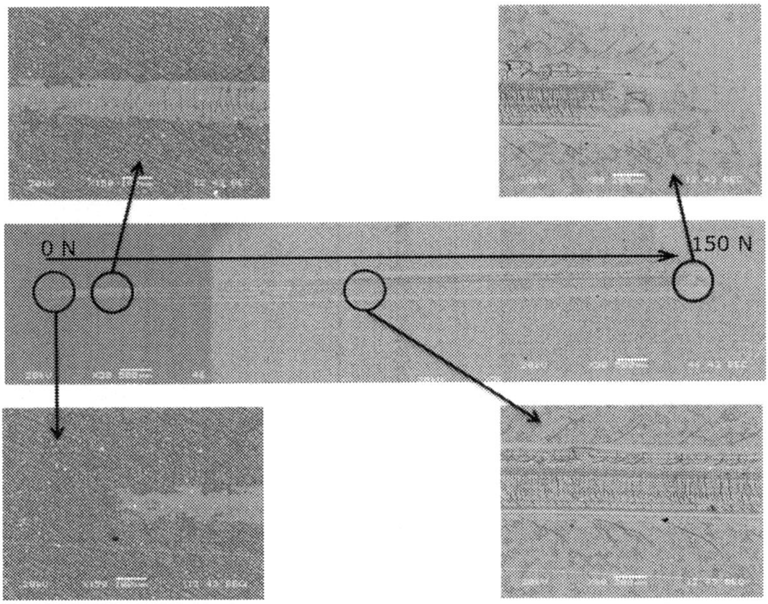

Figure 7.25 SEM images of scratch test on 25 μm thick as deposited nanostructured electroless nickel coating.
Source: Vitry et al. (2011), reproduced with permission.

As can be seen on Figure 7.25, damage induced by scratch testing up to 150 N on as-deposited coating (thickness = 25 μm) occurs as transversal intercolumnar cracks from 10 N (0.6 mm from the beginning of scratch), then as parallel intercolumnar cracks from 55 N (3.6 mm), which are cohesive failures in the coating (due to the low cohesion between adjacent columns that grew simultaneously but independently).

This is confirmed by Figure 7.26 where details of the end point of a scratch carried out up to 150 N on a 15 μm thick coating is shown: there is no evidence of adhesive failure.

Complete failure of the coating does not occur for 25 μm thick coatings but it can be observed for thinner coatings. In this case, the failure mechanism is ductile perforation, without evidence of adhesive failure (Vitry et al., 2011). This is in concordance with results carried out on other soft substrate–hard coating systems by Sander et al. (2011).

After heat treatment at 400°C (see Figure 7.27), damage following scratch testing with an increasing load up to 150 N on nanocrystalline electroless nickel-boron coatings takes the following form: intercolumnar chevron cracks at the edge of the scratch, then at its bottom, from 16 N (1.15 mm from the start of the scratch), followed by the appearance of much deeper cracks at the very end of the testing (from 143 N, or 9.7 mm). However, no adhesive failure can be observed. Those failure modes (plastic deformation and through thickness coating cracking) are in accordance with S.J. Bull's model (Bull, 1997) for a soft substrate/hard coating system.

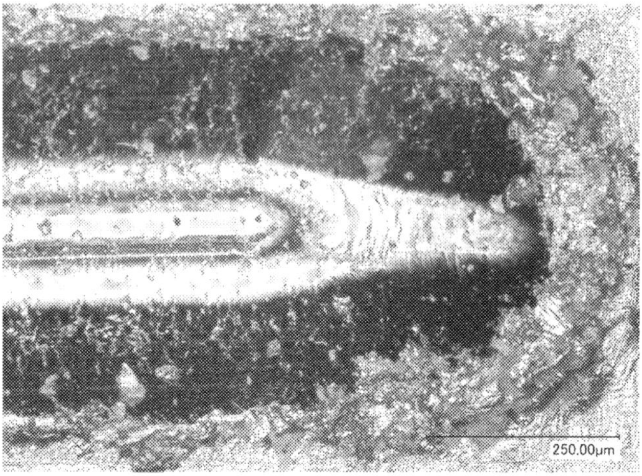

Figure 7.26 Digital image of detail (end point) of scratch test on 25 μm thick as-deposited nanostructured electroless nickel coating.

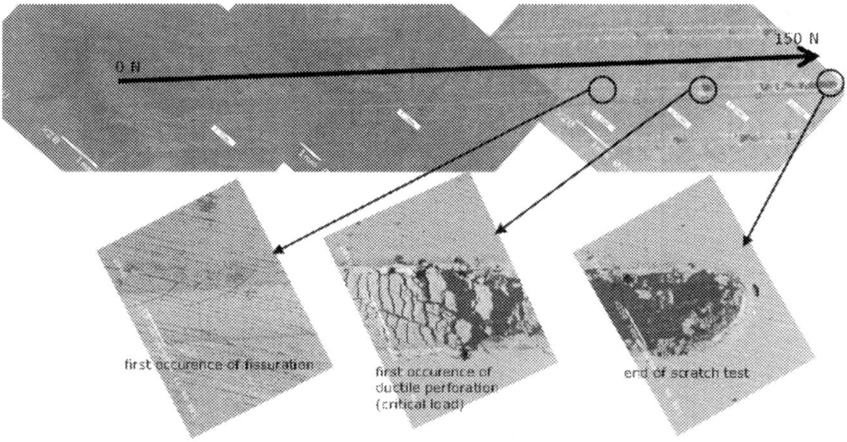

Figure 7.27 SEM images of scratch test on 25 μm thick heat treated (400°C, 1 h) nanostructured electroless nickel coating.
Source: Vitry et al. (2011), reproduced with permission.

7.4.4 Wear resistance

Nanocrystalline electroless nickel-boron coatings are usually found more suitable for wear application than their phosphorous containing counterparts because their hardness is already high in the as-deposited condition but can still be improved by a suitable heat treatment (Sahoo and Das, 2011). Moreover, their columnar structure allows retention of lubricants in adhesive wear conditions (Srinivasan et al., 2010) and the smoothness and cauliflower-like texture are favourable because they allow for reduced friction coefficient. It is suggested that Ni-B coated aluminium or magnesium could replace similarly plated steel in applications where only wear resistance is needed (Correa et al., 2013; Kanta et al., 2009; Sahoo and Das, 2011).

7.4.4.1 *Friction coefficient*

The friction coefficient of nanocrystalline electroless nickel-boron coatings is low compared to other hard coating materials such hard chrome of electroless nickel-phosphorous. The stability of the friction coefficient during sliding wear test has been observed by Krishnaveni et al. (2005).

Riddle (2005) measured friction coefficients by the ASTM-D2714 test, under lubrication and reported a value of 0.09 for nickel-boron, 0.14 for electroless nickel-phosphorous and 0.16 for hard chrome.

Shaffer and Rogers (2007) compared the friction coefficient of various coatings in pin on side geometry and self-mated contact. They obtained friction coefficients between 0.2 and 0.26 for electroless nickel-boron and at moderate speed (37.8 cm/s) and close to 0.45 for high speed (358 cm/s).

Friction coefficienst were recorded during sliding wear tests, with varying loads and heat treatments (Krishnaveni et al., 2005) and a value of 0.74 was obtained for a load of 20 N in the as-deposited condition, with a decrease down to 0.68 after heat treatment.

Das and Sahoo (2009) optimized the plating parameters to reduce the friction coefficient of nanocrystalline electroless nickel-boron coatings in dry, non-lubricated conditions. They obtained in their initial conditions values of 0.396 and were able to reduce the friction coefficient down to 0.129. They were able to identify concentration of the reducing agent as the most important parameter.

7.4.4.2 *Abrasive wear resistance*

Abrasive wear resistance of electroless nickel-boron has been investigated by several authors, using the Taber abrasion test. Taber Wear Index values are given in Table 7.5.

Taber wear test results can be difficult to compare because the results are influenced as much (if not more) by test conditions, such as the type of abrasive wheels that were used and the test load, than by the nature of the tested substrate.

For example, results obtained by Vitry's group on similar coatings but with different wheels (Vitry et al., 2011, 2012b) provided very different results, with a TWI twice as high with CS-17 wheels as was obtained with CS-10 wheels.

It is likewise very difficult to compare results obtained for electroless nickel-boron with other coatings because they are rarely carried out in similar conditions. However, Hari Krishnan et al.'s (2006) study helps with this as they tested both kind of coatings and observed a much better behaviour for electroless nickel-boron. Nevertheless, this study does not include information about the chemistry of coatings, which is an important parameter.

Other tests were used to characterize abrasive wear of electroless nickel-boron coatings. For instance, Riddle and Bailer (2005) used the dry sand, rubber wheel

Table 7.5 **Abrasion resistance (Taber Wear Index), in as-plated and heat treated conditions, of various electroless nickel-boron coatings**

Reference	Vitry et al., 2011	Sudagar et al., 2013	Srinivasan et al., 2010	Vitry et al., 2012d	Vitry et al., 2012d	Hari Krishnan et al., 2006	Hari Krishnan et al., 2006	Morcos and Barnstead, 2010	Morcos and Barnstead, 2010
Test conditions	CS-17 wheels, 72 rpm, 1 kg, 10 000 cycles		1 kg, 2000 cycles	CS-10 wheels, 72 rpm, 1 kg, 10 000 cycles	CS-10 wheels, 72 rpm, 1 kg, 10 000 cycles	CS-10 wheels, 10 N	CS-10 wheels, 10 N	CS-10 wheels, 100 g	CS-10 wheels, 100 g
As-deposited TWI	28.5 ± 3.5	8	5.45	11.4 ± 0.5	23.7 ± 0.7	9.0	17.0	7.0–9.0	3.0–10.0
Heat treatment	400°C, 1 h		400°C, 1 h	400°C, 1 h	400°C, 1 h	400°C, 1 h	500°C, 1 h		
Heat treated TWI	13 ± 1.5	3	3.1	6.0 ± 0.2	13.9 ± 0.9	3.0	6.0		
Boron content	6 wt%–1 wt% Pb	Medium B	4.94% B	6 wt%–1 wt% Pb	Ni-B-Tl		Ni–P	Low boron	High boron

test, as described in ASTM G65 procedure B and obtained an adjusted mass loss of 0.387 for electroless nickel-boron against 1.02 for hard chrome and 1.025 for low-phosphorous electroless nickel.

In all cases, researchers agree that electroless nickel-boron deposits have a very good abrasive wear resistance.

7.4.4.3 Sliding wear resistance

Apart from abrasive wear, which was described in the previous section, several groups submitted electroless nickel-boron to various other wear tests, mainly to measure sliding wear.

Das and Sahoo (2009) used plate-on roller test and observed limited wear of the surface, mainly in the form of abrasive wear. Riddle and Bailer (2005) used the Falex test and observed a good resistance of the coating.

Krishnaveni et al. (2005) used pin-on-disc tests and observed torn patches in as-deposited coatings related to loss of adhesion of the coating. However, on heat-treated coatings, they observed very different behaviour: a bright and smooth finish with the presence of fine grooves along the sliding direction. This was accompanied by a decrease of specific wear rate by a factor of 2. They attributed those differences to variations in adhesion between the tested sample (pin) and the mild steel counter disc and to the hardening of the coating brought about by heat treatment that decreases the true area of contact.

Ahmadkhaniha and Mahboubi (2012) obtained similar results and managed to further reduce the specific wear rate through the use of plasma nitriding. Similarly, Sankara Narayanan et al. (2003) showed a good resistance of electroless nickel-boron, improved by heat treatment. They further reduced the wear rate by using duplex Ni–P/Ni–B coatings and showed that the sliding wear resistance of electroless nickel-boron was better than that of electroless nickel-phosphorous.

Riddle and Bailer (2005) also used the pin-on-disc test but they compared electroless nickel-boron to other materials. They were able to show a better performance of nickel-boron in reciprocal wear than that of hard chrome and electroless nickel-phophorous.

Correa et al. (2013) used the ball-on-disc test to measure the sliding wear resistance of nanocrystalline electroless nickel-boron coatings on magnesium alloys and obtained wear rates close to 7.34×10^{-11} kg N^{-1} m^{-1}, with little visible evidence of wear.

7.5 Corrosion resistance

Corrosion resistance is not the main objective of nanocrystalline electroless nickel-boron coatings. However, as parts are often submitted to aggressive environments, it is interesting to get an overview of their corrosion resistance properties.

Several authors investigated the corrosion resistance of electroless nickel-boron by potentiodynamic polarization, mostly in 3.5% NaCl solution. Results from recent investigations are presented in Table 7.6. The best results were obtained by Vitry (2010) and Das and Sahoo (2011a), with a corrosion potential close to −150 mV Vs SHE. Those results are better than reported values for electroless nickel-phosphorous coatings. However, most authors obtain corrosion resistance lower than that of electroless nickel-phosphorous coatings (Sahoo and Das, 2011; Sankara Narayanan and Seshadri, 2004; Sudaghar et al., 2013). Some authors also used electrochemical impedance spectroscopy studies (Das and Sahoo, 2011b; Kanta et al., 2009, 2010a,b; Sankara Narayanan and Seshadri, 2004; Srinivasan et al., 2010).

Dervos et al. (2004) measured corrosion rates in 3.5% NaCl and obtained 0.058 mm/year for as-deposited coatings and 0.135 mm/year for coatings submitted to a special vacuum heat treatment.

The effect of heat treatment on the corrosion resistance of nanocrystalline electroless nickel-boron coatings is usually accepted as being detrimental (Anik et al., 2008; Contreras et al., 2006; Sankara Narayanan and Seshadri, 2004) due to the creation of grain-boundaries in the coating. However, some studies have shown an improvement of corrosion resistance of electroless nickel-boron deposits after heat treatment (Das and Sahoo, 2011b; Hamid et al., 2010; Kanta et al., 2009), as can be seen on Figure 7.28.

Some researchers sought to optimize corrosion resistance, with success (Das and Sahoo, 2011a,b) and others studied the effect of various parameters on this property (Anik et al., 2008).

The following corrosion mechanisms were proposed: corrosion of the intercolumnar zones, which have a lesser density and a slightly different chemistry compared to the bulk of the coating (Das and Sahoo, 2011b; Kanta et al., 2010a) and formation of galvanic coupling due to chemical heterogeneities in the coating (Crobu et al., 2008).

7.6 Conclusion

The properties of nanostructured electroless nickel-boron coatings have been observed, mainly in view of their mechanical applications. The most relevant values are summarized in Table 7.7.

Those coatings present an amorphous to nanocrystalline structure in their as-deposited state but are always nanocrystalline (with variable grain size) after heat treatment at 400°C. They present interesting properties in view of abrasion resistance: high hardness (up to 1200 hk_{25}), low Taber Wear Index and favourable surface texture (with a typical cauliflower-like shape).

Table 7.6 **Corrosion potential and corrosion current measured by potentiodynamic polarization on various as-plated and heat treated electroless nickel-boron coatings**

Reference	Material	Reference electrode	Medium	E_{corr} (mV)	E_{corr} (mV agains SHE)	I_{corr} ($\mu A/cm^2$)
Sankara Narayanan and Seshadri, 2004	Ni–B as-plated	SCE	3.5% NaCl	−516	−292	9.15
Sankara Narayanan and Seshadri, 2004	Ni–B heat treated (400°C)	SCE	3.5% NaCl	−520	−296	14.60
Sankara Narayanan and Seshadri, 2004	Ni–P as-plated (9% P)	SCE	3.5% NaCl	−354	−130	3.62
Sankara Narayanan and Seshadri, 2004	Ni–P heat treated (9% P)	SCE	3.5% NaCl	−492	−268	6.89
Dervos et al., 2004	Ni–B as-plated	SCE	3.5% NaCl	−536	−312	4,858
Dervos et al., 2004	Ni–B vacuum heated at 850°C	SCE	3.5% NaCl	−483	−259	11.27
Vitry, 2010	Ni–B as plated	Ag/AgCl (KCl Sat'd)	3.5% NaCl	−335	−138	0.09
Vitry, 2010	Ni–B heat treated (400°C)	Ag/AgCl (KCl Sat'd)	3.5% NaCl	−240	−43	0.178
Srinivasan et al., 2010	Ni–B as-plated	SCE	3.5% NaCl	−568	−344	13.2
Srinivasan et al., 2010	Ni–B heat treated (400°C)	SCE	3.5% NaCl	−576	−352	14.7
Das and Sahoo, 2011a	Ni–B initial	SCE	3.5% NaCl	−381	−157	5.04
Das and Sahoo, 2011a	Ni–B optimized	SCE	3.5% NaCl	−275	−51	0.11

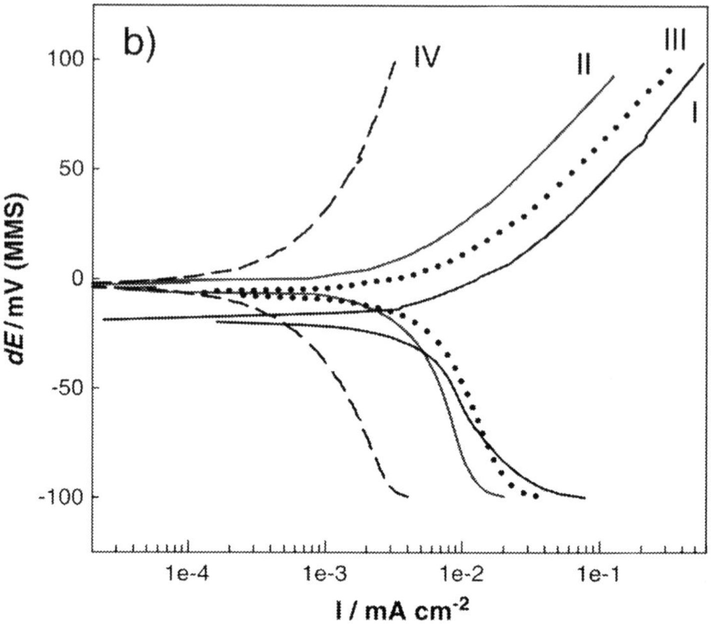

Figure 7.28 Polarization Tafel plot of steel coated with electroless nickel-boron: (I) substrate; (II) as-plated coated sample; (III) coated sample after heat treatment at 250°C; (IV) coated sample after heat treatment at 400°C.
Source: Hamid et al. (2010) reproduced with permission.

Table 7.7 Summary of properties of electroless nickel-boron coatings

	As plated	**After heat treatment at 400°C**
Structure	Nanocrystalline (grain size decreases with increasing B content)	Nanocrystalline
Phases observed	Nickel	Ni, Ni_3B, Ni_2B
Grain size (6 wt% B)	1 nm or less	~40 nm
Hardness	700 to 900 hk_{25}	1100 to 1200 hk_{25}
TWI (CS-10 wheels, 1 kg)	23.7 (Ni-B-Tl) to 11.4 (Ni-B-Pb)	13.9 (Ni-B-Tl) to 6 (Ni-B-Pb)
Friction coefficient (lubricated, ASTM-D2714)	0.09	
Friction coefficient (unlubricated)	0.12 to 0.8 according to conditions	
Specific wear rate (ball on disc)	7.34×10^{-11} kg N^{-1} m^{-1}	
Corrosion potential	−50 to −350 Vs SHE	−50 to −350 Vs SHE

References

Agarwala, R.C., Agarwala, V., 2003. Electroless alloy/composite coatings: a review. Sadhana. 28 (3–4), 475–493.

Ahmadkhaniha, D., Mahboubi, F., 2012. Effects of plasma nitriding on properties of electroless Ni-B coating. Surf. Eng. 28 (3), 195–198.

Alexis, J., 2000. Substitution des traitements de surface des alliages d'aluminium à base de chrome hexavalent en solution: Adhérence de revêtments d'alliages Zn-Ni, PhD thesis, Ecole Nationale d'Ingénieurs de Tarbes.

Anik, M., Korpe, E., Sen, E., 2008. Effect of coating bath composition on the properties of electroless nickel-boron films. Surf. Coat. Technol. 202, 1718–1727.

Baskaran, I., Kumar, R.S., Sankara Narayanan, T.S.N., Stephen, A., 2006. Formation of electroless Ni–B coatings using low temperature bath and evaluation of their characteristic properties. Surf. Coat. Technol. 200, 6888–6894.

Bielinski, J., Krol, J., Bielinska, A., 1990. Inorganic additives in the solution for electroless nickel-boron alloy deposition. Bull. Electrochem. 6, 828–831.

Brenner, A., Riddel, G., 1956. Nickel plating on steel by chemical reduction. J. Res. Natl. Bur. Stand. 37, 31–34.

Bulbul, F., Altun, H., Kucuk, O., Ezirmik, V., 2012. Tribological and corrosion behaviour of electroless Ni-B coating possessing a blackberry like structure. Met. Mater. Int. 18 (4), 631–637.

Bull, S.J., 1997. Failure mode maps in the thin film scratch adhesion test. Tribol. Int. 30 (7), 491–498.

Chang, Q.H., Chang, Z.Z., Bai, Y.T., Chen, X.Y., Peng, D., 2011. The effect of NaOH on the surface of magnesium alloy Ni-B alloy plating properties. Adv. Mater. Res. 189–193, 347–350.

Charbonnier, M., Romand, M., Goepfert, Y., 2006. Ni direct electroless metallization of polymers by a new palladium-free process. Surf. Coat. Technol. 200, 5028–5036.

Cheong, W.J., Luan, B.L., Shoesmith, D.W., 2004. The effects of stabilizers on the bath stability of electroless Ni deposition and the deposit. Appl. Surf. Sci. 229, 282–300.

Clerc, M.A., 1986. Morphologie, propriétés tribologiques et rhéologiques des alliages nickel-bore amorphes obtenus par voie chimique. PhD thesis, Université de Besançon.

Collin, R., Murphy, B., Zabrocky, S., 2010. Surface conditioning of ferrous alloys prior to electroless. In: Proceedings of En conference 2010, 8–9 November, Charleston, SC.

Contreras, A., León, C., Jimenez, O., Sosa, E., Pérez, R., 2006. Electrochemical behavior and microstructural characterization of 1026 Ni-B coated steel. Appl. Surf. Sci. 253 (2), 592–599.

Correa, E., Zuleta, A.A., Sepúlveda, M., Guerra, L., Castaño, J.G., et al., 2012. Nickel–boron plating on magnesium and AZ91D alloy by a chromium-free electroless process. Surf. Coat. Technol. 206, 3088–3093.

Correa, E., Zuleta, A.A., Guerra, L., Gomez, M.A., Castaño, J.G., et al., 2013. Tribological behavior of electroless Ni-B coatings on magnesium and AZ91D alloy. Wear. 305 (1–2), 115–123.

Crobu, M., Scorciapino, A., Elsener, B., Rossi, A., 2008. The corrosion resistance of electroless deposited nano-crystalline Ni-P alloys. Electrochim. Acta. 53 (8), 3364–3370.

Das, S.K., Sahoo, P., 2009. Optimization of electroless NiB coatings based on mutiple roughness characteristics. In: Proceedings of the International Conference on Mechanical Engineering 2009 (ICME2009), 26–28 December 2009, Dhaka, Bangladesh.

Das, S.K., Sahoo, P., 2011a. Study of potentiodynamic polarization behaviour of electroless Ni-B coatings and optimization using Taguchi method and grey relational analysis. J. Miner. Mater. Charact. Eng. 10 (14), 1307–1327.

Das, S.K., Sahoo, P., 2011b. Electrochemical impedance spectroscopy of Ni-B coatings and optimization by Taguchi method and grey relational analysis. Port. Electrochim. Acta. 29 (4), 211–231.

Das, S.K., Sahoo, P., 2012. Influence of process parameters on microhardness of electroless Ni-B coatings. Adv. Mech. Eng.art. no. 703168.

Delaunois, F., 2002. Les dépôts chimiques de nickel-bore sur alliages d'aluminium. PhD thesis, Faculté polytechnique de Mons, Belgium.

Delaunois, F., Lienard, P., 2002. Heat treatments for electroless nickel-boron plating on aluminium alloys. Surf. Coat. Technol. 160, 139–148.

Delaunois, F., Petitjean, J.P., Lienard, P., Jacob-Duliere, M., 2000. Autocatalytic electroless nickel-boron plating on light alloys. Surf. Coat. Technol. 124, 201–209.

Dervos, C.T., Novakovic, J., Vassiliou, P., 2004. Vacuum heat treatment of electroless nickel-boron coatings. Mater. Lett. 58, 619–623.

Dervos, C.T., Vassiliou, P., Novakovic, J., 2005. Electroless Ni-B plating for electrical contact applications. Rev. Metal. Madrid. Extr, 232–238.

Evans, W.T., Schlesinger, M., 1994. Effect of solution pH and heat-treatment on the properties of electroless nickel boron films. J. Electrochem. Soc. 141, 78–83.

Fan, N., Huang, M., Wang, L., 2010. Electroless nickel-boron plating on magnesium alloy. In: Proceedings 2010 11th International Conference on Electronic Packaging Technology & High Density Packaging (ICEPT-HDP), pp. 288–292.

Gao, M.S., Bai, S.L., 1987. A study on the electroless nickel-boron alloy plating. Plat. Finish. 9 (1), 12–14.

Gervasio, D., Xu, M., Thomas E., 2005. Properties of aqueous alkaline sodium borohydride solutions and by-products formed during hydrolysis. In: Proceedings of Fuel Cell Design, Fabrication and Materials Selection Workshop, University of Illinois at Urbana-Champaign, IL, 26–27 July.

Gray, J.E., Luan, B., 2002. Protective coatings on magnesium and its alloys – a critical review. J. Alloys. Compd. 336 (1–2), 88–113.

Gu, C., Lian, J., Li, G., Niu, L., Jiang, Z., 2005. Electroless Ni-P plating on AZ91D magnesium alloy from a sulfate solution. J. Alloys. Compd. 391 (1–2), 104–109.

Hajdu, J., 1990. Surface preparation for electroless nickel plating. In: Hajdu, J., Mallory, G. O., Hajdu, J.B. (Eds.), Electroless Plating: Fundamentals and Applications. Noyes Publications, Park Ridge, NJ (Chapter 7).

Hamid, Z.A., Hassan, H.B., Attyia, A.M., 2010. Influence of deposition temperature and heat treatment on the performance of electroless Ni-B films. Surf. Coat. Technol. 205, 2348–2354.

Hari Krishnan, K., John, S., Srinivasan, K.N., Praveen., J., Ganesan., M., et al., 2006. An overall aspect of electroless Ni-P depositions–a review article. Metall. Mater. Trans. 37A, 1917–1926.

Henry, J.R., 2006. Electroless (autocatalytic) plating. Met. Finish. 104, 354–365.

Huang, Y.S., Cui, F.Z., 2007. Effect of complexing agent on the morphology and microstructure of electroless deposited Ni–P alloy. Surf. Coat. Technol. 201 (9–11), 5416–5418.

Kanta, A.F., Vitry, V., Delaunois, F., 2009. Wear and corrosion resistance behaviours of autocatalytic electroless plating. J. Alloys. Compd. 486, L21–L23.

Kanta, A.F., Poelman, M., Vitry, V., Delaunois, F., 2010a. Nickel-boron electrochemical properties investigations. J. Alloys. Compd. 505 (1), 151–156.

Kanta, A.F., Vitry, V., Delaunois, F., 2010b. Wear and corrosion resistance of electroless nickel-boron coated mild steel. Mater. Sci. Forum. 638–642, 846–851.

Khoperia, T.N., 2003. Investigation of the substrate activation mechanism and electroless NiP coating ductility and adhesion. Microelectron. Eng. 69, 391–398.

Krishnaveni, K., Sankara Narayanan, T.S.N., Seshadri, S.K., 2005. Electroless nickel-boron coatings: preparation and evaluation of hardness and wear resistance. Surf. Coat. Technol. 190, 115–121.

Kumar, P.S., Nair, P.K., 1994. X-ray diffraction studies on the relative proportion and decomposition of amorphous phase in electroless nickel-boron deposits. Nanostruct. Mater. 5, 183–198.

Lacourcelle, L., 1979. Nickelage. Techniques de l'Ingénieur, m1565.

Lee, C.K., 2008. Corrosion and wear-corrosion resistance properties of electroless Ni-P coatings on GFRP composite in wind turbine blades. Surf. Coat. Technol. 202 (19), 4868–4874.

Liu, Z., Gao, W., 2006. Electroless nickel plating on AZ91 Mg alloy substrate. Surf. Coat. Technol. 200 (16–17), 5087–5093.

Matsubara, H., Yonekawa, T., Ishino, Y., Saito, N., Nishiyama, H., et al., 2002. Observation of initial deposition process of electroless nickel plating by quartz crystal microbalance method and microscopy. Electrochim. Acta. 47, 4011–4018.

Matsubara, H., Yonekawa, T., Ishino, Y., Saito, N., Nishiyama, H., et al., 2006. The observation of the nucleation and growth of electrolessly plated nickel deposited from different bath pH by TEM and QCM method. Electrochim. Acta. 52, 402–407.

McComas, E., 2004. Nodular Nickel Boron Coating. US patent 20040111947 A1, Universal Chemical Technologies Inc., Florida.

Meyers, M.A., Mishra, A., Benson, D.J., 2006. Mechanical properties of nanocrystalline materials. Prog. Mater. Sci. 51 (4), 427–456.

Morcos, B., Barnstead, M., 2010. Electroless nickel plating. Prod. Finish. 75 (3), 126–131.

Ohno, I., 1991. Electrochemistry of electroless plating. Mater. Sci. Eng. A146, 33–49.

Ohno, I., Wakabayashi, O., Haruyama, S., 1985. Anodic oxidation of reductants in electroless plating. J. Electrochem. Soc. 132, 2323–2330.

Oraon, B., Majumdar, G., Ghosh, B., 2008. Improving hardness of electroless Ni-B coatings using optimized deposition conditions and annealing. Mater. Des. 29 (7), 1412–1418.

Pal, S., Verma, N., Jayaram, V., Biswas, S.K., Riddle, Y., 2011. Characterization of phase transformation behaviour and microstructural development of electroless Ni-B coating. Mater. Sci. Eng. A528 (28), 8269–8276.

Paunovic, M., 1983. Electrochemical aspects of electroless nickel deposition. Plat. Surf. Finish. 70, 62–66.

Peixoto Barbosa, D., Knörnschild, G., 2009. Anodization of Mg-alloy AZ91 in NaOH solutions. Surf. Coat. Technol. 203 (12), 1629–1636.

Rao, Q.L., Bi, G., Lu, Q.H., Wang, H.W., Fan, X.I., 2005. Microstructure evolution of electroless Ni-B film during its depositing process. Appl. Surf. Sci. 240, 28–33.

Riddle, Y.W., Bailer, T.O., 2005. Friction and wear reduction via an Ni-B electroless bath coating for metal alloys. JOM. 4, 40–45.

Riedel, W., 1991. Electroless Nickel Plating. Finishing Publications, Stevenage (UK).

Sahoo, P., Das, S.K., 2011. Tribology of electroless nickel coatings – a review. Mater. Des. 32, 1760–1775.

Saito, T., Sato, E., Matsuoka, M., Iwakura, C., 1998. Electroless deposition of Ni-B, Co-B, and Ni-Co-B alloys using dimethylamineborane as a reducing agent. J. Appl. Electrochem. 28, 559–563.

Sander, T., Tremmel, S., Wartzack, S., 2011. A modified scratch test for the mechanical characterization of scratch resistance and adhesion of thin hard coatings on soft substrates. Surf. Coat. Technol. 206 (7), 1873–1878.

Sankara Narayanan, T.S.N., Seshadri, S.K., 2004. Formation and characterization of borohydride reduced electroless nickel deposits. J. Alloys. Compd. 365, 197–205.

Sankara Narayanan, T.S.N., Krishnaveni, K., Seshadri, S.K., 2003. Electroless Ni-P/Ni-B duplex coatings: preparation and evaluation of microhardness, wear and corrosion resistance. Mater. Chem. Phys. 82 (3), 771–779.

Schlessinger, H.I., Brown, H.C., 1949. Preparation of Alkali Metal Borohydrides, US patent 2,461,663.

Shaffer, S.J., Rogers, M.J., 2007. Tribological performance of various coatings in unlubricated sliding for use in small arms action components – a case study. Wear. 263 (7–12), 1281–1290 (special issue).

Srinivasan, K.N., Meenakshi, R., Santhi, A., Thangavelu, P.R., John, S., 2010. Studies on development of electroless Ni–B bath for corrosion resistance and wear resistance applications. Surf. Eng. 26 (3), 153–158.

Sudagar, J., Lian, J., Sha, W., 2013. Electroless nickel, alloy, composite and nano coatings – a critical review. J. Alloys. Compd. 571, 183–204.

Suryanarayana, C., Koch, C.C., 2000. Nanocrystalline materials – current research and future directions. Hyperfine Interact. 130, 5–44.

Tjong, S.C., Chen, H., 2004. Nanocrystalline materials and coatings. Mater. Sci. Eng., R. 45 (1), 1–88.

Vitry, V., 2010. Electroless Nickel-Boron Deposits: Synthesis, Formation and Characterization; Effect of Heat Treatments; Analytical Modeling of the Structural State. PhD thesis, University of Mons, Belgium.

Vitry, V., Delaunois, F., Dumortier, C., 2008. Mechanical properties and scratch test resistance of Nickel-boron coated aluminium alloy after heat treatments. Surf. Coat. Technol. 202 (14), 3316–3324.

Vitry, V., Kanta, A.F., Delaunois, F., 2010. Initiation and formation of electroless nickel-boron coatings on mild steel: effect of substrate roughness. Mater. Sci. Eng., B. 175 (3), 266–273.

Vitry, V., Kanta, A.-F., Delaunois, F., 2011. Mechanical and wear characterization of electroless nickel-boron coatings. Surf. Coat. Technol. 206 (7), 1879–1885.

Vitry, V., Kanta, A.F., Delaunois, F., 2012a. Evolution of reactive concentration during Borohydride-reduced electroless nickel-boron plating and design of a replenishment procedure. Ind. Eng. Chem. Res. 51 (27), 9227–9234.

Vitry, V., Sens, A., Delaunois, F., 2012b. Abrasion Resistance of Steel Coated with Various Chemical Nickel Deposits, paper presented at Junior Euromat 2012.

Vitry, V., Sens, A., Kanta, A.-F., Delaunois, F., 2012c. Experimental study on the formation and growth of electroless nickel-boron coatings from borohydride-reduced bath on mild steel. Appl. Surf. Sci. 263, 640–647.

Vitry, V., Kanta, A.-F., Dille, J., Delaunois, F., 2012d. Structural state of electroless nickel-boron deposits (5wt.% B): characterization by XRD and TEM. Surf. Coat. Technol. 206 (16), 3444–3449.

Wang, Z.C., Jia, F., Yu, L., Qi, Z.B., Tang, Y., Song, G.-L., 2012. Direct electroless nickel–boron plating on AZ91D magnesium alloy. Surf. Coat. Technol. 206 (17), 3676–3685.

Warf, J.C., Schaltegger, K.H., 1955. Method of Coating Steel with Nickel-boron. US patent 2.726.170, (Superweld Corporation).

Watanabe, T., Tanabe, Y., 1983. The lattice images of amorphous-like Ni-B alloy films prepared by electroless plating method. Trans. Jpn. Inst. Met. 24, 396–404.

Williams, R.S., Feist, W.C., 1994. Effect of preweathering, surface roughness, and wood species on the performance of paint and stains. J. Coat. Technol. 66 (828), 109–121.

Wu, L.-P., Zhao, J.-J., Xie, Y.-P., Yan,g, Z.-D., 2010. Progress of electroplating and electroless plating on magnesium alloy. Trans. Nonferrous Met. Soc. 20 (0), s630–s637.

Zhang, W.X., Jiang, Z.H., Li, G.Y., Jiang, Q., Lian, J.S., 2008. Electroless Ni-P/Ni-B duplex coatings for improving the hardness and the corrosion resistance of AZ91D magnesium alloy. Appl. Surf. Sci. 254 (16), 4949–4955.

Ziyuan, S., Deqing, W., Zhimin, D., 2004. Surface strengthening of pure copper by Ni-B coating. Appl. Surf. Sci. 221, 62–68.

Wear resistance of nanocomposite coatings

8

*Vahe Vardanyan**, Bouddah Poaty*, Véronic Landry†, Gregory Chauve†, Tigran Galstian*, and Bernard Riedl**
*Université Laval, Québec, Canada; †FPInnovations, Québec, Canada

8.1 Introduction

Coatings are generally employed for two main reasons: to modify the appearance (color, texture, etc.) of materials (decorative coatings) or to protect them against numerous stresses (UV light, mechanical, chemical, or thermal stresses). Coatings have to be selected according to the intended application.

According to Hosker (1993) and Bulian et al. (2009), coatings used for exterior applications (garden furniture, sidings, patios, etc.) must be formulated to achieve good resistance to UV rays, changes in temperature and humidity, insects, and rot and mildew. In contrast, floor coatings must be resilient to mechanical stresses. In fact, finishes used for interior applications such as furniture, kitchen cabinets, and floors must have good mechanical (abrasion, scratches, impact, and hardness), thermal (fire and flame, thermal stability), and chemical resistance (detergent, food, etc.) (Bulian, 2009). Coatings selection must also be made according to the substrate to be coated. Plastic materials are known as low surface energy materials and, for this reason, coating adhesion is one of the major concerns associated with plastic. On the other side, coatings applied to metals have to be corrosion resistant if intended for exterior applications. Wood coatings are no exception; coatings must be adapted to its numerous particularities.

8.1.1 The importance of coatings in the wood industry

Wood is a naturally durable material, long recognized for its versatility as well as its exceptional structural properties (Feist, 1983). However, to maintain its appearance as well as its mechanical properties, it must be kept under favorable conditions. According to Feist (1983), like any biological material, wood is subject to degradation caused by light, moisture, and oxidation. For this reason, products made from wood should generally be treated in order to preserve their properties. The use of finishing products is often the preferred avenue. Among the most desirable characteristics for interior wood coatings, abrasion or wear resistance is certainly among the first. Abrasion is a degradation due to mechanical wear by hard and rough objects. It generally occurs when paint or varnish is exposed to mechanical erosion, rubbing, or friction, such as in floor finishes. A low abrasion resistant coating will lead to many

Anti-Abrasive Nanocoatings.

surface defects after a short period of time (Surface Modification Center, 2013). Resistance to abrasion is related to the coating's toughness and depends on its ability to dissipate the mechanical energy. As wood is a soft material subject to indentation, the coating system must have a good mechanical performance.

The abrasion resistance of a coating will depend on the type of coating used and the fillers added. Different types of coating can be used for interior wood products, the most frequently used being nitrocellulose, pre- and post-catalyzed lacquer, conversion varnish, catalyzed polyurethane, polyester, and UV-curable coatings (high solids and water-based) (Gustafson, 2003). Besides performance criteria, drying time, costs per volume, application methods, drying methods, and equipment required are important factors to consider when selecting a coating. As it is not a decision based on a single criterion, one might choose a coating that is less efficient in terms of mechanical performance but more suitable taking into consideration the facilities of the industry or the coating costs per volume. For this reason, reinforcing fillers are often added to the coating formulation in order to achieve better mechanical performance. Water-based coatings, which are increasingly used by wood industries, are no exception.

8.1.1.1 Water-based coatings

Year after year, water-based coatings gain market share due to new regulations on volatile organic compound (VOC) emissions and the numerous voluntary certification programs to which specifiers (architects, engineers, etc.) subscribe. Organic solvents used in most coatings generate issues with respect to the environment and human health, not to mention handling and storage. As a result, regulations on VOCs are becoming tighter in several countries. In fact, European, American, and Canadian architectural regulations have recently been modified to restrict usage and encourage less hazardous alternatives. On the basis of their broad experience with solvent-borne coatings over several decades, the woodworking industries and consumers expect to meet these new VOC requirements while maintaining coating quality (performance and aspect) achieved with solvent-borne coatings.

The proposed alternatives to solvent-based coatings include water-based coatings. Even if they are gaining significant market shares, they entail significant disadvantages, and these still restrict their adoption by a portion of the wood industry. One major disadvantage relates to the mechanical performance of water-based coatings, which typically differs from that of solvent-based coatings. One way to overcome the lack of performance of water-based coatings is to add to the coating formulation different fillers that will improve the mechanical performance and more specifically the abrasion resistance.

8.1.1.2 Reinforcing fillers

Reinforcing fillers are often used in order to increase the performance thus the service life of coatings. In the wood industry, silica (SiO_2) and alumina (Al_2O_3) are used at the micrometer scale to get a good abrasion and scratch resistance. These metal oxides are relatively cheap and abundant, and they possess a relatively low refractive index compared, for example, with titanium and zinc oxides which are used as UV absorbers. Silica and alumina are hard materials, silica being around 6–7 on the Mohs scale and alumina (corundum) being 9. However, the addition of micrometric particles, even if their refractive index is relatively low and near that of the resin, diffracts light, which leads to a reduced transparency of the coating. For UV-curable coatings, it can also interfere with the curing process. Light scattering is one of the reasons among others explaining why over the last few years several coating suppliers have switched to nanosize fillers.

8.1.2 The special case of nanosize fillers

Nanocomposites can be defined as multiphase materials comprising minimally a phase size of less than 100 nm. This well-accepted definition is not based on an arbitrary choice. In fact, as the size scale of matter becomes smaller, down to the nanoscale, the principles which govern macroscopic materials become nonscalable. This could be explained on the basis of intermolecular forces that govern material behavior (Fernando and Sung, 2009). Although gels, colloids, and copolymers can be included in the nanocomposite definition, the term nanocomposite generally refers to the addition of a nano-sized phase in a matrix, both with different properties as well as a different structure. In general, nanoparticles are used for their optical, magnetic, or conductive properties, rather than for their quality of mechanical reinforcement. However, some nanoparticles such as aluminum oxide (Al_2O_3) and silica (SiO_2) are used, similarly to their micrometric counterparts, to improve the mechanical properties of polymer matrices. Coatings containing well-dispersed nanosize inorganic materials can be considered as nanocomposites (Ajayan et al., 2003; Nalwa, 2003; Ray and Bousmina, 2006). Table 8.1 summarizes studies performed to improve different properties using nanoparticles.

Table 8.1 Summary of the most frequently used nanoparticle in coatings

Nanoparticle		Property achieved	References
Type	**Geometry**		
Clay	Lamellar	Barrier properties	Asif et al. (2010); Park et al. (2012)
Silica (SiO_2) Aluminum oxide (Al_2O_3)	Spherical	Hardness Wear resistance Impact resistance Scratch resistance	Jalili et al. (2007); Amerio et al. (2008); Ranjbar and Rastegar (2009); Sow et al. (2011); Sangermano et al. (2008); Jalili and Moradian (2009)
Zinc oxide Titanium dioxide	Spherical	Resistance to UV light	Cristea et al. (2011); Auclair et al. (2011)

A number of studies have already been conducted on UV-curable acrylate coatings with nano-size aluminum oxide and silicon oxide. One of the major issues related to nanocomposite coatings is to achieve a good dispersion of the nanoparticles in the polymer matrix. For this reason, Bauer et al. (2002) have grafted different silanes at the surface of nanoparticles. 3-(methacryloxy) propyl-triethoxysilane (MEMO) was grafted at the surface of silica nanoparticles and nano alumina. The addition of 35% by weight of the acrylate monomer-modified nanoparticles led to a significant increase in scratch resistance. The polymer nanocomposite with and without nanoparticles has also been subjected to wear resistance experiments (steel wool, 60 cycles). The nanocomposite coating was found to have a mechanical strength much greater than the one of the coating without nanoparticles. The nanocomposite remained intact while the unmodified coating was severely affected by these tests.

In another paper Bauer et al. (2004) have expanded the range of nanosize fillers and silane-coupling agents added to polyacrylates. They compared the nanocomposites prepared with MEMO with nanocomposite prepared with two other coupling agents: vinyl trimethoxysilane (VTMO) and propyl trimethoxysilane (PTMO). These coupling agents were grafted onto the same silica nanoparticles and also onto aluminum oxide (Al_2O_3) nanoparticles, zirconium oxide (ZrO_2), and titanium oxide (TiO_2). Taber rotary platform experiments were performed and mass losses were measured. These tests were performed with the use of the coupling agent VTMO in the resin. For all nanocomposites, it was possible to note a significant increase in the wear resistance. The results were closely related to the hardness of the nanofiller as well as its refractive index. The high refractive indices of the zirconium and titanium oxide led to a decrease in curing efficiency. Ultraviolet penetration decreases because of light scattering. Bauer et al. (2006) also published results on the improvement of polyacrylates by adding both nanoparticles and microparticles. Abrasion resistance tests have been carried out and showed that the mixture of two types of particles leads to a significant increase in resistance to abrasion. Indeed, the abrasion resistance is much higher than for the formulation with the same concentration of microparticles or nanoparticles. In a second article on the subject (Bauer et al., 2007), they concluded that the micrometer aluminum oxide size is important. Using the same proportions of silica and alumina as in the previous article, they varied the size of the aluminum oxide and determined the abrasion resistance. The results obtained show that the 16 microns aluminum oxide combined with nanoscale silica leads to excellent abrasion resistance. The work realized by Bauer demonstrated the importance of the chemical composition, the size of nanoparticles and microparticles chosen, and the nature of the coupling agent. Sow et al. (2011) have also demonstrated in their work on UV-curable water-based nanocomposites the importance of good nanoparticle dispersion on the final mechanical properties. Aggregates were found to decrease the hardness of the coating films. Dispersing nanoparticles is a real challenge. Their high surface area, promoting agglomeration and aggregation, makes it necessary to modify the surfaces of the inorganic nanoparticles in order to make them compatible with the organic matrix. Using an organic filler could help to limit the dispersion issues.

8.1.2.1 Cellulose nanocrystals: a potential reinforcing filler for coatings

Cellulose is the most abundant organic material on earth (more than 50% of the biomass). It is the main component of wood and other woody plants. Associated cellulose macromolecules form microfibrils, which themselves are associated in layers, which form the cell walls of plant fiber. Strong hydrogen bonds are present between the anhydroglucose molecules of the different cellulose chains.

Many products in everyday life are cellulose derivatives, including a new product, cellulose nanocrystals (CNC), also called "whiskers." CNC particles are extracted from native cellulose sources such as wood pulp by controlled acid hydrolysis (Dong et al., 1998). The resulting rod-like shape and negative surface charge of CNCs give rise to electrostatically stable colloidal suspensions. Wood is not the only source of CNC; other sources include: cotton (Revol et al., 1994), green algae (Sugiyama et al., 1994), beet sugar (Heux et al., 1999). Bacterial CNC can also be produced (Berndt et al., 2013).

There are very few studies in the literature on water-based coatings with the CNC because it is a new product. Some work has been done by the Dufresne group, which uses "whiskers" of cellulose (Helbert et al., 1996). Several particles were mixed with waterborne coatings for wood-derived products in order to improve their performance. CNC was found to be a good reinforcing filler when mixed with aqueous latex matrices (Favier et al., 1995). Similar to the inorganic fillers, the main challenge of mixing CNC particles into a host matrix is to achieve a homogeneous dispersion in order to maximize the macroscopic effects on the mechanical reinforcement. Landry et al. (2010, 2011a) have shown that the addition of small concentrations of CNC particles to clear water-based coatings could help improve the mechanical properties (abrasion and scratch resistance, hardness).

As it is difficult to incorporate inorganic nanoparticles in a polymer matrix (organic), it will possibly be easier to use organic nanoparticles. For this purpose, we investigated the replacement of the inorganic nanoparticles, studied previously, with CNC particles. These have a higher aspect ratio than inorganic nanoparticles, which are usually spherical and are also bio-based. The research hypothesis assumes that CNC, possibly with appropriate modification of its surface, will show high dispersion and stability of nanoparticles in nanocomposite coatings.

The atomic force microscope (AFM) has been used to study the quality of pigment dispersions and reinforcements. Karakas and Celik (2012) have shown the extent of the distribution and particle aggregation in seven different formulations in relation to the surface of paint films. The experimental results showed that the addition of TiO_2 can be reduced by 4% by optimizing the calcite optimum size. Farrokhpay (2012) conducted a literature review on the new assessment techniques of painted surfaces and obtained quantitative results on the surface roughness characterization by AFM. Thometzek et al. (1999) also used transmission electron microscopy (TEM) and AFM to quantify the surface quality of painting and correlated the degree of hydrophobicity of the particles with the quality of the dispersion.

This chapter aims to qualify the dispersion of CNC in varnish/paint coatings, determine the surface roughness of the nanocomposite coatings by AFM and link these results with the abrasion resistance performance.

8.2 Materials and methods

UV-curable clear and opaque coatings were prepared minimally from a water-based emulsified polyurethane acrylate resin, a defoaming agent, a surfactant, a dispersant, a thickener, CNC and a photoinitiator. Opaque formulations also include TiO_2 (Table 8.2).

The resin selected is the Bayhydrol UV 2282 (Bayer Material Science), a polyurethane acrylate (PUA) oligomer emulsified in water that was primarily developed for applications on wood. The photoinitiator used is a bis-acyl phosphine oxide (Irgacure 819DW, BASF Resins – Inks and OPV) dispersed in water (45 wt%). TiO_2 was used as a pigment. After trials, it was found that ~30% (dry weight) of TiO_2 gives an opacity of 80%.

Table 8.2 summarizes the weight of each chemical used to prepare the different formulations. CNC concentrations were set to 0.5, 1, 1.5, 2, and 3% (dry weight).

In Table 8.2, the initial concentration of each ingredient can be described as follows:

$$I = \frac{M_{\text{dry}}}{M_{\text{liquid}}} \times 100\% \tag{8.1}$$

where M_{dry} and M_{liquid} are the dry and liquid weight each ingredient when added to the formulation. The CNC used in this research is in gel state (14% w/w in water). Final liquid weight of formulations is different, approximately 310.5 g for varnishes and 265.5 g for paints. In both cases, the dry weight of coatings equals 114.28 g. This was done to have comparative formulations. So, overall initial concentration, as solids content, is about 37% for varnishes and about 43% for paints.

8.2.1 Cellulose nanocrystals

Aqueous CNC suspensions were prepared in FPInnovations' pilot plant by sulfuric acid hydrolysis of a commercial bleached softwood kraft pulp according to a procedure modified from the literature (Klemm et al., 2011). Samples of the suspension of CNC were sonicated using a Sonics Vibra-cell 130W, 20 kHz ultrasonic processor with a 6 mm diameter probe. Typically, 15 ml of a 2–3% w/w CNC suspension was placed in a plastic tube of 50 ml and sonicated at 60% of the maximum power. This sonication treatment (corresponding to a total energy input of ~3600 J/g CNC) was performed in an ice bath to prevent the CNC degradation caused by the heating. Following this process, retrieved nanofibers have the following properties: specific surface area of 600 m^2/g, a Young's modulus of 150 GPa and a tensile

Table 8.2 **Initial concentration of coating components**

	%CNC dry w/w	Binder	Antifoaming agent	Surfactant	Dispersant	Pigment	Solvent	CNC	Thickener	Photoinitiator	Total
		Liquid mass (g)									
Varnish	0	274.73	1.86	1.15	3.72	–	25.50	–	1.24	2.30	310.50
	0.5	273.36	1.85	1.14	3.70	–	22.84	4.08	1.23	2.29	310.49
	1	271.98	1.84	1.14	3.68	–	20.19	8.16	1.23	2.28	310.50
	1.5	270.61	1.83	1.13	3.66	–	17.53	12.24	1.22	2.27	310.49
	2	269.24	1.82	1.13	3.65	–	14.87	16.33	1.21	2.26	310.51
	3	266.49	1.80	1.12	3.61	–	9.56	24.49	1.20	2.23	310.50
Paint	0	197.23	1.86	1.15	3.72	31	26.5	–	1.24	2.3	265.00
	0.5	196.24	1.85	1.14	3.70	30.85	23.61	4.08	1.23	2.29	264.99
	1	195.26	1.84	1.14	3.68	30.69	20.72	8.16	1.23	2.28	264.99
	1.5	194.27	1.83	1.13	3.66	30.54	17.83	12.24	1.22	2.27	265.01
	2	193.29	1.82	1.13	3.65	30.38	14.94	16.33	1.22	2.25	264.99
	Initial concentration (%)	40	25.5	97.8	40.5	100	–	14	20.1	45	

strength of 10 GPa (Klemm et al., 2011). Cellulose fibers that remain after this treatment are almost entirely crystalline and as such are called "crystalline." The precise physical dimensions of the crystallites depend on several factors, including the source of cellulose hydrolysis conditions accurate, and ionic strength.

The aqueous CNC suspension was then concentrated in a Labconco RapidVap (Labconco, Kansas City, MO) down to 14% w/w. Such high CNC concentration is required to adjust the CNC loading within the formulation and avoid excessive dilution.

8.2.2 CNC particle size measurement by light scattering

Particle size or aggregate size was determined by photon correlation spectroscopy (PCS) using a Malvern Zetasizer Nano ZS (Malvern Instruments Ltd, Malvern, UK). To avoid excess light scattering in the spectrometer, the samples were diluted to 0.05 wt% CNC in the presence of 5 mM NaCl. Samples were filtered with 0.45 μm nylon or 0.7 μm GF/F Whatman syringe filters prior to measurement. No significant loss of CNC occurred as proven by gravimetry measurements after filtration. Measurements were performed in triplicate at 25°C.

Because PCS is a light-scattering method, the measured CNC particle size values quoted are the *z*-average (intensity mean) hydrodynamic diameters of equivalent spheres and do not represent actual physical dimensions of the rodlike CNC particles. However, they are valid for comparison purposes.

8.2.3 Coating formulations preparation: dispersion method

The choice of the dispersion method plays an important role in the final dispersion state of nanoparticles (Landry et al., 2010). Several types of apparatus are commonly used in the industry, for example: high-speed mixer, three-cylinder mill, and ball mill. Although the last two methods lead to very high shear rates and a high degree of dispersion, high-speed mixer (Dispermat, VMA-Getzmann GMBH D-51580 Reichshof) was preferred as it is most commonly encountered in the industry. The mixer used can reproduce shear rates comparable to those of industrial mixers.

To achieve good TiO_2 dispersion, ultrasonication was used after high-speed mixing, the formulation was transferred to a 1 L Erlenmeyer flask and placed in a cooling bath containing ethylene glycol maintained at 5°C. An ultrasonic probe of 750 Watts (Ultrasonic processor, Cole Palmer) and a thermometer were immersed in the aqueous formulation. The ultrasonic frequency used for the dispersion was set to 20 kHz. The ultrasonic treatment was carried out for 2 minutes by setting the maximum temperature attainable in the formulation at 40°C (Sow et al., 2011).

8.2.4 Preparation of the formulations

The protocols followed to prepare the paint and varnish formulations are presented in Figure 8.1. To ensure reproducible results, the order of addition of the different ingredients was closely followed. The stirring speed was set to 400 rpm. After the

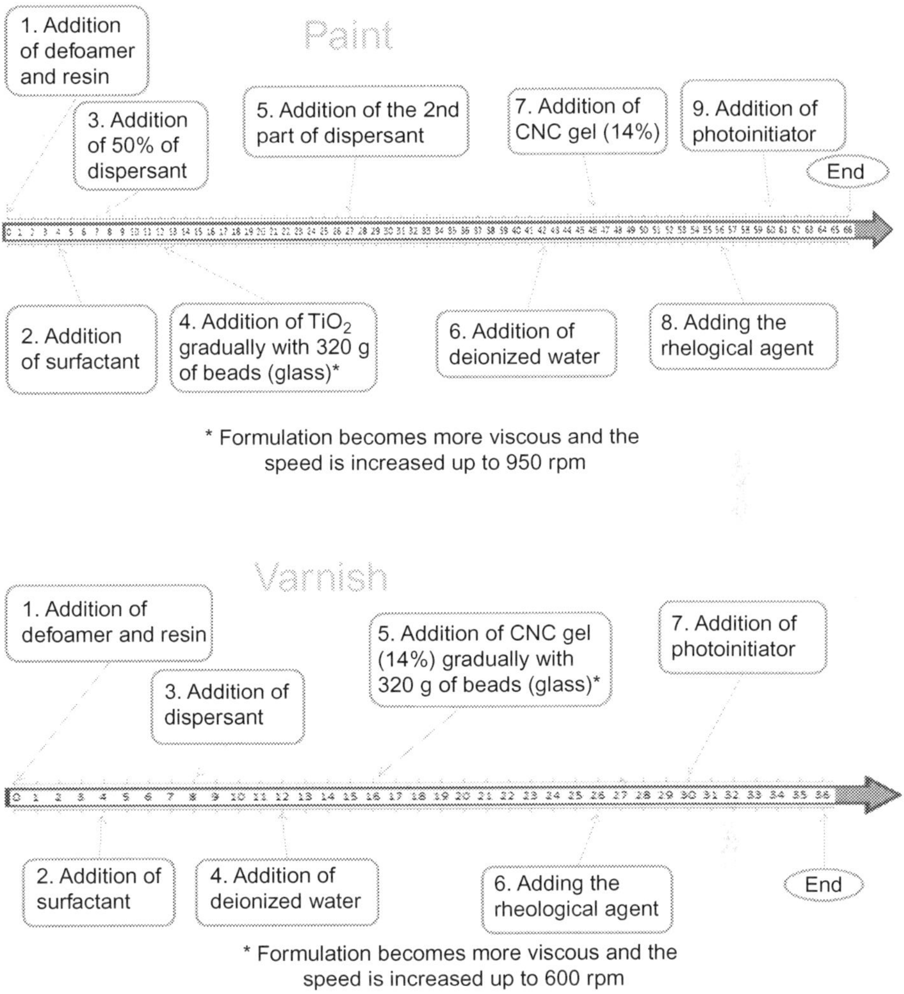

Figure 8.1 Formulations preparation protocol: paint and varnish.

dispersion of the nanoparticles and the TiO_2, formulations were filtered to remove the glass beads.

8.2.5 Wood samples preparation

The dimensions of the sanded Sugar Maple (*Acer saccharum*) wood samples used in this project are 96 mm × 96 mm × 15 mm (Figure 8.3). Samples were conditioned at 60% relative humidity prior to coating application. Paint and varnish formulations were applied by pulverization. After application (dry film thickness of $127 \pm 20\ \mu m$ for varnish and $152 \pm 20\ \mu m$ for paint formulations), samples were placed in an oven for 10 minutes at 60°C to evaporate the water. After complete water evaporation, the UV curing was carried out using a UV oven equipped with a medium pressure mercury lamp (600 W/cm). The intensity of the UV light, with complete UV range beginning at 184.9 nm, was 570 mJ/cm^2 and the perceived temperature during curing ranges between 25 and 30°C. These steps gave a final dry coating thickness (solid film) of about 100 microns.

Figure 8.2 illustrates the appearance of the coatings discussed in this chapter. At the top are uncoated wood surfaces. In the middle, varnishes are presented: the surface is shiny, and there is no discernible difference between left and right, which are, respectively, the varnish without CNC and the varnish with 1% CNC. At the bottom, painted samples are presented. They are white and opaque thanks to the TiO_2 particles. Again, there is no major difference between coatings with and without CNC.

8.2.6 Atomic force microscopy

AFM observations were carried out using a NanoScope V, fitted with a Hybrid XYZ scanner (Veeco Instruments Inc., Santa Barbara, CA). AFM measurements were performed under ambient air conditions in tapping mode (tip radius is 8 nm). The resolution was set to 256 lines by 256 pixels for all observations. Surface roughness was calculated in 10 μm × 10 μm scan areas, using the classical mean surface roughness parameter. The parameters were calculated by the Research Nanoscope 7.2 software using this equation:

$$R_a = \frac{1}{n}\sum_{i=1}^{n} |Z_i - Z_{\text{ave}}| \tag{8.2}$$

Figure 8.2 Samples with and without coatings (without CNC (left) and with CNC (right)).

where R_a is the mean roughness, the arithmetic average of the absolute values of the surface height deviations, Z_i is the current Z value, Z_{ave} is the average of the Z values within the given area, and n is the number of points within the given area (256×256).

8.2.7 Scanning electron microscopy and elemental analysis

Scanning electron microscopy (SEM) observations were carried out using a Quanta 3D FEG high-resolution SEM (FEI Company, Hillsboro, OR). UV-cured waterborne films were analyzed to investigate the dispersion of TiO_2 and CNC in the films.

The elemental analysis of the UV-cured waterborne films was studied using an EDAX's Sapphire Si (Li) Energy Dispersive Spectroscopy (EDS) Detector for the SEM with classic 10-Liter Dewar.

8.2.8 Abrasion resistance

Abrasion resistance of the paint and varnish formulations was studied using a Taber Abraser (Taber Abraser 5135, TABER Industries, North Tonawanda, NY) following the ASTM D 4060 standard. Mass loss was recorded after 100 rotations, with 500 g discs, with type S-42 sanded paper.

8.2.9 Scratch resistance

To measure the scratch resistance, the surface must be scraped with a standard load for all samples. This test was performed according to the ASTM D7027-05 standard, using a Maluti-finger Scratch/Mar Tester, model 710, (TABER Industries, North Tonawanda, NY). The samples were scratched with a 10 N load, followed by depth profiling with the profilometer.

8.3 Results and discussion

8.3.1 CNC samples characterization

Prior to investigation of the surface roughness of the nanocomposite coatings, the average size of the CNC particles in water was measured. Figure 8.3 displays the particle size distribution profile by intensity, with logarithmic X-axis and linear Y-axis. These experiments were performed in water, as it is not possible to perform the particle size analysis with this equipment in the presence of resin.

This technique, as mentioned previously, does not recognize the fiber-like nature of CNC and gives a value equivalent to a hydrodynamic volume size. The characteristic dimension of CNC, assimilated to a hydrodynamic volume size, as measured in aqueous suspension gives an averaged particle diameter of 110 nm ($\pm$5 nm). According to measurements performed by FPInnovations, the dimensions of CNC particles are 6–10 nm (width) and 100–130 nm (length).

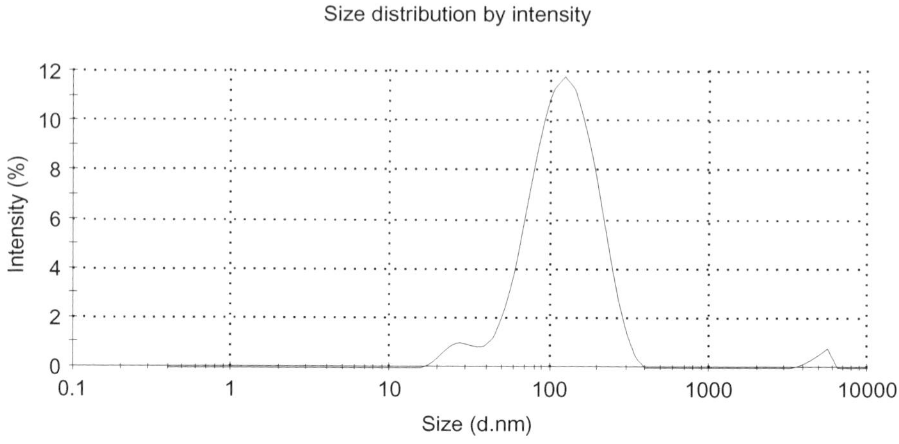

Figure 8.3 Size distribution profile of the 0.05% w/w CNC suspension in water.

8.3.2 Polymerized coating formulations characterization

The most widely used method for analyzing the dispersion state of nanoparticles in dry films is electron microscopy – scanning (SEM) or transmission (TEM). However, in this project, it is difficult to use them, as the polymer matrix (resin) and the CNC particles are both organic, which leads to a very low contrast between the resin and the particles. SEM images of the clear coatings (varnish) with or without CNC (Figure 8.4(a), (b)) show no significant differences. In fact, varnish surfaces are very homogeneous, and the presence of CNC does not impact the surface. Other research has shown that wood surfaces are rather rough at the nanoscale, since wood is a porous substrate, but once covered with a coating, even with only 30 microns thickness, it becomes highly even at the nanoscale (Cool and Hernandez, 2011). In the case of paint formulations, a good contrast between the resin and the titanium oxide was observed, but not for the CNC particles (Figure 8.4(c), (d)). For the paint formulations, TiO_2 particles clearly appear as white dots on the micrographs. To make sure that the white dots are indeed microparticles of TiO_2, elemental analysis was performed. Two points were selected and studied on Figure 8.5: (a) where aggregates were observed and (b) on a smooth part of the film without aggregates.

Thanks to this analysis, the presence of TiO_2 microparticles has been shown in the aggregates. The small peaks (Figure 8.5(b)) are due to the presence of TiO_2 nanoparticles, invisible in this micrograph at this scale, but probably present in all film volume. It is not possible to perform a similar analysis for the varnish films as no aggregates can be observed. To overcome this issue, AFM experiments were performed and results are presented in Figure 8.6.

Nanoscopic roughness is a reflection of the quality of the CNC dispersion within the formulations. If the surface is smooth at the nanometer level, CNC particles are well dispersed and aggregates are small. As presented in Figure 8.6(a),

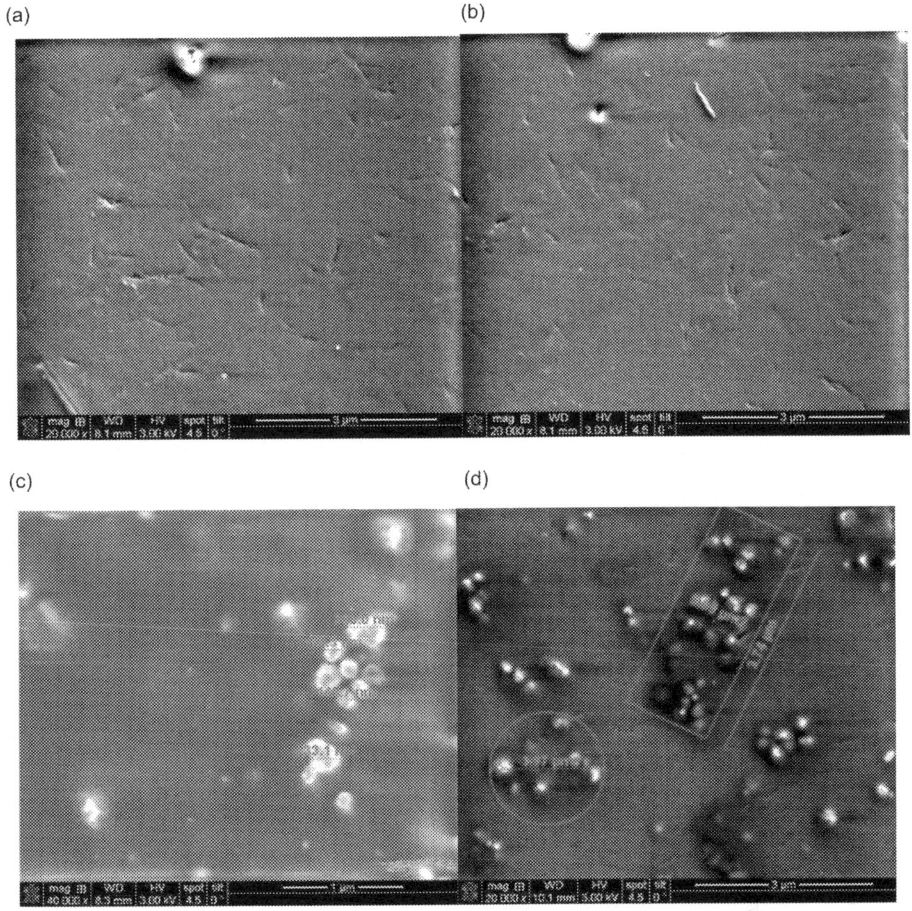

Figure 8.4 SEM imaging (a) varnish without CNC; (b) varnish with 1.5% CNC; (c) paint without CNC; (d) paint with 1.5% CNC.

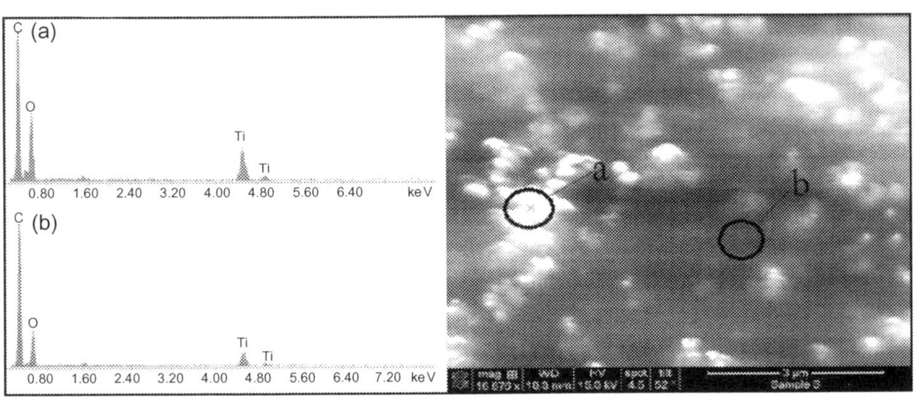

Figure 8.5 Elemental analysis of paint surface: (a) where there are aggregates; (b) where there are no aggregates.

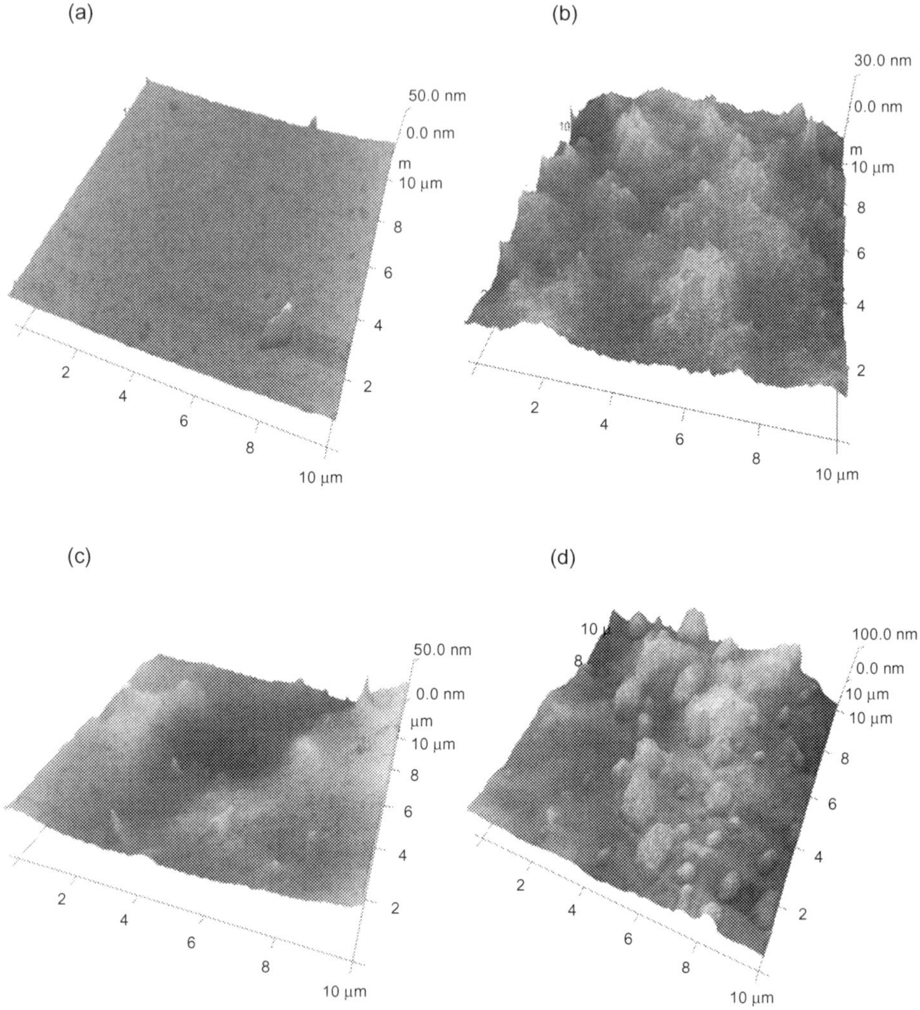

Figure 8.6 AFM profile images of: (a) varnish without CNC; (b) varnish with 1.5% CNC; (c) paint without CNC; (d) paint with 1.5% CNC.

which represents the varnish without CNC, the roughness is low, even at the submicron scale. On the other hand, as shown in Figure 8.6(b), the addition of 1.5% of CNC to a varnish formulation leads to a sharp increase in roughness. This reveals strong aggregation of CNC particles, as the length of the cellulose nanoparticles is only about 100 nm and the approximate aggregate size for this image is about 1 micron. Figure 8.6(c) presents the surface of a paint formulation without CNC particles. As can be seen, roughness increases, but not as much as for the varnish formulation with CNC particles. Finally in Figure 8.6(d), the addition of both CNC and TiO_2 was found to significantly increase the size of the

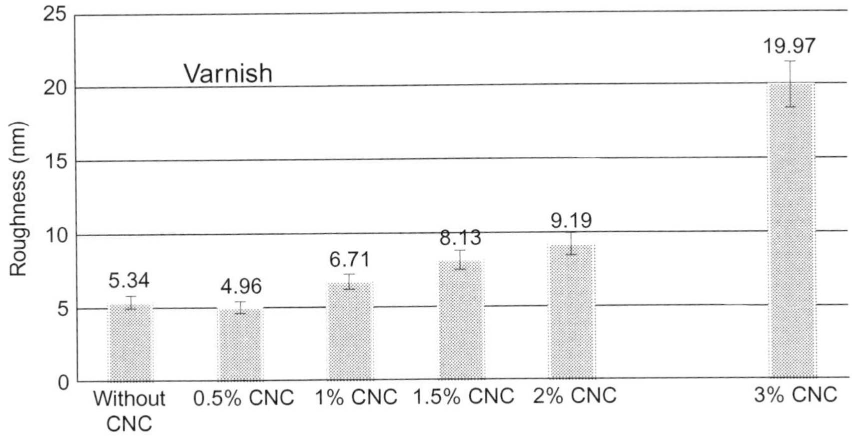

Figure 8.7 Varnish average roughness (Eq. 8.2) as a function of the CNC content.
Source: Reproduced with permission from Vardanyan et al. (2014).

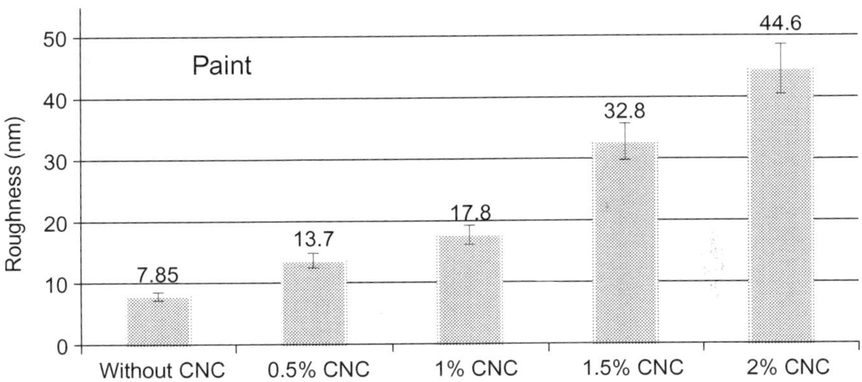

Figure 8.8 Paint average roughness (Eq. 8.2) as a function of the CNC content.
Source: Reproduced with permission from Vardanyan et al. (2014).

aggregates, much more defined than in Figure 8.6(b) and Figure 8.6(c), indicating an interaction between TiO_2 and the CNC. It should be noted that it is very important to measure the surface roughness for quite small ($10 \times 10\,\mu m$ in our case) areas, so that the surface irregularities at larger scales do not disturb the measurement of the degree of dispersion of nanoparticles. Thus, $10 \times 10\,\mu m$ images were chosen for characterization of the CNC dispersion in both varnish and paint formulations.

Four AFM characterizations ($10 \times 10\,\mu m$) were performed on 10 samples for concentrations of CNC varying between 0 and 3% (0, 0.5, 1, 1.5, 2, and 3%) (Figure 8.7) for varnish formulations, and between 0 and 2% (0, 0.5, 1, 1.5, and 2%) for paint formulations (Figure 8.8). The reason why CNC concentration was limited to 2% for paint formulations is that, above 2% in CNC content, the paint becomes highly viscous and it is impossible to spray the coating on the substrate without adding water.

The average roughness measurements as a function of the CNC content were measured for the varnish (Figure 8.7) and the paint formulations (Figure 8.8). As mentioned previously, it is not possible to perform the same characterizations by SEM/TEM on sections of coatings due to the lack of contrast between the varnish and the CNC, both organic. Thus, AFM roughness measurements are considered as measures of dispersion quality of TiO_2 and CNC particles.

It is important to mention that even if the size of the CNC particles is about 100 nm (Figure 8.3) the surface roughness is much smaller, as the CNC particles are not stacked at the surface of the coating but disperse in the coating film. Only part of the CNC sticks stand out of the surface. Figure 8.9 shows a schematic representation of the CNC and TiO_2 particle dispersion in the films. With no added micro- or nanoparticles, the varnish film is quite even and average roughness is about 5 nm, while with addition of 2% CNC it almost doubles (Figure 8.8), a little over 9 nm. The length of the CNC nanoparticles, or rods is about 100 nm (Figure 8.4), so it could be argued that the CNC particles on the surface slightly protrude from the surface individually or in aggregates. The presence of TiO_2 particles seems to increase the roughness, but as for the CNC particles, the roughness is not of the magnitude of the particle size of the TiO_2 particles.

8.3.3 Influence of the sonication on the CNC dispersion

The coatings analyzed so far revealed that the presence of CNC drastically changes the surface roughness and topology. This can be attributed to CNC instability in the medium (water) caused by the presence of the other ingredients and the mixing to disperse it within the formulation. Indeed, the CNC aqueous suspensions are very sensitive to pH and ionic strength, especially with regard to their liquid crystal states, in concentrated form (Klemm et al., 2011). Changes in these two parameters would probably affect the stability and the agglomeration state of the CNC suspensions. Moreover, the chemical affinity and surface characteristics of the CNC would have to be changed to better interact with the other components. For now, the

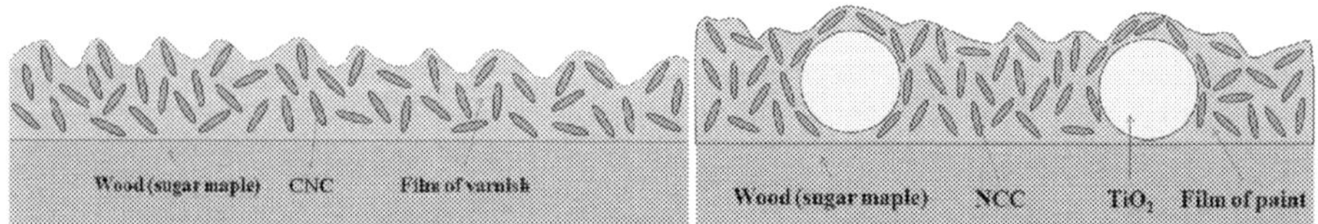

Figure 8.9 Schematic (not to scale) representation of the (a) CNC-containing varnish layer; and (b) CNC-containing paint layer.

mixing aspect only is addressed. However, to optimize the dispersion of the CNC particles, CNC chemical surface modification is being addressed, as of now, in our laboratory.

Sonication impacts on the agglomeration state of the CNC and TiO_2 particles and should help to make the formulation more homogeneous in terms of surface roughness (Beck et al., 2011). SEM imaging was performed before and after sonication (Figure 8.10), for paints at 1% CNC content.

As presented in the SEM images in Figure 8.10, the aggregates are almost twice as small in the sonicated media than in the nonsonicated media. The roughness, after coating polymerization, was also measured by AFM. For sonicated samples, the roughness was found to be significantly lower ($R_a = 7.11$ nm) than for the nonsonicated samples (Figure 8.10, 1% CNC), which is 17.8 nm. This experiment clearly shows that this roughness parameter is closely related to the degree of dispersion.

By comparing results found for the SEM and the AFM experiments for the paint formulations, it is possible to characterize the dispersion of CNC in the varnish, therefore the dispersion of organic nanoparticles in an organic matrix as organic, which is very difficult to do using SEM, because there is no contrast between the two materials.

8.3.4 Abrasion resistance

Figure 8.11 presents the mass loss found for the paint and varnish formulations after 100 rotations. For the varnish, the abrasion resistance is improved by the addition of CNC particles. The maximum abrasion resistance is obtained for a CNC concentration of 2%. For a formulation with more CNC – 3% – the abrasion resistance was found to decrease slightly. As it was possible to see on the AFM images, roughness was found to increase significantly in this range of concentration (2–3%) for the varnish formulations. For these results, it is possible to conclude that the aggregation interferes with the mechanical properties but can be tolerated

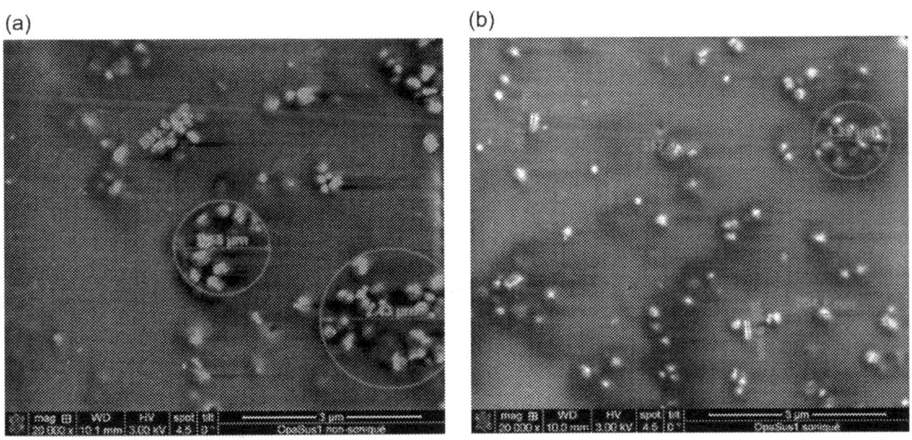

Figure 8.10 SEM imaging for paints with 1% CNC (a) nonsonicated; and (b) sonicated.

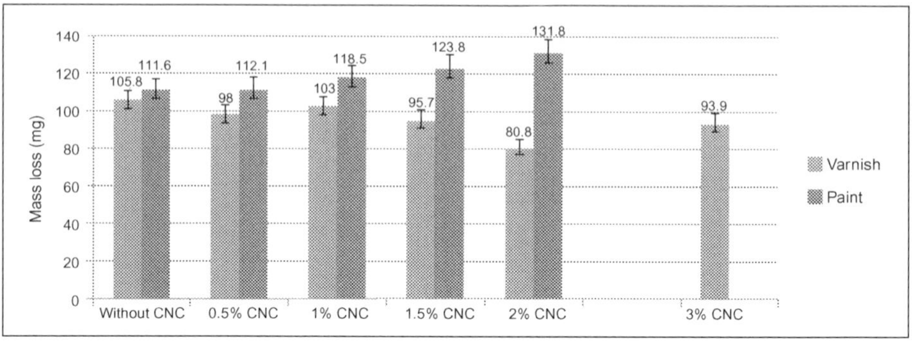

Figure 8.11 Mass loss of the paint and varnish coatings with different CNC concentrations.

to a certain extent. As CNC particles possess a refractive index close to the one of most acrylic-acrylate resins (Landry et al., 2011b), at low concentration it should not diffract light significantly, nor impede the polymerization process. Paint formulations show a totally different behavior. Even at low concentration, the addition of CNC decreases the abrasion resistance (higher mass loss). There is no optimum concentration such as for the varnish formulations. There could be many reasons for this negative result. First, as seen on the AFM images, roughness is important even at low CNC concentration for the paint formulations. High roughness increases light scattering decreasing the UV light penetration and the curing efficiency. Paint formulations already contain high particle content (TiO_2), adding CNC, as present on the AFM images, seems to create a special aggregation phenomenon. Studying the interactions between the TiO_2 particles and the CNC particles would be beneficial for gaining a better understanding of this phenomenon.

8.3.5 Scratch resistance

Figure 8.12 presents the scratch depth for the paint and varnish formulations after scratching by 10 N load. There is some improvement for varnishes by the addition of CNC particles. The minimum depth (maximum resistance) is obtained for a CNC concentration of 2%, like the abrasion resistance, but, at CNC content over 2%, its decreases, because of the bad dispersion of particles in the varnish (Figure 8.7). In the case of paints, the depth almost did not change up to 1% of CNC particles, but at over 1% CNC content, it increases abruptly, like the roughness (Figure 8.8). So it can be concluded, that for both test (abrasion and scratch resistances) we have improvement by the addition of 2% CNC particles in the varnishes, while properties remain at same level for the paints.

8.3.6 Multilayer coatings

We need opaque paint for some applications (kitchen furniture, etc.). Addition of CNC particles can't improve mechanical resistance because of aggregations with

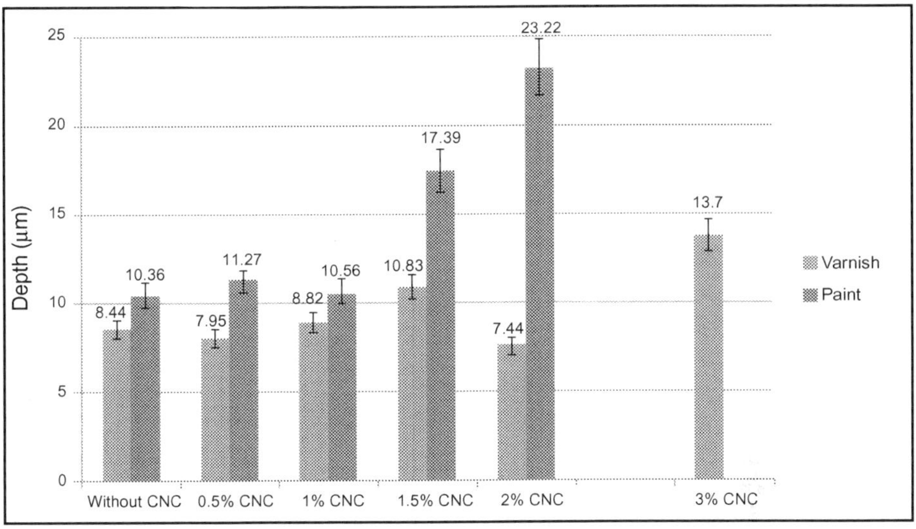

Figure 8.12 Depth after scratching of the paint and varnish formulations with different CNC concentrations.

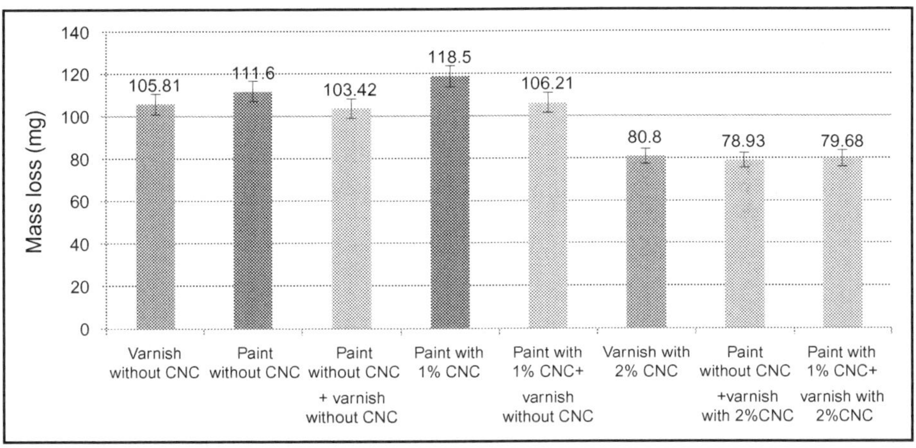

Figure 8.13 Mass loss of the multilayer coatings with different CNC concentrations, compared to that of single layer coatings.

nonorganic pigment (TiO_2). Pigment can be changed, but CNC could again create aggregation with new pigment too. One way to increase wear resistance is to make a multilayer coating: first layer with paint on the wood for obtaining opacity and white (or other) color, followed with one varnish layer to improve mechanical resistance. This technique is well known in car industry. Figure 8.13 presents the mass loss found for the multilayer coating after 100 rotations (all four green bars), as well as some other monolayer results. The mass loss of multilayer coatings is nearly equal to those of varnishes, so that the second (varnish) layer adheres to the first, and protects the wood.

There is no reason to repeat the scratch resistance test for the multilayer coatings, because the maximum depth which registered for the varnishes is 13.7 μm (Figure 8.13, varnish with 3% CNC) and the varnish layer thickness is about 50 μm.

These multilayer coatings retain the white color of the paint because of the highly transparent varnishes (Figure 8.2).

8.4 Conclusions

In this chapter, UV-curable varnish and paint formulations containing CNC particles have been prepared. The surface roughness was measured by AFM and gives good indications of the degree of CNC dispersion and agglomeration for paints. It also gives indications of the degree of dispersion for varnish formulations. This is an efficient method of evaluating the dispersion in cases where SEM cannot be employed, as there is no contrast between the particles and matrix (organic composition). This method can easily be used to characterize the dispersion of other nanoparticles in other matrices and give not only qualitative results but also quantitative results. In fact, the average of *N* measurements of roughness parameters, as obtained thanks to the AFM software, can easily and precisely be used to compare roughness parameters.

The average roughness of the painted surfaces is small, considering that wood is a rather rough substrate. This may be due to the high surface tension of the water, which, while evaporating, flattens all irregularities.

Still the presence of CNC can be detected by the topography of the varnish or paint surface. This is due to the sensitivity of the AFM technique, but it was found to have no effect on the optical properties of the coating.

Finally, the impact of the CNC addition in the paint and varnish formulations was assessed. Varnish and paint formulations behaviors differ significantly in terms of abrasion and scratch resistance. For the varnish formulations, an optimum concentration (2%) was found. In comparison, paint formulations were all negatively impacted by the addition of CNC particles. When the reinforced varnish was applied to an opaque paint layer, rather than the wood, the same reinforcement took place.

This work was performed with unmodified CNC. Surface-modified CNC and its influence on dispersion will be introduced in another publication.

Acknowledgments

The authors thank Arboranano, Nanoquebec, FRNTQ (Québec, Canada) for funding this project; the FPInnovations' labs (Québec and Pointe-Claire, QC, Canada) for providing the CNC samples and helping with the coating formulation; as well as Professors Carmel Jolicoeur and Bohuslav Kokta for help with funding application.

References

Ajayan, P.M., Schadler, L.S., Braun, P.V., 2003. Nanocomposite Science and Technology. Wiley, New York.

Amerio, E., Fabbri, P., Malucelli, G., Messori, M., Sangermano, M., Taurino, R., 2008. Scratch resistance of nano-silica reinforced acrylic coatings. Prog. Org. Coat. 62 (2), 129–133.

Asif, A.A., John, B., Rao, V.L., Ninan, K.N., 2010. Surface morphology, thermomechanical and barrierproperties of poly(ether sulfone)-toughened epoxy clay ternary nanocomposites. Polym. Int. 59 (7), 986–997.

Auclair, N., Riedl, B., Blanchard, V., Blanchet, P., 2011. Improvement of photoprotection of wood coatings by using inorganic nanoparticles as ultraviolet absorbers. For. Prod. J. 61, 20–27.

Bauer, F., Sauerland, V., Glasel, H.J., Ernst, H., Findeisen, M., et al., 2002. Preparation of scratch and abrasion resistant polymeric nanocomposites by monomer grafting onto nanoparticles, 3. Effect of filler particles and grafting agents. Macromol. Mater. Eng. 287, 546–552.

Bauer, F., Ernst, H., Hirsch, D., Naumov, S., Pelzing, M., et al., 2004. Preparation of scratch and abrasion resistant polymeric nanocomposites by monomer grafting onto nanoparticles, 5(a) – application of mass spectroscopy and atomic force microscopy to the characterization of silane-modified silica surface. Macromol. Chem. Phys. 205, 1587–1593.

Bauer, F., Flyunt, R., Czihal, K., 2006. Nano/micro particle hybrid composites for scratch and abrasion resistant polyacrylate coatings. Macromol. Mater. Eng. 291, 493–498.

Bauer, F., Flyunt, R., Czijal, K., Langguth, H., Mehnert, R., et al., 2007. UV curing and matting of acrylate coatings reinforced by nano-silica and micro-corundum particles. Prog. Org. Coat. 60 (2), 121–126.

Beck, S., Bouchard, J., Berry, R., 2011. Controlling the reflection wavelength of iridescent solid films of cellulose nanocrystals. Biomacromolecules. 12, 167–172.

Berndt, S., Wersarg, F., Wiegand, C., Kralish, D., Müller, F.A., 2013. Antimicrobial porous hydrid consisting of bacterial nanocellulose and silver nanoparticles. Cellulose. 20, 771–783.

Bulian, F., Graystone, J.A., 2009. Wood coatings: theory and practice. Elsevier, Amsterdam.

Cool, J., Hernandez, R.E., 2011. Improving the sanding process of black spruce wood for surface quality and water-based coating adhesion. For. Prod. J. 61, 372–380.

Cristea, M.V., Riedl, B., Blanchet., P., 2011. Effect of addition of nanosized UV absorbers on the physico-mechanical and thermal properties of an exterior waterborne stain for wood. Prog. Org. Coat. 72, 755–762.

Dong, X.M., Revol, J.F., Gray, D.G., 1998. Effect of microcrystallite preparation conditions on the formation of colloid crystals of cellulose. Cellulose. 5, 19–32.

Farrokhpay, S., 2012. Application of spectroscopy and microscopy techniques in surface coatings evaluation: a review. Appl. Spectrosc. Rev. 47, 233–243.

Favier, V., Chanzy, H., Cavaille, J.Y., 1995. Polymer nanocomposites reinforced by cellulose whiskers. Macromolecules. 28, 6365–6367.

Feist, W.C., 1983. Weathering and protection of wood. Am. Wood-Preserv. Assoc. 79, 195–205.

Fernando, R.H., Sung, L.P. (Eds.), 2009. Nanotechnology Applications in Coatings. American Chemical Society, Oxford University Press, Washington, DC, ACS Symposium Series 1008.

Gustafson, S.R., 2003. Architectural Woodwork Quality Standards. first ed. Woodwork Institute, West Sacramento, CA.

Helbert, W., Cavaille, J.Y., Dufresne, A., 1996. Thermoplastic nanocomposites filled with wheat straw cellulose whiskers. 1. Processing and mechanical behavior. Polym. Compos. 17, 604–611.

Heux, L., Dinand, E., Vignon, M.R., 1999. Structural aspects in ultrathin cellulose microfibrils followed by C-13 CP-MAS NMR. Carbohydr. Polym. 40, 115–124.

Hosker, I., 1993. Complete Wood Finishing. Guild of Master Craftsman Publications, London.

Jalili, M.M., Moradian, S., 2009. Deterministic performance parameters for an automotive polyurethaneclearcoat loaded with hydrophilic or hydrophobic nano-silica. Prog. Org. Coat. 66 (4), 359–366.

Jalili, M.M., Moradian, S., Dastmalchian, H., Karbasi, A., 2007. Investigating the variations in properties of 2-pack polyurethane clear coat through separate incorporation of hydrophilic and hydrophobic nano-silica. Prog. Org. Coat. 59 (1), 81–87.

Karakas, F., Celik, M.S., 2012. Effect of quantity and size distribution of calcite filler on the quality of water borne paints. Prog. Org. Coat. 74, 555–563.

Klemm, D., Kramer, F., Moritz, S., Lindström, T., Ankerfors, M., et al., 2011. Nanocelluloses: a new family of nature-based materials. Angew. Chem. Int. Ed. 50, 5438–5466.

Landry, V., Blanchet, P., Riedl, B., 2010. Mechanical and optical properties of clay-based nanocomposites coatings for wood flooring. Prog. Org. Coat. 67, 381–388.

Landry, V., Blanchet, P., FPInnovations, 2011a. Coatings Containing Cellulose Nanocrystals, Processes for Preparation and Use Thereof. Patent WO 2011075837A1.

Landry, V., Alemdar, A., Blanchet, P., 2011b. Cellulose nanocrystals: morphological, physical and mechanical properties. For. Prod. J. 61, 104–112.

Nalwa, H.S., 2003. Handbook of Organic-Inorganic Hybrid Materials and Nanocomposites, vol. 2. American Scientific Publishers, Stevenson Ranch, CA.

Park, S.H., Lee, H.S., Choi, J.H., Jeong, C.M., Sung, M.H., et al., 2012. Improvements in barrier properties of poly(lactic acid) films coated with chitosan or chitosan/clay nanocomposite. J. Appl. Polym. Sci. 125, E675–E680.

Ranjbar, Z., Rastegar, S., 2009. The influence of surface chemistry of nano-silica on microstructure, optical and mechanical properties of the nano-silica containing clear-coats. Prog. Org. Coat. 65 (1), 125–130.

Ray, S., Bousmina, M., 2006. Polymer Nanocomposites and their Applications. Department of Chemical Engineering, Université Laval, Québec, Canada.

Revol, J.F., Godbout, L., Dong, X.M., Gray, D.G., Chanzy, H., et al., 1994. Chiral nematic suspensions of cellulose crystallites – phase separation and magnetic field orientation. Liq. Cryst. 16, 127–134.

Sangermano, M., Lak, N., Malucelli, G., Samakande, A., Sanderson, R.D., 2008. UV-curing and characterization of polymer-clay nanocoatings by dispersion of acrylate-funtionalizedorganoclays. Prog. Org. Coat. 61 (1), 89–94.

Sow, C., Riedl, B., Blanchet, P., 2011. UV-waterborne polyurethane-acrylate nanocomposite coatings containing alumina and silica nanoparticles for wood: mechanical, optical, and thermal properties assessment. J. Coat. Technol. Res. 8, 211–221.

Sugiyama, J., Chanzy, H., Revol, J.F., 1994. On the polarity of cellulose in the cell-wall of Valonia. Planta. 193, 260–265.

Surface Modification Center, 2013. SpecialChem. Available from: <http://www.specialchem4coatings.com> (accessed 12.06.13.).

Thometzek, P., Ludwig, A., Karbach, A., Kohler, K., 1999. Effects of morphology and surface treatment of inorganic pigments on waterborne coating properties. Prog. Org. Coat. 36, 201–209.

Vardanyan, V., Poaty, B., Chauve, G., Landry, V., Galstian, T., et al., 2014. Mechanical properties of UV-waterborne varnishes reinforced by cellulose nanocrystals. J. Coat. Technol. Res. 11 (6), 841–852.

Machining medical grade titanium alloys using nonabrasive nanolayered cutting tools

9

Mark J. Jackson, Rodney G. Handy*, Michael D. Whitfield*, John W. Burgess*, Grant M. Robinson*, Tamara Novakov*, Waqar Ahmed**, Ahmed M. Elhissi**, St John J. Crean**, David A. Phoenix**, Álisson Rocha Machado†, and Marcio Bacci da Silva†*

*Purdue University, Indiana, USA; **University of Central Lancashire, UK; †Federal University of Uberlandia, Brazil

9.1 Metallurgical Aspects

9.1.1 Introduction

This section focuses on the metallurgical structure of titanium alloys so that the reader fully understands how cutting tools respond to various types of titanium materials as a function of machining conditions and tools coatings. In summary, the following explanation of the nature of titanium and its metallurgy was initially published by Jackson and colleagues and is reproduced here for completeness and brevity. The re-publication was authorized by Inderscience publishers (see "Acknowledgments" below). We have updated the text and provided supplementary data on titanium machining with coated tools. A full list of references to works shown in the original article is shown at the end of the chapter so that readers can refer to the original works published that characterize the nature of titanium alloy metallurgy. For the benefit of readers of this chapter, Table 9.1 shows the classification of types of titanium and resulting mechanical property ranges. The ranges are dependent not only on alloying elements introduced into the pure metal, but also on subsequent heat treatment. In essence, information on the nature of α-, $\alpha+\beta$-, and β-type titanium alloys can be found in references (Freese et al., 2001; Froes and Bomberger, 1985; Karasevskaya et al., 2003; Lin et al., 2002).

9.1.2 Basic aspects of titanium metallurgy

Titanium is a transition metal and it occurs in several minerals including rutile and ilmenite, which are well dispersed over the Earth's crust. Even though titanium is as strong as some steels, its density is only half that of steel. Titanium is broadly

Anti-Abrasive Nanocoatings.

used in a number of fields, including aerospace, power generation, automotive, chemical and petrochemical, sporting goods, and dental and medical industries (Sibum, 2003; Wang, 1996; Rack and Qazi, 2006). The large variety of applications is due to its desirable properties, mainly the relative high strength combined with low density and enhanced corrosion resistance Niinomi (2002). Among metallic materials, titanium and its alloys are considered the most suitable materials in medical applications because they satisfy the property requirements better than any other competing materials, like stainless steels, Cr-Co alloys, commercially pure (CP) Nb, and CP Ta (Lütjering and Williams, 2003; Long and Rack, 1998). In terms of biomedical applications, the properties of interest are biocompatibility, corrosion behavior, mechanical behavior, processing ability, and availability (Katti, 2004; Disegi, 2000; He and Hagiwara, 2006).

Titanium may be considered as being a relatively new engineering material. It was discovered much later than the other commonly used metals, its commercial application starting in the late 1940s, mainly as structural material. Its usage as an implant material began in the 1960s (Bannon and Mild, 1983). Despite the fact that titanium exhibits superior corrosion resistance and tissue acceptance when compared with stainless steels and Cr-Co-based alloys, its mechanical properties and tribological behavior restrain its use as biomaterial in some cases. This is particularly true when high mechanical strength is necessary, like in hard tissue replacement or under intensive wear use (Oliveira et al., 1998). To overcome such restrictions, CP titanium was substituted by titanium alloys, particularly, the classic grade 5, that is, Ti-6Al-4V alloy. The Ti-6Al-4V $\alpha + \beta$-type alloy, the most utilized titanium alloy worldwide, was initially developed for aerospace applications. (Boyer, 1996; Ferrero, 2005). Although this type of alloy is considered a good material for surgically implanted parts, recent studies have found that vanadium may react with the tissue of the human body (Wang, 1996). In addition, aluminum may be related with neurological disorders and Alzheimer's disease (Wang, 1996). To overcome the potential vanadium toxicity, two new vanadium-free $\alpha + \beta$-type alloys were developed in the 1980s. Vanadium, a β-stabilizer element, was replaced by niobium and iron, leading to Ti-6Al-7Nb and Ti-5Al-2.5Fe $\alpha + \beta$-type alloys (Niinomi, 2002; Long and Rack, 1998; Semlitsch et al., 1985). While both alloys show mechanical and metallurgical behavior comparable to those of Ti-6Al-4V, a disadvantage is that they all contain aluminum in their compositions.

In recent years, several studies have shown that the elastic behavior of $\alpha + \beta$-type alloys is not fully suitable for orthopedic applications (Vail et al., 1998; Niinomi et al., 2005; Kikuchi et al., 2006; Kim et al., 2006). A number of studies suggest that unsatisfactory load transfer from the implant device to the neighboring bone may result in its degradation (He and Hagiwara, 2006). Also, numerical analyses of hip implants using finite element method indicate that the use of biomaterials with elastic behavior similar to cortical bones improves the distribution of stress around the implanted bone (Gross and Abel, 2001). While the elastic modulus of a cortical bone is close to 18 GPa (Katti, 2004), the modulus of Ti-6Al-4V alloy is 110 GPa (Katti, 2004). In such a case, the high elastic modulus of the implant material may lead to bone resorption and possible unsuccessful implantation procedure.

The elastic behavior mismatch between the implant and the adjacent bone is named "stress shielding effect" (Gross and Abel, 2001).

Since CP titanium and some specific $\alpha + \beta$-type titanium alloys do not completely meet the demands of medical applications, especially concerning mechanical behavior and toxicity to the human body, a new class of alloys has been investigated for biomedical applications in the last decade, the β-type alloys. After proper heat treatments this type of alloy may exhibit low elastic modulus, very good corrosion resistance, and suitable good biocompatible behavior, as they may be obtained by adding biocompatible alloying elements like Nb, Ta, and Zr to titanium (Hao et al., 2003; Hao et al., 2002; Gunawarman et al., 2005; Sakaguchi et al., 2005; Kuroda et al., 1998).

The microstructure diversity of titanium alloys is a result of an allotropic phenomenon. Titanium undergoes an allotropic transformation at 882°C. Below this temperature, it exhibits a hexagonal close-packed (HCP) crystal structure, known as α phase, while at higher temperature it has a body-centered cubic (BCC) structure, β phase. The latter remains stable up to the melting point at 1670°C (Lütjering and Williams, 2003). As titanium is a transition metal, with an incomplete d shell, it may form solid solutions with a number of elements and hence, α and β phase equilibrium temperature may be modified by allowing titanium with interstitial and substitutional elements.

Titanium alloying elements fall into three classes: α-stabilizers, β-stabilizers, and neutral. While elements defined as α-stabilizers lead to an increase in the allotropic transformation temperature, other elements, described as β-stabilizers, provoke a decrease in such a temperature (Peters et al., 2003). When a eutectoid transformation takes place, this β-stabilizer is termed a eutectoid β-stabilizer, otherwise, it is called an isomorphous β-stabilizer. If no significant change in the allotropic transformation temperature is observed, the alloying element is defined as a neutral element. Figure 9.1 shows a schematic representation of types of phase diagram between titanium and its alloy elements (Lütjering and Williams, 2003; Peters et al., 2003).

As a result, titanium alloys with an enormous diversity of compositions are possible. Among α-stabilizer elements are the metals of IIIA and IVA groups (Al and Ga) and the interstitials C, N, and O. On the contrary, β-stabilizer elements include the transition elements (V, Ta, Nb, Mo, Mg, Cu, Cr, and Fe) and the noble metals.

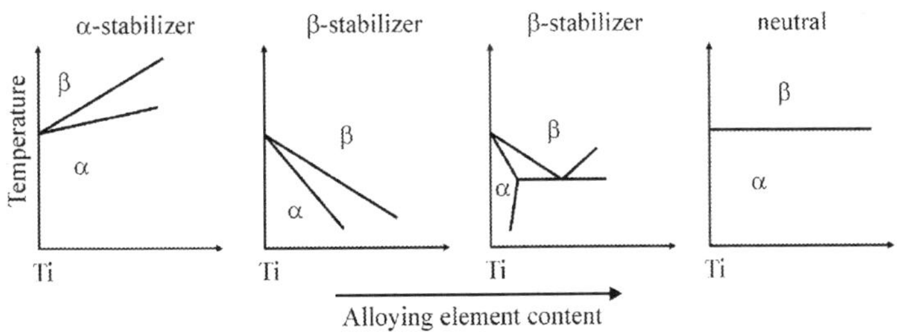

Figure 9.1 Schematic representation of types of phase diagram between titanium and its alloying elements (Lütjering and Williams, 2003; Peters et al., 2003).
Source: Re-printed with permission from Inderscience Publishers. Copyright is retained by Inderscience.

Addition of α- and β-stabilizer elements to titanium gives rise to a field in the corresponding phase diagram where both α and β phase may coexist. Titanium alloys exhibit a variety of properties, which are connected to chemical composition and metallurgical processing. (Ari-Gur and Semiatin, 1998; Lütjering, 1999; Prasad and Seshacharyulu, 1998). According to the nature of their microstructure, titanium alloys may be divided as α alloys, β alloys, or α + β alloys (Freese et al., 2001). Beta alloys may be further classified into near β and metastable β alloys.

Alpha titanium alloys are especially formed by CP titanium and alloys with α-stabilizer elements, which present only α phase at room temperature. Such alloys show high creep resistance and are thus suitable for high-temperature service. Since no metastable phase remains after cooling from a high temperature, no major modification in terms of microstructure and mechanical properties is possible using heat treatments. Finally, as α phase is not subjected to ductile-brittle transition, these alloys are proper for very low temperature applications. Regarding mechanical and metallurgical properties, α alloys present a reasonable level of mechanical strength, high elastic modulus, good fracture toughness, and low ability to be forged, which is due to their HCP crystal structure.

Beta titanium alloys are obtained when a high amount of β-stabilizer elements are added to titanium, which decreases the temperature of the allotropic transformation (α/β transition) of titanium (Froes and Bomberger, 1985). If the β-stabilizer content is high enough to reduce the martensitic start temperature (M_S) to temperatures below the room temperature, nucleation and growth of α phase will be very restricted, and hence, metastable β is retained at room temperature under rapid cooling, as depicted in Figure 9.2. This type of titanium alloy may be hardened by using heat treatment procedures (Karasevskaya et al., 2003). In some cases, depending upon composition and heat treatment parameters, precipitation of ω phase is possible. However, ω phase may cause

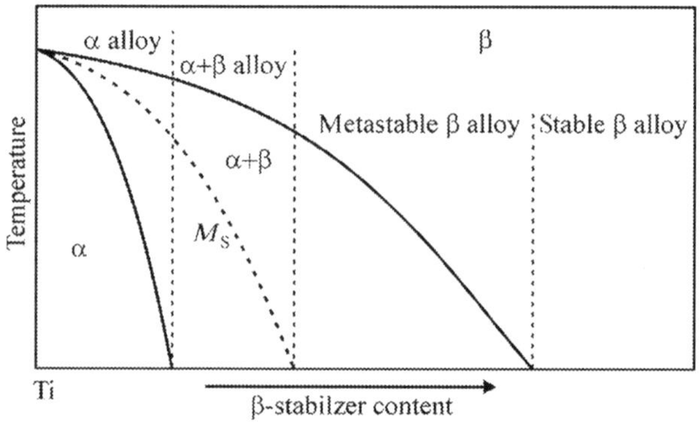

Figure 9.2 Partial phase diagram of titanium and a β-stabilizer element (Lütjering and Williams, 2003; Long and Rack, 1998).
Source: Re-printed with permission from Inderscience Publishers. Copyright is retained by Inderscience.

embrittlement of a titanium alloy and, in general, its precipitation must be avoided (Lin et al., 2002). β-titanium alloys are very brittle at cryogenic temperatures and are not meant to be applied at high temperatures, as they show low creep resistance.

Finally, *α + β alloys* include alloys with enough α- and β-stabilizers to expand the α + β field to room temperature (Lütjering and Williams, 2003; Peters et al., 2003). The α- and β-phase combination allows one to obtain an optimum balance of properties. The characteristics of both α and β phases may be tailored by applying proper heat treatments and thermomechanical processing. A significant assortment of microstructures may be obtained when compared to α-type alloys. The Ti-6Al-4V alloy is an example of α + β-type alloy. Due to its wide availability, very good workability, and enhanced mechanical behavior at low temperatures, such an alloy is the most common composition among the titanium alloys and based on these characteristics it is still largely applied as a biomaterial, mainly in orthopedic implant devices. Figure 9.3 depicts the microstructures of β and α + β titanium alloys.

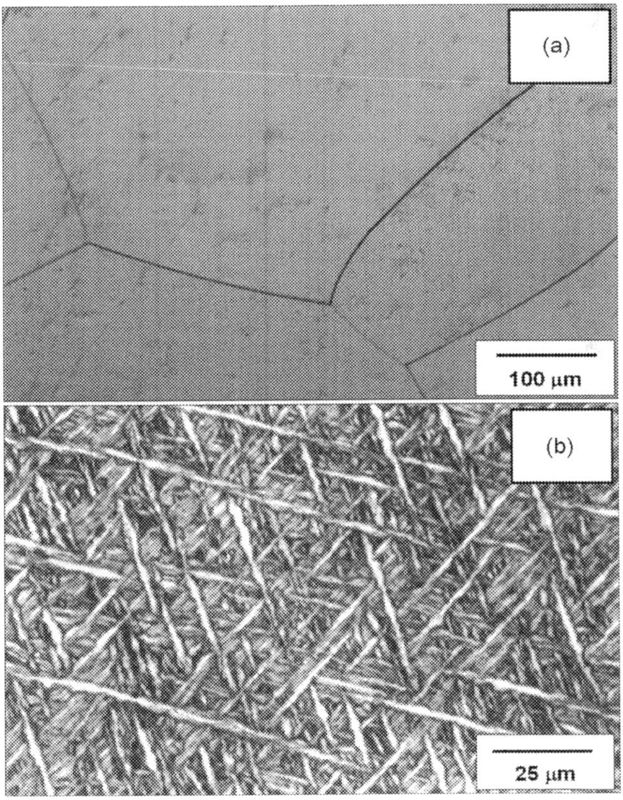

Figure 9.3 Microstructures of (a) β Ti-35Nb (wt%) and (b) α + β Ti-6Al-7Nb (wt%) alloys cooled in air.
Source: Re-printed with permission from Inderscience Publishers. Copyright is retained by Inderscience.

As in the case of iron (steels), allotropic transformation is the main reason for the enormous variety of microstructure in titanium alloys. Titanium alloy microstructures are formed by stable and metastable phases (Moffat and Larbalestier, 1988; Flower et al., 1974). In general, for limited β-stabilizer content and depending on cooling conditions, titanium alloys show only α and β phases.

However, if the thermodynamic equilibrium is not reached, metastable phases may be retained at room temperature, mainly martensitic and ω phases. According to several authors (Tang et al., 2000; Dobromyslov and Elkin, 2003; Dobromyslov and Elkin, 2006), titanium alloys with β-stabilizer elements such as Mo, Nb, Ta, and V may form two types of martensitic structures. If the β-stabilizer content is considered low, rapid cooling leads to formation of hexagonal martensite, termed α'. When this content is high, α' martensite undergoes a distortion, loses its symmetry and is substituted by orthorhombic martensite, defined as α'' (Dobromyslov and Elkin, 2006).

When titanium alloys with β-stabilizer elements are submitted to rapid cooling from a high temperature, the β-phase may transform either in martensitic structures or, eventually, in metastable ω phase (Tang et al., 2000). Figure 9.4 presents a microstructure of the Ti-25Nb (wt%) after cooling in water and in air showing α'' and ω formation.

Precipitation of ω phase occurs only in a limited range of the alloy element and may arise during the quenching from high temperature (β phase), forming a thermal ω phase. However, ω phase may also form after aging of a rapidly quenched structure at medium temperatures, resulting in isothermal ω phase, as indicated in Figure 9.5.

9.1.3 Mechanical behavior

Concerning mechanical behavior, biomedical titanium alloys applied as biomaterial mainly in hard tissue replacement must exhibit a low elastic modulus combined with enhanced strength, good fatigue resistance and good workability. Mechanical behavior of titanium alloys is directly related to composition and mainly, thermomechanical processing. Some mechanical properties of selected titanium-based materials applied as biomaterials are shown in Table 9.1 (Niinomi, 1998).

Mechanical strength may be increased by adding alloying elements, which may lead to solid-solution strengthening, or even precipitation of second phases. Also, by using aging processes, metastable structures obtained by rapid quenching from a β field may give rise to fine precipitates, which considerably increases mechanical strength.

Titanium alloys present a high strength-to-weight ratio, which is higher than with most of steels. While CP titanium has yield strength between 170 (grade 1) and 485 MPa (grade 4), titanium alloys may present values higher than 1500 MPa (Peters et al., 2003).

The elastic modulus or Young's modulus corresponds to the stiffness of a material and is associated to the way interatomic forces vary with distance between atoms in the crystal structure. A comparison between both crystal structures of titanium has led to the conclusion that the HCP structure presents higher values of

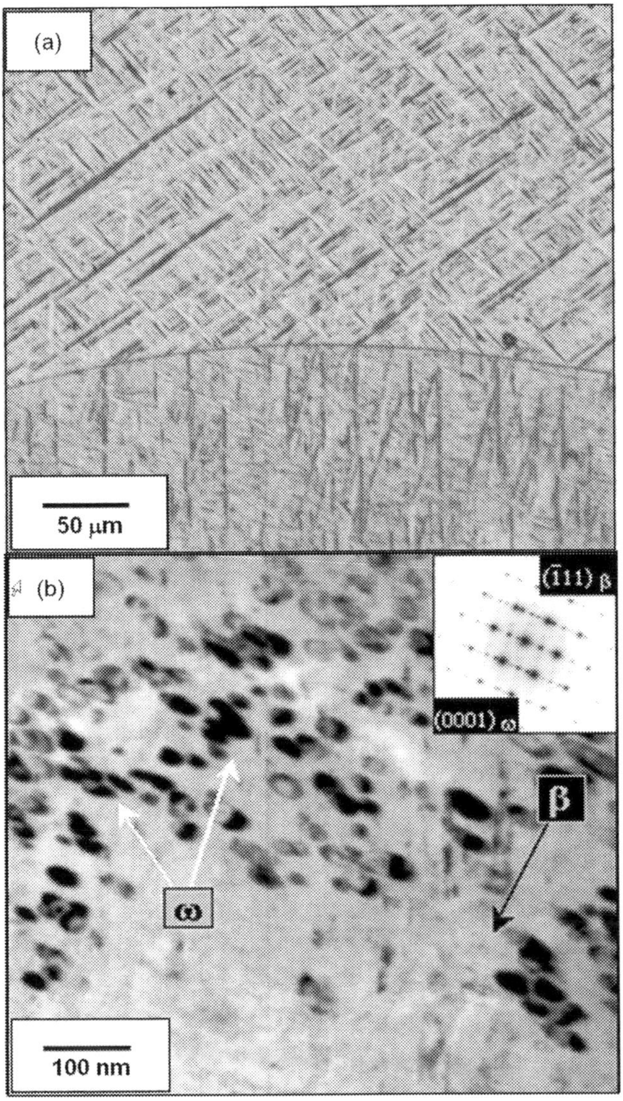

Figure 9.4 Microstructure of the Ti-25Nb (wt%) alloy: (a) water cooled sample showing martensitic structure (MO analysis) and (b) air cooled sample showing ω phase dispersed in β matrix and respective SADP showing ω and β phases microstructure (TEM analysis). *Source:* Re-printed with permission from Inderscience Publishers. Copyright is retained by Inderscience.

elastic modulus than the BCC structure. Hence, the addition of β-stabilizer elements allows β-phase stabilization and, hence, low elastic modulus alloys. While CP titanium shows elastic modulus values close to 105 GPa and Ti-6Al-4V type α + β alloy presents values between 101 and 110 GPa, β-type titanium alloys may present

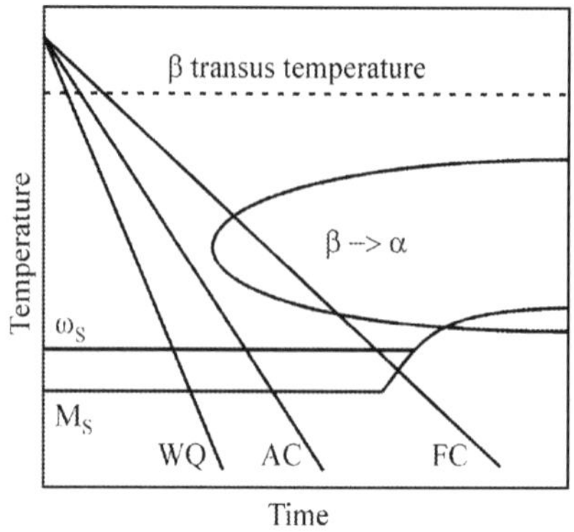

Figure 9.5 A schematic TTT diagram for β-phase transformation in titanium alloys with β-stabilizer elements (Tang et al., 2000).
Source: Re-printed with permission from Inderscience Publishers. Copyright is retained by Inderscience.

Table 9.1 Selected Ti-base materials developed for medical applications (Niinomi, 1998)

Material	Tensile strength (MPa)	Yield strength (MPa)	Elongation (%)	Elastic modulus (GPa)
α type (Lütjering and Williams, 2003; Long and Rack, 1998; Katti, 2004; Disegi, 2000; He and Hagiwara, 2006)				
Pure Ti grade 1	240	170	24	102.7
Pure Ti grade 2	345	275	20	102.7
Pure Ti grade 3	450	380	18	103.4
Pure Ti grade 4	550	485	15	104.1
α + β type (Lütjering and Williams, 2003; Bannon and Mild, 1983; Oliveira et al., 1998; Boyer, 1996; Ferrero, 2005; Semlitsch et al., 1985; Vail et al., 1998; Niinomi et al., 2005; Kikuchi et al., 2006; Kim et al., 2006; Peters et al., 2003)				
Ti-6Al-4V	895–930	825–869	6–10	110–114
Ti-6Al-4V ELI	860–965	795–875	10–15	101–110
Ti-6Al-7Nb	900–1050	880–950	8.1–15	114
Ti-5Al-2.5Fe	1020	895	15	112
β type (Hao et al., 2003; Hao et al., 2002; Gunawarman et al., 2005; Sakaguchi et al., 2005; Kuroda et al., 1998; Freese et al., 2001; Froes and Bomberger, 1985; Karasevskaya et al., 2003; Lin et al., 2002; Tang et al., 2000; Dobromyslov and Elkin, 2003; Dobromyslov and Elkin, 2006)				
Ti-13Nb-13Zr	973–1037	836–908	10–16	79–84
Ti-12Mo-6Zr-2Fe	1060–1100	1000–1060	18–22	74–85
Ti-15Mo	874	544	21	78
Ti-15Mo-5Zr-3Al	852–1100	838–1060	18–25	80
Ti-15Mo-2.8Nb-0.2Si	979–999	945–987	16–18	83
Ti-35.3Nb-5.1Ta-7.1Zr	596.7	547.1	19	55
Ti-29Nb-13Ta-4.6Zr	911	864	13.2	80

Source: Re-printed with permission from Inderscience Publishers. Copyright is retained by Inderscience.

values as low as 55 GPa Niinomi (1998). When compared with common alloys used as biomaterials, such 316L stainless steel (190 GPa), Co-Cr alloys (210–253 GPa), and low elastic modulus titanium alloys display a more compatible elastic behavior to that of the human bone. Brunski (2004). In general, as the elastic modulus decreases so does mechanical strength, and vice versa.

Analysis of slip systems in different crystal structures reveals that plastic deformation is easier in a BCC crystal structure than in an HCP structure. It explains the enhanced ductility of β phase when compared to α phase. In an HCP structure the number of slip systems is only three, while this number increases to 12 in the case of a BCC structure. In addition, the ease of plastic deformation facility is directly connected to the minimum slip distance, b_{min} (Peters et al., 2003), which is given by the interatomic distance divided by the respective lattice parameter. Since an HCP structure exhibits a higher slip distance than a BCC structure, it is possible to conclude that the atomic planes slip or the plastic deformation is easier in a BCC structure than an HCP structure. Hence, β-type alloys present the best formability among the titanium alloys.

9.2 Machining of titanium alloys

Titanium alloys are among the most widely used and promising materials for medical implants. Selection of titanium alloy for implementation is determined by a combination of the most desirable characteristics including immunity to corrosion, biocompatibility, shear strength, density, and osteointegration (Rack and Qazi, 2006). The excellent chemical and corrosion resistance of titanium is to a large extent due to the chemical stability of its solid oxide surface layer to a depth of 10 nm (Tengvall and Lundstrom, 1992). Under in vivo conditions the titanium oxide (TiO_2) is the only stable reaction product whose surface acts as catalyst for a number of chemical reactions. However, micromotion at the cement-prosthesis and cement-bone are inevitable and, consequently, titanium oxide and titanium alloy particles are released in the cemented joint prosthesis. Sometimes this wear debris accumulates as periprosthetic fluid collections and triggers giant cell response around the implants (Henrich and Cox, 1994). TiO_2 film, such as the ones anodically formed in aqueous electrolytes, consists mainly of anatase and is an n-type semiconductor with low electronic charge conductivity and a high resistance to anodic current (Thull and Grant, 2001).

Processes of machining titanium alloys involve conventional machining operations (turning, face milling, high-speed cutting (HSC), milling, drilling), forming operations (cold and hot forming; hydroforming, forging), and alternative machining operations (laser cutting, waterjet cutting, direct metal laser sintering). Machining operations of titanium alloys are considered to be difficult, due to its relatively high tensile strength, low ductile yield, 50% lower modulus of elasticity (104 GPa), and approximately 80% lower thermal conductivity than that of steel (Klocke, 2001). The lower modulus of elasticity may cause a greater "spring back" and deflection effect of the workpiece. Therefore, more rigid set-ups and greater clearances for tools are required. In the tool contact zones high pressures and temperatures occur (the tool-to-workpiece interface). The amount of heat removed by laminar chips is only approximately 25%, the rest is removed via the tool. Owing to this phenomenon titanium alloys can be machined at comparatively low cutting speeds. At higher temperatures caused by friction the titanium becomes more chemically reactive and there is a tendency for titanium to weld to tool bits during machining operations. Overheating of the surface can result in interstitial pickup of oxygen and nitrogen, which will produce a hard and brittle alpha case. Carbides with high WC-Co content (K-grades) and high-speed steels with high cobalt content are suitable for use as cutting materials in titanium machining operations (Klocke, 2001). Turning operations of titanium alloys should have cutting depths as large as possible, and cutting speeds V_C from 12 to 80 m/min and approximately 50% lower when high-speed steel (HSS) tools are used. The heat generated should be removed via large volumes of cooling lubricant. Chlorinated cutting fluids are not recommended because titanium can be susceptible to stress corrosion failures in the presence of chlorine. Any type of hot working or forging operation should be carried out below 925°C due to the high level of titanium reactivity at high temperature. Some medical implants are produced modularly, using different materials and processing techniques. For example, the femoral stem as part of the hip endoprosthesis is produced in a combination of casting, forging, and milling. The final machining operation is performed on a CNC machine using CAD–CAM principles. Good alternatives to conventional machining techniques are alternative techniques such as waterjet cutting, sintering, or direct metal laser sintering. The latter is a rapid prototyping technique enabling prompt modeling of metal parts with high bulk density on the basis of individual three-dimensional data, including computer tomography models of anatomical structures (Hollander et al., 2006). The concept of layer-by-layer building rather than removing waste material to achieve the desired geometry of a component opens up endless possibilities of alternative manufacture of medical devices and is more environment friendly.

9.3 Machining with coated cutting tools: a case study

In 2013, it was estimated that well over half of the tooling used in industry consisted of physical vapor deposition (PVD) and chemical vapor deposition (CVD) coated carbides (Grzesik, 2003). Coatings are a few microns in thickness, which does not affect the geometry of the coated tool. PVD and CVD coatings can provide many different functions to minimize the wear of the tool and extend its life. It is unknown which coating is most effective at extending the tool life for a given set of machining parameters and substrate materials. The proper coating for a given application must be determined experimentally. This section of the chapter focuses on the experimental turning of a well-known titanium alloy (Ti6Al4V) using TiCN-, TiN-, and $Ti_{0.5}Al_{0.5}N$-coated carbide tools. An uncoated carbide insert will be used as a control for this experimental work. The inserts used for the turning process consist of a simple geometry 80-degree diamond-shaped insert. There is no chip breaking geometry present on these inserts to help ensure that the coating is the only variable that can enhance the performance of the tool. The coatings were supplied by a well-established vendor (OerlikonBalzers, 2013).

9.3.1 Material analysis

The workpiece material used in this study is Ti6Al4V alloy; 0.875-inch and 0.75-inch rods were used in the turning procedure. The mechanical properties and material components used for Ti6Al4V alloys are taken from MatWeb (MatWeb, 2012) and are shown on that particular website.

9.3.2 Coating analysis

The coatings used on the turning inserts were TiN, TiCN, and $Ti_{0.5}Al_{0.5}N$. The coatings were applied using a PVD process. Three layers of coating were applied to each insert. The applied coating layers are 3 μm thick. These coatings are highly wear resistant and chemically inert. Due to the extremely thin coatings, the alteration to the production tolerances of the inserts is negligible. The $Ti_{0.5}Al_{0.5}N$ has a maximum service temperature of 900°C and has a microhardness of 3300 (HV 0.05) (OerlikonBalzers, 2013). The TiN has a maximum service temperature of 600°C and has a microhardness of 2300 (HV 0.05) (OerlikonBalzers, 2013). The TiCN has a maximum service temperature of 400°C and has a microhardness of 3000 (HV 0.05) (OerlikonBalzers, 2013). The coatings may increase tool life of the insert and reduce tooling costs (Figure 9.6).

Figure 9.6 Carbide inserts used in Ti machinability study. From left to right: TiCN, TiN, uncoated, and $(Ti_{0.5}Al_{0.5})N$.

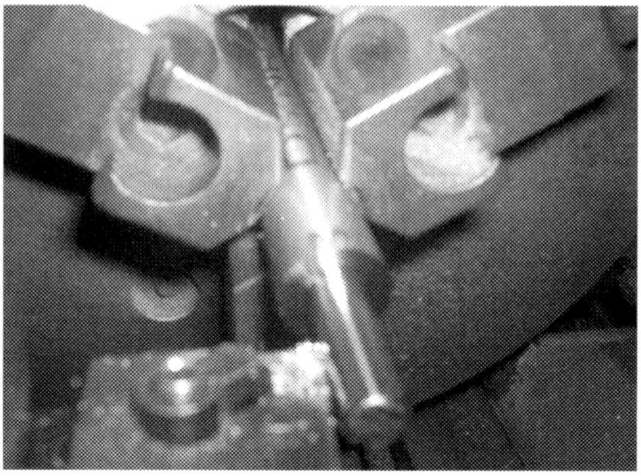

Figure 9.7 Image of the turning process.

9.3.3 Experimental parameters and procedure

The experimental turning of Ti6Al4V alloy was done on a Monarch metal cutting lathe. The workpiece was turned from 0.875 inches diameter down to 0.25 inches diameter through a series of 1-inch length passes. Chips from these passes were collected and analyzed to calculate strain rate. Surface roughness readings were taken using a Fowler Pocket Surf Profilometer. Flank wear was measured after each set of passes using a microscope fitted with a PaxCam camera and Paxit Software (Figure 9.7).

The cutting parameters were as follows:

- 0.0025 in, (0.0635 mm) depth of cut;
- 0.0025 in/rev, (0.066 mm/rev) feed rate;
- spindle speed 319 rpm; and
- cutting speed 73.03 ft/min, (22.3 m/min).

9.3.4 Experimental data

9.3.4.1 Surface roughness

Figures 9.8–9.11 show surface roughness data collected from samples of Ti6Al4V turned using a $Ti_{0.5}Al_{0.5}N$, uncoated, TiN, and TiCN-coated tools. The parameters for this turning process were 0.066 mm/rev, 0.0635 mm depth of cut, at 319 rpm, 21.3 m/min. The surface roughness results are in μin.

Surface roughness measurements were conducted in a controlled environment and procedures used are standard throughout the medical device industry.

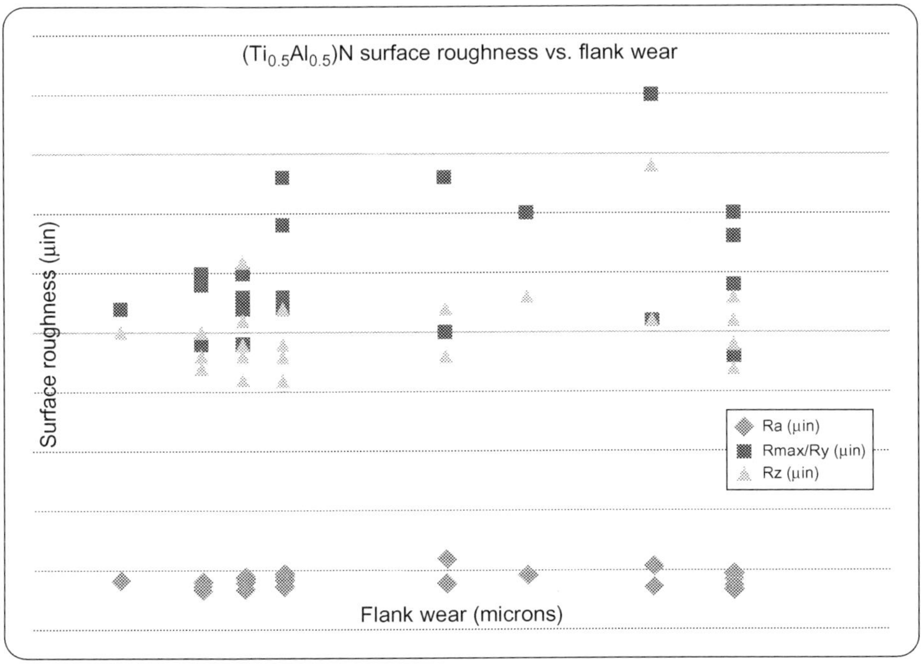

Figure 9.8 Experimental surface roughness for $(Ti_{0.5}Al_{0.5})N$ coated tool.

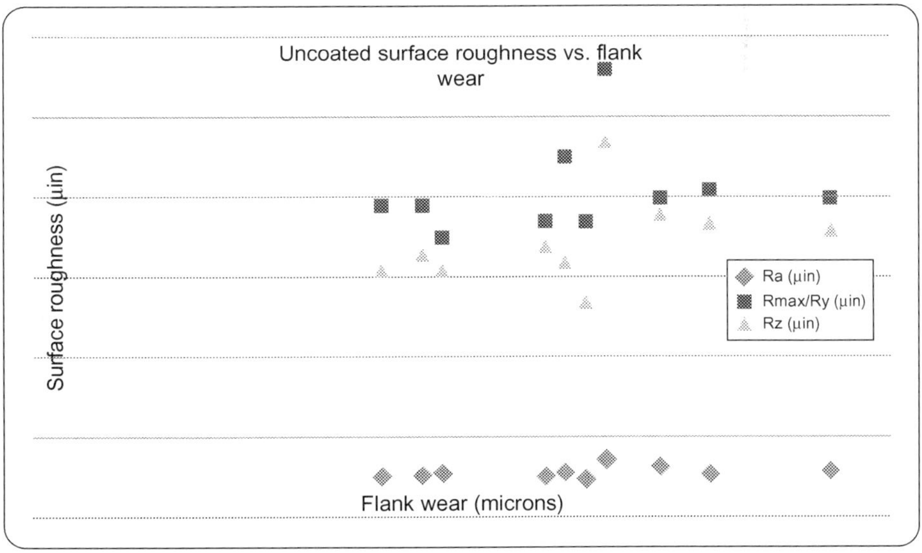

Figure 9.9 Experimental surface roughness for uncoated tool.

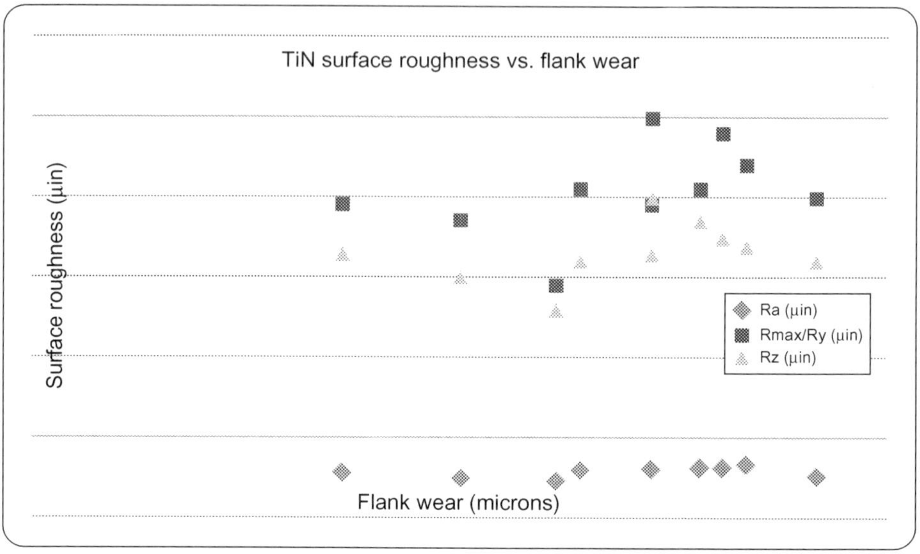

Figure 9.10 Experimental surface roughness for TiN-coated tool.

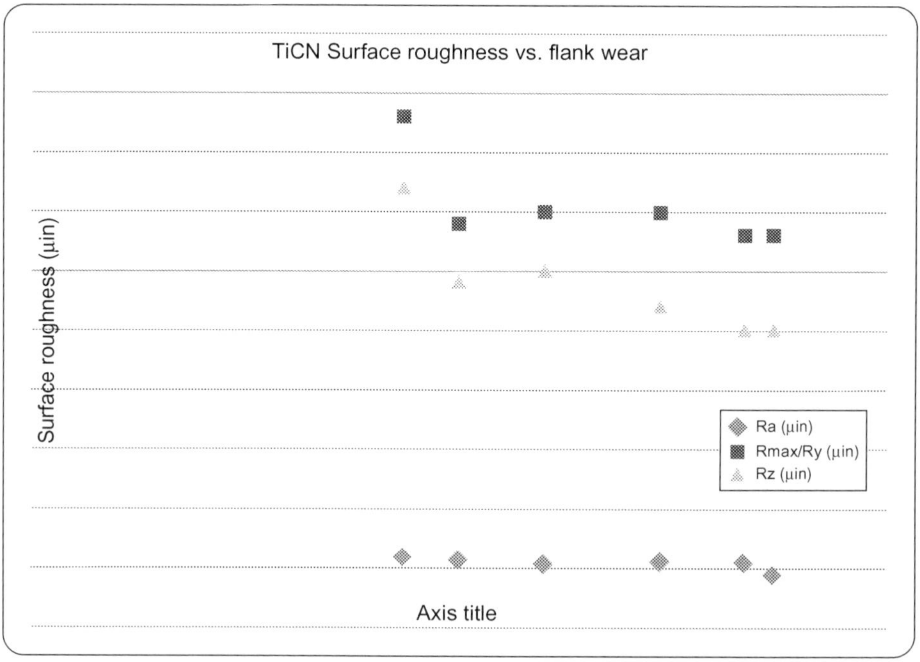

Figure 9.11 Experimental surface roughness for TiCN-coated tool.

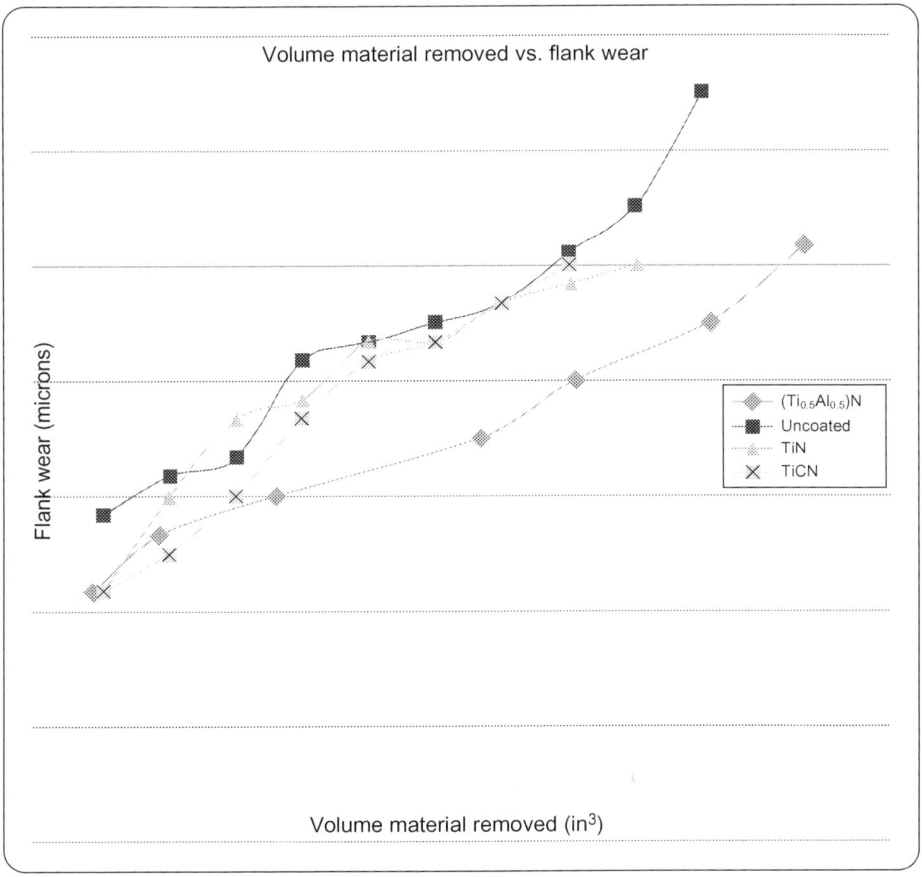

Figure 9.12 Flank wear vs. volume of material removed.

9.3.4.2 Flank wear of tool inserts

Figure 9.12 shows the experimental flank wear versus volume of material removed for the turning of Ti6Al4V using carbide inserts with $Ti_{0.5}Al_{0.5}N$, TiN, TiCN coating, and uncoated carbide. One can see that flank wear is progressive for all tools, but the rate of progression is clearly influenced by the type of coating used in the study.

9.3.4.3 Wear images of tool inserts

Figures 9.13–9.18 show the progressive wear of tool inserts when machining Ti6Al4V titanium alloy using a variety of uncoated and coated cutting tools. The wear criterion identified is flank wear.

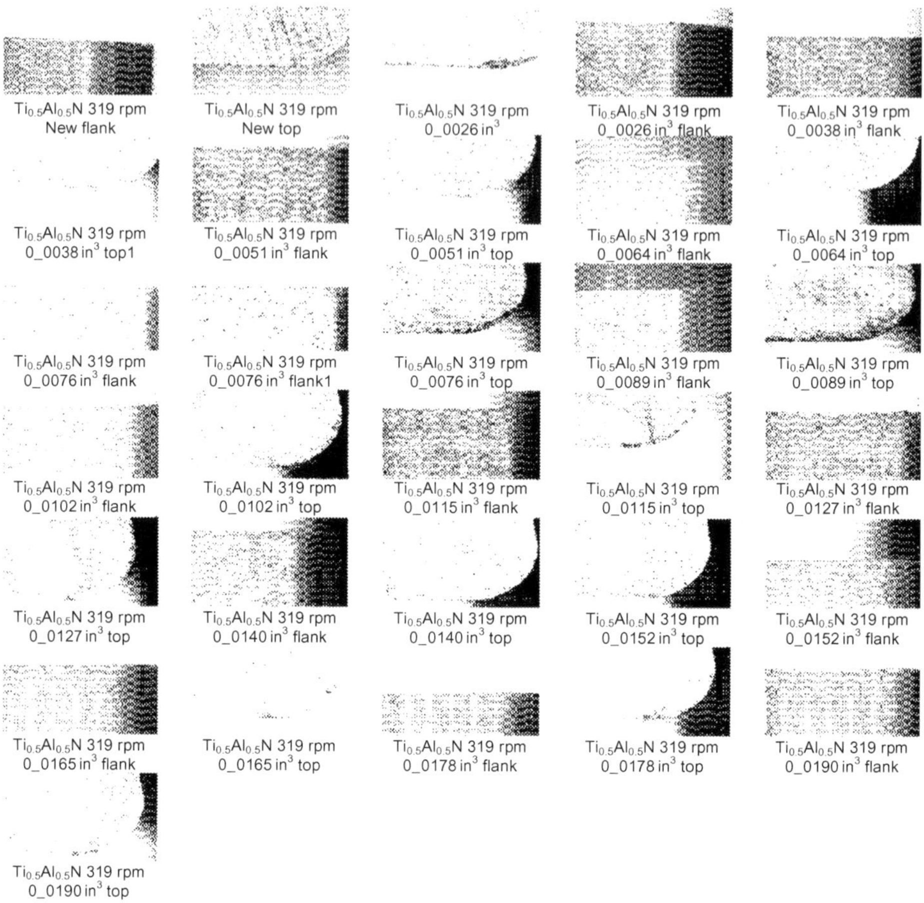

Figure 9.13 Wear scars on a virgin $Ti_{0.5}Al_{0.5}N$ tool that has machined 0.0190 in^3 of material.

The observed stages of wear of coated inserts shows that flank wear is intermittently influenced by galling of the workpiece as the thermal source of heat is concentrated at the immediate contact zone between the tool edge and workpiece. This is not unusual for titanium materials and further studies are being conducted to understand the precise removal mechanisms for these materials and coating types.

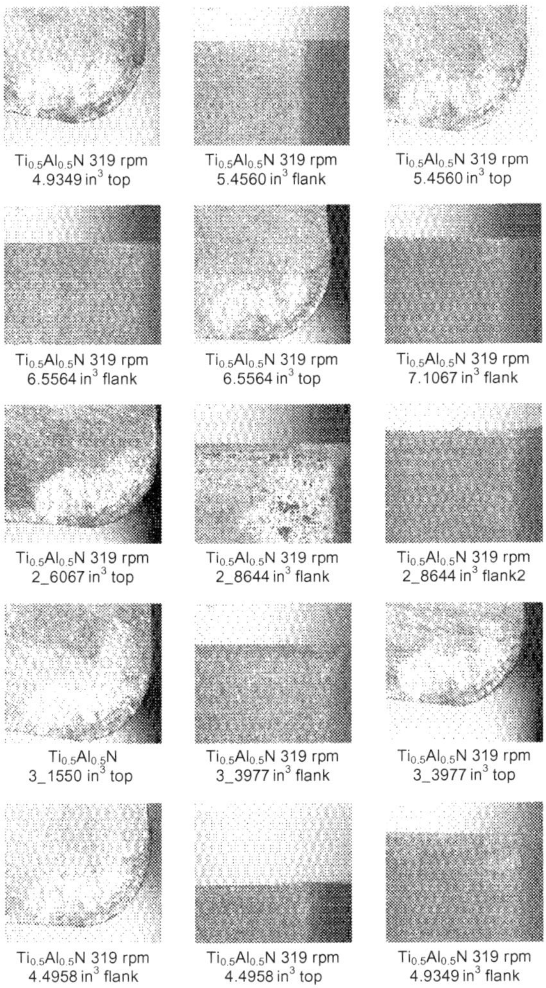

Figure 9.14 $Ti_{0.5}Al_{0.5}N$ tool machining 2.6067 in^3 of material to a volume of 7.6569 in^3.

9.3.4.4 Economic analysis

The economic analysis presented uses the following nomenclature to calculate the cost per component for each coating (The values used in this chapter are based on research findings recently conducted at Purdue University.):

$$C_{pr} = Mt_l + Mt_m + (t_r/t_p) * (M * t_{ct} + C_t)$$

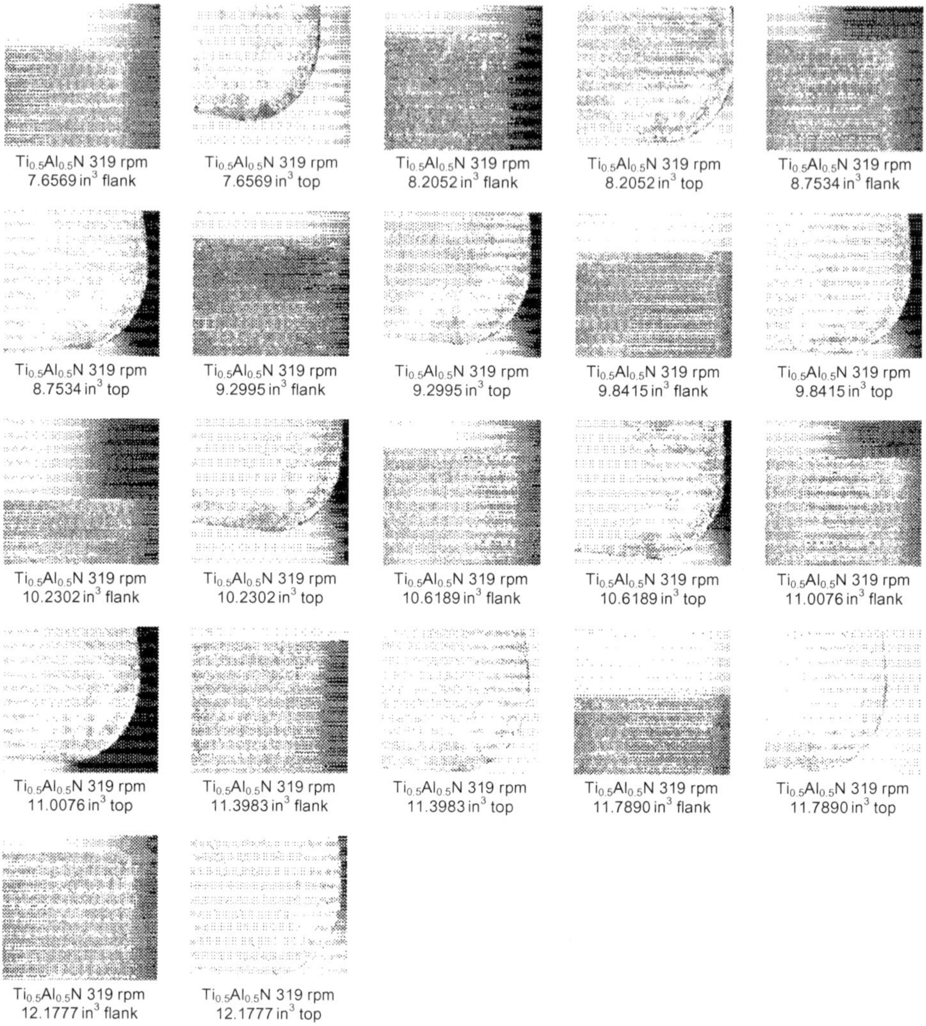

Figure 9.15 $Ti_{0.5}Al_{0.5}N$-coated tool machining 7.6569 in^3 of material to a final volume of 12.1777 in^3.

Where:

C_{pr} is the cost of tooling per component – \$;
M is total machine and operator rate including overhead – \$/min;
Mt_l is machine load time – min;
Mt_m is machine reference time – min;
t_r is reference time – min;
t_p is time predicted based on tool life – min;
t_{ct} is tool change time – min; and
C_t is tool cost – \$.

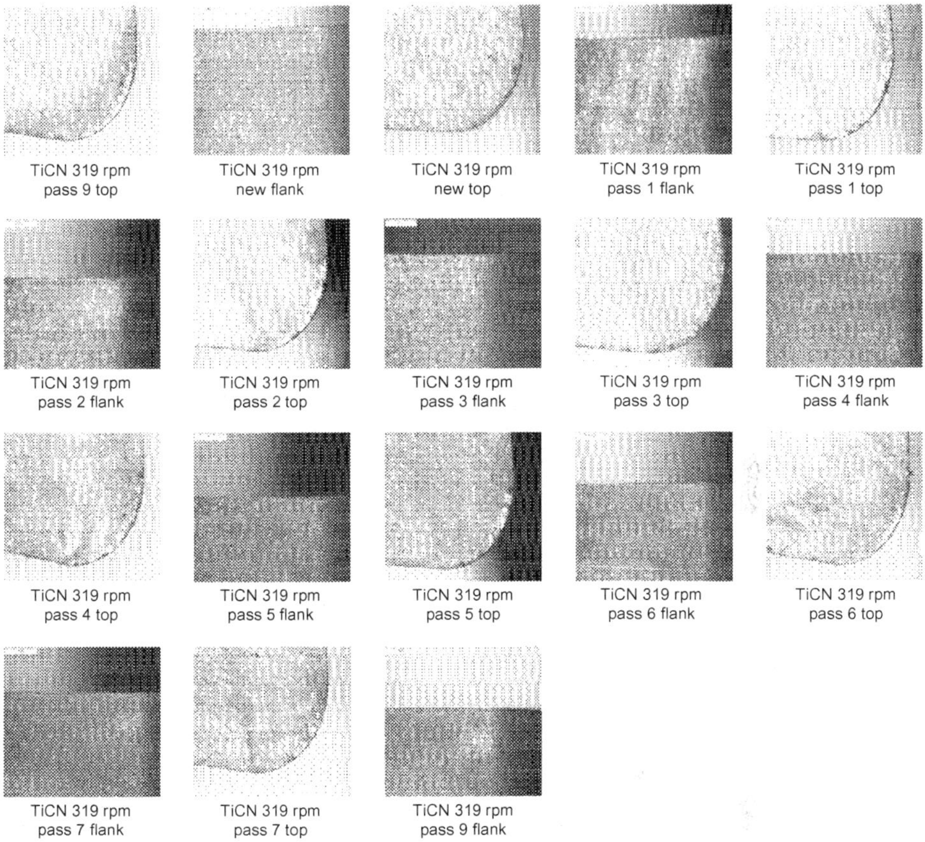

Figure 9.16 TiCN-coated tool showing 3.1237 in^3 of material removed.

It should be noted that $108/hr is the shop rate used in this investigation to establish the most economic coating to use in this particular machining application.

Using the notation shown above, the relative economic performance of using coated tools is shown in Table 9.2.

The coatings are ranked in the order of increase in productivity versus an uncoated tool. For this particular case study, it is shown that the $Ti_{0.5}Al_{0.5}N$-coated WC-Co insert was the most economic tool coating using the machining parameters specified. Of course, this may not be the best coating for other specific machining operations. A thorough investigation of how coatings affect economic performance of machining is currently being conducted at Purdue University's Center for Advanced Manufacturing.

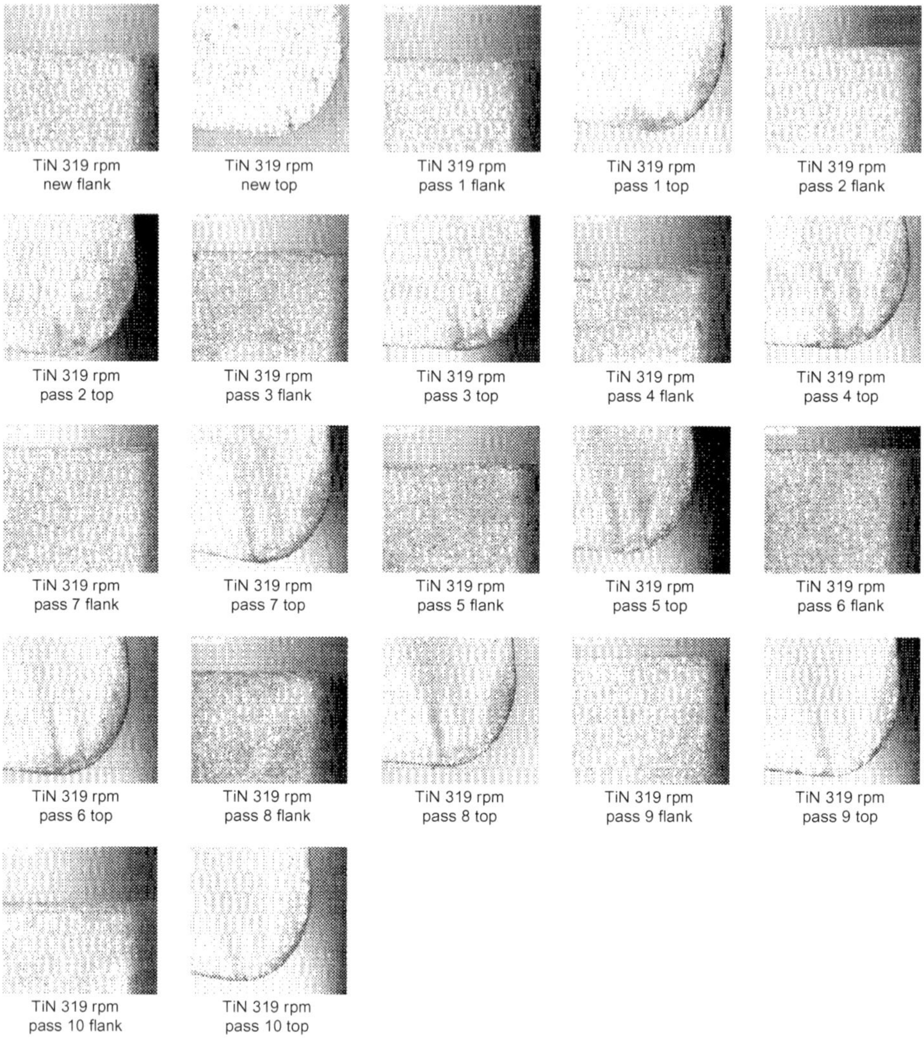

Figure 9.17 TiN-coated machining 3.5243 in^3 of material from the virgin state.

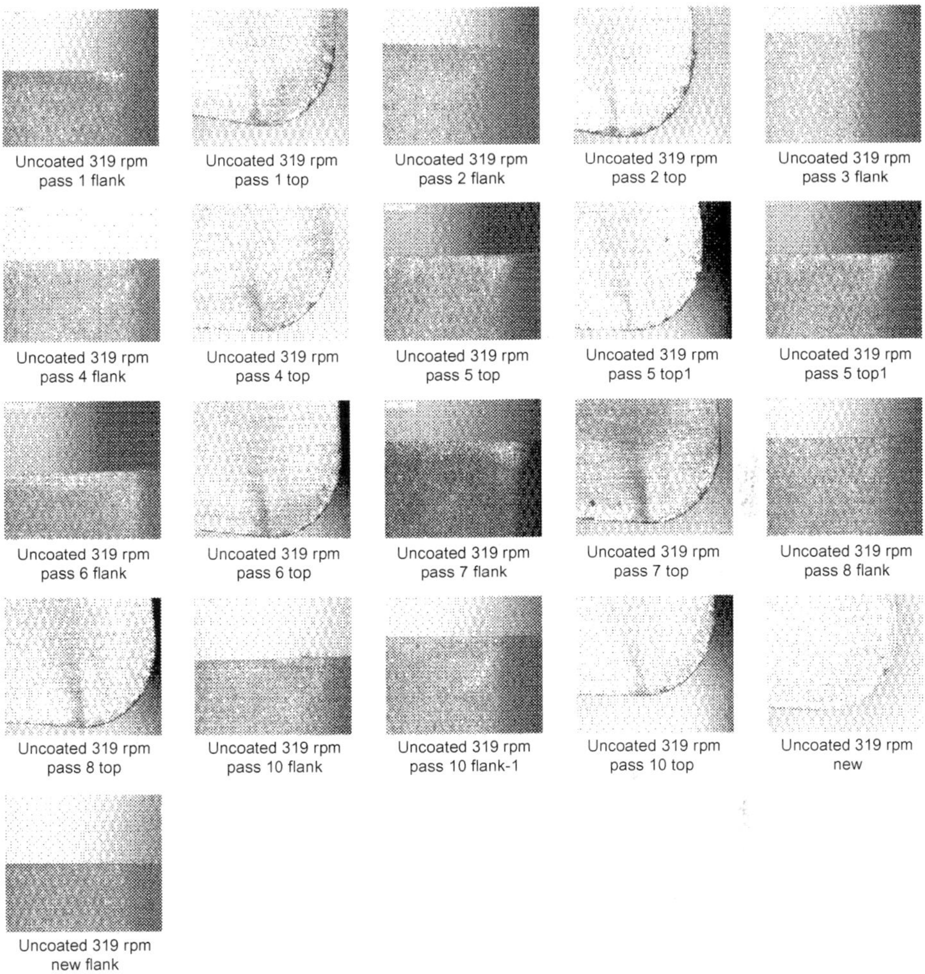

Figure 9.18 Uncoated tool machining 3.9011 in^3 of material from the virgin state.

Table 9.2 **Economic analysis of tool coatings**

Coating	C_t	Mt_l	Mtm	t_r	t_p	M	t_{ct}	C_{pr}	%	*n*
$Ti_{0.5}Al_{0.5}N$	18.01	1.2	18	900	510.8108108	0.03	300	66.78904762	6.38	0.37
TiCN	18.01	1.2	18	900	489.4736842	0.03	300	68.86354839	3.47	0.38
TiN	17.23	1.2	18	900	469.2307692	0.03	300	69.51	2.56	0.39
Uncoated	15	1.2	18	900	414.2857143	0.03	300	71.33793103	0	0.42

9.4 Conclusions

This report has provided an outline of the effects of using a variety of coatings on milling tools machining a specific titanium alloy, Ti6Al4V. The report explains how different coatings affect the economics of the machining process under fixed conditions. It should be noted that the selection of the coating might change when machining conditions change. Only a thorough determination of machining conditions will determine the best coating to use for a particular application. The conclusions drawn from this study are:

- The tool wear shown is consistently on the flank face and is known as flank wear.
- The coated tools performed well under the cutting parameters described above, when cutting Ti6Al4V alloy.
- The coating is able to withstand the heat generated in the cutting process to prevent thermal breakdown of the coating and/or tool.
- From the economic analysis it is cost effective to use coated tooling.
- From the initial study, $Ti_{0.5}Al_{0.5}N$ exceeds the performance of other coatings, followed by TiCN, TiN that outperformed uncoated tools.
- From the initial study, the wear of the cutting tool does not affect the surface finish at low levels of flank wear.

Acknowledgments

The authors thank Inderscience for permission to reproduce (copyright retained by Inderscience) the article that forms the first part of this chapter. The authors have updated and provided supplementary data on titanium machining with coated tools. The re-publication was authorized by Inderscience publishers and originally appeared in the *International Journal of Nanomanufacturing* as: M. Blazic, J. Kopac, M.J. Jackson, and W. Ahmed (2007) "Review of titanium and titanium alloy applications in medicine," *International Journal of Nanomanufacturing* 1(1).

References

Ari-Gur, P., Semiatin, S.L., 1998. Evolution of microstructure, macrotexture and microtexture during hot rolling of Ti-6Al-4V. Mater. Sci. Eng. A. 257, 118.

Bannon, B.P., Mild, E.E., 1983. Titanium alloys for biomaterial application: an overview. In: Luckey, H.A., Kubli Jr, F. (Eds.), Titanium Alloys in Surgical Implants, ASTM STP 796. American Society for Testing and Materials, New York, pp. 7–15.

Boyer, R.R., 1996. An overview on the use of titanium in the aerospace industry. Mater. Sci. Eng. A. 213, 103.

Brunski, J.B., 2004. In: Ratner, B.D., Hoffman, A.S., Schoen, F.J., Lemons, J.E. (Eds.), Biomaterials Science – An Introduction to Materials in Medicine. Elsevier Academic Press, San Diego, CA, pp. 137–153.

Disegi, J.A., 2000. Titanium alloys for fracture fixation implants. Injury. 31, S-D14.

Dobromyslov, A.V., Elkin, V.A., 2003. Martensitic transformation and metastable b-phase in binary titanium alloys with d-metals of 4–6 periods. Mater. Sci. Eng. A. 354, 121.

Dobromyslov, A.V., Elkin, V.A., 2006. The orthorhombic α'-phase in binary titanium base alloys with d-metals of V–VIII groups. Mater. Sci. Eng. A. 438–440, 324–326.

Ferrero, J.G., 2005. Candidate materials for high-strength fastener applications in both the aerospace and automotive industries. J. Mater. Eng. Perform. 14, 691.

Flower, H.M., Henry, S.D., West, D.R.F., 1974. The $\beta \rightleftharpoons \alpha$ transformation in dilute Ti-Mo alloys. J. Mater. Sci. 9, 57.

Freese, H.L., Volas, M.G., Wood, J.R., 2001. Metallurgy and technological properties of *titanium* and *titanium* alloys. In: Brunette, D.M., Tengvall, P., Textor, M., Thomsen, P. (Eds.), Titanium in Medicine. Springer, New York; Heidelberg, Germany, pp. 25–51.

Froes, F.H., Bomberger, H.B., 1985. The beta titanium alloys. J. Met. 37, 28.

Gross, S., Abel, E.W., 2001. A finite element analysis of hollow stemmed hip prostheses as a means of reducing stress shielding of the femur. J. Biomech. 34, 995.

Grzesik, W., 2003. Advanced Protective Coatings For Manufacturing and Engineering. Hanser Gardner Publications, Cincinnati, OH.

Gunawarman, B., Niinomi, M., Akahori, T., Souma, T., Ikeda, M., et al., 2005. Mechanical properties and microstructures of low cost β titanium alloys for healthcare applications. Mater. Sci. Eng. C. 25, 304.

Hao, Y.L., Niinomi, M., Kuroda, D., Fukunaga, K., Zhou, Y.L., et al., 2002. Young's modulus and mechanical properties of Ti-29Nb-13Ta-4.6Zr in relation to α'' martensite. Metall. Mater. Trans. 33A, 3137.

Hao, Y.L., Niinomi, M., Kuroda, D., Fukunaga, K., Zhou, Y.L., et al., 2003. Aging response of the Young's modulus and mechanical properties of Ti-29Nb-13Ta-4.6Zr. Metall. Mater. Trans. 34A, 1007.

He, G., Hagiwara, M., 2006. Ti alloy design strategy for biomedical applications. Mater. Sci. Eng. C. 26, 14.

Henrich, V.E., Cox, P.A., 1994. The Surface Science of Metal Oxides. Cambridge University Press, Cambridge.

Hollander, D.A., von Walter, M., Wirtz, T., Sellei, R., Rohlfing, B.S., et al., 2006. Structural, mechanical and in vitro characterization of individually structured Ti–6Al–4V produced by direct laser forming. Biomaterials. 27, 955–963.

Karasevskaya, O.P., Ivasishin, O.M., Semiatin, S.L., Matviychuk, Y.V., 2003. Deformation behavior of beta-titanium alloys. Mater. Sci. Eng. A. 354, 121.

Katti, K.S., 2004. Biomaterials in total joint replacement. Colloids Surf. B. 39, 133.

Kikuchi, M., Takahashi, M., Okuno, O., 2006. Elastic moduli of cast Ti-Au, Ti-Ag, and Ti-Cu alloys. Dent. Mater. 22, 641.

Kim, H.-S., Kim, W.-Y., Lim, S.-H., 2006. Microstructure and elastic modulus of Ti-Nb-Si ternary alloys for biomedical applications. Scr. Mater. 54, 887.

Klocke, F., 2001. Manufacturing Technology I. WZL-RWTH, Aachen, Germany.

Kuroda, D., Niinomi, M., Morinaga, M., Kato, Y., Yashiro, T., 1998. Design and mechanical properties of new β-type titanium alloys for implant materials. Mater. Sci. Eng. A. 243, 244.

Lin, D.J., Chern, J.H., Ju, C.P., 2002. Effect of omega phase on deformation behavior of Ti-7.5Mo-xFe alloys. Mater. Chem. Phys. 76, 191.

Long, M., Rack, H.J., 1998. Titanium alloys in total joint replacement – a materials science perspective. Biomaterials. 19, 1621.

Lütjering, G., 1999. Property optimization through microstructural control in titanium and aluminum alloys. Mater. Sci. Eng. A. 263, 117.

Lütjering, G., Williams, J.C., 2003. Titanium. Springer-Verlag, Berlin, Germany.

MatWeb. Stratasys® ArcamTi6Al4V Titanium Alloy (2012) Available from: <http://www.matweb.com/search/DataSheet.aspx?MatGUID=237df255c8324fe6a06dc004e86d86b6&ckck=1> (accessed 29.06.12.).

Moffat, D.L., Larbalestier, D.C., 1988. The competition between the alpha and omega phases in aged Ti-Nb alloys. Metall. Trans. 19A, 1687.

Niinomi, M., 1998. Mechanical properties of biomedical titanium alloys. Mater. Sci. Eng. A. 243, 231.

Niinomi, M., 2002. Recent metallic materials for biomedical applications. Metall. Mater. Trans. 33A, 477.

Niinomi, M., Akahori, T., Takeuchi, T., Katsura, S., Fukui, H., et al., 2005. Mechanical properties and cyto-toxicity of new beta type titanium alloy with low melting points for dental applications. Mater. Sci. Eng. C. 25, 417.

OerlikonBalzers. Balinit® coatings (2013) Available from: <http://www.oerlikon.com/ecomaXL/index.php?site=BALZERS_EN_balinit_coatings> (accessed 20.07.13.).

Oliveira, V., Chaves, R.R., Bertazzoli, R., Caram, R., 1998. Preparation and characterization of Ti-Al-Nb orthopedic implants. Braz. J. Chem. Eng. 17, 326.

Peters, M., Hemptenmacher, H., Kumpfert, J., Leyens, C., 2003. Review – titanium and titanium alloy applications in medicine. In: Leyens, C., Peters, M. (Eds.), Titanium and Titanium Alloys. Wiley-VCH, Switzerland, pp. 1–57.

Prasad, Y.V.R.K., Seshacharyulu, T., 1998. Processing maps for hot working of titanium alloys. Mater. Sci. Eng. A. 243, 82.

Rack, H.J., Qazi, J.I., 2006. Titanium alloys for biomedical applications. Mater. Sci. Eng. C. 26, 1269.

Sakaguchi, N., Niinomi, M., Akahori, T., Takeda, J., Toda, H., 2005. Relationship between tensile deformation behavior and microstructure in Ti-Nb-Ta-Zr. Mater. Sci. Eng. C. 25, 363.

Semlitsch, M., Staub, F., Weber, H., 1985. Titanium-aluminum-niobium alloy, development for biocompatible, high-strength surgical implants. Biomed. Tech. 30, 334.

Sibum, H., 2003. Titanium and titanium alloys – from raw material to semi-finished products. Adv. Eng. Mater. 5 (6), 393.

Tang, X., Ahmed, T., Rack, H.J., 2000. Phase transformations in Ti-Nb-Ta and Ti-Nb-Ta-Zr alloys. J. Mater. Sci. 35, 1805.

Tengvall, P., Lundstrom, I., 1992. Physico-chemical considerations of titanium as a biomaterial. Clin. Mater. 9, 115–134.

Thull, R., Grant, D., 2001. Physical and chemical vapor deposition and plasma-assisted techniques for coating titanium. In: Brunette, D.M., Tengvall, P., Textor, M., Thomsen, P. (Eds.), Titanium in Medicine. Springer-Verlag, Berlin; Heidelberg, Germany, pp. 284–335.

Vail, T.P., Glisson, R.R., Koukoubis, T.D., Guilak, F., 1998. The effect of hip stem material modulus on surface strain in human femora. J. Biomech. 31, 619.

Wang, K., 1996. The use of titanium for medical applications in the USA. Mater. Sci. Eng. A. 213, 134.

Functional nanostructured coatings via layer-by-layer self-assembly

10

*Katalin Halasz**, George Grozdits*†, and Levente Csóka***
*Institute of Wood and Paper Technologies, University of West Hungary, Sopron, Hungary; †Louisiana Tech University, Ruston, LA, USA

10.1 Introduction

Layer-by-layer (LbL) self-assembly, which means a spontaneous adsorption of polyelectrolytes and/or nanoparticles, was first introduced by Decher and Hong (1991). The LbL provides a surface coating method that allows formation of single molecular layers or multilayers on different substrates. LbL has many advantages compared to other coating methods such as spin coating, thermal deposition, or solution casting. The process is inexpensive, simple, and quick and it can be carried out in water media avoiding the use of harmful and toxic solvents. It offers ultrathin film creation with desired composition and properties on several kinds of substrates with different sizes, irregular shapes, and 3D forms. A wide variety of materials can be used to perform the LbL deposition including inorganic or organic polycations, polyanions and nanoparticles which have surface charges when solved or dispersed in water (or other processing media). LbL self-assembly is based on electrostatic attractions between the oppositely charged constituents, although hydrogen-bonding, van der Waals forces, or charge-transfer interactions can also play a role in the thickness or stability of the self-assembled layers. With LbL deposition ultrathin mono-, bi-, or multilayer coating can be achieved with more than 1 nm precision (Ai et al., 2003). The properties of the films can be finely tuned by varying the process parameters (components, concentration, pH, ionic strength, immersion time).

LbL self-assembled coatings not only show complexity in their structure but also in their properties. Using the well-adjusted parameters the LbL method can result in well-adsorbed, strong, anti-abrasive nanostructured coatings with multifunctional properties. Abrasion resistance is a prime requirement in the case of functional coatings. If the coating is not resistant to abrasion wear it cannot provide other functionalities either or it will lose them during the usage. When the interfacial interactions between the substrate and the layers are strong enough and the coating is abrasion wear-resistant (in the environment and at the level at which the coated product is used) the coating can give several other valuable properties to the product such as anti-corrosion properties; flame-resistance; resistance to fungi/bacteria;

or improved barrier properties to protect the surface, the product itself, or its environment.

Although in most of the studies only one or two functionalities of the coatings are examined, and there is usually no information about the anti-abrasive properties or wear resistance, for instance, in the case of antimicrobial coatings, there are a few authors who report coatings with different functions having good mechanical properties, and scratch and abrasion wear resistance. According to the literature, with LbL electrostatic self-assembly robust, abrasion-resistant films with controlled thickness and properties can be created (Rogach et al., 2000; DeLongchamp et al., 2003; Pastoriza-Santos et al., 2000; Croll et al., 2006; Paloniemi et al., 2006; Chunder et al., 2009).

The importance of LbL self-assembled functional coatings is verified by the large number of published scientific papers. In this chapter we will present a review of these studies and show the industrial justification for using the LbL technique, which can offer abrasion, scratch and corrosion resistance, anti-flammability, self-healing, microbial resistance, and improved barrier properties.

10.2 LbL process

In a dipping LbL process (Figure 10.1) the first step in creating LbL layers is the immersion of the substrate (which has negative or positive charge) into an oppositely charged colloid solution. The first monolayer is adsorbed into the substrate, which changes the surface charge and also the charge distribution of the surface. Immersion is followed by washing (to remove the unbounded materials) and, if required, drying. The second layer (or bilayer) is formed when the substrate is dipped into the colloidal solution which contains the other constituent and has the opposite charge to the previous solution. The alternating deposition of

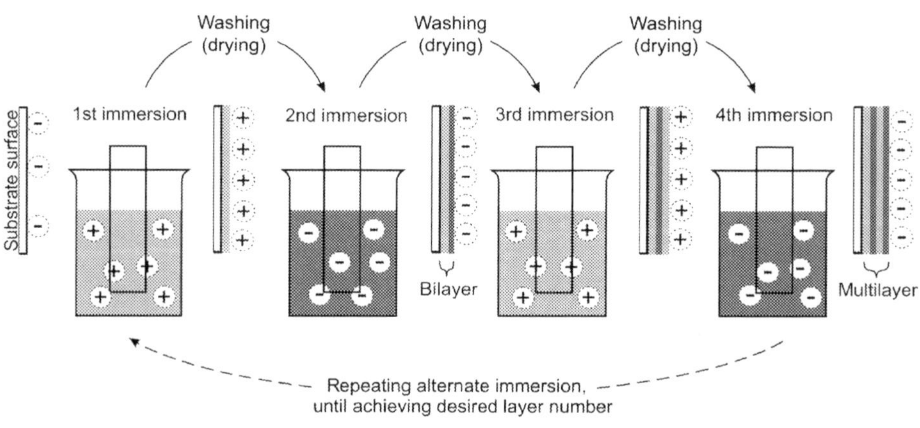

Figure 10.1 Dipping LbL process.

polyelectrolyte chains and nanoparticles involves resaturation of the oppositely charged polyelectrolyte and nanoparticle adsorption, which results in the reversal of the terminal surface charge after deposition of each LbL layer (Lu et al., 2007). After the alternate immersions a multilayer is created on the surface. The electrostatic attraction between the charged LbL deposition components results in strong interlayer adhesion without any activation steps (Liu et al., 1999). Cross-sectional TEM images of the LbL-deposited multilayer structure of chitosan (CH) and montmorillonite (MMT) on polystyrene substrate are shown in Figure 10.2, where Laufer et al. (2012b) demonstrate the effect of the pH on the deposition as well.

The ideal self-assembly takes place in single molecule layers; therefore it is a nanotechnology method (10^{-9} molecular dimensions). When only a single molecule layer is formed, the adjoin, the subsequent layer directly affects the other layer on a one-on-one molecular basis. When this occurs without self-assembly, it is referred to as a physical–chemical reaction forming a new product. That is, while the properties of molecules in subsequent layers are different, when measured in bulk, but in single-molecule thick layers the interaction is one-on-one, just like forming a new compound without self-assembly. Therefore the multilayered nanoassemblies have different properties compared to the properties of individual layers. Researchers of nano-self-assembly methods often report that the properties of

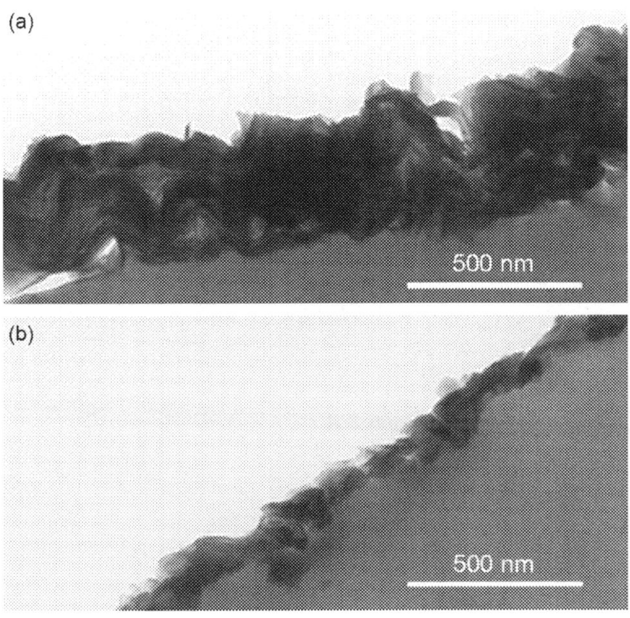

Figure 10.2 TEM cross section of LbL-deposited (by dipping process) 100 bilayers of CH/pH 6-MMT and 100 bilayers of CH/pH 3-MMT deposited on polystyrene (the pH was varied only in the case of MMT).

Source: Reprinted with permission from Laufer et al. (2012b). © 2013 American Chemical Society.

multilayered nanolayers differ, and are often synergistically better than the properties of the individual layers. With parameters such as concentration, immersion time, pH, or ionic strength the thickness and the properties of the layers can be modified remarkably.

The dip-LbL process is well-established and useful in many cases; however, it is commonly observed that a long time is required to achieve adsorption equilibrium. Because of this limitation some alternative methods have been developed such as spin- or spray-assisted LbL, which can provide high-speed processing using the same materials and allowing electrostatic self-assembly.

10.2.1 Spin-assisted LbL nanoassembly

A spin coater is required to perform spin-assisted LbL nanoassembly (Figure 10.3). After inserted onto the substrate a desired amount of charged colloidal solution, the substrate is rotated at high speed. The fast spinning causes centrifugal force, which enables the first molecular layer to develop quickly (in a few seconds) on the full substrate area. Usually, after the first layer is spin-coated onto the substrate a rinsing cycle is carried out to remove the surplus coating solution, then the process is followed by sequential adsorption of the next layers.

The advantage of this method is not only the fast coating process but also that with the spin-assisted LbL method, a more highly ordered coating structure, a uniform smooth surface can be achieved (Vozar et al., 2009). Moreover, spin coating can also be used in the case of hydrogen- or covalent-bonded LbL (Li et al., 2012). The spin-assisted LbL thin films, created with polyelectrolytes, show less interpenetration between the layers (Hammond, 2011), thus a linear growth of the thickness can be observed instead of the exponential growth which is common in case of dip-LbL. Multilayer films built up with polyelectrolytes and charged nanoparticles through spin-assisted LbL can exhibit remarkable biaxial strength and toughness (Hammond, 2011). The main disadvantage of the spin-assisted LbL is that only

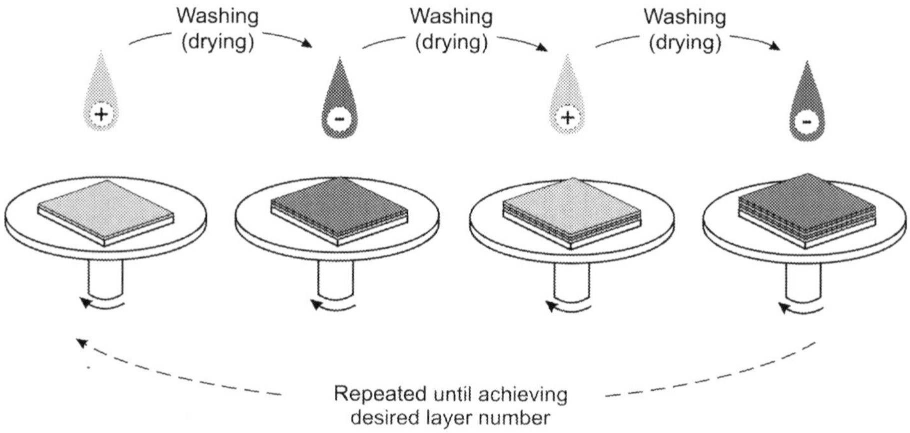

Figure 10.3 Spin-assisted LbL process.

planar, nonporous substrates are suitable for the method. Furthermore, producing thin, uniform films on substrates with large areas, especially using aqueous solutions, is difficult to achieve (Li et al., 2012).

10.2.2 Spray-assisted LbL assembly

LbL assembly of polyelectrolytes and nanoparticles can be achieved with alternate spraying of the colloid solutions onto the substrate. In spray-assisted LbL deposition (Figure 10.4) the first step is the spraying of the first oppositely charged solution containing polyelectrolytes or nanoparticles onto the substrate. The excess solution is continually draining from the surface. To remove the remaining solution the surface is sprayed with water. After the surface is washed, the alternate adsorption of the layer is carried out by spraying the charged solutions onto the substrate. The process can be optimized by varying parameters such as spraying and draining time, or solution flow-rate. The time required to assemble a bilayer, which is the basic building block of multifilm, can be reduced from minutes to seconds (Nogueira et al., 2011). Another big advantage of spraying compared to the spin coating is that objects with different sizes and shapes can be covered with thin multilayers.

10.2.3 Substrates, polyelectrolytes, nanoparticles

A minimal surface charge is enough to cover a surface (e.g., plastic, metal, or ceramic) with an ultrathin layer via the LbL technique. A little electrostatic attraction is enough between the polyelectrolyte, the nanoparticles, and the surface because both the coating substance and the surface provide many attachment points (Bolto and Gregory, 2007). By pretreating the surface with chemicals or a plasma treatment (especially in the case of plastics) the charge density and charge distribution can be improved. Modification of the surface roughness can also help to enhance the adsorption of the first monolayer on the substrate. A wide variety of substrates are suitable for covering by LbL nanoassembly including metals and

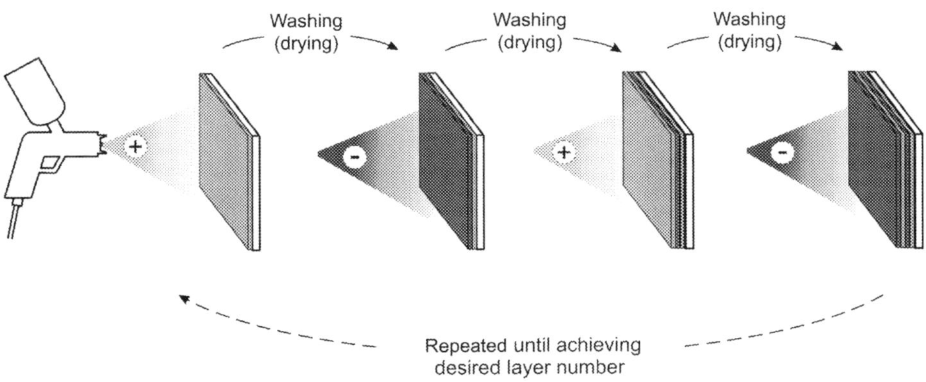

Figure 10.4 Spray-assisted LbL process.

metal oxides (such as Au, TiO_2, SiO_2, mica), ceramics, and natural or synthetic polymers (such as cellulose fibers, cotton fabrics, poly(methyl methacrylate) (PMMA), poly(ethylene terephthalate) (PET), or surface-treated polyethylene (PE), polypropylene (PP) other polymeric materials).

To perform the electrostatic LbL bottom-up nanofabrication technique on a substrate several kinds of polyelectrolytes and nanoparticles are available. The most common polycations are: branched poly(ethylenimine) (PEI or BPEI), linear poly (dimethyldiallyl ammonium chloride) (PDDA), poly(allylamine hydrochloride) (PAH), and poly(*N*-octyl-4-vinyl pyridinium iodide) (PNOVP). Among the natural-based polycations, the most important are CH and starch. Widely used cationic polymers are poly(styrenesulfonate) (PSS); poly(3,4-ethylenedioxythiophene)-poly (styrenesulfonate) (PEDOT-PSS); poly(acrylic acid) (PAA); and the natural-based alginate, heparine, poly(L-arginine), gelatin B, carboxymethyl cellulose (CMC), and lignin sulfonate. According to Andreeve at al. (2010) the conformation of polyelectrolytes depends on their nature and adsorption conditions and rather than on the substrate type and charge density of the substrate surface. Using polyelectrolytes many kinds of surfaces including nonionic or polar ones (Andreeva et al., 2010) can be covered by the LbL technique.

The multilayers can be built up with alternate adsorption of oppositely charged polyelectrolytes and/or with alternate adsorption of oppositely charged polyelectrolytes and nanoparticles (or only with nanoparticles). Many kinds of organic and inorganic nanoparticles, which show surface charge in aqueous media, can been used to form nanoassemblies including metal oxide or metal nanoparticles such as SiO_2, TiO_2, ZrO_2, CeO_2, Fe_2O_3, Co, Au, or organic particles such as globular protein, cellulose nanocrystals (CNCs), polymeric nanospheres, as well as different types of naturally occurring clay or graphene nanoplatelets, carbon and ceramic nanotubes.

The ultimate thickness and properties of the layers strongly depend also on the processing parameters. Varying the pH, concentration, ionic strength, media, and immersion (spraying or spinning) time the final multilayer structure can be precisely tuned.

10.2.4 Parameters

10.2.4.1 Ionic strength and pH

The ionic strength has a strong effect on the expansion of the polyelectrolytes. Highly charged polyelectrolytes can be shielded by varying ionic strength, thus modifying the charge density and adsorption behavior (Shiratori and Rubner, 2000). At low ionic strength the polymer coil can expand to a high extent, thus the formed layer is thin and the adsorbed amount is low. Increasing the ionic strength, the repulsion between the charged polymer segments is screened by ions in the solution involving less-expanded polymer coil (Bolto and Gregory, 2007; Villiers et al., 2011) and a thicker deposited layer (with loops and tails) with a higher adsorbed amount of mass. Increasing further the salt content the adsorption of the polyelectrolyte becomes less effective and after exceeding a critical salt concentration the salt ion can even displace the polyelectrolytes which were already adsorbed on the surface (Villiers et al., 2011). Adsorption is usually maximal at intermediate salt concentration (Van Tassel, 2012). In the case of weak polyelectrolytes the charge density, the relative amount of charge along the backbone, can be altered by varying the pH (Hammond, 2000; Shiratori and Rubner, 2000). When the charges of both polyelectrolyte and surface are pH dependent, a maximum in adsorption is generally observed at a pH between the isoelectric points of the surface and the polymer (Van Tassel, 2012).

For instance, Findenig et al. (2012) examined the effect of salt content on thickness in the case of four types of polyelectrolytes on two types of substrates (Figure 10.5). According to their results only little differences occurred in the thickness comparing the cellulose film and silicon wafer substrates. However the influence of the salt content can be clearly seen in Figure 10.5.

Ariga et al. (1999) have found that the presence of salt can affect the adsorption mechanism in the case of nanoparticles too. According to their results, the addition of NaCl to the aqueous SiO_2 improved the quality of the obtained film.

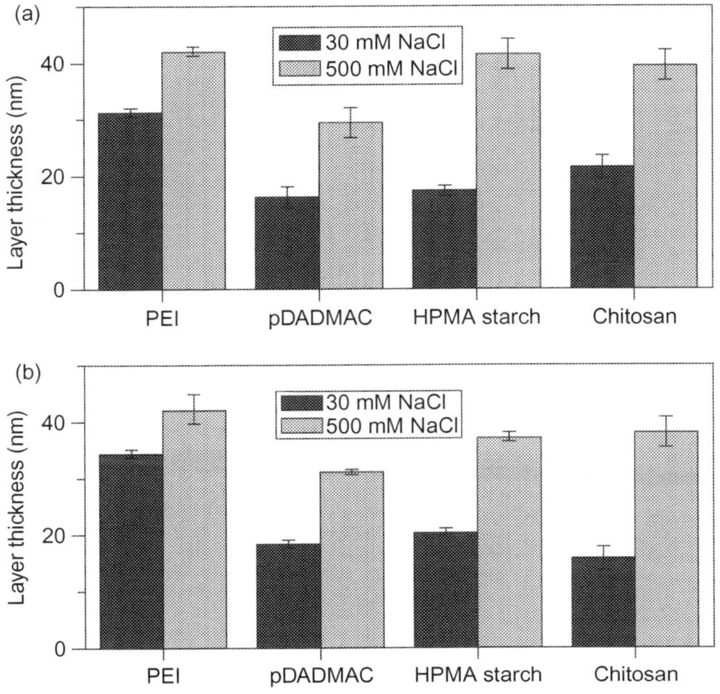

Figure 10.5 Influence of ionic strength on the layer thickness of different polyelectrolyte/clay systems (five bilayers) on (a) cellulose films and (b) SiO_2 wafers.
Source: Reprinted with permission from Findenig et al. (2012). © 2012 American Chemical Society.

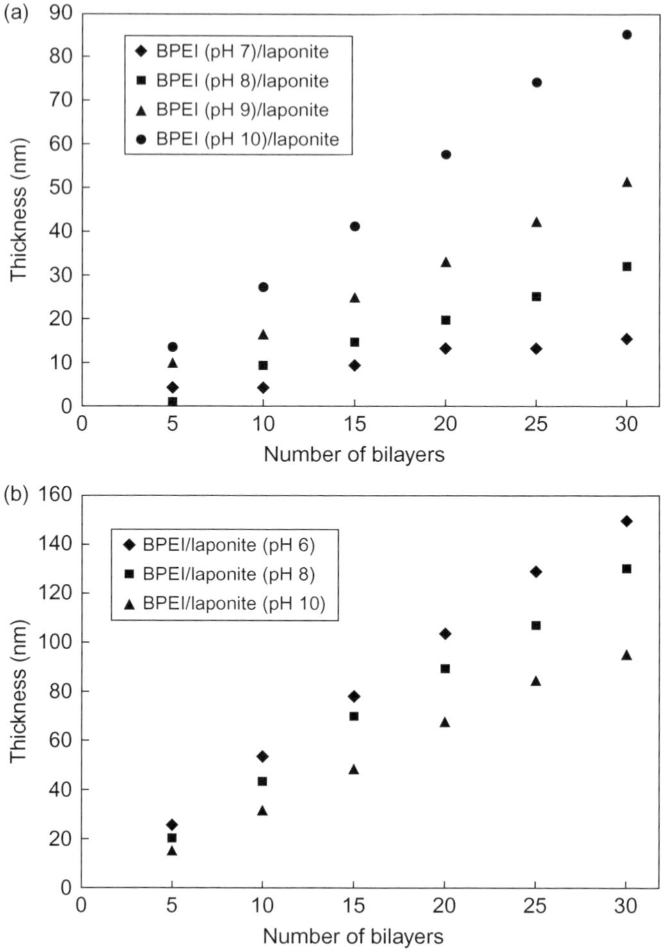

Figure 10.6 Film thickness as a function of number of bilayers deposited for LbL assemblies made with varous BPEI (a) and laponite (b) deposition mixture pHs. *Source:* Reprinted with permission from Li et al. (2009). © 2009 American Chemical Society.

Li et al. (2009) investigated how the pH affects the adsorption of PEI and the layered silicate, laponite. According to the ellipsometric measurements the nanoclay is also sensitive to the pH – the higher the pH was, the lower the thickness became. In contrast, the higher the pH was, the higher the thickness formed on the substrate in case of the polyelectrolyte (Figure 10.6).

10.2.4.2 Media, concentration, immersion time

The electrostatic LbL nanoassembly is usually performed in aqueous solutions. However, the thin films can be assembled in other mediums such as alcohol

(Beyer et al., 2007), or even in nonpolar solvents like dichlorometane (Borodina et al., 2009), formamide (Kamineni et al., 2007), formamide/water mixture (Zhang et al., 2008), or toluene (Tettey et al., 2010). Any kinds of media can be used which enable the ionization of the polyelectrolytes or the nanoparticles (Villiers et al., 2011).

Solutions with the proper concentrations are required to achieve the desired adsorption mechanism. The solution should be adjusted to a value that can prevent depletion during the LbL process (Villiers et al., 2011). Lvov et al. (1997) have found that the higher the concentration of the PDDA solution, the higher the adsorbed amount onto the substrate (Figure 10.7). However, above a threshold no differences in the thickness or adsorbed amount can be observed (Chen and McCarthy, 1997). Usual concentration of the polyelectrolytes are 1 mg/ml (Ye et al., 2004; Hirsjävi et al., 2006; Shi et al., 2008; Ali et al., 2010) but it can go up to 5 mg/ml too (Hua et al., 2004; Lee et al., 2008; Lvov et al., 1998). The optimal concentration of the nanoparticle colloids are usually higher, 5–10 mg/ml (Hua et al., 2004; Wu et al., 2008; Ariga et al., 1999), although in the case of single-walled carbon nanotubes Nepal et al. (2008) successfully used a colloid solution with only a 45 mg/l concentration.

Times required for the self-assembly vary for the different kinds of materials. For instance, in the case of polyelectrolytes, first only a few segments of the chain

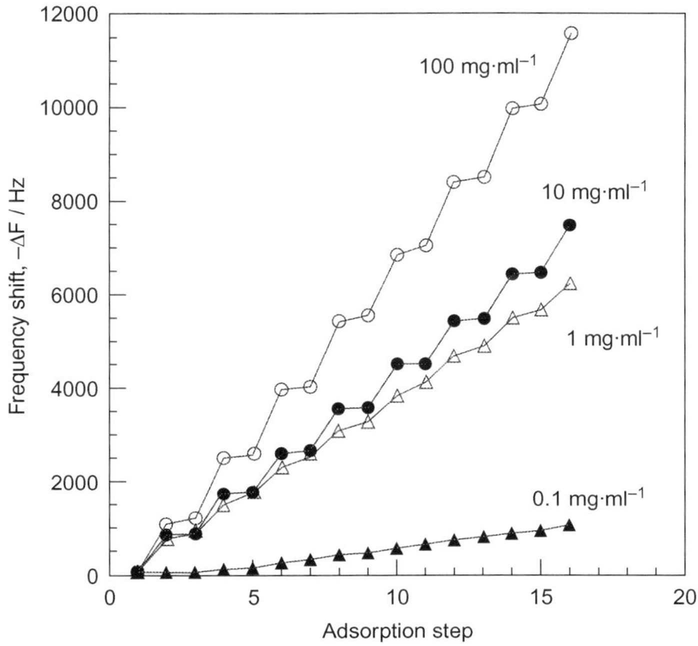

Figure 10.7 Effect of adsorption steps and concentration of PDDA on the frequency shift measured by QCM.
Source: Reprinted with permission from Lvov et al. (1997). © 1997 American Chemical Society.

are adsorbed on the substrate but after a given time that depends on a number of factors which are not well understood yet, the polymer chain reaches its equilibrium configuration on the substrate (Bolto and Gregory, 2007). Linear polyions usually need 8–15 minutes (Lvov et al., 1999; Ai et al., 2003), PSS and PAH 5 minutes (Ramsden et al., 1995), and PDDA 40 seconds (Lvov et al., 1998) to reach the equilibrium arrangement; polymers with high-molecular weight and long polymer chains might need more time (Bolto and Gregory, 2007). Particles in nanometer range in all the three dimensions generally require only a few seconds to reach the equilibrium. Required adsorption times for the SiO_2 nanoparticles, for example, is 10 seconds (Lvov et al., 1998) and for microfibrillated cellulose 100 seconds (in the case of spray coating) (Aulin et al., 2010). Nanoplates, like layered silicates, show similarity to the adsorption of polyelectrolytes. A quick adsorption can be observed when the platelets contact with the substrate, which is followed by a longer rearrangement process. The rearrangement of the nanoclay can take 5–6 minutes.

10.3 LbL-deposited nanostructured coatings with different functions

10.3.1 Hard, anti-abrasive, and anti-scratching coatings via LbL

The mechanical properties and stability of the coatings on a surface depend greatly on the interfacial interactions between the coating and the substrate. As the electrostatic attraction between the LbL assembly components is a strong interlayer adhesion, LbL nanocomposite coatings can perform abrasion and scratch resistance (Liu et al., 1999; Zhou et al., 2009) (Table 10.1).

In the case of commercial coatings the scratch resistance is usually achieved by incorporating cross-links in the binder of the coatings; however, a highly cross-linked hard coating prepared in this way, because of its reduced flexibility, has poor impact resistance. If the coating contains less cross-linkings, the film is more soft and flexible but is less resistant to scratching or abrasion (Ghosh, 2006). LbL coatings containing organic and inorganic constituents can combine the advantages of softness of the organic phase and the rigidity of the inorganic phase (Zhou et al., 2009). The most used nanoparticles creating hard coatings are ZrO_2, $CaCO_3$, SiO_2, SWCNT, Al_2O_3; however, the oriented organic rod-like CNC can also provide enhanced surface mechanical strength and wear resistance (Hoeger et al., 2011). According to Liu et al. (1999) the improved microhardness is due to high molecular packing density of the adsorbed layers and the ultrasmall size of the nanoparticles. Thermal treatment (burning out the polyelectrolyte layers) can enhance even more the hardness and the strength of the multilayers (Rosidian et al., 1998; Liu et al., 1999; Han et al., 2005; Chunder et al., 2009).

Rosidian et al. (1998) created hard nanocomposite thin films of ZrO_2 nanoparticles, PAH and poly(styrene sulfonate) (PSS) on quartz and silicon wafers. The first four PAH/PS119(molecular dye) bilayers were deposited on the substrates to

Table 10.1 Hard, anti-abrasive, and anti-scratching coatings via LbL

Multilayer composition	Further treatment	Properties	Reference
ZrO_2/PAH, ZrO_2/PSS	400°C 2 h 900°C 1 h	Hard coating, increased hardness with TT (without treatment 2.242 GPa, with treatments 19.311 GPa and 25.129 GPa, respectively)	Rosidian et al. (1998)
PDDA/PS + ZrO_2/PSS	900°C 2 h	Hard, very dense, homogenous thin-film coatings (after TT microhardness of 25.1 GPa with Young's modulus of 285 GPa)	Liu et al. (1999)
PDDA/PS + Al_2O_3/PSS or + Al_2O_3/PSS/ZrO_2/PSS	900°C 2 h	Slighter increase both in hardness and Young's modulus after TT	Liu et al. (1999)
PAH/PAA + PAH/PAA-coated ZrO_2 + SiO_2/PAH	215°C and fluorination	Stable, hard, and super hydrophobic coating	Han et al. (2005)
PDDA/PSS + PDDA/SWCNT	–	Flexible and strong multilayer coating	Xue and Cui (2007)
PEI/PSS + PAH/PSS + ZrO_2	500°C and 900°C	hard coating after TT, TT caused large shrinkage and cracking, but the coating remained stable	Zlotnikov et al. (2008)
PAH/silica	Functionalization of polycarbonate (PC) surface with APTS, sol-gel treatment on 100°C	Hard, abrasion-resistant, anti-reflective coating	Chunder et al. (2009)

(*Continued*)

Table 10.1 **(Continued)**

Multilayer composition	Further treatment	Properties	Reference
PAH/PAA-coated $CaCO_3$	180°C for 5 h	Strong, scratch-resistant, and transparent coating before and after a thermal cross-linking process	Liu et al. (2010)
PEI/CNC	–	High friction coefficient, elastic transversal modulus, and hardness of 8.3 and 0.38 GPa	Hoeger et al. (2011)
TiO_2/SiO_2	Atomic layer deposition of Al_2O_3	Stable, abrasion-resistant coating on PC with atomic layer deposition (ALD) coating prepared on the LbL coating	Dafinone et al. (2011)

Note: TT is thermal treatment.

enhance the adhesion of the first ZrO_2 layer on the surface. This was followed by the deposition of 35 bilayers of ZrO_2/PAH or ZrO_2/PSS on silicon substrates respectively. The film adsorbed onto the silicon wafer was treated at 400°C for 2 hours and at 900°C for 1 hour. It was found that applying thermal treatment can further enhance the hardness of the deposited multilayer. The Vickers microhardness (applying 10 and 100 nm maximum indentation) of the coatings was increased with the thermal treatment (without treatment 2.242 GPa, with thermal treatments 19.311 and 25.129 GPa, respectively).

Hard, very dense, homogenous thin-film coatings with highly ordered and uniform multilayer structures could be successfully prepared by electrostatic self-assembly (Liu et al., 1999). To achieve a strong interaction between the hard multilayers and the substrate, there were PAH/PS119 or PDDA/PSS bilayers deposited onto the substrate. Afterwards 75 ZrO_2/PSS and Al_2O_3/PSS bilayers, and Al_2O_3/PSS/ZrO_2/PSS quadlayers were created on the pre-coated substrate. Without thermal treatment the microhardness value, which was measured by nanoindentation, of the ZrO_2/PSS coating, was 2.2 GPa with Young's modulus of 52 GPa. After thermal treatment at 900°C for 2 hours it became 25.1 GPa and for Young's modulus 285 GPa. In the case of Al_2O_3/PSS and Al_2O_3/PSS/ZrO_2/PSS multilayers, the thermal treatment had no adverse influence but a slight increase both in hardness and Young's modulus could be observed.

Stable, hard, and super hydrophobic coating can be prepared by electrostatic LbL self-assembly using PAH, PAA-coated ZrO_2 (5 and 100 nm), and silica nanoparticles (Han et al., 2005). In the research of Han et al. (2005) the coating was prepared on a silicon wafer. First five bilayers of PAH/PAA were deposited (in order to improve the density of the deposited PAA-coated ZrO_2 nanoparticles), then 10, 20, 30, 40, and 60 bilayers of PAH/PAA-coated ZrO_2. According to the AFM image (Figure 10.8(a)) of the 5 nm ZrO_2 containing system the authors concluded that it gave flat, well-assembled multilayer film, while scanning electron microscopy (SEM) images of 100 nm ZrO_2 containing films (Figure 10.8(b–d)) showed porous net structures even after 10 deposition cycles, which are presumably caused by the mismatch of the electrostatic force between the weak polyelectrolytes and the large nanoparticles.

In order to obtain the superhydrophobic surface 1.5 bilayers of SiO_2/PAH were assembled onto the multilayers. After the deposition process, thermal treatment at 215°C and simple fluorination was carried out. It was found that 10 bilayers of PAH/PAA-coated ZrO_2 is enough to enhance the water contact angle from 48° to 95° and

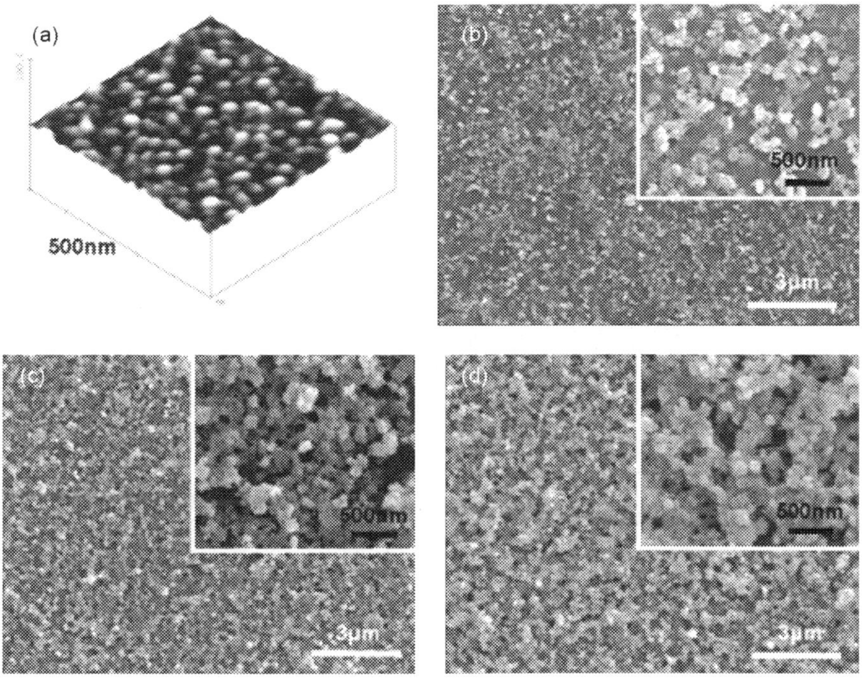

Figure 10.8 Atomic force microscopic image of (PAH/PAA-coated ZrO_2) multilayer film with 5 nm ZrO_2 nanoparticles on Si wafer (a), SEM images of organic–inorganic PAH/PAA-coated ZrO_2 (100 nm) films deposited onto Si wafers (b–d), where deposition number is 5 (b), 10 (c), and 20 (d).
Source: Reprinted with permission from Han et al. (2005). © 2005 American Chemical Society.

20 bilayers to 139°. The nanoindentation test showed the hardness of PAH/PAA polyelectrolyte multilayer film could be increased by heat-induced cross-linking. Sixty bilayers of PAH/PAA noncross-linked coating (642.8 nm) had a hardness of 0.75 GPa, while cross-linked (576.0 nm) had 1.36 GPa. Incorporating ZrO_2 nanoparticles into the system resulted in higher improvement in hardness: in the case of 60 bilayers of cross-linked (712.9 nm) PAH/PAA-coated ZrO_2 coating, it was 2.15 GPa.

In contrast to the other papers Zlotnikov et al. (2008) deposited only one layer of ZrO_2 to obtain a hard coating. The ZrO_2 was biomimetically deposited onto a previously organofilized (with two PEI/PSS and six PAH/PSS bilayers) Si wafer. The substrate coated with polyelectrolytes was kept in the $Zr(SO_4)_2X\ 4H_2O$ and hydrochloric acid solution for 24 hours at 70°C. After the immersion the substrate was cleaned ultrasonically. The authors found that after 8 hours the film completely covered the surface and the layer had grown with time, reaching a thickness of 110 nm after 24 hours. The coating was stable – after ultrasonic cleaning it did not detach from the surface and it remained crack-free; after the adhesive tape test the coating did not peel off either. As the deposited coating was amorphous and consisted of a mixture of ZrO_2 and $Zr(SO_4)_2$, it had low mechanical properties. To improve the hardness and the Young's modulus of the coating, thermal treatment was carried out at 500°C and 900°C. After the thermal treatments, although the treatment at 900°C led to large shrinkage and thus cracking, both the coatings remained adherent to the substrate and could have not been removed by the tape test. The authors reported that the improvement of mechanical properties was due to the densification, crystallization, and purification of the zirconium oxide.

Flexible and strong multilayer coating without any post-treatment was prepared by Xue and Cui (2007) using PDDA, PSS, and single-walled carbon nanotube (SWCNT). First, two PDDA–PSS bilayers were deposited on a silicon substrate followed by the six PDDA–SWCNT bilayers (Figure 10.9). According to the

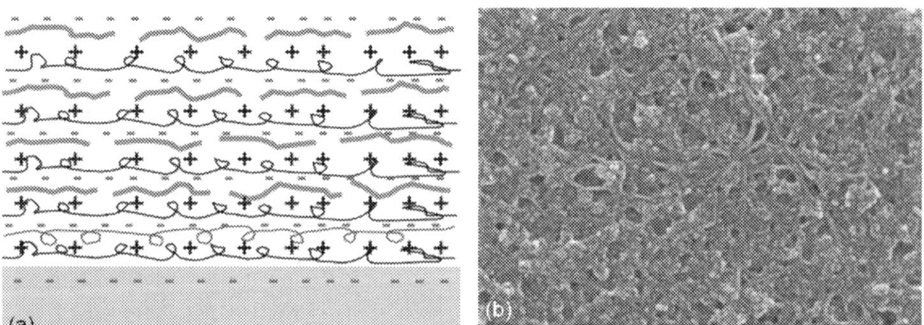

Figure 10.9 The structure of LbL self-assembled (PDDA/PSS)*2 + (PDDA/SWNT)*5 multilayer thin film on Si substrate (a) and SEM image of assembled multilayer containing SWNT (b).
Source: After Xue and Cui (2007) © IOP Publishing. Reproduced by permission of IOP Publishing. All rights reserved.

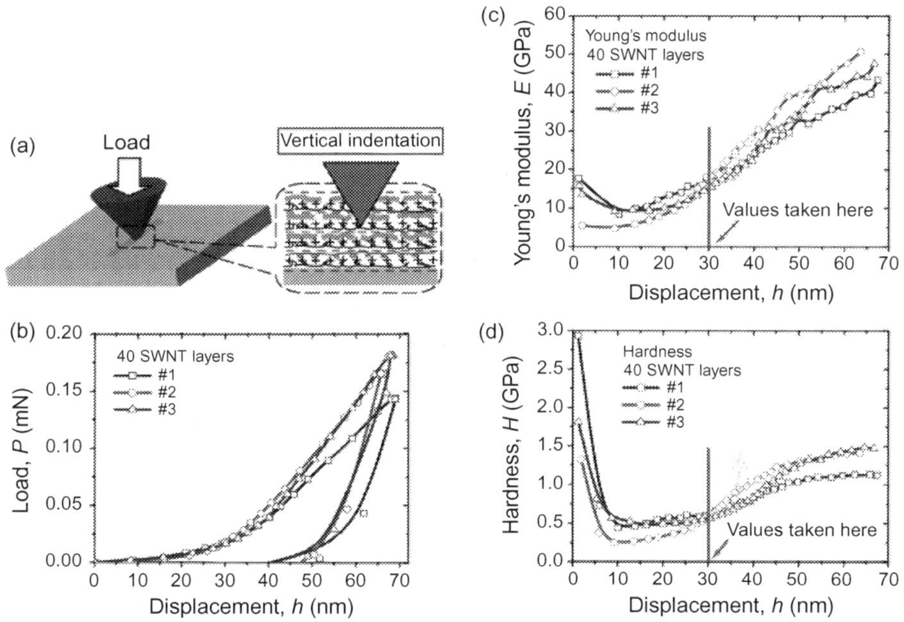

Figure 10.10 (a) Scheme of nanoindentation test on assembled SWNT multilayer thin films, (b) load–displacement curves, (c) measured Young's moduli, and (d) hardness of (PDDA/SWNT) multilayer thin films on silicon substrates.
Source: After Xue and Cui (2007) © IOP Publishing. Reproduced by permission of IOP Publishing. All rights reserved.

quartz crystal microbalance measurement the six bilayers of polymer and SWCNT were less than 50 nm. Despite the small thickness the mechanical properties of the coating was between those of silicon and hard polymers (Figure 10.10). Conductivity could be achieved by thermal treatment and could be tuned by changing the temperature of the annealing of PDDA–SWCNT-coated silicon.

Abrasion-resistant anti-reflective coating was deposited via spray-assisted LbL methods on PC substrate with PAH–silica bilayers by Chunder et al. (2009). The abrasion resistance was improved by functionalization of the PC surface with aminopropyltrimethoxylsilane (APTS) and the sol-gel treatment of the coating. After the coated PC was thermally treated at 100°C, the SiO_2 nanoparticles were fused through the reaction with tetrahydroxylsilane in the sol-gel solution which led to a hard, abrasion-resistant coating where the transmittance was not affected by the treatment.

Calcination is a high temperature (>500°C) thermal treatment that can improve the mechanical properties of nanoparticle thin films but usually not in the case of polymer substrates or thin films containing organic materials (Dafinone et al., 2011). Treatments such as sorption and cross-linking of a polymerizable sol-gel precursor, chemical vapor deposition of inorganic precursors, and hydrothermal calcination at lower temperatures can be performed too; however, these methods can change the initial properties of the thin film (Definone et al., 2011). According to Dafinone et al. (2011) the atomic layer deposition (ALD) of Al_2O_3 is a suitable method which does not drastically affect the LbL coating but improves the thin film mechanical properties. By applying a nanoindentation test they found that the modulus and hardness of the coating on silicon wafers increased drastically and systematically with increasing ALD cycles. After the abrasion test the scanning electron microscopic images showed that the LbL films without the ALD coating totally abraded and delaminated while the films with ALD coating showed good adhesion; they did not undergo delamination, and only small numbers of circular scratches were visible, which were due to the flattened nanoparticle multilayer. In the case of PC substrate where the adhesion between the substrate and the first layer was weak, the ultrathin Al_2O_3 could prevent abrasion and the LbL coating was removed completely; however, deposition of the Al_2O_3 layer onto the PC substrate improved the interactions and, after LbL deposition and the covering (10 cycles) layer, a stable, abrasion-resistant coating was created.

It was found by Liu et al. (2010) that low loading content of homogeneously dispersed $\sim$2 nm $CaCO_3$ nanoparticles can greatly improve the mechanical strength of the PAH/PAA-coated $CaCO_3$ coatings, and can provide a strong, scratch-resistant, and transparent coating before and after a thermal cross-linking process (at 180°C for 5 hours) as well. To characterize the adhesion between the substrate and the (PAH/PAA-coated $CaCO_3$)*20 coating, a cross-cut tape test was applied, where the results showed the highest level (ASTM class 5B) of adhesion. In contrast, the thermally cross-linked (PAA/PAH)*30 coating was totally removed from the substrate after the test, which indicates that the presence of $CaCO_3$ nanoparticles led to excellent adhesion to the substrate. The nanoindentation test proved improved mechanical properties as well. The hardness of the untreated, $CaCO_3$ containing coating was more than double the coating containing only a (PAA/PAH)*30 multilayer (0.70 and 0.30 GPa, respectively). The modulus was improved too (from 11.18 to 14.52 GPa) in the case of the presence of $CaCO_3$. Cross-linking improved the hardness and the modulus further, where the hardness and the modulus of the coating with $CaCO_3$ was 1.06 GPa and 15.33 GPa, without $CaCO_3$ 0.48 GPa and 12.62 GPa, respectively. The authors highlight that the hardness enhancement caused by the addition of $CaCO_3$ nanoparticles is more significant than thermal cross-linking. According to rubbing tests after 80 cycles the cross-linked (PAA/PAH)*30 coating was strongly damaged, while no scratches could be observed on the cross-linked (PAH/PAA-coated $CaCO_3$)*20 coating even after 1100 rubbing cycles. The well-dispersed nanoparticles and the strong coordination interaction between $CaCO_3$ and PAA/PAH matrix coatings led to a significantly enhanced hardness and scratch-resistance of the (PAH/PAA-coated $CaCO_3$)*20 coating.

Hoeger et al. (2011) created multilayers built up with rod-like CNCs and PEI. The nanoassembly process was not strictly an LbL process since the adsorption of CNC layers was carried out using convective and shear forces in order to achieve a well-ordered CNC structure in the multilayer. According to the nanoindentation measurement the PEI/aligned-CNC multilayers provided an elastic transversal modulus and hardness of 8.3 and 0.38 GPa, respectively. Furthermore, Hoeger et al. (2011) found that the friction coefficient of CNC films was higher than the desired in applications with high tribological demands.

10.3.2 Self-healing, anti-corrosive coatings

Sometimes anti-abrasion and hardness alone are not enough to protect a surface. Corrosion, which is mostly caused by chemical attack, can cause high industrial losses in case of metallic (and also plastic, concrete, or wood) products. Protecting the surface is a huge industrial demand. In the case of metals usually polymer coatings (deposited by sol-gel method or plasma polymerization) are applied to prevent corrosion (Ghosh, 2006; Shchukin et al., 2006). The effectiveness of the coating depends mostly on the thickness, adhesion, and permeability of corrosive species (Ghosh, 2006). If the protective coating is damaged, the corrosive agents are able to penetrate to the surface causing quality loss (Shchukin et al., 2006). Using LbL self-assembly technique in corrosion protection can offer coatings with tuned properties (thickness, permeability) with good adhesion to the substrate; furthermore, it can provide a "smart" corrosion protection system with self-healing effect (Table 10.2).

Although the self-healing properties were not examined, Farhat and Schlenhoff (2002) reported the effectiveness of ultrathin (70 nm) LbL-deposited multilayers built up of polyelectrolytes (PDDA/PSS and PNOVP/PSS) inhibiting corrosion on abraded stainless steel wires in saltwater. The anti-corrosive effect was attributed to the polyelectrolyte multilayer's ability to exclude small ions and to the molecular contact of the multilayer with the surface, which prevent the occlusion of pockets of electrolyte at the steel/coating interface.

Shchukin et al. (2006) combined the sol-gel method with the LbL deposition to prepare self-healing, corrosion protective coating on aluminum alloy. LbL-assembled nanoreservoirs (70 nm SiO_2 particles coated with PEI/PSS polyelectrolyte multilayer) were embedded in hybrid epoxy-functionalized ZrO_2/SiO_2 sol-gel coatings. The benzotriazole inhibitor was entrapped within the polyelectrolyte thin multilayer film during the LbL deposition. According to the authors the nanoreservoirs increase the long-term corrosion protection of the coated Al substrate and

Table 10.2 Self-healing, anti-corrosive coatings

Multilayer composition	Properties	Reference
PDDA/PSS, PNOVP/PSS	Anti-corrosive coating on stainless steel wires in saltwater	Farhat and Schlenhoff (2002)
SiO_2 particles coated with PEI/PSS + ZrO_2/SiO_2 sol-gel coatings	Self-healing, corrosion-protective coating on aluminum alloy with LbL nanoreservoirs	Shchukin et al. (2006)
PEI/PSS, PEI/PAA, PDDA/PSS, with corrosion inhibitor	Anti-corrosive, scratch-resistant, self-healing multicomponent nanonetwork on Al plate	Andreeva et al. (2010)

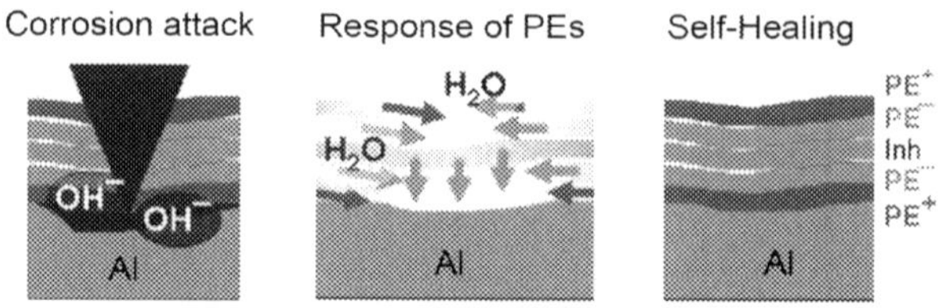

Figure 10.11 Schematic mechanism of self-healing action of a "smart" polyelectrolyte anticorrosion coating.
Source: Reprinted with permission from Andreeva et al. (2010). © 2010 American Chemical Society.

offer a good storage of the inhibitor and prolonged release to damaged zones with self-healing ability.

Anti-corrosive, scratch-resistant, self-healing multicomponent nanonetwork coating can be formed without nanoparticles by polyelectrolyte multilayers with a corrosion inhibitor (quinoline) incorporated into the polyelectrolytes, since polyelectrolytes exhibit very good adhesion to the substrate surface and are able to seal surface defects (Andreeva et al., 2010). Preparing a protective coating Andreeva et al. (2010) applied the fact that polyelectrolyte films are sensitive to the physical and chemical conditions of the surrounding media. The pH sifts or mechanical impacts allow regulated release of the inhibitor species entrapped into multilayers as can be seen in Figure 10.11.

In their study, Andreeva et al. (2010) built up 10 bilayers of PEI/PSS (80 nm thick), PEI/PAA (800 nm thick), and PDDA/PSS (200 nm thick) on the surface of aluminum plates by dip-, spin-, and spray-coating methods (all the coating methods provided successful modification of the surface). According to the authors the multilayers of polyelectrolytes showed high corrosion protection because of the nature and versatility of the PE complex. Andreeva et al. (2010) gave three explanations for the action of the polyelectrolyte multilayer system in the case of corrosion attack: the polyelectrolytes have pH-buffering activity and thus they are able to stabilize the pH at the metal surface in corrosive media; the inhibitors in the polyelectrolyte multilayer are released from the multilayer after the start of the corrosion process and directly into the rusted area, which prevents corrosion propagation; mobility of the polyelectrolytes enables sealing and eliminating surface damage caused by mechanical effects (Figure 10.12).

10.3.3 Flame-retardant nanocoatings

Anti-flammability is a required property, especially for fabrics. More than 94% of the fire fatalities in homes are related to upholstered furniture fires (Carosio et al., 2011). There are several techniques for creating flame-resistant fabrics such as surface treating, blending the material with fire-retardant additives (like inorganic metal hydroxides, halogenated compounds, phosphorous materials), or one of the most recent developments – the incorporation of nanoreinforcements (nanoclay)

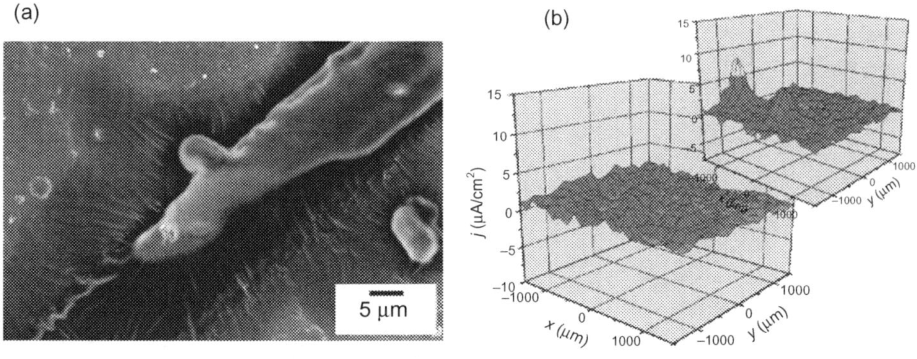

Figure 10.12 (a) SEM image of the surface of the PEI/PAA coating after 12 hours' immersion in 0.1 M NaCl, (b) scanning vibrating electrode measurements of the ionic currents above the surface in 12 hours (inset, in 10 hours 30 minutes) shows that the corrosion peak disappeared because of the mobility of the PE system.
Source: Reprinted with permission from Andreeva et al. (2010). © 2010 American Chemical Society.

Table 10.3 Flame-retardant nanocoatings

Multilayer composition	Properties	Reference
BPEI/laponite	Cotton fabric with improved thermal properties and flame-resistance	Li et al. (2009)
Alumina coated SiO_2/SiO_2	Flame-retardant thin film on PET fabric	Carosio et al. (2011)
MMT	Flame-resistant and thermally more stable polyimide fabric	Xu et al. (2012)
Alumina coated SiO_2/SiO_2	Flame-retardant thin film on PC films, significant improvement in fire retardancy and in time to ignition, good transparency	Carosio et al. (2013)
CH/PA	Intumescent multilayer nanocoating on cotton fabric	Laufer et al. (2012b)
PSP/PAA	Intumescent nanocoating on cotton fabric, lowered flammability	Apaydin et al. (2013)

into the bulk material. Current research shows that water-based environmentally friendly LbL nanoassembly is another useful method for making synthetic and natural fabrics or films anti-flammable (Table 10.3). It was reported that LbL flame-retardant coating with excellent barrier properties does not affect the mechanical properties or structure of the fabric: it does not close the surface, but covers only the individual fibers (as in the case of antimicrobial coatings) (Li et al., 2009, 2011; Laufer et al., 2012b). Furthermore, the LbL-coated fibers can keep their original

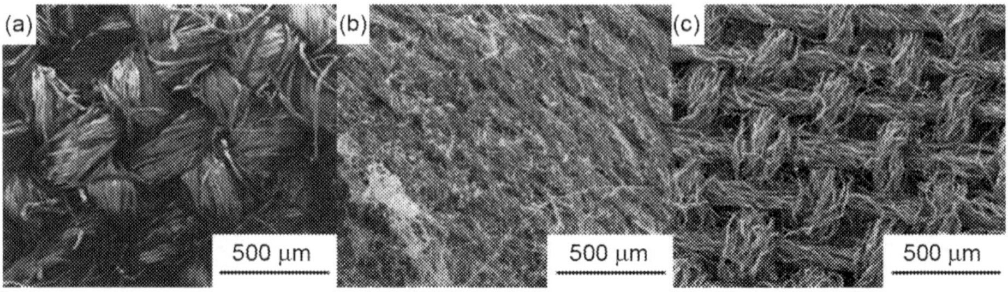

Figure 10.13 SEM images of virgin fabric (a) before, and (b) after flame testing; (c) BPEI/laponite-coated fabric after flame testing.
Source: Reprinted with permission from Li et al. (2009). © 2009 American Chemical Society.

shape and integrity; no shrinkage of the fabric can be observed after burning (Li et al., 2009; Laufer et al., 2012b). Several authors such as Li et al. (2011) and Laufer et al. (2012b) proved that it is possible to create intumescent multilayer nanocoating on fabrics with the alternate adsorption of cationic and anionic polyelectrolytes.

Li et al. (2009) were the first to create cotton fabric with improved thermal and flame-resistance properties using LbL assembly of branched poly(ethylene imine) (BPEI) and laponite nanoplatelets. According to the thermogravimetric results, at 500°C under atmosphere the weight loss in the case of the uncoated fabric was more than 99%, while 95% and 96% with coating. Twenty bilayers of BPEI/laponite preserved the cotton fabric structure (Figure 10.13) and caused stronger char formation. Although after flame testing a significant degradation also occurred in the case of the coated fabric, the authors consider that clay-based assemblies might be an interesting alternative to current flame-suppression technologies for fibers and fabrics.

Xu et al. (2012) used also layered silicate, but montmorillonite nanoplatelets (without polyelectrolyte) to cover the polyimide fabric and make it flame-resistant and thermally more stable. After heating up to 700°C the coated fabric left 15% more char than the uncoated fabric. Furthermore, the LbL-coated fabric showed reduced heat release and resistance to thermal degradation from direct flame. After the combustion test with SEM they observed the same as Li et al. (2009), that the weave structure and the fiber shape of the MMT-coated fabric was preserved.

Flame-retardant thin film was deposited onto PET fabric using positively charged colloidal alumina-coated silica and negatively charged silica nanoparticles by Carosio et al. (2011). The results showed that the coating with five bilayers provided increased (45%) time to ignition and decreased (20%) heat release rate peak. Furthermore, it reduced the burn time and prevented the incandescent melt dipping of PET fabric. In their next study, Carosio et al. (2013) examined the flame-resistant effect of the same LbL composition on 0.2 and 1 mm thick plasma-treated PC films. After the measurements it was found that in the case of PC, significant improvement in fire retardancy and time to ignition can be achieved as well, with a good transparency. In the case of 0.2 mm thick samples the coating hindered the incandescent melt dipping during the flammability test. However, it had been reported that the low surface-to-bulk ratio and the poor stability of the coating during the combustion could limit the flammability properties of the 1 mm thick, coated PC films.

A PAH and MMT multilayer was deposited in order to prepare flame-retardant, thermally more stable polyamide-6 (PA6) film by Apaydin et al. (2013). The MMT nanoplatelets were aligned parallel to the substrate, which could provide good mechanical and barrier properties to the LbL-assembled coating. Twenty bilayers (with 5 μm thickness) of the polyelectrolyte and MMT nanoplatelets decreased the peak of heat release rate by more than 60%. After combustion, the thickness of the coating was doubled and a continuous, heat and mass transfer limiting charred layer occurred. The stabilizing effect observed was attributed to the barrier effect of the nanoclay and also to the char formation induced by the MMT platelets.

Li et al. (2011) showed, first, that LbL deposition of polyelectrolytes is a promising method for achieving intumescent nanocoating. They deposited poly(sodium phosphate) and polyallylamine (PAA) (as blowing agent) onto cotton fabric. When the PAA degraded due to heating, low-molecular weight components were released which acted as blowing gas. According to the vertical flame testing, while the uncoated fabric was completely consumed, the coated one, even with just five bilayers, had 19.3 wt% residue left and was preserved as a complete piece. Using 10 and 20 bilayers of PSP/PAA the unburned area of the fabric increased, where the amount of residue after the burning was 41 and 95 wt%, respectively. The horizontal test showed that in case of five bilayers the flame spreading speed was 247.4 mm/min^{-1} while in case of 10 and 20 bilayers the speed could not been measured. The PSP/PAA intumescent nanocoating greatly reduced the flammability of the cotton fabric under a 35 kW m^{-2} heat flux.

Intumescent multilayer nanocoating (having high phosphorous content) can be prepared on cotton fabric using fully renewable polyelectrolytes. CH and pytic acid (PA) were deposited onto cotton fabric for the first time in order to enhance flame resistance by Laufer et al. (2012b). The vertical flame test showed that the CH–PA coating completely extinguished the flame, while the uncoated cotton was totally consumed. In the case of the coated fabrics, a reduction in peak heat release rate (pkHRR) of at least 50% was observed. Coating deposited at pH 4 provided the highest reduction (of 60% and 76%) in pkHRR and in total heat release. The greater improvement in flame resistance of the coating deposited at lower pH was attributed to the higher phosphorous content and to the formation of a thin multilayer on the fibers, which delayed the interfiber linking, that preserved the coating and eliminated the gaps in the protective coating. Laufer et al. (2012b) reported that thin coating LbL deposited at pH 4 (66 wt% PA) left as much as 95% preserved fabric (completely unburned material with a small amount of char) after burning (Figure 10.14). According to the authors this intumescent coating did not expand to the same extent as most conventional intumescent coatings; it remained able to slow down, and in some cases stop the flame propagation.

Figure 10.14 Images of uncoated cotton fabric and fabric coated with 30 bilayers of CH–PA (top) and 18 wt% CH–PA (bottom) deposited at varying pH level.
Source: Reprinted with permission from Laufer et al. (2012b). © 2012 American Chemical Society.

10.3.4 Barrier coatings

Barrier coatings are very important in industrial fields such as packaging. It is a strong requirement in many cases to cover the product with a transparent high barrier material. In order to reduce the permeability of gases such as O_2 or CO_2 and water vapor, several kinds of methods and materials have been developed to create barrier coatings on different substrates, especially on plastic surfaces. The most popular coating consists of SiO_x, which can provide high oxygen barrier properties, transparency, and microwaveability and which can be performed by sputtering, electronbeam deposition, and the most common plasma-enhanced chemical vapor deposition (PECVD) (Erlat and Spontak, 1999; Singh et al., 2007). Although these methods give coatings with good barrier properties, coatings like SiO_x are susceptible to cracking, bending, and often exhibit poor adhesion to plastic substrates (Jang et al., 2008). Incorporating nanoparticles and especially nanoplatelets into the polymeric matrix can also improve remarkably the barrier properties, however producing a well-dispersed (intercalated, exfoliated), well-distributed structured nanocomposite is a major challenge because of the agglomeration tendency. According to Jang et al. (2008), LbL self-assembled nanostructured coatings provide both transparency and flexibility, making them a good alternatives in packaging applications (Table 10.4).

The most effective barrier improvers are layered silicate nanoclays, which are not only well-studied in the context of nanocomposites, but recently also in the context of LbL deposition. The well-known tortuous path model (Figure 10.15) is valid for LbL thin films too, because they can be considered as thin nanocomposite layers (if they consist of nanoclay and polyelectrolyte). Furthermore, these nanoclays in the assembled system are highly ordered, highly transparent and act like impermeable nanobricks (Priolo et al., 2010; Findenig et al., 2012). Jang et al. (2008) and Findenig et al. (2012) showed that these systems with nanobricks are stable, scratch-resistant and do not peel off when the substrate is bent. The success of these barrier coatings comes from the brick-wall structure and the molecular interactions through electrostatic forces and hydrogen bonding (Svagan et al., 2012).

Table 10.4 Barrier coatings

Multilayer composition	Properties	Reference
PEI/MMT	Oxygen barrier coating on PET films, high transparency	Priolo et al. (2010)
PEI/MMT	Barrier coating on PET film, highest barrier effect with 2.0 wt% MMT suspension concentration	Priolo et al. (2011)
CH/MMT	Highly reduced oxygen transmission rate in case of PLA film	Svagan et al. (2012)
PEI/MMT, HPMA starch/MMT, PDDA/MMT, CH/MMT	Stable, scratch-resistant water vapor barrier coating on cellulose acetate and PVDC-coated cellulose films, PEI, and HPMA starch are the most suitable for transparent barrier coatings	Findenig et al. (2012)
CH/MMT	Highly effective oxygen barrier coating on PLA film	Laufer et al. (2012a)
PVP/MMT	Improved oxygen barrier properties in case of PET film	Holder et al. (2012)
CH/PAA/CH/MMT	Oxygen barrier coatings with quadlayers on PET and PLA films	Laufer et al. (2012c)
CH/MMT/CR, CH/CR/CH/MMT	Reduced OTR where quadlayers were more effective	Laufer et al. (2013)

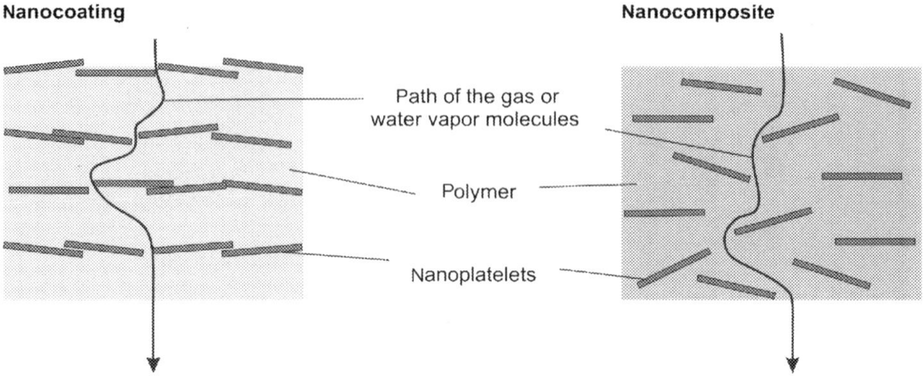

Figure 10.15 Tortuous path model in case of nanocomposites and nanostructured multilayers.

The most extensively studied layered silicate is MMT, which can be combined with different kinds of polyelectrolytes. PEI–MMT bilayers were built up on PET films by Priolo et al. (2010) and Priolo et al. (2011) to improve the oxygen barrier properties, and on cellulose films by Findenig et al. (2012) to improve the water vapor barrier properties. Priolo et al. (2010) studied how the pH of the PEI solution affects the barrier properties. Their result showed that at pH 7 the oxygen transmission rate (OTR) of the film is 8.42 cc/m^2 · day · atm while at pH 10 the OTR is only 0.34 cc/m^2·day·atm. This is obviously because, at higher pH, the PEI chains can be extended more than at lower pH, which causes thicker layers. As a function of number of bilayers at pH 10 the OTR shows an exponential reduction. With 70 bilayers the OTR was below the detection limit of commercial instrumentation (the OTR was less than 0.005 cc/m^2·day·atm). Besides the improved barrier properties the researchers showed that even with high clay concentration (up to 84.3 wt%) the PET kept its transparency up to 97.5% in the visible radiation range.

In their next study Priolo et al. (2011) investigated the effect of MMT suspension concentration on the barrier properties of the PET film. The higher the concentration, the stronger the barrier effect is. They found that there was an optimal clay suspension concentration to be used when creating gas barrier films, which is 2.0 wt% (the higher concentration the researchers used). This is because the nanoplatelets pack tighter and more platelets overlap. In this paper they reported also having highly transparent film; although the concentration of the MMT was as high as 93.3 wt% the film showed an average 95.1% transmittance in the range of 390–750 nm.

According to the research of Findenig et al. (2012) the PEI–MMT bilayers are useful for reducing the water vapor transmission rate (WVTR) too if they are deposited onto cellulose acetate and PVDC-coated cellulose films. Findenig et al. (2012) also examined using 2-hydroxy-3-trimethylammonium propyl chloride (HPMA) starch, PDDA, and CH as polyelectrolytes. It was concluded that it is possible to improve the water vapor barrier properties with hydrophilic components,

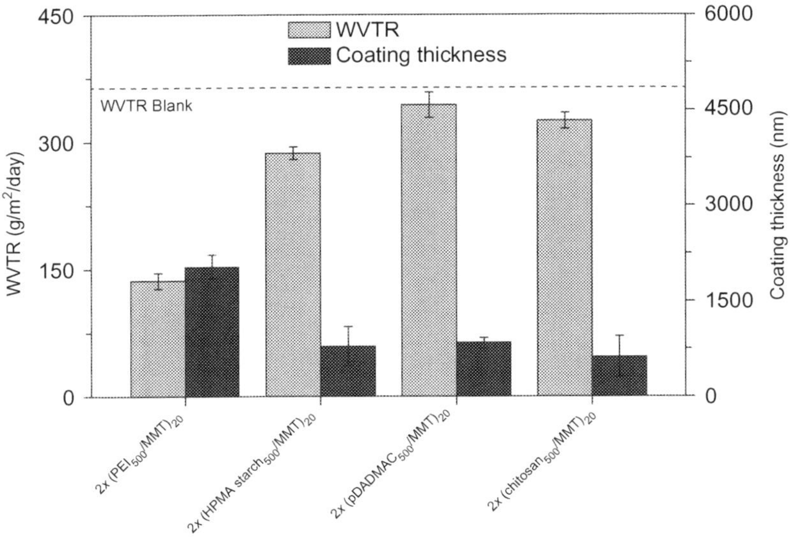

Figure 10.16 Influence of different polyelectrolytes on the water vapor barrier properties of commercial cellulose films (dashed line shows the WVTR of the uncoated cellulose film). *Source:* Reprinted with permission from Findenig et al. (2012). © 2012 American Chemical Society.

and PEI and HPMA starch are the most suitable for the preparation of transparent barrier coatings. The barrier properties were increased most in case of PEI with higher ionic strength due to the greater thickness of the polyelectrolyte layers. The five bilayers of PEI–MMT already led to a 22% improvement, and the 40 (2*20) bilayers a 68% improvement in water vapor barrier. The different water vapor transmission rates and coating thickness are shown in Figure 10.16, where coatings with 2 × 20 bilayers containing PEI, pDADMAC, HPMA starch and CH are compared. The researchers showed that after a certain number of bilayers there is no further improvement in the barrier properties. They explain this with the phenomenon that the coating is not well-structured after a certain number of bilayers. Furthermore, Findenig et al. found that there is a limit in the number of bilayers which can be built up. It was not possible to create a stable 2 × 40 bilayer coating on the substrate because the coating becomes too thick in a wet state and the adsorbed layers are removed from the surface.

Although CH was not as effective as PEI or starch in reducing WVTR with MMT it is suitable for reducing oxygen transmission as reported by Svagan et al. (2012) and Laufer et al. (2012a). CH is more acceptable for food packaging applications because compared to PEI, CH is proven to be non cytotoxic (Svagan et al., 2012). The coatings were assembled onto poly(lactic acid) films in both cases. After building up 40 bilayers of MMT–CH Svagan et al. (2012) observed 95% and 90% reduction in oxygen permeability at 20% and 50% RH, respectively (LbL assembly was performed with 0.2 wt% MMT and CH solution, where the pH of the CH solution was adjusted to pH 6). Seventy bilayers of MMT–CH caused 20% reduction in water vapor transmission. According to the authors, the lack of larger water vapor barrier improvement can be attributed to the hydrophilic nature of the CH and reduced CH–clay interaction at the interface due to water. Laufer et al. (2012a) reported high oxygen barrier effect of the MMT–CH multilayer too (where the concentration of MMT and CH solution (pH 6) was 1 wt%, and 0.1 wt% respectively). With 10 bilayers the oxygen permeability of the PLA film was equivalent to bearing PET film and after the deposition of the 30th bilayer the OTR became less than 0.005 cc/m^2·day·atm.

Swelling of the polyelectrolyte can improve the oxygen barrier properties according to Holder et al. (2012). The authors showed a reduced OTR of PET films with increasing relative humidity by creating hydrogen-bonded LbL multilayers of polyvinylpyrolidone (PVP) and MMT nanoplatelets. A 40 bilayer of PVP–MMT film showed an 11.1% decrease in OTR at 100% RH. The improved barrier properties were attributed to the increase in tortuous pathway length due to the swelling of PVP.

Preparing quadlayers instead of bilayers can enhance further the barrier effect of the LbL films, as shown by Laufer et al. (2012c and 2013). Laufer et al. (2012c) created 10 quadlayers of CH-PAA-CH-MMT (<100 nm thick) on corona-treated PLA and PET substrates and found that the coating reduced the OTR by two orders of magnitude under dry conditions and more than one order of magnitude at 38°C and 90% RH in both cases. This can be attributed to the near perfectly oriented clay platelets that creates an extremely tortuous path for diffusing oxygen molecules (Laufer et al., 2013). In their following study Laufer et al. (2013) replaced PAA to carrageenan (CR) and examined the effect of 10 trilayers of CH-MMT-CR (<40 nm thick) and 10 quadlayers of CH-CR-CH-MMT (<60 nm thick) and proved that quadlayers are more effective in improving barrier properties. In the case of trilayers the OTR was reduced by an order of magnitude and in the case of quadlayers by two orders of magnitude under the same conditions, which can be ascribed to the greater clay spacing in the quadlayers. Due to the high exfoliation and orientation of the MMT nanoplatelets the transparencies of the films were higher than 98%. LbL nanoassembly with a relatively small number of layers offers a flexible, transparent "green" barrier coating technique that can compete with the commercial coating techniques used in the packaging industry.

LbL-deposited barrier coatings are useful for making plastics as good gas and water vapor barrier materials but they can be also suitable to prevent corrosion of the surface in metals.

10.3.5 Antimicrobial coatings

Antimicrobial coating can inhibit the adhesion and growth of bacterium or fungi on a substrate. The coating thus can protect a surface from discoloration, microbial corrosion and, most importantly, it can prevent health risks. A wide variety of nanoparticles are suitable for creating antimicrobial coating through LbL (Table 10.5). Silver is one of the most powerful antimicrobial agents, with strong activity, even

Table 10.5 Antimicrobial coatings

Multilayer composition	Properties	Reference
CH/heparin	Antimicrobial coating (against *Escherichia coli* (*E. coli*)) on PET film (where heparin was used as anti-adhesive agent)	Fu et al. (2005)
Ag-PMA/PDDA	Antimicrobial coating on nylon and silk fibers	Dubas et al. (2006)
PDDA/PAA + PDDA/MMT, PDDA/PAA + PDDA/Ag, PDDA/PAA + PDDA/ MMT/PDD/Ag	Antimicrobial nacre-like coating (against *E. coli*) with good mechanical properties	Podsaido et al. (2007)
ALG/CH-organic rectorite	Stronger antimicrobial activity against *E. coli* compared to CH/ALG nanocoating	Deng et al. (2011)
CH/ALG	Durable nanocoating with antimicrobial effect against both Gram-positive and Gram-negative bacteria on cotton fibers	Gomes et al. (2012)

in the LbL multilayer, toward a broad range of microorganisms and simultaneously remarkably low human toxicity (Csóka et al., 2012). The antimicrobial mechanism of silver nanoparticles is still not fully understood. According to Morones et al. (2005) the silver nanoparticles can directly damage the microbial cell membranes by attaching themselves to the sulfur-containing proteins of cell membranes, leading to an increase in permeability of the cell membrane, which causes the death of the bacteria. Silver nanoparticles can also disrupt the adenosine triphosphate (ATP) production and DNA replication or, according to Lok et al. (2007) and Smetana et al. (2008), the reactive oxygen species created by the nanoparticles can be harmful to the microbes.

Dubas et al. (2006) used silver nanoparticle capped with poly(methacrylic acid) (PMA) and PDDA to create LbL coating on nylon and silk fibers. The results showed that both synthetic and natural fibers can be coated with layer-by-layer nanoassembly; however, the fiber surface played an important role in the growth of the film, the assembly on silk surface was better, and the coating was smoother and more uniform. The coated silk fibers had higher antimicrobial activity as well: with 10 bilayers the number of attached bacteria was reduced by 41% and with 20 bilayers by 80%. In the case of nylon fiber the effect was weaker: 10 bilayers did not cause any change, 20 bilayers showed a 53% decrease in bacteria growth. The created fibers can be useful in applications such as water sanitation or antimicrobial fabrics.

Antimicrobial nacre-like coating was created with PDDA, PAA, silver nanoparticles, and MMT nanolayers by Podsaido et al. (2007) on glass substrate. The surface was pretreated with three bilayers of PDDA/PAA, then the 10 PDDA/MMT, the 10

PDDA/Ag, or the 10 PDDA/MMT/PDDA/Ag layers were assembled onto the surface. According to the results a nanostructured, hybrid, and multifunctional composite layer was created with good mechanical properties, strong antimicrobial characteristic toward *E. coli*, and compatibility with human osteoblasts. The researchers found that the inhibition of the bacteria growth could be mostly attributed to the presence of silver and not to any of the polyions or clay.

Although silver nanoparticles are effective antimicrobial additives, in the case of LbL nanoassembly the most studied antimicrobial coating constituent is CH. In case of polyelectrolytes, like CH, although the antimicrobial activity is proven, the exact mechanism of the effect is less known, according to Baur et al. (1999) the interaction between the positively charged polyelectrolyte and the negatively charged microbial cell membranes may lead to a leakage of proteinaceous and other intracellular constituents.

Fu et al. (2005) applied CH (weak polycation) as an antibacterial agent and heparin (strong polyanion) as an anti-adhesive agent to create functional coating on PET films. The in vitro bacterial test presented that the CH/heparin multilayer can successfully kill the *E. coli* bacteria. Furthermore, the researchers showed that the pH of the dipping solution strongly affects the effectiveness of the antibacterial coating. The number of the *E. coli* bacteria which could be adsorbed on the surface to the PET film decreased by 68%, 58%, and 46% when using pH = 3.8, 2.9, and 6.0, respectively. The pH value defines the thickness of the outermost CH layer and thus the ability of heparin penetration through the polyactionic layer. The thinner the CH layer, the higher the penetration degree of the heparin, and the higher the hydrophilicity of the coating which can reduce the attachment of the bacteria to the surface.

Joshi et al. (2010) first deposited CH/poly(sodium-4-styrene sulfonate) (PSS) antimicrobial multilayers onto cotton fabric, with a new, ultrasound-assisted technique which provides a coating on the textile surface without affecting its flexibility, feel, or breathability. Ultrasonication was applied during the washing steps to ensure the complete removal of loosely adhering polyelectrolyte layers. According to the results, the ultrasonic treatment helped the deposition of more uniform bilayers on the individual fiber surfaces. In addition, with treatment the retention of antimicrobial activity was much higher than without the treatment.

Gomes et al. (2012) also used CH for antimicrobial coating on cotton fibers. In this research the surface of the cotton fibers were pretreated with TEMPO oxidation, which ensured the success of the assembly of the CH/alginate (ALG) on the fibers. The coating had excellent durability to washing, which indicated a good adhesion between the LbL-assembled layers and the cotton surface. The thus prepared CH/ALG multilayers provided stable antimicrobial effect against both Gram-positive and Gram-negative bacteria.

To perform LbL nanoassembly on nanofibrous cellulose mats, negatively charged sodium alginate and positively charged CH–organicrectorite composites (with intercalated structure) were applied by Deng et al. (2011). Using organic rectorite with CH led to a thicker bilayer and a better antimicrobial activity against *E. coli* compared to those coatings which contained simple CH/ALG bilayers.

10.4 Conclusions

The LbL nanoassembly method is a versatile, inexpensive, simple and quick, automatable method for the modification of the material surface. Any kind of material, with any shapes and dimensions, which has got a minimal surface charge can be used as a substrate. There is a wide variety of coating materials suitable for the LbL technique, such as cationic and anionic polyelectrolytes (also from renewable, natural sources) and nanoparticles (such as metal, metal oxide nanoparticles, naturally occurring nanocrystals, and nanoclays or nanotubes). The thickness and properties of the multilayers prepared through the alternate adsorption of polyelectrolytes and/or nanoparticles can be finely tuned by varying the processing parameters such as pH, ionic strength, concentration, media, or immersion time. Using the optimal parameters and coating materials the LbL can provide formation of complex, nanostructured functional coatings that combine several properties (anti-abrasive; scratch-, corrosion-, and flame-resistance; self-healing; antimicrobial; barrier; anti-reflective; or photovoltaic properties) is high demanded from industry.

Acknowledgment

This study was supported by the environment-conscious energy efficient building TAMOP-4.2.2.A−11/1/KONV-2012-0068 project sponsored by the EU and European Social Foundation.

References

Ai, H., Jones, S.A., Lvov, M.Y., 2003. Biomedical applications of electrostatic layer-by-layer nano-assembly of polymers, enzymes, and nanoparticles. Cell Biochem. Biophys. 39, 23−43.

Ali, M., Yameen, B., Cervera, J., Ramírez, P., Neumann, R., et al., 2010. Layer-by-layer assembly of polyelectrolytes into ionic current rectifying solid-state nanophores: insight from theory and experiment. J. Am. Chem. Soc. 132, 8338−8348.

Andreeva, D.V., Skorb, E.V., Shchukin, D.G., 2010. Layer-by-layer polyelectrolyte/inhibitor nanostructures for metal corrosion protection. Appl. Mater. Interfaces. 2 (7), 1954−1962.

Apaydin, K., Lachachi, A., Ball, V., Jimenez, M., Bourbigot, S., et al., 2013. Polyallylamine-montmorillonite as super flame retardant coating assemblies by layer-by-layer deposition on polyamide. Polym. Degrad. Stab. 98, 627−634.

Ariga, K., Lvov, Y., Ichinose, I., Kunitake, T., 1999. Ultrathin films of inorganic materials (SiO2 nanoparticle, montmorillonite microplate, and molybdenum oxide) prepared by alternate layer-by-layer assembly with organic polyions. Appl. Clay Sci. 15, 137−152.

Aulin, C., Johansson, E., Wagberg, L., Lindstrom, T., 2010. Self-organized films from cellulose I nanofibrils using the layer-by-layer technique. Biomacromolecules. 11, 872−882.

Baur, J.W., Rubner, M.F., Reynolds, J.R., Kim, S., 1999. Förster energy transfer studies of polyelectrolyte heterostructures coating conjugated polymers: a means to estimate layer interpenetration. Langmuir. 15 (19), 6460–6469.

Beyer, S., Mak, W.C., Trau, D., 2007. Reverse-phase LbL-encapsulation of highly water soluble materials by layer-by-layer polyelectrolyte self-assembly. Langmuir. 23, 8827–8832.

Bolto, B., Gregory, J., 2007. Organic polyelectrolytes in water treatment. Water Res. 41 (11), 2301–2324.

Borodina, T.N., Grigoriev, D.O., Andreeva, D.V., Möhwald, H., Shchukin, D.G., 2009. Polyelectrolyte multilayered nanofilms as a novel approach for the protection of hydrogen storage materials. Appl. Mater. Interfaces. 1 (5), 996–1001.

Carosio, F., Laufer, G., Alongi, J., Camno, G., Grunlan, J.C., 2011. Layer-by-layer assembly of silica-based flame retardant thin film on PET fabric. Polym. Degrad. Stab. 96, 745–750.

Carosio, F., Di Blasio, A., Alongi, J., Malucelli, G., 2013. Layer by layer nanoarchitectures for the surface protection of polycarbonate. Eur. Polym. J. 49, 397–404.

Chen, W., McCarthy, T.J., 1997. Layer-by-layer deposition: a tool for polymer surface modification. Macromolecules. 30, 78–86.

Chunder, A., Etcheverry, K., Wadsworth, S., Boreman, G.D., Zhai, L., 2009. Fabrication of anti-reflection coatings on plastics using the spraying layer-by-layer self-assembly. J. SID. 17 (4), 389–395.

Croll, T.I., O'Connor, J.A., Stevens, W.G., Copper-White, J.J., 2006. A blank state? Layer-by-layer deposition of hyaluronic acid and chitosan onto various surfaces. Biomarcomolecules. 7, 1610–1622.

Csóka, L., Bozanic, D., Nagy, V., Dimitrijevics-Brankovic, S., Luyt, A.S., et al., 2012. Viscoelastic properties andantimicrobial activity of cellulose fiber sheets impregnated with Ag nanoparticles. Carbohydr. Polym. 90, 1139–1146.

Dafinone, M., Feng, G., Brugarolas, T., Tettey, E., Lee, D., 2011. Mechanical reinforcement of nanoparticle thin fims using atomic layer deposition. ACS Nano. 5 (6), 5078–5087.

Decher, G., Hong, J.D., 1991. Buildup of ultra-thin multilayer films by a selfassembly process: II. Consecutive adsorption of anionic and cationic bipolar amphiphiles and polyelectrolytes on charged surfaces. Berichte Der Bunsengesellschaft für Physikalische Chemie. 95, 1430–1434.

DeLongchamp, M.D., Kastantin, M., Hammond, P.T., 2003. High-contrast electrochromism from layer-by-layer polymer films. Chem. Mater. 15, 1575–1586.

Deng, H., Wang, X., Liu, P., Ding, B., Du, Y., et al., 2011. Enhanced bacterial inhibition activity of layer-by-layer structured polysaccharide film-coated cellulose nanofibrous mats via addition of layered silicate. Carbohydr. Polym. 83, 239–245.

Dubas, S.T., Kumlangdudsana, P., Pityaraj, P., 2006. Layer-by-layer deposition of antimicrobial silver nanoparticles on textile fibers. Colloids Surf. A: Physocochem. Eng. Aspects. 289, 105–109.

Erlat, A.G., Spontak, R.J., 1999. SiOx gas barrier coatings on polymer substrates: morphology and gas transport considerations. J. Phys. Chem. B. 103, 6047–6055.

Farhat, T.R., Schlenhoff, J.B., 2002. Corrosion control using polyelectrolyte multilayers. Electrochem. Solid-State Lett. 5 (4), B13–B15.

Findenig, G., Leimgruber, S., Kargl, R., Sprik, S., Stana-Kleinschenk, K., et al., 2012. Creating water vapor barrier coatings from hydrophilic components. Appl. Mater. Interfaces. 4, 3199–3206.

Fu, J., Ji, J., Yuan, W., Shen, J., 2005. Construction of anti-adhesive and antibacterial multilayer films via layer-by-layer assembly of heparin and chitosan. Biomaterials. 26, 6684–6692.

Ghosh, S.K., 2006. Functionl coatings and microencapsulation: a general perspective. In: Ghosh, S.K. (Ed.), Functional Coatings. Wiley-VCH, Weinheim, Germany, pp. 1–28.

Gomes, A.P., Mano, J.F., Queiroz, J.A., Gouveia, I.C., 2012. Layer-by-layer deposition of antibacterial polyelectrolytes on cotton fibers. J. Polym. Environ. 20, 1084–1094.

Hammond, P.T., 2000. Recent explorations in electrostatic multilayer thin film assembly. Curr. Opin. Colloid Interface Sci. 4, 430–442.

Hammond, P.T., 2011. Engineering materials layer-by-layer: challenges and opportunities in multilayer assembly. Am. Inst. Chem. Eng. J. 57 (11), 2928–2940.

Han, J.T., Zheng, Y., Cho, J.H., Xu, X., Cho, K., 2005. Stable superhydrophobic organic-inorganic hybrid films by electrostatic self-assembly. J. Phys. Chem. B. 109, 20773–20778.

Hirsjävi, S., Peltonen, L., Hirvonen, J., 2006. Layer-by-layer polyelectrolyte coating of low molecular weight poly(lactic acid) nanoparticles. Colloids Surf. B: Biointerfaces. 49, 93–99.

Hoeger, I., Orlando, R.J., Efimenko, K., Velev, O.D., Kelly, S.S., 2011. Ultrathin film coatings of aligned cellulose nanocrystals from convective-shear assembly system and their surface mechnaical properties. Soft Matter. 7, 1957–1967.

Holder, K.M., Priolo, M.A., Secri, K.E., Greenlee, M., Nolte, A.J., et al., 2012. Humidity-responsive as barrier of hydrogen-bonded polymer-clay multilayer thin films. J. Phys. Chem. 116, 19851–19856.

Hua, F., Cui, T., Lvov, Y.M., 2004. Ultrathin cantilevers based polymer-ceramic nanocomposite assembled through layer-by-layer adsorption. Nano Lett. 4 (5), 823–825.

Jang, W.-S., Rawson, I., Grunlan, J.C., 2008. Layer-by-layer assembly of thin film oxygen barrier. Thin Solid Films. 516, 4819–4825.

Joshi, M., Khanna, R., Shekhhar, R., Jha, K., 2010. Chitosan nanocoating on cotton textile substrate using layer-by-layer self-assembly technique. J. Appl. Polym. Sci. 119, 2793–2799.

Kamineni, K.V., Lvov, M.Y., Dobbins, T.A., 2007. Layer-by-layer nano-assembly of polyelectrolytes using formamide as the working medium. Langmuir. 23, 7423–7427.

Laufer, G., Kirkland, C., Cain, A.A., Grunland, J.C., 2012a. Clay-chitosan nanobrick walls: completely renewable gas barrier and flame-retardant nanocoatings. Am. Chem. Soc. Appl. Mater. Interfaces. 4, 1643–1649.

Laufer, G., Kirkland, C., Morgan, A.B., Grunlan, J.C., 2012b. Intumescent multilayer nanocoating, made with renewble polyelectrolytes, for flame-retardant cotton. Biomacromolecules. 13, 2843–2848.

Laufer, G., Priolo, M.A., Kirland, C., Grunlan, J.C., 2012c. High oxygen barrier, clay and chitosan-based multilayer thin films: an environmentally friendly foil replacement. Green Mater. 1 (1), 4–10.

Laufer, G., Kirkland, C., Cain, A.A., Grunlan, J.C., 2013. Oxygen barrier of multilayer thin films comprised of polysaccharides and clay. Carbohydr. Polym. 95, 299–302.

Lee, H., Lee, Y., Statz, A.R., Rho, J., Park, T.G., et al., 2008. Substrat-independent layer-by-layer assembly by using mussel-adhesive inspired polymers. Adv. Mater. 20 (9), 1619–1623.

Li, Y., Wang, X., Sun, J., 2012. Layer-by-layer assembly for rapid fabrication of thick polymeric films. Chem. Soc. Rev. 41, 5998–6009.

Li, Y.-C., Schulz, J., Grunlan, J.C., 2009. Polyelectrolyte/nanosilicate thin film assemblies: influence of pH on growth, mechanical behavior, and flammability. Appl. Mater. Interfaces. 1 (10), 2338–2347.

Li, Y.-C., Mannen, S., Morgan, A.B., Chang, S., Yang, Y.-H., et al., 2011. Intumescent all-polymer multilayer nanocoating capable of extinguishing flame on fabric. Adv. Mater. 23, 3926–3931.

Liu, X., Zhou, L., Liu, F., Ji, M., Tang, W., et al., 2010. Exponential growth of layer-by-layer assembled coatings with well-dispersed ultrafine nanofillers: a facile route to scratch-resistant and transparent hybrid coatings. J. Mater. Chem. 20, 7721–7727.

Liu, Y., Claus, R.O., Rosidian, A., Zeng, T., 1999. A new route to prepare hard and anti-scratching coatings at room temperature. Mater. Res. Soc. Symp. Proc. 576, 213–219.

Lok, C.N., Ho, C.M., Chen, R., He, Q.Y., Yu, W.Y., et al., 2007. Silver nanoparticles: partial oxidation and antibacterial activities. J. Biol. Inorg. Chem. 12, 527–534.

Lu, Z., Eudala, S., Zheng, Z., Xu, K., Grozdits, G., et al., 2007. Layer-by-layer nanoparticle coatings on lignocellulose wood microfibers. Colloids Surf. A: Physicochem. Eng. Aspects. 292, 56–62.

Lvov, Y., Ariga, K., Onda, M., Ichinose, I., Kunitake, T., 1997. Alternate assembly of ordered multilayers of SiO2 and other nanoparticles and polyions. Langmuir. 13 (23), 6195–6203.

Lvov, Y., Rusling, J.F., Thomsen, D.L., Papadimitrakopoulos, F., Kawakami, T., et al., 1998. High-speed multilayer film assembly by alternate adsorption of silica nanoparticles and linear polycation. Chem. Commun. 11, 1229–1230.

Lvov, Y., Ariga, K., Ichinose, I., Kunitake, T., 1999. Molecular film assembly via layer-by-layer adsorption of oppositely charged molecules (linear polymer, protein and clay) and concanavalin A and glycogene. Thin Solid Films. 284–285, 797–801.

Morones, J.R., Elechiguerra, J.L., Camacho, A., Holt, K., Kouri, J.B., et al., 2005. The bactericidal effect of silver nanoparticles. Nanotechnology. 16, 2346–2353.

Nepal, D., Balasubramanian, S., Simonian, A.L., Davis, V.A., 2008. Strong antimicrobial coatings: single-walled carbon nanotubes armored with biopolymers. Nano Lett. 8 (7), 1896–1901.

Nogueira, G.M., Banerjee, D., Chine, R.E., Rubner, F., 2011. Spray-layer-by-layer assembly can more rapidly produce optical-quality multistack heterostructures. Langmuir. 27, 7860–7867.

Paloniemi, H., Lukkarinen, M., Aaritalo, T., Areva, S., Leiro, J., et al., 2006. Layer-by-layer electrostatic self-assembly of single wall carbon nanotube polyelectrolytes. Langmuir. 22, 74–83.

Pastoriza-Santos, I., Koktysh, S.D., Mamedov, A.A., Gierrsig, M., Kotov, A.N., et al., 2000. One-pot synthesis of Ag@TiO2 core-shell nanoparticles and their layer-by-layer assembly. Langmuir. 16, 2731–2735.

Podsaido, P., Paternel, S., Rouillard, J.-M., Zhang, Z., Lee, J., et al., 2007. Layer-by-layer assembly of nacre-like nanostructured composites with antimicrobial properties. Langmuir. 23, 7901–7906.

Priolo, M.A., Gamboa, D., Grunlan, C.J., 2010. Transparent clay-polymer nano brick wall assemblies with tailor able oxygen barrier. Appl. Mater. Interfaces. 2 (1), 312–320.

Priolo, M.A., Holder, K.M., Gamboa, D., Grunlan, J.C., 2011. Influence of clay concentration on the gas barrier of clay-polymer nanobrick wall thin film assemblies. Langmuir. 27, 12106–12114.

Ramsden, J.J., Lvov, Y.M., Decher, G., 1995. Determination of optical constants of molecular films assembled via alternate polyion adsorption. Thin Solid Films. 254, 246–251.

Rogach, A.L., Koktysh, D.S., Harrison, M., Kotov, N.A., 2000. Layer-by-layer assembled films of HgTe nanocrystals with strong infrared emission. Chem. Mater. 12, 1526–1528.
Rosidian, A., Liu, Y., Claus, R.O., 1998. Ionic self-assembly of ultrathin ZrO2/polymer nanocomposite thin films. Adv. Mater. 10 (4), 1087–1091.
Shchukin, D.G., Zheludkevics, M., Yasakau, K., Lamaka, S., Ferreira, M.G.S., et al., 2006. Layer-by-layer assembled nanoconationers for self-healing corrosion protection. Adv. Mater. 18, 1672–1678.
Shi, F., Wang, Z., Zhang, X., 2008. Combining a layer-by-layer assembly technique with electrochemical deposition of gold aggregates to mimic the legs of water striders. Adv. Mater. 17 (8), 1005–1009.
Shiratori, S.S., Rubner, M.F., 2000. pH-dependent thickness behavior of sequentially adsorbed layers of weak polyelectrolytes. Macromolecules. 33, 4213–4219.
Singh, B., Bouchet, J., Rochat, G., Leterrier, Y., Manson, J.-A.E., et al., 2007. Ultra-thin hybrid organic/inorganic gas barrier coatings on polymers. Surf. Coat. Technol. 201, 7107–7114.
Smetana, A., Klabunde, K., Marchin, G., Sorensen, C., 2008. Biocidal activity of nanocrystalline silver powers and particles. Langmuir. 24, 7457–7464.
Svagan, A.J., Akesson, A., Cardenas, M., Bulut, S., Knudsen, J.C., et al., 2012. Transparent films based on PLA and montmorillonite with tunable oxygen barrier properties. Biomolecules. 13, 397–405.
Tettey, E.K., Yee, Q.M., Lee, D., 2010. Layer-by-layer assembly of charged particles in nonpolar media. Langmuir. 26, 9974–9980.
Van Tassel, P.R., 2012. Polyelectrolyte adsorption and layer-by-layer assembly: electrochemical control. Curr. Opin. Colloid Interface Sci. 17, 106–113.
Villiers, M.M., Otto, D.P., Strydom, S.J., Lvov, Y.M., 2011. Introduction to nanocoatings produced by layer-by-layer (LbL) self-assembly. Adv. Drug Deliv. Rev. 63, 701–715.
Vozar, S., Poh, Y.-C., Serbowicz, T., Bacner, M., Podsiadlo, P., Qin, M., et al., 2009. Automated spin-assisted layer-by-layer assembly of nanocomposites. Rev. Sci. Instrum. 80, 023903-1–023903-5.
Wu, Q., Wang, Z., Kong, X., Gu, X., Xue, G., 2008. A facilate strategy for controlling the self-assembly of nanocomposite particles based on colloidal steric stabilization theory. Langmuir. 24, 7778–7784.
Xu, T., Zhong, Y., Liu, Y., Yu, H., Mao, Z., 2012. Flammability properties of PI fabric coated with montmorillonite. J. Therm. Anal. Calorim. 111, 27–33.
Xue, W., Cui, T., 2007. Characterization of layer-by-layer self-assembled carbon nanotube multilayer thin films. Nanotechnology. 18, 1–7.
Ye, S., Wang, C., Liu, X., Tong, Z., 2004. Multilayer nanocapsules of polysaccharide chitosan and alginate through layer-by-layer assembly directly on PS nanoparticles for release. J. Biomater. Sci. Polym. Ed. 16 (7), 909–923.
Zhang, P., Qian, J., An, Q., Du, B., Liu, X., et al., 2008. Influences of solution property and charge density on the self-assembly behavior of water-soluble polyelectrolyte sulfonated poly(sulphone) sodium salts. Langmuir. 24 (5), 2110–2117.
Zhou, S., Wu, L., You, B., Gu, G., 2009. Preparation, structure and properties of organic-inorganic nanocomposite coatings. In: Provder, T., Baghdaci, J. (Eds.), Smart Coatings II, ACS Symphosium Series. American Chemical Society, Washington, DC, pp. 193–219.
Zlotnikov, I., Gotman, I., Gutmanas, E.Y., 2008. Characterization and nanoindentation testing of thin ZrO2 films synthesized using layer-by-layer (LbL) deposited organic templates. Appl. Surf. Sci. 235, 3447–3453.

Theoretical study on an influence of fabrication parameters on the quality of smart nanomaterials

11

Velaphi Msomi
Cape Peninsula University of Technology, South Africa

11.1 Introduction

Materials application depends on their behaviour at different environmental conditions. In general, most of the materials change their properties due to a change in environmental conditions, which normally affect the performance of materials. There are some materials that change colour due to external factors; those materials are known as chromic materials.

There are materials which change colour due to temperature change, known as thermochromic materials, while others change colour due to light, known as photochromic, and others change colour due to pressure, known as piezochromic and so on. These materials are applied in various fields; for example, thermochromic materials are applied in window coating, laser protection and optical and electrical switching. Thermochromism occurs by means of rearrangements of the molecular structure of a material, as a result of a change in temperature (Tampere University of Technology, n.d.).

The general overview on window coating application is that a thin film of a thermochromic material on an exterior window could modify its reflectance properties, depending on outside ambient temperature. The solar radiation that is not able to pass through the window when it is in its darkened state must be either reflected or absorbed. In a cooling-dominated application, the window would pass all or part of the visible radiation incident on the window and reflect the majority of the sun's near-infrared radiation. Incident solar radiation that is not transmitted is absorbed or reflected. Absorption will cause significant heating of the window if left to stagnate in a coloured state under conditions of high irradiance (Mott, 1990; Kucharczyk and Niklewski, 1979).

An increase in temperature in the window will give rise to a radiant heat source adjacent to the room, potentially leading to thermal discomfort, and will impose additional demands on the temperature stability of the materials used in the smart window. Effective thermochromic window coatings would respond to this heating by increasing their reflectance and compensating for the increased heating by reflecting more heat away. Such intelligent coatings could be used in countless applications, including automobile windscreens, sunscreens and greenhouses. The development of

Table 11.1 **Outline of work done on the application of VO_2 thin films**

Authors	Application of VO_2
J.M. Wu and L.B. Liou (2011)	Ultrafast optical switches
Li et al. (2011)	Field-effect transistors
Goldflam et al. (2011)	Memory metamaterials
A.K. Prasad et al. (2014)	Methane sensing technology
X. Chen et al. (2012)	Smart window coating
Z. Chen et al. (2011)	
E. Syrrakou et al. (2006)	

such coatings will lead to large savings in energy costs (e.g. power to air conditioning units) and environmental benefits (e.g. reduced CO_2 emissions). This work focuses mainly on the quality of thermochromic material called vanadium dioxide (VO_2) synthesized with two different techniques: inverted cylindrical magnetron sputtering (ICMS) and pulsed laser deposition (PLD).

This material is defined as a material which undergoes a reversible semiconductor-to-metal transition at 68°C (Chain, 1986). VO_2 has a monoclinic structure below transition temperature and becomes tetragonal above transition temperature (Morin, 1959; Case, 1987; Remka et al., 1979). The transition of VO_2 is accompanied by abrupt change in the electrical resistivity and optical transmittance (Greenberg, 1983; University of Nebraska–Lincoln, n.d.; University of Nebraska–Lincoln, n.d.). VO_2 has a very low transmittance in the infrared (IR) region during the metallic phase but has higher transmittance in the IR region during the semiconducting phase. The transition temperature (68°C) of this material can be lowered or increased using dopants (e.g. tungsten molybdenum) depending on the purpose of the application. Table 11.1 outlines previous and the present work done around the analysis and application of VO_2.

11.2 Literature survey on VO_2

The thermochromism of VO_2 was discovered by F.J. Morin in 1959 (Morin, 1959). VO_2 has a monoclinic structure at temperatures below transition temperature and it is IR transparent. In the semiconducting phase, VO_2 has a high transmittance due to the low free carrier concentration. The free carrier concentration increases in the metallic phase, therefore, the transmittance in the IR decreases.

VO_2 thin film is reddish-brown in the semiconducting state and becomes greenish-blue in the metallic state. At temperatures above the critical temperature, each vanadium atom is located at the centre of an oxygen tetrahedron. At lower temperatures, the structure is distorted as vanadium atoms pair up and are alternately displaced from the centre of oxygen (Goodenough and Hong, 1973).

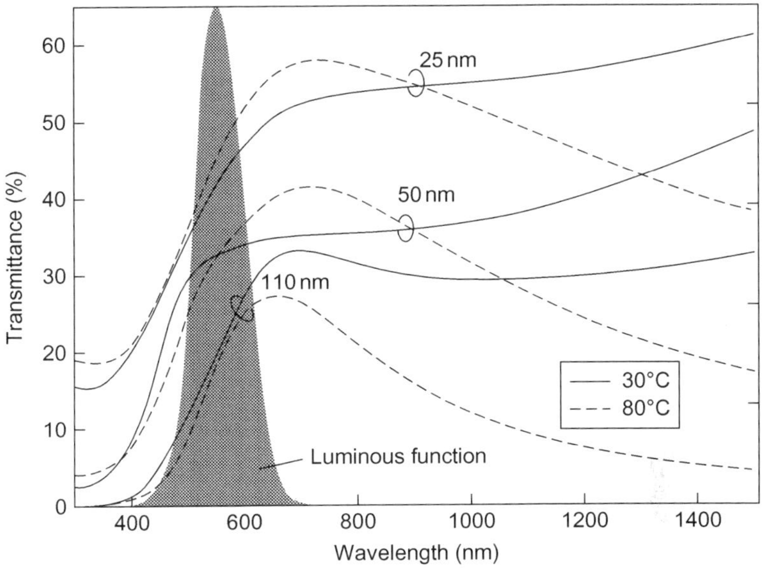

Figure 11.1 Transmittance of 25−110 nm of VO_2, respectively (Xu et al., 2005).

The transformation in VO_2 is of shear type, with a slight atomic displacement and a small dilation (0.044%) accompanying the phase change (Golan et al., 2003).

The electrical and optical properties of the film depend mostly on the stoichiometry and the microstructure, as well as the levels of residual stress caused by the deposition technique (Golan et al., 2003). The dependence phenomena are due to crystallinity of the film. The film morphology is affected by the thickness of the film. The increase in film surface roughness due to an increase in thickness leads to a decrease in transition temperature (Xu et al., 2005). An increased number of atoms distributed randomly at grain boundaries and/or large surface/interface ratio for a very thin VO_2 film may damage the zigzag chains of V−V pairs characteristic of the low temperature phase, resulting in the destabilization of the low temperature phase and therefore a decrease in transition temperature (Xu et al., 2005). The film thickness also affects the transmittance: as the thickness increases, transmittance decreases.

When the film thickness is less than 100 nm, the transmittance of the metallic phase exceeds that of a semiconducting phase near the IR region, while for those films thicker than 100 nm, the transmittance of the semiconducting phase exceeds that of the metal phase (see Figure 11.1).

Various deposition methods have been developed to produce VO_2 thin films. These include physical vapour deposition (reactive sputtering, reactive evaporation and laser ablation), chemical vapour deposition (CVD), sol-gel techniques and controlled oxidation and reduction (Kucharczyk and Niklewski, 1979; Chain, 1986; Morin, 1959; Case, 1987; Remka et al., 1979; Greenberg, 1983).

During the semiconductor-to-metal-transition, the optical properties of VO_2 are characterized by a sharp decrease in optical transmission in the IR spectrum. Because of this anomalous behaviour, VO_2 has been presented as an attractive thin film material for electrical and optical switches, optical storage, laser protection and solar energy control for the windows of space satellites (Morin, 1959; Case, 1987).

VO_2 has been studied in various ways using various deposition techniques as mentioned above. It has been doped with various dopants, for example tungsten, molybdenum, lithium and hydrogen, in order to increase or decrease the transition temperature. According to the Tang et al. (1985) model, the incorporation of dopant atoms leads to the loss of $V^{4+}-V^{4+}$ pairs, which are essential for the crystal structure of the semiconducting phase.

The loss of these homopolar $V^{4+}-V^{4+}$ bonds are caused by the incorporation of the dopants, as electrons are injected from the donors into the vanadium 3d valence band, resulting in the transformation of V^{4+} to V^{3+} ions via charge transfer. Thus doping elements destabilize the semiconducting phase, and consequently lower the semiconducting-to-metal transition temperature (Burkhardt et al., 2002). VO_2 has been doped with tungsten and fluorine in order to decrease the transition temperature to room temperature, with the purpose of applying it to smart windows.

The pure VO_2 thermochromism was compared with the W- and F-doped VO_2. The atomic force microscopy (AFM) results show the effect of dopants in the surface of VO_2 thin films (see Figure 11.2). The introduction of 1.8 at.% W on VO_2 thin film decreases the crystal size of the film. The surface roughness decreases

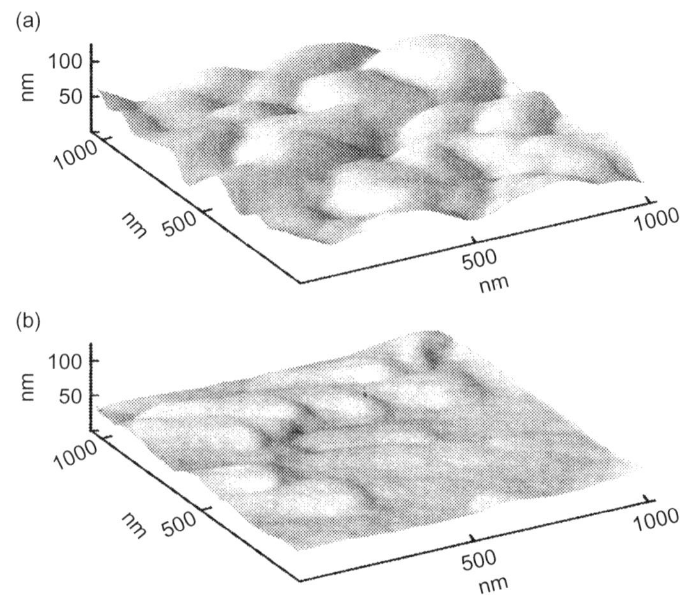

Figure 11.2 AFM images of $1000 \times 1000\text{-nm}^2$ area of VO_2 films: (a) undoped grown at 800 K substrate temperature; (b) doped with 1.8 at.% W grown at 800 K (Burkhardt et al., 2002).

with the increase of tungsten as a dopant hence the VO_2 quality decreases. Burkhardt et al. reported that doping VO_2 with tungsten only affects the transition temperature, whereas doping with fluorine affects the transition temperature and the absorption edge. To achieve a colourless appearance for VO_2 films, the optical absorption edge has to be shifted towards shorter wavelengths in order to sustain the blue part of the visible light (Burkhardt et al., 2002).

The higher the concentration of fluorine, the higher the absorption edge shifts. The effect of F doping depends on the presence of tungsten in VO_2. If the film is only doped with F, the transmittance of the metallic phase converges towards the high transmittance of the semiconducting phase as the F content is increased (Burkhardt et al., 2002).

Concerning the consequences for the VO_2 film, the two effects are contrasting: F and W dopants. Doping with fluorine implies the stabilisation of the semiconducting phase, whereas doping with tungsten implies the stabilisation of the metallic phase.

VO_2 has been doped with lithium so as to increase the transmittance and the blue shifts the absorption edge. The transmittance successfully increased but the layers changed colour from brown to bluish due to the lithium content. The switching efficiency for VO_2 doped with lithium become poor and the luminous transmittance increases (Fu, 2007). VO_2 lithium doped causes higher transmittance but lower transition temperature. The increase in lithium content in the film of VO_2 leads to the increase in transmittance but the thermochromism decreases.

VO_2 thin films have been synthesized using many techniques, as mentioned above. Each technique has its own effect on the synthesized film, such as stress. PLD, radio frequency (RF), magnetron sputtering, CVD and so on, have been used in several cases for synthesizing VO_2. There is a recent technique that has been developed in synthesizing VO_2, known as rf-ICMS (Kana et al., 2008). It is defined as the supplementary exceptional way to engineer nanosystems at high deposition rates onto complex shaped substrates.

This technique has been used to synthesize thermochromic VO_2. J.B.K. Kana et al. (2008) have deposited VO_2 at different deposition times using the ICMS. The VO_2 film thickness increases with the increase in deposition time, hence the average grain size increases with time. The increase in peaks intensity (Figure 11.3) with time indicates a volume decrease and hence an increase in film grain size with time (Kana et al., 2008).

The transition temperature also decreases with the increase in film thickness (see Figure 11.4).

The transition temperature decreases with the increase in deposition time, which indicates the film thickness increment (see Figure 11.5). From the literature, it is apparent that the optical transmittance for VO_2 could be decreased or increased by the dopants. It was reported from the literature that the deposition time affected the transition temperature. The increase in deposition time leads to an increase in film thickness (Kana et al., 2008). It was also reported that oxygen pressure is the critical parameter to obtain pure VO_2 (Kim et al., 2003). It has been reported from literature that VO_2 thin film gives the first strongest peak at (011) plane (Goodenough and Hong, 1973; Xu et al., 2005; Vernardou et al., 2004; Maaza et al., 2000; Golan et al., 2003; Kana et al., 2008).

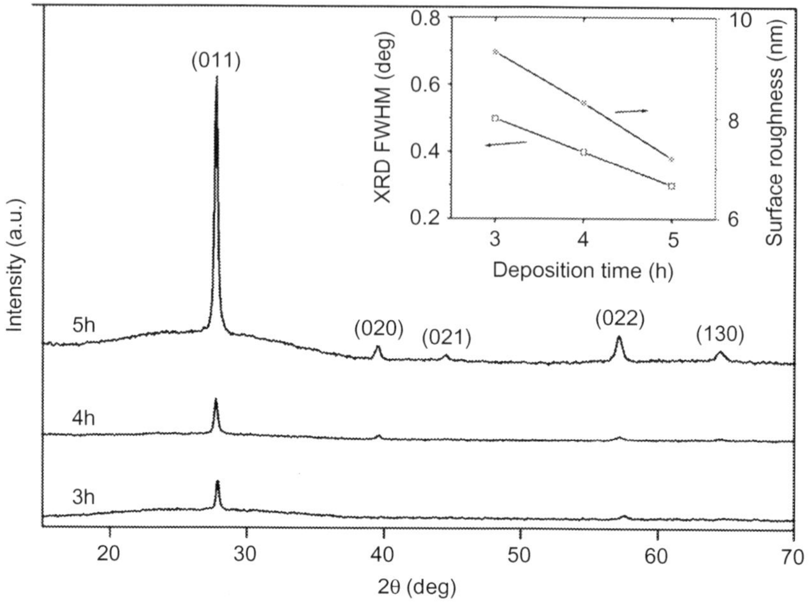

Figure 11.3 X-ray diffraction (XRD) patterns of VO_2 thin films at different deposition times (Kana et al., 2008).

11.3 Synthesis techniques description

In this section, two depositing techniques are reviewed and the comments are based on work reported by Msomi in 2008 (Msomi, 2008). The two techniques to be reviewed are PLD and ICMS (see Figures 11.5 and 11.6).

PLD is a thin film deposition technique where a high-power pulsed laser beam is focused inside a vacuum chamber to strike a target of the desired composition. ICMS is a supplementary exceptional way to engineer nanosystems at high deposition rates onto complex shaped substrates. These techniques were used to synthesize VO_2 thin films which were studied comparatively. The comparative study focuses on the quality of this film, that is, microstructure crystallography, morphological structure and so on. The sodalime glass substrate was used for deposition. The substrate temperature was varied so as to study its effect on the quality of the film.

11.3.1 Comparative discussion

The comparative discussion is based on the results from the work done by V. Msomi (Msomi, 2008). It should be noted that the other deposition parameters were kept constant while varying the substrate temperature. Similar effect of the substrate temperature on the film deposited by PLD and ICMS was observed. Films deposited by PLD show a drastic change in film microstructure with a substrate temperature increase, whereas films deposited by ICMS show gradual change in

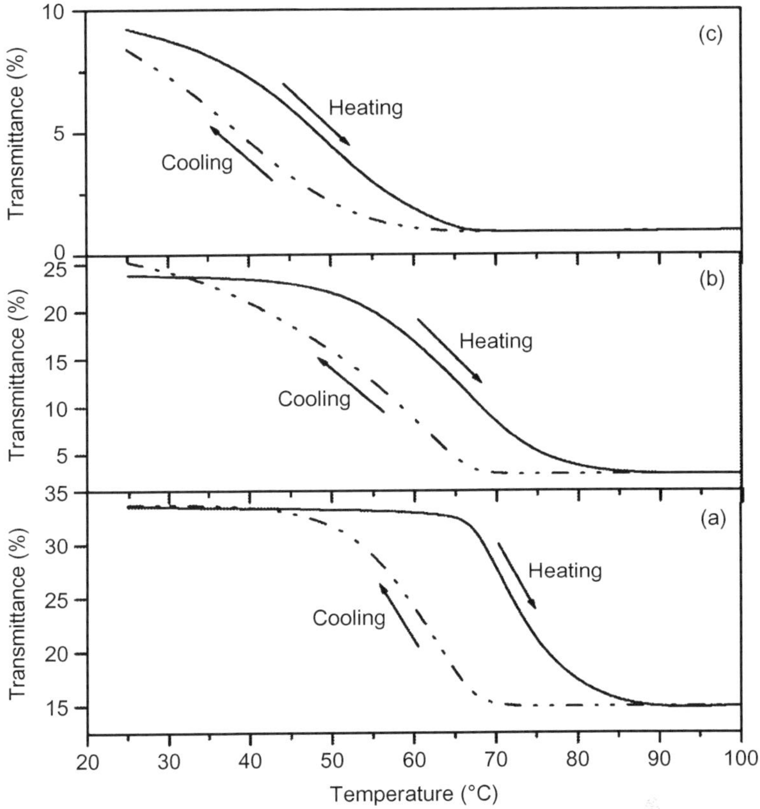

Figure 11.4 Transmittance at wavelength of 1100 nm for VO_2 thin film as a function of temperature at different deposition time: (a) 3 hours; (b) 4 hours; (c) 5 hours (Kana et al., 2008).

microstructure. The X-ray diffraction results from both techniques showed clearly that the film crystallinity evolves with substrate temperature.

The films from both techniques show an increase in surface roughness with the increase in substrate temperature, which indicates an increase in the grain size of the film. The films from both techniques show almost constant thickness with the increase in substrate temperature. The films synthesized by the PLD show a decrease in optical transmittance with the increase in substrate temperature, whereas the films deposited by ICMS show a fluctuation in optical transmittance with the increase in substrate temperature.

The films deposited by ICMS show good electrical properties, whereas the films synthesized by PLD show poor electrical properties with the increase in substrate temperature. We observed an obvious correlation between the microstructure, electrical and optical properties of the film. It is suggested that one can use ICMS to find good-quality VO_2. The uniform surface roughness can be obtained using ICMS

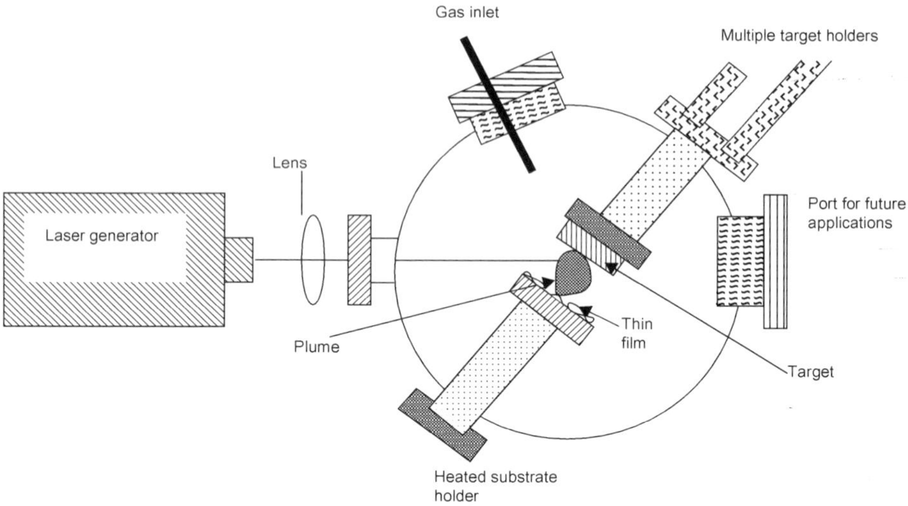

Figure 11.5 Schematic diagram for an PLD system.

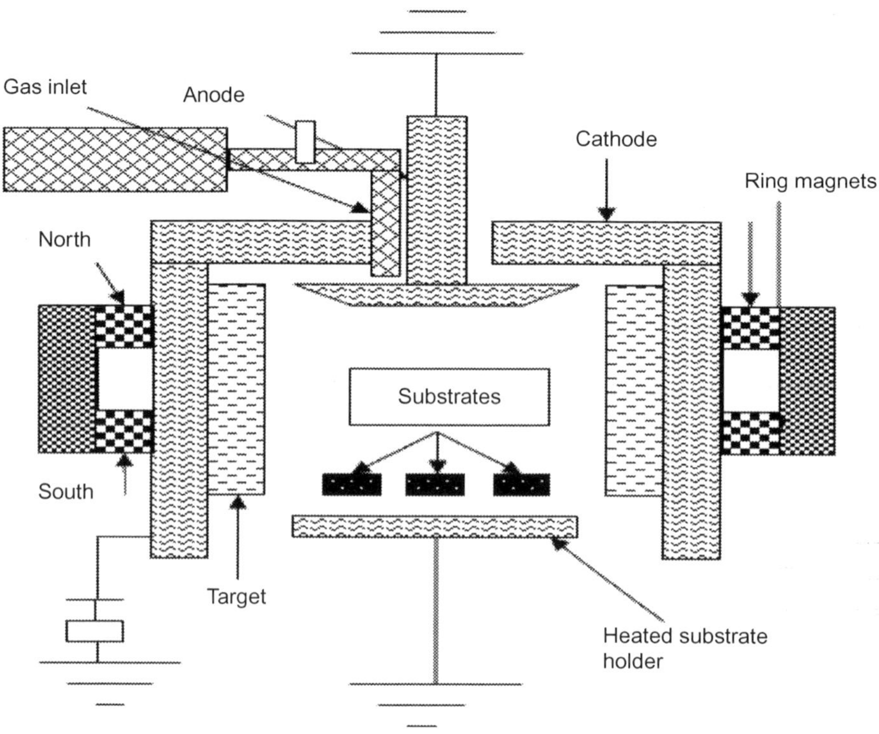

Figure 11.6 Schematic diagram for an ICMS system.

when the substrate temperature is 400°C. This indicates that the ICMS technique among other techniques can be considered one of the best techniques to be used for smooth nanocoatings.

11.4 Conclusion

A comparative study was conducted so as to find the recommended technique for synthesizing a good VO_2 thin film. Based on the results reported on Msomi's work, it was then recommended that ICMS can be a good technique to synthesize good quality of thin films or nanostructures. This technique (ICMS) can be used for smooth nanocoating applications.

References

Burkhardt, W., Christmann, T., Meyer, B.K., Niessner, W., Schalch, D., et al., 2002. W- and F-doped VO2 films studied by photoelectron spectroscopy. Thin Solid Films. 402, 226–231.

Case, F.C., 1987. Influence of ion beam parameters on the electrical and optical properties of ion-assisted reactively evaporated vanadium dioxide thin films. J. Vac. Sci. Technol. A. 5, 1762.

Chain, E.E., 1986. The influence of deposition temperature on the structure and optical properties of vanadium oxide films. J. Vac. Sci. Technol. A. 4, 432.

Chen, X., Qiang, L.V., Yi, X., 2012. Smart window coating based on nanostructured VO2 thin film. Optik. 123, 1187–1189.

Chen, Z., Gao, Y., Kang, L., Dua, J., Zhang, Z., et al., 2011. VO2-based double-layered films for smart windows: optical design, all-solution preparation and improved properties. Sol. Energy Mater. Sol. Cells. 95, 2677–2684.

Fu, G., 2007. Metal-semiconductor Transition Materials: FeS and VO2 Thin Films by RF Reactive Sputtering. Doctoral dissertation, I. Physikalisches Institut Justus-Liebig-Universität Gießen.

Golan, G., Axelevitch, A., Sigalov, B., Gorenstein, B., 2003. Metal insulator phase transition in vanadium oxides films. Microelectron. J. 34, 255–258.

Goldflam, M.D., Driscoll, T., Chapler, B., Khatib, O., Jokerst, N.M., et al., 2011. Reconfigurable gradient index using VO2 memory metamaterials. Appl. Phys. Lett. 99, 044103.

Goodenough, J.B., Hong, H.Y-P., 1973. Structures and a two-band model for the system V1-xCrxO2. Phys. Rev. B. 8, 1323.

Greenberg, C.B., 1983. Optically switchable thin films: a review. Thin Solid Films. 110, 73.

Kana, J.B.K., Ndjaka, J.M., Owono Ateba, P., Ngom, B.D., Manyala, N., et al., 2008. Thermochromic VO2 thin films synthesized by RF-inverted cylindrical magnetron sputtering. Appl. Surf. Sci. 254, 3959–3963.

Kim, H.T., Chae, B.G., Youn, D.H., Maeng, S.Y., Kang, K.Y., 2003. 'Gate-induced Mott transition', Cond-mat/0305632.

Kucharczyk, D., Niklewski, T., 1979. Accurate X-ray determination of the lattice parameters and the thermal expansion coefficients of VO2 near the transition temperature. J. Appl. Crystallogr. 12, 370.

Li, W.W., Zhu, J.J., Liang, J.R., Hu, Z.G., Liu, J., et al., 2011. External electric field manipulations on structural phase transition of vanadium dioxide nanoparticles and its application in field effect transistor. J. Phys. Chem. C. 47, 23558.

Maaza, M., Bouziane, K., Maritz, J., McLachlan, D.S., Swanepool, R., et al., 2000. Direct production of thermochromic VO2 thin film coatings by pulsed laser ablation. Opt. Mater. 15, 41–45.

Morin, F.J., 1959. Oxides which show a metal-to-insulator transition at the Neel temperature. Phys. Rev. Lett. 3, 34–36.

Mott, N.F., 1990. Metal-Insulator Transitions. Taylor and Francis, London; New York, pp. 185–189.

Msomi, V., 2008. Vanadium Dioxide Thermochromic Thin Films Correlation Between Microstructure, Electrical and Optical Properties. Master's thesis, University of Zululand.

Prasada, A.K., Amirthapandiana, S., Dharaa, S., Dasha, S., Muralib, N., et al., 2014. Novel single phase vanadium dioxide nanostructured films for methane sensing near room temperature. Sens. Actuators, B. 191, 252–256.

Remka, R.L., Walser, R.M., Bene, R.V., 1979. Transition layers between VO2 films and oxide substrates. Thin Solid Films. 61, 73.

Syrrakou, E., Papaefthimiou, S., Yianoulis, P., 2006. Eco-efficiency evaluation of a smart window prototype. Sci. Total Environ. 359, 267–282.

Tampere University of Technology, (n.d.) <http://www.tut.fi/units/ms/teva/projects/intelligenttextiles/chromic.htm> (accessed 28.01.08.).

Tang, C., Georgopoulos, P., Fine, M.E., Cohen, J.B., Nygren, M., et al., 1985. Local atomic and electronic arrangements in WxV1-xO2. Phys. Rev. B. 31, 1000–1011.

University of Nebraska–Lincoln, (n.d.) <http://www.unl.edu/CMRAcfem/interact.htm> (accessed 16.07.13.).

University of Nebraska–Lincoln, (n.d.) <http://www.unl.edu/CMRAcfem/gun.htm> (accessed 16.07.13.).

Vernardou, D., Manning, T.D., Parkin, I.P., Pemble, M.E., Sheel, D., 2004. Intelligent window coatings: atmospheric pressure chemical vapour deposition of tungsten-doped vanadium dioxide. Surf. Coat. Technol. 188–189, 250.

Wu, J.M., Liou, L.B., 2011. Room temperature photo-induced phase transitions of VO2 nanodevice. J. Mater. Chem. 21, 5499.

Xu, G., Jin, P., Tazawa, M., Yoshimura, K., 2005. Thickness dependence of optical properties of VO2 thin films epitaxially grown on sapphire (0 0 0 1). Appl. Surf. Sci. 244, 449–452.

Formation of dense nanostructured coatings by microarc oxidation method

12

Vladimir N. Malyshev
Gubkin Russian State University of Oil and Gas, Russia

12.1 Introduction

The microarc oxidation (MAO) method (Anon, 1981; Markov et al., 1992), also known as plasma electrolytic oxidation (Rudnev et al., 1990; Yerokhin et al., 1999; Yerokhin et al., 2003), is a modern method for forming high-strength and wear-resistant ceramic coatings (e.g., Hradgovsky, 1977; Kurze, Krysmann et al., 1980; Kurze, Marx, et al., 1980; Yamada and Mita, 1982; Malyshev, 1995, 2006). It significantly expands the range of aluminum alloys used in manufacturing by creating a surface layer of aluminum oxide with a high-strength and wear-resistant phase – α-Al_2O_3 (corundum) (Petrosyants et al., 1984; Malyshev et al., 1985). The high-strength properties of the coatings enable them to be used in a variety of industries, in particular in heavily loaded friction units of machines and mechanisms, such as bearings, mechanical seals, rotating shafts, valves, and regulating equipment.

12.2 Phenomena of MAO-coating formation

When submitting voltage to electrodes, the voltage increases in the first few seconds on the anode due to the valve properties of the oxide film existing on the surface of the sample with thickness in some Angstroms (Young, 1967). The alternative voltage grows in such a way that an anodic and cathode voltage are installed asymmetrically to the zero line $U_a \sim 460-560$ V and $U_k \sim 60-120$ V, depending on the concrete conditions of oxidation (electrolyte composition, processed material, density of current, etc.). The current oscillogram looks virtually symmetrical to the zero line, with the relations of $I_k/I_a = 1$ (Figure 12.1). The sample is covered completely by the microarc discharges that are burning on the entire surface.

In approximately 8–10 minutes on the ascending branch of the current pulse in the anode half-cycle a small spade appears (the local maximum), which in 15 minutes of oxidation is seen clearly on the diagram. For this time the surface of the sample covers the oxide layer with a thickness of $\sim$10–15 μm and looks like a film that may be produced by the usual oxidation. However, in some places, especially on the periphery of the sample, the small dots forming a new layer with a

Anti-Abrasive Nanocoatings.

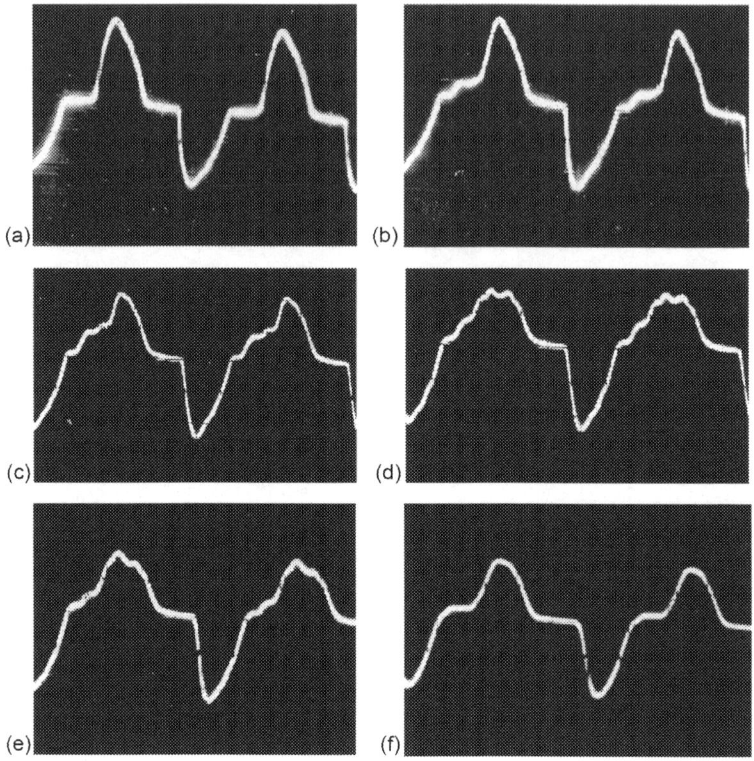

Figure 12.1 Oscillograms of the current pulses for different moments of formation (min): (a) 1–2, (b) 15, (c) 30, (d) 60, (e) 120, and (f) 150–180.

dense structure can be seen already. After 30 minutes, the local maximum on the ascending branch of current pulse is rising. The amount of island embryos of the new layer increases in the oxide film at this time and their growth occurs. With the rise of the local maximum on the ascending branch of current pulse after about 60 minutes the island embryos cover practically the whole surface of the sample. The thickness of the coating at this time is ~60–80 μm and is quite uneven. The current's signal changes in the anode half-cycle to look like a sinusoid, but more broad in the base and with less value on the amplitude.

Further evolution of the current's signal leads to alignment of the individual local maximums in the anodic half-cycle. The coating thickness at this time continues to grow and becomes more even. The relations between the thicknesses of the upper porous layer and dense inner layer are charged. The latter begins to exceed the upper layer. The intensity of combustion of microarc discharges at this time changes again: they become more localized and reduced in quantity.

After the next 30–60 minutes the shape of the current's signal is similar to the beginning of the process (see Figure 12.1); however, in comparison to the initial curve, the first current's pause is little higher than the zero line and the maximum of the signal in the anodic half-cycle is lower and narrower than usual maximum. It must be said that this form of the current's signal in the definite mode remains constant from this point onward (even if the process lasts 4–6 hours or more), when the main layer of great thickness is needed.

The appearance in the anodic half-cycle on the diagram of the second spade (another local maximum) on the ascending branch must be related, in the first line, to the qualitative changes that happen at this time in the coating, and particularly to the changes in its effective resistance.

The key moments which lead to the formation of a dense main layer in the MAO coating can be analyzed. The current's registration parameters in chronological oscillograms (Malyshev, 1996) at the coating formation and the changes of its characteristics are shown as follows.

The "insular" growth of the coating main layer (see Figure 12.2) is a particular characteristic of forming the MAO coating in anodic-cathode mode. The evolution of the current's signal shape (Malyshev, 1996) leads to stabilization of the individual local maximums of current in the anodic half-cycle.

It is important at this time not to allow any of the single and strong discharges to start burning, because this leads to irreversible damage – they can burn through to the metal of the formed coating. The anode voltage already approximates to the threshold voltage at this time, which is characteristic of a move to the next stage: arc discharge. To continue with the microarc discharge stage and increase the thickness of the main layer, it is necessary to reduce the current density slightly, or change the relation between the cathode and anode currents I_k/I_a, up to the value 1.05–1.2; the latter is more convenient.

At the final stage of the coating formation the conditions for creating a so-called "soft" MAO-mode are favorable. In this case, the microarc discharges are localized in some spot in the field with an area of 3–4 cm^2, which begins to wander over the whole processed area in the auto fluctuating mode. The crashing noise of the microarc discharges is noticeably reduced, and the coating forms very evenly and with a high quality.

12.3 Voltage–current characteristics

Even more clearly, the appearance of the evolution's changes in the "anode–coating–electrolyte–cathode" system can be seen on the voltage–current characteristics (VCC). The voltage and current data were registered by oscillograph. On Figure 12.3 the VCC curves are presented for the corresponding moments of MAO-coating formation.

The appearance and the presence of the second maximum (spade) on the ascending branch of the current pulse of VCC is noted with the deviations (see Figure 12.3, curves (b), (c), (d)) from the initial curve (Figure 12.3, curve (a)). These changes in the VCC must be connected in the first line with the structure changes in the coating and its effective resistance R_{eff}. It is intended that the area below the VCC curve (see

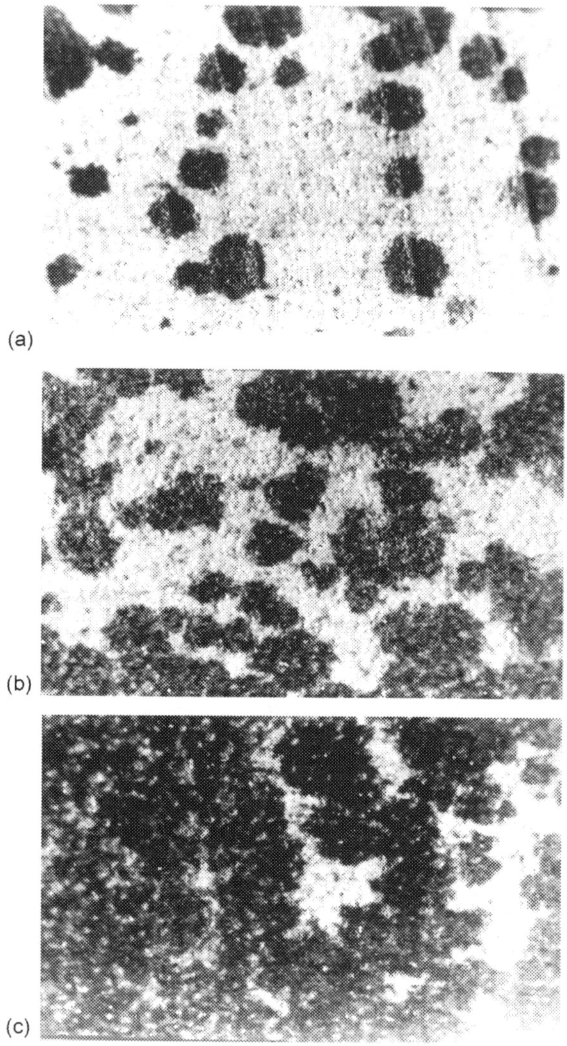

Figure 12.2 The "insular" growth of MAO coating at different moments of formation (min): (a) 15; (b) 30; (c) 60.

Figure 12.3(III)) provides an energy performance and shows how much power was used for the formation of the layer in the anode and cathode half-cycles. The instantaneous value of the power can be considered in the anode half cycle:

$$P_a = \int_0^{I^a_{\max}} U_{ai} dI - \int_{I^a_{\max}}^0 U'_{ai} dI = \int_0^{I^a_{\max}} (U_{ai} - U'_{ai}) dI$$

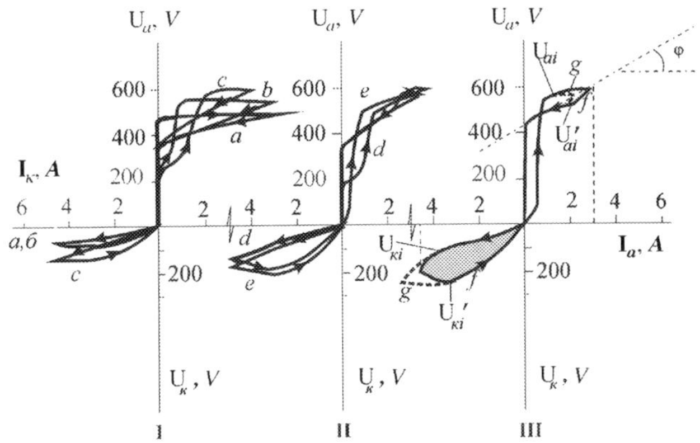

Figure 12.3 The voltage–current characteristics of microarc oxidation aluminum alloy 2024 (Д16) in anodic-cathode mode at the moments of formation (min): (a) 1–2, (b) 15, (c) 30, (d) 60, (e) 90, (f) 120–150, (g) when the current feedback is applied (the ratio change between I_k/I_a)

In the cathode half-cycle it will be:

$$P_k = \int_0^{I^k_{\max}} U_{ki} dI - \int_{I^k_{\max}}^0 U'_{ki} dI = \int_0^{I^k_{\max}} (U_{ki} - U'_{ki}) dI$$

Then the overall power used for the coating formation is:

$$P = P_a + P_k = \sum_{i=1}^{n} \left[\left| \int_0^{I^a_{\max}} (U_{ai} - U'_{ai}) dI \right| + \left| \int_0^{I^k_{\max}} (U_{ki} - U'_{ki}) dI \right| \right]$$

It is characteristic that with the growth of coating thickness λ the share of power increases in the cathode half-cycle and decreases in the anode half-cycle (see Figures 12.3(I) and 12.3(III)). This confirms again the importance of the cathode current pulses in the coating formation, and agrees with the conclusions of the research authors (Markov et al., 1992).

With regard to the resistance change at the coating formation it is possible to judge on the change in the angle φ (see Figure 12.3(III)): the more the angle φ, the more the resistance and the thickness of the coating λ.

12.4 Discussion about growth mechanism of MAO coating

The beginning of MAO coating growth in usual mode occurs as shown in Figure 12.2. It represents an "insular" growth mechanism, when the main layer of coating begins from some defects or inhomogeneities in the surface and then expands, increasing in size and covering the whole treated surface. A description of such coating growth mechanisms can be obtained from the standpoints of fractal theory and synergetics.

Indeed, the achievement of the above-mentioned so-called "soft" mode of MAO-coating formation can be connected with the emergence of some self-organization of the system, "sample (the anode)—coating—electrolyte—bath (the cathode)," which is expressed in the transition to such dissipative structures (Haken, 1985; Prigozhin and Glensdorf, 1973; Haken, 1988; Nicolis and Prigozhin, 1989) in the kind of microarc discharge spots that migrate on the whole surface of the sample and ensure the smallest production of entropy in such conditions.

As is known (Nicolis and Prigozhin, 1984; Haken, 1988), in open systems, and the process of MAO-coatings formation can be attributed to such a system, in certain conditions, the processes of ordering of matter occur through the interaction of the flow of energy and matter, which leads to a reduction in entropy, and to the formation of self-organized dissipative structures. In nonequilibrium conditions the evolution of a system is connected with the self-organization of dissipative structures due to the exchange of energy and matter with an environmental media (Nicolis and Prigozhin, 1989).

According to the Prigozhin—Glensdorf principle (Prigozhin and Glensdorf, 1973), the driving force of this process is the minimization of entropy production, defining the behavior of a system as far distant from thermodynamic balance.

The creation of nonequilibrium conditions in the microarc process is ensured by unchanging power delivery (electric potential) and substance (anions of the electrolyte), and is regulated by the main control parameters: the applied current density d_I and the relation between the cathode and anode currents I_k/I_a. They are used in this manner as feedback. The implementation of this feedback in the microarc oxidation process leads to the manifestation of a hierarchical sequence of energy dissipation mechanisms, namely the formation of macrostructure coatings, structural changes in the layers at the micro level, and nonequilibrium phase transitions. The account of this hierarchy is of fundamental importance for the development of technological modes of coatings production with a structure corresponding to the specified properties.

Previously it was shown that the area under the curve VCC (see Figure 12.3(III)) is, in fact, power energy expended in coating layer formating in the anodic (P_a) and cathode (P_k) half-cycles. In this case, the total power of the coating formation is $P = P_a + P_k$.

The area under the curve VCC (U_a) in the anode half-cycle until an inflection (maximum currents amplitude value) is the power that goes to the breakdown of the film and its oxide thermal dissociation. The area under the descending curve (U_a') is obviously the power consumed by the thermal conversion in the coating layers, chemical interaction of the electrolyte, and the electrode elements as well as phenomena associated with light and sound effects (because the anodic current oscillations are observed most on the descending branch of the waveform). One should not forget also the power that is consumed in the heating of a coating in the location of breakdown, namely the sample and the electrolyte.

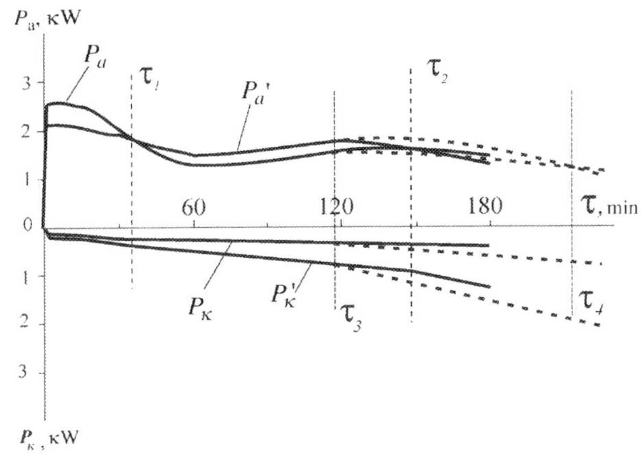

Figure 12.4 Dependencies of instantaneous power changes in the anodic and cathode half-cycles (solid lines) and when the current feedback is applied (dotted lines).

On the basis of data from Figure 12.3, a graph of power changes in the anodic P_a and cathode P_k half-cycles in time was plotted (see Figure 12.4), on which it is clearly seen at what time in the formation process this or other components prevailed.

The power expended on the breakdown of coating P_a prevails in the beginning of the process when there is an intensive growth of the primary dielectric film. With the growth of the film thickness λ and its resistance, at some point (τ_1 on Figure 12.4) the components of capacity become equal. At this point, the coating is thick enough that the power consumed for its handling of P_a' (structural changes, compaction, etc.) prevails over the power of the breakdown P_a. It should be said that this time τ_1 corresponds approximately to the time (about 30 minutes) when the conditions for the beginning of the main working layer formation (dense structure) are created in the coating. Furthermore, from this moment the value of the cathode current component becomes more significant (the power in cathode half-cycle begins to increase significantly).

Next is the formation and growth of the main layer under the prevailing operating power of P_a'. This process ends at the time marked on Figure 12.4 as point τ_2 (about 140–150 minutes). At this time, the components of P_a and P_a' are again equalized. The total thickness λ is 130 ... 160 μm at the time τ_2, and the thickness of the base working layer is 80 ... 100 μm.

As the process continues, the component of power P_a begins to prevail over P_a' again. A new layer begins to form in the coating, which, when the process continues, soon begins to be broken down by the powerful arc discharges, that is, microarc mode spontaneously moves to arc mode, which in this case causes an irreversible defect – a breakdown through the coating to the metal.

As mentioned above, to prevent spontaneous transition to arc mode, it is necessary either to reduce the current density, or to change the ratio of the cathode and anode currents I_k/I_a. An increase of I_k/I_a allows the system to return to the area of relations between P_a and P'_a as in the area between points τ_1 and τ_2 (see Figure 12.4, the solid curves), that is, when P'_a prevails over P_a. It is in this case that the system goes to the above-mentioned "soft" MAO mode with oscillatory wandering of discharge spots, which may increase the thickness of the coating up to 300–400 μm with high quality. This state of the system marked by dotted lines (see Figures 12.3 and 12.4, curve (g)) is characterized by the transition points, τ_3, because of imposing the feedback, and τ_4, the moment of transition to arc mode.

If we express graphically the changes in the effective resistance R_{eff} of coating during this same time period, the curve $\Omega - \tau$ will appear as shown in Figure 12.5. In this case, with the change of the current's pulse ratio I_k/I_a the curve $\Omega - \tau$ changes its course and shifts the boundary of transition into arc mode, when the breakdown of the coating up to the metal is possible.

In this sense, the parameter Ω – effective resistance R_{eff} – is the order parameter, responsible ultimately for the quality of the formed coating. With sufficient availability of its measurements and the presence of sensors that control this parameter during the formation process, high-quality coatings can be guaranteed by introducing a flow diagram of the formation with the feedback element in the form of a controlling parameter – the ratio I_k/I_a, in which the change agrees with the change in effective resistance. Moreover, the system automatically selects the most appropriate setting of the control parameter (I_k/I_a) at a particular formation time, providing a self-oscillating mode of wandering spots of discharges. Thus, the MAO technology can be considered as technology that creates a self-adjusting system or so-called "smart" technology.

The component intersection points of P_a and P'_a (τ_1 и τ_2, τ_3 и τ_4, Figures 12.3 and 12.4) in terms of synergetics (Nicolis and Prigozhin, 1989) are characteristic of the phase transition points or the points of bifurcation, when passing through which

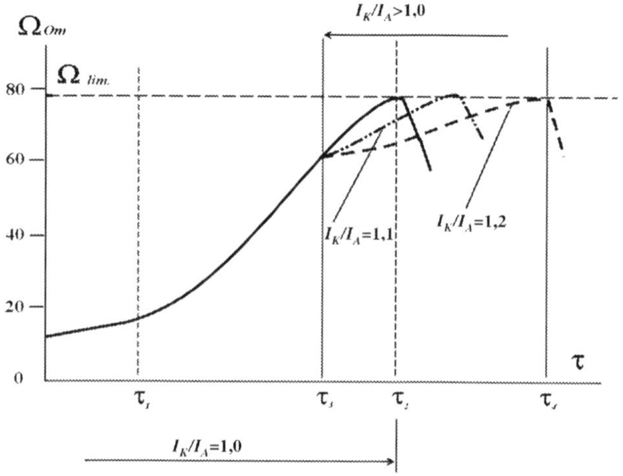

Figure 12.5 Changes of effective resistance Ω within the formation process of MAO coating.

the "anode–coating–electrolyte–cathode" system forms new dissipative structures with a spontaneous change in the properties of the medium.

Indeed, at point τ_1 the system proceeds itself spontaneously to the formation of a coating structure of another type – with dense structure. At points τ_2 and τ_4, there is a spontaneous transition to arc mode, or, with the application of the feedback at point τ_3, to "soft" MAO mode with the formation of dissipative structures – the self-oscillation wandering spots of microarc discharges.

Thus, the specificity of MAO-coating formation in accordance with anodic-cathode mode lies in the fact that the formation of its dense structure with high physical and mechanical properties occurs not immediately but after some time. Moreover, most often the formation of the coating has a so-called "insular" character (this largely depends on the scattering power of the electrolyte and the applied current density), that is, the embryonic elements of dense structure are initially being formed on the surface. Then they increase in size gradually and are spliced with each other, eventually covering the entire surface. The thickness of the main layer is small initially, then rapidly increases and becomes comparable to the thickness of a top layer, and at the end of the process begins to exceed it by about 2.5–3.1.

This feature of the MAO coating formation of the main working layer causes a specificity in the operating procedure that is expressed in the need for strict adherence to the duration of the process formation. The minimum time to form a working layer formation with high quality is determined by the achievement of the bifurcation point τ_2 (Figures 12.3 and 12.4), which is 2.2–2.5 hours, or at the application of feedback at point τ_4 (about 3.0–4.0 hours). Of course, this time may be reduced by the application of higher current density. However, here a reasonable compromise is necessary, as the excessively high current density leads to a considerable increase in the uneven thickness of the main working layer, particularly when the coating will be formed from the details of complicated forms.

12.5 Model of fractal growth of the dense wear-resistant layer

The MAO method, by way of its energy action on the processed material, can be attributed to the methods of surface hardening by concentrated streams of energy. The physics and chemistry of the MAO process, especially in its anodic-cathode mode, are quite complex due to the simultaneous action of complex physical, plasma-chemical, thermal, and other processes. These are quite problematic when fully taken into account in developing models.

By considering the microarc oxidation process in terms of synergetics and fractal theory, that is, as a physical and chemical system which leads to some transient phenomena, large-scale ordering and symmetry breaking away from the state of equilibrium, it is possible to understand and analyze this complex process, using as a prototype the model for other complex systems, for which the laws of the evolution of relevant variables have been studied already. This new view of the microarc oxidation process allows us to take the simulation of complex dynamic "anode–coating–electrolyte–cathode" systems to another level and to formulate a specific task on a computer, using numerical modeling.

To understand some of the limitations that any model inevitably suffers from the physical point of view, we consider first the phenomenological description of the discharge in the electrolyte, using the model of elemental plasma-chemical breakdown (Gogish-Klushin et al., 1993).

After filing the next voltage pulse in the electrolyte the current begins to flow and the redistribution of the discharge occurs. Due to the formation of an electrical double layer (EDL) on the surface of the existing dielectric film, the bulk of the voltage falls directly to the oxide coating. When the tension of the electric field is sufficient for the thermal breakdown of the dielectric film, a conductive channel is formed in the oxide coating, which is heated by passing the current density about 10^7 A/m^2 to a temperature of several thousand degrees ($t \sim 10^4$ K). Thus, according to the phenomenological model (Gogish-Klushin et al., 1993), when almost all the applied voltage on level EDL have been on the dielectric oxide film, we can apply the dielectric breakdown model described by the Laplace equation (Murat, 1988).

The local impact of the discharge on the surface of metal is determined by the electrical characteristics of microarc discharge, the parameters of the medium (electrolyte) and the characteristics of the metal itself. Because the place where discharges burn (at least in the second stage of the process, see Figure 12.2) is determined mainly by fluctuation phenomena (Markov et al., 1992), the microarc oxidation process can be described as a multifactor stochastic process. Formation of the dense structure of such coatings in the anodic-cathode mode, based on the experience of numerous experiments and analysis of the process, as thermodynamically non-equilibrium, occurs so that we can apply the percolation model to it (Nicolis and Prigozhin, 1989).

Fractal theory uses the concept of a cluster to describe the object consisting of a large number of solid particles rigidly connected to each other and having a loose and branching structure. The cluster is called a system consisting of large number of atoms and molecules which retain their individuality within this system (Smirnov, 1991). Fractal clusters are distinguished from non fractal in that they have the property of self-similarity. The concept of fractal clusters is universal and therefore applicable to systems of a different physical nature. Let us briefly characterize existing models of growth of fractal clusters.

The nature of the process depends on the actual physical conditions. In some cases, the process of cluster growth occurs as the result of a merger of individual particles; in other cases, at the first stage of the process, there are many particles that are combined in small clusters, and they in turn are combined in large clusters. So in the end it is going to be a fractal cluster. Moreover, the compactness and physical properties of the fractal cluster depend on the nature and movement of particles and clusters in the space, and also on the probability of adhesion of the particles and clusters within their contact.

Historically, the first computer model for the assembly of the cluster was proposed by Witten and Sander in 1981 (Anon, 1986; Gould and Tobochnik,

1990/1986: Part I, p. 349, Part II, p. 400; Smirnov, 1991), and named as the DLA (Diffusion Limited Aggregation) model. In this model it is assumed that the particles performing the Brownian motion in space adjoining the cluster, with a certain probability adhere to it. In this way the growth of the cluster is exercised.

This model allows us to describe the evolution of the cluster and to follow the details of its growth. The Witten–Sander model has played an important role in the study of fractal clusters, because it offered an easy way to build the cluster. The creation of this model has attracted attention to the process of formation and growth of fractal clusters and was the beginning of a comprehensive modeling of these processes and the creation of new models and algorithms.

Quite another set of physical conditions that are fundamentally different from the case of DLA models make the CCA (Cluster–Cluster Aggregation) model (Anon, 1986; Smirnov, 1991). In this case, the particles which are moving along a certain path make the initial step in the given region of space, collide and then coalesce into clusters. Large numbers of clusters of small dimensions are produced in the first stage of the process, which are then combined into larger clusters. Over time the number of clusters in the space decreases as their size increases. Clearly, in this case the formed clusters are looser than in the DLA model, because in this method of cluster assembly it is more difficult to fill the voids.

In the above mentioned models it is assumed that the contact of the particles leads to their coalescence with the probability of the order of unity. Another physical situation corresponds to the case when the probability of sticking by the contact of particles is smaller (i.e., less than 1). Moreover, in the CCA model the clusters penetrate deeper into each other and as a result a more compact cluster is formed. The model that is characterized by a low probability of sticking cluster–cluster aggregation is called the RLCA (Reaction Limited Cluster Aggregation) model (Anon, 1986; Smirnov, 1991).

When analyzing the first two models, it was thought that the particles make associating diffusive motion in space, that is, the mean free path of the particle is small compared to the characteristic dimensions of the region responsible for the growth of the cluster. Another situation is possible physically when the mean free path of the particles is larger than the size of the space. In this case the pattern amounts to replacing a rectilinear trajectory on the way, and the cluster formed is more compact. This cluster model has a long history and is called a model of Eden (Gould and Tobochnik, 1990/1986: Part I, p. 349, Part II, p. 400; Smirnov, 1991) or ballistic model.

In the present cursory description of existing models for cluster assembly, the details of individual models were not considered; only their basic features were described.

It should be said that the model of forming densely structured MAO coating developed in this work does not fully meet the known models of clusters considered above – DLA, CCA, RCLA, and so on (Anon, 1986), because has its own very specific physical conditions, namely, the process of cluster growth (dense structure) followed by sequential assembly of particles formed by oxidation of the starting material under the applied voltage depending on the stochastic resistance change of the oxide film.

12.5.1 Algorithm for generating densely structured MAO coating

The process of growth of individual clusters (the number of seed cells *n*) begins in the lattice model of the form *LxL* broken into elementary cells, depending on the amount of seed (embryo) *n* cells, that is, in the space in which the cluster growth occurs.

Moreover, the experiments show that the number of the original cells particles (cells) is dependent, in particular, on the applied current density, the characteristics of the electrolyte (its scattering ability).

The formation of a dense layer MAO coating in the anodic-cathode mode, as shown earlier, does not begin immediately, but only after the establishment of certain specific conditions. The birth of the first cell in the formation of the dense structure (for the model – cell) is, obviously, in those parts of the surface in which, because of various defects in the pre-formed film, there is a some critical (i.e., the lowest) current density d_{cr}.

Since, as a rule, the distribution of the electric field lines is such that on the perimeter it is greater than in the central portion of the surface (Ivanov et al., 1986: 216), the value of d_{cr} in the zone close to periphery will obviously be sooner than in the central zone of the specimen, and hence the emergence of initial cell units (cells) should be started sooner from the peripheral zone. This is consistent with experimental observations.

In the electrolytes with better scattering ability the distribution of the electric field lines is more even (Ivanov et al., 1986: 216). Therefore the achievement of the value d_{cr} at the same time for more points on the surface is more likely. That is why the coatings formed in such electrolytes are more uniform. Consequently, the characteristics of the electrolyte, in particular, its scattering ability and conductivity, affect the number of seed cells that appear simultaneously.

For nucleation of seed cells the following are necessary:

1. a corresponding potential difference between the anode and the cathode;
2. the critical current density d_{cr}, which should have been achieved at this point for the breakdown;
3. minimal film resistance at this point at this moment.

Further cluster growth occurs so that the breakdowns and the increase in coating is most likely implemented in those places where the resistance of the oxide film is the least. The growth of individual clusters in the first stage of the model goes by way of the method described above. Moreover, as the motion (i.e., such as a Brownian motion of particles) of clusters in the model does not occur, at the second growth stage the individual clusters are joined to the one large cluster in accordance with the CCA model (Anon, 1986) by being fused to each other and filling all the lattice space.

The program on the computer to implement the growth of the cluster within the lattice space *LxL* is developed using the algorithm below.

The growing cluster is an object with the following properties:

1. thickness λ – integer (determined by the number of coating layers at a given point);
2. is a logical variable such as: "Processed" or "Unprocessed." "Truth is," if the cluster has been processed for the current iteration, and "Lie" otherwise.

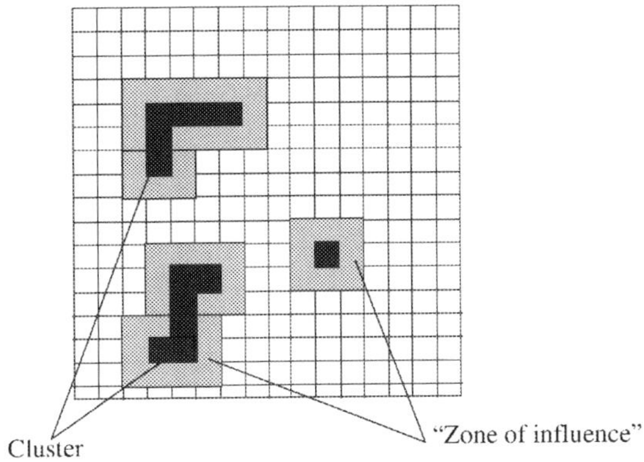

Figure 12.6 Schematic of the fractal cluster formation.

3. has a "Zone of influence" (see Figure 12.6), which also represents a logical variable. And, the "Truth is," if the cluster is a "Zone of influence" (i.e., the thickness $\lambda = 0$, but one of the neighboring cells of the cluster is not empty, and therefore a cluster breakdown occurs).

To realize the given algorithm the following input parameters should be set:

- the width of the window;
- an indicator of the probability of breakdown η;
- the size of the lattice space LxL;
- the number of seed cells n_{embr}.;
- iteration accuracy, (accuracy) in per cent (shows the percentage of clusters that will be processed before the transition to the next iteration).

12.5.2 Algorithm

12.5.2.1 Initialization

1. Each cluster is assigned to the thickness $\lambda = 0$.
2. The amount of seed cells n_{embr}. is selected, to each of which is assigned a "Zone of influence": = Yes.

12.5.2.2 Running the program

1. Each cluster is assigned to: "Processed": = No
2. Start of iteration: The cluster for which the "Processed" is selected: = No, and/or the thickness $\lambda > 0$ or "Zone of influence": = Yes;

   If these clusters are absent, or are less than the required "Accuracy," go to 1.
3. The selected cluster is assigned to: "Processed": = Yes;
4. Calculated random integer in the range of ($N = 1 \ldots\ldots \lambda^{\eta}$). If $N \neq 1$, go to 2., otherwise there is a breakdown.
5. The thickness is set to $\lambda := \lambda + 1$;
6. For each of the neighboring clusters is assigned: "Zone of influence": = Yes;
7. Go to 2 (end of iteration).

To calculate the fractal dimension d_f of a growing cluster, according to Ivanova et al. (1994: 383), the conditional "mass" M of the cluster (from the center of lattice space LxL) should be calculated depending on the distance r and plotted in double logarithmic scale (see Figure 12.7). Next, using the method of least squares the obtained curve should be approximated by a straight line and the angle of this line to the horizontal axis – lgr calculated. Since, in accordance with the expression (Ivanova et al., 1994: 383; Feder, 1988):

$$M(r) \approx r^{d_f}, \text{ then } d_f = \frac{\lg M(r)}{\lg r}$$

the exponent in the formula is a slope of the straight line, that is, the fractal dimension d_f. This algorithm is implemented in programs written in Turbo Pascal.

The analysis of models of dense structure formation vary depending on different formation conditions which are defined by the input parameters of the models (the number of primary nucleation points – n, which are correlated with the current density, an indicator of the probability of microarc discharge, that is proportional to the scattering ability of the electrolyte and parameter characterizing the degree of continuity of treatment, that is showing what percentage of the surface are processed by modeling "microarc discharges" at their moving from iteration to iteration). There are different examples and performed calculations, which together with the value of the fractal dimensions are shown in Figures 12.8 and 12.9.

Numerous experiments on the computer were conducted using the developed programs, allowing us to find the fractal dimension based on certain input parameters of the models. These results were the basis for the assumption of certain dependencies of the fractal dimension from the microarc oxidation process parameters and allowed us to develop practical recommendations for evidence-based modes of coatings formation in order to obtain their high wear resistance.

In particular, the most dense and wear-resistant coatings correspond to the value of fractal dimension in the range d_f= 1.85–1.98 with the structure in accordance

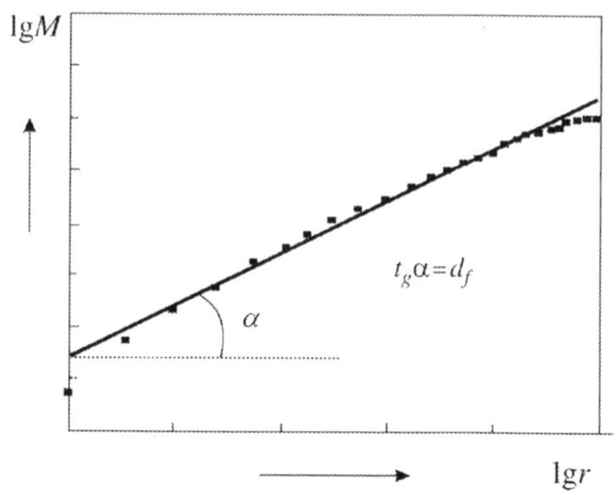

Figure 12.7 Determination of fractal dimension on graph in double logarithmic coordinates.

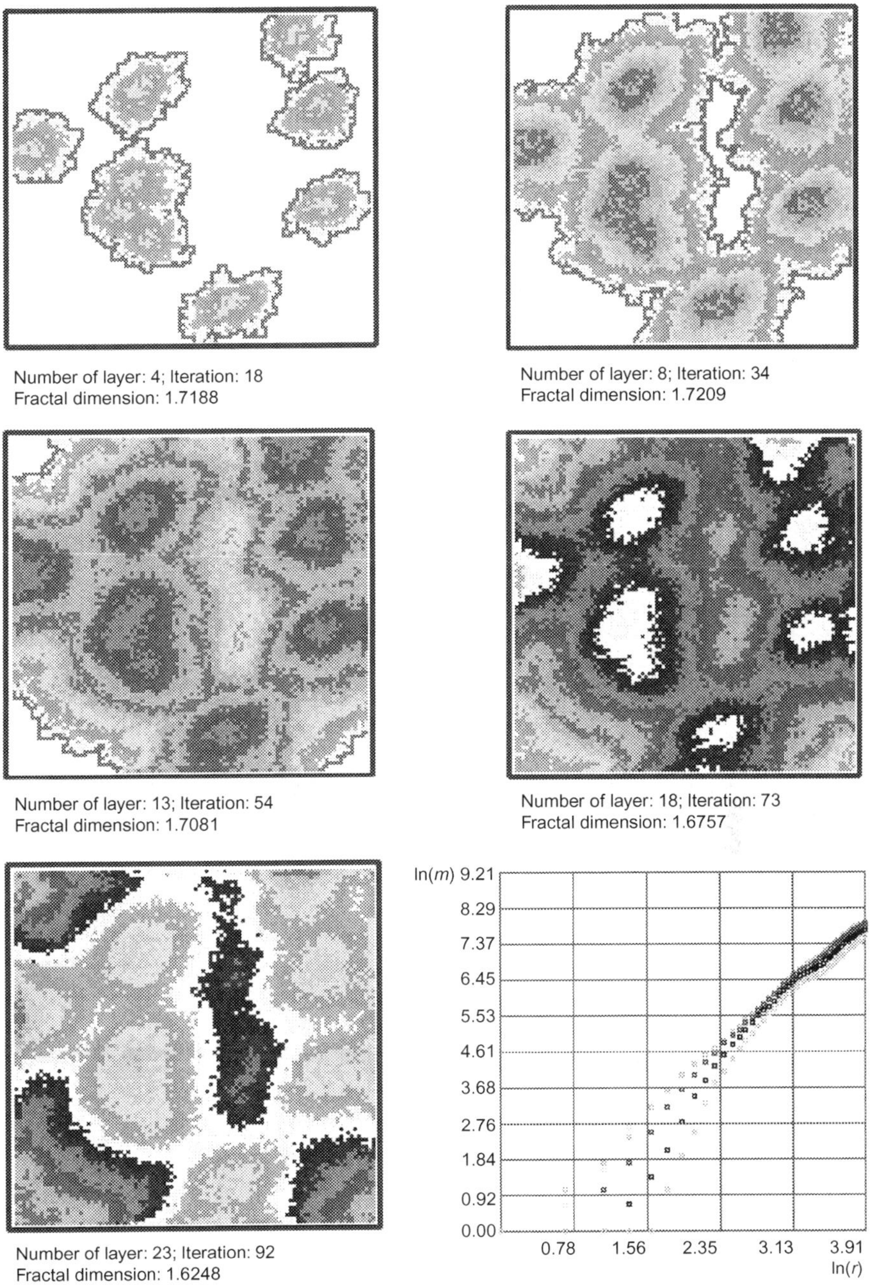

Figure 12.8 The evolution of the fractal structure of dense MAO coating on the various process stages of the model formation with a probability exponent of breakdown $\eta = 0.1$ (the number of embryos $n = 10$) and the graph of fractal dimension.

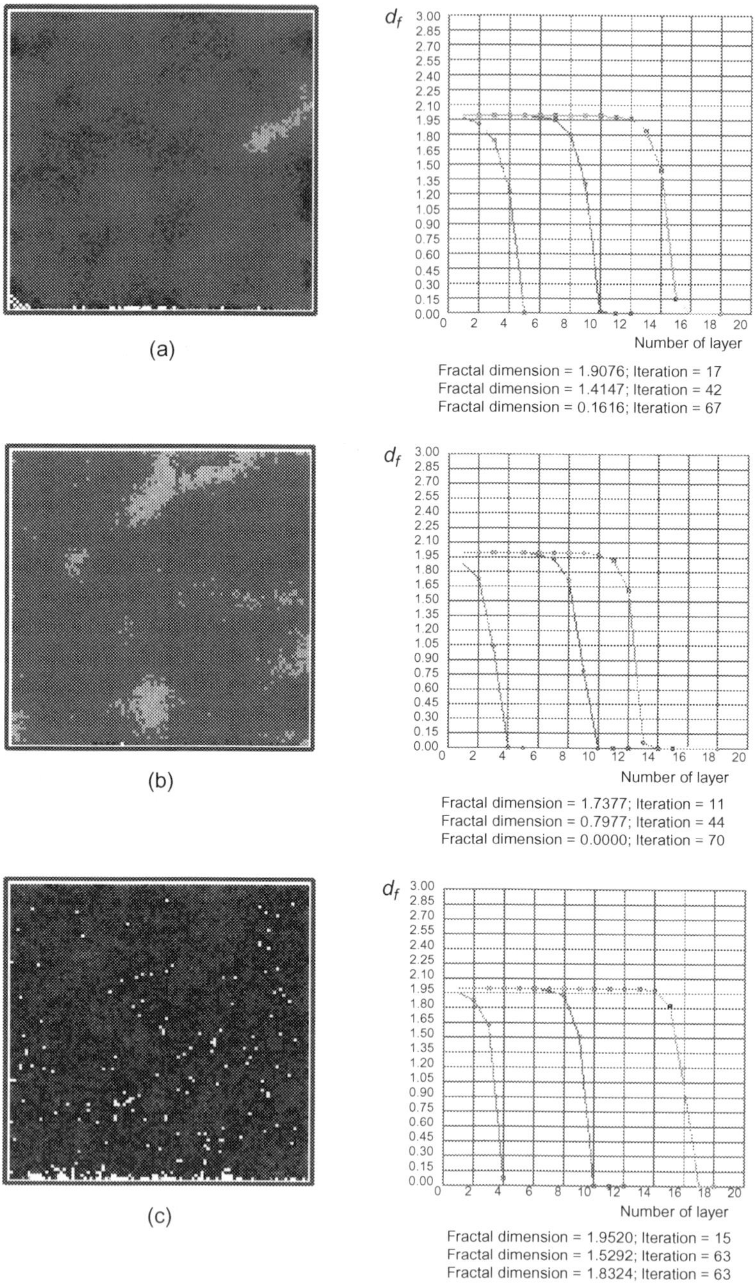

Figure 12.9 Fractal structure of the models and the change of fractal dimension of the coatings layers, formed under the following conditions: $n = 100$, accuracy = 70%, the index of breakdown probability: (a) $\eta = 0.5$; (b) $\eta = 1.0$; (c) $\eta = 1.5$.

with Figure 12.2. This in turn suggests that in order to produce such coatings the best value of current density is in the range of 13–18 A/dm^2, which corresponds to the experimental data, and is confirmed by the high values of microhardness (up to 20 GPa) and wear resistance.

12.6 Macro- and microstructure of MAO coatings

The application of a structural approach explicated recently in the physics of toughness and materials technology gives us a fuller understanding of the mechanisms of hardening at all levels regarding hierarchies of deformation structural levels, to prognosticate coatings work capacity under different conditions. On the basis of the results of spectral analyses it is possible to claim that the coating consists, basically, of oxidized phases of elements contained in a metal. To confirm this, the X-ray diffraction phase analysis on diffraction devices DRON (ДРОН-2.0 and ДРОН - 3.0) were conducted. The filming of coatings samples was conducted using X-ray tubes with $Co_{K\alpha}$- and $Cu_{K\alpha}$- radiation and applying nickel and manganese filters. The measurement ranges of devices were selected depending on the intensity of diffraction lines within limits (0.1–0.4) 10^3 Imp/s. The roentgenograms were taken off on the angles $2\theta = 15-11°$. Identification of roentgenograms was conducted on interplanar distances using the handbook (Anon, 1967). The most representative diffraction patterns received from MAO coatings are shown in Figure 12.10.

The data of roentgenograms and Table 12.1 (Malyshev, 2008) make it possible to note that MAO coatings consist of the crystallized phases of elements contained in aluminum alloys. Moreover, unlike anode oxide films formed by normal anodizing and composed of roentgen-amorphous aluminum oxide, the availability of diffraction peaks on a narrow basis are characteristic for MAO coatings. This specifies that these coatings have a brightly expressed crystalline structure. The lines of α- and γ-Al_2O_3 are revealed legibly on all three roentgenograms, identification of which (see Table 12.1, for example) testifies that these phases are the basic structural components of MAO coatings.

On the basis of X-ray investigations of MAO coatings formed on a wide gamut of aluminum alloys of different systems of doping, Fedorov and Velikoselskaya (1990) specify that besides oxide phases of aluminum (α- and γ-Al_2O_3), mullite ($3Al_2O_3 \cdot 2SiO_2$) was also found as a basic phase. The presence in the coating of this phase is identified because of the occurrence of lines with $d/n = 3.38$; 2.20; 3.41 and spinel Al_3O_4. Besides the lines α- and γ-Al_2O_3, the data of roentgenograms (Figure 12.10) indicate the presence on a background of γ-Al_2O_3, the δ-Al_2O_3, identification of which may be seen due to the occurrence of a line with $d/n = 2.61$ and by the splitting of wide reflex $d/n = 1.52$ of γ-Al_2O_3 on $d/n = 1.545$ and $d/n = 1.51$. However, the indicated phases, as moreover some other modifications of alumina, such as: ρ-, χ-, θ-, and β-Al_2O_3 should be considered as the intermediate forms discovered during thermal transmutations of aluminum hydroxides.

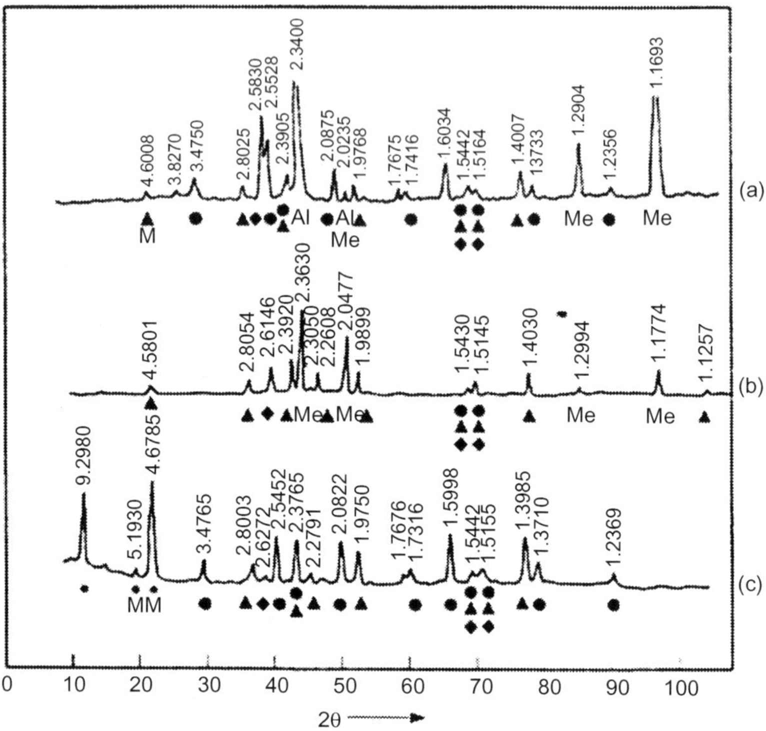

Figure 12.10 The diffraction patterns of MAO coatings: (a) the coating separated from the substrate, (b) the coating on the alloy 2024 (Д16Т), and (c) the coating on the alloy 5086 (АМг6).

Note: ● – α-Al_2O_3; ▲– γ-Al_2O_3; ♦ – δ-Al_2O_3; **Al** – aluminum; **M** – mullite $3Al_2O_32SiO_2$; **Me** – substrate material (compounds of type Al_2MgCu, Al_6Mg_4Cu, $Al_7Mg_6Cu_3$); *complex oxides compound $KAl_2Si_3AlO_{10}(OH)_2$.

These are present on coatings roentgenograms obtained on alloys 2024 (Д16Т) (Al-Cu-Mg system) and 5086 (АМг6) (Al-Mg-Mn system) (Figure 12.10(b) and (c)), besides lines for aluminum oxides, the diffraction peaks of which identification allows them to be referred to as the lines of substrate materials. For alloy 2024 (Д16Т) the presence of the lines of aluminum and phases $CuAlMg_2$, $CuMg_4Al_6$, and also Al_3Mg_3SiFe and $Al_7Cu_3Mg_6$ are characteristic (Anon, 1967).

The fullest information on distribution of elements in transition area can be obtained using a method of X-ray spectral analysis (Mayer, 1979: 447). On Figures 12.11 and 12.12 the outcomes of qualitative and quantitative X-ray spectral analysis of MAO coatings on alloy 2024 (Д16Т) obtained on the device "Cameca" MS-46 are shown. The basic elements which are found in the structure of alloy and in the coating – Al, Mg, Cu, Mn, Si, K, and so on. were exposed to the analysis. On all pictures, the border of transition of analyzed elements concentration lines are legibly indicated. The boundary zone is shown in X-ray $Al_{K\alpha}$-radiation (Figure 12.11(b))

Table 12.1 Diffraction pattern of MAO coating separated from the substrate

2θ°	*d/n*, A°	*I/I*$_o$	Reference data			Identified phases
			d/n, A°	*I/I*$_o$	Hkl	
11.06	9.29	90	9.9	100	001	
19.85	5.19	20	5.04	27	002	$KAl_2Si_3AlO_{10}(OH)_2$ –
22.06	4.67	100	4.49	90	020	Muscovite
29.84	3.47	40	3.47	50	101	α-Al_2O_3
37.28	2.80	30	2.80	20	220	γ-Al_2O_3
39.84	2.62	10	2.61	20	–	δ-Al_2O_3
41.18	2.54	70	2.55	75	112	α-Al_2O_3
44.25	2.37	60	2.37	30	311,101	α-,γ-Al_2O_3
46.25	2.27	20	2.27	13	222	γ-Al_2O_3
50.77	2.08	80	2.08	80	102	α-Al_2O_3
53.92	1.975	60	1.975	72	400	γ-Al_2O_3
60.81	1.76	10	–	–	–	–
62.34	1.73	30	1.74	50	202	α-Al_2O_3
68.04	1.59	80	1.59	100	123	α-Al_2O_3
70.84	1.54	10	1.54	5	–	α-,γ-,δ-Al_2O_3
72.40	1.51	10	1.50	5	233	α-,γ-,δ-Al_2O_3
79.58	1.398	80	1.396	100	440	γ-Al_2O_3
81.50	1.37	40	1.37	50	112	α-Al_2O_3
92.82	1.235	20	1.233	20	234	α-Al_2O_3

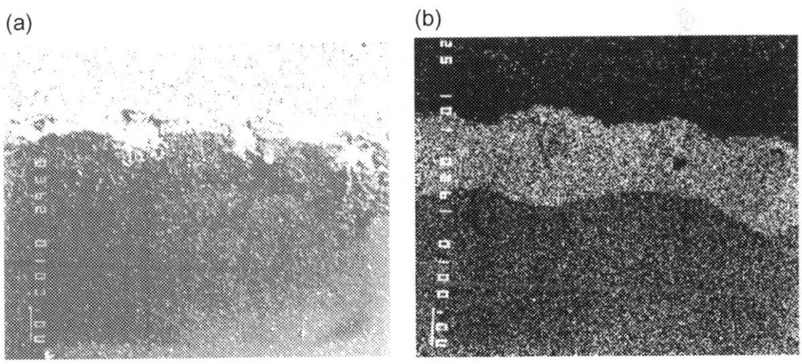

Figure 12.11 The transverse section view of coating (a) and distribution of aluminum in the characteristic $Al_{K\alpha}$ X-ray (b) (magnification −1000×, the device Cameca MS-46)

and also by concentration curves of Al distribution through a border (Figure 12.12). It is represented a little washed out, which testifies to the possibility of a diffusion zone. It should be noted that for practically all concentration lines some increase (positive gradient) of concentrations at the approach to the border of the transition area from the side of the substrate is characteristic. A possible explanation is that

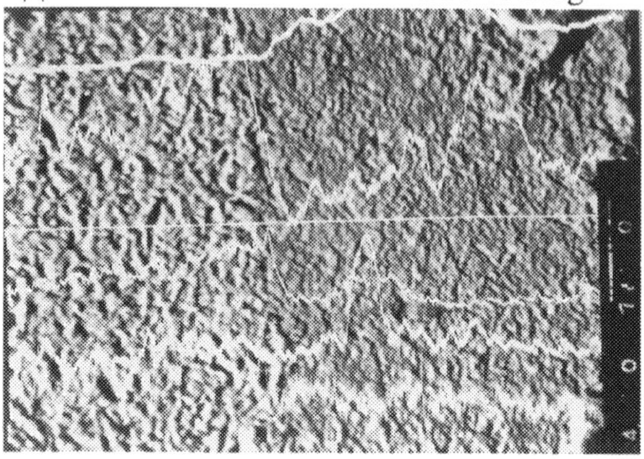

Figure 12.12 The qualitative distribution of the main components of the MAO coating: Al, K, Cu, Mg, Si in the transition area 'substrate – coating', microsection was etched, magnification – 300×.

during processing there is a thermal influence on the frontier area of a substrate. As a result of the diffusion of elements caused by the alignment of differences in the kinetic energies of atoms, this zone appears a little enriched in contrast to areas remote from the boundary zone. The generalized information about the structure of MAO coatings is represented in Table 12.2.

High wear resistance of microarc oxidation coatings is substantiated by their high level of physical and mechanical characteristics and structural features (Malyshev et al., 1985). The investigations carried out (Alekhin et al., 1991) showed a capability of MAO coatings formation depending on an elemental composition of aluminum alloys in three kinds of hyperthermal α-Al_2O_3, namely: like bundles of separate single crystals (pure Al), bundles of threadlike monocrystals (Al-Cu-Mg alloy), and in fine-crystalline shape (Al-Mg-Mn alloy). The highest mechanical and tribotechnical characteristics have MAO coatings, the main working layer of which consists of bundles of threadlike single crystals α-Al_2O_3 and mullite ($3Al_2O_3 \cdot 2SiO_2$). The MAO coatings have this microstructure on duralumin alloys. On metallographic microsections (Figure 12.13) it is possible to see these threadlike single crystals having about 4–8 μm or less in size with a ratio of length to thickness about $l/d > 15$. The bundles of threadlike single crystals are sharply curved and together with an isotropic matrix from γ-Al_2O_3 form separate globules–skeins. Such globular and cellular structure looks like the carapace of a turtle (see Figure 12.14) and represents the most effective structure for ensuring high wear-resistance. The increase in wear resistance in all aspects of abrasive outwearing is provided first of all by magnification of the plasticity resistance of coating material or, in other words, by a heightening of its toughness properties (HV, σ, E, etc.). However, the fragility is increased at excessive hardening, and breakdowns occur through intensive shearing of the surface layer (low value of K_{Ic}). Therefore, coating material with a large value of σ will be worn less

Table 12.2 **Results of studies of composition MAO coatings by different methods**

Elements	Methods of research							
	Emission spectroscopy (ДФС-8)		Micro-X-ray spectroscopy (Cameca, MS-46)		X-ray analysis (ДРОН-2.0, 3.0)		Chemical analysis	
	Volume content %	Sensitivity %	Volume content, %		Identified phases	Volume content %	Identified phases	Volume content %
			Substrate	Coating				
Al	>>1	$1 \cdot 10^{-3}$	92–94	24–34	α-Al_2O_3 γ-Al_2O_3 δ-Al_2O_3	28 44 3	Al_2O_3	60.8
Mg	>0.1	$1 \cdot 10^{-3}$	0.8–1.5	0.3–1.3	Compounds of type:		–	–
Cu	>0.1	$5 \cdot 10^{-4}$	1.9–2.6	0.5–1.6	$3Al_2O_32SiO_2$		–	–
Si	>0.1	$1 \cdot 10^{-3}$	0.1	0.6–3.2	Al_3Mg_3SiFe,		SiO_2	9.4
Mn	$>3 \cdot 10^{-2}$	$4 \cdot 10^{-3}$	0.4	0.05–0.2	$CuMgAl_2$,	Up to 24	–	–
Fe	>0.1	$1 \cdot 10^{-3}$	0.04–0.1	0.15–0.4	$CuMg_4Al_6$,		Fe_2O_3	1.95
Na	–	–	No	<0.1	$Al_7Cu_3Mg_6$		Na_2O_2	2.14
K	–	–	No	<0.1	and others		K_2O_2	2.25
Others	–	–	–	–			–	22.7

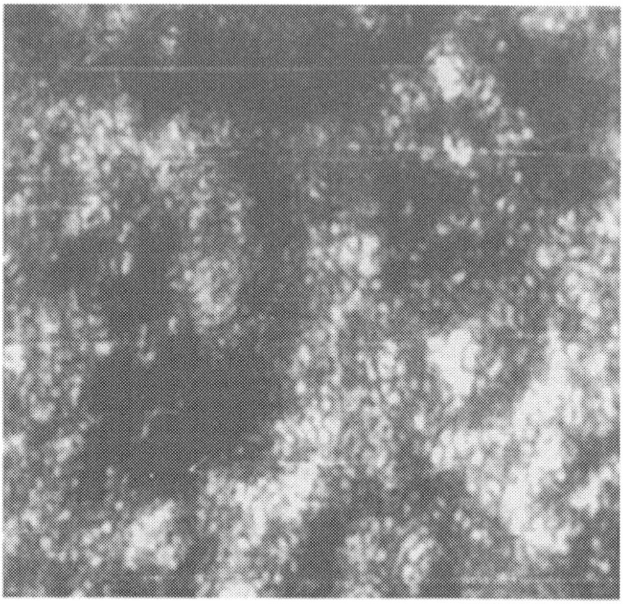

Figure 12.13 Microstructure of threadlike single crystals of α-Al_2O_3, magnification – 5000×.

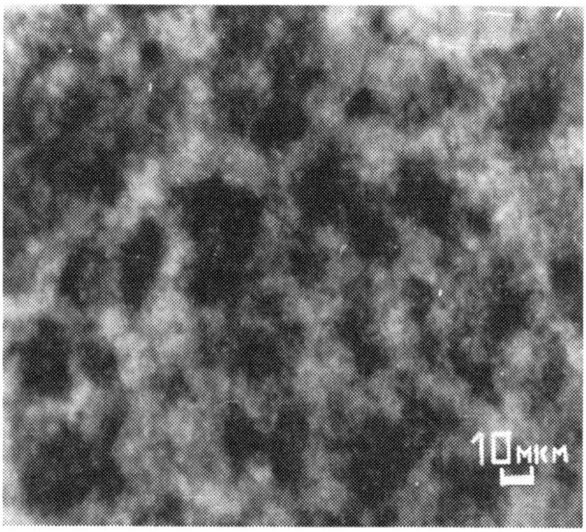

Figure 12.14 Morphology of cellular structure of MAO coating that is similar to the carapace of a turtle, magnification – 500×.

intensively at identical fracture resistance, and at the equal level of σ, while the higher the value of K_{Ic}, the more wear resistant the coating will be.

12.7 Wear-resistant properties

The method of tests on abrasive outwearing is most usable (Khrutshev and Babichev, 1970: 252) for an estimation of the wear resistance of MAO coating, because they have rather high mechanical characteristics (microhardness, elastic modulus, etc.). The estimation of relative wear resistance of MAO coatings was conducted on an installation (Malyshev et al., 1983), designed by the author, using the friction schema "pin on disk" at the outwearing of coatings samples on fastened abrasive particles (abrasive wheel of green silicon carbide, with a graininess of <0.070 mm). The friction was always applied to a fresh abrasive surface, due to the radial motion of the sample on the rotating abrasive wheel with a velocity of 0.3 mm/rev (path of friction – Archimedes spiral). The stripping of the circle surface with the help of a small diamond abrasive wheel was simultaneously carried out. The products of wear were deleted from the surface by submission of water in a zone of friction. The specific load on a sample can be changed from 0.25 up to 1.25 MPa. Each test was conducted during 30 seconds that corresponded to 22 m of friction path. The velocities of sliding were changed in an interval 0.6–1.0 m/s. The wear was evaluated on a change of mass (on analytical balance W21 with an exactitude ±0.0001 g) and linear also by means of an indicator with an exactitude ±0.001 mm. The application of two methods of wear estimation allowed us to control gross errors in measurements and to increase the level of reliability of the data obtained.

For comparison of data on an abrasive outwearing of MAO coatings in the same conditions a number of other materials were tested. The choice of such materials was carried out based on a comparison of data of MAO coating wear resistance with these materials out of which one could judge the place of the coatings among a number of wear-resistant materials.

Plenty of materials and coatings that are applied to increase wear resistance to different friction knots and links are now known. Among them it is possible to mark the composites on the basis of tungsten carbides with ligaments from copper–nickel alloys; different surface hardening materials, such as ПГ-ХН80Ср4, ВСНГН, and so on; siliconized graphite of different kinds such as СГ-К, СГ-М, and СГ-Т (Vinogradov, 1977), mineral ceramics ЦМ-332 on the basis of Al_2O_3 (99.0–95.5%), cemented-carbide compositions, such as ВК-6, ВК-8, and Т15К6 (Khebda et al., 1989: 1, 400), and others. The wear resistance of MAO coatings was compared to the following materials: a composite on the basis of tungsten carbides WC-Cu-Ni; siliconized graphite СГ-К; and also cast iron 180 BS 1452 (СЧ18), steel 1045 or 1.1191(45), hardened up to HRC 44, and steel USt 37-2 or A283 Grade C (3) as the etalon. Table 12.3 (Malyshev, 2008) presents the basic physical and mechanical characteristics of the tested materials.

Table 12.3 **Basic physical and mechanical properties of materials**

No.	Material	Density g/cm^3	Ultimate strength, MPa			Hardness		Young's modulus, GPa	Thermal conductivity Coeff.Ccal/ m·h	Linear expansion Coeff.1/ grad
			Tensile	Compression	Bending	HV, GPa	HRC, HRA, HB			
1	WC-Cu-Ni	14.3–14.5	50–70	1030–1100	–	–	HRA 80	3–5	–	12–14
2	Siliconized graphite	2.4	50–60	420–450	100–120	–	HRC 75	127	130–150	4.2–5.0
3	ПГ-ХН80Ср4	7.6	–	–	–	–	HRC 50	–	–	–
4	Steel 45	7.85	610	–	–	–	HRC44	200	1	11
5	Cast iron СЧ18	7.2	176	–	353	–	HB200	95–160	–	–
6	Steel 3	7.8	370–480	–	–	1.5	–	–	–	–
7	MAO coatings	3.65	–	–	–	11–25	–	190–290	–	–

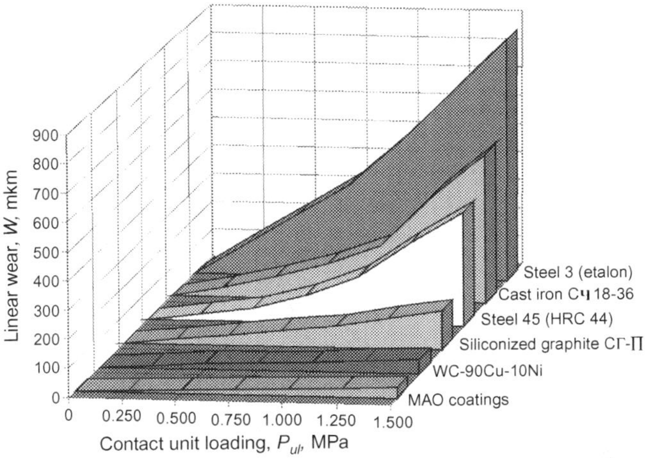

Figure 12.15 Comparable linear wear of tested materials and coatings.

The dependences of linear wear W (Figure 12.15), intensity of outwearing I_h (Figure 12.16) and relative wear resistance ε (Figure 12.17) from the unit load on contact – P_{ul} – at the friction on fastened abrasive particles are represented. Very high wear resistance of MAO coatings can be seen, which is comparable to the wear resistance of materials traditionally used against an abrasive wear – composites on the basis of tungsten carbides. Moreover, it is characteristic that the intensity of abrasive outwearing grows proportionally with an increase in unit load on contact up to P_{ul} = 1.5 MPa (Figure 12.16), while for cast iron, steel 3 and steel 45, at a loading higher than 1.00 MPa more intensive outwearing is seen.

Of great interest also is the comparison of the outcomes on wear resistance of MAO coatings with data of relative wear resistance investigations of materials conducted by other investigators (Khrutshev and Babichev, 1970: 252). Such comparison obviously allows the presentation of the level of relative wear resistance of ceramic coatings in contrast to a wide range of materials. To do this, the data points of MAO coatings' magnitudes of relative wear resistance were added to the graph of relative wear resistance ε from the ratio of hardness of abrasion to the tested materials H_a/H_m.

Steel 3 was chosen as the etalon in our tests. The MAO coatings' relative wear resistance ε was evaluated in relation to wear of this material. The magnitudes of relative wear resistance ε were put on the diagram (Figure 12.18) forming the area of MAO coatings that described their place among other materials. Additionally, to expand the comparison spectrum, on the same diagram were also added the data of relative wear resistance for different hardening coatings, surface hardening, welding and modified layers, achieved by modern methods of materials processing such as PVD technologies (Semenov et al., 1992: 404), laser hardening (Stafford, 1991: 238), and so on.

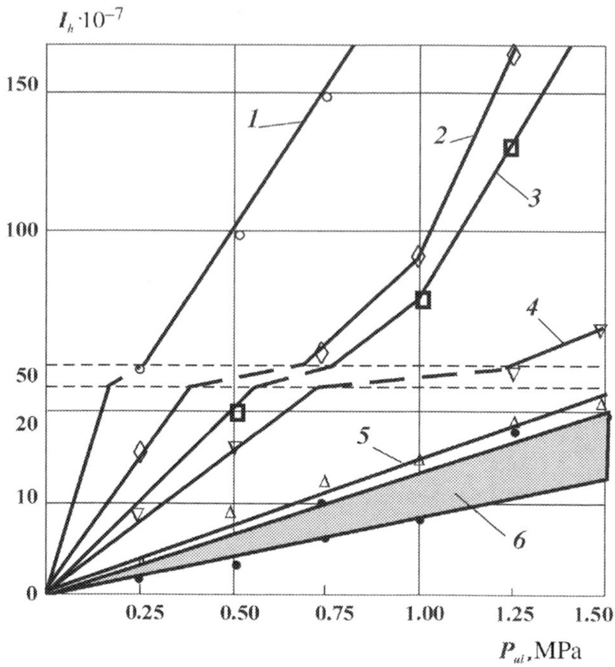

Figure 12.16 Dependence of abrasive outwearing intensity I_h from unit contact load P_{ul} for materials: (1) Steel 3 (etalon), (2) Cast iron СЧ 18, (3) Steel 45 (HRC 44), (4) siliconized graphite СГ-П, (5) composite WC-90Cu-10Ni, and (6) MAO coatings.

From Figure 12.18 you can see that the MAO coatings have a very high wear resistance comparable to wear resistance of materials and coatings traditionally used against abrasive wear: WC-Cu-Ni, boron layers (FeB, Fe_2B), ionic-vacuum coatings. The reason for the high wear resistance of MAO coatings is, apparently, more to do with their structural features than with their toughness characteristics (microhardness), because the structural modifications of alumina in MAO coatings (α-, γ-, δ-Al_2O_3) during the treatment form among themselves the strongest type of bond corresponding to the possible interparticle connections in the material – metalic, covalent, ionic and molecular (Katsheev, 1978: 213).

Some authors (Alekhin et al., 1991) have pointed out that a considerable increase in materials wear resistance is achieved by introducing high-tensile inserts. The big difference in the properties of α- and γ-Al_2O_3 means that the phase of α-Al_2O_3 can be considered as these high-tensile inserts in MAO coating, due to which their high wear resistance is ensured. Thus, the aggregate of components of alumina phases in MAO coating represents some dispersed system that is composed of a dispersed phase (α-Al_2O_3), distributed in a certain way in the dispersion medium (matrix), or, in fact, the composite material.

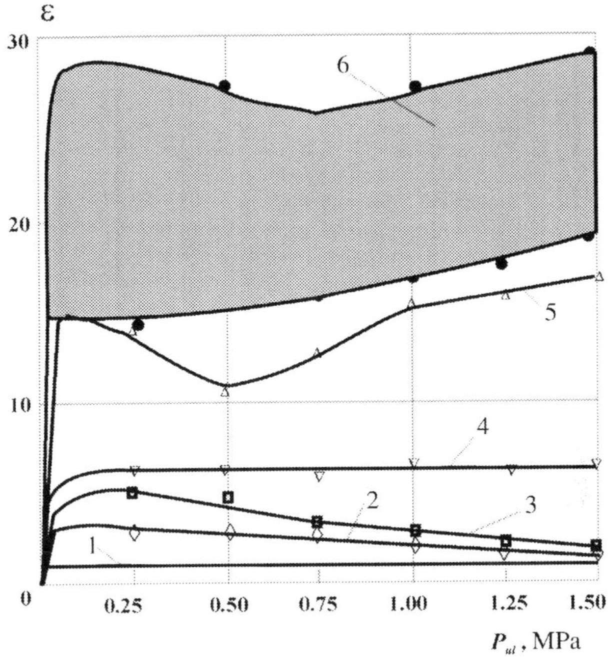

Figure 12.17 Dependence of relative wear resistance ε from unit contact load P_{ul}: (1) Steel 3 (etalon), (2) Cast iron СЧ 18, (3) Steel 45 (HRC 44), (4) siliconized graphite СГ-П, (5) composite WC-90Cu-10Ni, and (6) MAO coatings.

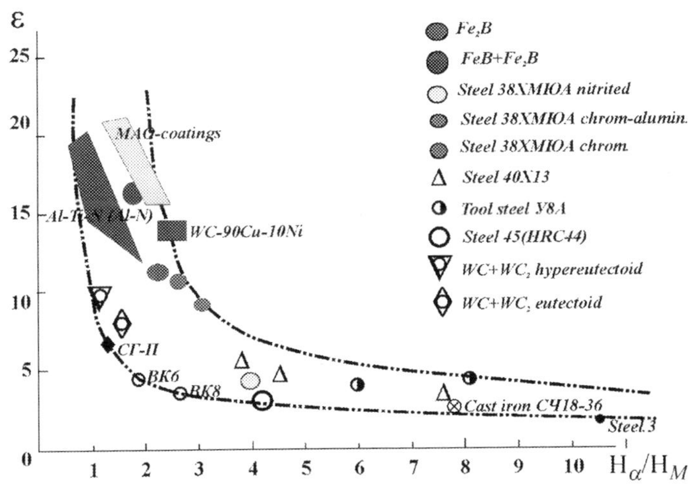

Figure 12.18 Dependence of relative wear resistance of different materials, hardened layers and coatings ε from ratio hardness of abrasive to material.

Deeper microstructure investigations of different phases of alumina forming at the MAO treatment have shown (Rudnev et al., 1990) that it really is a fact. In conditions of burning microarc discharges at the expense of high-speed crystallization, changes in thermal fields and the influence of internal stresses, there is a growth in oxide phase crystal formations in different structural states. As noted above, the basic hardening phase – α-Al_2O_3 – can be formed in the dense working layer in three forms: bundles of separate single crystals, bundles of threadlike single crystals, and microcrystalline form. This explains that observable essential differences in the toughness characteristics of MAO coatings have practically identical phase structures.

The essential differences in the properties of aluminum oxide phases can be seen in Table 12.4.

The coatings in which α-Al_2O_3 consists of skeins of threadlike monocrystals built-in in a microcrystalline matrix from γ-Al_2O_3 and mullite ($3Al_2O_32SiO_2$) have the greatest wear resistance and toughness.

The MAO coatings on duralumin alloys have such microstructures – 2024, 2017, and 2117 (Д16, Д1, Д18, Д20 etc.). The bundles of threadlike crystals of corundum (α-Al_2O_3) twisted in skeins (see Figure 12.12), together with the microcrystalline matrix from γ-Al_2O_3 are practically deprived of the possibility of deformation to shifts at the expense of the strain of the matrix only.

In this is hidden the reason for the high wear resistance of MAO coatings on duralumin alloy 2024 (Д16). On the other hand, at the deformations caused by friction the localization of the centers of destruction happens on surfaces with a cellular structure.

The specificity of MAO coatings formation in an anodic-cathode mode consisting in the separation of the coating on layers in the defined stage of the process stipulates apparently the difference in the properties of these layers. This difference is of greatest interest in the estimation of MAO coatings wear resistance, because the change in the properties of the coatings layers will be reflected in their wear resistance. It is through this link that the experiments that have allowed the character of wear resistance changes in coatings layers to be discovered have been carried out (Petrosyants et al., 1984). The investigations were conducted on installation of the friction schema "pin on disk." The tests were accelerated to reveal the kinetics of MAO coatings' outwearing using the "level-by-level" method, through the friction of the fastened abrasive particles on the coating samples, that is, an abrasive wheel of green silicon carbide was used as a disk-counterbody. The test parameters were: unit load −0.25 MPa, duration of each experience 60 seconds.

On a Figure 12.19 the outcomes of this experiment are represented. Here the curves of linear wear *W* (1), microhardness *HV* (2), intensity of outwearing I_h (3), and porosity *P* (4) are the integral curves constructed by the results of tests on five to seven samples of MAO coatings.

The microhardness was measured by a standard method on a PMT-3 (ПМТ-3) device by way of layer by layer on transversal microsections of samples with coating at the load on indenter (Vickers diamond pyramid) 1.96 N. The measurements were conducted beginning from the substrate 2024 (Д16) through the coating to the

Table 12.4 **Physical and mechanical properties of different aluminum oxide phases**

Phases	Density, $kg/m^3 \cdot 10^3$	Physical and mechanical properties					
		Type of lattice	Temper. of form., °C	Temper. of melt., °C	Thermal expansion coeff., $\alpha \cdot 10^6$, $°C^{-1}$	Microhardness, GPa	Young's modulus, GPa
α-Al_2O_3	3.96–4.01	Rhombohedral	1000–1500	2320	8.90	25.5	400
γ-Al_2O_3	3.29–3.42	Cubic	500–600	1050	6.8	12.5	250
β-Al_2O_3	3.5	Hexagonal	1000	1700	–	–	–
δ-Al_2O_3	2.4	Hexagonal		–	–	–	–
θ-Al_2O_3	–	Cubic	1000	–	–	–	–

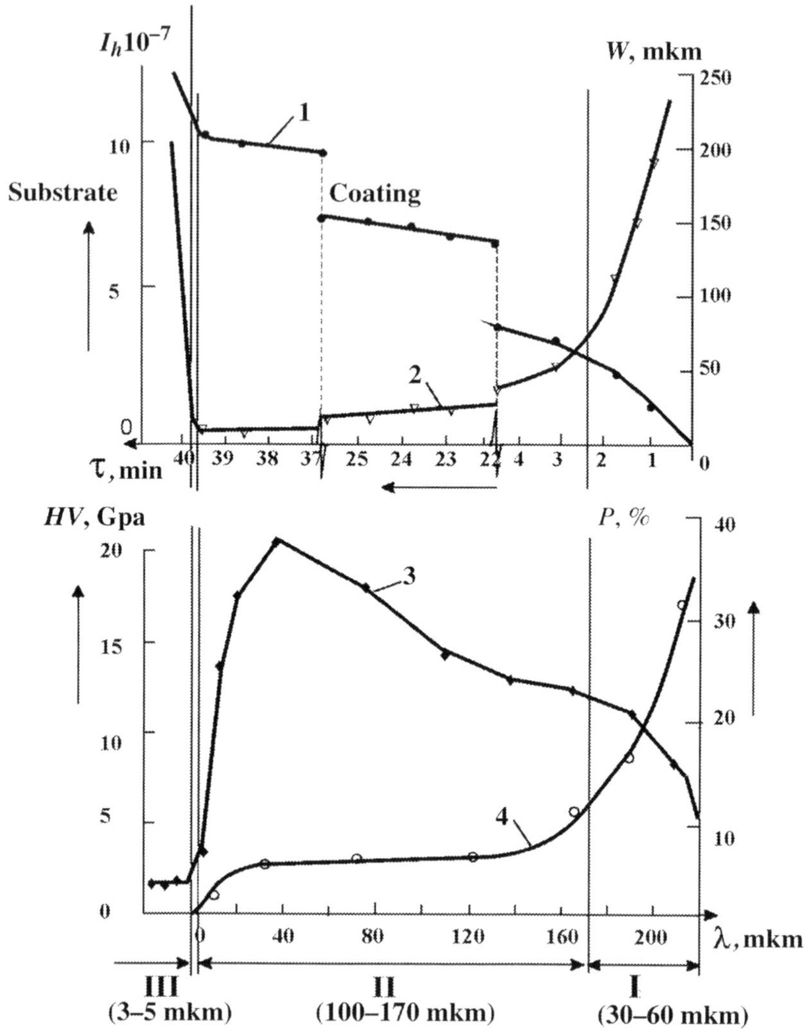

Figure 12.19 Dependences of changes linear wear W (1), outwearing intensity I_h (2), microhardness HV (3), and porosity P (4) over the thickness of MAO coatings.

surface layer with a pitch in the average 10–20 ± 5 μm. The generalized dependence $HV = f(\lambda)$ shown on Figure 12.19 reflects the general character of the microhardness change in thickness of the formed coating and can be expressed as follows:

$$HV = \exp\left(3.6 - 0.0074\lambda - \frac{11,1}{\lambda}\right)$$

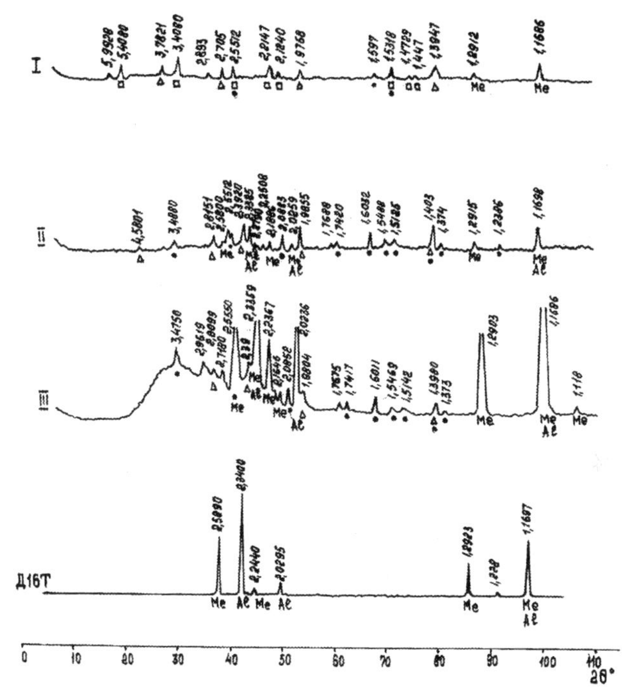

Figure 12.20 Diffractograms of MAO coatings layers: (I) surface layer, (II) basic working layer, (III) frontier zone.
Note: ● – α-Al_2O_3; ▲ – γ-Al_2O_3; ■ – compounds like $Al_2O_3 \cdot SiO_2$ or $3Al_2O_3 \cdot 2SiO_2$ (mullite); **Al** – aluminum; **Me** – compounds like Al_3Mg_3SiFe, Al_6Mg_4Cu, and others.

where λ is the thickness of coating beginning from the substrate.

It is necessary to note the general character of this dependence that is reasonable for MAO coatings formed on any aluminum alloys. It is confirmed by the outcomes of other contributors (Fedorov and Velikoselskaya, 1990) and can be explained only by the specificity of ceramic coating formation in anodic-cathode mode.

The porosity *P* was defined by the stereometrical metallography method by applying the analysis texture system of images TAS Leitz. The identification of the obtained diffractograms (Figure 12.20) was conducted in accordance with Anon (1967).

All of the investigations carried out have allowed installations in which the MAO coatings are inhomogeneous in thickness and are changed both in structure and on the constitution. Three zones (see Figure 12.19) are distinguished in coatings legibly: zone I – surface layer of coating with the greatest intensity of outwearing I_h, rather low magnitude of microhardness *HV* and highest porosity *P* (up to 40%); zone II – basic layer of coating has the best mechanical characteristics: a big magnitude of microhardness, the least intensity of outwearing, and with a stationary value of porosity (within of 1–10%); zone III – the frontier zone between coating and substrate material.

The roentgenograms taken off in each of the isolated coating zones and in the substrate on Figure 12.20 are represented. The level-by-level filming was conducted with allowance for recommendations (Rudnev et al., 1990). The decoding of outcomes has allowed a pattern of structure changes in the coating's thickness to be presented. In particular, from Figure 12.20 you can see that approaching the substrate, beginning from the surface layer, crystallization of the structure phases increases considerably, and there is an essential influence of the substrate phase's structure on the structural constitution of the coating. So, if zone I in the surface layer predominantly consists of α-, γ-Al_2O_3, and a spinel compound such as $3Al_2O_32SiO_2$ – mullite – the basic layer (zone II), apart from aluminum oxides, is distinguished by phases composed of elements from substrate material (Al and its compounds with Cu, Mg, and others). To such combinations, first of all, it is necessary to attribute Al_3Mg_3SiFe and Al_6Mg_4Cu, and others. Some shift in diffraction peaks from their tabulated magnitudes (see Table 20.4), which characterizes the inter-influence of these compounds against each other (Anon, 1967). In zone III the influence of the substrate is most essential, because the characteristic of aluminum alloy 2024 (Д16) phases Al_6Mg_4Cu and Al_2MgCu is that they have considerable share here (more than 50%). The basis of coating consists of α- and γ-Al_2O_3 (up to 70%), which ensures its high microhardness can reach up to 22 GPa (Figure 12.19). The structural heterogeneity and the nonuniform distribution in coating thickness reduces to change the microhardness in accordance with the modification of coatings phase structure and in the greatest degree depending on the distribution of high-hard phase α-Al_2O_3 (Figure 12.21). The microhardness in this zone depending on the distance from the substrate is in an interval of 20–15 GPa.

In reality, if the ratio between α- and γ-Al_2O_3 in the surface layer does not exceed 0.5, in the basic layer of coating this relation already comes nearer to 1.0; and in the layer adjoining the substrate it becomes equal to 2.5. And this is the reason that the distribution of microhardness on thickness of coating has the character of the curve shown on Figure 12.19 (or Figure 12.21).

The fact that in the coating formation process, especially in the zone III boundary, it is detected that the phases have great affinity with the structure of metal explains the high adhesion of the coating to the substrate and characterizes the presence of chemically bonded compounds between the substrate material and the coating. The greatest wear and accordingly intensity of outwearing of the coating (Figure 12.19) is noted in the surface layer, which has the maximum amount of micro- and macrodefects (pores, friability), and rather low microhardness. If the surface layer is taken off, the microhardness is increased up to magnitudes about 13–17 GPa, the intensity of outwearing is changed accordingly diminishing to $(2.5–4) \cdot 10^{-7}$. On approaching the substrate at a distance of 30–60 μm the greatest significance of microhardness is noticed – 20 to 22 GPa (see Table 12.5), in some cases up to 25 GPa – and the least intensity of outwearing $(1–2) \cdot 10^{-7}$. It must be emphasized that the area with the highest microhardness, almost always, irrespective of MAO coatings thickness (for $\lambda > 100$ μm), is at the indicated distance from the substrate.

This can be explained by the fact that, with the growth of the basic (dense) MAO coating layer, there is a shift in the zone of maximal temperature of microarc

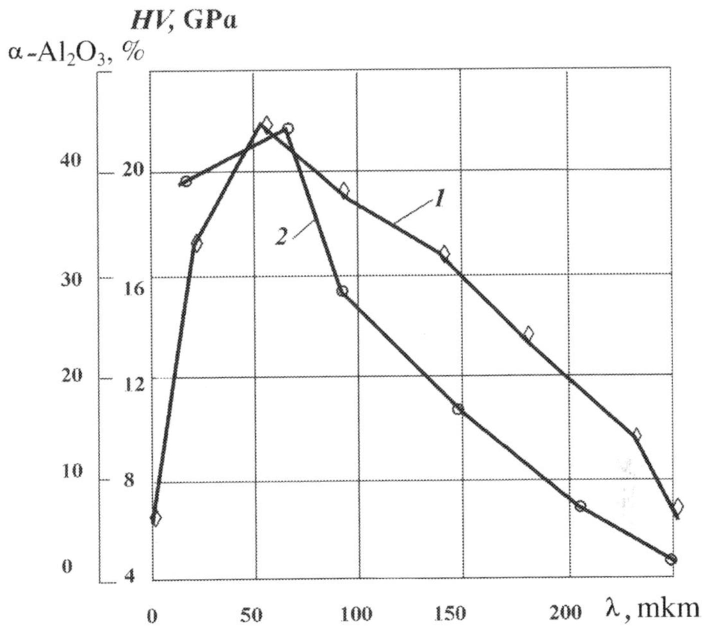

Figure 12.21 Generalized dependences of changes microhardness *HV* (1) and contents α-Al_2O_3 (2) over the thickness of MAO coatings on alloy 2024 (Д16).

discharges deep into the oxide, at the expense of a change in its electrophysical characteristics. Indirectly it is confirmed by distribution of coatings components and elements (see Figure 12.12) at the thickness where the peaks (maxima in a zone 30–60 μm from the substrate) are seen. Namely, in this area the maximum temperature is concentrated and consequently the maximum heating of the coatings material happens. Accordingly, in this area the maximum content of high-temperature phase – α-Al_2O_3 – is located. As is known (Mayer, 1979: 447), the diffusion of elements in melted mediums flows more actively than in solids, therefore increasing the content of alloying elements in a coating (the maxima on the distribution curves of Figure 12.12) specifies uniquely on an area with the maximum gradient of temperature.

The indicated feature allows the least coating thickness sufficient to guarantee the high wear resistance of the friction knots to be established. With allowance for subsequent grinding and removal of the surface layer during running-in it should be not less than 100–120 μm. The magnification of film thickness more than 300 μm is not rational from the point of view of electric power costs necessary for its formation, because considerable time for item processing is required. Best from the technological point of view is a thickness 200–250 μm, which after grinding and running-in will be 140–200 μm. Such coating thickness will ensure a sufficiently long life for the friction pair, even in the presence of abrasive particles.

Table 12.5 **Fiberwise microhardness measurements of MAO coatings**

Place of indenter Injection	Microhardness $HV_{1.96N}$, GPa					
	MAO modes					
	4 g/l KOH + 3 g/l Na_2SiO_3, $I_k/I_a = 1{,}0$			4 g/l KOH + 4 g/l Na_2SiO_3, $I_k/I_a = 0{,}6$	4 g/l KOH, $I_k/I_a = 0{,}8$	4 g/l KOH, $I_k/I_a = 1{,}0$
	1	2	3	4	5	6
Metal of substrate (Д16)	2.18	1.07	1.23	1.91	1.96	1.83
	1.74	1.01	1.25	1.48	1.72	1.77
	1.96	1.03	1.24	1.85	1.83	1.64
	1.94			1.88	1.72	
	1.91			1.83		
	1.88					
	1.70					
Metal coating border	3.60	1.95	1.49	2.86	3.34	3.24
	3.60		1.45	2.06	2.32	2.77
	4.58					
Basic working layer of coating	12.70	20.35	7.52	15.85	20.35	6.12
	23.30	25.75	16.48	19.50	15.85	6.08
	21.30	16.48	18.65	17.90	22.28	8.92
	14.50	18.15	19.47	12.70	16.50	8.42
	10.70	18.48	13.14	10.40	17.88	6.60
	10.10	15.23	11.83		13.65	
	10.40				12.70	
	10.10				11.44	
	11.85				10.70	
Upper surface layer of coating	9.16	8.91	6.43	9.47	8.66	3.60
	7.90	8.40	3.53		7.28	
					6.00	

12.8 Conclusion

The basic difference between the MAO process and traditional anodizing is the use of energy from electric discharges migrating onto the treated surface immersed in electrolyte. These discharges render, in particular, thermal, plasma chemical, and hydrodynamic influences on the basic metal coating and electrolyte, as a result of which a ceramic coating is formed, which can be widely adjusted by elements and phase composition, structure, and properties.

Among the characteristic features of the MAO process are high temperatures in the discharge channels and, as the corollary, the formation of high-temperature phases in coatings (for aluminum alloys it is corundum α-Al_2O_3); a splitting of water with the formation of atomic and ionization oxygen; local magnification of electrolyte concentration; and particular plasma chemical reactions in a zone of discharges.

The multifunctionality of MAO coatings makes them ideal for application in a variety of industries and the spectrum of treatment detail is constantly growing. Nevertheless, research into the possibilities of the MAO method has not been exhausted. Work on perfecting the process must be continued, by way of developing new electrolytes, selecting optimal modes, and developing and creating new technological current sources.

References

Alekhin, V.P., Fedorov, V.A., Bulychev, S.I., Tyurpenko, O.A., 1991. Features of microstructure of hardened surface layers obtained by microarc oxidation. Phys. Chem. Treat. Mater. 5, 121–126 (in Russian).

Anon, 1967. INDEX (inorganic) to the Powder Diffraction File. ASTM, Pennsylvania, PA, ASTM Publication, PDIS-171.

Anon, 1981. Microarc oxidation. Science and Humanity. Znanie, Moscow, Russia, p. 341 (in Russian).

Anon, 1986. Fractals in Physics. Proceedings of the Sixth International Symposium on Fractals in Physics (Trieste, Italy, July 9–12). Elsevier Science Publishers BV, North-Holland, the Netherlands, p. 627.

Feder, J., 1988. Fractals. Plenum Press, New York, London.

Fedorov, V.A., Velikoselskaya, N.D., 1990. Influence of microarc oxidation on the wear resistance of aluminum alloys. Phys. Chem. Mater. Treat. 4, 57–62 (in Russian).

Gogish-Klushin, S.Y., Markeshin, A.V., Fridman, A.A., Kharitonov, D.Y., 1993. Plasma-chemical model of elemental breakdown at the electrolytic-sparking oxidation. Proceedings of 'Interanod-93'. KAI, Kazan, p. 59 (in Russian).

Gould, H., Tobochnik, J., 1990. Computer Simulations in Physics. Part I and Part II (translated from English). Mir, Moscow, Russia (in Russian).

Haken, H. (Ed.), 1985. Synergetics (translated from English). Mir, Moscow, Russia (in Russian).

Haken, H., 1988. Information and Self-organization: A Macroscopic Approach to Complex Systems. Springer-Verlag, Berlin, Germany.

Hradgovsky, R., 1977. Patent USA 4082626. Process for forming a silicate Coatings on metals/Hradgovsky R. -Заявл.17.12.76, опубл. 04.04.78; МКИ C25d 11/02, 11/06.

Ivanov, V.T., Gusev, V.G., Fokin, A.N., 1986. Optimization of Electric Fields, Control and Automation of Galvanic Technology. Mashinostroenie, Moscow, Russia (in Russian).

Ivanova, V.S., Balankin, A.S., Bunin, I.Z., Oksogoev, A.A., 1994. Synergetics and Fractals in Materials Science. Nauka, Moscow, Russia (in Russian).

Katsheev, V.N., 1978. Processes in a Zone of Metals Friction Contact. Mashinostroenie, Moscow, Russia (in Russian).

Khebda, M., Chichinadze, A.V., Brown, E.D., Evdokimov, Y.A., Luzhnov, Y.M., et al., 1989. Handbook on Tribotechnic. Theoretical Basis. Mashinostroenie, Moscow, Russia (in Russian).

Khrutshev, M.M., Babichev, M.A., 1970. Abrasive Outwearing. Nauka, Moscow, Russia (in Russian).

Kurze, P., Krysmann, W., Marx, G., 1980. Verfahren zur Erzeugung α-Al2O3-haltiger Schichten auf Aluminiummetallen. Declared. 07.03.79, Published. 18.06.80; МКИ C25d 11/04. Patent DDR 142360.

Kurze, P., Marx, G., Krysmann, W., Ditrich K., 1980. Verfahren zur Oberflaechenbehandlung von Titanium und Legierungen. Declared. 23.09.80, Published. 21.07.82; МКИ C25 d 9/06. Patent DDR 156003.

Malyshev, V.N., 1995. Mikrolichtbogen-oxidation – Ein neuartiges Verfahren zur Verfestigung von Aluminiumoberflaechen. Metalloberflaeche. 8, 606–608.

Malyshev, V.N., 1996. Features of coating formation by anodic-cathode microarc oxidation method. Prot. Met. 32 (6), 662–667 (in Russian).

Malyshev, V.N., 2006. Neue Anwendungsmoeglichkeiten fuer aluminium. Metalloberflaeche. 1–2, 28–29.

Malyshev, V.N., 2008. Effective electrochemical transformation method of valve metals surface layer into high-tensile ceramic coating. In: Duncan, J.I., Klein, A.B. (Eds.), Chemical Reactions on Surfaces. Nova Science Publishers, Hauppauge, NY, pp. 211–262.

Malyshev, V.N., Golub, M.V., Kharlamenko, V.I., 1983. Investigation of composite materials relative wear resistance, VNIIOENG. Mach. Oil Equip. 4, 6–8 (in Russian).

Malyshev, V.N., Bulychev, S.I., Markov, G.A., Fedorov, V.A., Petrosjants, A.A., et al., 1985. Physical and mechanical characteristics and wear resistance of coatings formed by microarc oxidation method, Physika i Khimija Obrab. Materials. 1, 82–87 (in Russian).

Markov, G.A., Belevantsev, V.I., Terleeva, O.P., Shulepko, E.K., Slonova, A.I., 1992. Microarc oxidation. Vestn. MSTU Ser. Mashinostr. 1, 34–42 (in Russian).

Mayer, K., 1979. Physical and Chemical Crystallography (translated from German). Metallurgy, Moscow, Russia (in Russian).

Murat, M., 1988. Two-dimensional dielectric breakdown between the parallel lines. Fractals in Physics. Proceedings of the Sixth International Symposium on Fractals. Mir, Moscow, Russia, pp. 234–237 (in Russian).

Nicolis, G., Prigogine, I., 1984. Selforganization. Nonequilibrium Systems (translated from English). Mir, Moscow, Russia (in Russian).

Nicolis, G., Prigogine, I., 1989. Exploring Complexity: An Introduction. Freeman, New York.

Petrosyants, A.A., Malyshev, V.N., Fedorov, V.A., Markov, G.A., 1984. Trenie I Iznos 5: 350–354 (for English translation see J. Friction & Wear, ISSN 1068–2666) (in Russian).

Prigogine, I., Glensdorf, P., 1973. Thermodynamic Theory of Structure, Stability and Fluctuations (translated from English). Mir, Moscow, Russia (in Russian).

Rudnev, V.S., Gordienko, P.S., Orlova, T.I., 1990. About one formation mechanism of MAO-coating on aluminum alloy. Electronnaya Obrabotka Materialov. 3, 48–50 (in Russian).

Semenov, A.P., Kovsh, I.B., Petrova, I.M., 1992. Methods and Means of Details Machines Surfaces Hardening by Concentrated Flows of Energies. Nauka, Moscow, Russia (in Russian).

Smirnov, B.M., 1991. Physics of Fractal Clusters. Nauka, Moscow, Russia (in Russian).

Stafford, K.N., 1991. Coatings and Surface Treatment for Protection from Corrosion and Wear. Metallurgy, Moscow, Russia (in Russian).

Vinorgadov, Y.M., 1977. Wear-resistant Materials in Chemical Machine Industry: Handbook. Mashinostroenie, Leningrad, Russia (in Russian).

Yamada, M., Mita, I., 1982. Formation of θ-alumina by anodic oxidation of aluminium. Chem. Lett. 5, 759–762.

Yerokhin, A.L., Matthews, A., Nie, X., Leyland, A., et al., 1999. Plasma electrolysis for surface engineering. Surf. Coat. Technol. 122 (2–3), 73–93.

Yerokhin, A.L., Snizhko, L.O., Gurevina, N.L., Leyland, A., Pilkington, A., et al., 2003. Discharge characterization in plasma electrolytic oxidation of aluminium. J. Phys. D Appl. Phys. 36, 2110–2120.

Young, L., 1967. Anode Oxidation Films. Energy, Leningrad, Russia (in Russian).

Current trends in molecular functional monolayers

13

Prashant K. Sarswat, Amarchand Sathyapalan, and Michael L. Free
University of Utah, USA

13.1 Introduction

Molecular monolayers or self-assembled monolayers (SAMs) are spontaneously formed on surfaces such as gold or silicon by adsorption of surfactants (Ulman, 1996). Figure 13.1 shows a typical SAM on gold substrate. The field of SAMs has seen high growth rates over past 30 years (Ulman, 1996). The field of SAMs came into existence more than 50 years ago, when an article was published discussing preparation of a monolayer on a metal surface by adsorption. Later, chlorosilane derivatives were applied to hydrophobize glass (Ulman, 1996). In addition, SAMs of alkanethiolates were also prepared by adsorption of di-n-alkyl disulfides on gold surfaces from dilute solutions.

Monolayers can be obtained by various means. For example, surfactant molecules contain a polar end and a non-polar end; they can form a monolayer at the junction between a polar substance and a nonpolar one. Such surfactant monolayers can also be found on the surface of soapy water, paints, oil droplets, and milk. It was observed that success of SAMs is much better when a crystalline gold substrate was used as well as when alkyl tri chloro-silane is discarded.

Molecular monolayers have unique insulating behavior, dielectric properties, and tunable thickness. They are also solution processable. These monolayers can be self-assembled monolayers, which can change into multilayers or even nanodielectrics, and are formed spontaneously or intermittently on various surfaces by adsorption, and they can result in ordered domains. These unique characteristics take place due to several important chemical processes including chemisorption, electrostatic interaction, hydrogen bonding, metal-coordination, and covalent bonding. Thus, the concept of SAMs with specific chemical characteristics accommodates the requirement of chemical functionalities for specific uses and has emerged as a promising platform for versatile applications. Such applications include nanoelectromechanical systems (NEMS) and microelectromechanical systems (MEMS) devices, chemical and electrical biosensors, biomedical and photovoltaics (Srinivasan et al., 1998).

Functional monolayers on gold, silicon, diamond, indium tin oxide (ITO), quartz, nanoparticles, and quantum dots are feasible with covalent and non-covalent interactions, selective activation, layer-by-layer deposition, and weak Van der Waals forces. Although many such SAMs systems have been investigated and reported, the SAMs of alkanethiolates on gold substrate are one of the most studied systems.

Anti-Abrasive Nanocoatings.

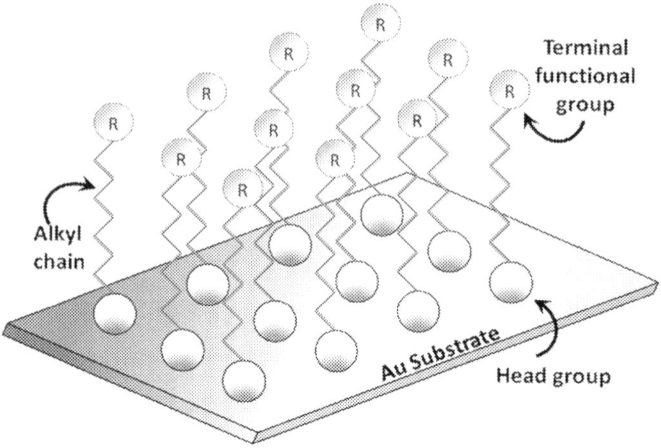

Figure 13.1 A schematic diagram showing a typical SAM.

The ability to adjust both head and tail ends of the constituent molecules in SAMs makes them versatile systems for a more fundamental understanding of phenomena which can be changed by forces such as intermolecular, molecular-substrates and molecule-solvent interactions. In many cases molecules of monolayer do not strongly bond with substrates. A typical example of this case is two-dimensional supramolecular networks of perylene-tetracarboxylicacid-dianhydride on gold substrates (Witte and Wöll, 2004). However, SAMs containing a head group of thiols, silanes, and phosphonates have strong binding affinity with substrates. Most of the time, SAMs are fabricated by the chemisorption of head groups onto a substrate from either the vapor or liquid phase and subsequent slow organization of a functional tail end is allowed. The adsorption process starts with a small molecular density on the surface. Over time, three-dimensional crystalline or semicrystalline structures on the substrate surface start forming.

Monolayer formation on silicon with Si-C bonding (without an intervening oxide) was introduced two decades ago (Linford and Chidsey, 1993). Such an invention is beneficial for better performing robust devices using flexible organic and supramolecular assemblies. The structural details of monolayers, molecular design–property relationships, duration of stable layer formation, and intramolecular rearrangement are some of the important key points to understand. These concepts will be discussed in this chapter. A concise summary of monolayer types and their applications is presented in Table 13.1.

13.2 Steps for self-assembly

Self-assembled monolayers are one of the important nanofabrication techniques. Often, SAMs follow a bottom-up fabrication approach. Most of the time adsorption

Table 13.1 **Different types of monolayers and their applications**

Substrate	Parent molecule	Film characteristics	Applications	References
AgO/Al_2O_3	Fatty acids	High ordering	Sensors, electronic devices	Ulman (1996)
Silica/gold	Methyl 2,3-(trichlorosilyl) tricosanoate, -hydroxyalkanethiolate monolayers	Thick film formation	Optical and molecular electronics	Ulman (1996)
Polycrystalline silicon microstructures	Octadecyltrichlorosilane	Thermal stability	MEMs	Srinivasan et al. (1998)
Silica	2H,2H-perfluorode-cyltrichlorosilane	Thermal stability	MEMs	Srinivasan et al. (1998)
Si	Planar molecules, Acenes, phtalocyanines	Large carrier mobility	Organic electronics	Witte and Wöll (2004)
Si/gold	Diacyl peroxide	C-Si covalent bonding	Sensor, electronics	Linford and Chidsey (1993)
Si	Dichlorodimethylsilane	Anti-stiction nature	MEMs	Ashurst et al. (2001)
Plastic/ silica nanoparticles	Poly (allylamine hydrochloride), aminopropyltrimethoxylsilane	High adhesion	Abrasion resistant coating	Chunder et al. (2009)
Si	Porphyrins	Thermal stability	Molecular memory	Liu et al. (2003)
Gold	Meso -Pyridyl Porphyrin Monolayer	Memory	Electronics	Sathyapalan et al. (2005)
Gold/graphene	Thiols and mixed molecules	High affinity	Molecular electronics and medical diagnosis	Newton et al. (2013)
Polyester matrix	Human retinal pigment epithelium (RPE) stem cells	Long-term functionality	Age related macular degeneration: treatment	Stanzel et al. (2014)

kinetics are defined for such self-assembly processes (Vericat et al., 2010). For self-assembly of thiol on gold substrates (Au (111)), adsorption is done using solutions of thiols, dialkyldisulfides or dithiols of concentration of ~100 μM. The duration of adsorption depends on the nature of the molecule and the type of solvent. For short chain alkanethiols, adsorption generally needs longer times ~24 hours, whereas longer chain molecules require 2–12 hours for good quality SAM formation. Adsorption starts by adsorption of the thiols onto an Au (111) substrate with their molecular axis parallel to the surface. Over the course of time, surface coverage increases to near saturation and a two-dimensional phase transition occurs; this transition causes thiols to change from a horizontal to a vertical configuration so that their molecular axis is oriented 30° from the surface normal. A schematic diagram is shown in Figure 13.2 that represents generalized steps of monolayer formation using thiols or their derivatives. Validating atomic force microscopy (AFM) imaging was performed and associated plots suggest that, at the initial stage, the height of topographic features was ~5.0 Å, which changed to 10.0–100 Å, when continuous exposure of thiols was maintained (Xu et al., 1998). It is also believed that during the initial stage of growth, the physisorbed phase (in contact with clean Au) saturates quickly, and therefore the coverage of these molecules is independent of pressure and temperature change. It is also believed that these physisorbed molecules subsequently convert to the chemisorbed state, resulting in the horizontal configuration. It is believed that the coverage of the vertical phase increases at the expense of the horizontal phase as the level of coverage increases. It is worth knowing that the whole self-assembly process in SAMs formation is not a single process as shown in Figure 13.2. It is a process which involves several steps, such as bulk solution transport, surface adsorption, and two-dimensional organizations on the

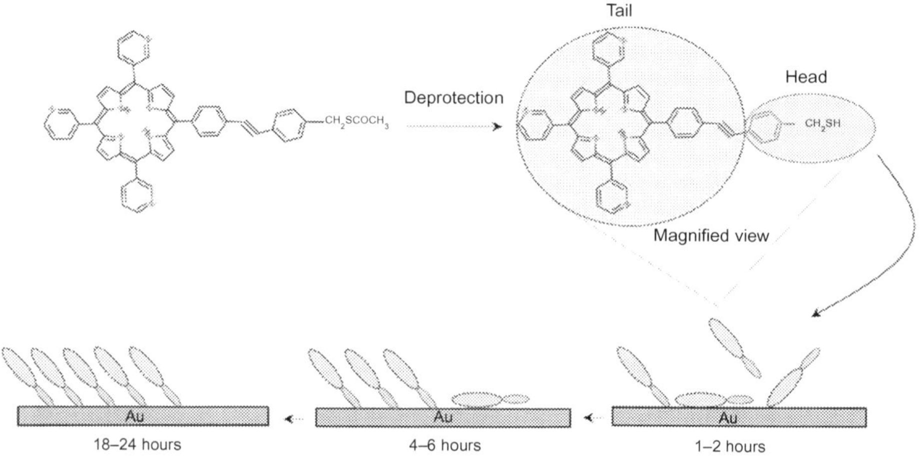

Figure 13.2 A schematic diagram showing different steps taking place during the self-assembly of alkanethiol on gold; a mesopyridyl porphyrine molecule is synthesized and used for making ordered SAMs via Au-S bonding (based on the work of Sathyapalan et al., 2005).

substrate of interest. Various nucleation and growth models are involved to explain SAMs growth. Schwartz et al. investigated the multi-stage mechanism based on phase diagram predictions, and one needs to consider all the phase transformation changes in the overall SAM formation. More details are discussed in the next section.

13.3 Mechanism

Stable SAMs are different from ordinary physisorbed surfactant monolayers, as one end of the molecule is generally designed to have specific interaction with a solid substrate (Schwartz, 2001). The specific interaction and associated bonding makes the SAMs more stable than surfactant monolayers that rely on typical Van der Waals and electrostatic forces for adsorption. Stable SAMs remain as intact monolayers of molecules even after removal of the substrate from the solution. SAM growth starts with solution phase transportation (diffusive and convective transport) of adsorbate molecules to the solid–liquid junction, followed by the adsorption of the molecules on the substrate. The adsorption dynamics are governed by diffusion controlled, adsorption controlled, or mixed (intermediate) controlled processes. During this state of self-assembly, the adsorption of surface-active molecules at the liquid–vapor interface is a very important step for the growth process (Schwartz, 2001). Hence, at this interface, surface tension is one of the important parameters determining growth. A summary of difference between surfactant monolayer and SAMs is shown in Table 13.2.

Two-dimensional (2D) molecular organization is important for SAMs' stability. Hence, molecular ordering is essential during adsorption and surface coverage growth. It is a stepwise process that follows an isothermal path through

Table 13.2 Key differences between surfactant monolayer and SAMs

Serial number	Self-assembled monolayer	Surfactant monolayer
1	Two-dimensional molecular organization is an important characteristic for SAM stability and function.	Primary function of surfactants is reduction of surface tension.
2	Head end of the molecule possesses a favorable and specific interaction with the solid substrate.	Amphiphilic molecules can adsorb at various interfaces such as liquid–liquid, liquid–vapor, and solid–liquid interfaces.
3	High coverage is obtained at substrate surface.	Relative low coverage is observed.

quasiequilibrium. The most common states in this process are a low-density 'vapor' phase (in this state isolated adsorbate molecules randomly deposit on the substrate), an intermediate-density phase involving disordered molecules and, finally, a high-density phase containing ordered-close packed molecules (which resembles a solid phase) that are oriented almost normal to the substrate. If the temperature is lower than the triple point, adsorbed molecules possess a tendency to form a dilute 2D-vapor phase (Schwartz, 2001). Under conditions of a relatively low surface concentration, there is a coexistence region between the vapor and the condensed (solid like) phase. It results in nucleation and growth of domains (islands) of solid like phase surrounded by isolated adsorbate molecules in the vapor phase. These domains will grow, over the course of time, to cover the entire substrate. However, the growth mechanisms of SAMs above the critical point temperature are different. When the vapor phase exceeds a certain surface concentration, domains of an intermediate, condensed phase will start nucleating and growth will proceed. Such a phase may be a disordered 2D-liquid or an ordered phase with relative lower density. The vapor phase is completely transformed to the low-density condensed phase as time proceeds. During the course of adsorption, a second transition also occurs which involves nucleation, growth, or coalescence of solid-phase domains, surrounded by the low-density condensed phase.

13.4 Characterization of SAMs

The surface topography, roughness, and thicknesses of SAMs can be measured using ellipsometry or AFM. Various spectroscopic techniques such as near edge X-ray absorption fine structure, second harmonic generation, sum frequency generation, surface-enhanced Raman scattering, as well as high-resolution electron energy loss spectroscopy are utilized to investigate the orientation of molecules and ordering in the film. The structures and grains of SAMs are generally examined using AFM and scanning tunneling microscopy (STM). Use of AFM is advantageous because it does not require SAMs to be conductive. These techniques are helpful to understand chemical functionality, surface electronics properties, structural features, surface stabilities, and frictional forces of SAMs. X-ray reflectivity and contact angle measurements techniques are also some of the techniques used for SAMs' quality evaluations. A larger angle formed using water drops on an alkylsilane monolayer indicates relatively more hydrophobic and closely packed film. Surface free energy can also be determined using contact angle data. Surface free energy can be correlated with average composition of the surface of the SAM. The kinetics of adsorption/desorption and time-dependent structure information can also be obtained using low energy ion scattering (LEIS) and time of flight direct recoil spectroscopy (TOFDRS). Some other techniques such as grazing incidence X-ray diffraction (GIXD) and helium atom reflectometry (HAR) are also utilized to investigate SAMs.

13.5 Use of SAMs for various applications

13.5.1 Antistiction coatings and MEMS devices

Micro- or nano-sized components, possessing a high surface area/volume ratio, are more prone to adhesion during physical contact (Baker and Li, 2006). Such adhesion is generally known as stiction. The most common stiction phenomena are in use in device- and release-related stiction. If stiction exceeds the restoring force, it will result in device failure. SAMs are used in many cases to minimize stiction in various miniature devices. For silicon-based MEMS devices, a monolayer of alkylsilane is utilized most of the time. It has been observed that wear reduces by a factor of two or more when SAMs are present (see Figure 13.3).

A comparison of wear properties of MEMS has been done for devices having SAMs coatings and devices without coatings (Baker and Li, 2006). It was observed that for most bare silicon surfaces as well as surfaces with SAMs coatings, nearly equal wear occurs below the normal load of 35 μN (Baker and Li, 2006). However, when the load exceeds 40 μN, the severity in wear was high for Si wafers. Polysilicon substrates show high wear resistance at higher load due to smaller grain size. 4,4-dihydroxybiphenyl SAMs monolayers also work well for polysilicon substrates and show similar reduction in wear (Baker and Li, 2006). It was observed that silicon nitride layers show better wear resistance compared to polysilicon. It was observed that at a normal load of ~60 μN, wear depth was approximately 13 times higher for the uncoated MEMS polysilicon compared to uncoated MEMS silicon nitride. This result may be due to high hardness of covalently bonded silicon nitride layers and a microstructure which makes it resistant to wear.

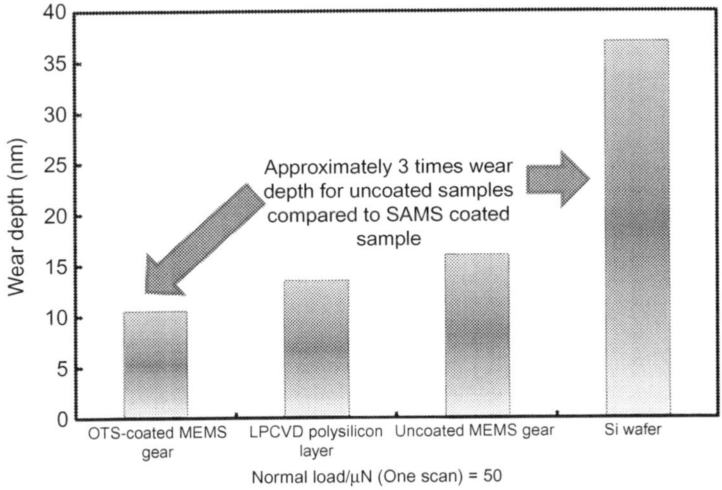

Figure 13.3 A schematic diagram showing normal load-wear depth relationships for various SAMs coated substrate.

Anti-stiction coatings based on primary alkene containing hydrogen terminated silicon was also reported (Ashurst et al., 2001). Such coatings do not produce hydrochloric acid during monolayer formation in contrast to other chlorosilane based coatings that do. This coating does not require an intervening oxide layer formation. Importantly, this coating is robust and more resistant to humidity. The large surface-area-to-volume ratios of surface and bulk micro-machined materials bring the role of stiction into the foreground, as adhesion to adjacent surfaces for such materials is a major failure mode for MEMS. Stiction is related to the unintentional adhesion of compliant microstructure surfaces when it is difficult for restoring forces to overcome Van der Waals, capillary, and electrostatic interfacial forces. It should be noted that a high static friction force initiates sliding.

Microstructure surfaces may come into contact through kinetic, static, or electrostatic forces. Permanent adhesion causes device failure due to in-use stiction. These adhesion problems can be mitigated by controlling surface topography and chemical composition of contacting surfaces. Octadecyltrichlorosilane (OTS) and perfluorodecyltrichlorosilane (FDTS) SAMs have been utilized for reducing adhesion in polycrystalline silicon microstructures (Ashurst et al., 2001). 1-octadecene monolayers and the OTS SAM-molecules contain a straight 18-carbon chain, and bind to the substrate at only one end. Hence, the films created with these particular molecules are directly comparable in terms of anti-stiction behavior.

Sacrificial layer etching is a common microstructure release procedure, performed at the beginning. This is generally conducted using concentrated hydrogen fluoride (HF) for a time that is dependent on the fabrication history of the structures being released. After this step, the silicon surfaces will be hydrogen terminated. However, it is essential to remove the etch products and solution. High-quality de-ionized water rinsing is generally applied for these purposes. However, during rinsing, the water may oxidize the hydrogen-terminated surface, which can cause damage to the integrity of the alkene coating that may be present. Thus, for hydrogen-terminated surface recovery, the HF is displaced with isopropyl alcohol (IPA), which does not oxidize a hydrogen-terminated silicon. IPA also serves as a useful intermediate between the organic 1-octadecene coating solution and aqueous HF. The structures are subsequently rinsed with the 1-octadecene. These structures can be transferred into a reactor containing the octadecene coating solution. This solution is prepared using hexadecane and 1-octadecene in a ratio of $\sim$9:1 v/v. Diluted octadecene can reduce the bulk polymerization of the di-functional alkene. After the structures are placed inside the sealed reactor, dry nitrogen is circulated throughout the heating process. The reactor is heated to 180°C for 30 minutes. Heating can generate free radicals; this is the step during which the monolayer forms. The reactor is then allowed to cool. The structures are then removed from the coating solution and through displacement rinsed in petroleum ether. The structures are then displaced into IPA and rinsed further. Next, the IPA is completely washed away with deionized (DI) water, while the liquid surface is aspirated. At this point, the structures can be removed from the water.

SAMs can also be deposited using vapor phase technique in order to improve the performance of MEMS devices as well as alleviate stiction failures. Vapor deposition offers many advantages of scalability and repeatability for manufacturing. One of the technical obstacles is a lack of chemical stability of these coatings in certain conditions such as application in liquid immersion for long durations. These coatings often peel off from the target surface after prolonged liquid exposure. This phenomenon can result in penetration of the polar molecules through the film and breakage of the film surface. Additionally, some SAM coatings do not withstand thermal exposure, such as in back-end semiconductor manufacturing where temperatures may exceed 250°C.

SAM coatings made using a 'sequential' vapor technique can be deposited on a variety of substrates and can withstand harsh conditions. This can be done using a 'layered' vapor deposition of two different films. First, an adhesion promoter layer is used to increase the binding sites for the subsequent functional SAM coating. Subsequently, the deposited SAM layer that forms is denser and is believed to be better attached to the surface depending on the substrate material. The resulting molecular coatings were observed to have improved stability to immersion application as a result of the improved surface coverage compared to SAMs. The vapor deposition chamber is evacuated and further purged with nitrogen. A temperature-controlled vapor delivery system is used to deliver SAM precursors and catalytic agents. An oxygen plasma treatment is used for surface conditioning prior to the organic layer deposition. The chamber walls and the delivery lines are mildly heated to eliminate condensation and residual vapor contamination. The reaction pressure is maintained in the range of 0.1 to 50 Torr.

13.5.2 Anti-reflection coatings

The important requirements for anti-reflection (AR) coatings are their strong adhesion to the substrate, adequate refractive index, abrasion resistance and precise thickness (Chunder et al., 2009). The most commonly used AR coating is magnesium fluoride. However, such coatings reduce reflectance by only ~1.5% to 4% when applied to plastic substrates. Hence, an approach such as formation of a layer of controlled porosity is adopted. Such an approach can reduce the refractive index of a layer by up to ~1.25. Other approaches include vacuum coating and chemical vapor deposition (CVD) coating techniques. A layer-by-layer adsorption of material method is another way to create conformal thin film coatings of adequate physical and chemical properties. This approach is based on interaction forces such as hydrogen bonds, acid-base interaction, and electrostatic attraction. There are a number of options for making AR coatings, such as dipping, sol-gel spin coating, or spraying techniques. AR coatings can also be made on polycarbonate (PC) substrates using layer-by-layer spraying of PAH/silica nanocrystals. The functionalization of PC surfaces and sol-gel treatments help to improve the mechanical strength of the film, and the corresponding reflectance reduces from 5% to 0.3%.

13.5.3 Injection molding industry

The injection molding industry generally uses prototype molds and mold inserts from melt spun aluminum for applications such as optics, photonics, microfluidics, and automobile industry parts such as connecting rods, cylinder sleeves, pistons, and valve retainers (Cech and Taboryski, 2012; Öveçoğlu et al., 2003). The use of aluminum saves cost and time. Most of the conventional mold coatings with thicknesses of a few microns can eliminate small features; hence, functional monolayer coatings on silicon/SiO_2 lithographic stamps are widely used.

Such coatings have antistiction properties as well as nanostructure reproducibility. A fluorinated trichloro-silane based thin coating was reported on aluminum or its alloys grown by molecular vapor deposition. It was observed that such coatings are sub-nanometer monolayers, which are controllable, covalently bonded, durable, affordable, and scalable to production. Such coatings are very suitable for rapid mold prototyping as well as mold geometry testing.

13.5.4 Organic monolayers and transistors

Organic monolayers and transistors, based on materials such as silicon, gold, and ITO have gained attention in recent years and undergone significant advancement (Salinas and Halik, 2013). In the transistor structure, tuning of the gate voltage, and consequently modulation of the charge density in the semiconductor channel, can also cause alteration of the electrical conductivity of the charge carrier channel. This modulation can be achieved by incorporating tunable molecular architecture on the surface, where the utility of SAMs plays a vital role. SAMs as part of a thin hybrid dielectric are potentially useful, and it is helpful for realization of low-power consuming, flexible, stable, organic thin-film transistors. SAMs of molecules with C_{60} head groups were grown on a thin aluminum oxide (AlO_x) dielectric in order to study the photoinduced charge transfer at low voltage in pentacene as well as α,ω-dihexylsexithiophene (DH6T) transistor devices. Fast charge transfer between an electron donor and acceptor and trapping of charges at the dielectric–semiconductor interface are key processes governing the optical response. SAMs' ability to alter the charge carrier density is useful in lowering and tuning the voltages required for optical switching of the transistors.

13.5.5 Monolayers on silicon with controlled levels of functionality

Patterned and functionalized silicon surfaces with organics and inorganics also have gained attention. Significant advances have been achieved in the preparation of patterned, scribed silicon surfaces with alkenes, amines, and epoxides. It was observed that epoxide reaction with scribed silicon produces monolayers of greater efficiency compared to 1-alkenes, 1-alkynes, 1-halo- and α,ω-dihalo- alkanes, and alcohols (Lua et al., 2003). Use of the various functionalization techniques enables the modification or alteration of silicon-based structures and devices. The monolayers of redox active porphyrin molecules attached to silicon have shown charge storage capabilities, which can be potentially useful for molecular scale memory applications (Liu et al., 2003). In sensing applications, Si microcantilevers have been treated by robust monolayers with quarternary ammonium tail ends, for sensing of chromate ions (Boiadjiev et al., 2005).

13.5.6 Modification of electronic properties of carbon nanotube

Electronic properties of carbon nanotube network transistors can be permanently modified by applying SAMs to the devices (Vosgueritchian et al., 2010). It is not easy to grow carbon nanotube (CNT) films onto most surfaces from a solution. This problem was overcome by using a 'dry' technique to perform such a transfer onto surfaces containing various SAMs. The SAMs modify the surface. Various SAMs containing a terminal group such as n-octadecyltrichlorosilane (OTS), 11-cyanoundecyltrimethoxysilane (CTS), and 11-bromoundecyltrimethoxysilane (BTS) are used. Electron withdrawing or donating properties of the molecules are responsible for charge transfer between the CNTs and functional groups.

13.6 Self-assembled monolayers on gold substrates

Ordered monolayers on gold substrates have received great scientific attention due to the strong Au–S chemical bonding and associated stability of monolayers on surfaces (Badia et al., 1997). Thiols are generally utilized for SAMs preparation over a noble substrate such as gold. It is essential to note that sulfur has a strong affinity to noble metals. Use of noble metal such as gold is advantageous because gold is relatively inert and biocompatible (Badia et al., 1997). Gold is also easy to pattern using various lithography techniques. It can be prepared by immersing a clean Au substrate into a solution of thiol. Gold substrate is generally prepared by e-beam deposition on the glass substrate. These SAMs can be useful as they can detect and immobilize molecules such as proteins and DNA. Use of different functional groups allows interfacial charge modification, which can act as an adaptor for material release and immobilization through electrostatic force. Most of the suggested bonding modes are gold thiolates (RS^-) and dialkyl disulfides. It was recently concluded by some researchers that chemisorbed species on gold nanocrystals are probably thiolates, based on NMR data.

Monolayers on gold surfaces with the thiolated molecules form well-ordered SAMs. The tail end having functional groups which possess the ability to bind other molecules and form derivitized functional monolayers as needed. Figure 13.4 shows typical functionalized SAMs on a gold substrate. A similar functionalized imine molecule with a tri phenyl amine core was used for photovoltaic applications (Sarswat et al., 2013). Multilayers can form onto surfaces by chemical interactions between molecules or by electrostatic interactions.

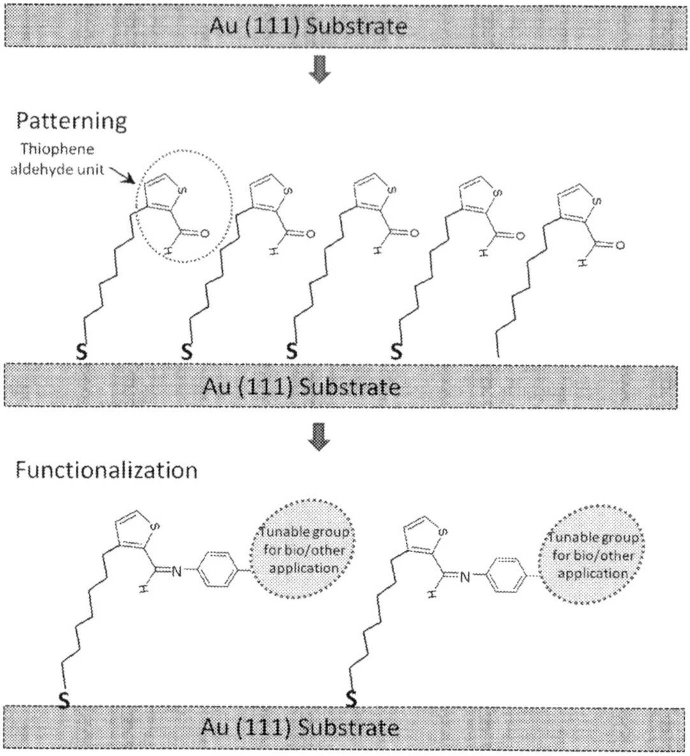

Figure 13.4 SAMs with Au-S bonding and its functionalization.

Polyelectrolyte multilayers (PEMS) are tunable and being used to manipulate topographical properties for biomaterials, tissue engineering, sequential assembly of nanocomposite materials, drug delivery, and nanomedicine. Electrostatic interactions play an important role in PEMS. Thus, by varying deposition parameters such as pH of solution, ionic strength, number of layers, and post-assembly modifications, multilayer properties can be modulated. A quartz/glass substrate of negatively charged surface can be modified with a positively charged poly-cation (e.g. diallyl dimethyl ammonium chloride (DADMAC)) and positively charged poly (sodium 4-styrene sulfonate (PSS)), followed by incorporation of required molecules, for specific applications.

13.7 Si-C monolayer formation and C-C bonding

Tribological properties of silicon are not very adequate (Bhushan, 2002). Techniques such as ion implantation and oxidation improve the tribological properties of silicon. It was observed that SiC exhibits significantly better tribological properties than conventional silicon-based materials and is a good candidate for MEMS/NEMS devices. Another advantage of SiC is its utility as a high temperature microsensor and microactuator and in other high-power devices. The modification of silicon to form required interfaces is typically accomplished via silane chemistry on oxidized silicon surfaces. The formation of monolayers on silicon without an intervening oxide layer is of great interest for many electronic and semiconductor applications.

Silica can be treated with sulfuric acid and hydrogen peroxide in a ratio of ~50:50 or 70:30 or other combinations (usually called Piranha cleaning) to create Si-O-H terminated layers. Subsequent ammonium fluoride treatment will generate hydrogen terminated surfaces as shown in the following equations:

$$SiO_2 \xrightarrow{H_2SO_4/H_2O_2} \text{Si-OH}$$

$$SiO_2 \xrightarrow{20\%NH_4F} \text{Si-H}$$

Oxide-free monolayers will be formed and strong Si-C covalent bond formation will take place due to the formation of atomically flat Si-H terminated surfaces.

Based on the Pauling electronegativity scale and the low electronegativity of Si with partially oxidized, H-terminated surfaces will create charge distribution and consequent hydrosilation (Figure 13.5). It explains the formation of a Si-C monolayer without an intervening oxide layer.

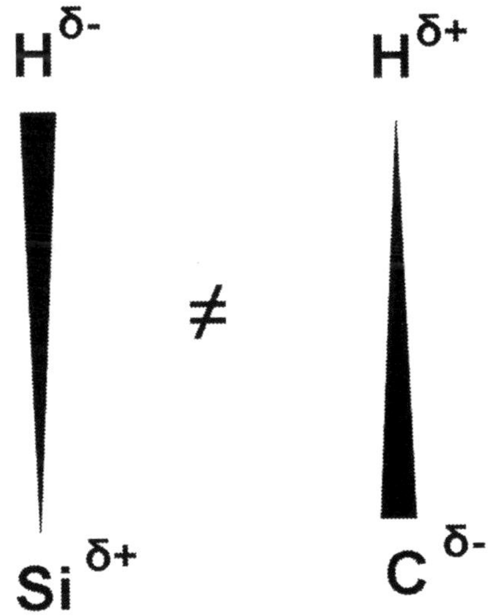

Figure 13.5 A schematic representation of Si-H charge distribution.

13.8 Supramolecular assembly on surface–host-guest interactions and other non-covalent bonding

Monolayer formation of supramolecules on surfaces usually occurs by weak Van der Waal forces due to host-guest chemistry. This non-covalent bonding is important in biological processes, where large molecules need to bind specifically to one another. Calixarenes, crown ethers, rotaxanes, porphyrins, and other proteins may be arranged via these non-covalent bonding or hydrophobic interactions, hydrogen bonding, and ionic bonding. However, though these types of bonding usually take place in monolayer formation with large molecules, modification of molecules like porphyrin (see Figure 13.2) may result in stronger binding (Sathyapalan et al., 2005). Among the supramolecules for monolayer formation, porphyrin-based molecules gained great attention due to their flexibility to adopt conformational changes and, hence, they are potentially useful for vast applications such as charge storage, light harvesting, electronic and photonic devices. Thiol and dithiol linkers attached to porphyrins and mesopyridyl porphyrins have been used for conductance switches in nanopores, and they exhibit tunneling behavior in junctions. The advantages of forming all these functional monolayers are their adequate strength and high-level molecular definition (Sathyapalan et al., 2005).

The self-assembled monolayers of large supramolecular structures in water were also investigated; it is one of the approaches for systems able to interact with biomolecules such as oligonucleotides, proteins, and cells (Pieters et al., 2013). A monolayer-protected gold cluster (Au MPC) with attractive intrinsic optical and electronic properties, when combined with high biocompatible molecules, becomes very useful for applications such as nanomedicine and diagnostics. The block copolymer self-assembly method is useful to create templates with nanoscale patterns. Such patterning can be achieved using a Langmuir-Blodgett (LB) technique/supramolecular approach.

13.9 Self-assembled monolayers on other surfaces such as titania nanotubes

Photocatalytic properties such as remote degradation are of specific relevance to SAMs attached to solid surfaces such as TiO_2. Good self-assembled monolayers can be formed on titanium nano-tubes for various applications, including sensors for volatile organic chemicals and explosive sensors (Paz, 2011). The same sensor platform modification can lead to reliable photoluminiscent and electrochemical sensor platforms. TiO_2 is among those oxides, which can form surface hydroxyl groups, and hence SAM formation on titania is feasible. It is interesting to note that the bond length of Ti-O is not very different from Si-O. Hence a similar compactness is expected. Although there are some similarities, there is a difference in electronegativity of Ti and Si. A typical illustration of such a monolayer is shown in Figure 13.6. Here the piranha treatment was applied to titania. A zinc or cobalt complexed molecule, amine(3-aminopropyl triethoxy silane)-hydroxyl aryl

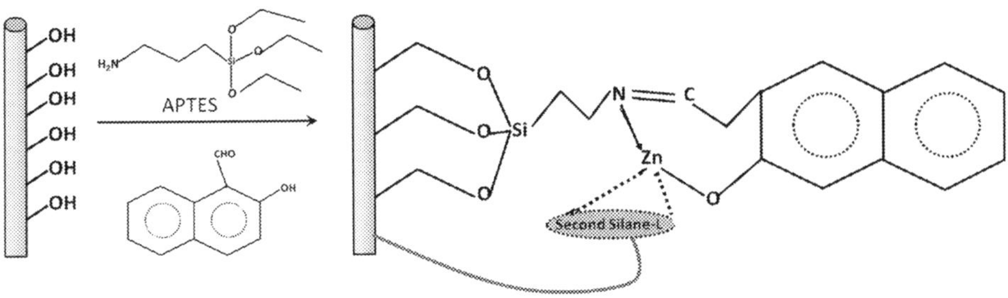

Figure 13.6 SAM formation on titanium nanotube.

aldehyde (such as salicylaldehyde or hydroxynaphthaldehyde), is immobilized onto a Ti-O-H surface (Sathyapalan et al., 2012). These successfully functionalized metal complexes on titania nanotube substrates showed good luminescent or electrochemical applications such as sensors.

13.10 Chemical and electrical biosensors

The above discussed monolayers are viable and suitable candidates for chemical and biosensors. Adequate linkers can be attached to the surface by chemisorption of carboxyl or amino derivatives such as dihydrolipoic acid or cysteamine. Molecular wires, which are linear molecules with conjugated π bonds, can be grown on gold substrates by chemisorption of thiols. π electrons can transport along the molecule, which enables electrical contact between redox species and metal within the sensing layer (Banica, 2012).

Recent advances in the areas of monolayers and devices and the need for high-performance integrated nano-multi-biosensing lead to the quest for molecular level detection. Such advancement is also due to the development of more sophisticated instrumentation techniques such as scanning probe microscopy and surface plasmon resonance. Enzyme immunosensors and DNA sequencing are still required to achieve the goal in bio-applications (Sathyapalan and Azar, 2012). Typical DNA nanopore sequencing is shown in Figure 13.7. While creating 1 nm size nanopores, the use of critical knowledge in monolayer formation is inevitable. The responses related to the passage of gold or single stranded DNA need to be noise free, as the signal to noise ratio in such devices may be low. The feedback from such devices/sensors amplifies the translocation and the passage (slowing down or speeding up) increases the displacement current and hence affects the sensor characteristics.

13.11 Quality improvement

There are various monolayer defects observed due to intrinsic as well as external factors. The cleanliness of the substrate prior to layer growth of a monolayer is

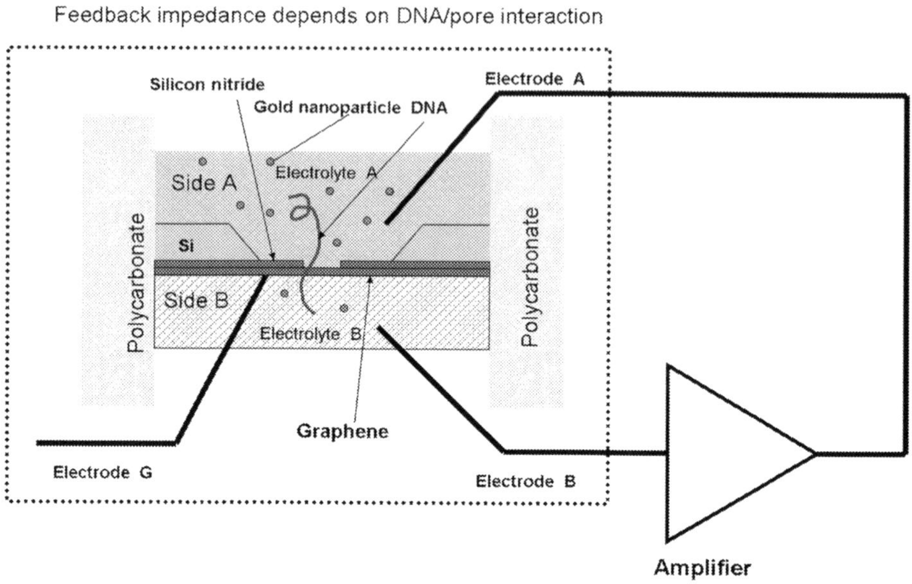

Figure 13.7 A schematic representation of a nanopore device in the feedback of a differential amplifier to affect the translocation of the nanoparticles.

very essential. Methods of preparation as well as purity of absorbates are also equally important for high-quality SAM growth. Proper control of the solvent and of the self-assembly conditions can lead to high-quality SAMs with better chain ordering. For example, self-assembly from hexane solutions under dark conditions and in the absence of oxygen yields dense and well-ordered SAMs of dithiols in vertical orientation.

13.12 Conclusions

In this chapter the use of the self-assembly concept on various surfaces as a promising platform for versatile applications such as devices, sensors, biochemical and nano-sensors is discussed and demonstrated. Appropriate self-assembled films can be achieved by selecting proper organic molecules with flexible tunability for controlled surface engineering. The versatile self-assembly processes and their basic understanding with suitable examples and illustrations are discussed. In addition, structural details of monolayers, molecular design–property relationships, duration of stable layer formation, and intramolecular rearrangement events with related examples and figures are presented. Some systematic ways of characterization of these self-assembled monolayers, current trends of using polymers, including immobilizing biomolecules for sensors, DNA sequencing, anti-adhesion coatings, and patterned SAMs, as well as device applications are also discussed in this chapter.

References

Ashurst, W.R., Yau, C., Carraro, C., Maboudian, R., Dugger, M.T., 2001. Dichlorodimethylsilane as an anti-stiction monolayer for MEMS: a comparison to the octadecyltrichlorosilane self-assembled monolayer. J. Microelectromech. Syst. 10 (1), 41–49.

Badia, A., Demers, L., Dickinson, L., Morin, F.G., Lennox, R.B., et al., 1997. Gold – sulfur interactions in alkylthiol self-assembled monolayers formed on gold nanoparticles studied by solid-state NMR. J. Am. Chem. Soc. 119 (45), 11104–11105.

Baker, M.A., Li, J., 2006. The influence of an OTS self-assembled monolayer on the wear-resistant properties of polysilicon based MEMS. Surf. Interface Anal. 38 (4), 863–867.

Banica, F.G., 2012. Chemical Sensors and Biosensors: Fundamentals and Applications. Wiley-Blackwell, Oxford.

Bhushan, B., 2002. Introduction to Tribology. Wiley, London, New York.

Boiadjiev, V.I., Brown, G.M., Pinnaduwage, L.A., Goretzki, G., Bonnesen, P.V., et al., 2005. Photochemical hydrosilylation of 11-undecenyltriethylammonium bromide with hydrogen-terminated si surfaces for the development of robust microcantilever sensors for Cr(VI). Langmuir. 21 (4), 1139–1142.

Cech, J., Taboryski, R., 2012. Stability of FDTS monolayer coating on aluminum injection molding tools. Appl. Surf. Sci. 259, 538–541.

Chunder, A., Etcheverry, K., Wadsworth, S., Boreman, G.D., Zhai, L., 2009. Fabrication of anti-reflection coatings on plastics using the spraying layer-by-layer self-assembly technique. J. SID. 17 (4), 389–395.

Linford, M.R., Chidsey, E.D.C., 1993. Alkyl monolayers covalently bonded to silicon surfaces. J. Am. Chem. Soc. 115 (26), 12631–12632.

Liu, Z., Yasseri, A.A., Lindsey, J.S., Bocian, D.F., 2003. Molecular memories that survive silicon device processing and real-world operation. Science. 302 (5650), 1543–1545.

Lua, Y.-Y., Lee, M.V., Fillmore, W.J.J., Matheson, R., Sathyapalan, A., et al., 2003. Amine-reactive monolayers on scribed silicon with controlled levels of functionality: reaction of a bare silicon surface with mono- and diepoxides. Angew. Chem. Int. Ed. 42, 4046–4049.

Newton, L., Slater, T., Clarka, N., Vijayaraghavan, A., 2013. Self assembled monolayers (SAMs) on metallic surfaces (gold and graphene) for electronic applications. J. Mater. Chem. C. 1, 376–393.

Öveçoğlu, M.L., Ünlü, N., Eruslu, N., Genç, A., 2003. Characterization investigations of a melt-spun ternary Al–8Si–5.1Cu (in wt.%) alloy. Mater. Lett. 57 (21), 3296–3301.

Paz, Y., 2011. Self-assembled monolayers and titanium dioxide: from surface patterning to potential applications. Beilstein J. Nanotechnol. 2, 845–861.

Pieters, G., Pezzato, C., Prins, L.J., 2013. Controlling supramolecular complex formation on the surface of a monolayer-protected gold nanoparticle in water. Langmuir. 29 (24), 7180–7185.

Salinas, M., Halik, M., 2013. Photoactive self-assembled monolayers for optically switchable organic thin-film transistors. Appl. Phys. Lett. 102, 203301.

Sarswat, P.K., Sathyapalan, A., Zhu, Y., Free, M.I., 2013. Design, synthesis, and characterization of TPA-thiophene-based amide or imine functionalized molecule for potential optoelectronic devices. J. Theor. Appl. Phys. 7, 4.

Sathyapalan, A., Azar, M.T., 2012. Selectively modified DNA bases for potential nano-pore sequencing. paper presented at *Nano biosensing-NORM*, Boise, ID.

Sathyapalan, A., Lohani, A., Santra, S., Goyal, S., Ravikanth, M., et al., 2005. Preparation, characterization, and electrical properties of a self-assembled meso -pyridyl porphyrin monolayer on gold surfaces. Aust. J. Chem. 58, 810–816.

Sathyapalan, A., Mohanty, S., Free, M.L., Misra, M., 2012. Surface functionization and immobilization of titania nanotubes/silica surfaces with zinc or cobalt using simple organic molecules as sensors. Paper presented at the American Chemical Society Rocky Mountain Regional Meeting, CO.

Schwartz, D.K., 2001. Mechanisms and kinetics of self assembled monolayer formation. Annu. Rev. Phys. Chem. 52, 107–137.

Srinivasan, U., Houston, M.R., Howe, R.T., Maboudian, R., 1998. Alkyltrichlorosilane-based self-assembled monolayer films for stiction reduction in silicon micromachines. J. Microelectromech. Syst. 7 (2), 252–260.

Stanzel, B.V., Liu, Z., Somboonthanakij, S., Wongsawad, W., Brinken, R., et al., 2014. Human RPE stem cells grown into polarized RPE monolayers on a polyester matrix are maintained after grafting into rabbit subretinal space. Stem Cell Rep. 2 (1), 64–77.

Ulman, A., 1996. Formation and structure of self-assembled monolayers. Chem. Rev. 96, 1533–1554.

Vericat, C., Vela, M.E., Benitez, G., Carro, P., Salvarezza, R.C., 2010. Self-assembled monolayers of thiols and dithiols on gold: new challenges for a well-known system. Chem. Soc. Rev. 39, 1805–1834.

Vosgueritchian, M., LeMieux, M.C., Dodge, D., Bao, Z., 2010. Effect of surface chemistry on electronic properties of carbon nanotube network thin film transistors. ACS Nano. 4, 6137–6145.

Witte, G., Wöll, C., 2004. Growth of aromatic molecules on solid substrates for applications in organic electronics. J. Mater. Res. 19, 1889–1916.

Xu, S., Cruchon-Dupeyrat, S.J.N., Garno, J.C., Liu, G.Y., Jennings, J.K., et al., 1998. In situ studies of thiol self-assembly on gold from solution using atomic force microscopy. J. Chem. Phys. 108, 5002.

Surface engineered nanostructures on metallic biomedical materials for anti-abrasion

14

Kelvii Wei Guo
City University of Hong Kong

14.1 Introduction

An increasing demand for reliable metallic biomedical materials is being observed as a result of an aging population with higher quality of life expectations. Improved metallic biomaterials and the relevant manufacturing techniques are being introduced to meet this demand. Apart from the biocompatibility requirement, metallic biomedical materials also have to meet high standards regarding performance (functionality, reliability) and selection criteria (cost, manufacturing ability).

Up to now, the most often used metallic biomedical materials include stainless steel, cobalt-based alloys, and commercially pure titanium or titanium-based alloys. However, the implant metallic biomaterials, like all mechanical components, are subject to degradation and have a limited lifetime. A damaged implant requires successive operations to replace worn components and so intensive efforts are made to increase durability of metallic biomaterials used for implants.

For instance, a titanium alloy such as Ti6Al4V has been widely used in implantable devices for several decades, because of its attractive mechanical and corrosion characteristics (Khan et al., 1996). However, failure often occurs, due to wear in the contact areas (Okazaki and Gotoh, 2005; Khan et al., 1999). Release of metallic ions and wear fragments due to wear corrosion lead to allergy and to tissue inflammation. Cytotoxicity concerns are mainly related to the presence of vanadium, but also high concentrations of titanium and aluminum ions can be dangerous (McKay et al., 1996; Eisenbarth et al., 2002; Hanawa, 2002).

Many titanium alloys under the co-joint action of corrosion and sliding wear have been studied (Khan et al., 1999). Under these conditions, Ti6Al4V is found to undergo the strongest degradation among all the alloys tested, but according to different tests it results in the best combination of both corrosion and wear resistance (Khan et al., 1996). The wear corrosion mechanism consists in the removal of the surface passive film in the sliding contact and its subsequent regeneration (Dearnley et al., 2004). It should be stressed that the two mechanisms of degradation, that is wear and corrosion, do not proceed separately but depend on each other

Anti-Abrasive Nanocoatings.

in a complex way. Normally corrosion is accelerated by wear and, similarly, wear may be affected by corrosion phenomena (Khan et al., 1996; Khan et al., 1999; Barril et al., 2002; Ponthiaux et al., 2004). Corrosion was found to be stimulated by wear, because of galvanic coupling between the anodic wear track and the cathodic region outside (Hanawa, 2002). The effect of wear on the electrochemical behavior is marked and depends on frictional load, applied potential, and pH of the solution (Okazaki, 2002). Duisabeau et al., (2004). studied the tribocorrosion behavior of a Ti6Al4V/316L stainless steel pairs in a Ringer's solution, under gross slip conditions, and have concluded that the dissipated mechanical energy and fretting regimes are strongly affected by the presence of a corrosive lubricant, because of the electrochemical phenomena caused by the electrolyte. Many investigations on titanium and its alloys are carried out under fretting corrosion. Degradation is accompanied by depassivation and repassivation phenomena, which can be investigated by the electrochemical noise technique (Vieira et al., 2006). At anodic potentials, it is found that the electrochemical response of the alloy to tribocorrosion is critically affected by the prevailing fretting regime (stick, mixed, or slip regime) and significant wear and enhancement of the anodic current occurs only in the presence of slip (Barril et al., 2004). The applied potential markedly stimulates the wear process because of continuous oxidation of the surface layer formed by mechanical mixing of plastically deformed metal and oxide debris (Barril et al., 2005).

Recently, nanotechnology has been defined as "the creation of functional materials, devices and systems through control of matter on the nanometer length scale (1−100 nm), and exploitation of novel phenomena and properties (physical, chemical, and biological) at that length scale" (Ashby et al., 2009a). Nanotechnology involves materials that have nanosized topography or are composed of nanosized materials. These materials have a size range between 1 and 100 nm (10^{-9} m). Nanostructured materials also have enormous surface areas per unit weight or volume, so that vastly more surface area is available for interactions with other materials around them. That is useful because many important chemical and electrical reactions occur only at surfaces and are sensitive to the shape and texture of a surface as well as its chemical composition.

The novel properties of nanomaterials offer great promise to provide new technological breakthroughs. Nanotechnology has been explored for creating lighter and stronger materials, for cleaning contaminated groundwater, for replacing toxic chemicals in various applications, for enhancing solar cell efficiency, and for targeted cancer treatment. Nanotechnology is already used in hundreds of products across various industries such as electronics, healthcare, chemicals, cosmetics, materials, and energy (Ashby et al., 2009a,b,c; Kostoff et al., 2007; Pitkethly, 2004; Van Hove, 2009; Zäch et al., 2006).

Nanotechnology has become a powerful tool in modern materials science. It is able to incorporate biomimicry on the nanoscale into materials engineering. Research on nanotechnology/nanostructured biomaterials is also one of the hottest topics in the biomaterials field and in fact, "nanotechnology/nanostructured biomaterials" were also very important keywords at WBC (World Biomaterial Congress) 2008. Nanoscale materials are thought to interact with some proteins more effectively than conventional materials. For instance, osteoblast functions are mediated due to their similar size and altered energetics. Balasundaram and Webster (2006) suggested that nanophase materials may be used as alternative orthopedic implant materials because of their ability to mimic the dimensions of the constituents and components in natural bone-like proteins and hydroxyapatite (HAP). Nanofibers with a shape resembling HAP crystals in bones can influence the conformation of typical adhesive proteins such as fibronectin and vitronectin as well as osteoblast behavior (Stevens and George, 2005; Wilson et al., 2005). Some other nanostructures that spur enhanced response from osteoblasts and subsequently more efficient deposition of calcium containing minerals include nanometals, carbon nanofibers and nanotubes, nanopolymers, and nanocomposites of ceramics and polymers. A number of studies have been conducted to evaluate the nanomaterials chemistry and preliminary results suggest that increased bone cell functions may be independent of the bulk materials chemistry but rather rely on the degree of the nanostructured surface roughness (Webster and Ejiofor, 2004; Price et al., 2003; Webster and Smith, 2005; Manjubala et al., 2006).

Many nanostructures can be found in nature (Aizenberg, 2005). For example, the surface of lotus leaves is composed of micro- and nanoscale hierarchical structures composed of fine-branched nanostructures (c. 120 nm) on top of micropapillae (5–9 mm). Scanning electron microscopy (SEM) reveals that on the lotus leaf surface, every epidermal cell forms a micrometer-scale papilla and has a dense layer of epicuticular waxes superimposed on it. Each of the papillae consists of branch-like nanostructures. The synergistic effects between these special double-scale surface structures and hydrophobic cuticular waxes are thought to be the reason for super-hydrophobicity as indicated by a high water contact angle and low sliding angle. As another example, butterflies (Wagner et al., 1996) and cicadae (Lee et al., 2004) can keep themselves uncontaminated by removing dust particles, dew, or water droplets easily from their wings spontaneously. Such properties also originate from the special structures on their wings. The cicada wing is composed of aligned nanocolumns with diameters of about 70 nm and a column-to-column distance of about 90 nm. These surface microstructures bestow the wing with its self-cleaning property. The relevant references cited are listed in Table 14.1.

Nanostructures also exist in the human body. For example, the size of ions, DNA, proteins, and viruses in the human body ranges from several nanometers to submicrometers. Human bones are composed of nanosized organic and mineral phases agglomerating to form a large macrostructure. Proteins present in the extracellular matrix of bones are nanostructured comparable to collagen fibrils. Furthermore, tiny mineral particles composed of calcium phosphate, an important constituent of the bone matrix, is compositionally and structurally nanostructured.

It is well known that surface engineering (SE) has been applied to metallic biomedical materials in order to improve their mechanical, chemical, and physical properties, such as anti-abrasion, corrosion resistance, biocompatibility, and surface energy, and understand the long-term impact on human health.

For enhancing mechanical retention, engineered nanostrucutres on the surface are normally taken to increase the effective surface area on implant surfaces (such as both dental and orthopedic applications) to exhibit biological, mechanical, and morphological compatibilities to receiving vital hard/soft tissue, resulting in promoting osseointegration.

Table 14.1 Focus of the cited references

References	Focus
Balasundaram and Webster (2006)	Nanophase materials for orthopedic implant applications.
Stevens and George (2005)	Approaches to control cell behavior through the nanoscale engineering of materials surfaces.
Wilson et al. (2005)	Mediation of cell responses to biomaterials coated with adsorbed proteins.
Webster and Ejiofor (2004)	Synthesization, characteristics, and evaluation of osteoblast adhesion on nanophase metals (specifically, Ti, Ti6Al4V, and CoCrMo alloys).
Price et al. (2003)	Determine effects of select properties of novel carbon nanofibers and polycarbonate urethane/carbon nanofiber (PCU/CNF) composites (specifically, dimension, surface energy, and chemistry) on osteoblast, fibroblast, chondrocyte, and smooth muscle cell adhesion.
Webster and Smith (2005)	Efficacy of nanophase materials as implants - the first evidence of greater functions of osteoblasts on poly (lactic-*co*-glycolic acid) (PLGA) scaffolds containing nanophase compared to conventional titania.
Manjubala et al. (2006)	Nucleate the HAP crystals onto three-dimensional porous polymeric scaffolds with nanoapatite formation.
Aizenberg (2005)	Bio-inspired approach to controlled crystallization at the nanoscale.
Wagner et al. (1996)	Wettability and contaminability of insect wings as a function of their surface sculptures.
Lee et al. (2004)	Simple and highly reproducible method for fabricating well-defined nanostructured polymeric surfaces with aligned nanoembosses or nanofibers of controllable aspect ratios similarity with interesting natural biostructures such as the wing surface of *Cicada orni* and the leaf surface of *Lotus*.

Surface-engineered nanostructures play a crucial role in anti-abrasion because (1) the surface of a biomaterial is the only part in contact with the bioenvironment; (2) the surface region of a biomaterial is almost always different in morphology and composition from the bulk; (3) for biomaterials that do not release or leak biologically active or toxic substance, the characteristics of the surface governs the biological response (foreign material vs. host tissue); and (4) surface properties such as topography affect the mechanical stability of the implant–tissue interface.

Nanostructured surfaces can be produced on conventional biomaterials by surface modification. Depending on the ways surfaces are modified, nanofunctionalization techniques can be categorized into two groups: nanocoating and nanofilm deposition, and in situ surface nanofunctionalization. These two types of techniques are often combined to produce surfaces with hybrid nanostructures such as a coating/film and a nanostructured zone.

In the past 10 years, much work has been performed on surface nanofunctionalization of titanium. Coatings and films with nanostructures such as nanograins, nanopores, nanotubes, nanopartices, nanowires, and nanorods have been synthesized on titanium and its alloys.

Moreover, such important surfaces can be fabricated in a favorable fashion to accommodate, facilitate, or promote more biofunctionality and bioactivity in mechanical, chemical, electrochemical, and thermal technologies, or any combination of these.

Technologies such as chemical/electrochemical action, mechanical indentations (sandblasting, shot peening, etc.), and physical or chemical depositing techniques (physical vapor deposition (PVD), chemical vapor deposition (CVD), plasma spraying, etc.) are normally used to increase or improve the effective surface area. In fact, such modifications can be divided into two categories: surface concave texturing and surface convex texturing. Surface concave textures can be achieved by chemical/electrochemical action and mechanical indentations (caused by sandblasting, shot peening, etc.). Surface convex textured surfaces can be formed by depositing certain types of particles by one of several physical or chemical depositing techniques (such as PVD, CVD, plasma spraying, etc.). If density and porosity of deposited particles can be appropriately controlled, a porous surface can be achieved, leading to successful bone ingrowth.

Biological survival, particularly longevity of biological adhesive joints, is often dependent on thin surface films. Surfaces and interfaces behave completely differently from bulk properties. The characteristics of a biomaterial surface govern the processes involved in biological response. Surface properties such as surface chemistry, surface energy, and surface morphology play a crucial role in biological interactions for four reasons: (1) the surface of a biomaterial is the only part in contact with the bioenvironment; (2) the surface region of a biomaterial is almost always different in morphology and composition from the bulk; (3) for biomaterials that do not release or leak a biologically active or toxic substance, the characteristics of the surface governs the biological response (foreign material vs. host tissue); and (4) some surface properties such as topography affect the mechanical stability of the implant–tissue interface. Like the interface, the surface has a certain characteristic thickness: (1) for the case when the interatomic reaction is dominant, such as wetting or adhesion, atoms within a depth of 100 nm (1000 Å) will be important; (2) for the case of the mechanical interaction, such as tribology and surface hardening, since the elasticity due to the surface contact and the plastically deformed layer will be a governing area, a thickness of about 0.1–10 μm will be important; and (3) for the case when mass transfer or corrosion is involved, the effective layer for preventing the diffusion will be within the range 1–100 μm.

14.2 Surface technologies on metallic biomedical materials for anti-abrasion

Tribology is the science dealing with the interaction of surfaces in tangential motion. Hence, tribology includes the nature of surfaces from both a chemical and physical point of view, including topography, the interaction of surfaces under load and the changes in the interaction when tangential motion is introduced.

Macroscopically, the interactions are manifested in the phenomena of friction and wear. Modification of the interaction through the interposition of liquid, gaseous, or solid films is known as the lubrication process. Hence, from a macroscopic point of view, tribology includes lubrication, friction, and wear. There are four major clinical reasons to remove the implants: (1) fracture; (2) infection; (3) wear; and (4) loosening. Among these, removal due to infection generally occurs in the relatively early stages after implantation, while the other three incidents typically increase gradually over the years, because they are somewhat related to biotribological reactions. Human joints are one example of natural joints, and show low wear and exceedingly low friction through efficient lubrication. Disease or accident can impair the function at a joint, and this can lead to the necessity for joint replacement.

Special tribological element problems arise in many biodevices used in various areas, for example, medical implants (prostheses, assists, and artificial organs). Medical implants, such as prostheses, should demonstrate very good tribological properties, as well as biological inertness. Their longevity depends significantly on the wear of the rubbing elements and must be extended as much as possible. The developments of total replacement hip, knee, ankle, elbow, shoulder, and hand joints (and the appropriate surgical techniques) has been the major success of orthopedic surgery, and would not have been possible without extensive in vitro and in vivo studies of the tribological problems, especially wear, associated with such artificial joints.

Most orthopedic implants are subjected to biotribological actions. Metal ions released from the implant surface are suspected to play some contributing role in the loosening of hip and knee prostheses, which are substantially subjected to biotribological environments.

Therefore, in order to achieve a structurization in the desired sub-100 nm scale range of the biomaterials (Webster et al., 1999), fabricated nanophase ceramics, and metals. In principle, ceramic or metal powders that show an average grain size distribution at the nanometer level are pressed and further sintered. Thus, surfaces of materials synthesized via this processing procedure can be used for in vitro response investigations. In the case of ceramics, alumina and titania powders have been used. By sintering the pressed powders in air for 2 hours at either 1000 or 600°C the ceramics were successfully produced. Depending on the temperature the starting average grain sizes of 23 nm and 32 nm can be enlarged to values up to 177 nm and 2.12 μm. In addition, the created surfaces of different grain sizes possess the same crystallinity and surface chemistry, altering only in degree of nanometer surface features. It is reported to significantly increase adhesion behavior on nanophase alumina surfaces compared to grain sizes in the range of 177 nm, and to cover slide glasses used as control. Comparable results could be found for nanophase grain sizes in case of titania surfaces. At the same time, an increase of adherent fibroblasts was reported on nanophase ceramics (Webster et al., 2001a). Supporting evidence of decreased fibroblast function on nanophase ceramics has been presented (Webster et al., 2001b; Webster et al., 2000; Mustafa et al., 2005).

The adhesion behavior of osteoblasts has been reported to show a comparable behavior in relation to the average grain size of biomedical alloys such as commercial pure titanium (cp-titanium; cp-Ti), Ti6Al4V and CoCrMo (Webster and Ejiofor, 2004). This elucidated that the adhesion behavior of cells in dependence of the average grain sizes of the used surfaces may be independent of any surface chemistry, as this topographical dependency could be found on either ceramics or metals and other materials (Price et al., 2003; Cai et al., 2006; Balasundaram and Webster, 2006; Popat et al., 2007).

Methods for fabricating surface topographies with high lateral resolution in the sub-100 nm range show limitations. Many of the presented approaches are not based on biomedical implant materials that can be used for hard tissue applications, or are not clinically relevant. Although this plays a major role in the case of transferring the identified dependencies or influencing factors of surface model topographies at the nanometer level to real biomedical material surfaces, the most important task is to have a material system that possesses a good adjustability of nanometer-scaled features in the sub-100 nm range in a highly reproducible manner. Moreover, most of the examples shown for structuring techniques are very time consuming, difficult to process and the generated surface structures are not sufficiently defined. The majority of the shown structures do not fit into the requirement for nanometer features in the sub-100 nm range as they are often much bigger than 100 nm. Although numerous investigations have been reported by using the presented surface structures and by using various cell types, up to now it has been somewhat difficult to compare the observations, as the investigations have proceeded on totally different material systems with nanofeatures of different sizes, making it hard to normalize the results. The desired successful adjustment of nanofeatures of different sizes in narrow alternations has not been described so far. A precise control of defined spacing between adhesive ligands of cells on interfaces at a length scale of 10–100 nm still remains a major challenge. Recently, two methods have been reported to possess the ability to structure nanofeatures at high and distinct lateral resolutions and in a highly reproducible and fairly simple manner. These techniques are either based on a non clinically relevant material system of gold nanodot arrays that are further coupled with cell adhesive sequences placed in distinct distances on wafers, or the growth of anodic nanoporous or nanotubular oxide structures on valve metals that can be tailored in size, for example, on biomedical titanium (Koegler et al., 2012; Md Jani et al., 2013).

The field of implantology steadily evolves as more is learned about the specific biological interrelations of the implant and its surroundings. Important factors from the viewpoint of SE include the impact of the surface chemistry, the topography at the micro- and nanometer level, physicochemical effects, and biological factors such as biochemically mediated cell differentiation, the unavoidable bacterial colonization of the implant, the biologic dimensions, and the histology of surrounding structures. Proper surface modification techniques not only retain the desired bulk attributes of biomedical materials, but also improve specific surface properties required by different clinical applications (Schwartz and Boyan, 1994; Puleo and Nanci, 1999).

Table 14.2 Surface technologies of metallic biomedical materials applications for anti-abrasion

Surface technology Chemical/electrochemical action	
Mechanical indentation	Sandblasting Shot peening
PVD CVD Plasma spraying	

The influence of surface roughness on the rate of osseointegration and biomechanical fixation of hard tissue implants has been identified as a key factor. Mainly surface topographies at the micron level were reported as important, and several surface modification techniques operating at this length scale were developed. In particular, the observation of more rapid and increased bone contact formation by micron-scale roughened surfaces manufactured through blasting and subsequent acid etching stimulated considerable activities. This observation also leads to the conclusion that hard tissue implants based on alloys, mainly titanium, were not only entirely bioinert or biocompatible, but proper surface conditioning could also influence protein adsorption, cellular activity, or tissue response and this in turn could be exploited to achieve a higher level of osseointegration. Morphological features at the micro level control the rate and quality of new tissue formation at the interface (Buser et al., 1991; Gotfredsen et al., 1990; Wennerberg, 1998; Deligianni et al., 2001; Degasne et al., 1999). Kieswetter et al. (1996) reported on the influence of surface roughness on titanium to affect the production of local factors involved in bone formation by osteoblasts, suggesting that the complement of autocrine and paracrine factors produced by cells at the bone–implant interface can be directed by altering the implant surface roughness to directly affect the type of interface that is formed at the implant site.

Optimal, mechanical interlocking of implants to the host tissue is required to achieve well-accepted and integrated implantations. Not only the geometric requirements and stress distribution factors of the host tissue guide the topographical demands on the implant surface but also the consideration that bone adapts to mechanical loading by osteocytes which act as mechanosensors (Hansson, 2006; Wong et al., 1995; Wennerberg et al., 2006). To meet the demands for an enhanced bone implant contact formation various methods are employed to create and establish such microstructural surface features. Such methods listed in Table 14.2 include chemical methods (such as acid-etching, anodization, micro arc oxidation), blasting, plasma-spraying, and so on.

14.2.1 Chemical/electrochemical action

Chemical or electrochemical action using acids, alkalis, and hydrogen peroxide involve chemical reactions at the interface between the materials and solution, which are considered enabling techniques from the viewpoint of the surface

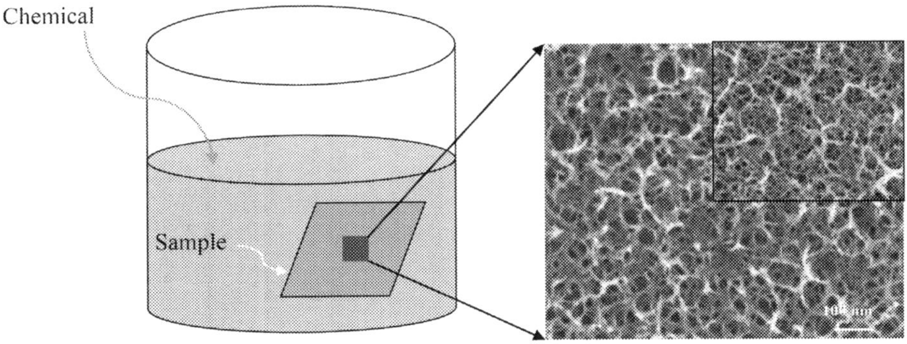

Figure 14.1 Schematic of chemical/electrochemical action for surface modification of biomaterials.

modification of biomaterials, especially metals, due to the simplicity and flexibility. The relevant diagram is shown in Figure 14.1. The inset SEM image is the higher magnification result of nanostructures on cp-Ti modified by electrochemical method. It shows that after chemical treatment, the nanostructures distribute on the cp-Ti surface uniformly resulted in better properties of succedent biomedical applications. It is well known that titanium and titanium (Ti) alloys are generally encapsulated by a native protective oxide film typically composed of three different layers, TiO (inner layer in contact with the metal), Ti_2O_3 (intermediate layer), and TiO_2 (outer layer) (McCafferty and Wightman, 1999). When titanium is immersed in an oxidizing solution, a chain of reactions occurs to form nanostructures on the surface (Variola et al., 2008). The suboxides, TiO and Ti_2O_3 are transformed into TiO_2 in the oxidizing medium while the etching solution penetrates the nanopits and reaches the underlying metal. Hence, passivation layers are naturally created. The native layered structure composed of TiO (in contact with the metal), Ti_2O_3 (intermediate), and TiO_2 (outer layer) is reestablished. The oxidation process increases the amount of TiO_2 to the extent that the underlying suboxides can no longer be detected but the layer organization is not altered.

There are several experimental results on chemical, electrochemical, thermal, and combinations of these with regard to altering the biomaterial surface to facilitate better surface chemical, mechanical, and biological reactions.

Zirconium has a good compatibility with human tissue and might thus be considered as a biomaterial for use in surgical implants (Okazaki et al., 1996; Davidson et al., 1994; Li et al., 2008). However, similar to titanium, its tribological properties might limit its application. For this reason several studies have been performed to improve the wear resistance of zirconium and titanium-based alloys. One possible approach is the formation of a nitride layer on the surface by thermal nitridation. Whereas in the case of titanium two nitrides can be formed – TiN and Ti_2N – only one nitride is known for the zirconium/nitrogen system (ZrN). Ti_2N has a narrower composition range than ZrN or TiN. As a result, the diffusion coefficient of nitrogen in Ti_2N is smaller and thus Ti_2N might act as a diffusion barrier, decreasing the nitridation rate. This suggests that the thermal formation of a nitride layer may be much easier on a Zr-based alloy. Preliminary studies using a pin-on-disc tribometer have shown a significant improvement in the mechanical properties of a zirconium alloy by thermal nitridation. The development of these wear-resistant coatings is hampered by the limited sensitivity of classical gravimetry to measure small wear losses. This leads to a need for long testing times. This problem could be overcome by using the thin-layer activation (TLA) technique, its benefits in the tribological testing of biomaterials having already been researched (Neumann et al., 1990).

However, ziconium tribological properties are rather poor. In order to overcome this drawback, Brugnoni et al. (1998) prepared a ZrN coating by thermal nitridation of zircalloy-2. Nitridation was performed by introducing the samples into the furnace containing 0.5 bar nitrogen at 850°C for 100 hours. Heating and cooling of the samples were performed in vacuum better than 2×10^{-5} mbar. The temperature was selected to be high, but still below the α/β phase transition, which might cause specimen distortion. The thickness of the ZrN layer formed was in the range of 5 μm. Results show that a ZrN coating had a very good wear resistance, indicating its potential use as a biomaterial for surgical implants. Also, TLA offers significant benefits for the evaluation of wear behavior of very resistant materials such as ZrN coatings. Of paramount importance is its high sensitivity, which is orders of magnitude higher than for gravimetry.

Nickel-titanium (NiTi) shape memory alloy (SMA) is well known for its unique physical and mechanical properties, such as shape memory effect, good strength, and superelasticity (Otsuka and Ren, 1999; Kasano and Morimitsu, 1997; Miyazaki, 1998; Ryhanen et al., 1999). In recent years, NiTi alloy has been used as a biomaterial because of its unique properties as well (Duerig et al., 1999; Yahia, 2000). However, the high content of Ni (about 50 at%) in biomedical NiTi SMA is of great health concern because it can trigger allergy and adverse reactions when its ions are released into the human body (McKay et al., 1996; Wever et al., 1998). Therefore, it is necessary to improve the corrosion resistance of the NiTi alloy and suppress the release of Ni ions by surface modification technique.

Zirconia (ZrO_2) is a ceramic and bioinert which shows good wear resistance, corrosion resistance and biocompatibility. In addition, it shows a high strength and toughness index; therefore, adding ZrO_2 into HAP coating (The mechanical strength of HA is too poor to be used in load-bearing applications.) is an effective method for improving the interfacial bonding strength of the matrix and coating (Fu et al., 2000; Wang et al., 2010).

Qiu et al. (2010) investigated the characterization and corrosion behavior of HAP/zirconia composite coating on NiTi fabricated by electrochemical deposition. The electrolyte solution contained 0.0167 M $Ca(NO_3)_2$, 0.01 M $NH_4H_2PO_4$, 0.1 M $NaNO_3$, 10 ml/L H_2O_2•$NaNO_3$ was added to improve the ionic strength of the electrolytes, and H_2O_2 was added to erase the evolution of H_2 gas and promote the deposition of dense coating. The pH of the electrolyte was adjusted to 6.0 at 25°C by ammonia water and nitric acid. For the fabrication of HAP/ZrO_2 composite coatings, 7 g/L ZrO_2 with a typical size of about 40–50 nm was added into the electrolyte. Electrodeposition was carried out in a cell consisting of NiTi as the anode and a graphite plate as the cathode. The deposition was carried out in galvanostatic mode and the current density was maintained at 0.5 mA/cm^2 for 40 minutes, keeping the temperature at 65°C. Pure HAP coating was designated as the control. The samples were washed with deionized water and dried at room temperature in air after deposition.

Results indicated that carbonate ions replaced phosphate ions into the HAP lattice in electrochemical deposition. The electrochemical and chemical reactions on the surface of the NiTi alloy were suggested as follows:

$$O_2 + 2H_2O + 4e- \rightarrow 4OH- \tag{14.1}$$

$$2H_2PO_4 - + 2e- \rightarrow 2HPO_4^{2-} + H_2\uparrow \tag{14.2}$$

$$2HPO_4^{2-} + 2e- \rightarrow 2PO_4^{3-} + H_2\uparrow \tag{14.3}$$

These reactions resulted in production of OH-; besides, the pH of the electrolyte had been adjusted to 6.0 before electrochemical deposition, so the pH around the surface of the NiTi alloy increased when a voltage was applied. The high concentration hydroxyl ions could promote the formation of HAP, as follows:

$$10Ca_2 + + 6PO_4^{3-} + 2OH \rightarrow Ca_{10}(PO_4)_6(OH)_2 \tag{14.4}$$

when zirconia was added to the electrolyte, it could combine with H_3O^+, and the zirconia is positively charged. ZrO_2 was deposited on the NiTi surface by electrophoretic deposition.

SEM results show that the pure HAP coating appeared as regular thin flake-flower-like crystals, and the HAP/ZrO_2 composite coating consisted of needle-flower-like crystals and some particles, the former being HAP and the latter ZrO_2. It was clearly seen that incorporation of ZrO_2 particles resulted in a change in the structure of the HAP. It indicated the presence of ZrO_2 particles prompted the HAP crystals to grain arranged in the [001] direction during electrochemical deposition. ZrO_2 particles dispersed uniformly between the HAP crystals, forming the dense coating.

Chen et al. (2012) performed the spark anodizing of titanium in a phosphoric acid solution producing a porous TiO_2 layer on its surface, and subsequent hot water treatment to modify the surface characteristics of the resultant oxide layer of commercially pure titanium. The titanium plate specimen and a platinum plate were connected to the anode and the cathode, respectively, which were 3 cm apart. The spark anodizing treatment was carried out at room temperature in 1 M phosphoric acid solution using a DC power supply and lasted for 2 minutes at a voltage of 200 V. A magnetic stirrer at a given rotation speed was employed to achieve a homogeneous electrolyte and to lift the bubbles generated during the treatment. After the treatment, anodized titanium plate specimens were gently washed with ultrapure water, and then dried in an oven at 40°C for 24 hours. For the subsequent hot water treatment, anodized titanium plate specimens were immersed in 60 ml of ultrapure water at 80°C for 48 hours, and then dried in an oven at 40°C for 24 hours.

Results show that for the TiO_2 layer before water treatment, a porous oxide surface with microprojections and craters can be observed, and a smooth oxide surface is around the crater. After water treatment, the surface morphology of the TiO_2 layer was significantly changed. A granular surface composed of numerous nanoparticles was formed on the water-treated oxide layer. Such a significant surface change was ascribed to the crystallization behavior of the TiO_2 layer during water treatment. It indicated that before water treatment, the TiO_2 layer consisted of amorphous or poorly crystallized oxides. However, after water treatment, crystalline anatase was formed with a high degree of crystallinity via a solid-state reaction. Generally, crystalline TiO_2 layers are preferable to amorphous ones for implants application because the latter easily dissolve in physiologic fluids. Furthermore, crystalline TiO_2 layers exhibit better apatite-forming ability than amorphous ones. In addition to the phase structure, the surface topography of biomaterials also plays an important role in implants application, especially in regulating cell response. Rougher biomaterial surfaces usually performed better bone fixation than smooth ones due to the promotion of osteoblast functions. Moreover, surfaces with nanoscale structures can enhance osteoblast adhesion greatly. The present hot water treatment of the TiO_2 layer resulted in a nanostructured surface, and in vitro biocompatibility study of oxide layers demonstrated better cell adhesion on the water-treated oxide layer than the untreated one. It elucidated that the significant fatigue damage was induced in the TiO_2 layer. After water treatment, the TiO_2 layer exhibited severe fatigue damage, and extensive cracking and delamination of the TiO_2 layer occurred at a lower maximum cyclic stress.

Recently, micro arc oxidation (MAO) followed by hydrothermal treatment (HT) were introduced for HA/TiO_2 coatings (Nie et al., 2000). MAO is one of the effective methods of forming porous, rough, and homogeneous oxide layers, which can be controlled by process parameters, on the surface. Furthermore, Ca and P elements, which are the main components of HAP, can be merged into the titania layer by changing the electrolyte. With this method, better adhesion of HA to the substrate was achieved along with the fact that improvement of the mechanical properties, corrosion and wear resistance, and biocompatibility were obtained by forming a HA/TiO_2 composite layer on the surface (Liu et al., 2005; Han et al., 2003).

Vangolu et al. (2011) studied the wear properties of micro arc oxidized and hydrothermally treated Ti6Al4V alloy in simulated body fluid (SBF) to improve the adhesion and the wear rate of the HA coatings. The MAO process was carried out in an electrolyte which consisted of 0.2 M $Ca(C_2H_3O_2)_2{\cdot}H_2O$ (calcium acetate monohydrate), 0.02 M $C_3H_7CaO_6P$.aq (calcium β-glycerophosphate), and distilled water. For the treatment, unipolar pulse voltage was applied by an AC pulse power supply while the specimens were the anode of the system and a stainless steel bath was the cathode. The system temperature was kept constant during the experiments via a circulation pump. The MAO was performed at a constant voltage, frequency, duty circle, and time of +450 V, 300 Hz, 6–6.5%, and 5 minutes, respectively. The coated samples were flushed with water after MAO treatment and then they were placed inside a 100 ml aqueous solution in an autoclave with the pH value of 11.0–11.5 for time intervals 5 hours and 10 hours and temperatures of 160°C, 190°C, and 220°C in order to form HAP. The pH was adjusted via NaOH addition. Wear experiments were performed in SBF solution at 37°C using a pin-on-disc tribotester. All wear experiments were conducted with 6 mm diameter tungsten carbide (WC) balls in sliding contact with a load of 1 N at a sliding velocity of 6.5 cm/s. Wear tests wear carried out for a sliding distance of 50 m.

Results show that a homogenous film layer which has a thickness of 10 μm was achieved after MAO and HT, and the surface morphology of the oxide layer after MAO treatment was rough and porous. The pores with miscellaneous dimensions, including micropores and submicropores, which are a natural result of MAO, are related to the frequency value of the process. After HT at various times and temperatures, HA crystals were observed on the surfaces of the substrates. HA crystals completely covered the surface. Even the columnar HA crystals grew in through the pores.

According to the wear results, it illustrated that the friction coefficients were decreased in every specimen after the surface treatments. The wear resistance of all specimens increased compared to that of Ti6Al4V. It is known that the wear mechanism of Ti6Al4V starts with oxidation and then adhesive and abrasive wear takes place (Budinski, 1991). Titanium has a native oxide layer on its surface because of its high affinity to the oxygen. If it is not stable and the applied normal and shear stresses are high enough to disrupt this layer, the abrasive effect takes place and the wear resistance decreases (Long and Rack, 1998). Deep wear tracks formed due to the abrasion of Ti6Al4V, and abrasive particles were observed on the inner surface of the wear tracks. The width of the wear track of the TiO_2 layer, which was formed by MAO, was narrow compared with the untreated Ti6Al4V. The wear mechanism for oxidized materials depends on the adhesion between the film layer and the substrate. So, the weak adhesion causes the disruption of the oxide layer. This layer may bond adhesively to the mating body or it also may modify a passive layer. The modified passive layer may be subjected to a continuous removal in the first situation. This continuous removal gradually depletes both the coating and the substrate and increases the roughness, causing a relatively high metal loss. Ultimately, the breakdown of the oxide layer leads to the potential for abrasive wear, where the hard oxide debris acts as third body abrasive components. The wear track width of HA/TiO_2 composite layers is almost close to that of the specimens coated only with a titania layer. MAO and HT allow homogenous HA crystals to form on the surface which improve the adhesion. Thus, the increased adhesion makes separation of the coating from the substrate difficult and the wear resistance increases. Consequently, the abrasive effect was relatively low on the surfaces which contained HA.

14.2.2 Sandblasting

Usually blasting is explained as the use of abrasive particles such as hard ceramics against another material under high pressure in order to make it smoother, remove surface contaminants, or to roughen the surface (Hignett et al., 1987; Kern and Thompson, 1994a,b). The dynamic contact between the forced particles and the surface leads to higher roughness values, increasing metal surface reactivity

(Wennerberg, 1998). The desired roughness can be set up by the particle size. Thus, for the blasting of biomedical materials the particles should be chemically stable and biocompatible. Usually alumina, titania, or HAP particles are applied for blasting treatments.

Sandblasting possesses three purposes: (1) cleaning surface contaminants; (2) roughening surfaces to increase effective surface area; and (3) producing beneficial surface compressive residual stress. As a result, such treated surfaces exhibit higher surface energy, indicating higher surface chemical and physical activities, and enhancing fatigue strength as well as fatigue life.

In order to obtain satisfactory fixation and biofunctionality of biotolerated and bioinert materials, some of the mechanical surface alternation such as threaded surface, grooved surface, pored surface, and rough surface have been produced that promote tissue and bone ingrowth. But so far, there is no report on suitable roughness to specific metallic biomaterials. In general, on the macroscopic level (>10 μm), roughness will influence the mechanical properties of the interface, the way stresses are distributed and transmitted, the mechanical interlocking of the interface and the biocompatibility of biomaterials. On a smaller scale, surface roughness in the range 10 nm–10 μm may influence the interface biology, because it is of the same order in size as cells and large biomolecules. Topographic variations of the order of 10 nm and less may become important because microroughness on this scale length consists of material defects such as grain boundaries, steps, and vacancies, which are known to be active sites for adsorption, and thus may influence the bonding of biomolecules to the implant surface. There is evidence that surface roughness on a micron scale allows cellular adhesion that alters the overall tissue response to biomaterials. Microrough surfaces allow early better adhesion of mineral ions or atoms, biomolecules, and cells, form stronger fixation of bone or connective tissue, result in a thinner tissue-reaction layer with inflammatory cells decreased or absent, and prevent microorganism adhesion and plaque accumulation, when compared with the smooth surfaces (Wen et al., 1996).

Although alumina (Al_2O_3) or silica (SiO_2) particles are most frequently used as a blasting media, there are several different types of powder particles utilized as media. This process may cause the release of cytotoxic silicon or aluminum ions in the peri-implant tissue. To generate a biocompatible roughened titanium surface, an innovative grid-blasting process using biphasic calcium phosphate (BCP) particles was developed. Ti6Al4V discs were either polished, BCP grid-blasted, or left as-machined. BCP grid-blasting created an average surface roughness of 1.57 μm compared to the original machined surface of 0.58 μm. X-ray photoelectron spectroscopy indicated traces of calcium and phosphorus and relatively less aluminum on the BCP grid-blasted surface than on the initial titanium specimen. It was reported that (1) SEM observations and measurement of mitochondrial activity (MTS assay) showed that osteoblastic MC3T3-E1 cells were viable in contact with the BCP grid-blasted titanium surface; (2) MC3T3-E1 cells expressed alkaline phosphate (ALP) activity and conserved their responsiveness to bone morphogenetic protein BMP-2; and (3) the calcium phosphate grid-blasting technique increased the roughness of titanium implants and provided a noncytotoxic surface with regard to mouse osteoblasts (Citeau et al., 2005). Tribo-chemical treatment has been proposed to enhance the bond strength between titanium crown and resin base (Goodacre et al., 2012). Using silica-coated alumina as a blasting media under relatively low pressure, the silica layer is expected to remain on the blasted surface so that retention force is enhanced by silancoupling treatment.

Although blasting shows several advantages, there is evidence of adverse effects: (1) surface contamination, depending on type of blasting media; and (2) distortion of blasted workpiece, depending on the blasting manner and intensity. Miyakawa et al. (1996) studied the surface contamination of abraded titanium. Despite low grinding speeds and water cooling, the abraded surfaces were found to be contaminated by abrasive constituent elements. Element analysis and chemical bond state analysis of the contaminants were performed using an electron probe microanalyzer. X-ray diffraction (XRD) of the abraded surface was performed to identify the contaminants. It was reported that (1) the contamination of titanium was related to its reactivity as well as its hardness; (2) in spite of water cooling and slow-speed abrading, titanium surfaces were obviously contaminated; (3) contaminant deposits with dimensions ranging from about 10 to 30 μm occurred throughout the surfaces; and (4) the contaminant of titanium, although related also to the hardness, resulted primarily from a reaction with abrasive materials, and such contamination could negatively influence titanium's resistance to corrosion and its biocompatibility (Miyakawa et al., 1996).

Normally, fine alumina particles (50 μm Al_2O_3) are recycled within the sandblasting machine. Ceramics such as alumina are brittle in nature, therefore some portions of recycled alumina might be brittle-fractured. If fractured sandblasting particles are involved in the recycling media, it might result in irregular surfaces, as well as potential contamination. Using fractal dimension analysis (Mandelbrot, 1983; Chesters et al., 1989; Sayles and Thomas, 1978), a sample plate surface was weakly analyzed in terms of topographic changes, as well as chemical analysis of sampled recycled Al_2O_3 particles. It was found that after accumulated use time exceeded 30 minutes, the fractal dimension (DF) (Mandelbrot, 1983) remained a constant value of about 1.4, prior to that it continuously increased from 1.25 to 1.4. By the electron probe microanalysis on collected blasting particles, unused Al_2O_3 contains 100% Al, whereas used (accumulated usage time was about 2400 s) particles contained Al (83.32 wt%), Ti (5.48), Ca (1.68), Ni (1.36), Mo (1.31), S (1.02), Si (0.65), P (0.55), Mn (0.49), K (0.29), Cl (0.26), and V (0.08), strongly indicating that the used alumina powder was heavily contaminated, and there was a high risk of the next material surface being contaminated. Such contaminants were from previously blasted materials having various chemical compositions, and investing materials as well (Oshida et al., 1993).

There is evidence of surface contamination due to mechanical abrasive actions in several cases. As a metallographic preparation, the surface needs to be mechanically polished with a metallographic paper (which is normally SiC-adhered paper) under running water (Park et al., 2012; Guo et al., 2010). It is worth mentioning here that polishing paper should be changed between different types of materials, and particularly when a dissimilar metal-couple is used for galvanic corrosion tests, the couple should not be polished prior to corrosion testing because both materials could become cross-contaminated (Guo et al., 2010). Hence, there are attempts to use TiO_2 powder for blasting onto titanium material surfaces. It was reported that titanium surfaces were sandblasted using TiO_2 powder (particle size ranging from 45, 45–63, and 63–90 μm) to produce the different surface textures prior to fibroblast cell attachment (Cooper et al., 2006).

Recently, it was reported that sandblasting using alumina as a media caused a remarkable distortion on a Co–Cr alloy and a noble alloy (Peutzfeld and Asmussen, 1996; Multigner et al., 2010). It was estimated that the stress causing the deflection exceeded the yield strength of the tested materials. It was also suggested that the sandblasting should be done using the lowest air pressure, duration of blasting period, and particle size alumina in order to minimize distortion of crowns and frameworks. To measure distortion, Co–Cr alloy plates (25 mm long, 5 mm wide, and 0.7 mm thick) were sandblasted with Al_2O_3 of 125 μm. Distortion was determined as the deflection of the plates as a distance of 20 mm from the surface. It was reported that (1) the mean deflections varied between 0.37 and 1.72 mm; and (2) deflection increased by an increase in duration of the blasting, pressure, particle size, and by a decrease in plate thickness (Peutzfeld and Asmussen, 1996).

14.2.3 Shot peening and laser shock peening

Shot peening (which is a similar process to sandblasting, but has more controlled peening power, intensity, and direction) is a cold working process in which the surface of a part is bombarded with small spherical media called shot. Each piece of shot striking the material acts as a tiny hammer, imparting to the surface small indentations or dimples. In order for a dimple to be created, the surface fibers of the material must be yielded in tension. Below the surface, the fibers try to restore the surface to its original shape, thereby producing below the dimple a hemisphere of cold-worked material highly stressed in compression. Overlapping dimples (which are sometimes called forged dimples) develop an even layer of metal in residual compressive stress. It is well known that cracks will not initiate or propagate in a compressively stressed zone due to a tendency to crack-closure. Since nearly all fatigue and stress corrosion cracking failures originate at the surface of a part, compressive stresses induced by shot peening provide considerable increases in part life, since advancing crack-opening is suppressed by pre-existing compressive residual stress. The maximum compressive residual stress produced at or under the surface of a part by shot peening is at least as great as half the yield strength of the material being peened. Many materials will also increase in surface hardness due to the cold-working effect of shot peening. Both compressive stresses and cold-working effects are used in the application of shot peening in forming metal parts, called "shot forming."

Hashemi et al. (2011) investigated the wear resistance of shot peened-nitrided 316 L austenitic stainless steel. The shot-peening treatment was performed on some of the 316 L stainless steel sheet samples (3 mm in thickness) for 15 minutes using a SBRT-ISO shot-peening machine containing steel balls with mean hardness of 40–45 RC, diameter of 1–2 mm, and shooting angle of 45°. Wear tests were carried out using a rotating pin-on-disc wear testing machine with a speed of 2 cm s^{-1}. The diameter of wearing circumferential trajectory was adjusted to be 7 mm. These tests were performed against a 5 mm diameter SAE 52100 pin (ball) and repeated at least three times per specimen. The sliding distance was 250 m with normal loads of 5 and 10 N. The weight loss value of the specimens and the traversed distance were the criterion for comparison of the wear resistance.

Results show that severe plastic deformations due to shot peening caused strain-induced martensite transformation in the surface layers, and the amounts of grain boundaries and other defects increased in the shot-peened specimens. The wear rate due to motion of pin on the specimen's surface at a constant load, for the shot peened-nitrided specimen, was less than this quantity for the nitride specimen, which was itself less than that of the untreated specimen. When 5 N load was exerted, the wear rate of specimens was eligible; however, exerting the 10 N load made the wear rate of untreated specimens increase intensely, while the wear rate for the other specimens did not show a significant increase. Therefore, surface treatments played an important role in increasing the hardness and wear resistance of 316 L stainless steel, and according to the surface hardness of the specimens, shot peening-nitriding treatment had the highest effect in decreasing wear. Signs of a mixture of adhesive (which was usually the predominant mechanism in wear of austenitic stainless steel) and abrasive wear mechanisms were observed on the surface of the untreated specimen. The crushed particles due to the high pressure of the pin were probably the reason for abrasive wear between the ball and the specimen.

It seems that however much the surface hardness increases, the abrasive mechanism becomes more controlling. The controlling mechanism for the untreated specimen is adhesive–abrasive, according to its low hardness value, while investigating shot-peened specimens showed that the controlling mechanism goes more and more to an abrasive mechanism, because of the considerable increase in the surface hardness.

Laser peening (laser shot peening) technology has recently been developed, claiming a non contact, nomedia, and contamination-free peening method. Before treatment, the workpiece was covered with a protective ablative layer (paint or tape) and a thin layer of water. High-intensity (5–15 GW/cm^2) nanosecond pulses (10–30 ns) of laser light beam (3–5 mm width) striking the ablative layer generated a short-lived plasma which caused a shock wave to travel into the workpiece. The shock wave-induced compressive residual stress that penetrated beneath the surface and strengthened the workpiece, resulting in improvements in fatigue life and retarding the occurrence of stress corrosion cracking.

Sealy and Guo (2011) researched surface integrity and process mechanics of laser shock peening of a novel biodegradable magnesium–calcium (Mg–Ca) alloy. An Nd:YAG laser (wavelength D 1064 nm, pulse width (τ) = 5–7 ns, frequency (f) = 30 Hz) with power range 0.1–10 W was used in the experiments. The estimated spot diameter was approximately 250 μm using a 100 mm focal length lens where the cross-sectional area was 0:0491 mm^2. The sample was oriented vertically so particulate could not accumulate in the laser's path. Each experiment was performed in a water-confined regime at a depth of 1–2 mm. Black tape with a thickness of 0.05 mm was used as the ablative medium. The average power measured 8 W.

As material behavior is characterized by stress/strain graphs along the peening or depth direction and radial directions, tensile pile-up is critical to tribological applications in implants. Results showed that due to the radial expansion of neighboring peens, the magnitude of the pile-up increased approximately 50%, which drastically affected the wear and fatigue performance of a surface.

14.2.4 Physical vapor deposition

Fretting is a wear phenomenon occurring when two contacting solids are subjected to a relative oscillatory tangential motion of small displacement amplitude. Even displacement amplitudes in the submicron range can result in wear damage. Fretting becomes an important aspect to designers when the functionality of a component deteriorates or its lifetime is reduced by induced material damage. In severe cases, material damage initiates or promotes other failure mechanisms, such as fretting fatigue, fretting corrosion, or stress-corrosion cracking.

Fretting fatigue is a form of adhesive wear damage due to small oscillatory movement between two contacting bodies under the action of uniform or non-uniform cyclic loads. Cyclic loads may be experienced due to vibration of one or both the bodies eventually leading to failure at the contact area. Fretting damage is also experienced by load-bearing implants within the body environment such as hip joints, knee joints, and bone plates. Damage characterization is important from the point of view of minimizing in vivo failures. Therefore, fretting of components under cyclic load leads to in-service failures without prior warning. Human joints such as hip, knee, or shoulder joints are highly prone to degeneration leading to acute pain and joint stiffness commonly termed as osteoarthritis and rheumatoid arthritis. These load-bearing joints necessitate arthroplastic surgery involving complete replacement of diseased joint surfaces with biocompatible devices. Titanium alloys are frequently used as bioimplants due to their excellent biocompatibility and low modulus of elasticity compared to stainless steel or Co–Cr–Mo alloys.

Fretting wear damage of load-bearing implants can be minimized through a suitable surface modification process. SE is one of the emerging areas in the field of material science or tribology aimed at modification of interacting surfaces of the components for improving their life and efficiency while retaining their original bulk properties. In biomedical devices, quality overrules the cost of the product. Therefore, it is necessary to maintain the superior quality of the device for better performance in the human body. PVD coating is a kind of popular surface modification process used for biomedical devices; a diagram of its process flow is shown in Figure 14.2. It favors the formation of a hard titanium nitride (TiN) layer on the surface. A TiN layer is characterized by high abrasion resistance, low coefficient of friction, high temperature stability, and high hardness. The quality of the film depends upon the control of the process. The formed TiN coating is shown in Figure 14.2. The inset image is the surface morphology of the TiN coating. It shows that the fine pore-free columnar structures are packed closely in cross-sectional

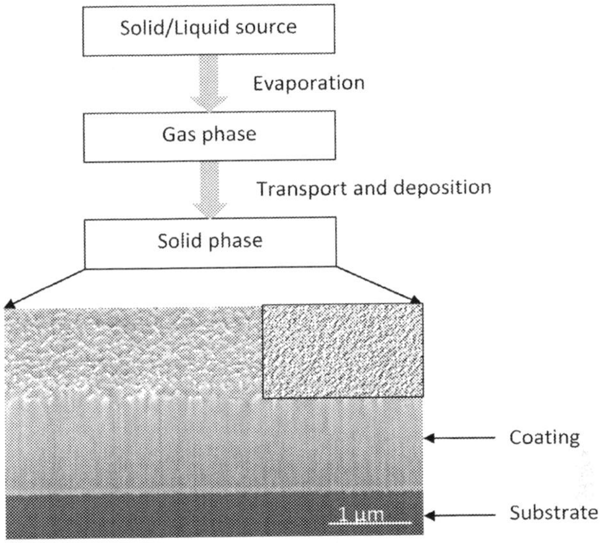

Figure 14.2 Diagram of physical vapor deposition (PVD) process flow.

view. In the meantime, the nanostructures are distributed on the film surface evenly, which are propitious to anti-abrasion for biomedical materials.

Vadiraj and Kamaraj (2006) studied the fretting fatigue behavior of 2 μm thick PVD TiN-coated titanium alloys. The fretting fatigue tests were conducted in an enclosed chamber with protein and saline liquid, which perfectly simulated the body environment. Ringer solution maintained at 37°C was continuously streamed on pads and collected in a container below. This flow was adjusted to obtain complete immersion of a contact area similar to the fretting action of hip joints surrounded by a continuous flow of the physiological medium. Cyclic loads of 3–7.5 kN at 5 Hz frequency were applied during the test.

Results show that fretting fatigue of uncoated Ti6Al4V failed within 3% of the plain fatigue life. Plain fatigue life of PVD TiN coated Ti6Al4V at 500 MPa cyclic load did not fail even after 106 cycles. This indicated TiN coatings prevented the formation of surface intrusions and extrusions, which were generally responsible for fatigue failures. Comparison of friction coefficients of the uncoated and TiN-coated alloys showed that higher friction was being experienced without TiN coatings. For uncoated alloys, the friction coefficient gradually increased and later decreased with ongoing fretting damage. Reduction of friction coefficient at the later stages of fretting of uncoated alloys could be attributed to a continuous flow of ringer solution providing a lubricating effect at the contact. Friction was minimized with TiN coatings due to its excellent tribological properties compared to uncoated alloys. But it could not withstand the fretting process for a long time. The film ruptured at the later stages of fretting leaving networks of cracked layers. Delaminated TiN particles acted as potential abrasives for inducing damage on both the surfaces. Their sizes varied from less than a micron to more than 10 μm.

Particles with sharper edges would be more harmful as they easily dig into the surface during fretting.

Carbon-based films are biocompatible with good bearing capacity, wear resistance, corrosion resistance, and have a low coefficient of friction. However, great intrinsic stress prevents their wider application, mainly as implant coatings. To reduce this undesirable effect special deposition procedures are under development and/or the films are doped with suitable elements. Diamond-like carbon (DLC) films, which present a metastable form of carbon in an amorphous structure with a mixture of sp^2 and sp^3 C—C bonds, have a wide range of potential applications.

Additionally, the mechanical properties of DLC films are directly related to the fraction of sp^3 C—C bonds. DLC is an excellent base coating to be alloyed with different other elements. The amorphous nature of DLC opens up the possibility of introducing certain amounts of additional elements, such as Si, F, N, O, W, V, Co, Mo, Ti, and their combinations, into the film and still maintaining the amorphous phase of the coating (Santavirta et al., 2006; Anttila et al., 1997). Understanding the wear mechanism of diamond coatings in the friction process is of great technological and scientific interest. Also characterization of their tribological properties is required. As a result, it is reasonable to investigate the performance of diamond coatings under fretting friction conditions.

Fretting wear combines most of the existing wear mechanisms, such as abrasive, adhesive, oxidative, and fatigue (Czichos, 1978). Moreover, fretting wear can be easily produced on hard surfaces with minimal energy input into the friction surfaces, because wear is concentrated in a small contact area between the two contacting solid surfaces. In addition, fretting conditions are dynamic and simulate the behavior of many mechanical systems.

In order to understand the relationships between the fretting wear behavior and mechanical properties of films based on carbon (Zdravecká et al., 2012) deposited as thin carbon films on the conventional bioimplant material CoCrMo (ISO 5832-4) by DC using the pulsed arc discharge PVD nitrogen-doped amorphous carbon (a-C) films with lower sp^3 content and the filtered pulsed arc discharge deposition system tetrahedral carbon films (ta-C) with a higher sp^3 (up to 85%), where the a-C films were deposited by pulsed arc discharge PVD vacuum equipment, and ta-C coatings were prepared using the filtered pulsed carbon plasma technique. Deposition parameters are shown in Table 14.3. Samples were rotated in order to obtain uniform thickness. Thickness of ta-C film was determined as 0.70 mm and for a-C film was 1.25 mm. Fretting wear experiments were conducted in humid air (RHE = 45%, $T = 21°C$) at constant oscillating frequency of 10 Hz, with $F = 5$ N, number of cycles = 10,000, total sliding distance = 100 mm. For ta-C samples both unlubricated and lubricated contact sliding against corundum ball (α-Al_2O_3) were investigated. Bovine serum diluted to 25% with deionized water, with penicillin, streptomycin, and sodium azide added, was used for lubrication.

This elucidates that the wear depends on the type of coating and on the use of a lubricant. Fretting tests performed using reciprocal movement and dry conditions (without lubricant) disclosed that the wear mechanism of ta-C was abrasive—adhesive wear. Results from the study of fretting and lubrication with bovine serum show that

Table 14.3 Deposition parameters of DLC films

Type of films	Ar_2 cm^3/min	N_2 cm^3/min	Arc voltage V	Pulse rate Hz	Pulse duration microsec	Deposition distance cm	Deposition time min	Deposition technique
a-C[a]	70	60	400	3	300–500	20	30	Pulsed arc discharge
ta-C*			500	10	100	10	30	Filtered pulsed arc discharge

*Deposition atmosphere: vacuum, $10e^{-6}$ mbar/100 μPa.
[a]*RCL* parameters: $R = 0.1$ Ohm, $C = 34.8$ μF, and $L = 12$ muH.

ta-C coatings, despite their high hardness, have very low friction coefficients and low volume losses. The value of the friction coefficient for dry sliding contact of a-C coating against a corundum ball was higher but showed a moderate increase.

Furthermore, development and applications of thin hard carbon layers are not yet finished and they represent an ongoing theme. A more detailed understanding of the correlation between requirements for biomaterial applications and coating properties will lead to realistic simulations for testing the performance of the coatings produced.

14.2.5 *Chemical vapor deposition*

As mentioned above, titanium is most commonly used as an orthopedic implant or bone substitute material because it possesses good biocompatibility and sufficient mechanical properties for medical applications (Aparicio et al., 2006; Aparicio et al., 2011; Subramani and Mathew, 2012; Ravelingien et al., 2010). However, the abrasion resistance of Ti is low and minute Ti abrasion powders may cause inflammatory reactions (Cassar et al., 2012; Cozza et al., 2011; Dalili et al., 2010; Cvijović-Alagić et al., 2011). Therefore it is desirable to develop implant materials with both biocompatibility and abrasion resistance. Dental implant abutment division (the mucosa penetration division) is exposed in the oral cavity. Plaque and dental calculus adheres more easily to the surface of the implant than natural teeth. Scaling treatment is the only method for removal of them, and it is a necessary treatment process to obtain a good prognosis throughout the long-term maintenance of the implant (Zhu and Watari, 2007). Therefore, for abutment division of an implant, it is important to possess high abrasion resistance to keep the smoothness of the implant surface after scaling treatment (Mengel et al., 1998; Quirynen and Bollen, 1995). Titanium carbide (TiC) is a useful material for biomedical instruments because it possesses a range of desirable properties. The combination of very high hardness, high melting temperature, and excellent thermal and chemical stabilities makes TiC suited to a number of commercial applications. For instance, TiC is often used in abrasives, cutting tools, grinding wheels, and coated cutting tips (Koc et al., 2000; Eskandarany, 2000). Usually, TiC films are produced by the techniques PVD, CVD, and plasma vapor deposition (PCVD). However, these methods need the substrates for deposition to be high temperature (>1000°C), which sometimes causes thermal damage to the deposited films (Rie et al., 1996). Therefore, it is desirable to develop a deposition technique using low-temperature (<900°C) substrates preventing thermal damage to the films.

Zhua et al. (2012) demonstrated TiC film deposition on Ti substrate by an ion-enhanced triode plasma CVD method. The plasma reactor used a triode system in which a substrate bias circuit with two diodes was employed to accelerate the deposition through ion bombardment processes. An ion-enhanced triode plasma CVD joined the discharge plasma in the CVD process to activate the deposition reaction, so that the film temperature was reduced by a considerable amount with PVD, and maintained the characteristics of the CVD. The base materials were Japanese Industrial Standards (JIS) type I pure Ti plates (10 mm × 10 mm × 0.5 mm). Before the deposition, the reactor is evacuated to about 1×10^{-5} Torr by a diffusion pump subsidized by a rotary pump. TiC film was deposited on Ti substrate at about 800°C by an ion-enhanced triode plasma CVD method using a $TiCl_4 + CH_4 + H_2$ gas mixture. The gases used are $TiCl_4$ (99.9%), CH_4 (99.9%), and H_2 (99.999%), and these are introduced into the reactor at flow rates of 10–15 sccm, 25 sccm, and 250 sccm, respectively. The production characteristics of the TiC films are investigated in the low frequency of power from 50 to 200 Hz and the bias voltage (V_b) ranging from −100 to −400 V with the triode technique.

XRD analysis showed that TiC was formed on the surface of Ti as a consequence of the basal substrate in addition to TiC coverage with the relative thin thickness. The pattern for TiC suggested that the film contained TiC crystal with TiC (111), TiC (200), and TiC (220) orientations. The peaks representing these orientations were in about 2.0° of FWHM (full-width at half maximum), while peaks for Ti were sticking out sharply. No peaks of impurity, such as C, were shown. Both peaks of Ti and TiC appeared, which suggested that the deposition of layer was totally transformed into carbide and the TiC layer was thin. These results showed the surface of TiC was significantly rougher than Ti. The results of SEM images of a cross section of TiC showed that the continuous TiC layer was formed on Ti and the thickness of the TiC layer was about 6 μm. The Hv of TiC was much higher than pure Ti at the same load and decreased along with the increase in load. The highest value of Hv was 2580 at load 0.98 N, and it quickly decreased to 1860 at load 1.96 N, and the decreasing amount was gradually less for loads from 1.96 N to 4.9 N. This was mainly because the indentation depth exceeded the thickness of the carbide layer and reached the softer matrix of Ti while with an increase in load to 4.9 N. However, the indentation depth of TiC was still smaller than that of pure Ti, because the penetration of the indenter was suppressed by the carbide layer.

There are many studies about preparation of TiC film on the substrate surface under different conditions (such as different substrate heating temperatures, different types of gas mixtures and gas ratio, duty cycle) by PACVD methods (Shanaghi et al., 2012a,b). However, studies on the abrasion resistance of implant abutment division material have rarely been attended. In dental clinics, the operation to remove dental calculus by ultrasonic scaler was carried out with a light pressure of about 0.49 N. In this study, Hv of TiC film was 2580 at load 0.98 N, therefore, for clinical use, the authors anticipate that there will be very limited damage on the surface of TiC film even if a stainless steel tip is used.

14.2.6 Plasma spraying

The number of patients requiring and receiving biomedical implants to correct skeletal defects and heal diseases is constantly increasing. Worldwide, large demand exists for loadbearing hip, knee, and dental endoprosthetic implants, but also for bone replacement parts in the maxillar–mandibular area, the ossicular chain of the inner ear, and alveolar ridge and iliac crest augmentation. To date, in the United States and in Europe, more than 1 million hip and knee arthroplasties are being performed annually, and this is increasing.

The growing demand for endoprostheses is the result of the wear and tear the hip and knee joints are subjected to during a human lifetime. Consider this: an average person walks about 1 million steps per year with a frequency of about 1 Hz. Using a conservative width of stride of 1.2 m, this amounts in an average lifespan to $7-10^7$ load changes during walking a distance of 84,000 km or about twice the equatorial circumference of the earth. The load on the joints is roughly 1 kN (body mass) during rest, 2 to 3 kN during normal walking, up to 5 kN while jogging, and up to 8 kN while jumping. Beyond that threshold the risk of damage to the joints increases dramatically. As people generally live longer and are increasingly overweight due to overeating and lack of exercise, eventually the protective tissue lining of the acetabular cup wears away; friction increases; and inflammation, pain, and finally immobilization will result. At this point a total hip replacement (THR) is the only reasonable option to maintain mobility, freedom from pain, and hence a rewarding life in old age. However, increasingly younger people require these operations because their lifestyle, including damaging sports activities, promotes premature wear on their joints. Here a problem arises because the younger patient will generally outlive the average lifetime of today's implants so that a remediation operation is required with substantial additional cost to the healthcare system of the country involved. Moreover, the early onset of pain and associated mental stress on the patient adds another dimension to the problem of the incompatibility of the lifespan between the patient and the implant.

Plasma spraying is another method used for the structuring of implant surfaces at the micro/nanoscale level, and the principle of plasma spraying is shown in Figure 14.3. This method is a version of thermal spraying, a technique for producing coatings using a plasma jet at high temperatures that projects particles onto the surface of the implant where they condense and fuse together. Usually, deposits produced with plasma spraying show thicknesses from micrometers to several millimeters and can be produced from a variety of materials. In biomedical applications bioinert ceramics with excellent mechanical properties, such as titania, zirconia, or

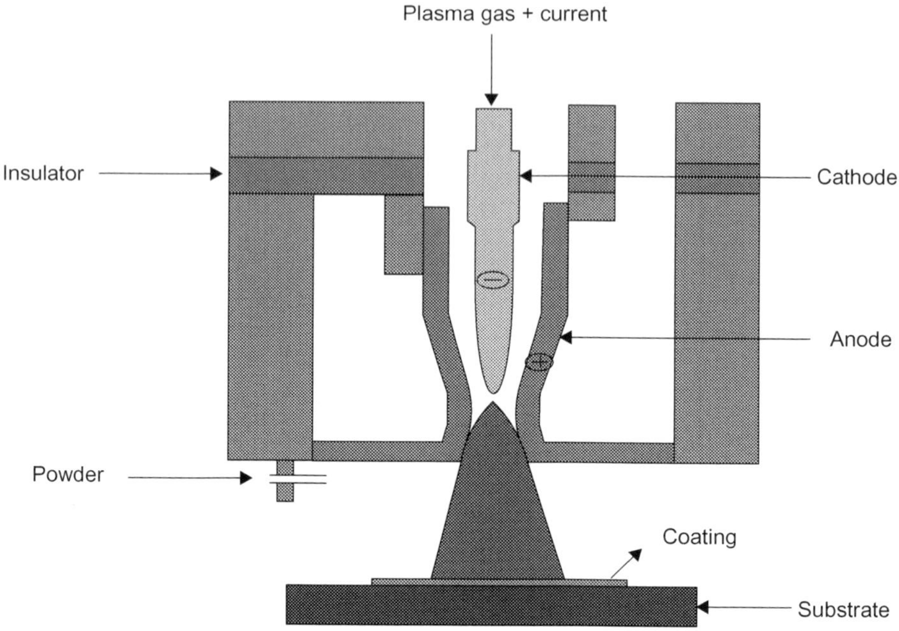

Figure 14.3 The principle of plasma spraying.

alumina are deposited mostly onto titanium-based alloys by plasma spraying to increase the surface roughness and at the same time to modify the surface chemistry. Alumina and zirconia coatings are being used clinically, mostly due to their higher wear resistance than titania. However, alumina and zirconia coatings cannot bind directly to bone tissues due to their bioinert nature, thus limiting their use in hard tissue applications (Liu et al., 2004). Plasma spraying of titania has been used to produce rough implant surfaces – such coatings may show an average roughness of around 20 μm (using high deposition thicknesses to achieve uniform deposits) (Martin et al., 1995). Buser et al. demonstrated that such coatings induce an increase in the bone-to-implant contact formation by measuring the tensile strength at the interface (Buser et al., 1991). However, there is a controversy on the binding strength and on particle release from plasma-sprayed coatings into the host tissue, caused by either dissolution or fretting (Urban et al., 2000). Plasma-sprayed surfaces are often used in combination with other modifications such as blasting or etching (Martin et al., 1995; Taba et al., 2003).

More recently, plasma spraying is increasingly used for HAP coatings, as, beside the structural influence, these coatings also show improved abilities in terms of bioactivation (Ong et al., 2004).

Such surfaces mostly show random topographies with surface structures in a wide range, from nanometers to millimeters. Zinger et al. (Zinger et al., 2004; Landolt et al., 2003) reported that combining surface patterning techniques at the submicrometer level with subsequent sandblasting and etching led to surface roughnesses with combined micrometer and nanometer structures, and thus showed an enhanced osteoblast activity. From such studies it was deduced that cell attachment in vitro and the bone implant interface in vivo may be influenced by both nanoscale and microscale parameters of topography, where osteoblasts showed an enhanced attachment on submicron-scale structures (Anselme et al., 2002; Zhu et al., 2004). The role of surface roughnesses at both length scales, either micron or nanometer level, requires the consideration that molecular interactions with the surface occur and, as a direct consequence, cell adhesion phenomena and local biomechanical features of the established interface are directly influenced by this length scale. Modifications of surfaces at the nanoscale will have an effect on the chemical reactivity of a biomedical material and thus affect ionic or biomolecular interactions of the surface with the host tissue. Such changes in surface properties altered by modifications at the nanoscale may change wetting properties, lead to a different protein adsorption, or have an influence on the mineralization of de novo bone formation. The importance of topography at the nanometer scale has been emphasized, where one interpretation of the sensitive reaction to nanotopography occurs because there are small differences in chemistry between one part of the topography and another. An opposite view is that even when small local differences in chemistry are made by techniques such as nanoprinting, there is also a small difference in topography (Curtis et al., 2004; Flemming et al., 1999; Teixeira et al., 2003; Turner et al., 1997).

Pure and crystalline HAP ($Ca_{10}(PO_4)_6(OH)_2$) coating is a good candidate for improving the biocompatibility properties of metallic implants such as Ti, Ti-alloys, and stainless steel. Currently, HA materials are used extensively as biomaterial because their structure is close to human bone and exhibits good compatibility with body fluid (Tan et al., 2012; Khandelwal et al., 2013; Wu et al., 2012; Sukjai et al., 2012). The most common method used to deposit HA coatings on metallic implants is thermal spraying. Thermal spraying processes can develop thick, adhered, and crystalline HA coatings in a short time compared with other plating methods (Gross et al., 2012; Heimann, 2006; Fu et al., 1998). The biological, chemical, and mechanical properties of thermal sprayed HA coatings applied to the stem of hip endoprostheses, dental root implants, and osteosynthetic bone replacement plates are crucially dependent on the phase composition, crystallinity, and microstructure of the coatings as well as on their adhesive strength to the metallic implant surface. In clinical application, coating failure by chipping, spalling, delamination, and dissolution is being observed on explanted endoprostheses, usually close to the Ti implant/coating interface (Drnovšek et al., 2012; Zhou et al., 2012; Ribeiro et al., 2012; Liu et al., 2013). This has frequently been attributed to a layer of amorphous calcium phosphate (ACP) formed by rapid quenching of molten or semi molten droplets of calcium phosphate with exceptionally high cooling rates at the coating/Ti interface. Such a continuous ACP layer is thought to act as a fracture path and, owing to its comparatively high solubility, will preferentially be dissolved in vivo, thus further weakening the mechanical integrity of the interface. Controlling the spraying parameters is the key to minimizing ACP and other phases such as tri-calcium phosphate and oxyhydroxyapatite, which have degradation effects for HA coatings (Saber-Samandari and Gross, 2012; Saber-Samandari and Gross, 2013; dos Santos et al., 2012; Xu et al., 2012; Barnes et al., 2012; Abdelkebir et al., 2012; Cizek and Khor, 2012).

In 2006, Morks and Kobayashi (2006) studied the influence of gas flow rate on the microstructure and mechanical properties of HAP coatings (HAP powder of particle size $+10-45\ \mu m$) on 304 stainless steel substrate fabricated by gas tunnel-type plasma spraying, and the abrasive wear resistance of HA coatings sprayed at different gas flow rates were investigated. The main advantage of using gas tunnel plasma spraying is its capability for spraying crystalline HA coatings without an APC phase. Gas flow rate affects the plasma power and velocity of the flight particles and hence the particle temperature at the moment of impinging on the substrate.

An abrasive wear test was done for HA coatings sprayed at different gas flow rates. The tested samples were cut from the coated samples with a surface area 20×20 mm. The samples were moved horizontally (oscillation mode) a distance of 1 cm under load 25 g on a slightly moved wheel (25 mm/minute) covered with SiC emery paper. The width of the wheel is 1 cm and the contacted area with the test sample is 1 cm^2. The abrasive wear test was performed in dry condition in air atmosphere at room temperature.

Results show that HA coatings sprayed at 120, 140, 150, and 160 l/minute had lower porosity, and some unmelted powders compared with HA coatings sprayed at 100 l/minute. HA coatings sprayed at 170 l/minute exhibited dense structure with few pores. The flying particles became semi molten during spraying at 160 l/minute and completely molten at 170 l/minute.

The abrasive wear rate of HA coatings sprayed at low gas flow rate of 100 l/minute was higher than that sprayed at higher gas flow rates mainly due to the presence of many pores and unmelted particles inside the coatings. The low hardness value of coatings sprayed at 100 l/minute also affected the abrasion resistance due to the fact that the cohesion bonds among the splats became very weak. The abrasion resistance of HA coatings became higher and the wear rate reached a minimum value when spraying at high gas flow rate of 170 l/minute. The coatings sprayed at 170 l/ minute exhibited high hardness, low porosity, and a dense structure and these properties improved the wear resistance of the coating. The abrasion resistance of HA coatings sprayed at 120 l/minute was slightly higher than those sprayed at 140 and 150 l/minute mainly due to their slightly higher hardness value. It indicated that the higher the plasma power, the higher the abrasion resistance of sprayed coatings.

Two years later, an attempt to improve the mechanical properties of HA coating, Morks (2008) investigated the microstructure and mechanical properties of plasma-sprayed HA/SiO_2 coatings for biomedical application. SUS 304 stainless steel was used as substrate material and pure HA and fused silica powders of particle size 10–45 and 10–25 μm were used as spraying materials.

Results indicated that the degree of crystallinity of the coatings decreased as the amount of silica increased. The presence of amorphous silica particles affected the crystallinity of HA–silica coatings mainly due to the increase of the melting degree of HA particles. The high melting point silica particles adjacent to or in contact with HA particles increased their melting temperature degree. The silica phase did not appear in XRD patterns because of its amorphous structure and low amount. By means of energy dispersive X-ray spectroscopy (EDX) analysis, it was possible to prove that many of these particles mainly consisted of deformed and unmelted silica. Spherical silica particles had been found inside the coatings, which indicated that the silica particle was melted and deformed from an angular to spherical shape. However, in the other photo the silica particle was still angular in shape. From the surface morphology of HA/SiO_2 coatings, it was concluded that small silica particles (<5 μm) were semi- or fully melted and large silica particles (>7 μm) were semi- or unmelted. However, the majority of HA particles were fully melted, flattened and mechanically adhered together in a dense manner with the silica particles.

It was found that the abrasion resistance was significantly improved for HA coatings reinforced with silica particles as compared with pure HA coatings, and the abrasion resistance of the HA/SiO_2 coatings increased with increasing silica content. The silica particles in the HA/SiO_2 coatings are removed and ground, by the effect of abrasion, into very fine particles and act as a thin solid lubricant layer that improved the abrasive wear properties of sprayed HA coatings.

14.3 Future prospects

Technologies should be developed for nanostructured surfaces of metallic biomaterials with/without masking the unwanted area. Through the choice of processing parameters, the width and depth of the nanaotextured features can be controlled over a wide range resulting in higher anti-abrasion in the bioenvironment. Nanotexturing prosthetic items such as hip implants, textures could be sized and shaped to favor the ingrowth of adjacent bone to anchor the implants with the suitable properties of tribology in the practical applications.

Proposing an ideal implant structure with nanomodifaction should be integrated by bioengineering, nanotechnology, and biomaterials science. The four important factors and requirements for successful and biofunctional implant systems should be satisfied: (1) biological compatibility (or, biocompatibility); (2) mechanical compatibility (or mechanocompatibility); (3) morphological compatibility; and (4) crystallographic compatibility (or micromorphological compatibility).

Currently, there is no absolute knowledge on the influence of nanostructure features on the biological environment, because of the absence of standardized surfaces with repetitive topography at the nanoscale level with highly controllable lateral resolution. The increasing availability of well-understood and standardized surface structures in the sub-100 nm scale will greatly help to understand the interactions between specific proteins and cells.

References

Abdelkebir, K., Morin-Grognet, S., Gaudière, F., Coquerel, G., Labat, B., et al., 2012. Biomimetic layer-by-layer templates for calcium phosphate biomineralization. Acta Biomater. 8 (9), 3419–3428.

Aizenberg, J., 2005. A bio-inspired approach to controlled crystallization at the nanoscale. Bell Labs Tech. J. 10 (3), 129–141.

Anselme, K., Bigerelle, M., Noel, B., Iost, A., Hardouin, P., 2002. Effect of grooved titanium substratum on human osteoblastic cell growth. J. Biomed. Mater. Res. 60, 529–540.

Anttila, A., Lappalainen, R., Tiainen, V.M., Hakovirta, M., 1997. Superior attachment of high-quality hydrogen-free amorphous diamond films to solid materials. Adv. Mater. 9, 1161–1164.

Aparicio, C., Padrós, A., Gil, F.J., 2006. In vivo evaluation of micro-rough and bioactive titanium dental implants using histometry and pull-out tests. J. Mech. Behav. Biomed. Mater. 4 (8), 1672–1682.

Aparicio, C., Rodriguez, D., Gil, F.J., 2011. Variation of roughness and adhesion strength of deposited apatite layers on titanium dental implants. Mater. Sci. Eng. C. 31 (2), 320–324.

Balasundaram, G., Webster, T.J., 2006. A perspective on nanophase materials for orthopedic implant applications. J. Mater. Chem. 16, 3737–3745.

Barnes, D., Johnson, S., Snell, R., Best, S., 2012. Using scratch testing to measure the adhesion strength of calcium phosphate coatings applied to poly(carbonate urethane) substrates. J. Mech. Behav. Biomed. Mater. 6, 128–138.

Barril, S., Debaud, S., Mischler, S., Landolt, D., 2002. A tribo-electrochemical apparatus for in vitro investigation of fretting-corrosion of metallic implant materials. Wear. 252, 744–754.

Barril, S., Mischler, S., Landolt, D., 2004. Influence of fretting regimes on the tribocorrosion behaviour of Ti6Al4V in 0.9 wt.% sodium chloride solution. Wear. 256, 963–972.

Barril, S., Mischler, S., Landolt, D., 2005. Electrochemical effects on the fretting corrosion behaviour of Ti6Al4V in 0.9% sodium chloride solution. Wear. 259, 282–291.

Brugnoni, C., Lanza, F., Macchi, G., Miiller, R., Parnisari, E., et al., 1998. Evaluation of the wear resistance of ZrN coatings using thin layer activation. Surf. Coat. Technol. 100–101, 23–26.

Budinski, K.G., 1991. Tribological properties of titanium-alloys. Wear. 151, 203–217.

Buser, D., Schenk, R.K., Steinemann, S., Fiorellini, J.P., Fox, C.H., et al., 1991. Influence of surface characteristics on bone integration of titanium implants – a histomorphometric study in miniature pigs. J. Biomed. Mater. Res. 25, 889–902.

Cai, K., Bossert, J., Jandt, K.D., 2006. Does the nanometre scale topography of titanium influence protein adsorption and cell proliferation? Colloids Surf. B. 49, 136–144.

Cassar, G., Banfield, S., Avelar-Batista Wilson, J.C., Housden, J., et al., 2012. Micro-abrasion wear testing of triode plasma diffusion and duplex treated Ti–6Al–4V alloy. Wear. 274–275, 377–387.

Chen, Z.X., Wang, W.X., Takao, Y., Matsubara, T., Ren, L.M., 2012. Characterization and fatigue damage of TiO2 layer on spark-anodized titanium before and after hot water treatment. Appl. Surf. Sci. 262, 2–7.

Chesters, S., Wen, H.Y., Lundin, M., Kasper, G., 1989. Fractal-based characterization of surface texture. Appl. Surf. Sci. 40, 185–192.

Citeau, A., Guicheux, J., Vinatier, C., Layrolle, P., Nguyen, T.P., et al., 2005. In vitro biological effects of titanium rough surface obtained by calcium phosphate grid blasting. Biomaterials. 26, 157–165.

Cizek, J., Khor, K.A., 2012. Role of in-flight temperature and velocity of powder particles on plasma sprayed hydroxyapatite coating characteristics. Surf. Coat. Technol. 206 (8–9), 2181–2191.

Cooper, L.F., Zhou, Y.S., Takebe, J., Guo, J.L., Abron, A., et al., 2006. Fluoride modification effects on osteoblast behavior and bone formation at TiO2 grit-blasted c.p. titanium endosseous implants. Biomaterials. 27 (6), 926–936.

Cozza, R.C., Tanaka, D.K., Souza, R.M., 2011. Friction coefficient and wear mode transition in micro-scale abrasion tests. Tribol. Int. 44 (12), 1878–1889.

Curtis, A.S.G., Gadegaard, N., Dalby, M.J., Riehle, M.O., Wilkinson, C.D.W., et al., 2004. Cells react to nanoscale order and symmetry in their surroundings. IEEE Transactions. Nanobioscience. 2004, 61–65.

Cvijović-Alagić, I., Cvijović, Z., Mitrović, S., Panić, V., Rakin, M., 2011. Wear and corrosion behaviour of Ti–13Nb–13Zr and Ti–6Al–4V alloys in simulated physiological solution. Corros. Sci. 53 (2), 796–808.

Czichos, H., 1978. Tribology – System Approach to the Science and Technology of Friction, Lubrication and Wear. Elsevier, Amsterdam, The Netherlands.

Dalili, N., Edrisy, A., Farokhzadeh, K., Li, J., Lo, J., et al., 2010. Improving the wear resistance of Ti–6Al–4V/TiC composites through thermal oxidation (TO). Wear. 269 (7–8), 590–601.

Davidson, J.A., Mishra, A.K., Kovacs, P., Poggie, R.A., 1994. New surface hardened, low-modulus, corrosion-resistant Ti-13Nb-13Zr alloy for total hip arthroplasty. Bio-Med. Mater. Eng. 4 (3), 231–243.

Dearnley, P.A., Dahm, K.L., Çimenoğlu, H., 2004. The corrosion-wear behaviour of thermally oxidised CP-Ti and Ti–6Al–4V. Wear. 256, 469–479.

Degasne, I., Basle, M.F., Demais, V., Hure, G., Lesourd, M., et al., 1999. Effects of roughness, fibronectin and vitronectin on attachment, spreading, and proliferation of human osteoblast-like cells (Saos-2) on titanium surfaces. Calcif. Tissue Int. 64, 499–507.

Deligianni, D.D., Katsala, N., Ladas, S., Sotiropoulou, D., Amedee, J., 2001. Effect of surface roughness of the titanium alloy Ti–6Al–4V on human bone marrow cell response and on protein adsorption. Biomaterials. 22, 1241–1251.

dos Santos, E.A., Moldovan, S., Mateescu, M., Faerber, J., Acosta, M., et al., 2012. Physical-chemical and biological behavior of an amorphous calcium phosphate thin film produced by RF-magnetron sputtering. Mater. Sci. Eng. C. 32 (7), 2086–2095.

Drnovšek, N., Rade, K., Milačič, R., Štrancar, J., Novak, S., 2012. The properties of bioactive TiO2 coatings on Ti-based implants. Surf. Coat. Technol. 209 (25), 177–183.

Duerig, T., Pelton, A., Stockel, D., 1999. An overview of nitinol medical applications. Mater. Sci. Eng. A. 273–275, 149–160.

Duisabeau, L., Combrade, P., Forest, B., 2004. Environmental effect on fretting of metallic materials for orthopaedic implants. Wear. 256, 805–816.

Eisenbarth, E., Velten, D., Schenk-Meuser, K., Linez, P., Biehl, V., et al., 2002. Interactions between cells and titanium surfaces. Biomol. Eng. 19, 243–249.

Eskandarany, M.S.E., 2000. Structure and properties of nanocrystalline TiC full-density bulk alloy consolidated from mechanically reacted powders. J. Alloys Compd. 305, 225–238.

Flemming, R.G., Murphy, C.J., Abrams, G.A., Goodman, S.L., Nealey, P.F., 1999. Effects of synthetic micro- and nano-structured surfaces on cell behavior. Biomaterials. 20, 573–588.

Fu, L., Khor, K.A., Lim, J.P., 2000. Yttria stabilized zirconia reinforced hydroxyapatite coatings. Surf. Coat. Technol. 127, 66–75.
Fu, Y.Q., Batchelor, A.W., Wang, Y., Khor, K.A., 1998. Fretting wear behaviors of thermal sprayed hydroxyapatite (HA) coating under unlubricated conditions. Wear. 217, 132–139.
Goodacre, C.J., Garbacea, A., Naylor, W.P., Daher, T., Marchack, C.B., et al., 2012. CAD/CAM fabricated complete dentures: Concepts and clinical methods of obtaining required morphological data. J. Prosthet. Dent. 107 (1), 34–46.
Gotfredsen, K., Hjorting-Hansen, E., Budtz-Jorgensen, E., 1990. Clinical and radiographic evaluation of submerged and nonsubmerged implants in monkeys. Int. J. Prosthodont. 3, 463–469.
Gross, K.A., Muller, D., Lucas, H., Haynes, D.R., 2012. Osteoclast resorption of thermal spray hydoxyapatite coatings is influenced by surface topography. Acta Biomater. 8 (5), 1948–1956.
Guo, Z.X., Wang, X.L., Huang, H., 2010. Plasma-mediated ablation of biofilm contamination. Appl. Surf. Sci. 257 (4), 1247–1253.
Han, Y., Hong, S.H., Xu, K.W., 2003. Structure and in vitro bioactivity of titania-based films by micro-arc oxidation. Surf. Coat. Technol. 168, 249–258.
Hanawa, T., 2002. Evaluation techniques of metallic biomaterials in vitro. Sci. Technol. Adv. Mater. 3, 289–295.
Hansson, S., 2006. The dental implant meets bone – a clash of two paradigms. Appl. Osseointegration Res. 1, 15–17.
Hashemi, B., Yazdi, M.R., Azar, V., 2011. The wear and corrosion resistance of shot peened-nitrided 316L austenitic stainless steel. Mater. Des. 32, 3287–3292.
Heimann, R.B., 2006. Thermal spraying of biomaterials. Surf. Coat. Technol. 201, 2012–2019.
Hignett, B., Andrew, T.C., Downing, W., Duwell, E.J., Belanger, J., et al., 1987. Surface cleaning, finishing and coating. In: Wood, W.G. (Ed.), Metals Handbook. American Society for Metals, Metals Park, OH, pp. 107–127.
Kasano, F., Morimitsu, T., 1997. Utilization of nickel-titanium shape memory alloy for stapes prosthesis. Auris, Nasus, Larynx. 24, 137–142.
Kern, M., Thompson, V.P., 1994a. Effects of sandblasting and silica-coating procedures on pure titanium. J. Dent. 22, 300–306.
Kern, M., Thompson, V.P., 1994b. Sandblasting and silica coating of a glass-infiltrated alumina ceramic – volume loss, morphology, and changes in the surface-composition. J. Prosthet. Dent. 71, 453–461.
Khan, M.A., Williams, R.L., Williams, D.F., 1996. In vitro corrosion and wear of titanium alloys in the biological environment. Biomaterials. 17, 2117–2126.
Khan, M.A., Williams, R.L., Williams, D.F., 1999. Conjoint corrosion and wear in titanium alloys. Biomaterials. 20, 765–772.
Khandelwal, H., Singh, G., Agrawal, K., Prakash, S., Agarwal, R.D., 2013. Characterization of hydroxyapatite coating by pulse laser deposition technique on stainless steel 316 L by varying laser energy. Appl. Surf. Sci. 265, 30–35.
Kieswetter, K., Schwartz, Z., Hummert, T.W., Cochran, D.L., Simpson, J., et al., 1996. Surface roughness modulates the local production of growth factors and cytokines by osteoblast-like MG-63 cells. J. Biomed. Mater. Res. 32, 55–63.
Koc, R., Meng, C., Swift, G.A., 2000. Sintering properties of submicron TiC powders from carbon coated titania precursor. J. Mater. Sci. 35, 3131–3135.

Koegler, P., Clayton, A., Thissen, H., Santos, G.N.C., Kingshott, P., 2012. The influence of nanostructured materials on biointerfacial interactions. Adv. Drug Deliv. Rev. 64 (15), 1820–1839.

Kostoff, R.N., Koytcheff, R.G., Lau, G.Y., 2007. Global nanotechnology research literature overview. Technol. Forecast. Soc. Change. 74, 1733–1747.

Landolt, D., Chauvy, P.F., Zinger, O., 2003. Electrochemical micromachining, polishing and surface structuring of metals: Fundamental aspects and new developments. Electrochim. Acta. 48, 3185–3201.

Lee, W., Jin, M.K., Yoo, W.C., Lee, J.K., 2004. Nanostructuring of a polymeric substrate with well-defined nanometer-scale topography and tailored surface wettability. Langmuir. 20 (18), 7665–7669.

Li, S.J., Yang, R., Li, S., Hao, Y.L., Cui, Y.Y., et al., 2008. Wear characteristics of Ti-Nb-Ta-Zr and Ti–6Al–4V alloys for biomedical applications. Wear. 257, 869–876.

Liu, F., Song, Y., Wang, F.P., Shimizu, T., Igarashi, K., et al., 2005. Formation characterization of hydroxyapatite on titanium by microarc oxidation and hydrothermal treatment. J. Biosci. Bioeng. 100, 100–104.

Liu, X., Chu, P.K., Ding, C., 2004. Surface modification of titanium, titanium alloys, and related materials for biomedical applications. Mater. Sci. Eng. R. 47, 49–121.

Liu, Z.F., Wang, W.Q., Liu, H.Y., Wang, T.Y., Qi, M., 2013. Formation and characterization of titania coatings with cortex-like slots formed on Ti by micro-arc oxidation treatment. Appl. Surf. Sci. 266 (1), 250–255.

Long, M., Rack, H.J., 1998. Titanium alloys in total joint replacement – a materials science perspective. Biomaterials. 19, 1621–1639.

Mandelbrot, B.B., 1983. The Fractal Geometry of Nature. Freeman, New York, p. 34.

Manjubala, I., Scheler, S., Bossert, J., Jandt, K.D., 2006. Mineralisation of chitosan scaffolds with nano-apatite formation by double diffusion technique. Acta Biomaterilia. 2 (1), 75–84.

Martin, J.Y., Schwartz, Z., Hummert, T.W., Schraub, D.M., Simpson, J., et al., 1995. Effect of titanium surface roughness on proliferation, differentiation, and protein synthesis of human osteoblast-like cells (MG63). J. Biomed. Mater. Res. 29 (3), 389–401.

McCafferty, E., Wightman, J.P., 1999. An X-ray photoelectron spectroscopy sputter profile study of the native air-formed oxide film on titanium. Appl. Surf. Sci. 143 (1), 92–100.

McKay, G.C., Macnair, R., MacDonald, C., Grant, M.H., 1996. Interactions of orthopaedic metals with an immortalized rat osteoblast cell line. Biomaterials. 17, 1339–1344.

Md Jani, A.M., Losic, D., Voelcker, N.H., 2013. Nanoporous anodic aluminium oxide: Advances in surface engineering and emerging applications. Prog. Mater. Sci. 58 (5), 636–704.

Mengel, R., Buns, C., Vidahl, T., 1998. An in vitro study of the treatment of implant surfaces with different instruments. Int. J. Oral Maxillofac. Implants. 67, 91–96.

Miyakawa, O., Watanabe, K., Okawa, S., Kanatani, M., Nakano, S., et al., 1996. Surface contamination of titanium by abrading treatment. Dent. Mater. 15, 11–21.

Miyazaki, S., 1998. Medical and dental applications of shape memory alloys. In: Otsuka, K., Wayman, C.M. (Eds.), Shape Memory Materials. Cambridge University Press, Cambridge, pp. 267–281.

Morks, M.F., 2008. Fabrication and characterization of plasma-sprayed HA/SiO2 coatings for biomedical application. J. Mech. Behav. Biomed. Mater. 1, 105–111.

Morks, M.F., Kobayashi, A., 2006. Influence of gas flow rate on the microstructure and mechanical properties of hydroxyapatite coatings fabricated by gas tunnel type plasma spraying. Surf. Coat. Technol. 201, 2560–2566.

Multigner, M., Ferreira-Barragáns, S., Frutos, E., Jaafar, M., Ibáñez, et al., 2010. Superficial severe plastic deformation of 316 LVM stainless steel through grit blasting: Effects on its microstructure and subsurface mechanical properties. Surf. Coat. Technol. 205 (7), 1830–1837.

Mustafa, K., Oden, A., Wennerberg, A., Hultenby, K., Arvidson, K., 2005. The influence of surface topography of ceramic abutments on the attachment and proliferation of human oral fibroblasts. Biomaterials. 26, 373–381.

Nanomaterials and nanotechnologies: an overview. In: Ashby, M.F., Ferreira, P.J., Schodek, D.L. (Eds.), Nanomaterials: Nanotechnology and Design. Butterworth-Heinemann, Elsevier, Oxford.

Nanomaterials synthesis and characterization. In: Ashby, M.F., Ferreira, P.J., Schodek, D.L. (Eds.), Nanomaterials: Nanotechnology and Design. Butterworth-Heinemann, Elsevier, Oxford.

Nanomaterials and nanotechnologies in health and the environment. In: Ashby, M.F., Ferreira, P.J., Schodek, D.L. (Eds.), Nanomaterials: Nanotechnology and Design. Butterworth-Heinemann, Elsevier, Oxford.

Neumann, W., Woelfli, W., Heimgartner, P., Streicher, R.M., 1990. Thin-layer activation of hip-joint prostheses for tribological tests. Nucl. Instrum. Methods B. 850, 57–61.

Nie, X., Leyland, A., Matthews, A., 2000. Deposition of layered bioceramic hydroxyapatite/TiO2 coatings on titanium alloys using a hybrid technique of micro-arc oxidation and electrophoresis. Surf. Coat. Technol. 125, 407–414.

Okazaki, Y., 2002. Effect of friction on anodic polarization properties of metallic biomaterials. Biomaterials. 23, 2071–2077.

Okazaki, Y., Gotoh, E., 2005. Comparison of metal release from various metallic biomaterials in vitro. Biomaterials. 26, 11–21.

Okazaki, Y., Ito, Y., Kyo, K., Tateishi, T., 1996. Corrosion resistance and corrosion fatigue strength of new titanium alloys for medical implants without V and Al. Mater. Sci. Eng. A. 213, 138.

Ong, J.L., Carnes, D.L., Bessho, K., 2004. Evaluation of titanium plasma-sprayed and plasma-sprayed hydroxyapatite implants in vivo. Biomaterials. 25, 4601–4606.

Oshida, Y., Munoz, C.A., Winkler, M.M., Hashem, A., Ito, M., 1993. Fractal dimension analysis of aluminum oxide particle for sandblasting dental use. J. Biomed. Mater. Eng. 3, 117–126.

Otsuka, K., Ren, X., 1999. Recent developments in the research of shape memory alloys. Intermetallics. 7, 511–528.

Park, J.H., Olivares-Navarrete, R., Baier, R.E., Meyer, A.E., Tannenbaum, R., et al., 2012. Effect of cleaning and sterilization on titanium implant surface properties and cellular response. Acta Biomater. 8 (5), 1966–1975.

Peutzfeld, A., Asmussen, E., 1996. Distortion of alloy by sandblasting. Am. J. Dent. 9, 65–66.

Pitkethly, M.J., 2004. Nanomaterials – the driving force. Mater. Today. 7, 20–29.

Ponthiaux, P., Wenger, F., Drees, D., Celis, J.P., 2004. Electrochemical techniques for studying tribocorrosion processes. Wear. 256, 459–468.

Popat, K.C., Chatvanichkul, K.-I., Barnes, G.L., Latempa, T.J., Grimes, C.A., et al., 2007. Osteogenic differentiation of marrow stromal cells cultured on nanoporous alumina surfaces. J. Biomed. Mater. Res. A. 80, 955–964.

Price, R.L., Waid, M.C., Haberstroh, K.M., Webster, T.J., 2003. Selective bone cell adhesion on formulations containing carbon nanofibers. Biomaterials. 24, 1877–1887.

Puleo, D.A., Nanci, A., 1999. Understanding and controlling the bone-implant interface. Biomaterials. 20, 2311–2321.

Qiu, D.L., Wang, A.P., Yin, Y.S., 2010. Characterization and corrosion behavior of hydroxyapatite/zirconia composite coating on NiTi fabricated by electrochemical deposition. Appl. Surf. Sci. 257, 1774–1778.

Quirynen, M., Bollen, C.M.L., 1995. The influence of surface roughness and surface-free energy on supra- and subgingival plaque formation in man. J. Clin. Periodontol. 22, 1–14.

Ravelingien, M., Hervent, A.S., Mullens, S., Luyten, J., Vervaet, C., et al., 2010. Influence of surface topography and pore architecture of alkali-treated titanium on in vitro apatite deposition. Appl. Surf. Sci. 256 (11), 3693–3697.

Ribeiro, A.A., Vaz, L.G., Guastaldi, A.C., Campos, J.S.C., 2012. Adhesion strength characterization of PVDF/HA coating on cp Ti surface modified by laser beam irradiation. Appl. Surf. Sci. 258 (24), 10110–10114.

Rie, K.T., Gebauer, A., Whole, J., 1996. Plasma assisted CVD for low temperature coatings to improve the wear and corrosion resistance. Surf. Coat. Technol. 86–87, 498–506.

Ryhanen, J., Kallioinen, M., Serlo, W., Peramaki, P., Junila, J., et al., 1999. Bone healing and mineralization, implant corrosion, and trace metals after nickel-titanium shape memory metal intramedullary fixation. J. Biomed. Mater. Res. 47, 472–480.

Saber-Samandari, S., Gross, K.A., 2012. Nano-indentation on amorphous calcium phosphate splats: Effect of droplet size on mechanical properties. J. Mech. Behav. Biomed. Mater. 16, 29–37.

Saber-Samandari, S., Gross, K.A., 2013. Contact nanofatigue shows crack growth in amorphous calcium phosphate on Ti, Co-Cr and Stainless steel. Acta Biomater. 9 (3), 5788–5794.

Santavirta, S.S., Lappalainen, R., Anttila, A., Pekko, P., Anttila, A., et al., 2006. The counterface, surface smoothness, tolerances, and coatings in total joint prostheses. Clin. Orthop. Relat. Res. 369, 92–102.

Sayles, R.S., Thomas, T.R., 1978. Surface topography as a non-stationary random process. Nature. 271, 431–434.

Schwartz, Z., Boyan, B.D., 1994. Underlying mechanisms at the bone-biomaterial interface. J. Cell. Biochem. 56, 340–347.

Sealy, M.P., Guo, Y.B., 2011. Surface integrity and process mechanics of laser shock peening of novel biodegradable magnesium-calcium (Mg-Ca) alloy. J. Mech. Behav. Biomed. Mater. 3, 488–496.

Shanaghi, A., Rouhaghdam, A.R.S., Ahangarani, S., Chu, P.K., 2012a. Effect of plasma CVD operating temperature on nanomechanical properties of TiC nanostructured coating investigated by atomic force microscopy. Mater. Res. Bull. 47 (9), 2200–2205.

Shanaghi, A., Rouhaghdam, A.R.S., Ahangarani, S., Chu, P.K., Farahani, T.S., 2012b. Effects of duty cycle on microstructure and corrosion behavior of TiC coatings prepared by DC pulsed plasma CVD. Appl. Surf. Sci. 258 (7), 3051–3057.

Stevens, M.M., George, J.H., 2005. Exploring and engineering the cell surface interface. Science. 310, 1135–1138.

Subramani, K., Mathew, R.T., 2012. Titanium surface modification techniques for dental implants – From microscale to nanoscale. In: Ahmed, W., Subramani, K. (Eds.), Emerging Nanotechnologies in Dentistry. William Andrew, Norwich, NY, pp. 85–102.

Sukjai, O., Asanithi, P., Limsuwan, P., Limsuwan, S., 2012. Growth of hydroxyapatite on silk fibers using simulated body fluid. Procedia Eng. 32, 1087–1093.

Taba Jr., M., Novaes Jr., A.B., Souza, S.L., Grisi, M.F., Palioto, D.B., et al., 2003. Radiographic evaluation of dental implants with different surface treatments: An experimental study in dogs. Implant Dent. 12, 252–258.

Tan, F., Naciri, M., Dowling, D., Al-Rubeai, M., 2012. In vitro and in vivo bioactivity of CoBlast hydroxyapatite coating and the effect of impaction on its osteoconductivity. Biotechnol. Adv. 30 (1), 352–362.

Teixeira, A.I., Abrams, G.A., Bertics, P.J., Murphy, C.J., Nealey, P.F., 2003. Epithelial contact guidance on well-defined micro- and nanostructured substrates. J. Cell Sci. 116, 1881–1892.

Turner, S., Kam, L., Isaacson, M., Craighead, H.G., Shain, W., et al., 1997. Cell Attachment on Silicon Nanostructures. sixth ed. AVS, Dana Point, CA, pp. 2848–2854.

Urban, R.M., Jacobs, J.J., Tomlinson, M.J., Gavrilovic, J., Black, J., et al., 2000. Dissemination of wear particles to the liver, spleen, and abdominal lymph nodes of patients with hip or knee replacement. J. Bone Joint Surg. Am. 82, 457–476.

Vadiraj, A., Kamaraj, M., 2006. Characterization of fretting fatigue damage of PVD TiN coated biomedical titanium alloys. Surf. Coat. Technol. 200, 4538–4542.

Van Hove, M.A., 2009. Atomic-scale structure: From surface to nanomaterials. Surf. Sci. 603, 13015.

Vangolu, Y., Alsaran, A., Yildirim, O.S., 2011. Wear properties of micro arc oxidized and hydrothermally treated Ti6Al4V alloy in simulated body fluid. Wear. 271, 2322–2327.

Variola, F., Yi, J.H., Richert, L., Wuest, J.D., Rosei, F., et al., 2008. Tailoring the surface properties of Ti6Al4V by controlled chemical oxidation. Biomaterials. 29, 1285–1298.

Vieira, A.C., Ribeiro, A.R., Rocha, L.A., Celis, J.P., 2006. Influence of pH and corrosion inhibitors on the tribocorrosion of titanium in artificial saliva. Wear. 261, 994–1001.

Wagner, T., Neinhuis, C., Barthlott, W., 1996. Wettability and contaminability of insect wings as a function of their surface sculptures. Acta Zool. 77, 213–225, <http://dx.doi.org/10.1111/j.1463-6395.1996.tb01265.x>.

Wang, J., Huang, C., Wan, Q., Chen, Y., Chao, Y., 2010. Characterization of fluoridated hydroxyapatite/zirconia nano-composite coating deposited by a modified electrocodeposition technique. Surf. Coat. Technol. 204, 2576–2582.

Webster, T.J., Ejiofor, J.U., 2004. Increased osteoblast adhesion on nanophase metals: Ti, Ti6Al4V, and CoCrMo. Biomaterials. 25, 4731–4739.

Webster, T.J., Smith, T.A., 2005. Increased osteoblast function on PLGA composites containing nanophase titania. J. Biomed. Mater. Res. A. 74, 677–686, <http://dx.doi.org/10.1002/jbm.a.30358>.

Webster, T.J., Siegel, R.W., Bizios, R., 1999. Osteoblast adhesion on nanophase ceramics. Biomaterials. 20, 1221–1227.

Webster, T.J., Ergun, C., Doremus, R.H., Siegel, R.W., Bizios, R., 2000. Specific proteins mediate enhanced osteoblast adhesion on nanophase ceramics. J. Biomed. Mater. Res. 51, 475–483.

Webster, T.J., Seigel, R.W., Bizios, R., 2001a. Design and evaluation of nanophase alumina for orthopedic/dental implant efficacy. Scr. Mater. 44, 1639–1642.

Webster, T.J., Schadler, L.S., Siegel, R.W., Bizios, R., 2001b. Mechanisms of enhanced osteoblast adhesion on nanophase alumina involve vitronectin. Tissue Eng. 7, 291–301.

Wen, X., Wang, X., Zhang, N., 1996. Microrough surface of metallic biomaterials: A literature review. J. Biomed. Mater. Eng. 6, 173–189.
Wennerberg, A., 1998. The importance of surface roughness for implant incorporation. Int. J. Mach. Tools Manuf. 38, 657–662.
Wennerberg, A., Ektessabi, A., Albrektsson, T., Johansson, C., Andersson, B., 2006. A 1-year follow-up of implants of differing surface roughness placed in rabbit bone. Int. J. Oral Maxillofac. Implants. 12, 486–494.
Wever, D.J., Veldhuizen, A.G., de Vries, J., Busscher, H.J., Uges, D.R.A., et al., 1998. Electrochemical and surface characterization of a nickel-titanium alloy. Biomaterials. 19, 761–769.
Wilson, C.J., Clegg, R.E., Leavesley, D.I., Pearcy, M.J., 2005. Mediation of biomaterial-cell interactions by adsorbed proteins: A review. Tissue Eng. 11 (1–2), 1–18.
Wong, M., Eulenberger, J., Schenk, R., Hunziker, E., 1995. Effect of surface-topology on the osseointegration of implant materials in trabecular bone. J. Biomed. Mater. Res. 29, 1567–1575.
Wu, X.D., Song, X.F., Li, D.S., Liu, J.G., Zhang, P.B., et al., 2012. Preparation of mesoporous nano-hydroxyapatite using a surfactant template method for protein delivery. J. Bionic Eng. 9 (2), 224–233.
Xu, J.L., Joguet, D., Cizek, J., Khor, K.A., Liao, H.L., et al., 2012. Synthesis and characterization on atomphospheric plasma sprayed amorphous silica doped hydrxoyapatite coatings. Surf. Coat. Technol. 206 (22), 4659–4665.
Yahia, L., 2000. Shape Memory Implants. Springer, Berlin, Germany.
Zäch, M., Hägglund, C., Chakarov, D., Kasemo, B., 2006. Nanoscience and nanotechnology for advanced energy systems. Curr. Opin. Solid State Mater. Sci. 10, 132–143.
Zdravecká, E., Tiainen, V.M., Konttinen, Y.T., Franta, L., Vojs, M., et al., 2012. Relationships between the fretting wear behavior and mechanical properties of thin carbon films. Vacuum. 86, 675–680.
Zhou, X., Siman, R., Lu, L., Mohanty, P., 2012. Argon atmospheric plasma sprayed hydroxyapatite/Ti composite coating for biomedical applications. Surf. Coat. Technol. 207 (25), 343–349.
Zhu, X., Chen, J., Scheideler, L., Altebaeumer, T., Geis-Gerstorfer, J., et al., 2004. Cellular reactions of osteoblasts to micron- and submicron-scale porous structures of titanium surfaces. Cells Tissue Organs. 178, 13–22.
Zhu, Y.H., Watari, F., 2007. Surface carbonization of titanium for abrasion resistant implant materials. Dent. Mater. J. 26, 244–252.
Zhua, Y.H., Wang, W., Jia, X.Y., Akasaka, T., Liao, S., et al., 2012. Deposition of TiC film on titanium for abrasion resistant implant material by ion-enhanced triode plasma CVD. Appl. Surf. Sci. 262, 156–158.
Zinger, O., Anselme, K., Denzer, A., Habersetzer, P., Wieland, M., et al., 2004. Time-dependent morphology and adhesion of osteoblastic cells on titanium model surfaces featuring scale-resolved topography. Biomaterials. 25, 2695–2711.

Theoretical modeling of friction and wear processes at atomic level

15

Yuri F. Migal
Southern Scientific Center of Russian Academy of Sciences, Rostov-on-Don, Russia

15.1 Introduction

The tribology which has arisen as a science concerning processes in macroscopic objects in recent decades extends to the micro- and nanolevels and is engaged more and more in studying atomic interactions on friction surfaces. To understand more deeply the origin of such characteristics of materials as a coefficient of friction, wear resistance and so on, it is necessary to reveal the mechanisms of the various phenomena at atomic level occurring on a surface, and estimate their influence on tribological processes. Advances in this direction are impossible without using of the newest experimental and theoretical methods of research.

Essential progress in studying friction mechanisms was made in the 1980s. Thanks to the emergence of new experimental methods, namely scanning tunneling, atomic force and frictional force microscopy (see Binnig et al., 1982, 1986; Mate et al., 1987), a new direction, nanotribology has arisen in tribology. This is the science studying friction, lubrication and wear at the nanostructure level. The main objects of research are the phenomena in nanotribocontacts: the effect of stick-slip sliding, adhesion in tribocontacts, the dependence of adhesion on structure of surfaces and chemical interactions and so on.

Along with new methods based on microscopy, the spectroscopic methods which have successfully proved themselves in solid physics – X-ray electron, Auger electron spectroscopy, X-ray absorption near-edge structure, electron energy loss spectroscopy, infrared spectroscopy, secondary ion mass spectroscopy, nuclear magnetic resonance and so on – are widely used for obtaining information on processes occurring on friction surfaces. These methods allows us to investigate, for example, the chemical composition and structure of small areas of a surface and subsurface layers (some micrometers in diameter), changes in distribution of chemical elements during friction (see Table 15.1).

Of course, when studying tribosystems an important role is played by chemical methods through which changes in chemical bonds on triboconjugated surfaces are investigated. The contribution of chemistry and especially surface chemistry in the solution of tribology problems is constantly growing (see Gellman and Spencer, 2002; Hsu, 2004). Chemistry at atomic level is the quantum chemistry which makes

Anti-Abrasive Nanocoatings.

Table 15.1 Methods of study of friction and wear processes at atomic level

	Methods	Objects
Nanotribology	Scanning tunneling, atomic force microscopy, frictional force microscopy	Effect of stick-slip sliding, adhesion in tribocontacts, dependence of adhesion on structure of surfaces, chemical interactions, etc.
Spectroscopy	X-ray electron spectroscopy, Auger electron spectroscopy, X-ray absorption near-edge structure, electron energy loss spectroscopy, infrared spectroscopy, secondary ion mass spectroscopy, nuclear magnetic resonance, etc.	Chemical composition and structure of small areas of a surface and subsurface layers (some micrometers in diameter), change in distribution of chemical elements during friction, etc.
Theoretical	Methods of molecular dynamics (MD)	Large systems of particles interacting classically
	Methods of quantum mechanics and quantum chemistry	Small systems of particles

it possible to determine types of chemical compounds, energy of bonds, direction of processes, and so on.

Currently there are two different theoretical approaches which are used in studying nanotribological processes. The first of them is classical, based on methods of molecular dynamics (MD), and the second is quantum chemical based on methods of quantum mechanics and quantum chemistry. The widely known MD method appears effective for the study of a large number of particles only if it is possible to neglect the reorganization of the electronic structure and change in chemical bonds in a system. Otherwise it is necessary to use principles of quantum mechanics. Approximate methods of quantum chemistry, which are now actively used in studying chemical interactions in the various phenomena (e.g. adsorption of molecules on a surface of solids, homogeneous and heterogeneous catalysis, biochemistry), are applicable for the solution of a number of problems of tribology. Nevertheless, it should be noted that wide application of quantum chemistry to tribology is restrained by the fact that the number of atoms in systems considered by quantum chemical methods is relatively small. It usually does not exceed several hundred – much less, than in systems studied by the MD method. However, the development of quantum chemical methods and exponential growth in the speed of calculations and memory size of modern computers do now allow us to carry out deep quantum chemical studies in the field of tribology.

Below both theoretical approaches, MD and quantum chemistry, are considered. There are many published reviews on the application of MD to tribological tasks currently (see, e.g. Braun and Naumovets, 2006; Mosey and Müser, 2007). For this reason the MD method is described briefly here, and the main focus of the chapter is the quantum chemical approach. Five types of tasks solved through methods of quantum chemistry are considered: calculation of the coefficient of friction; interaction of grease molecules with a metal surface; solid coatings; tribochemical reactions; and the influence of boundary grain segregation on wear resistance of polycrystalline materials.

15.2 MD method

Today the MD method is the most developed theoretical approach in nanotribology. In spite of the fact that the interaction of atoms is not subjected to laws of classical mechanics and, strictly speaking, when studying nanotribological processes it is necessary to use methods of quantum mechanics, in the case of big systems consisting of tens of thousands of atoms and more, as practice shows, a classical approach is applied. In fact, the MD method represents a scheme of calculation of equilibrium and transport properties of systems of many bodies with given potentials of interaction for which laws of classical mechanics are valid.

One of the important tribological tasks solved by means of MD is the study of boundary lubrication. In this case the following model (Figure 15.1a) is typical. Two contacting surfaces are divided by a thin layer of lubricant. The bottom surface

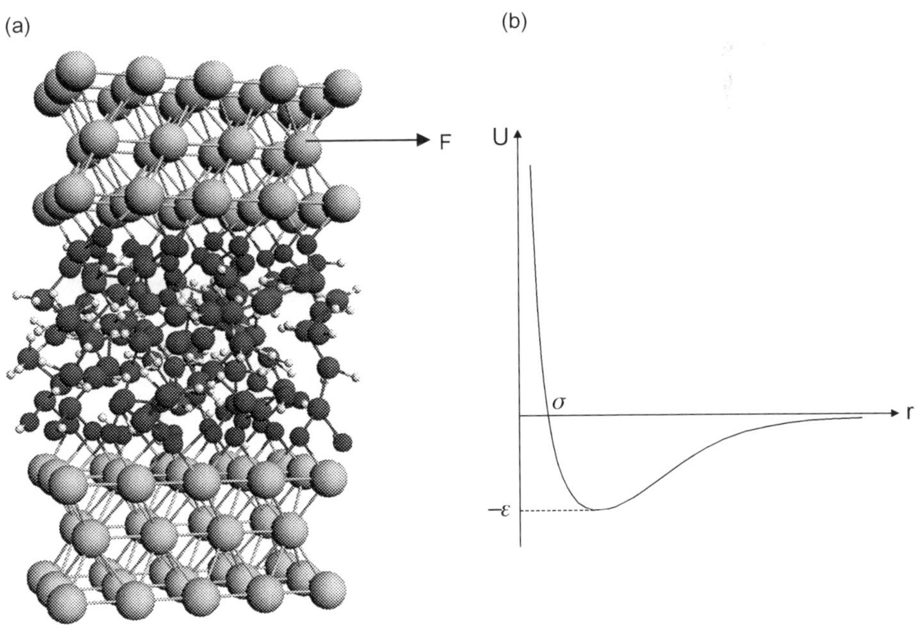

Figure 15.1 Model of sliding (a) and Lennard–Jones potential (b).

is fixed, and a shifting force F acts upon the top surface. Both the surfaces and the layer of lubricant are imitated by a system from a large number of atoms, interaction between which is described by Lennard–Jones potential:

$$U(r) = 4\varepsilon\left[\left(\frac{\sigma}{r}\right)^{12} - \left(\frac{\sigma}{r}\right)^{6}\right]$$

(or its modifications), where r is the distance between centers of particles, ε is the depth of the potential well and σ is the distance at which the interaction is equal to zero (Figure 15.1b).

At the solution of the equations of movement in the directions parallel to the plane of sliding, periodic boundary conditions are usually used. It is supposed that at shift along the system on the $\vec{a}$ vector (a lattice constant) the distribution of deformations, stresses and values of speed of particles remain invariable. In particular, the following ratio is valid:

$$v(\vec{r} + \vec{a}) = v(\vec{r}) \tag{15.1}$$

Under such conditions, in order to study the whole system it is enough to consider only the elementary cell.

Let's adduce some important results received by means of the MD method. If atomic interaction within a lubricant is weaker than the interaction of the atoms of the lubricant with the surfaces, the tribosystem is considered a soft system (see Braun and Naumovets, 2006). The lubricant in such a system melts right away after motion starts. In the opposite case of a hard tribosystem the interaction of atoms of a lubricant among themselves is stronger than the interaction with surfaces, the lubricant can remain solid in the course of sliding. Within the framework of the model it is established that the temperature T_c of melting of a lubricant film can be significantly higher than the temperature T_v of melting of lubricant bulk. For a monolayer film the relation T_c/T_v can be equal to 3. The melted lubricant consisting of several layers has a pronounced layered structure.

If both surfaces and a lubricant have ideal crystal structure and the periods of their lattices do not coincide, the friction in such a system is very small. For example, low coefficients of friction of graphite lubricants and other layered materials (of MoS_2 or Ti_3SiC_2 type) are explained by this fact. However, these advantages of lubricants disappear if the contacting surfaces are not ideal. In the case of defective (amorphous or glassy) film structures the coefficient of friction becomes big, even bigger than for liquid lubricants, typical of soft tribosystems.

Several ideas permitting the coefficient of friction to be lowered and the wear of sliding surfaces to be reduced are studied by means of the MD method. In particular, the use of fullerenes and nanotubes as lubricants allowing us to pass from sliding friction to rolling friction is considered. Methods of control and purposeful management of friction by means of additives to greases are also studied.

15.3 Quantum chemistry methods

As it was mentioned above, the MD method is effective for the study of relative movement only when it is possible to neglect the rebuilding of electronic structure and change in chemical bonds in a system. Otherwise it is necessary to use methods of quantum mechanics and quantum chemistry.

In the first work where the quantum-chemical approach was applied, rather simple semi-empirical methods were used. As an example we can refer to a work (see its description in Kolesnikov et al., 2010) in which the structure of transfer film of a polytetrafluroethylene (PTFE)-based composite was investigated. The analysis of X-ray electron spectra of the film and calculations carried out by using the Extended Hückel method (Hoffmann, 1963) allowed us to study the process of formation of the film over time and determine types of chemical compounds in the film.

Modern quantum-chemical methods, including Hartree-Fock, configuration interaction (see Cramer, 2004; Prasad, 2006), and density functional theory (DFT) which is most often used today (Kohn and Sham, 1965; Jones and Gunnarsson, 1989; Kohn et al., 1996), give us the possibility to rather precisely estimate the energy of atomic interactions, determine equilibrium distances between atoms, establish the direction of chemical processes, and so on. It should be noted that the exact quantum-chemical analysis of even elementary interactions (atom + atom) in the case where one of atoms is the atom of a heavy element, is fraught with major technical difficulties. Therefore modeling is an obligatory stage in the research into atomic interactions on a surface.

Below we will consider at first basic idea of the DFT method, and then ways of modeling which are applied to the quantum-chemical study of tribological problems.

15.3.1 *Basic ideas of DFT*

The main notion of quantum chemistry is the wave function which is the solution of the Schrödinger equation. If the wave function of a system is known we can determine, in principle, all characteristics of the system. However, the determination of the wave function itself for polyatomic systems with an acceptable accuracy represents an extremely difficult task.

In many simplified approaches, instead of wave function one uses the notion of electronic density n(r). This quantity can be found by minimization of the function of the total energy:

$$E[n] = T[n] + V[n] + U[n]$$

where $T[n]$ describes kinetic energy of particles, $V[n]$ describes interaction of electrons with nuclei and $U[n]$ describes interaction between electrons. In the DFT method based on the work of Kohn and Sham (1965), it is supposed that electrons move independently from each other and their potential energy is equal to

$$V_{KS}(x) = V_{ext}(r) + V_C(r) + V_{xc}(x) \tag{15.2}$$

where x is the total space-spin coordinate of one electron including space coordinate r and spin coordinate σ, V_{ext} is potential energy in a field created by external sources, $V_C(r)$ is the energy of Coulomb interactions in a system and V_{xc} is so-called exchange-correlation energy. Two approaches most often applied to calculate the two last terms in Eq. 15.2 are LDA (local density approximation) and GGA (generalized gradient approximation). Within LDA the electron energy depends only on density at a given point of a field, and within GGA the density gradient is also taken into account.

Work in which the basic ideas of DFT are stated and its applicability to problems of tribology is considered are presented in Table 15.2.

Table 15.2 Development of DFT and its application in tribology

References	Description
Jones and Gunnarsson (1989)	Considering a scheme that reduces the calculations of ground-state properties of systems of interacting particles exactly to the solution of single-particle Hartree-type equations. The authors survey the formalism and some of its applications (in particular to atoms and small molecules).
Becke (1993)	Testing a semi-empirical exchange-correlation functional containing local-spin-density, gradient, and exact-exchange terms on atomic energies of first- and second-row systems.
Stephens et al. (1994)	The unpolarized absorption and circular dichroism spectra of the fundamental vibrational transitions of the chiral molecule 4-methyl-2-oxetanone are calculated *ab initio* with different functionals and compared with experimental data. It was found that B3LYP functionals give excellent results.
Kohn et al. (1996)	Aim of the paper is to convey the basic concepts of DFT and breadth of its applications: the current status and trends of approximation methods (local density and generalized gradient approximations, hybrid methods) and the new light which DFT has been shedding on important concepts like electronegativity, hardness and chemical reactivity index.
Adamson et al. (1999)	An efficient algorithm for the calculation of short-range Coulomb energies is examined. The algorithm uses a boxing scheme and a prescreening for negligible integrals.
Hsu (2004)	Monomolecular thin films have been developed to investigate the fundamental mechanism of boundary lubricating films. Results provide additional insights of how antiwear films work in the lubrication system.

This approach was successful in calculations of molecules and solids, as chemical interaction between atoms is prevalent (the electron shells of interacting atoms are overlapping). If physical interaction is prevalent (electron shells are not overlapping), this approach becomes significantly less exact. In this case it is necessary to consider long-range van der Waals forces. A similar situation arises at studying compounds with layered structure, and it will be considered below in Section 15.4.3.

15.3.2 Cluster model

When studying the interaction of molecules in a tribological system (consisting, e.g. of two bodies sliding from each other) it is necessary to take into account that the full quantum-chemical analysis of such a sharply nonuniform system is practically impossible because of a huge number of atomic interactions in a system. The way out of the situation is as follows. From the surface layers of two bodies we can select some microscopic volume containing atoms of both bodies. These atoms in total represent a polyatomic cluster which allows us to imitate a phenomenon occurring in a system. The size of a cluster has to be big enough for the qualitative and quantitative regularities of the phenomenon to be reproduced by the cluster more precisely. On the other hand, the cluster should be not so big that the time of calculation of its main characteristics exceeds reasonable values. Thus, the choice of cluster size represents a compromise between two inconsistent requirements, the accuracy of reproduction of the main characteristics of system and the limitations in calculation time.

The practice of quantum-chemical calculations shows that in many cases clusters of rather small sizes can reproduce features of the studied phenomenon at qualitative and semi-quantitative levels (see, e.g. Lebedev, 2006; Montiel et al., 2011). If in a system there is a translation symmetry, then it is enough to consider, as a cluster, an elementary cell characterizing this system, and demand that the system meet periodic boundary conditions (see Eq. 15.1). The size of a model cluster, type of boundary conditions and choice of a method of solution depend on the problem posed.

15.4 Basic types of problems

Below we consider five types of tasks solved by means of quantum chemistry: calculation of the coefficient of friction, interaction of grease molecules with a metal surface, solid coverings, tribochemical reactions and influence of grain boundary segregation on wear resistance of polycrystalline materials. This list is not exhaustive, but limited by the brevity of the review.

15.4.1 Calculation of coefficient of friction

For theoretical determination of this quantity it is expedient to use periodic boundary conditions. First one needs to find the energy of interaction of two surfaces sliding from each other, and its dependence on the relative arrangement of the surfaces. Some work relating to this problem are presented in Table 15.3.

Values of coefficient of friction μ are determined by the following formula (see Zhong and Tománek, 1990; Tománek et al., 1991; Overney et al., 1992; Matsuzawa and Kishii, 1997):

$$\mu = \frac{\Delta U_{\max}}{f_{\text{ext}} \Delta x},$$

where $\Delta U_{\max}$ is the maximum change of potential energy in the course of movement; Δx is the distance between points of maximum and minimum energy.

By means of this approach the coefficient of friction was calculated by Koskilinna et al. (2007) for sliding surfaces of two types of boron nitride, cubic and hexagonal. The number of atoms in model clusters was some tens and varied over a

Table 15.3 Calculation of coefficient of friction

References	Description
Zhong and Tománek (1990)	The atomic-scale friction associated with a layer of Pd atoms moving across a graphite substrate is determined from *ab initio* total-energy calculations. The friction energy is caused by variations of the chemical bond strength and work against an external force.
Tománek et al. (1991)	The microscopic mechanism of energy dissipation in the friction force microscope (FFM) is investigated. Based on *ab initio* results for the interaction between Pd and graphite, the authors determine the atomic-scale modulation of the friction force and the corresponding stick-slip motion at the interface during the relative motion between these solids.
Overney et al. (1992)	The authors present a first-principles theory of atomic force microscopy (AFM) on layered elastic surfaces. The effect of a finite size tip on the substrate distortions and the resulting AFM image is discussed.
Matsuzawa and Kishii (1997)	*Ab initio* molecular orbital calculations were performed to predict the friction force and the coefficient of friction between surfaces contacted by van der Waals forces. A theory to account for the calculated microscale friction is presented, and the derived results are compared to the calculated results.
Koskilinna et al. (2007)	Friction coefficients for cubic boron nitride were determined with the use of hybrid density functional B3LYP calculations. Cluster models and an infinite periodic model were applied.
Wang et al. (2012)	First-principles calculations within DFT were performed to predict the friction properties between two single-side hydrogenated graphene sheets at the atomic scale. The study suggests that atomic-scale friction may be controlled by adjusting the electronic structure.

wide range. Calculations were carried out by the hybrid B3LYP method (see Becke, 1993; Stephens et al., 1994), which is a modification of the DFT method. A trajectory of relative movement of two contacting surfaces was chosen to correspond to the smallest coefficient of friction. The results of calculations of μ closest to experimental values were obtained for clusters with periodic boundary conditions. In particular, the calculated value of coefficient of friction between two c-BN surfaces was equal to 0.21–0.29 while experimental value is equal to 0.1–0.2. The corresponding values in case of h-BN are equal to 0.226–0.265 and 0.23–0.25, respectively. At loads of several nanonewtons, the calculated coefficients of friction appear reliable. At much higher loads resulting in structural deformation of the surfaces, the applied methodology would likely to fail due to the frozen structure of the surfaces.

Wang et al. (2012) performed first-principles calculations within DFT to predict the friction properties between two single-side hydrogenated graphene sheets at the atomic scale. The coefficients of the friction along two sliding paths were calculated under normal loads ranging from 1 to 9 nN. The calculated results show a general increase in the coefficient of friction with increasing normal load, and the coefficients of friction for this system exhibit an isotropic feature. Values of coefficients of friction in the range of 0.01–0.05 were predicted.

15.4.2 Interaction of grease molecules with a metal surface

A significant amount of work (e.g. Adam et al., 2001; Wang et al., 2006) has been devoted to this problem.

The main questions in the quantum-chemical study of greases are the change in structure of grease molecules, first of all, the molecules of additives near a metal surface, and determining the energy of adsorption. Kolesnikov et al. (2009) considered from these points of view heteropolyphosphates of alkali metals, M^IPMoO_6 where M^I = Li, Na, K, Rb, Cs. These inorganic oleophilic substances are promising anti-wear and anti-scoring additives to greases. Molecules of these compounds received by crystallization from fusion have a cyclical shape with alternating blocks which contain atoms of phosphorus and molybdenum (Figure 15.2). But after interaction with an iron surface the prevailing shape of the molecules is linear. For such a change in shape it is necessary, as calculations show, to provide two conditions: (1) emergence in the molecules of a negative electric charge (that is provided with transfer of electrons from an iron surface); and (2) the presence of oxygen in the environment. In this case a reaction occurs:

$$M^I_nP_nMo_nO_{6n} + m\bar{e} + \tfrac{1}{2}O_2 \rightarrow (M^I_nP_nMo_nO_{6n+1})^{m-},$$

where $n = 3$ (in the case of M^I = Li) or 4 (in the case of M^I = Na, K, Rb, Cs), m is the number of the electrons captured by a molecule. Emergence of a linear shape of heteropolyphosphates of alkali metals further provides their strong binding with a surface and, as a result, effective protection of the surface against wear.

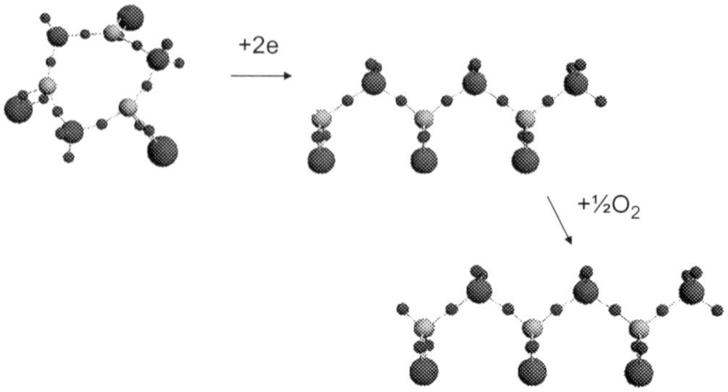

Figure 15.2 Transformation of a cyclic molecule $Li_3P_3Mo_3O_{18}$ into a radical $(Li_3P_3Mo_3O_{18})^-$ and then into a steady linear chain $Li_3P_3Mo_3O_{19}$.

The most difficult question for the quantum-chemical analysis of the interaction in a system additive + surface is the calculation of adsorption energy. Even in the absence of chemical reactions on a surface this question is usually considered approximately enough. Other cases of research into the interaction of lubricants with a surface are considered below.

15.4.3 Solid coverings

Solid coverings are widely applied for protection of surfaces against wear, including abrasive wear. These substances, as a rule, wear themselves out, and the task of researchers is to develop solid coverings (lubricants) with a long life. The most important properties of these substances are the low coefficient of friction and strong coupling with a protected surface. The best known solid lubricants are layered inorganic and organic substances: graphite, molybdenum disulfide, tungsten disulfide, polytetrafluoroethylene, carbon films, self-lubricated composite materials and so on. Research relating to the study of coverings are listed in Table 15.4.

The goal of quantum-chemical analysis of such substances is to help to look into the detail of interaction between coverings and a surface. A distinctive feature of layered materials is a prevalence of a rather weak physical interaction between the layers. In this case electron shells of atoms are not overlapping and the long-range van der Waals forces are prevalent. For the traditional DFT method accounting for these forces represents a difficult problem. In this connection, modifications of DFT are actively developed, in which van der Waals forces are considered rather precisely.

One of the most successful schemes is based on the nonlocal exchange-correlation functional of Dion et al. (2004). This functional is known as vdW-DF and is defined as:

$$E_{xc}[n] = E_{revPBE(x)}[n] + E_{LDA(c)}[n] + E_{nl(c)}[n]$$

Table 15.4 Development of DFT modifications and the study of coverings

References	Description
Dion et al. (2004)	A scheme within density functional theory is proposed that provides a practical way to generalize to unrestricted geometries the method for layered geometries. It includes van der Waals forces in a seamless fashion.
Lee et al. (2010)	A second version of the van der Waals density functional, employing a more accurate semi-local exchange functional and the use of a large-N asymptote gradient correction in determining the vdW kernel is proposed.
Birowska et al. (2011)	Novel functionals accounting for van der Waals forces are employed to the test cases of graphite and graphene layers.
Graziano et al. (2012)	The authors report results on the structural and energetic properties of graphite and hexagonal boron nitride where interlayer binding is dominated by van der Waals forces. An analysis of the physical nature of the interlayer binding is also reported.
Martin et al. (2001)	The nature and properties of polyphosphate glasses formed in the anti-wear action of zinc dithiophosphate (Zndtp) is investigated. Special attention is paid to the advantage of coupling three surface analytical techniques on the same Zndtp tribofilm: Auger (AES), XPS and XANES spectroscopies.
McCook et al. (2005)	It is found that a composite coating of polytetrafluoroethylene and epoxy shows 100 × improvements in wear resistance as compared to either of its constituents alone and a reduced friction coefficient under testing on a pin-on-disc tribometer.
Wang and Yan (2006)	PTFE-based composites containing 15 vol.% MoS_2, graphite, aluminum and bronze powder were respectively prepared. It was found all these fillers improved the wear-resistant capability of the composites.
Montiel et al. (2011)	The adsorption capability of furan derivatives with different iron clusters was studied by DFT. The results show that since the compounds possess suitable structural and electronic parameters for metal adhesion, it is observed that the functional group (NH_2, OH and SH) of furan derivatives strongly adsorb over the metal clusters.
Onodera et al. (2009)	The friction mechanism of hexagonal MoS_2 is investigated by using a hybrid quantum chemical/classical MD method. Friction for the oxygen-containing MoS_2 sheets was influenced by not only the Coulomb repulsive interaction but also the atomic-scale roughness of the MoS_2/MoS_2 sliding interface.
Onodera et al. (2010)	The authors review a quantum chemical MD tribochemical simulator, developed by them on the basis of a hybrid tight-binding quantum chemical/classical MD method. Also its applications to the tribochemical reaction dynamics of a few lubricant additives are considered.

(*Continued*)

Table 15.4 (Continued)

References	Description
Khedkara et al. (2002)	The tribological behavior of PTFE and PTFE composites with filler materials such as carbon, graphite, E glass fibers, MoS_2 and poly-*p*-phenyleneterephthalamide (PPDT) fibers was studied.
Wang et al. (2013)	The van der Waals corrected first-principles approach within DFT has been performed to calculate the interaction energies and friction coefficients for hydrogen- and fluorine-terminated diamond film systems. It was found that the fluorine-terminated diamond film is more stable and chemically inert than that of hydrogen termination.
Menezes et al. (2012)	The tribological behavior of metal matrix composites reinforced with graphite particles (aluminum–graphite, magnesium–graphite, copper–graphite, silver–graphite and nickel–graphite composites) were studied. The influence of various variables on the friction coefficient and wear rate is discussed.
Vasil'ev et al. (2011)	It is shown that the coefficient of friction of polycrystalline graphite materials is governed by the crystallite size and adsorption heat of the molecules making up the lubricating layer.
Wang et al. (2006)	The molecular geometries optimization and electronic structures of 14 disulfide compounds were investigated by DFT at 6–31G basis set level. The calculated results indicated that the S–S bond and C–S bond of the compounds tend to be broken when organic disulfide compounds interact with a metal.

where $E_{\mathrm{revPBE(x)}}[n]$ is the exchange energy received by means of the revPBE-functional (Adamson et al., 1999), $E_{\mathrm{LDA(c)}}[n]$ is the correction to the local density approximation LDA and $E_{\mathrm{nl(c)}}[n]$ is nonlocal correction with an approximate accounting of van der Waals interaction.

Graziano et al. (2012) applied the functional vdW-DF and its modifications to calculations of graphite and hexagonal boron nitride (h-BN). These layered substances, in spite of their different chemical composition, have close geometrical and energy characteristics, including interlayer distances. This has stimulated numerous theoretical studies into these substances thanks to the fact that they have become model benchmark systems for research. The purpose of calculations at this stage is the best adjustment of calculated parameters to their experimental values. The main purpose of such research is to develop the optimum scheme of calculations suitable for the whole class of objects.

Graziano et al. (2012) used the periodic boundary conditions and various approximations to test the considered approach. At the beginning, the C–C bond lengths and atomization energies for a single layer of graphite were carried out and compared with experimental values and results of calculations by traditional DFT. It was shown that for all considered functionals the C–C bond length is approximately identical and equal to 1.41–1.42 Å, in agreement with the experiment.

Atomization energy E_{atom} within the LDA approach was overestimated essentially (by almost 20%). The other functionals at a calculation of E_{atom} offer much better agreement with experiment.

Then the following interlayer characteristics of graphite, for which experimental values are also known, were calculated: the distance b, binding energy E_{inter} and the elastic constant C_{33} in the direction between layers. Experimental values are: $b = 3.33$ Å, $E_{inter} = -52 \pm 5$ (meV/atom) and $C_{33} = 36 \pm 1$ GPa. It was found that the simplest LDA approach in which long-range interaction in explicit form are not considered, reproduces well interlayer distance (3.31 Å); however, binding energy is significantly underestimated (−20 meV/atom). Within the vdW-DF2 approach, described in detail in Lee et al. (2010), the following results are achieved: $b = 3.54$ Å, $E_{inter} = -48$ (meV/atom) and $C_{33} = 33$ GPa. These can be accepted as good results.

Comparing the results of calculations of distance and binding energy between layers in graphite and hexagonal boron nitride, it was discovered that almost all considered functionals give close values of calculated quantities for both substances. The performed analysis reveals the reason for such proximity. It was shown that in h-BN, thanks to the polar nature of this compound, the contribution of electrostatic interaction to binding energy between layers is slightly greater than in unpolar graphite. On the other hand, the contribution of van der Waals forces in graphite is greater, and the total effect from two types of interaction in both substances is approximately identical.

Birowska et al. (2011) also compared the various functionals, including functional vdW-DF, which take into account van der Waals forces, through studying bulk graphite and layered graphene. Here the efficiency of the functionals investigated was also confirmed.

In Wang et al. (2013) the van der Waals corrected first-principles approach within DFT was performed to calculate the interaction energies and coefficients of friction for hydrogen- and fluorine-terminated diamond film systems. Quantitative calculations showed that the hydrogen-terminated diamond film system has larger potential barriers and hence larger coefficients of friction of about 0.15 under the loads of 0.5−3 nN, which are almost twice as big as a fluorine-terminated system (about 0.08). The results are in agreement with experiments and other theoretical predictions.

Onodera et al. (2009) proposed a new approach for studying the friction mechanism of hexagonal MoS_2. The authors used a hybrid method combining quantum chemical and classical MD methods. First, several parameters for MD simulations via accurate quantum-chemistry calculations were determined and MoS_2 and $MoS_{2-x}O_x$ structures were successfully reproduced. The atomic interactions between MoS_2 sheets were studied by using the hybrid method. It was found that the predominant interaction between two sulfur layers in different MoS_2 sheets is Coulomb repulsion, which directly affects the MoS_2 lubrication. MoS_2 sheets adsorbed on a nascent iron substrate reduced friction further due to much larger Coulomb repulsive interactions. Friction for the oxygen-containing MoS_2 sheets was influenced by not only the Coulomb repulsive interaction but also the atomic-scale roughness of the

MoS_2/MoS_2-sliding interface. This hybrid approach was developed further by Onodera et al. (2010), who also applied it to the tribochemical reaction dynamics of a few lubricant additives.

The important direction in the theory of solid coverings is also the study of composite materials. It is known, for example, that PTFE has a low coefficient of friction on polished steel, lying in the range of 0.05–0.10. However, pure PTFE is suitable for covering a small amount because it is subject to creep. For this reason the PTFE has to be used in combination with stronger materials providing greater wear resistance (see Khedkara et al., 2002; McCook et al., 2005; Wang and Yan, 2006). Thus, there is a compatibility problem for mixed components. It is necessary to select components with great adhesion to each other than a lubricant that is not stratified in use. The quantum-chemical analysis could allow us to estimate the adhesion factor between components of composite materials and predict their wear resistance.

One more type of tribological task which can be investigated by quantum chemistry is research into interphase boundary interactions in the systems representing inclusion of soft materials (like graphite) in metal matrix composites (see Menezes et al., 2012) or, for example, ultraphosphates and chalcogens in pores of graphite (see Vasil'ev et al., 2011).

15.4.4 Tribochemical reactions

Many reactions which are impossible under normal conditions can occur on a surface of solids during friction. This is caused by the energy emitted at local sites (on the whole at the tops of the roughness) exceeding the energy of activation of any reaction (see, e.g. Bulgarevich et al., 2012).

It is known that high pressures and temperatures, which appear in tribocontacts in the course of friction, promote specific chemical reactions as a result of which the properties of lubricants and films on sliding surfaces can change significantly. As an example let us consider change of the shape of polymer chains of zinc phosphide (ZnP). This compound forms the basis of anti-wear films derived from zinc dialkyldithiophosphate (ZDDP), which is an additive to engine oils (see Mosey and Müser, 2007). Modeling showed that the flexible linear molecules of ZnP that occur in the course of sliding at the tops of roughness of metal surfaces (where pressure at friction exceeds 5 GPA), can be sewn in the cross directions, creating a rigid net structure. This process is irreversible, and as a result the surface of the metal becomes covered by a film whose hardness is close to the hardness of iron. If such an additive is on an iron surface, the wear resistance of the surface increases. In the case of an aluminum surface the net structure from the molecules of ZnP works as an abrasive destroying a surface.

Monasse and Montmitonnet (2008) shown that sulfur-containing molecules, which are used as extreme pressure additives for low-alloy steel surfaces, react with ferrous surfaces. They form FeS_x compounds which strengthen the surface and reduce scuffing. The stable conformation and the chemical stability and reactivity of an alkyl sulfur molecule (DTDP) with oxygen and iron surface are predicted by the DFT model.

Hayashi et al. (2012) studied the chemical reaction occurring in layered diamond-like carbon (DLC) in the presence of methanol. It became clear that the availability of methanol vapors improves the tribological characteristics of DLC films. By using the tight-binding quantum chemical MD method developed by the authors, they constructed a simulation model in which one methanol molecule is sandwiched between two hydrogen-terminated DLC films. By means of this model it was shown that the methanol molecule decomposed and then OH-termination of the DLC was realized and the CH_3 species was incorporated into the DLC film. Such a phenomenon leads to good low-friction properties.

15.4.5 Influence of grain boundary segregation on wear resistance of polycrystalline materials

One of the causes of the destruction of surface layers of polycrystals, mentioned in the literature, is grain boundary segregation during which the atoms of alloying and impure elements go out from the bulk of the grains on their boundaries. There are various points of view on how segregated atoms contribute to this destruction. One of the hypotheses is based on the fact that in process of grain boundary segregation the metal–metal bond is substituted by a weaker or stronger bond of metal with segregated atoms.

From practice it is known that some elements, for example sulfur and phosphorus, embrittle steel. By means of simple estimates Seah (1980) has shown that atoms of embrittling elements are capable of weakening the coupling between grains. To achieve this result only the data of the sublimation enthalpy and sizes of atoms of studied substances were used. Methods of quantum chemistry allow for more accurate (than in Seah, 1980) analysis of atomic interactions affecting the destruction of metals. In Kolesnikov et al. (2008, 2010) and Migal et al. (2012) a study of the chemical bond of atoms of various elements with a surface of iron grains was performed by means of cluster modeling. Previously it was shown that clusters with around 18 or more atoms allow us to investigate at qualitative and semi-quantitative levels changes of energy of coupling between the segregated atoms and a surface of iron grains, occurring in the course of segregation. Calculations were carried out by means of the software package ADF (see te Velde et al., 2001) based on the DFT method. This package makes it possible to study compounds of transitional metals, which is essential for the analysis of phenomena on a surface of steel. It was established that the going out of an element X on a grain boundary increases wear resistance of steel if the energy of the Fe–X coupling modulo is more than the energy of Fe–Fe coupling. If the energy of Fe–X coupling is less than the energy of Fe–Fe coupling, wear resistance decreases. Thus, the degree of strength of the coupling of segregated atoms with iron atoms in a surface layer of metal is the essential factor influencing wear resistance.

Migal et al. (2012) carried out calculations of binding energy of 18 atomic clusters $Fe_2Fe_6X_2Fe_2Fe_6$ which contain adsorbed and segregated atoms of

Table 15.5 Energy necessary for dividing segregate clusters $Fe_2Fe_6X_2Fe_2Fe_6$ (eV)

	I	II	III	IV	V	VI	VII	VIII			0
1	H **9.9**										He **1.7**
2	Li **3.4**	Be **6.8**	B **11.7**	C **12.3**	N **8.7**	O **7.8**	F **2.7**				Ne **1.0**
3	Na **2.0**	Mg **4.6**	Al **6.3**	Si **7.5**	P **7.7**	S **6.4**	Cl **1.6**				Ar **0.7**
4	K **1.6**	Ca **3.9**	Sc **5.8**	Ti **8.1**	V **9.7**	Cr **10.9**	Mn **11.1**	Fe **10.4**	Co **9.0**	Ni **8.1**	
	Cu **6.2**	Zn **5.3**	Ga **5.8**	Ge **5.8**	As **5.4**	Se **3.8**	Br **0.8**				Kr **0.3**

chemical elements of the first four periods of the periodic table, with atomic numbers from 1 to 36 (from hydrogen to krypton). Using the data obtained, the values of energy, which are necessary to break clusters into two parts imitating grains in model clusters, were calculated. The results of the calculations are presented in Table 15.5 constructed by analogy to the periodic table. This table actually reflects compatibility of various elements with iron on its grain boundary surfaces. From this it follows that atoms of boron, carbon and manganese sew together strongly the iron grains because the binding energy of the clusters containing these atoms appears to be lower than the binding energy of the same-sized clusters consisting of pure iron. On the contrary, atoms of lithium, silicon, phosphorus, sulfur, copper and nickel, though entering into chemical bonds with iron, weakens the bond between the grains because the binding energy of the corresponding clusters is less than the binding energy of a pure iron cluster. Obviously, the weakening of a bond between grains in the surface layer reduces steel wear resistance. From Table 15.5 it follows that the energy necessary for dividing segregate clusters is in periodic dependence on the charge of atomic nuclei, which corresponds to Mendeleyev's periodic law.

In spite of the approximate character of the calculations, the results obtained conform to known experimental facts. The received dependence of the energy of clusters on an atomic number of segregate can be a theoretical base for the prediction of strength properties of polycrystalline materials with different alloying components.

15.5 Lubrication and one-electron transfers

Today the most detailed theoretical information on processes occurring on a surface of solids can be obtained from quantum-chemical calculations of a polyatomic system modeling a real object. However, some estimates of required quantities can be achieved approximately, without the full calculation of a system. One of

Table 15.6 Lubrication chemistry viewed from electronic structural principles

References	Description
Pearson (1986)	The concepts of absolute electronegativety and absolute hardness are incorporated into molecular orbital theory. A graphic and concise definition of hardness is given as twice the energy gap between the highest occupied molecular orbital and the lowest unoccupied molecular orbital.
Lee (1991)	The acid–base interaction in terms of the molecular interaction is discussed from a broader background. On the basis of DFT, two chemical parameters, that is chemical potential μ and absolute hardness η, are shown to govern an acid-based interaction. From these, the number of transferred electrons ΔN can be estimated.
Martin (1999)	A set of tribochemical reactions between polyphosphates and oxides is proposed on the basis of the hard and soft acids and bases (HSAB) principle. In the case of ZDDP the model predicts a layered tribofilm with the presence of polymer-like zinc metaphosphate overlying a mixed-transition-metal phosphate glass.
Li et al. (2004)	Fundamental molecular issues in lubrication chemistry were reviewed under the categories of solution chemistry, contact chemistry and tribochemistry. By introducing the DFT-derived chemical reactivity parameters and related electronic structural principles, their relevancy to lubrication chemistry was explored.

the new directions of the analysis of friction processes is based on studying of electronic structure of interacting bodies. Some work with this theme is presented in Table 15.6.

It is supposed that information on the highest occupied molecular orbital (HOMO) and lowest unoccupied molecular orbital (LUMO) or gap between the conduction and valence energy bands of solids (E_g) for various components of the system are sufficient for predicting the properties of the whole system (see Li et al., 2004). Such a point of view can be explained by the fact that character of tribological processes is determined first of all by transitions of electrons between the occupied and unoccupied states.

Two basic parameters are introduced: electronic chemical potential (μ), and absolute or chemical hardness (η) (Pearson, 1986; Lee, 1991):

$$\mu = \left(\frac{\partial E}{\partial N}\right)_{V(r)}, \quad \eta = \frac{1}{2}\left(\frac{\partial^2 E}{\partial N^2}\right)_{V(r)},$$

where E is the electronic energy of the chemical system, N is the number of electrons and $V(r)$ is the sum of the potential created by nuclei and any external potentials. Chemical potential μ measures the tendency of electrons to escape from a subsystem, while chemical hardness η determines the resistance of a subsystem to lose electrons. Both μ and η are global properties in the sense that they characterize a subsystem as a whole.

Within the validity of Koopmans' theorem for closed shell objects one can write:

$$-\mu = \chi = (\varepsilon_{\mathrm{LUMO}} + \varepsilon_{\mathrm{HOMO}})/2;$$
$$\eta = (\varepsilon_{\mathrm{LUMO}} - \varepsilon_{\mathrm{HOMO}}/2.$$

If the difference of energies $\varepsilon_{\mathrm{LUMO}} - \varepsilon_{\mathrm{HOMO}}$ is small, it is easy to excite the electron transition from HOMO to LUMO. Thus, a subsystem with small value of η is easily excitable, and it is called soft. Otherwise, at great value of η it is necessary to transmit a large amount of energy to a subsystem, and this type of subsystem is called hard.

At the equilibrium, chemical potentials within a chemical system are equal everywhere, while absolute hardness is a function of position. According to Sanderson's electronegativity equalization principle, when two subsystems, *A* and *B*, are brought together, electrons will flow from that of lower χ to that of higher χ until the chemical potential reaches an equilibrium. As a first approximation, the fractional number of electrons transferred, ΔN, is given by:

$$\Delta N \approx \frac{\mu_B - \mu_A}{2(\eta_A + \eta_B)}$$

It can be seen from this expression that a big difference of chemical potentials promotes transfer of electrons between subsystems, and a large sum of chemical hardnesses opposes such a transfer. In the case of two hard subsystems the value of $\Sigma\eta_i$ can be very big, then ΔN becomes too small, and the interaction of the subsystems is determined by electrostatic interaction. On the other hand, for soft–soft interactions the value of $\Sigma\eta_i$ can be rather small, ΔN will be big, and the main effect is electron (charge) transfer. However, it should be noted, that molecular interactions involve other interactions besides electron transfer, and ΔN is not the total change of electrons, but is still useful in serving as an approximation for the bond strength.

The approach based on concepts of electronic chemical potential and chemical hardness was repeatedly applied to the analysis of tribosystems (see the reviews in Li et al., 2004 and Erdemir et al., 2005). Let us adduce some questions studied in Martin (1999) and Martin et al. (2001). Martin et al. explained the role of ZDTP in eliminating abrasive wear due to the presence of transition metal oxide species in boundary lubrication. Another example is the investigation of the two-layer structure of the ZDTP film, in which a chemical hardness model was developed to predict the formation of the layered tribofilm.

15.6 Conclusion

The two considered approaches, the methods of MD and quantum chemistry, allow us to solve a wide range of problems of tribology at an atomic level. Obviously, these methods mutually supplement each other. As it was noted above, the MD method allows us to consider systems from a large number of atoms and makes it possible to study collective effects in nanotribosystems, including the influence of ordering in an arrangement of atoms on a course of processes, heat transfer between various parts of system and so on. Quantum chemistry methods allow us to investigate quantum and chemical effects in rather small systems: tribochemical reactions, change of shape of molecules as a result of tribological processes and so on. Currently the amount of work in which these methods are used simultaneously at the solution to tribological tasks is small, and these only consider a limited number of questions. The generally accepted scheme of association of these methods has not yet been created, but, clearly, the emergence of such an association will lead in the near future to a deeper understanding of the mechanisms of the friction process.

References

Adam, J.B., Hector, L.G., Siegel, D.J., Yu, H., Zhong, J., 2001. Adhesion, lubrication, and wear on the atomic scale. Surf. Interface Anal. 31, 619–626.

Adamson, R.D., Dombroski, J.P., Gill, P.M.W., 1999. Efficient calculation of short-range Coulomb energies. J. Comput. Chem. 20 (9), 921–927.

Becke, A.D., 1993. Density-functional thermochemistry. III. The role of exact exchange. J. Chem. Phys. 98, 5648–5652.

Binnig, G., Rohrer, H., Gerber, C., Weibel, E., 1982. Surface studies by scanning tunneling microscopy. Phys. Rev. Lett. 49, 57–61.

Binnig, G., Quate, C.F., Gerber, C., 1986. Atomic force microscope. Phys. Rev. Lett. 56, 930–933.

Birowska, M., Milowska, K., Majewski, J.A., 2011. Van der Waals density functionals for graphene layers and graphite. Acta Phys. Pol. A. 120, 845–848.

Braun, O.M., Naumovets, A.G., 2006. Nanotribology: microscopic mechanisms of friction. Surf. Sci. Rep. 60, 79–158.

Bulgarevich, S.B., Boiko, M.V., Tarasova, E.N., Feizova, V.A., Lebedinskii, K.S., 2012. Kinetics of mechanoactivation of tribochemical processes. J. Frict. Wear. 33 (5), 345–353.

Cramer, C.J., 2004. Essentials of Computational Chemistry: Theories and Models. John Wiley, New York.

Dion, M., Rydberg, H., Schröder, E., Langreth, D.C., Lundqvist, B.I., 2004. Van der Waals density functional for general geometries. Phys. Rev. Lett. 92, 246401.

Erdemir, A., Li, S., Jin, Y., 2005. Relation of certain quantum chemical parameters to lubrication behavior of solid oxides. Int. J. Mol. Sci. 6, 203–218.

Gellman, A.G., Spencer, N.D., 2002. Surface chemistry in tribology. Proc. Inst. Mech. Engrs. J. Eng. Tribol. 216, 443–461.

Graziano, G., Klimeš, J., Fernandez-Alonso, F., Michaelides, A., 2012. Improved description of soft layered materials with van der Waals density functional theory. J. Phys. Condens. Matter. 24, 424216–424224.

Hayashi, K., Sato, S., Bai, S., Higuchi, Y., Ozawa, N., et al., 2012. Fate of methanol molecule sandwiched between hydrogen-terminated diamond-like carbon films by tribochemical reactions: tight-binding quantum chemical molecular dynamics study. Faraday Discuss. 156, 137–146.

Hoffmann, R., 1963. An extended Hückel theory. I. Hydrocarbons. J. Chem. Phys. 39, 1397–1412.

Hsu, S.M., 2004. Molecular basis of lubrication. Tribol. Int. 37, 553–559.

Jones, R.O., Gunnarsson, O., 1989. The density functional formalism, its applications and prospects. Rev. Mod. Phys. 61, 689–746.

Khedkara, J., Negulescub, I., Meletisa, E.I., 2002. Sliding wear behavior of PTFE composites. Wear. 252, 361–369.

Kohn, W., Sham, L.J., 1965. Self-consistent equations including exchange and correlation effects. Phys. Rev. A. 140, 1133–1138.

Kohn, W., Becke, A.D., Parr, R.G., 1996. Density functional theory of electronic structure. J. Phys. Chem. 100, 12974–12980.

Kolesnikov, V.I., Migal, Y.F., Mizhiritskaya, S.N., Doronkin, V.N., 2008. Quantum-chemical investigation of the effect of grain boundary segregation on wear resistance of steel. J. Frict. Wear. 29 (2), 99–107.

Kolesnikov, V.I., Migal, Y.F., Savenkova, M.A., Mizhiritskaya, S.N., 2009. Adsorption of heteropolyphosphates of alkaline metals on iron surface. J. Frict. Wear. 30 (6), 404–410.

Kolesnikov, V.I., Kozakov, A.T., Migal, Y.F., 2010. Study of friction and wear in the wheel-rail system by X-ray electron and auger-electron spectroscopy and quantum chemistry. J. Frict. Wear. 31 (1), 11–22.

Koskilinna, J.O., Linnolahti, M., Pakkanen, T.A., 2007. Friction paths for cubic boron nitride: an ab initio study. Tribol. Lett. 27, 145–154.

Lebedev, M.V., 2006. Mechanism of H_2S molecule adsorption on the GaAs(100) surface: ab initio quantum-chemical analysis. Phys. Solid State. 48, 164–171.

Lee, K., Murray, E.D., Kong, L., Lundqvist, B.I., Langreth, D.C., 2010. Higher-accuracy van der Waals density functional. Phys. Rev. B. 82, 081101–081105.

Lee, L.-H., 1991. Relevance of the density-functional theory to acid-base interactions and adhesion in solids. In: Mittal, K.L., Anderson Jr., H.R. (Eds.), Acid-Base Interactions: Relevance to Adhesion Science and Technology. VSP, Utrecht, The Netherlands, pp. 25–46.

Li, S., Yang, H., Jin, Y., 2004. Lubrication chemistry viewed from DFT-based concepts and electronic structural principles. Int. J. Mol. Sci. 5, 13–34.

Martin, J.M., 1999. Antiwear mechanisms of zinc dithiophosphate: a chemical hardness approach. Trib. Lett. 6, 1–8.

Martin, J.M., Grossiord, C., Le Mogne, T., Bec, S., Tonck, A., 2001. The two-layer structure of zndtp tribofilms. Part 1: AES, XPS and XANES analyses. Tribol. Int. 34, 523–530.

Mate, C.M., McClelland, G.M., Erlandsson, R., Chang, S., 1987. Atomic-scale friction of a tungsten tip on a graphite surface. Phys. Rev. Lett. 59, 1942–1945.

Matsuzawa, N.N., Kishii, N.J., 1997. Theoretical calculations of coefficients of friction between weakly interacting surfaces. J. Phys. Chem. A. 101, 10045–10052.

McCook, N.L., Burris, D.L., Bourne, G.R., Steffens, J., Hanrahan, J.R., et al., 2005. Wear resistant solid lubricant coating made from PTFE and epoxy. Tribol. Lett. 18 (1), 119–124.

Menezes, P.L., Rohatgi, P.K., Lovell, M.R., 2012. Self-lubricating behavior of graphite reinforced metal matrix composites. In: Nosonovsky, M., Bhushan, B. (Eds.), Green Tribology, Green Energy and Technology. Springer-Verlag, Berlin, Germany, pp. 445–480.

Migal, Y.F., Kolesnikov, V.I., Doronkin, V.N., Novikov, E.S., 2012. Interaction of atoms with grain surfaces in steel: periodic dependence of binding energy on atomic number and influence on wear resistance. Adv. Mater. Phys. Chem. 2, 201–207.

Monasse, B., Montmitonnet, P., 2008. DFT – modeling of the reaction of a polysulfur extreme-pressure lubricant additive on iron surface. Int. J. Mater. Form. 1 (Suppl. 1), 1251–1254.

Montiel, E., Jayanthi, N., Veloz, M.A., Pandiyan, T., Cruz, J., 2011. Theoretical studies on iron surface coating: adsorption of furan derivatives over Fe(n) clusters (n = 1–4). J. Nanosci. Nanotechnol. 11 (6), 5483–5490.

Mosey, N.J., Müser, M.H., 2007. Atomistic modeling of friction. In: Lipkowitz, K.B., Cundari, T.R. (Eds.), Reviews in Computational Chemistry, vol. 25. John Wiley, New York, pp. 67–124.

Onodera, T., Morita, Y., Suzuki, A., Koyama, M., Tsuboi, H., et al., 2009. A computational chemistry study on friction of h-MoS_2. Part I. Mechanism of single sheet lubrication. J. Phys. Chem. B. 113 (52), 16526–16536.

Onodera, T., Miura, R., Suzuki, A., Tsuboi, H., Hatakeyama, N., et al., 2010. Development of a quantum chemical molecular dynamics tribochemical simulator and its application to tribochemical reaction dynamics of lubricant additives. Modell. Simul. Mater. Sci. Eng. 18, 034009.

Overney, G., Tomanek, D., Zhong, W., Sun, Z., Miyazaki, H., et al., 1992. Theory for the atomic force microscopy of layered elastic surfaces. J. Phys. Condens. Matter. 4, 4233.

Pearson, R.G., 1986. Absolute electronegativity and hardness correlated with molecular orbital theory. Proc. Natl. Acad. Sci. U. S. A. 83, 8440–8441.

Prasad, R.K., 2006. Quantum Chemistry. New Age International (P), New Delhi, India.

Seah, M.P., 1980. Adsorption-induced interface decohesion. Acta Metall. 28, 955–962.

Stephens, P.J., Devlin, F.J., Chabalowski, C.F., Frisch, M.J., 1994. Ab initio calculation of vibrational absorption and circular dichroism spectra using density functional force fields. J. Phys. Chem. 98, 11623–11627.

te Velde, G., Bickelhaupt, F.M., van Gisbergen, S.J.A., Fonseca Guerra, C., Baerends, E.J., Snijders, J.G., et al., 2001. Chemistry with ADF. J. Comput. Chem. 22, 931–967.

Tománek, D., Zhong, W., Thomas, H., 1991. Calculation of an atomically modulated friction force in atomic–force microscopy. Europhys. Lett. 15, 887–892.

Vasil'ev, Y.N., Kolyaev, I.A., Fugol', V.A., 2011. On the friction mechanism of solid lubricants. J. Frict. Wear. 32 (5), 324–332.

Wang, J., Wang, F., Li, J., Wang, S., Song, Y., et al., 2012. Theoretical study of superlow friction between two single-side hydrogenated graphene sheets. Tribol. Lett. 48, 255–261.

Wang, J., Wang, F., Li, J., Sun, Q., Yuan, P., et al., 2013. Comparative study of friction properties for hydrogen- and fluorine-modified diamond surfaces: a first-principles investigation. Surf. Sci. 608, 74–79.

Wang, X.-Y., Li, X.-F., Wen, X.-H., Tan, Y.-Q., 2006. DFT method study on lubricant properties of disulfide compounds as extreme pressure additive. J. Theor. Comput. Chem. 5, 1–10.

Wang, Y., Yan, F., 2006. Tribological properties of transfer films of PTFE-based composites. Wear. 261, 1359–1366.

Zhong, W., Tománek, D., 1990. First-principles theory of atomic-scale friction. Phys. Rev. Lett. 64, 3054–3057.

Mechanical characterization of thin films by depth-sensing indentation

16

Jorge M. Antunes[*,†], *José V. Fernandes*[*], *and Natalyia A. Sakharova*[*]
[*]University of Coimbra, Portugal; [†]Instituto Politécnico de Tomar, Portugal

16.1 Introduction

Since a long time ago, thin films have been used in decoration and corrosion protection. However, more recently this technology has started to be studied in depth due to the growing need to acquire materials with particular features and mechanical properties. For example, the development of thin films applied in the coating of low-cost classic materials makes it possible to obtain composite materials with high performance and versatility. Currently, thin films are used in a number of fields such as cutting tools, medicine, computing, telecommunications and optics.

In this context, the development of methodologies to evaluate thin films' mechanical properties, in order to predict their performance in service, is required. Depth-sensing indentation (DSI) is the test most widely used for this propose. In fact, the non-destructive character of the test and the absence of specific requirements with regard to the size and geometry of the sample make DSI appropriate for the mechanical characterization of thin films. The mechanical properties usually evaluated by DSI tests are hardness and Young's modulus. Furthermore, other mechanical properties of the materials such as yield stress, strain hardening parameters and residual stresses can be also evaluated (see, e.g. Tabor, 1951; Johnson, 1970; Dao et al., 2001; Swaddiwudhipong et al., 2005; Antunes et al., 2007b; Tsui et al., 1996, 1997; Bolshakov et al., 1996; Suresh and Giannakopoulos, 1998; Chen et al., 2006; Jang, 2009).

Nevertheless, during indentation tests of composites with the aim of characterising thin films some difficulties can appear, specifically for increasing indentation depths the substrate can deform elastically and plastically. As consequence, the mechanical properties obtained for the composite are a complex function of the properties of the film and substrate (see, e.g. Doerner and Nix, 1986; Fabes et al., 1992; Page and Hainsworth, 1993; Chen and Vlassak, 2001; Chechenin et al., 1995; Tsui et al., 1999; Saha and Nix, 2002). In this context, the obvious solution for determining the mechanical properties of thin films is the use of low indentation depths in order to avoid as much as possible the contribution of the substrate on the results obtained by DSI. With this purpose, a rule that limit the maximum indentation depth to around 10−15% of the film's thickness

Anti-Abrasive Nanocoatings.

(see, e.g. Lichinchi et al., 1998; Fisher-Cripps, 2004) is usually used. This critical value for the indentation depth depends on the film and substrate mechanical properties, namely the ratio between both hardnesses. Furthermore, Manika and Maniks (2008) established that the critical indentation depth is sensitive to the film structure (crystal, polycrystalline or amorphous), which determines the plastic deformation mechanisms and, consequently, the size of the region under the indenter that deforms plastically. This shows the important role that this region plays in composite behaviour for relatively low indentation depths, as stated by Fernandes et al. (1998).

Obviously, the experimental procedure associated with the use of DSI tests requires careful preparation of the sample surface. This is particularly important when using very low indentation depths, which involves the increased risk of reducing the accuracy of the evaluated mechanical properties. However, uncertainties in the DSI data due to the roughness of the sample surface and the indenter's geometrical imperfections are inevitable to a certain extent (see, e.g. Antunes et al., 2002, 2006, 2007a).

As alternative to the use of very low indentation depths, methodologies that consider the substrate and film contribution to the composite mechanical properties were proposed. Several experimental and theoretical studies have been done to analyse the problem of extracting the mechanical properties of thin films from the results obtained for the composite (see, e.g. Burnett and Rickerby, 1987; Joslin and Oliver, 1990; Menčík et al., 1997; Korsunsky et al., 1998; Antunes et al., 2007b, 2008).

In the specific case of hardness and Young's modulus, these methodologies mainly consist of analytical and semi-empirical models. In both of these cases, the models are usually very simple to implement and intended to separate the contributions of the film and the substrate from the results obtained for the composite.

16.2 Hardness

Hardness is an important mechanical property, easily determined in the case of bulk and thin films. The hardness, H_{IT}, is expressed by the ratio between the maximum applied load, P_{max}, and the indentation contact area, A_c (ISO 14577, 2002):

$$H_{IT} = \frac{P_{max}}{A_c}. \tag{16.1}$$

For the most frequently used indenters in DSI tests, the Vickers and Berkovich indenters (both have pyramidal geometry, with square and triangular bases, respectively), the indentation contact area is given by:

$$A_c = 24.5h_c^2\,, \tag{16.2}$$

where the contact indentation depth, h_c, is evaluated from the unloading part of the indentation curve, using the following equation (Oliver and Pharr, 1992):

$$h_c = h_{max} - h_s = h_{max} - \varepsilon C P_{max}, \quad (16.3)$$

where C represents the compliance, that is, the inverse of the stiffness of the unloading curve, S, at the maximum load ($C = 1/S = (dh/dP)P_{max}$), and ε is a parameter, depending on the indenter geometry, which defines the value of $h_s = \varepsilon C P_{max}$, (see Figure 16.1). The value of the geometrical parameter, ε, is between 0.72 and 1 (see Figure 16.1(b)). For the Vickers and Berkovich indenters, the ε parameter is generally considered equal to 0.75 (Pharr and Bolshakov, 2002).

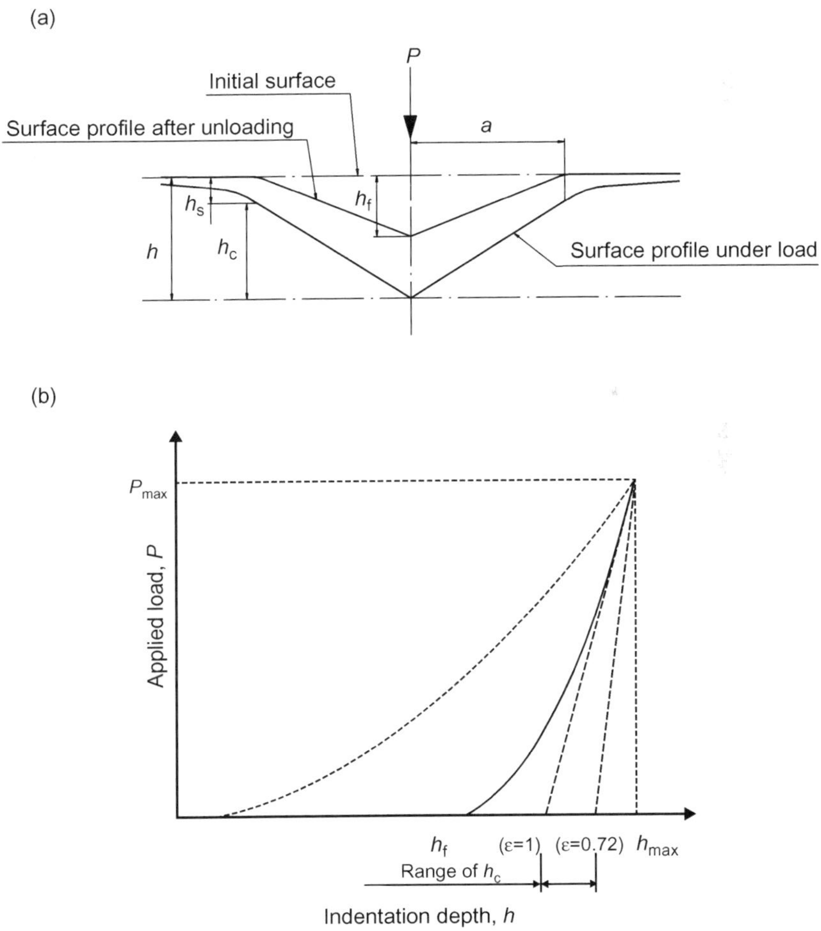

Figure 16.1 Schematic representation of: (a) geometrical parameters of the indentation; (b) load–indentation depth curve.
Source: Adapted from Antunes et al. (2002).

During the hardness test of composite materials, elastic and plastic deformations can occur in the film and substrate. For separating the contributions of the film and substrate to the composite's hardness results, a significant number of models were proposed (see, e.g. Bückle, 1973; Jönsson and Hogmark, 1984; Chicot and Lesage, 1995; Korsunsky et al., 1998; Ichinmura et al., 2000; Tuck et al., 2001; Puchi-Cabrera et al., 2002; Sakharova et al., 2013). In general, the models to extract the film's hardness from the overall hardness of the composite can be summarized by the formulation:

$$\frac{H - H_s}{H_f - H_s} = F\left(\frac{h_c}{t}\right), \tag{16.4}$$

where H, H_s and H_f are the composite, substrate and film hardness, respectively, t is the film thickness and h_c the contact indentation depth. The function $F(h_c/t)$ assumes different forms, more or less complex, depending on the model. This function varies in the range [0, 1], where $F(h_c/t) = 1$ corresponds to a critical value of the relative indentation depth h_c/t, below which plastic deformation only occurs in the film, and $F(h_c/t) = 0$ corresponds to deep indentations ($h_c \to \infty$). With regard to the function $F(h_c/t)$, the parameters involved depend on factors such as the geometry and size of the indentation and the mechanical properties of the materials of the film and the substrate. However, the understanding of these correlations is not always achieved, and often the function $F(h_c/t)$ is formulated empirically.

The feature that mainly influences the behaviour of the composite film substrate, when both deform plastically, is the size of the plastically deformed zone (see, e.g. Korsunsky et al., 1998). Furthermore, the composite behaviour is also affected by the ratio between the yield stresses of the film and substrate and their strain hardening parameters (see, e.g. Fernandes et al., 1998). So, it is difficult to achieve accurate results for the film hardness with models for which the parameters of the function $F(h_c/t)$, in Eq. 16.4, do not depend on the mechanical properties of both materials of the composite. This is, for example the case of the function proposed by Jönsson and Hogmark (1984) that uses a parameter, D, whose value does not depend on the film and substrate mechanical properties:

$$F\left(\frac{t}{d}\right) = \left(\frac{t}{d}\right)2D + \left(\frac{t}{d}\right)^2 D^2, \tag{16.5}$$

where d is the diagonal of the Vickers indentation ($d = 7h_c$) and D is a predefined parameter: $D \approx 1$ if, during indentation, only plastic strain occurs; and $D \approx 0.5$ if cracking also occurs in the film.

However, for applying this model, it is only necessary to perform a single hardness measurement on the composite, for a given h_c/t ratio, which is of major importance.

Burnett and Rickerby (1987) proposed a mixing law similar to the Jönsson and Hogmark model that considers the volumes of the plastically deformed regions:

$$H_c = \frac{v_f}{v} H_f + \frac{v_s}{v} \chi^3 H_s \quad \text{for} \quad H_s < H_f$$
$$H_c = \frac{v_f}{v} \chi^3 H_f + \frac{v_s}{v} H_s \quad \text{for} \quad H_s < H_f \,, \tag{16.6}$$

where V_f, V_s and V are the plastic deformation volume of the film, substrate and total, respectively. χ is a corrective factor, designated as an interface parameter, to take into account the interaction between film and substrate.

Following by this type of formulation, Chicot and Lesage (1995) developed a model that presumes continuity between the film and the substrate. In this case, the volumes of the deformed plastic regions are approached by cylinder geometry in the film and by hemispherical geometry in the substrate:

$$F\left(\frac{h_c}{t}\right) = \left\{ \frac{3t}{2d} \left[\left(\frac{H_f}{E_f}\right)^{1/2} + \left(\frac{H_s}{E_s}\right)^{1/2} \right] \tan^{1/3} \zeta \right\}, \tag{16.7}$$

where ζ is the indenter semi-angle, and E_s and E_f are the substrate and film's Young's modulus, respectively.

In this case, the value of the Young's modulus of the film, E_f, necessary to input in the equation, becomes an additional difficulty. It should be noted that, as will be shown below, the accurate determination of the film's Young's modulus can itself be challenging.

Another model, which makes use of dimensionless parameters, was proposed by Korsunsky et al. (1998):

$$F\left(\frac{h_c}{t}\right) = \frac{1}{1 + k\left(\frac{h_c}{t}\right)^2}, \tag{16.8}$$

where k is a dimensionless parameter, given by the ratio: $k = t/\gamma$; in which γ, which has the dimension of length, is linked to the fracture toughness. For a fracture-dominated case (when a facture occurs), γ is proportional to the film thickness, t; while for a plastically dominated case (when fracture does not occur), γ mainly depends on the ratio H_f/H_s and is weakly dependent on t.

The quality of the fit used in this model depends on the experimental hardness data obtained over a wide range of h_c/t values, which implies that nanoindentation data must be taken into account. Otherwise, there are insufficient data points for values close to zero to provide a good quality of fit. Therefore, the main difficulty associated with this type of model is to obtain the correct extrapolation of h_c/t to zero.

A slightly different model proposed by Puchi-Cabrera et al. (2002) considers two material dependent parameters. This model assumes the following formulation:

$$F\left(\frac{h_c}{t}\right) = \exp(-a(t/h_c)^n), \quad (16.9)$$

where a and n represent material parameters that characterize the change in hardness as the indenter passes from the film to the substrate.

In the indentation tests of composite materials, the evolution of the hardness as function of the relative indentation depth, h_c/t shows three different regimes (see, e.g. Iost and Bigot, 1996; Fernandes et al., 2000). Figure 16.2 presents the evolution of the composite hardness in the case of a hard film on a soft substrate. In the inverse case, a soft film on a hard substrate, region I presents the lowest hardness values, which appear at low indentation depths.

Figure 16.2 shows that at low indentation depths, region I, the response of the composite only depends on the film. As the indentation depth increases (region II), the hardness decreases gradually due to the complex response of the film and the substrate. In the central part of region II, a quasi linear evolution between the composite hardness and the inverse of the relative indentation depth h_c/t is observed (see, e.g. Wang et al., 2001; Iost et al., 2003). In region III, corresponding to high indentation depths, the substrate entirely controls the response of the composite.

Based on the linear evolution during region II, Sakharova et al. (2013) recently proposed, for the case of the film harder than the substrate (typical case), a simple linear model which considers the hardness behaviour of the composite described by a linear $F(h_c/t)$ function, which parameters depend on the film/substrate system, avoiding the use of indentation data obtained for very low

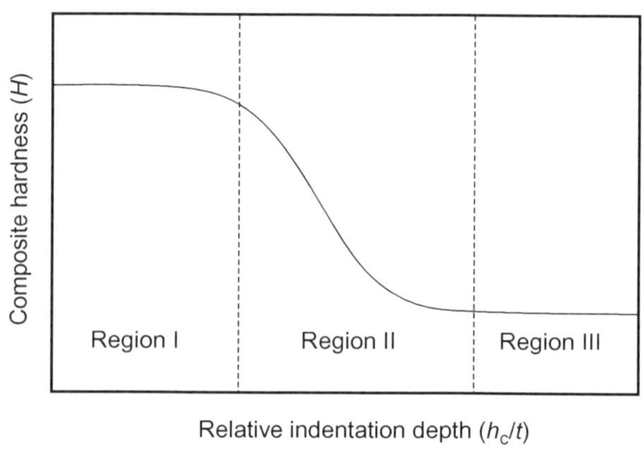

Figure 16.2 Schematic diagram of the evolution of composite hardness as function of the relative indentation depth, h_c/t, in a hard film on a soft substrate.

indentation depths. So, the linearity observed in region II is described by a general formulation of the type:

$$F\left(\frac{h_c}{t}\right) = \frac{H - H_s}{H_f - H_s} = A\left(\frac{t}{h_c}\right) + B, \tag{16.10}$$

where A and B are parameters obtained by fitting the equation to the composite hardness data at different values of the relative indentation depth, t/h_c. The parameter A characterises the slope of the straight lines from the plot $(H - H_s)/(H_f - H_s)$ as a function of t/h_c. From Eq. 16.10, the ratio H/H_s can be written as follows:

$$\frac{H}{H_s} = k_1\left(\frac{t}{h_c}\right) + k_2, \tag{16.11}$$

where k_1 and k_2 are parameters given by:

$$k_1 = \left(\frac{H_f}{H_s} - 1\right)A, \tag{16.12}$$

and

$$k_2 = 1 + \left(\frac{H_f}{H_s} - 1\right)B. \tag{16.13}$$

The slope, k_1, of the linear evolution described by Eq. 16.11 increases with the ratio H_f/H_s and so k_1 can be considered as a measurement of the relative film hardness, that is, of the ratio between the film and substrate hardness, H_f/H_s. Since the ratio H_f/H_s is also nearly a linear function of k_1 (Eq. 16.12), it is possible to write:

$$\frac{H_f}{H_s} = ak_1 + 1, \tag{16.14}$$

where the coefficient a is the inverse of the A parameter: $a = 1/A$ and is equal to 7.40 (Sakharova et al., 2013).

Using Eq. 16.14 it is possible to extract the relative film hardness, H_f/H_s, where the k_1 value can be determined from the experimental normalized composite hardness, H/H_s, obtained at different indentation depths, in region II, as can be seen in Figure 16.3.

Moreover, the description of the evolution of the ratio H_f/H_s as a linear function of k_1 is somewhat simplified due to the fact that the parameter A decreases slightly with the increase of the ratio H_f/H_s. Consequently, the coefficient a (Eq. 16.14) also evolves with the ratio H_f/H_s and, therefore, depends on k_1. In order to describe the ratio H_f/H_s as a function of k_1 more accurately, a second-order polynomial was used to fit a wide range of composite hardness results. In this way, the a coefficient is expressed by a linear function of k_1: $a = C_1k_1 + C_2$, where C_1 and C_2 are linearity

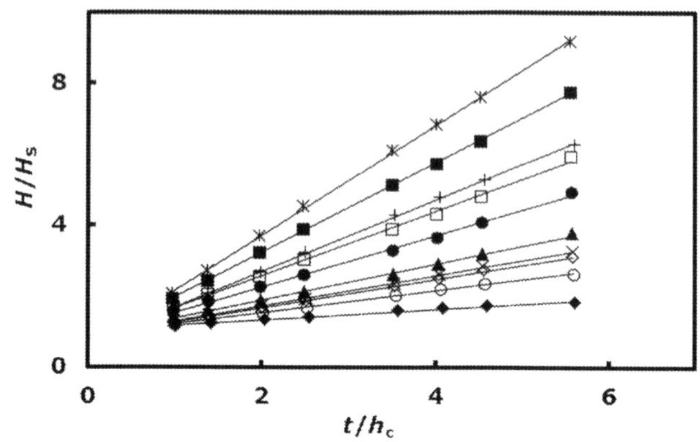

Figure 16.3 Illustration of the linear fitting of the numerical data during region II: Normalized composite hardness, H/H_s as a function of t/h_c for a hard film on 10 different soft substrates, with ratios H_f/H_s from 1.96 (lower line) to 13.77 (upper line).
Source: Adapted from Sakharova et al. (2013).

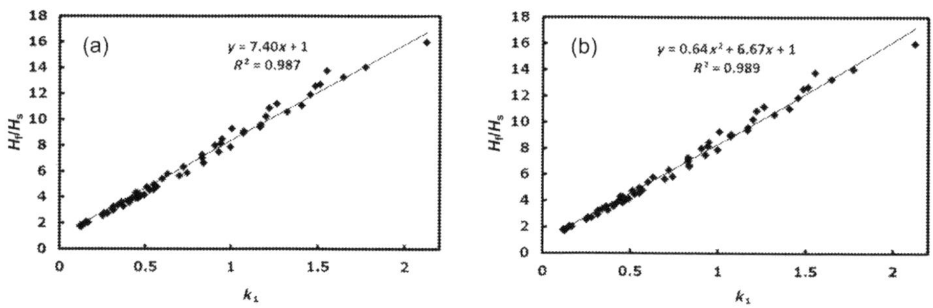

Figure 16.4 Fittings of numerical data during stage II for the ratio H_f/H_s versus k_1: (a) linear and (b) second-order polynomial.
Source: Adapted from Sakharova et al. (2013).

constants. Similar to the procedure used in Eq. 16.14, it is possible to extract the relative value of the film's hardness, H_f/H_s, using the following equation:

$$\frac{H_f}{H_s} = 0.64k_1^2 + 6.67k_1 + 1, \tag{16.15}$$

Eqs. 16.15 and 16.14 are valid within region II, namely in the range $1 < t/h_c < 6$ (Sakharova et al., 2013). The evolutions obtained resorting to Eqs. 16.14 and 16.15 are graphically represented in Figure 16.4.

16.3 Young's modulus

Sneddon (1965) has proposed an analytical solution to the problem of the contact between a rigid flat cylindrical indenter and the surface of a body with elastic behaviour. This solution relates the applied load, P, with the elastic deflection of the sample surface, h_e (see Figure 16.1(a)):

$$P = \frac{2E}{(1-\nu^2)} a h_e, \tag{16.16}$$

where E is the Young's modulus of the elastic body, a is the radius of the flat punch, h_e is the elastic deflection of the sample surface and ν is the Poisson ratio.

Similarly, in the case of indentation tests of elastic–plastic materials, the slope of the unloading curve, $S = dP/dh$, evaluated at the maximum load, is the derivative of Eq. 16.16 with respect to the indentation depth (see Figure 16.1(b)). With A_c being the indentation contact area ($A_c = \pi a^2$, in the cases of circular flat or conical indenters), the following relation between the inverse of the slope of the unloading curve at maximum load ($C = 1/S$, where S is the slope) and the Young's modulus is obtained (ISO 14577, 2002):

$$C = C_0 + \frac{\sqrt{\pi}}{2E_{SI}} \frac{1}{\sqrt{A_c}} \frac{1}{\beta} \tag{16.17}$$

where $C = (1/S)$ is called the compliance of the system (specimen + indenter) and C_0 is added to this equation to take into account the compliance of the equipment; A_c is the contact indentation area, β is a correction coefficient that depends on the indenter geometry. Among the references for this chapter, several values of β for different indenter geometries can be found (see, e.g. Oliver and Pharr, 2004; Antunes et al., 2006; Sakharova et al., 2009). The 'specimen + indenter' modulus, E_{SI}, only depends on the elastic parameters of the tested material if the indenter is considered perfectly rigid. However, in real cases, Young's modulus also depends on the elastic parameters of the indenter:

$$\frac{1}{E_{SI}} = \frac{1}{E^*} + \frac{1}{E_i^*} \tag{16.18}$$

with

$$E^* = E/(1-\nu^2) \text{ and } E_i^* = E_i/(1-\nu_i^2) \tag{16.19}$$

where E^* and E_i^* are the reduced Young's modulus, E and E_i the Young's modulus and ν and ν_i the Poisson ratios of the specimen and of the indenter, respectively.

In the case of elastically homogeneous composites, that is, the film and the substrate have the same reduced Young's modulus, $E_f^* = E_s^*$ and so E^* represents the value of the reduced elastic modulus of the specimen: $E^* = E_f^* = E_s^*$. If the film and the substrate are not elastically homogeneous, the reduced modulus evaluated with the help of Eqs. 16.18 and 16.19 does not correspond only to the film, especially if the film is very thin and/or the ratio between the Young's modulus of the film and the substrate is very different from 1. When the reduced Young's modulus of the film, E_f^*, and the substrate, E_s^*, differ, the evaluated reduced indentation modulus of the specimen, E^*, depends on the indentation depth. By increasing the indentation depth, the reduced Young's modulus, E^*, changes gradually from the value of the film to that of the substrate. As consequence, the evaluated Young's modulus depends on the film and the substrate, whatever maximum indentation depths are used. Similar to the hardness case, weight functions have been used for evaluating the reduced Young's modulus of thin films, E_f^*, which depends on the value of measured reduced Young's modulus of the composite, the reduced Young's modulus of the substrate, E_s^*, and the ratio h_c/t. The easiest and most common way of determining the value of E_f^*, by fitting the experimental results and applying a weight function, is to know in advance the value of E_s^* (see, e.g. Doerner and Nix, 1986; King, 1987; Menčík et al., 1997; Saha and Nix, 2002). However, in a study using different weight function to the experimental indentation data of several composites, Antunes et al. (2007b) showed that considering the substrate Young's modulus, E_s^*, as unknown, can present advantages.

Next, the main weight functions used to determine the film's Young's moduli are described. Menčík et al. (1997) proposed the following equation:

$$\frac{E^* - E_s^*}{E_f^* - E_s^*} = e^{\left(-a_1 \frac{h_c}{t}\right)}, \tag{16.20}$$

where E^*, E_s^*, e and E_f^* are the reduced Young's modulus of the composite, substrate and film respectively, t is the film thickness, h_c is the contact indentation depth and α_1 is a constant that depends on the composite. This function can be used for both cases $E_f^* > E_s^*$ and $E_f^* < E_s^*$, as illustrated in Figure 16.5. Eq. 16.20 can be developed into a linear form: $y = A + B(h_c/t)$, where $y = \ln|E^* - E_s^*|$, $A = \ln|E_f^* - E_s^*|$ and $B = -\alpha_1$. The parameters A and B can be determined using, for example, the least squares method. Knowing the constant A and the substrate reduced modulus, E_s^*, the film–reduced modulus, E_f^*, can be determined by: $E_f^* = E_s^* + Ke^A$, where $K = 1$ if $E_f^* > E_s^*$, and $K = -1$ if $E_f^* < E_s^*$ (Menčík et al., 1997).

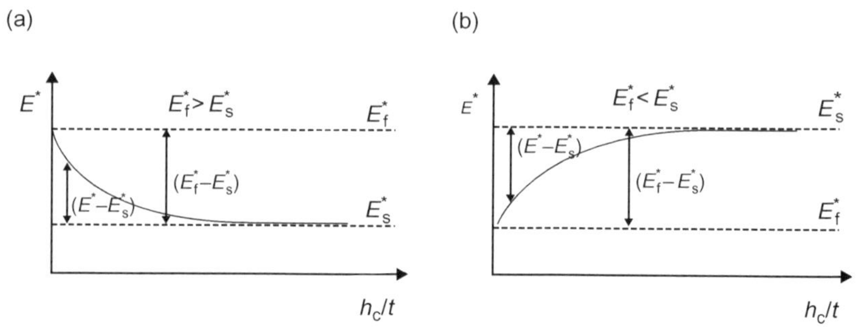

Figure 16.5 Illustration of the exponential function, expressed by Eq. 16.20. (a) $E_f^* > E_s^*$; (b) $E_f^* < E_s^*$.
Source: Adapted from Antunes et al. (2007b).

The reciprocal exponential function that considers, in the first term of Eq. 16.20, the inverse of the reduced Young's modulus instead of the reduced modulus was also proposed (Menčík et al., 1997):

$$\frac{1/E^* - 1/E_s^*}{1/E_f^* - 1/E_s^*} = e^{\left(-\alpha_2 \frac{h_c}{t}\right)}, \tag{16.21}$$

where α_2 is a constant. This function can also be used in two cases, $E_f^* > E_s^*$ and $E_f^* < E_s^*$, as shown in Figure 16.6. The determination procedure for the reduced Young's modulus is identical to the case of the exponential function.

Doerner and Nix (1986), also proposed a model of the exponential type, but t/h_c is used instead of h_c/t. The proposed equation can be written in the form (Doerner and Nix, 1986):

$$\frac{1/E^* - 1/E_f^*}{1/E_s^* - 1/E_f^*} = e^{\left(-\alpha_3 \frac{t}{h_c}\right)}, \tag{16.22}$$

where α_3 is a constant. The representation of Eq. 16.22 for the composites $E_f^* > E_s^*$ and $E_f^* < E_s^*$ is illustrated in Figure 16.7. Unlike the two previous functions, the linearisation of Eq. 16.22 does not make it possible to determine the unknowns, α_3 and E_f^*, by fitting a set of results of the composite reduced Young's modulus, E^*, measured at different maximum indentation depths. In this case, it is necessary to use an optimisation procedure to obtain the combination of α_3 and E_f^*. An optimization procedure, as for example, the minimization of the sum of the squared difference between the values of the reduced Young's modulus of the composite evaluated by hardness tests and those obtained with Eq. 16.22 for different combinations of α_3 and E_f^*, can be used (Menčík et al., 1997).

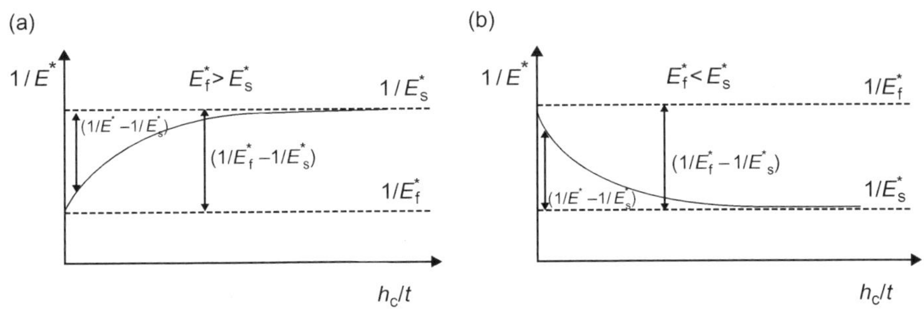

Figure 16.6 Illustration of the exponential function, expressed by Eq. 16.21. (a) $E_f^* > E_s^*$; (b) $E_f^* < E_s^*$.
Source: Adapted from Antunes et al. (2007b).

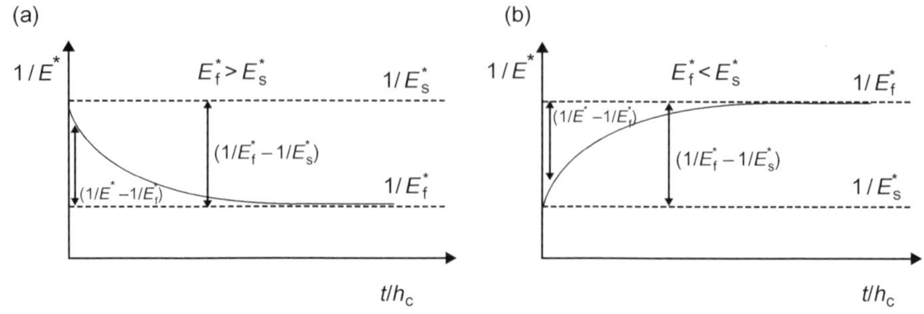

Figure 16.7 Illustration of the Doerner and Nix function, expressed by Eq. 16.22. (a) $E_f^* > E_s^*$; (b) $E_f^* < E_s^*$.
Source: Adapted from Antunes et al. (2007b).

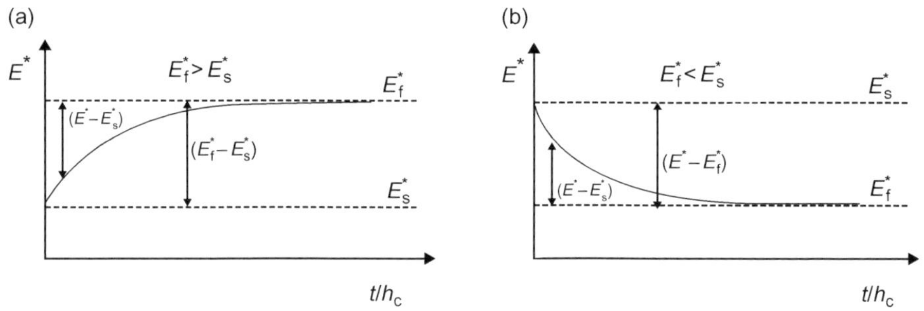

Figure 16.8 Illustration of the function expressed by Eq. 16.23. (a) $E_f^* > E_s^*$; (b) $E_f^* < E_s^*$.
Source: Adapted from Antunes et al. (2007b).

Another possibility for the exponential function, proposed by Antunes et al. (2007b), is expressed by:

$$\frac{E^* - E_f^*}{E_f^*/E_s^*} = e^{\left(-\alpha_4 \frac{t}{h_c}\right)}, \tag{16.23}$$

where α_4 is a constant. The representation of Eq. 16.23 is shown in Figure 16.8. Similar to Eq. 16.22, the linearization of the equation is not possible, and it is necessary to use an optimization procedure to determine the values of α_4 and E_f^*.

A non-exponential function was proposed by Gao et al. (1992). This function takes into account the analytical solution for the contact between a rigid circular flat indenter with radius, a, and a composite with elastic behaviour:

$$\frac{E^* - E_s^*}{E_f^* - E_s^*} = \Phi, \tag{16.24}$$

where Φ, is a weight function expressed by:

$$\Phi = \frac{2}{\pi}\text{arctg}\xi + \frac{(1+2\nu)\xi}{2\pi(1-\nu)}\ln\left(\frac{1+\xi^2}{\xi^2}\right) - \frac{1}{2\pi(1-\nu)}\frac{\xi}{1+\xi^2}, \tag{16.25}$$

with $\xi = \text{t}/\text{a} = \text{t}/Vh_c$, where t is the film thickness and V is a constant that relates the indention depth, h_c, with the contact radius of the indentation, a. For equivalence between the circular flat indenter and pyramidal (Vickers and Berkovich) indenters, the constant V is equal to 2.79 $\left(a = h_c\sqrt{24.5/\pi} = 2.79h_c\right)$.

The procedure for determining the film-reduced Young's modulus uses the results of the modulus of the composite, obtained at different maximum indentation depths. As proposed by Menčík et al. (1997), Eq. 16.24 is transformed into a straight line ($y = Bx$), where $y = |E^* - E_s^*|$, and x corresponds to the result of Eq. 16.25 for different maximum indentation depths. Accordingly, the fitting parameter $B = |E_f^* - E_s^*|$ is determined using, for example, the least squares method. The film-reduced Young's modulus is then determined by: $E_f^* = E_s^* + B$. This function, such as the above one can be used for both type of composites, $E_f^* > E_s^*$ and $E_f^* < E_s^*$.

Antunes et al. (2007b), based on the fact that the linear behaviour is better estimated by the evolutions that consider the inverse of the reduced composite Young's modulus, propose an alternative function deduced from Eqs. 16.24 and 16.25, called reciprocal of Gao function:

$$\frac{1/E^* - 1/E_s^*}{1/E_f^* - 1/E_s^*} = \Phi, \tag{16.26}$$

The parameters in Eq. 16.26 have the same meaning as indicated for Eqs. 16.24 and 16.25. Also, the procedures used in the determination of the reduced Young's modulus of the films are identical.

Antunes et al. (2007b) explored the above six models using two different procedures for determining the film-reduced Young's modulus: (a) the substrate modulus as a known parameter; and (b) the substrate modulus considered to be unknown. They evaluate the reduced modulus of the film for several composites with different mechanical properties, and conclude that the suitable procedure for determining the Young's modulus of the films must consider the substrate Young's modulus as unknown. Moreover, they conclude that in order to obtain accurate results of the Young's modulus of the films, the function that must be used is reciprocal of exponential (Eq. 16.21), for composites with $E_f^* > E_s^*$, and exponential (Eq. 16.20) for $E_f^* < E_s^*$. For both types of composites ($E_f^* > E_s^*$ and $E_f^* < E_s^*$) the reciprocal of the Gao function can be also used (Eq. 16.26). Furthermore, this last function has the advantage that the results do not depend on the ratio between the Young's modulus of the film and the substrate, E_f^*/E_s^* (Antunes et al., 2007b).

As an alternative to the use of models, a reverse analysis methodology for evaluating the Young's modulus of thin films was proposed by Antunes et al. (2008).

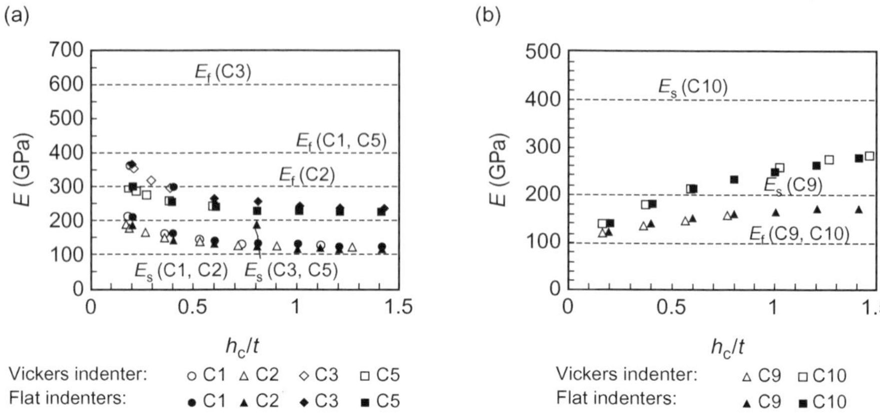

Figure 16.9 Composite Young's modulus, E, versus the normalized contact indentation depth, h_c/t (C1, C2, C3, C4, C5, C6, C7, C8, C9 and C10 concern composites with values of E_f/E_s of 4.00, 1.50, 3.00, 2.00, 0.50 and 0.25, respectively). Results obtained with Vickers and squared flat punch indenters (in this case, the contact indentation depth is equal to the one of the Vickers indenter with the same contact area). (a) Composites C1, C2, C3 and C5, (b) composites C9 and C10.
Source: Adapted from Antunes et al. (2008).

This methodology is based on the finding that the composite Young's modulus results do not depend on the geometry of the indenter (flat punch or pyramidal), but only on the contact area as shown in Figure 16.9 (see, e.g. Antunes et al., 2007b, 2008). It consists of the comparison of the composite Young's modulus, evaluated by the experimental hardness tests using a pyramidal indenter with the ones obtained by the numerical simulation with flat punch indenters, for the same values of the equivalent relative contact indentation depth h_c/t. The contact indentation depth of the flat punch that is equivalent to the pyramidal indenter is $h_c = \sqrt{A/24.5}$, where A is the flat punch area (i.e. the contact area is the same for both tests, experimental and numerical).

The reverse analyses methodology consists of the experimental determination, by using a pyramidal indenter, of the contact area and of the reduced Young's modulus of the composite, E^*, for a given relative contact indentation depth, h_c/t. Subsequently, numerical simulations of the indentation test with a flat punch indenter are performed. For these simulations, the flat punch indenter must have the same equivalent relative contact area as for an experimental test on the composite, with the pyramidal indenter; and the value of the substrate reduced Young's modulus is equal to that experimentally determined for the substrate, E^*, which must be done in advance. For the first simulation, a value for the film's Young's modulus, E_f^*(punch), is chosen: E_f^*(punch) must be higher than E^* in the case of $E^* > E_s^*$, or lower than E^* in the case of $E^* < E_s^*$. The value of the composite's Young's modulus, E^*(punch), numerically determined with the flat punch indenter is now compared with the one, E^*, experimentally determined with the pyramidal indenter. If two values of the reduced modulus, E^*(punch) and E^*, are similar enough (within

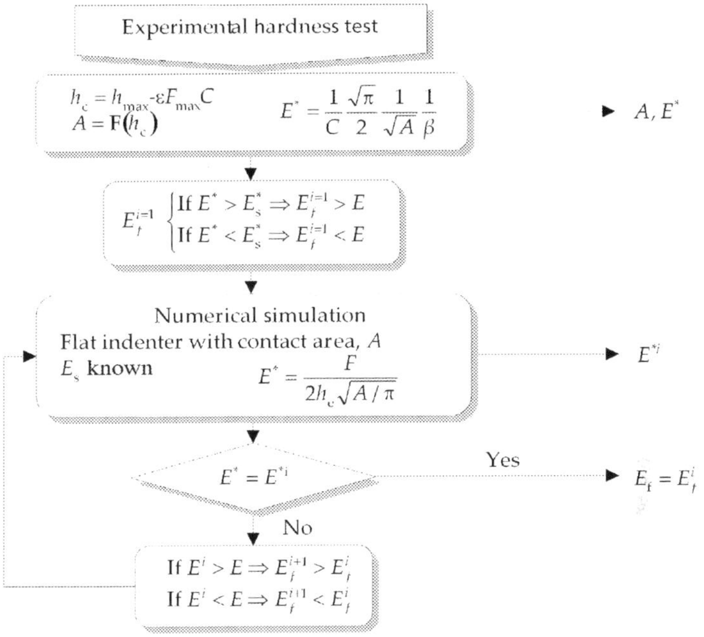

Figure 16.10 Reverse analyses algorithm for thin film Young's modulus determination. *Source:* Adapted from Antunes et al. (2008).

a predefined range of accuracy), then the reduced elastic modulus, E_f^*, can be estimated as being equal to E^*(punch). If there is a not negligible difference between the experimental (pyramidal) and numerical (flat punch) composite reduced elastic modulus (this is generally the case), then an iterative method will be used for optimization. That is to say, new values for E_f^*(punch) are chosen as input in the numerical simulations with the flat punch indenter until the numerical and experimental values of the composite reduced modulus, E^*(punch) and E^*, are within the delimited range of accuracy (Antunes et al., 2008). Figure 16.10 shows the algorithm of the described reverse analysis methodology.

The use of a flat punch indenter ensures that no plastic deformation occurs in the composite for small indentation depths, which means the plastic properties of the film and the substrate do not need to be known to perform the numerical simulations (Antunes et al., 2008). In fact, the numerical simulation results of indentation tests with flat punch indenters show a linear relationship between the load, P, and the elastic deflexion of the sample surface, h_e, as described by Eq. 16.16 and shown in Figure 16.11.

The application of a reverse analysis algorithm to several composites, with different values of the ratio E_f/E_s, shows that this reverse analysis methodology is generally more accurate than the above models, and easily performed (Antunes et al., 2008). In order to take maximum advantage of this methodology, the authors suggest performing experimental indentation tests in the range of relative indentation depths, h_c/t, between about 0.2 and 0.3.

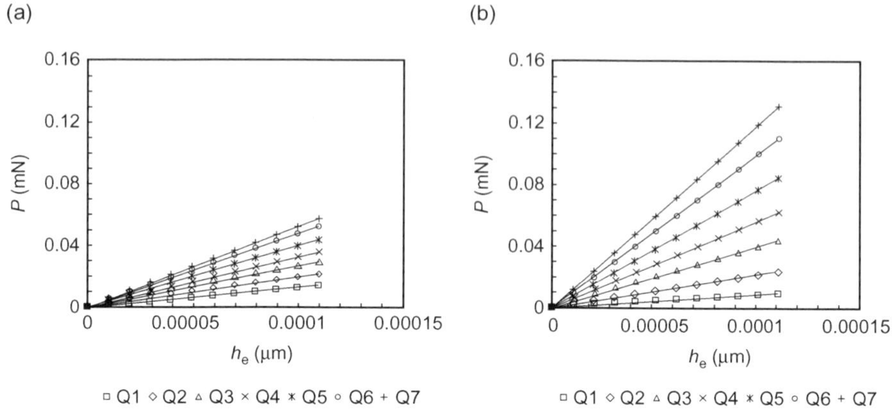

Figure 16.11 Load, P, versus the elastic deflexion of the sample surface, h_e. Results obtained by numerical simulation, using squared flat punch indenters, with different areas (Q1 = 0.25 μm^2; Q2 = 1.00 μm^2; Q3 = 2.25 μm^2; Q4 = 4.00 μm^2; Q5 = 6.25 μm^2; Q6 = 9.00 μm^2; Q7 = 12.25 μm^2): (a) $E_f = 400$ GPa e $E_s = 100$ GPa; (b) $E_f = 100$ GPa e $E_s = 400$ GPa.
Source: Adapted from Antunes et al. (2008).

16.4 Conclusion

The increasing application of thin films in various fields in industry is essentially due to their versatility and high performance in service. The hardness test, due to ease of application, is suitable for evaluating the mechanical properties of such materials. DSI equipment allows the evaluation of not only the hardness, but also other mechanical properties, such as Young's modulus. As consequence, considerable work has been done by several researchers to develop methodologies for the mechanical characterization of thin films based on results obtained by DSI on film/substrate composites.

In this context, this chapter reviewed the main methodologies proposed in the literature for the assessment of the hardness and the Young's modulus of thin films. In addition to a detailed description of the methods, their main advantages and limitations were analysed.

Acknowledgements

This research work is sponsored by national funds from the Portuguese Foundation for Science and Technology (FCT) via the projects PEst-C/EME/UI0285/2013 and PTDC/EME-TME/122472/2010 and by FEDER funds via the 'Programa Operacional Factores de Competitividade' – COMPETE.

References

Antunes, J.M., Cavaleiro, A., Menezes, L.F., Simões, M.I., Fernandes, J.V., 2002. Ultra-microhardness testing procedure with Vickers indenter. Surf. Coat. Tech. 149, 27–35.

Antunes, J.M., Fernandes, J.V., Menezes, L.F., 2006. Three-dimensional numerical simulation of Vickers indentation tests. Int. J. Solid Struct. 43, 784–806.

Antunes, J.M., Menezes, L.F., Fernandes, J.V., 2007a. Influence of Vickers tip imperfection on depth sensing indentation tests. Int. J. Solid Struct. 44, 2732–2747.

Antunes, J.M., Fernandes, J.V., Menezes, L.F., Sakharova, N.A., 2007b. On the determination of the Young's modulus of thin films using indentation tests. Int. J. Solid Struct. 44, 8313–8334.

Antunes, J.M., Fernandes, J.V., Sakharova, N.A., Menezes, L.F., 2008. Reverse analysis in depth-sensing indentation for evaluation of the Young's modulus of thin films. Philos. Mag. 88 (3), 313–325.

Bolshakov, A., Oliver, W.C., Pharr, G.M., 1996. Influences of stress on the measurement of mechanical properties using nanoindentation. 2. Finite element simulations. J. Mater. Res. 11 (3), 760–768.

Bückle, H., 1973. In: Westbrook, J.H., Conrad, H. (Eds.), The Science of Hardness Testing and its Research Applications. American Society for Metals, Metals Park, OH, pp. 453–494.

Burnett, J.P., Rickerby, D.S., 1987. The mechanical properties of wear-resistant coatings. I. Modelling of hardness behaviour. Thin Solid Films. 148, 41–50.

Chechenin, N.G., Bottiger, J., Krog, J.P., 1995. Nanoindentation of amorphous aluminium-oxide films. 1. The influence of the substrate on the plastic properties. Thin Solid Films. 261 (1–2), 219–227.

Chen, X., Vlassak, J.J., 2001. Numerical study on the measurement of thin film mechanical properties by means of nanoindentation. J. Mater. Res. 16 (10), 2974–2982.

Chen, X., Yan, J., Karlsson, A.M., 2006. On determination of residual stress and mechanical properties by indentation. Mater. Sci. Eng. A. 416 (1–2), 139–149.

Chicot, D., Lesage, J., 1995. Absolute hardness of films and coatings. Thin Solid Films. 254 (1–2), 123–130.

Dao, M., Chollacoop, N., Van Vliet, K.J., Venkatesh, T.A., Suresh, S., 2001. Computational modelling of the forward and reverse problems in instrumented sharp indentation. Acta Mater. 49 (19), 3899–3918.

Doerner, M.F., Nix, W.D., 1986. A method for interpreting the data from depth-sensing indentation instruments. J. Mater. Res. 1 (4), 601–609.

Fabes, B.D., Oliver, W.C., McKee, R.A., Walker, F.J., 1992. The determination of film hardness from the composite response of film and substrate to nanometer scale indentations. J. Mater. Res. 7 (11), 3056–3064.

Fernandes, J.V., Menezes, L.F., Trindade, A.C., 1998. The coated surface hardness: a kinematic model. Thin Solid Films. 335, 153–159.

Fernandes, J.V., Trindade, A.C., Menezes, L.F., Cavaleiro, A., 2000. Influence of substrate hardness on the response of W–C–Co-coated samples to depth-sensing indentation. J. Mater. Res. 15, 1766–1772.

Fisher-Cripps, A.-C., 2004. Nanoindentation. Springer-Verlag, New York.

Gao, H., Cheng-Hsin, C., Jin, L., 1992. Elastic contact versus indentation modelling of multi-layered materials. Int. J. Solid Struct. 29 (20), 2471–2492.

Ichimura, H., Rodriguez, F.M., Rodrigo, A., 2000. The composite and film hardness of TiN coatings prepared by cathodic arc evaporation. Surf. Coat. Tech. 127 (2–3), 138–143.
Iost, A., Bigot, R., 1996. Hardness of coatings. Surf. Coat. Tech. 165, 117–120.
Iost, A., Najjar, D., Hellouin, R., 2003. Modelling of the Vickers hardness of paint coatings deposited on metallic substrates. Surf. Coat. Tech. 165, 126–132.
ISO 14577, 2002. Metallic Materials – Instrumented Indentation Tests for Hardness and Materials Parameters. ISO Central Secretariat, Geneva, Switzerland.
Jang, J.I., 2009. Estimation of residual stress by instrumented indentation: a review. J. Ceram. Proc. Res. 10 (3), 391–400.
Johnson, K.L., 1970. The correlation of indentation experiments. J. Mech. Phys. Solids. 18 (2), 115–126.
Jönsson, B., Hogmark, S., 1984. Hardness measurements of thin-films. Thin Solid Films. 114 (3), 257–269.
Joslin, D.L., Oliver, W.C., 1990. New method for analysing data from continuous depth-sensing microindentation tests. J. Mater. Res. 5 (1), 123–126.
King, R.B., 1987. Elastic analysis of some punch problems for a layered medium. Int. J. Solid Struct. 23 (12), 1657–1664.
Korsunsky, A.M., McGurk, M.R., Bull, S.J., Page, T.F., 1998. On the hardness of coated systems. Surf. Coat. Tech. 99 (1–2), 171–183.
Lichinchi, M., Lenardi, C., Haupt, J., Vitali, R., 1998. Simulation of Berkovich nanoindentation experiments on thin films using finite element method. Thin Solid Films. 312, 240–248.
Manika, I., Maniks, J., 2008. Effect of substrate hardness and film structure on indentation depth criteria for film hardness testing. J. Phys. D Appl. Phys. 41, 740101–740106.
Menčík, J., Munz, D., Quandt, E., Weppelmann, E.R., 1997. Determination of elastic modulus of thin layers using nanoindentation. J. Mater. Res. 12 (9), 2475–2484.
Oliver, W.C., Pharr, G.M., 1992. An improved technique for determining hardness and elastic-modulus using load and displacement sensing indentation experiments. J. Mater. Res. 7 (6), 1564–1583.
Oliver, W.C., Pharr, G.M., 2004. Measurement of hardness and elastic modulus by instrumented indentation: advances in understanding and refinements to methodology. J. Mater. Res. 19 (1), 3–20.
Page, T.F., Hainsworth, S.V., 1993. Using nanoindentation techniques for the characterization of coated systems – a critique. Surf. Coat. Tech. 61 (1–3), 201–208.
Pharr, G.M., Bolshakov, A., 2002. Understanding nanoindentation unloading curves. J. Mater. Res. 17 (10), 2660–2671.
Puchi-Cabrera, E.S., Berrios, J.A., Teer, D., 2002. On the computation of the absolute hardness of thin solid films. Surf. Coat. Tech. 157 (2–3), 185–196.
Saha, R., Nix, W.D., 2002. Effects of the substrate on the determination of thin film mechanical properties by nanoindentation. Acta Mater. 50 (1), 23–38.
Sakharova, N.A., Fernandes, J.V., Antunes, J.M., Menezes, L.F., 2009. Comparison between Vickers, Berkovich and conical indentation: three-dimensional numerical simulation. Int. J. Solid Struct. 46, 1095–1104.
Sakharova, N.A., Fernandes, J.V., Oliveira, M.C., Antunes, J.M., 2013. On the determination of hardness of the film in hard film/substrate composites using depth-sensing indentation. Ceram. Int. 39, 6.
Sneddon, I.N., 1965. The relation between load and penetration in the axisymmetric Boussinesq problem for a punch of arbitrary profile. Int. J. Eng. Sci. 3, 47–57.

Suresh, S., Giannakopoulos, A.E., 1998. A new method for estimating residual stresses by instrumented sharp indentation. Acta Mater. 46 (16), 5755–5767.

Swaddiwudhipong, S., Tho, K.K., Lui, Z.S., Zeng, K., 2005. Material characterization based on dual indenters. Int. J. Solid Struct. 42 (1), 69–83.

Tabor, D., 1951. The Hardness of Metals. Clarendon Press, Oxford.

Tsui, T.Y., Oliver, W.C., Pharr, G.M., 1996. Influences of stress on the measurement of mechanical properties using nanoindentation. 1. Experimental studies in an aluminum alloy. J. Mater. Res. 11 (3), 752–759.

Tsui, T.Y., Ross, C.A., Pharr, G.M., 1997. Nanoindentation of soft films on hard substrates – the importance of pile-up. Proc. Mater. Res. Soc. Symp. 473, 207–212.

Tsui, T.Y., Vlassak, J., Nix, W.D., 1999. Indentation plastic displacement field: part I. The case of soft films on hard substrates. J. Mater. Res. 14 (6), 2196–2203.

Tuck, J.R., Korsunsky, A.M., Bhat, D.G., Bull, S.J., 2001. Indentation hardness evaluation of cathodic arc deposited thin hard coatings. Surf. Coat. Tech. 139 (1), 63–74.

Wang, H.L., Chiang, M.J., Hon, M.H., 2001. Determination of thin film hardness for a film/substrate system. Ceram. Int. 27, 385–389.

Part Three

Advanced bulk and thin film materials for harsh environment MEMS applications

17

Cinzia Caliendo and Fabio Lo Castro[1]
Istituto di Acustica e Sensoristica Corbino, Italy

17.1 Introduction

A considerable number of both civil and military applications – the aeronautics, metallurgy, and automotive fields, to name just a few – are in need of small and lightweight electronic devices able to survive directly inside an extreme environment, and to withstand temperatures of several degrees centigrade with lifetimes of hundreds of hours. Table 17.1 lists some possible applications of high-temperature, harsh-environment devices by industry. The device mounting and packaging, but most of all the device materials, must be stable with the working temperature, otherwise temperature-induced stress may result in device failures (Neudeck et al., 2002). Conventional-temperature-range electronics can withstand high temperatures if protected from the environment by using precautions that can be so complex as to make it preferable to use *special* electronics suitable for reliable operation for an extended time period at high temperatures. Such a solution requires the investigation of the temperature effects on the properties of the materials comprising the device and on the device performance.

Electroacoustic devices based on surface acoustic wave (SAW) technology, such as delay lines, filters, or resonators, are implemented on bulk single crystal piezoelectric substrates or on multilayered structures that include a piezoelectric film. The SAW excitation and detection is performed by thin film interdigitated transducers (IDTs) that are fabricated by lithographic processing of a metal film, usually 100–150 nm thick, deposited by sputtering or thermal evaporation on the surface of the piezoelectric substrate, or at the substrate/piezoelectric film interface. Figure 17.1 shows the schematic of a SAW delay line implemented on a bulk single crystal piezoelectric substrate, whose thickness H is greater than the acoustic wavelength λ, and on a piezoelectric film grown onto a nonpiezoelectric substrate, being the film thickness $h \leq \lambda$. The operating frequency f of the SAW device is determined by the IDT period p, which sets the SAW λ. A simple IDT is typically composed of several finger pairs of length $\leq 100\ \lambda$ with a finger width $p/4$; the IDT

[1] Cinzia Caliendo is the primary investigator, first author, and corresponding author. Fabio Lo Castro is a collaborating author.

Anti-Abrasive Nanocoatings.

Table 17.1 Sensors for high-temperature, harsh-environments: industrial applications

Field	Application
Energy and manufacturing industries	Testing and evaluation of structural/material integrity of turbine critical engine components; electronic controls placed directly inside jet engines.
Aerospace and aircraft industries	Monitoring aerospace propulsion component conditions; combustion sensors; pressure sensing in turbine engines and compressors; in aircraft engines for measuring the air pressure as it reaches the combustor for controlling the air/fuel mixture in order to reduce the risk of stalling.
Automotive electronics	Measuring temperature, oxygen pressure, and preignition to improve the efficiency and reliability of internal combustion engines; sensors for monitoring combustion pressure to optimize combustion timing and reduce the exhaust pollution; strain sensors.
Industrial applications	Monitoring of process variables, such as viscosity of melts; dynamic pressure monitors, vibration detectors, proximity detectors, etc.; gas sensors for controlling industrial emission and for in-line monitoring of production processes.
Nuclear power industry	Monitoring the internal state of the materials or structures; monitoring of furnace and reactor systems; level, flow, and temperature sensors for nuclear power utility applications.
Manufacturing plants	Health monitoring coal-fired electric-generation plants and nuclear plants; evaluation of structural/material integrity in harsh environments; sensors for controlling the manufacturing processes of power plant components; performing nondestructive testing (NDT) on welds during manufacture.

works at frequency $f = v/\lambda$, v being the substrate SAW phase velocity that depends on the crystal cut and SAW propagation direction. f can be increased by using an expensive submicrometer lithographic technique to reduce λ, or by increasing v (e.g., by choosing fast SAW propagating media). For example, an operation frequency of about 2.5 GHz is allowed for 1 μm IDT's line widths and spacing, by using a diamond substrate whose SAW velocity $v \sim 1 \cdot 10^4$ m s^{-1}. Unfortunately diamond is not piezoelectric, thus a piezoelectric overlayer, typically aluminum nitride (AlN), is required in order to excite the SAW propagation along the diamond substrate.

The physical properties of the propagating medium, such as the mass density, the elastic–piezoelectric–dielectric constants are temperature dependent, thus the

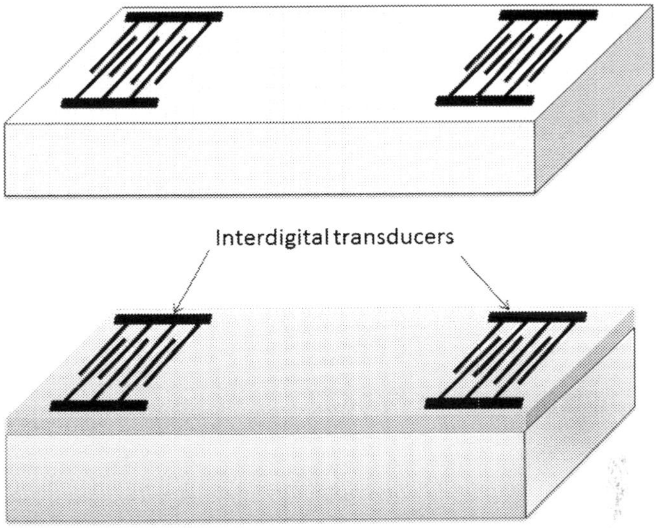

Figure 17.1 SAW delay line implemented on a bulk piezoelectric substrate and onto a layered structure.

SAW velocity is affected by the temperature as well as the operating frequency of the SAW device. The SAW device high-temperature stability demand can be met through an appropriate choice of temperature-stable piezoelectric crystals or by combining suitable overlays and substrate materials with opposite temperature coefficients of delay (TCDs). The SAW device high-operation frequency demand can be met through an appropriate choice of dispersive multilayered configurations that allow the propagation of higher order modes.

The most popular piezoelectric single crystal and thin film materials which can be surveyed at high temperature will be briefly reviewed and their applications will be described. Examples of nonpiezoelectric substrate materials that are suitable for high-frequency operation and high-temperature applications will be presented together with thin metal films suitable for IDT implementation.

17.2 Piezoelectric substrates

When considering piezoelectric materials for high-temperature applications (Zhang and Yu, 2011; Jaffe et al., 1971) a requirement is that the Curie transition temperature must be well above the operating temperature, otherwise the piezoelectric material may depolarize when heated. Table 17.2 lists typical piezoelectric materials along with reported maximum operating temperature. Commonly used piezoelectric crystals are quartz, $LiNbO_3$, $LiTaO_3$, $Li_2B_4O_7$, and $Bi_{12}Ge_4O_{20}$: among these materials quartz offers the highest thermal stability, as it shows some zero TCD cuts, but at a price of low K^2 and velocity. Quartz has an $\alpha-\beta$ transition at

Table 17.2 **Conventional piezoelectric materials**

Material	Maximum temperature of operation	K^2	Thermal stability	References
Quartz	573°C (α–β transition) 350°C (high losses)	Low	Zero TCD cuts (22.5°C TOT)	Darnjanovic (1998), Turner et al. (1994), Jiang et al. (2014), Heywang et al. (2008), Hartmann et al. (1973)
$AlPO_4$	587°C (α–β transition)	High	Zero TCD cuts	O'Connell and Carr (1977)
$GaPO_4$	930°C (α–β transition)	Good	High	Darnjanovic (1998), Jiang et al. (2014), Giurgiutiu et al. (2010)
$LiNbO_3$	600°C (chemical decomposition, increased attenuation)	High	Low	Darnjanovic (1998), Turner et al. (1994), Jiang et al. (2014), Heywang et al. (2008), Hartmann et al. (1973)
$LiTaO_3$,	~610°C	Moderate	Moderate	Darnjanovic (1998), Heywang et al. (2008)
$Li_2B_4O_7$	917°C (melting temp.)	Good	Less than ST-quartz (22.5°C TOT)	Darnjanovic (1998), Heywang et al. (2008)
AlN	1150°C (it loses piezoelectricity)	Moderate	Low	Jiang et al. (2014)
$Bi_{12}GeO_{20}$	918°C (melting point)	Good	Low: −140 pm/°C for (110) <001>	Heywang et al. (2008), Hartmann et al. (1973)
$Bi_4Ti_3O_{12}$	685°C (transition temp.)	Good-high	–	Turner et al. (1994), Kitanaka et al. (2012), Megriche et al. (1999)
$BaTiO_3$	115°C (Curie point)	$d_{33} = 191–419$ pC/N	Low	Heywang et al. (2008), Hartmann et al. (1973), Zheng et al. (2012)
$La_3Ga_5SiO_{14}$	1470°C (melting temp.)	Low	High	Darnjanovic (1998), Jiang et al. (2014) Heywang et al. (2008)
Sc-AlN thin film	500°C (56 hours life)	Good	Low	Akiyama et al. (2009), Umeda et al. (2013)
AlN/sapphire	700–1000°C (20 hours life)	Good	Good	Turner et al. (1994), Aubert et al. (2012)
Soft PZT	350°C (recommended temp.)	High	Moderate	PI Piezo Technology (n.d.)

573°C, which causes the loss of the piezoelectricity. Traditional microbalance systems based on AT-cut quartz crystal sensors are restricted to thin film monitoring below ≈300°C as the piezoelectric constant of quartz drops sharply above 300°C. For this reason quartz crystal sensors must be water cooled for accurate thickness measurements under typical thin film growth conditions. The temperature stability of the quartz is affected by the crystal cut angle. For example, the AT and SC cuts show thermal stability over narrow temperature ranges around 20°C and 90°C.

$LiNbO_3$ (LNO) has a high Curie temperature (1210°C), remarkable K^2 (up to 17% for 42° y-cut x propagation), quite high SAW velocity (4792 m/s) but a high temperature sensitivity (50 ppm/°C).

$LiTaO_3$ (LTO), like LNO, is pyroelectric and the high Li vacuum pressure limits the operation at elevated temperature. It has a quite low Curie temperature (610°C), remarkable K^2 (up to 7.8% for 42° y-cut x propagation), quite high SAW velocity (4022 m/s) but high temperature sensitivity (40 ppm/°C) (Heywang et al., 2008).

$Li_2B_4O_7$ (LBO) melts at 917°C and has high velocity, large coupling, several orientations with zero TCD but excessive ionic conductivity limits its maximum operating temperature to 230°C. Examples of temperature-compensated leaky SAW orientations expressed as Euler angles directions are (0°, 75°, 75°) and (45°, 90°, 70°), with phase velocity of 4225 and 3680 m/s and K^2 of 1.8 and 2.3%, respectively (Adachi et al., 1994).

As a consequence of the inadequacy of conventional piezoelectric materials to high-temperature applications, the research focused on the development of piezoelectric materials that preserve their piezoelectricity at elevated temperatures and show temperature stable cuts, such as gallium orthophosphate ($GaPO_4$), langasite (LGS, $La_3Ga_5SiO_{14}$), langanite (LGN, $La_3Ga_{11/2}Nb_{1/2}O_{14}$), and langatate (LGT, $La_3Ga_{11/2}Ta_{1/2}O_{14}$).

Douchet et al. (2010) compared the temperature behavior of vibrating beam resonators in length-extensional, flexural, and cylindrical-torsional modes with the same structures built in quartz, $GaPO_4$, LGN, and LGT crystals. All three kinds of modes exhibit temperature-compensated orientations at room temperature for singly rotated cuts, temperature-compensated cuts for flexural modes at high temperatures, and temperature-compensated cuts for cylinders in torsional modes are theoretically predicted.

LGS has no phase change below its melting point (1470°C) and shows some zero TCD cuts: for example, LGS 48.5° rotated y cut, 26.6° rotated x propagation direction shows TCD = 1.1 ppm/°C, but also low propagation velocity (2400 m/s) and moderate K^2 (0.32%). The performances of a SAW resonator implemented on LGS substrate oriented along the Euler angles (0°, 138.5°, 26.6°) have been tested from 250–750°C for time periods up to three weeks; the device shows just a quite low degradations in thel amplitude of the S_{21} scattering parameter up to 7 dB (Thiele and Pereira da Cunha, 2006). SAW properties (such as phase velocity, K^2, and power flow angle) of LGS in singly and doubly rotated cuts have been investigated at 250°C (Ji et al., 2005). Three regions for crystal cut orientations and SAW propagation directions at high temperature are put forward, defined by the Euler angles (0°, 20° to 50°, 35° to 45°), (0°, 85° to 110°, 0° to 5°), and (0°, 138° to

145°, 20° to 23°), respectively. The first region includes zero or very low TCD (<2 ppm) but small K^2 (0.2%–0.35%); the second one exhibits higher K^2 (0.35%–0.45%) and moderate TCD (<5 ppm); and the last region shows the highest K^2 (>0.45%) and larger TCD (25–30 ppm/°C). For some orientations within the previously mentioned regions, the temperature dependency of SAW characteristics is discussed up to 1000°C. The orientation (0°, 90°, 0°) has better SAW characteristics and is more suitable for high-temperature applications. Few crystallographic orientations are listed in Ji et al. (2005) showing turn over temperature (TOT) from 220 to 261°C. Some LGS cuts suitable for the propagation of Rayleigh SAWs (R-SAWs) and shear-horizontal SAWs (SH-SAWs), showing high K^2 and zero power flow angle, have been theoretically and experimentally studied up to 700°C (Bardong et al., 2013). The LGS cuts identified by the Euler angles (0°, 22°, 90°), (0°, 138.5°, 26.7°), and (0°, 22°, 31°) have been experimentally investigated between room temperature and 900°C, via in situ electrical measurements. Both (0°, 138.5°, 26.7°) and (0°, 22°, 31) cuts show reliable results up to 700°C; beyond this temperature the Pt IDTs degrade and the crystal shows decomposition. The (0°, 22°, 31°) cut has a TOT close to 300°C. A wireless passive SAW device operating at 434 MHz has been implemented on LGS and interrogated at 700°C with a 30 cm range interrogation distance (Francois et al., 2012). After 30 hours annealing at 700°C, the device shows increased dynamic admittance, constant K^2 and resonance frequency decreased by 50 kHz.

The SAW propagation along LGT substrates with Pt/Rh/ZrO_2-based IDTs is experimentally investigated up to 900°C for the Euler angle orientations (90°, 23°, Ψ) with $\Psi = 0°$, 13°, 48°, 77°, 119°, and 123°. Some crystal orientations are identified showing both zero TCD (in the temperature range from 400 to 500°C) and quite low TCD (Peter et al., 2011). The measured and theoretically predicted operation frequencies agree within experimental and prediction uncertainties in the temperature range 400 to 900°C.

$GaPO_4$, is a quartz homeotypic crystal but it has higher temperature stability, higher K^2 (twice the coupling as quartz and about the same as LGS) and lower damping than quartz. It is thus a promising material for the implementation of piezoelectric devices able to work in the temperature range up to 900°C (Krempl et al., 1997). $GaPO_4$ offers temperature compensated cuts for many operating temperatures, with a temperature sensitivity lower than quartz. One-inch diameter $GaPO_4$ wafers are commercially available for several crystal cuts; $GaPO_4$ microbalances thermally compensated at 500°C and 240°C, with Q of 70,000, are also commercially available from Piezocryst. The rotated Y-11° cut $GaPO_4$ is about 100 times less temperature sensitive than conventional AT-cut quartz crystals above 400°C. $GaPO_4$ sensors have been proven to be compatible with standard quartz crystal microbalance (QCM) instrumentation, and found to be an excellent replacement for quartz at temperatures above 300°C and below 720°C (Tichy et al., 2010). Singly rotated $GaPO_4$ Y-cut resonators have been investigated (Krispel et al.) for different rotating angles: Y-84° resonators showed Q factors in the order of millions but a low K^2, while Y + 27° resonators showed a very linear frequency dependence on temperature from room temperature up to 600°C.

The potential of the piezoelectric $Ca_3TaGa_3Si_2O_{14}$ (CTGS) crystals for applications over the temperature range from 60°C to ~700°C is demonstrated in Yu et al. (2011). The highest K^2 (18.9%) and piezoelectric coefficient d_{26} (11.5 pC/N) are obtained for (YXl)-25° cuts. Zero TCF cuts, with tunable TOTs from 20°C to 330°C, are explored as the rotation angle Θ varied from −22.5° to −35°. Bulk acoustic wave (BAW) resonators on Y(−30°) monolithic disks with a fundamental frequency 2.87 MHz show a mechanical quality factor Q ~24,000 and 10,000 at 20°C and 700°C, respectively.

The dielectric, piezoelectric, and electromechanical properties of disordered and ordered langasite-type piezoelectric single crystals at room temperature and 500°C are investigated for LGS, LTG, LNG, LTGA ($La_3Ta_{0.5}Ga_{5.3}Al_{0.2}O_{14}$), LNGA ($La_3Nb_{0.5}Ga_{5.3}Al_{0.2}O_{14}$), SNGS ($Sr_3NbGa_3Si_2O_{14}$), STGS ($Sr_3TaGa_3Si_2O_{14}$), CNGS ($Ca_3NbGa_3Si_2O_{14}$), CTGS($Ca_3TaGa_3Si_2O_{14}$), and CTAS ($Ca_3TaAl_3Si_2O_{14}$), showing stable temperature-dependent behavior. The ordered crystals possess lower dielectric permittivity and piezoelectric coefficients when compared to the disordered LGS crystal and its isomorphs; however, the totally ordered structure promises much higher mechanical Q and electrical resistivity at elevated temperatures. The higher acoustic velocity makes the fabrication of devices in the high frequency range easier. Meanwhile, the large time constant of the ordered LGS-type crystals is desirable for the potential of sensors, which operate at low frequency range and elevated temperature (Heywang et al., 2008).

Yttrium calcium oxyborate, $YCa_4O(BO_3)_3$ (YCOB), is a recently discovered single crystal piezoelectric material that doesn't undergo any phase transition up to its melting temperature of 1500°C. The resistivity, K^2, and piezoelectric coefficients of YCOB crystals have been investigated as a function of temperature for different crystal cuts (Yu et al., 2010). In particular, the coupling coefficient (22%) of the (ZXw) − 30° cut YCOB remains constant from room temperature up to 950°C, while the piezoelectric constant d_{26} varies from 10 to 12 pC/N. The high-temperature sensitivity make this material suitable as temperature sensor. In Zhang et al. (2008) the resonance frequency, K^2, and dielectric permittivity are studied in the temperature range of 30–950°C for different orientations and the most thermally stable cuts are found. A monolithic compression-mode piezoelectric accelerometer is fabricated using *(XYlw)* − 15°/45° cut YCOB single crystals; the sensor is tested up to 1000°C in the 100–600 Hz frequency range, showing a sensitivity of 2.4 ± 0.4 pC/g (Zhang et al., 2010b). A shear-mode piezoelectric accelerometer on YCOB is fabricated and tested from room temperature to 1000°C and at frequencies from 50 Hz to 350 Hz. The sensitivity of the prototype is found to be 5.9 ± 0.06 pC/g throughout the tested frequency and temperature ranges (Kima et al., 2012).

PiezoStar® KI100 is the tradename of a new piezoelectric material developed by Kistler in collaboration with the Leibniz Institute of Crystal Growth, IKZ, Berlin, Germany, and showing a structure similar to LGS or LGT (Claudio et al., 2011). It is not pyroelectric, and no twinning occurs up to its melting point. High temperature stability tests (from 500°C to 700°C) revealed that KI100 has remarkable features, such as outstanding insulation properties, very high resistivity (that remains stable in reducing atmosphere up to at least 600°C), and a small temperature coefficient that make this new crystal suitable for pressure sensors or accelerometers.

17.3 Non-piezoelectric substrates

SiC has potential for use in high-power, high-frequency, and radiation-resistant applications. The 6H-SiC high thermal conductivity (about 3.3 times that of Si at 300°K), high electric field breakdown strength (about 10 times that of Si), and wide bandgap (about 3 times that of Si) make it a material ideally suited for high-temperature, high-power, high-frequency, and high-radiation environments. The 6H-SiC lattice mismatch with AlN and gallium nitride (GaN) (as low as 1% and 3%) intimately links SiC substrates with AlN and GaN thin films for electroacoustic and optical device fabrication. Single-crystal wafers of 6H-SiC have been commercially available only since 1991 from Cree Research, Inc., Durham, NC, while 4H-SiC wafers have been available from Cree since 1994 and from ATMI since 1995 respectively. 3C-SiC and 15R-SiC wafers are produced in research environments, but not commercially. SiC-based electronics and sensors can operate in hostile environments where conventional silicon-based electronics cannot function. SiC has excellent mechanical properties for use in a harsh environment, is chemically inert and therefore difficult to micromachine.

Sapphire (Al_2O_3) has many outstanding properties: its hardness is next to diamond, it is chemically inert and is totally unaffected by all chemicals except some very hot caustics. It has high thermal conductivity (0.1 W/cm°K at 1000°C) and high electrical resistance, high biological resistance (it is unaffected by in vivo exposure), extremely high melt temperature (2040°C), and very wide optical transmission band from UV to near-IR. Its high acoustic velocity and relatively low loss make it suitable for high-frequency SAW device applications. Al_2O_3 wafers are commercially available in large diameter size, thickness, and several plane orientations, such as C, R, A, and M plane. C-Al_2O_3 has a large in-plane lattice mismatch with c-GaN, c-AlN, and c-ZnO, thus requiring a suitable buffer layer to optimize the nucleation and growth of these layer materials; in contrast, the lattice mismatch between GaN and *A*-plane sapphire is less than 2% (Doppalapudi et al., 1999). Epitaxial AlN and GaN films are grown on c-Al_2O_3, being the high mismatch compensated by dislocations confined at the layer/substrate interface (Ambacher, 1998). The growth of zinc oxide (ZnO) film with 38° tilted c axis on c-Al_2O_3 reduces to 2% value of the lattice mismatch (Wang et al., 2008).

Diamond is a suitable substrate for high-power, high-frequency electronic devices able to operate in harsh environments. Its hardness is larger than that of Si, Si_3N_4 and SiC (Kohn et al., 1999); its extreme chemical stability and bio-inertness (Specht et al., 2004) make it an ideal material for sensors operating in harsh or biologic environments. Although diamond is not piezoelectric, its high acoustic velocity makes it a desirable substrate for SAW devices when covered by a thin piezoelectric film (Chalker et al., 1999; Mortet et al., 2008). In addition to its wide band gap, high density, and relatively high thermal conductivity, diamond offers a very high SAW velocity, thus allowing the fabrication of SAW filters operating in the GHz range via viable feature sizes.

17.4 Thin piezoelectric films

Scientific and industrial communities have expressed a real need for chemical and physical sensors able to operate at elevated temperatures. SAW devices are small, robust, passive (batteryless), remotely requestable (wireless), cheap (if fabricated on a large scale), and able to operate in extreme conditions (if the materials constituting the device can withstand harsh conditions, such as high temperature and corrosive ambient) (Elmazria, 2011). The SAW devices implemented on *conventional* piezoelectric substrates, such as quartz, $LiNbO_3$, or $LiTaO_3$, are suitable to operate at ambient or near ambient temperature. SAW devices implemented on *nonconventional* piezoelectric materials, such as the LGS-family or $GaPO_4$, are suitable for harsh environment applications, but the relatively high acoustic propagation losses, which increase with frequency and temperature, and/or their quite low phase velocity limit the operating frequency below 1 GHz. As a result, research and development (R&D) is focused on a new generation of piezoelectric materials that are integrated circuit (IC) compatible, suitable for high-frequency operation and stable at those temperature values where the other materials fail. Multilayered structures based on thin film piezoelectric materials and nonpiezoelectric substrates have the potential for new types of sensors and signal processing devices, in which the optical, electronic, and acoustic features are hybridized. Moreover, in the layered structures the piezoelectric layer can recover the double role of electroacoustic transducer material and protective layer with respect to the metal IDTs when located at the substrate/film interface.

Among the piezoelectric materials able to withstand to harsh environment, the following are worthy of note: aluminum nitride (AlN), gallium nitride (GaN), and zinc oxide (ZnO):

- AlN is a promising piezoelectric material that has remarkable features: it can be grown at low temperature in thin film form (Naik et al., 1999; Tait and Mirfazli, 2001) with a high structural/morphological quality; it has high thermal conductivity (2.85 W cm^{-1}°C^{-1} at 300 °K) and resistance to caustic chemicals and mechanical stability at high temperature. In inert atmospheres AlN maintains its piezoelectricity up to 1200°C; in a vacuum, it decomposes at ~1800°C. In air, surface oxidation occurs above 700°C, but even at room temperature surface oxide layers of 5–10 nm have been detected (Kim et al., 1999). Above this temperature, bulk oxidation occurs (Kang et al., 2003). AlN exhibits high SAW phase velocity (5607 m/s in the c-plane) and quite good K^2 (0.3%). AlN-based multilayered structures look promising as substrates for the design of enhanced K^2 (with respect to that of the single piezoelectric material), near-zero TCD, high-frequency SAW devices (Caliendo, 2008). High-frequency SAW devices have been fabricated by growing highly c-axis oriented AlN films on diamond

(Assouar et al., 2007; Fatemi et al., 2013), SiC (Kim et al., 2007), and Al_2O_3 (Caliendo and Imperatori, 2004; Tsubouchi and Mikoshiba, 1985).

- GaN is a piezoelectric semiconductor suitable for application in high-power electronics; it exhibits a SAW phase velocity of 3701 m/s and $K^2 = 0.2\%$ in the c-plane, thermal conductivity (1.3 W/cm K), wide bandgap energy (3.39 eV for w-GaN), and a high breakdown field ($\sim 5 \cdot 10^6$ V cm^{-1}) (Strite and Morkoc, 1992). GaN is also chemically quite unreactive (very difficult to etch), with the N-face material being chemically more active than the Ga-face material. GaN-based structures and their alloys are piezoelectric semiconductors that belong to the III–V group; they show wide energy gaps suitable for applications in optoelectronic devices operating in the ultraviolet spectrum region (Khan et al., 1999; Ciplys et al., 2002a,b; Rimeika et al., 2000). SAW devices based on GaN films are especially attractive for developing visible–ultraviolet spectral range acousto-optic devices, such as modulators, deflectors, UV photodetectors (Calle et al., 2005), and UV sensors based on a SAW oscillator (Ciplys et al., 2001). High-velocity SAW devices based on high-quality GaN thin films are successfully fabricated on Al_2O_3 (Choi et al., 2003; Kim et al., 2002), SiC (Takagaki et al., 2002a), and diamond (Zhang et al., 2011; Zhang et al., 2010a) substrates. Other materials, including metals, metal oxides, metal nitrides, and semiconductors, have been investigated as possible substrates for the deposition of high-quality GaN layers (Kukushkin et al., 2008), and different kinds of buffer layers have been suggested for each substrate, in order to suppress crack formation in the GaN layer.
- ZnO is a multifunctional material whose electrical, optical, acoustic, and mechanical properties make it suitable in optoelectronic (blue and ultraviolet light emitters and detectors) and piezoelectric MEMS actuators to be used in the communications and sensors fields. ZnO has a SAW phase velocity of 2681 m/s in the c-plane and large K^2 (1%), a wide direct bandgap of $\sim$3.3 eV at room temperature and thermal conductivity in the range from 0.6 to 1 $Wcm^{-1} K^{-1}$. It can be chemically etched and epitaxially deposited in thin film form by reactive sputtering technique at low temperature (<400°C) on many different substrates. Through proper doping, it can be made as piezoelectric, ferromagnetic, transparent, and conducting. ZnO-based multilayered structures look promising as substrates for the design of enhanced K^2 (with respect to that of the single piezoelectric material), near-zero TCD, and high-frequency SAW devices (Naumenko, 2011). When ZnO is grown on high-velocity and low-loss substrates (such as Al_2O_3 (Mitsuyu et al., 1980; Kadota and Minakata, 1998), SiC (Didenko et al., 2000) and diamond (Dreifus et al., 1997; Hachigo et al., 1999)), on semiconductors (such as Si (Panwar, 2002; Moeller et al., 1994), GaAs (Matteo et al., 1996), and InP (Chang et al., 1995)), or on low K^2 piezoelectric materials (such as quartz (Tomar et al., 2005)). It is a promising candidate for high-frequency radio frequency (RF) devices, power electronics, monolithic acousto-optic light modulators, and high-temperature sensors.

17.4.1 High coupling efficiency

The SAW propagation along a dispersive structure can be designed to satisfy different electrical boundary conditions, resulting in the four coupling configurations schematically depicted in Figure 17.2. In the substrate/film/transducer (SFT) and substrate/transducer/film (STF) configurations the IDTs are located on the piezoelectric film-free surface and at the substrate/film interface. When a metal ground electrode is positioned opposite the piezoelectric layer side where the IDTs are, the substrate/metal/film/transducer (SMFT) and substrate/transducer/film/metal

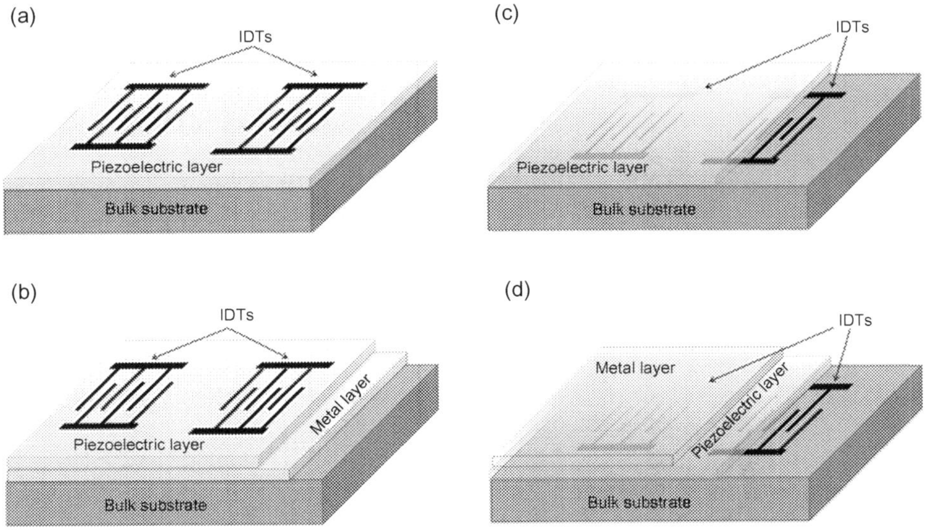

Figure 17.2 The four SAW coupling configurations: (a) SFT, (b) SMFT, (c) STF, and (d) STFM.

(STFM) configurations are obtained. The K^2, as well as the SAW phase velocity of these configurations, is frequency dispersive and can be larger than that of the individual piezoelectric material; it depends on the crystal cut and SAW propagation direction, and it is drastically affected by the electrical boundary conditions. As an example, Figure 17.3 shows the K^2 of the four configurations on 3C-SiC (001) <100> /c-AlN and 3C-SiC(001) <110> /c-AlN (Caliendo et al., 2013): a K^2 as high as 1.2% can be reached at the proper AlN film thickness and under the proper electrical boundary conditions.

17.4.2 TCD

The thermal stability of SAW devices implemented on bulk single-crystal piezoelectric substrates is typically achieved by selecting the zero-TCD cuts, often at the price of low K^2 and/or phase velocity. The TCD of a substrate/film multilayer is frequency dispersive and its value starts from the TCD of the substrate for $h/\lambda << 1$, then it asymptotically reaches the TCD of the film for $h/\lambda \geq 1$. A composite arrangement of materials with opposite TCDs enables achieving a thermally stable SAW device because the TCD of the structure crosses the zero TCD value for a certain film thickness (the temperature-compensated point). Si, GaAs, and Al_2O_3 substrates have positive TCDs (30, 50, and 70 ppm/°C (Cambon et al., 1981; Tsubuochi et al., 1981), respectively) and are suitable for implementing thermally stable SAW devices in combination with the proper AlN film thickness (TCD = −30 ppm/°C) (Tsubuochi et al., 1981). On the contrary, a compensation layer with negative TCD, such as SiO_2 (with TCD = −70 ppm/°C) is required if ZnO (Wu et al., 2001; Yu et al., 2007) or GaN films (Nishimura

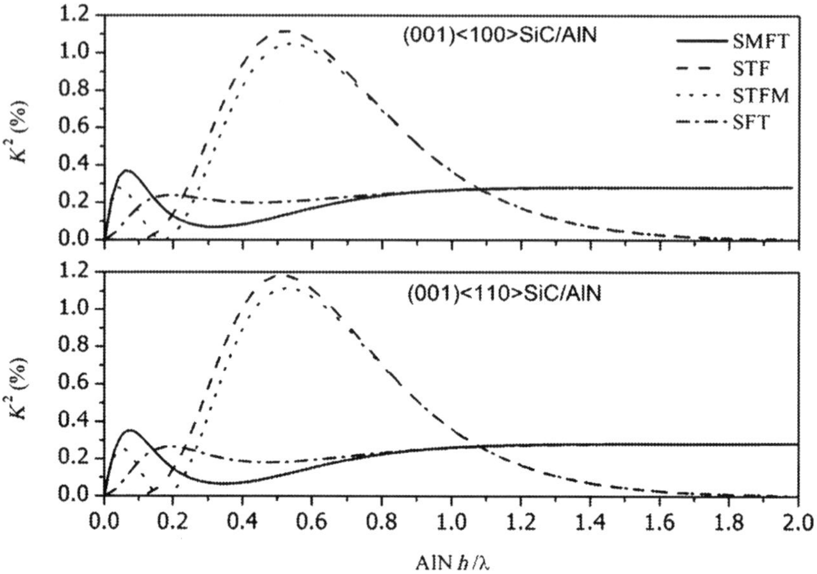

Figure 17.3 K^2 dispersion curves of the four coupling configurations on 3C-SiC (001) < 100 > /c-AlN and 3C-SiC(001) <110> /c-AlN.

et al., 2005) (with positive TCD = 35 and 50 ppm/°C) are preferred to AlN. A SiO_2 compensation layer is used to thermally compensate the positive TCD of the ZnO/Si multilayer at a specific frequency. By adding the proper thickness of a SiO_2 compensation layer, the TOT of an AlN-based device can be designed for a specific temperature value. In Lin et al. (2010), the temperature compensation of a Lamb wave resonator at 214, 430, and 512°C is described. The effects of the presence of the SiO_2 compensating layer have also been studied for high-velocity substrates and enhanced coupling configurations based on the propagation of higher order modes, and resulting in low TCD, GHz range SiO_2/ZnO/diamond/Si (Nakahata et al., 2003) structures.

An attractive alternative material to SiO_2 is the TeO_2 thin films whose negative TCD can be tuned by varying the oxygen background pressure during its growth by the RF sputtering technique. By increasing the oxygen percentage from 25% to 100% in the processing gas mixture, the TCD value increases continuously from $-1.862 \cdot 10^3$ ppm/°C to $-9.883 \cdot 10^3$ ppm/°C (Dewan et al., 2006). The positive TCD of ZnO/diamond-based SAW devices can be changed after integrating with a TeO_2 layer which enhances the coupling coefficient, and makes the layered TeO_2/ZnO/diamond structure temperature stable (zero TCD) (Dewan et al., 2008). Moreover, improved thermal stability of SAW devices implemented on 128° y-x LNO (Dewan et al., 2006; Dewan et al., 2005) can be obtained by depositing a thin TeO_2 film onto the piezoelectric substrate surface.

When the thermal stability of a SAW device implemented on bulk piezoelectric substrates is achieved at the expense of the coupling efficiency, one solution is to select two different bulk materials (one to be used for the implementation of the high K^2 SAW device and a second to ensure the thermal stability) and bond them together. After the bonding, the piezoelectric plate can then be thinned to a few tens of μm; being the base substrate much thicker than the thin piezoelectric plate where the SAW propagates, the bonding suppresses the coefficient of thermal expansion of the thin piezoelectric plate (Hashimoto et al., 2011).

17.4.3 High operation frequency

The dispersive film/substrate structure can allow the propagation of Rayleigh-like waves, higher order modes, pseudo-SAWs, and so on, depending on the SAW velocity of the two materials, the propagation direction, crystal cut and films thickness. On the contrary, the velocity of the SAW propagating along a piezoelectric semi-infinite substrate depends only on the crystal cut and SAW propagation direction. Expensive submicron lithography is required to increase the operation frequency of SAW devices implemented on bulk piezoelectric crystal substrates, while high-frequency SAW devices can be achieved by depositing thin piezoelectric films on high acoustic velocity substrates such as sapphire, SiC, and diamond. As an example, the SiO_2/IDT/ZnO/diamond/Si multilayer structure, which can provide a second Sezawa mode travelling at ~10000 m/s, a zero TCD, and a K^2 of ~1.2%, is promising for high-frequency and narrow-band devices (Nakahata et al., 2000).

The characteristics (such as phase velocity, loss, and K^2) of the SAW propagation along ZnO/6H-SiC has been theoretically and experimentally characterized in the film thickness-to-acoustic wavelength ratio range up to 0.12, for two acoustic surface modes, a generalized SAW (GSAW) mode with velocities between 6000 and 7000 m/s, and high-velocity pseudo-SAWs (HVPSAWs) with velocities between 8500 and 12500 m/s (Ashrafi et al., 2004). c-axis oriented ZnO films have been grown by metalorganic chemical vapor deposition on 6H-SiC substrates revealing an higher structural quality than the ZnO layers grown on conventional Al_2O_3 substrate. This higher ZnO quality might be attributed to the smaller lattice mismatch of −5% as well as the +c surface orientation in ZnO/SiC sample geometry (Ashrafi et al., 2004).

Extremely high-velocity SAW modes, showing moderate K^2, can be fabricated on AlN and GaN thin films deposited on SiC, diamond, or Al_2O_3 substrates (Takagaki et al., 2002a,b; Pastureaud et al., 2000). Large coupling higher-order GSAWs and HVPSAWs can propagate along ZnO films deposited on Al_2O_3, diamond, and SiC. A K^2 as high as 6% can be achieved for the Sezawa wave propagating along ZnO/r-Al_2O_3 (Ieki and Kadota, 1999); low-loss ZnO/r-Al_2O_3-based SAW filters with operation frequency as high as 2.5 GHz can be obtained for 0.53 μm IDTs line and space width (Emanetoglu et al., 2001).

The velocity and K^2 dispersion of GSAW, PSAW, and HVPSAW modes propagating along layered structures based on GaN/sapphire have been investigated for different Al_2O_3 cut planes (Kim et al., 2002). SAW devices operating in GHz range have been investigated on GaN/Si (with f = 5.69 GHz) (Müller et al., 2010) and for GaN/AlGaN (Lalinský et al., 2010). The frequency response of SAW delay lines implemented on GaN thin films grown on Al_2O_3 substrates has been measured at the fundamental and harmonic frequencies up to the 11th; operating frequencies from 230 MHz up to 2.1 GHz have been obtained for IDTs fingers width of 5 μm (Justice et al., 2012).

High-quality GaN films with small surface roughness of 4.9 nm and high c-axis orientation have been deposited on freestanding diamond substrates, confirming the great potential of GaN/diamond substrate for the development of SAW devices with high frequencies and low loss (Zhang et al., 2011).

Thus, combinations of a piezoelectric film and a nonpiezoelectric substrate can represent a good solution for high-frequency, enhanced-coupling, temperature-stable, and IC-compatible SAW devices.

17.4.4 High temperature measurements

The reliability of the piezoelectric thin film technology to the high temperature SAW devices requires the investigation of the device performances other than the thermal stability of the piezoelectric film, IDTs, and substrate materials. In Song et al. (1995) it has been observed that sputtered oxygen-rich ZnO films are thermally stable up to 700°C and the out-diffusion of zinc from the films heated at 450°C is negligible.

Fast and highly sensitive NO_2 and H_2 SAW sensors have been fabricated on *xz*-$LiNbO_3$ covered by a 1.2 μm ZnO layer and tested from 100 and 273°C (Ippolito et al., 2005). Improved sensor selectivity and sensitivity have been achieved by adding an InO_x layer on the device surface.

An O_2 SAW sensor has been fabricated on bulk LGS substrate covered by a thin (200 nm) ZnO film that works as the gas-sensitive layer. The gas concentration changes are revealed through the ZnO film conductivity change, thus producing the SAW sensor response (operating frequency and propagation loss shifts) via the electroacoustic effect. The sensor is able to work in the 500°C to 700°C range and its sensitivity has a peak at about 625°C (Zheng et al., 2011).

Shear horizontal SAW delay lines have been fabricated on ZnO/36° yx-$LiTaO_3$ substrates, being the device surface covered by an H_2-sensitive Pt and Au catalyst-activated WO_3 layer. The sensor sensitivity towards H_2 concentrations below 1% have been tested from 30 to 290°C. The Pt−WO_3 sensor shows the maximum sensitivity at operating temperatures between 100 and 270°C for the lower (from 0.06% to 0.25%) and higher (from 0.25% to 1%) H_2 concentrations; the Au−WO_3 sensor shows its best performance at 270°C for the whole H_2 concentration range (from 0.06% to 1%) (Ippolitoa et al., 2005).

The frequency response of SAW delay lines based on AlN/sapphire substrates and with Ir IDTs have been tested between 20°C and 1050°C in a vacuum (Aubert et al., 2013). Burnout is observed after 60 hours at 1050°C and is mainly attributed to the Ir transducers' agglomeration phenomena. A significant oxidation of the AlN film is also observed at such extreme temperatures. In Bruckner et al. (2010) the time and frequency responses of SAW delay lines on AlN/Al_2O_3 and operating at 680 MHz have been recorded during annealing treatments at temperatures up to 630°C; the observed device failure at 858°C is attributed to degradation of the Pt electrodes.

In Hou et al. (2012) c-axis oriented AlN thin film ultrasonic transducers have been fabricated on aluminum alloy and carbon steel substrates. High-temperature performance and durability of the transducers have been investigated using pulse-echo experiments at elevated temperatures, and the transducer failure modes are characterized. The AlN films' crystalline structure and orientation remained stable up to 600°C, while the transducers' performance at high temperature is limited by the deterioration of substrate properties.

In Goericke et al. (2011) AlN double-ended tuning forks (DETF) are presented as a transduction element to measure mechanical loads by measuring the resonance frequency shift. Device operation is demonstrated for strain sensing up to 91°C; transducer operation without applied strain is demonstrated up to 570°C and it is expected that higher temperatures can be achieved with appropriate circuitry. Strain sensing experiments indicate a strain sensitivity of 192 ppm/με at 22°C that increases with the temperature up to 243 ppm/με at 91°C. The frequency response of SAW delay lines implemented on highly textured c-AlN thin film sputtered at moderate temperature on Al_2O_3 substrates have been characterized before and after an air annealing at 800°C for 90 minutes (Aubert et al., 2011a). Both the acoustic measurements and the structural investigations demonstrated that the AlN films retain their piezoelectric properties after the thermal annealing.

A SAW delay line implemented on commercial monocrystalline AlN/sapphire plate has been investigated at high temperatures up to 950°C in air (Aubert et al., 2010). Frequency characterizations of the device indicate a slight increase of 2 dB in the insertion losses after 30 min annealing at 900°C. Scanning electron and atomic force microscopy as well as X-ray diffraction measurements suggest that these losses are due to the deterioration of the Pt/Ta electrodes and to a slight oxidation of the AlN film.

Highly c-axis oriented AlN films sputtered on Pt/Si and on (0001)Al_2O_3 substrates have been heated at 900°C in air up to 32 hours to test their resistance to high temperature (Caliendo and Latino, 2011; Caliendo, 2011). The thermal annealing improves the crystal quality of the AlN films sandwiched between the Pt films, as confirmed by the decreased full width at half maximum (FWHM) of the rocking curves, and the film piezoelectric coefficient d_{33} results unaffected by the temperature. Scanning electron microscopy, four points probe, and piezoelectricity tests reveal that the morphology and the sheet resistivity (in the range from 0.6 to 0.5 Ω/sq) of the Pt film, as well as the AlN piezoelectric constants d_{33} (in the range from 6.2 and $7.4 \cdot 10^{-12}$ C/N) results are unaffected by the high temperature annealing even after 32 hours' treatment. The AlN films on sapphire show a lattice elongation, perpendicular to the growth plane, that is usually associated with the compressive stress caused in the growth plane by the lattice mismatch between the film and the substrate. After the first few hours of annealing, the FWHM of the (002) AlN peak decreases but further annealings result in the FWHM broadening, depending on the film thickness. The structural investigations of thin (0.26 μm) and thick (from 2 to 4.7 μm) annealed AlN films reveals that the film behavior in a harsh environment is strongly affected by the film properties.

Thick films, whose structure is more relaxed than thin films, are able to survive to high temperature without suffering significant deterioration for longer annealing times than thin films.

Schottky diodes based on GaN layer on sapphire substrates and Pt catalytic gate are able to sense a large variety of gases (H_2, C_2H_4, C_4H_{10}, CO, NO) at operation temperature ranging from 200 to 600°C (Schalwig et al., 2001). An alternative to a Pt–GaN Schottky diodes-based gas sensor is the Pt–HEMTs (High Electron Mobility Transistors) devices formed from GaN/AlGaN heterostructures whose gas sensing properties have been tested at 300°C for H_2, C_2H_4, and CO gases (Stutzmann et al., 2002). SAW devices designed to operate at the fifth and seventh harmonics have been fabricated on GaN thin films on silicon and sapphire substrates and annealed at temperatures ranging from 450 to 1000°C. The device performance is measured through insertion loss, signal-to-noise ratio, operating frequency, and quality factors. SAW devices fabricated in this work have operating frequencies above 1 GHz, and quality factors up to and higher than 2000, depending on the harmonic mode. SAW devices on GaN showed good chemical stability and measured changes in device performance after exhaust exposure are negligible (Justice et al., 2013). The SAW propagation characteristics along GaN/Al_2O_3(0001)<01-10> and <2-1-10> directions are studied between −50 and 200°C for different GaN thicknesses. The TCD of bulk GaN is estimated to be 28.3 ppm/°C. The thermal stability of SAW filters on GaN/(0001)Al_2O_3 has been assessed up to 620°C by the Nippon Telegraph and Telephone Corporation (Nishimura et al., 2005). A GaN-based Lamb-wave oscillator fully integrated on Si substrate has been characterized from room temperature up to 250°C (Lu et al., 2013). The device oscillation frequency exhibits a linear dependence on temperature with a small temperature coefficient of frequency, TCF, of −47.5 ppm/°C.

17.5 Metal electrodes

The metal IDTs serve to convert electrical energy into acoustic energy on the piezoelectric substrate, and vice versa. Thus, the high-temperature performance of the electroacoustic devices is limited by the stability of the metal electrodes other than on that of the single layer materials. This limitation is especially serious for the case of SAW resonators, whose IDTs' thickness should be about 100–150 nm, as a consequence of the influence of the electrodes mass loading on the SAW propagation characteristics. The electrodes should exhibit good adhesion to the substrate and a high conductivity under harsh conditions. Noble metals, such as silver, gold and the platinum group (ruthenium, platinum, rhodium, palladium, osmium, and iridium) are suitable for high-temperature applications because they remain conductive in highly oxidizing conditions; Pt is highly resistant to oxidation, whereas Ir and Ru form conductive oxides (IrO_2 and RuO_2) when they become oxidized. Sometimes an intermediate layer, grown between the substrate and the thin metal electrode, is required to improve the electrode adhesion and to increase the long-term stability of the electrode. Table 17.3 summarizes the properties of the metals suitable for SAW

Table 17.3 **Properties of metals (Çiftyürek et al., 2013) that are suitable as IDTs and metal electrodes**

Metal	Melting point (°C)	Density at 298°K (kgm^{-3})	Resistivity at 293°K (10^{-8} Ωm)
Al	660	2700	2.65
Pt	1768	21500	10.5
W	3422	19300	5.28
Ru	2334	12100	7.1
Ag	962	10500	1.59
Au	1064	19300	2.214
Pd	1555	12000	10.54
Ta	3017	16400	13.1
Mo	2623	10200	5.34
Ti	1668	4506	39
Zr	1855	6520	42.1
Os	3033	22587	8.1
Rh	1964	12400	4.3
Ir	2446	22562	4.7

electrode implementation. Pt is an excellent choice for electrodes because it is inert and highly conductive; it has a high melting point and its electrical resistance changes linearly over a wide temperature range. Despite its chemical inertness, Pt can be wet etched in hot aqua regia in a fast and easy way (Kollensperger et al., 2012). Pt and other noble metals have poor adhesion to oxide surfaces such as quartz or LGS, thus a thin (10 to 20 nm) intermediate layer (such as Ti, Zr, Ta, etc.) between the substrate and the metal film is required to improve the thermal stability over prolonged exposure to high temperature. Various types of metal/Pt layer compositions have been studied to improve the noble metal adhesion and thermal stability. In Çiftyürek et al. (2013) the room temperature electrical resistivity of Pt films with various adhesion layers (Ti, Ta, Zr, and Hf), deposited on alumina ceramic substrates, are studied after annealing up to 1200°C in N_2 atmosphere, for 1 to 48 hours. The Hf–Zr–Pt multilayer remains conductive (resistivity $<62.5 \cdot 10^{-9}$ Ωm) after being processed at 1200°C for 48 hours, while the Ti/Pt bilayers show a complete loss of electrical conductivity after 5 hours' annealing treatment; both Ta/Pt and Zr/Pt bilayers lose their conductivity after 15 hours' annealing.

The thermal stability of Ti/Pt- and Zr/Pt-based IDTs on $GaPO_4$ substrates is tested for different Pt thicknesses. The thicker Pt electrode (135 nm) shows an improved long-term stability (192 h at 600°C with a relative frequency shift of 10 ppm/h) in comparison to the thinner ones (~74 nm) due to Au wire diffusion and Ti under layer oxidation. On the contrary, devices with a Zr underlayer are more stable (relative frequency shift of 2 ppm/h after 300 h at 600°C) (Hamidon et al., 2005).

The NASA Glenn Research Center, Cleveland, OH, widely investigated the long term thermal stability of metal contacts on SiC at high temperature (Neudeck et al.,

2007; Neudeck et al., 2010). This research center was the first to surpass 3000 hours of stable electrical operation at 500°C of a Ti/TaSi/Pt contact structure. Moreover, the W/Ni ohmic contact to commercially grown 2 μm-thick homoepitaxial 4H-SiC and 6H-SiC epilayers of n- and p-type conductivities on high-resistivity p-type substrates shows a stable contact resistivity even after rapid thermal annealing at 1000°C for 5 seconds in Ar and a subsequent thermal treatment at 900°C for 1 hour in argon ambient (Okojie et al., 2010). A Ta adhesion layer for Pt electrodes has been used in high-temperature SAW devices on LGS substrates (Aubert et al., 2011b). Ta exhibits a great adhesive strength and a very low mobility through the Pt film, ensuring about 1 hour and at least 20 hours' device lifetime at 900°C in air and vacuum; after that period, the Pt film continuity is broken. Thiele, J.A. and Pereira da Cunha (Thiele and Pereira da Cunha, 2005) describes the design and fabrication of an original high-temperature RF testing chamber (including packaging, test materials, test fixture, coaxial cables, coplanar transmission line, and Pt wire bonds to the IDTs' pads) suitable for the characterization of the Zr/Pt- and Zr/Pd-based IDTs on LGS substrate and capable of withstanding temperature values up to 1000°C. The Pt transducers on LGS SAW devices are able to operate at 750°C with a 7 dB decrease in the scattering parameter $|S_{21}|$ over a period of three weeks. The Pd transducers are able to operate at 250°C with a 2 dB decrease in $|S_{21}|$, over a period of six weeks. The transducer electrodes show no metal surface degradation, as confirmed by the resistivity measurements and the XPS data recorded.

It has been observed that the Ti adhesion layer material for Pt film grown on r-Al_2O_3 substrates migrates into the Pt film after prolonged exposure to temperatures between 200 and 700°C, thus causing the loss of the electrode adhesion (Bernhardt et al., 2001). Also the quality and reliability of the wire bond connections to the Pt bond pads is affected by the metal adhesion layer. Pt bond pad delamination occur in the 800–850°C range for both Zr and Ti adhesion layer, and the weakening begin sooner for samples with Ti adhesion layer than for samples with Zr adhesion layer.

Pd, Ti/Pt, Ti/Pt/Rh, Ir electrodes, and ceramic electrodes have been tested up to 1000°C for SAW devices on LGS and their resistance and life time have been studied for different thicknesses up to 1000°C (Richter and Fritze, 2011). The Ti/Pt electrode with a Al_2O_3 protection layer shows a stable resistance with no visible degradation up to temperatures of 900°C for 6 hours. It seems a suitable electrode material for applications up to 700°C, while Pd undergoes a significant degradation below 600°C. Pure $La_{0.65}Sr_{0.35}MnO_3$ (LSM) ceramic electrodes and Pt/LSM undergo a significant increased resistance at 1000°C and 900°C respectively.

Promising alternatives to metal electrodes are conductive ceramic electrodes, which may offer a good long-term stability at high temperatures. Common materials already used as cathode layers in high-temperature solid oxide fuel cells are, for example, LSM and LSCF ($La_{0.6}Sr_{0.4}Co_{0.2}Fe_{0.8}O_3$) (Lee et al., 2010). General disadvantages of this material group compared to metal electrodes are the lower conductivity of the materials in the low-temperature range. Furthermore, problems may be caused by changes in stoichiometry in reducing or oxidizing atmospheres.

17.6 Conclusion

The focus of this brief review is on the recent studies on SAW electroacoustic devices able to survive at high temperatures, and based on single-crystal piezoelectric substrates and dispersive layered structures. The review has covered thin piezoelectric films, nonpiezoelectric substrates, and metal electrode materials that are suitable for the implementation of high-frequency, enhanced coupling electroacoustic devices able to operate at high temperatures.

References

Adachi, M., Shiosaki, T., Ohtsuka, K., Kawabata, A., 1994. Leaky SAW propagation properties on Li2B407 substrates. In: IEEE Int. Freq. Control Symposium, pp. 296–300.

Akiyama, M., Kamohara, T., Kano, K., Teshigahara, A., Takeuchi, Y., et al., 2009. Enhancement of piezoelectric response in scandium aluminum nitride alloy thin films prepared by dual reactive cosputtering. Adv. Mater. 21, 593–596.

Ambacher, O., 1998. Growth and applications of Group III-nitrides. J. Phys. D: Appl. Phys. 3, 2653–2710.

Ashrafi, A.B.M.A., Zhang, B.P., Binh, N.T., Wakatsuki, K., Segawa, Y., 2004. Highquality ZnO layers grown on 6H-SiC substrates by metalorganic chemical vapor deposition. Jpn. J. Appl. Phys. 43 (3), 1114–1117.

Assouar, M.B., Elmazria, O., Kirsch, P., Alnot, P., 2007. High-frequency surface acoustic wave devices based on AlN/diamond layered structure realized using e-beam lithography. J. Appl. Phys. 101, 114507.

Aubert, T., Elmazria, O., Assouar, B., Bouvot, L., Oudich, M., 2010. Surface acoustic wave devices based on AlN/sapphire structure for high temperature applications. Appl. Phys. Lett. 96, 203503.

Aubert, T., Assouar, M.B., Legrani, O., Elmazria, O., Tiusan, C., et al., 2011a. Highly textured growth of AlN films on sapphire by magnetron sputtering for high temperature surface acoustic wave applications. J. Vac. Sci. Technol. A. 29, 021010–021010-6.

Aubert, T., Elmazria, O., Assouar, B., Bouvot, L., Hehn, M., et al., 2011b. Behavior of platinum/tantalum as interdigital transducers for SAW devices in high-temperature environments. IEEE Trans. Ultrason. Ferroelectr. Freq. Control. 58 (3), 603–610.

Aubert, T., Elmazria, O., Assouar, B., Blampain, E., Hamdan, A., et al., 2012. Investigations on AlN/sapphire piezoelectric bilayer structure for high-temperature SAW applications. IEEE Trans. Ultrason. Ferroelectr. Freq. Control. 59 (5), 999–1005.

Aubert, T., Bardong, J., Legrani, O., Elmazria, O., Assouar, M.B., et al., 2013. In situ high-temperature characterization of AlN-based surface acoustic wave devices. J. Appl. Phys. 114, 014505.

Bardong, J., Aubert, T., Naumenko, N., Bruckner, G., Salzmann, S., et al., 2013. Investigations of some useful langasite cuts for high-temperature SAW applications. IEEE Trans. Ultrason. Ferroelectr. Freq. Control. 60 (4), 814. IEEE Experimental and Theoretical.

Bernhardt, G., Silvestre, C., Le Cursi, N., Moulzolf, S.C., Frankel, D.J., et al., 2001. Performance of Zr and Ti adhesion layers for bonding of platinum metallization to sapphire substrates. Sens. Actuators B. 77, 368–374.

Bruckner, G., Bardong, J., Fachberger, R., Forsén, E., Eisele, D., 2010. Investigations of SAW delay lines on c-plane AlN/sapphire at elevated temperatures. In: Frequency Control Symposium (FCS), 2010 IEEE International, pp. 499–502.

Caliendo, C., 2008. Analysis of dispersive electroacoustic coupling configurations for application to gigahertz-band, temperature-compensated AlN-based acoustic devices. Appl. Phys. Lett. 92, 103501.

Caliendo, C., 2011. Surface acoustic wave devices for harsh environment. In: Acoustic Waves–From Microdevices to Helioseismology, InTech, Croatia, pp. 579–594.

Caliendo, C., Imperatori, P., 2004. Structural, optical, and acoustic characterization of high-quality AlN thick films sputtered on Al_2O_3(0001) at low temperature for GHz-band electroacoustic devices applications. J. Appl. Phys. 96 (5), 2610–2615.

Caliendo, C., Latino, P.M., 2011. Characterization of Pt/AlN/Pt-based structures for high temperature, microwave electroacoustic devices applications. Thin Solid Films. 519 (19), 6326–6329.

Caliendo, C., D'Amico, A., Lo Castro, F., 2013. Lamb waves propagation along 3C-SiC/AlN membranes for application in temperature-compensated, high-sensitivity gravimetric sensors. Sensors (Basel). 13, 550–564.

Calle, F., Pedrós, J., Palacios, T., Grajal, J., 2005. Nitride-based surface acoustic wave devices and applications. Phys. Stat. Sol. (c). 2 (3), 976–983.

Cambon, G., Lassale, J., Saurel, J., Attal, J., 1981. Design and realization of temperature stable surface acoustic wave delay lines on silicon and gallium arsenide. IEEE Ultrason. Symp. Proc. 364.

Chalker, P.R., Joyce, T.B., Johnston, C., Crossley, J.A.A., Huddlestone, J., et al., 1999. Fabrication of aluminium nitride/diamond and gallium nitride/diamond SAW devices. Diamond Relat. Mater. 8 (2–5), 309–313.

Chang, S.J., Su, Y.K., Shei, Y.P., 1995. High quality ZnO thin films on InP substrates prepared by radio frequency magnetron sputtering. II. Surface acoustic wave device fabrication. J. Vac. Sci. Technol. A. 13 (2), 385.

Choi, K.H., Joon, H., Chung, S.J., Kim, J.Y., Lee, T.K., et al., 2003. Experimental and theoretical characterization of the surface acoustic wave propagation properties of GaN epitaxial layers on c-plane sapphire. J. Mater. Res. 18 (5), 1157–1161.

Çiftyürek, E., Sabolsky, K., Sabolsky, E.M., 2013. Platinum thin film electrodes for high-temperature chemical sensor applications. Sens. Actuators B. 181, 702–714.

Ciplys, D., Rimeika, R., Sereika, A., Gaska, R., Shur, M.S., et al., 2001. GaN-based SAW delay-line oscillator. Electron. Lett. 37 (8), 545–546.

Ciplys, D., Rimeika, R., Shur, M.S., Gaska, R., Deng, J., et al., 2002a. Acousto-optic diffraction of blue and red light in GaN. Citation. Appl. Phys. Lett. 80, 1701.

Ciplys, D., Shur, M.S., Gaska, R., Rimeika, R., Sereika, A., et al., 2002b. GaN based surface acoustic wave devices for optoelectronics applications. Electrochem. Soc. Proc. 4, 119–138.

Claudio, C., Roland, S., Max, W., 2011. High temperature stability of new single crystal piezoelectric sensors. In: SENSOR + TEST Conferences 2011, SENSOR Proceedings, pp. 520–525.

Darnjanovic, D., 1998. Materials for high temperature piezoelectric transducers. Curr. Opin. Solid State Mater. Sci. 3 (5), 469–473.

Dewan, N., Tomar, M., Gupta, V., Sreenivas, K., 2005. Temperature stable $LiNbO_3$ surface acoustic wave device with diodes puttered amorphous TeO2 over-layer. Appl. Phys. Lett. 86, 223508.

Dewan, N., Sreenivas, K., Gupta, V., 2006. Temperature-compensated devices using thin TeO_2 layer with negative TCD. IEEE Electron Device Lett. 27 (9), 752–754.

Dewan, N., Sreenivas, K., Gupta, V., 2008. Theoretical studies on a TeO_2/ZnO/diamond-layered structure for zero TCD SAW devices. Semicond. Sci. Technol. 23, 085002.

Didenko, I.S., Hickernell, F.S., Naumenko, N.F., 2000. The experimental and theoretical characterization of the SAW propagation properties for zinc oxide films on silicon carbide. IEEE Trans. Ultrason. Ferroelectr. Freq. Control. 47 (1), 179.

Doppalapudi, D., Iliopoulos, E., Basu, S.N., Moustakas, T.D., 1999. Epitaxial growth of gallium nitride thin films on A-plane sapphire by molecular beam epitaxy. J. Appl. Phys. 85 (7), 3582–3589.

Douchet, G., Sthal, F., Bigler, E., Bourquin, R., 2010. A comparison of vibrating beam resonators in quartz, $GaPO_4$, LGS and LGT. Solid State Sci. 12, 325–332.

Dreifus, D.L., Higgins, R.J., Henard, R.B., Almar, R., Solie, L.P., 1997. Experimental observation of high velocity pseudo-SAWS in ZnO/diamond/Si multilayers. In: Proceedings of the IEEE Ultrasonics Symposium, p. 191.

Elmazria, O., 2011. Wireless SAW sensor for high temperature applications: material point of view. In: Proceedings of SPIE Smart Sensors, Actuators, and MEMS, Prague, Czech Republic, Vol. 8066, Session 1, pp. 18–20.

Emanetoglu, N.W., Patounakis, G., Liang, S., Gorla, C.R., Wittstruck, R.H., et al., 2001. Analysis of SAW properties of epitaxial ZnO films grown on r-Al_2O_3 substrates. IEEE Trans. Ultrason. Ferroelectr. Freq. Control. 48 (5), 1389.

Fatemi, H., Zeng, H., Carlisle, J.A., Abdolvand, R., 2013. High-frequency thin-film AlN-on-diamond lateral–extensional resonators. J. Microelectromech. Syst. 22 (3), 678–686.

Francois, B., Richter, D., Fritze, H., Davis, Z.J., Droit, C., et al., 2012. Wireless and passive sensors for high temperature measurements. In: SENSORDEVICES 2012 – The Third International Conference on Sensor Device Technologies and Applications, pp. 46–51.

Giurgiutiu, V., Xu, B., Liu, W., 2010. Development and testing of high-temperature piezoelectric wafer active sensors for extreme environments. Struct. Health Monit. 9, 513.

Goericke, F.T., Chan, M.W., Vigevani, G., Izyumin, I., Boser, B.E., et al., 2011. High temperature compatible aluminum nitride resonating strain sensor transducers. In: Proceedings of Solid-State Sensors, Actuators and Microsystems Conference (TRANSDUCERS), 2011 16th International, Beijing, China, June 5–9, pp. 1994–1997.

Hachigo, A., Nakahata, H., Itakura, K., Fujii, S., Shikata, S., 1999. 10 GHz narrow band SAW filters using diamond. In: Proceedings of the IEEE International Ultrasonics Symposium, p. 325.

Hamidon, M.N., Skarda, V., White, N.M., Krispel, F., et al., 2005. Fabrication of high temperature surface acoustic wave devices for sensor applications. Sens. Actuators A. 123–124, 403–407.

Hartmann, C.S., Bell Jr, D.T., Rosenfeld, R.C., 1973. Impulse model design of acoustic surface-wave filters. IEEE Trans. Microw. Theor. Tech. MTT-21, 162–175.

Hashimoto, K.-Y., Kadota, M., Nakao, T., Ueda, M., Miura, M., et al., 2011. Recent development of temperature compensated SAW devices. In: IEEE International Ultrasonics Symposium Proceedings, pp. 79–86.

Heywang, W., Lubitz, K., Wersing, W., 2008. Piezoelectricity: Evolution and Future of a Technology. Springer Series in Materials Science, Vol. 114. Springer-Verlag, Berlin; Heidelberg, Germany.

Hou, R., Hutson, D., Kirk, K.J., Qing Fu, Y., 2012. AlN thin film transducers for high temperature non-destructive testing applications. J. Appl. Phys. 111, 074510.

Ieki, H., Kadota, M., 1999. ZnO thin films for high frequency SAW devices. In: Proceedings of the IEEE Ultrasonic Symposium, p. 281.

Ippolito, S.J., Kandasamya, S., Kalantar-Zadeh, K., Wlodarski, W., Galatsis, K., et al., 2005. Highly sensitive layered $ZnO/LiNbO_3$ SAW device with InOx selective layer for NO2 and H2 gas sensing. Sens. Actuators B. 111−112, 207−212.

Ippolitoa, S.J., Kandasamya, S., Kalantar-Zadehab, K., Wlodarskia, W., 2005. Layered SAW hydrogen sensor with modified tungsten trioxide selective layer. Sens. Actuators B. 108, 553−557.

Jaffe, B., Cook Jr., W.R., Jaffe, H., 1971. In: Roberts, J.P., Popper., P. (Eds.), Piezoelectric Ceramics, Non Metallic Solids: A Series of Monographs, Number 3. Academic Press, London.

Ji, X., Han, T., Shi, W., Zhang, G., 2005. Investigation on SAW properties of LGS and optimal cuts for high-temperature applications. IEEE Trans. Ultrason. Ferroelectr. Freq. Control. 52 (11), 2075−2080.

Jiang, X., Kim, K., Zhang, S., 2014. Joseph Johnson and Giovanni Salazar, high-temperature piezoelectric sensing. Sensors. 14, 144−169.

Justice, J., Lee, K., Korakakis, D., 2012. Harmonic surface acoustic waves on gallium nitride thin films. IEEE Trans. Ultrason. Ferroelectr. Freq. Control. 59 (8), 1806−1811.

Justice, J., Elbaz, M., Rodak, L.E., Korakakis, D., 2013. High harmonic surface acoustic wave devices for harsh environment sensor applications. MRS Proc. 1519, 60−65.

Kadota, M., Minakata, M., 1998. Piezoelectric properties of ZnO films on a sapphire substrate deposited by an RF-magnetron-mode ECR sputtering system. Jpn. J. Appl. Phys. Part 1: Reg. Papers, Short Notes & Review Papers. 37 (5B), 2923.

Kang, H.C., Seo, S.H., Jang, H.W., Kim, D.H., Kim, J.W., et al., 2003. Synthesis of epitaxial gamma Al_2O_3 thin films by thermal oxidation of AlN/sapphire(0001) thin films. Appl. Phys. A. 77, 627−632.

Khan, A., Rimeika, R., Ciplys, D., Gaska, R., Shur, M.S., 1999. Optical guided modes and surface acoustic waves in GaN grown on (0001) sapphire substrates. Phys. Stat. Sol. (b). 216, 477.

Kim, J.W., Lee, S.-J., Kang, K.-Y., Kim, S.H., Kim, H.K., 1999. Critical thickness and surface oxidation of epitaxial AIN thin films. Integr. Ferroelectr. Int. J. 24 (1−4), 129−137.

Kim, M., Ohta, J., Kobayashi, A., Fujioka, H., Oshima, M., 2007. Epitaxial growth mechanisms of AlN on SiC substrates at room temperature. Appl. Phys. Lett. 91, 151903.

Kim, Y.-J., Lim, G.H., Choi, K.H., Chung, S.J. Kim, H.J., et al., 2002. SAW characteristics of GaN epitaxial films deposited on different plane sapphire substrates using MOCVD. In: Proceedings of the IEEE Ultrasonics Symposium, Vol. 1, pp. 427−430.

Kima, K., Zhangb, S., Salazara, G., Jianga, X., 2012. Design, fabrication and characterization of high temperature piezoelectric vibration sensor using YCOB crystals. Sens. Actuators A. 178, 40−48.

Kitanaka, Y., Noguchi, Y., Miyayama, M., Kagawa, Y., 2012. Elastic and piezoelectric properties of high-quality ferroelectric $Bi_4Ti_3O_{12}$ single crystals. Jpn. J. Appl. Phys. 51, 09LD08.

Kohn, E., Gluche, P., Adamschik, M., 1999. Diamond MEMS − a new emerging technology. Diam. Relat. Mater. 8, 934−940.

Kollensperger, P.A., Karl, W.J., Ahmad, M.M., Pike, W.T., Green, M., 2012. Patterning of platinum (Pt) thin films by chemical wet etching in aqua regia. J. Micromech. Microeng. 22, 067001.

Krempl, P., Schleinzer, G., Wallnofer, W., 1997. Gallium phosphate, $GaPO_4$: a new piezoelectric crystal material for high-temperature sensories. Sens. Actuators A. 61, 361−363.

Krispel, F., Reiter, C., Neubig, J., Lenzenhuber, F., Krempl, P.W., et al., Properties and applications of singly rotated $GaPO_4$ resonators, paper presented at IEE Int. Frequency Control Symposium & 17th European Frequency and Time Forum. Tampa, USA.

Kukushkin, S.A., Osipov, A.V., Bessolov, V.N., Medvedev, B.K., Nevolin, V.K., et al., 2008. Substrates for epitaxy of gallium nitride: new materials and techniques. Rev. Adv. Mater. Sci. 17, 1–32.

Lalinský, T., Rýger, I., Vanko, G., Tomáška, M., Kostiþ, I., et al., 2010. AlGaN/GaN based SAW-HEMT structures for chemical gas sensors. Procedia Eng. 5, 152–155.

Lee, S., Chu, C.-L., Tsai, M.-J., Lee, J., 2010. High temperature oxidation behavior of interconnect coated with LSCF and LSM for solid oxide fuel cell by screen printing. Appl. Surf. Sci. 256, 1817–1824.

Lin, C.-M., Yen, T.-T., Felmetsger, V.V., Hopcroft, M.A., Kuypers, J.H., Pisano, A.P., 2010. Thermally compensated aluminum nitride lamb wave resonators for high temperature applications. Appl. Phys. Lett. 97, 083501.

Lu, X., Ma, J., Yue, C.P., Lau, K.M., 2013. A GaN-based lamb-wave oscillator on silicon for high-temperature integrated sensors. IEEE Microwave Wireless Compon. Lett. 23 (6), 318–320.

Matteo, M., Passaro, V.M.N., Armenise, M.N., 1996. High-performance guided-wave acoustooptic bragg cells in $LiNbO_{3-}$ and GaAs-based structures. IEEE Trans. Ultrason. Ferroelectr. Freq. Control. 43 (2), 270.

Megriche, A., Lebrun, L., Troccaz, M., 1999. Materials of $Bi_4Ti_3O_{12}$ type for high temperature acoustic piezo-sensors. Sens. Actuators. 78, 88–91.

Mitsuyu, T., Ono, S., Wasa, 1980. Structures and SAW properties of RF-sputtered single-crystal films of ZnO on sapphire. J. Appl. Phys. 51 (5), 2464.

Moeller, F., Vandahl, T., Malocha, D.C., Schwesinger, N., Buff, W., 1994. Properties of thick ZnO layers on oxidized silicon. Proc. IEEE Ultrason. Symp. 403.

Mortet, V., Williams, O.A., Haenen, K., 2008. Diamond: a material for acoustic devices. Phys. Stat. Sol. (a). 205 (5), 1009–1020.

Müller, A., Neculoiu, D., Konstantinidis, G., Deligeorgis, G., Dinescu, A., et al., 2010. SAW devices manufactured on GaN/Si for frequencies beyond 5 GHz. IEEE Electr. Dev. Lett. 31 (12), 1398–1400.

Naik, R.S., Reif, R., Lutsky, J.J., Sodini, C.G., 1999. Low temperature deposition of highly textured aluminum nitride by direct current magnetron sputtering for applications in thin film resonators. Electrochem. Soc. 146 (2), 691–696.

Nakahata, H., Hachigo, A., Itakura, K., Shikata, S., 2000. Fabrication of high frequency SAW filters from 5 to 10 GHz using SiO_2/ZnO/diamond structure. In: Proceedings of the 2000 IEEE Ultrasonics Symposium, p. 349.

Nakahata, H., Fujii, S., Higaki, K., Hachigo, A., Kitabayashi, H., et al., 2003. Diamond-based surface acoustic wave devices. Semicond. Sci. Technol. 18, S96–S104.

Naumenko, N., 2011. Multilayered structure as a novel material for surface acoustic wave devices: physical insight. In: Beghi, Marco G. (Ed.), Acoustic Waves – From Microdevices to Helioseismology. InTech, Croatia, pp. 421–442.

Neudeck, P.G., Okojie, R.S., Chen, L.-Y., 2002. High-temperature electronics—a role for wide bandgap semiconductors? Proc. IEEE. 90 (6), 1065–1076.

Neudeck, P.G., Spry, D.J., Chen, L.-Y., Okojie, R.S., Beheim, G.M., et al., 2007. SiC field effect transistor technology demonstrating prolonged stable operation at 500°C. Mater. Sci. Forum. 556–557, 831–834.

Neudeck, P.G., Krasowski, M.J., Chen, L.-Y., Prokop, N.F., 2010. Characterization of 6H-SiC JFET integrated circuits over a broad temperature range from −150°C to +500°C. Mater. Sci. Forum. 645–648, 1135–1138.

Nishimura, K., Shigekawa, N., Yokoyama, H., Hohkawa, K., 2005. Temperature dependence of surface acoustic wave characteristics of GaN layers on sapphire substrates. Jpn. J. Appl. Phys. 44, L564–L565.

O'Connell, R.M., Carr, P.H., 1977. High piezoelectric coupling temperature-compensated cuts of berlinite ($AlPO_4$) for SAW applications. IEEE Trans. Sonics Ultrason. SU-24 (6), 376–383.

Okojie, R.S., Evans, L.J., Lukco, D., Morris, J.P., 2010. A novel tungsten–nickel alloy ohmic contact to SiC at 900°C. IEEE Electron Device Lett. 31 (8), 791–793.

Panwar, B.S., 2002. Characteristics of surface acoustic wave convolver in the monolithic metal–zinc oxide–silicon nitride–silicon dioxide–silicon structure. Appl. Phys. Lett. 80 (10), 1832.

Pastureaud, T., Soufyane, A., Camou, S., Ballandras, S., Schenck, D., et al., 2000. AlN and GaN layers deposited on sapphires or silicon substrates: theory and experiment. In: Proceedings of the IEEE Ultrasonics Symposium, p. 293.

Peter, M., Pereira da Cunha, D., Pereira da Cunha, M., 2011. Determination and experimental verification of high-temperature SAW Orientations on langatate. In: Frequency Control and the European Frequency and Time Forum (FCS), Joint Conference of the IEEE International, 2–5 May. San Fransisco, CA, pp. 1–6.

PI Piezo Technology, n.d. <http://www.piceramic.com> (accessed 28.08.14.).

Richter, D., Fritze, H., 2011. High-temperature stable electrodes for langasite based surface acoustic wave devices. In: Sensor TEST Conference Proceedings, pp. 532–537.

Rimeika, R., Ciplys, D., Gaska, R., Yang, J.W., Khan, M.A., et al., 2000. Diffraction of guided optical waves by surface acoustic waves in GaN. Appl. Phys. Lett. 77, 480.

Schalwig, J., Muller, G., Ambacher, O., Stutzmann, M., 2001. Group-III-nitride based gas sensing devices. Phys. Stat. Sol. (a). 185 (1), 39–45.

Song, Y., Kim, E.S., Kapila, A., 1995. Thermal stability of sputter-deposited ZnO thin films. J. Electron. Mater. 24 (2), 83–86.

Specht, C.G., Williams, O.A., Jackman, R.B., Schoepfer, R., 2004. Ordered growth of neurons on diamond. Biomaterials. 25, 4073–4078.

Strite, S., Morkoc, H., 1992. GaN, AIN, and InN: a review. J. Vac. Sci. Technol. B. 10 (4), 1237–1266.

Stutzmann, M., Steinhoff, G., Eickhoff, M., Ambacher, O., Nebel, C.E., et al., 2002. GaN-based heterostructures for sensor applications. Diamond Relat. Mater. 11, 886–891.

Tait, R.N., Mirfazli, A., 2001. Low temperature aluminum nitride deposition on aluminum by rf reactive sputtering. J. Vac. Sci. Technol. A. 19, 1586.

Takagaki, Y., Santos, P.V., Wiebicke, E., Brandt, O., Schönherr, H.P., et al., 2002a. Guided propagation of surface acoustic waves in AlN and GaN films grown on 4H-SiC(0001) substrates. Phys. Rev. B. 66, 155439.

Takagaki, Y., Santos, P.V., Wiebicke, E., Brandt, O., Schönherr, H.P., et al., 2002b. Superhigh-frequency surface-acoustic-wave transducers using AlN layers grown on SiC substrates. Appl. Phys. Lett. 81, 2538.

Thiele, J.A., Pereira da Cunha, M., 2005. Platinum and palladium high-temperature transducers on langasite. IEEE Trans. Ultrason. Ferroelectr. Freq. Control. 52 (4), 545–549.

Thiele, J.A., Pereira da Cunha, M., 2006. High temperature LGS SAW gas sensor. Sens. Actuators B. 113, 816–822.

Tichy, J., Erhart, J., Kittinger, E., 2010. Fundamentals of Piezoelectric Sensorics: Mechanical, Dielectric, and Thermodynamical Properties of Piezoelectric Materials. Springer-Verlag, Berlin; Heidelberg, Germany.

Tomar, M., Gupta, V., Sreenivas, K., Mansingh, A., 2005. Temperature stability of ZnO thin film SAW device on fused quartz. Device Mater. Reliab. IEEE Trans. 5 (3), 494–500.

Tsubouchi, K., Mikoshiba, N., 1985. Zero-temperature coefficient SAW devices on AlN epitaxial films. IEEE Trans. Sonics Ultrason. SU-32 (5), 634–644.

Tsubuochi, K., Sugai, K., Mikoshiba, N., 1981. AlN material constants evaluation and SAW properties on AlN/Al_2O_3 and AlN/Si. In: Proceedings of the IEEE Ultrasonics Symposium, Chicago, IL, pp. 375–380.

Turner, R.C., Fuierer, P.A., Newnham, R.E., Shrout, T.R., 1994. Materials for high temperature acoustic and vibration sensors: a review. Appl. Acoust. 41, 299–324.

Umeda, K., Kawai, H., Honda, A., Akiyama, M., Kato T., et al., 2013. Piezoelectric properties of ScAlN thin films for piezo-mems devices, paper presented at MEMS 2013, Taipei, Taiwan, 20–24 January.

Wang, C.M., Saraf, L.V., Qiang, Y., 2008. Microstructures of ZnO films deposited on (0001) and r-cut α-Al_2O_3 using metal organic chemical vapor deposition. Thin Solid Films. 516, 8337–8342.

Wu, P., Emanetoglu, N.W., Tong, X., Lu, Y., 2001. Temperature compensation of SAW in ZnO/SiO_2 /Si structure. 2001 IEEE Ultrasonic Symposium Proceedings, pp. 211–214.

Yu, F., Zhang, S., Zhao, X., Yuan, D., Wang, Q., et al., 2010. High temperature piezoelectric properties of yttrium calcium oxyborate single crystals. Phys. Status Solidi RRL. 4 (5–6), 103–105.

Yu, F., Zhang, S., Zhao, X., Yuan, D., Qin, L., et al., 2011. Investigation of $Ca_3TaGa_3Si_2O_{14}$ piezoelectric crystals for high temperature sensors. J. Appl. Phys. 109, 114103.

Yu, H., Pang, W., Zhang, H., Kim, E.S., 2007. Ultra temperature-stable bulk-acoustic-wave resonators with SiO_2 compensation layer. IEEE Trans. Ultrason. Ferroelectr. Freq. Control. 54 (10), 2102–2109.

Zhang, D., Bai, Y.Z., Qin, F.W., Bian, J.M., 2010a. Deposition of GaN films on freestanding CVD thick diamond films. Mater. Sci. Forum. 654–656, 1740–1743.

Zhang, D., Bai, Y., Qin, F., Jia, F., Wang, J., et al., 2011. Deposition and properties of highly c-oriented GaN films on diamond substrates. Appl. Phys. A. 102, 353–358.

Zhang, S., Yu, F., 2011. Piezoelectric materials for high temperature sensors. J. Am. Ceram. Soc. 94 (10), 3153–3170.

Zhang, S., Fei, Y., Chai, B.H.T., Frantz, E., Snyder, D.W., et al., 2008. Characterization of piezoelectric single crystal $YCa_4O(BO_3)_3$ for high temperature applications. Appl. Phys. Lett. 92, 202905.

Zhang, S., Jiang, X., Lapsley, M., Moses, P., Shrout, T.R., 2010b. Piezoelectric accelerometers for ultrahigh temperature application. Appl. Phys. Lett. 96, 013506.

Zheng, P., Chin, T.-L., Greve, D., Oppenheim, I., Malone, V., et al., 2011. High-temperature langasite SAW oxygen sensor. IEEE Trans. Ultrason. Ferroelectr. Freq. Control. 58 (8), 1538–1540.

Zheng, P., Zhang, J.L., Tan, Y.Q., Wang, C.L., 2012. Grain-size effects on dielectric and piezoelectric properties of poled $BaTiO_3$ ceramics. Acta Mater. 60, 5022–5030.

Plasma-assisted techniques for growing hard nanostructured coatings: an overview

18

*R.S. Pessoa**,†, *M.A. Fraga*†,‡, *L.V. Santos**,†, *N.K.A.M. Galvão*†, *H.S. Maciel**,†, and *M. Massi***
*IP&D, University of Paraiba Valley (UniVap), São José dos Campos/SP, Brazil;
†Technological Institute of Aeronautics (ITA-DCTA), São José dos Campos/SP, Brazil;
‡Faculty of Technology of São Paulo, São Paulo/SP, Brazil; **Federal University of São Paulo-ICT, São José dos Campos/SP, Brazil

18.1 Introduction

Nanostructured materials are a new class of materials that are distinguished from conventional materials by their extremely fine grain size from 100 nm down to a just few nm. This is due to the small grain and the significant contribution of the grain boundary regions causing the "nanoscale effect," in which the properties are optimized or exhibit completely new and/or different behavior compared to conventional materials (Baer et al., 2003; Musil, 2012). Several papers refer to the properties of nanostructured materials as "unique properties," for example, superior hardness, toughness, chemical stability, low friction and wear-resistance, and these unique properties allow then to be applied in various technological sectors (Lukaszkowicz, 2011; Baer et al., 2003; Lu et al., 2006; Pogrebnjak and Beresnev, 2012). Table 18.1 presents some nanostructured materials and respective technological applications.

The thin (nanometer-sized) coating of a bulk material is a crescent category in the field of nanostructured materials. The highly sophisticated surface-related properties of nanocoatings, such as super-plasticity, and optical, magnetic, electronic and catalytic properties, make them attractive for industrial applications in high-speed machining, tooling, biomedical, automotive, optical applications and magnetic storage devices (Musil, 2012; Baer et al., 2003).

There are many types of design models for nanostructured coatings, such as nanocomposite coatings (Pogrebnjak and Beresnev, 2012), nanoscale multilayer coatings (Hovsepian and Munz, 2003), superlattice coatings (Hovsepian and Munz, 2003; Münz et al., 2001), and nanocrystalline coatings (Malshe et al., 2002; Hernández-Torres et al., 2012). Moreover, there are different combinations for nanocoating formation, namely metal–metal, metal–ceramic, ceramic–ceramic, and solid–lubricant. Thus, by engineering these composite materials at the nanometer scale, it is possible to enhance the materials properties such as hardness, making it possible to obtain hard (with hardness above 20 GPa), super-hard (hardness above

Anti-Abrasive Nanocoatings.

Table 18.1 Nanostructured materials and respective technological applications in different fields of science and industry

Nanostructured materials	Some technological applications in different fields	References
Al-N based coatings	Optical protective coatings	Li et al. (2005)
	Optoelectronic (field emission devices, flat panel displays), heterostructure diodes for UV light emitters and homojunction light emitting diodes (LED)	Kasu and Kobayashi (2000), Miskys et al. (2003)
	Protective layers against corrosion	Miao et al. (1997)
	Electromechanical/microelectromechanical systems (MEMS) (piezoelectric sensing layers)	Perooly et al. (2005)
B-N based coatings	Field passivation and capping layers for thermal annealing	Bathe et al. (2001)
	Protective layers for metallic filaments	Bathe et al. (2001), Rudolph (1993)
	Electrical insulation	Rudolph (1993)
Si-C based coatings	Optical coatings	Fraga et al. (2012)
	Optoelectronic (LED and heterojunction diodes)	Kortright and Windt (1988)
	Piezoresistive sensing layers	Fraga et al. (2011)
Si-N based coatings	Microelectronics (sacrificial layers)	Williams et al. (2003)
	Electronic passivation of silicon surfaces and antireflection coating on silicon solar cells	Nagel et al. (1999)
Ti-C based coatings	Industrial (high-speed machining, tooling, and wear)	Steinmann and Hintermann (2009)
	Hard disk drives	Sangkla et al. (2011), Vitu et al. (2008)
	Biocompatible implants	Vitu et al. (2008)
Carbon based coatings	Biomedical	Takeno et al. (2010)
	Tribological	Bull (1995)

40 GPa), or even ultra-hard (80 GPa) materials, that rival diamond performance (Zhang et al., 2003). As is known, diamond demonstrates both high thermal conductivity and electrically insulating properties and much attention has been put into finding practical applications for this material. However, diamond has several limitations for mass industrial applications, including its high cost and oxidation at temperatures above 800°C (John et al., 2002). Therefore, recent research into super- and ultra-hard materials has focused on compounds that would be thermally and chemically more stable than pure diamond. Some examples of super-hard materials are cubic boron nitride (c-BN), carbon nitrides, silicon carbide (SiC), and ternary compounds such as boron-carbon-nitride (B-N-C). Sapphire and zirconia are examples of ultra-hard materials.

All these designs and manipulations of materials with unique properties that are often unachievable in bulk materials are now possible due to recent advances in techniques for deposition and processing of thin films. Additionally, the design of hard nanostructured coatings needs many factors to be taken into consideration, for example, the interface volume, grain size, single layer thickness, surface and interfacial energy, texture, epitaxial stress and strain, all of which depend significantly on materials selection, deposition methods, and processing parameters (Baer et al., 2003; Malshe et al., 2002).

Physical vapor deposition (PVD), chemical vapor deposition (CVD), ion implantation, and laser beam treatments are the most widely applied procedures for modifying the chemical composition and/or atomic structure of solid surfaces on a nanometer scale. In this chapter an overview is presented of the hard nanocoatings and the main plasma-assisted techniques for their synthesis. Among these techniques, plasma enhanced chemical vapor deposition (PECVD), magnetron sputtering, and atomic layer deposition (ALD) are highlighted. Characteristics of some hard coatings such as DLC, SiC, c-BN, and aluminum oxide (Al_2O_3) as well as heterostructures obtained by these techniques are presented.

18.2 Hard nanocoatings: from history to designs and properties

The concepts that seeded nanotechnology were first discussed in 1959 by renowned physicist Richard Feynman in his talk "There's Plenty of Room at the Bottom," in which he described the possibility of synthesis via direct manipulation of atoms (Drexler, 1992). Only in 1974 was the term "nanotechnology" defined by Tokyo Science University Professor Norio Taniguchi: "Nanotechnology mainly consists of the processing of, separation, consolidation, and deformation of materials by one atom or by one molecule" (Sandhu, 2006). The ideas contained in this definition were further explored and developed by K. Eric Drexler in the late 1970s and 1980s, culminating in the release of his book *Engines of Creation: The Coming Era of Nanotechnology* in 1986, which is considered the first book on the topic of nanotechnology (Drexler, 1986). Hence, emergence of nanotechnology as a field in the 1980s occurred through convergence of Drexler's theoretical and public work, which developed and popularized a conceptual framework for nanotechnology, and high-visibility experimental advances that drew additional wide-scale attention to the prospects of atomic control of matter.

At the same time, the German physicist Herbert Gleiter contributed significantly and at various levels, to the development of nanostructured materials (Nordmann, 2009). Several papers cite H. Gleiter as the first to introduce the field of nanostructured materials with his proposal to synthesize a new solid state structure – an interfacial material – which differs from the crystalline and amorphous counterparts of the same chemical composition (Nordmann, 2009; Gleiter, 2000; Meyers et al., 2006; Langdon, 2010). Gleiter's proposal to achieve such a new structure by introducing very large volume fractions of grain boundary or interface atoms by crystal size reduction was based on the understanding of grain boundary structure. Thus, by controlling the type and/or volume fraction of grain boundaries of conventional engineering materials, unique material performance characteristics in bulk materials as well as films and coatings can be achieved (Gleiter, 2000). After this work, the interest in nanostructured materials greatly increased alongside the development of techniques for engineering of these materials.

As mentioned in the Section 18.1, there are different design models for nanostructured coatings where the main classes are: nanoscale multilayer coatings and nanocomposite coatings (Musil, 2000; Lukaszkowicz, 2011; Pogrebnjak et al., 2013). The multilayer coatings exhibit composition and phase modulation in one direction of space (the growth direction), whereas the nanocomposite coatings are modulated in all three directions of space. Due to the size effect, these new classes of materials exhibit improved mechanical, electronic, magnetic and optical properties, and, of all the exotic properties, hardness is one of the most important for applications in manufacturing industry (Pogrebnjak and Beresnev, 2012). For example, nanocomposite coatings can present hardness two times greater than that of its harder component (Pogrebnjak and Beresnev, 2012; Musil et al., 2001). From knowledge of nanostructured materials emerged the following hardness classification: materials with hardness greater than 20 GPa and less than 40 GPa are called hard materials, super-hard materials and the ultra-hard materials refer to materials with hardness ≥40 GPa and ≥80 GPa respectively (Zhang et al., 2002).

The hardness of nanostructured coatings as well as conventional bulk materials is related to the material grain size (Lukaszkowicz, 2011; Zhang et al., 2003). The relationship of the hardness and the grain size is illustrated in Figure 18.1; from this

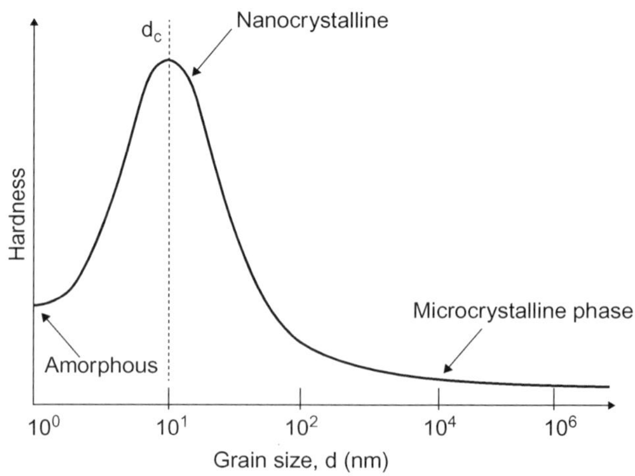

Figure 18.1 Behavior of the hardness of materials as a function of grain size.
Source: Adapted from Lukaszkowicz, 2011; Pogrebnjak and Beresnev, 2012; Zhang et al., 2003.

it can be seen that there is a critical value of the grain size $d_c \approx 10$ nm at which a maximum value of hardness, H_{max}, of the coating is achieved.

When the grain size is decreased to about 10 nm, the mobility of dislocations is restricted. This occurs because the refined grain and the boundary regions suppress the dislocation, causing the onset of incoherent deformations and increasing the hardness of the material accord with the "Hall-Petch" relationship (Musil, 2012; Zhang et al., 2003; Schiotz, 2001; Holleck, 1999):

$$H(d) = H_0 + K \cdot d^{-1/2} \tag{18.1}$$

where, H is the material hardness, K is strengthening coefficient and d is the grain size. This effect is especially prominent for grain size down to tens of nanometers (Lukaszkowicz, 2011; Zhang et al., 2003). In grain size of approximately 10 nm there is a critical value where the maximal hardness is achieved. For grain size less than d_c, due to the increase in grain boundary a new deformation mechanism (i.e. the Hall-Petch relation is inverted), called grain boundary sliding, replaces the dislocation activity that dominates deformation process in conventional materials, thus decreasing the hardness (Pogrebnjak and Beresnev, 2012; Zhang et al., 2003). Softening caused by grain boundary sliding is mainly attributed to the large amount of defects in the grain boundaries, which allows fast diffusion of atoms and vacancies with the applying of stress. A further increase in the strength and hardness with decreasing crystallite size can be achieved only if grain boundary sliding is blocked by appropriate coating design and materials selection.

18.2.1 Classifications of hard nanocoatings

There are many design ideas for nanocomposite coatings, and embedding nanocrystalline phases in the amorphous phase matrix is quite often applied which can be fulfilled easily by PVD and CVD methods (Zhang et al., 2002). According to Musil there are at least four types of nanostructures (Figure 18.2) that result in the enhanced hardness of nanocomposite coatings: (1) bilayers with nanosize period λ; (2) the columnar nanostructure; (3) nanograins surrounded by a very thin tissue phase, w ($\sim$1 to 2 monolayers); and (4) the mixture of nanograins with different crystallographic orientations and/or different phases; here $\lambda = h_1 + h_2$, h_1 and h_2 are the thickness of first and second layer of the bilayer, respectively (Musil, 2012).

Individual nanostructures are formed under different conditions using either a sequential deposition of individual layers in the nanosize bilayers or in transition regions where the coating structure changes from crystalline through nanocrystalline to amorphous. There are three transition regions: (1) the transition from the crystalline to the amorphous material; (2) the transition between two crystalline phases of different materials; and (3) the transition between two crystallographic orientations of grains of the same material (Pogrebnjak and Beresnev, 2012).

Additionally, in 1999, Holleck et al. proposed a classification of hard coatings in three groups according to their bonds (see Figure 18.3): (1) covalent hard materials; (2) metallic hard materials; and (3) ionic hard materials (Zhang et al., 2003; Holleck, 1999; Zhang et al., 2005; Stüber et al., 2011).

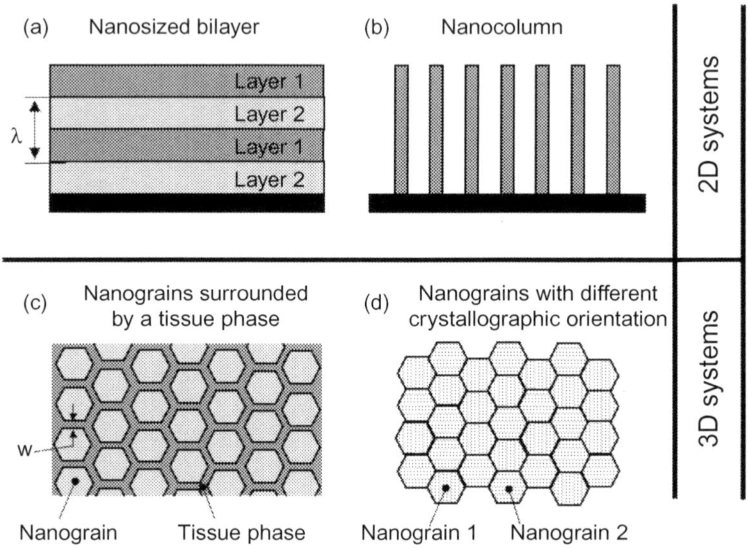

Figure 18.2 Schematic illustration of four nanostructures of the nanocomposite coating with enhanced hardness: (a) nanosize bilayers, (b) columnar nanostructure, (c) nanograins surrounded by a tissue phase, and (d) mixture of nanograins with different crystallographic orientation (adapted from Musil, 2012; Pogrebnjak and Beresnev, 2012). Where w is the tissue phase thickness.

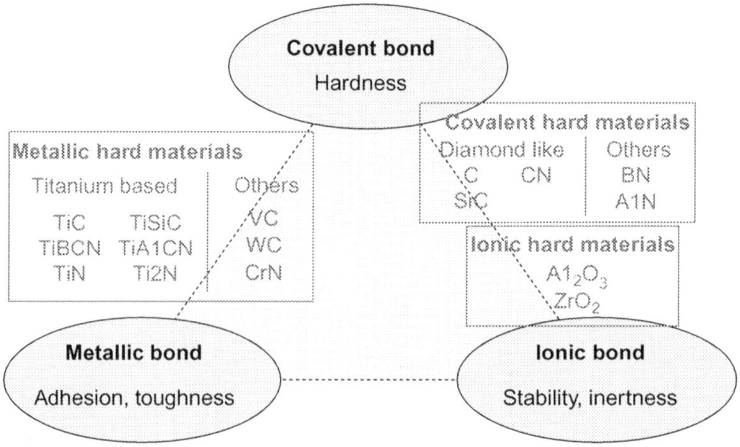

Figure 18.3. Classification of some hard coatings with relation to their bond nature. *Source:* Adapted from Zhang et al., 2003; Holleck, 1999; Zhang et al., 2005.

The specific properties of each hard material group are related to the bond type between the elements. For example, covalent coatings such as diamond and boron nitride are the hardest materials and have a high temperature strength, whereas the ionic (ceramic) coatings exhibit the best chemical inertness and higher brittleness. Metallic coatings have a good adhesion and toughness.

From Figure 18.3, S. Zhang et al. (2003) conclude that the best to design a nanocomposite coating with both high hardness and high toughness is probably to use ternary, quaternary or even more complex systems (for example, multilayer structure Zhang et al., 2005), with a high-strength amorphous phase as matrix (such as a-SiN_x, a-BN, a-C) and hard transition metal-nitride nanocrystals (such as TiN, BN) as the nanocrystalline phase to increase grain boundary complexity and strength. These nanocrystalline phases should be refractory and immiscible with each other, and could result in compositional modulation, segregation, and high thermal stability of the nanostructure. The aim is to maximize interfaces and form well-defined spinodal structure at interfaces.

18.2.2 Hardness mechanisms of crystalline nanostructured materials

There are few intrinsic super-hard materials highlighting diamond, c-BN and some binary and ternary B-C-N compounds. Studies show that the high intrinsic hardness of these materials is related to the following characteristics: a short interatomic distance, a high binding energy, and a high degree of covalent bonding (Gao et al., 2003).

On the other hand, there is a group of hard materials that have what is called extrinsic hardness. These crystalline materials have the hardness determined by their microstructure, that is, it is possible obtain hard and super-hard nanocoatings from these materials only by hindering or retarding the deformation generated by dislocation movements (Arzt, 1998).

The four main mechanisms that can modify the hardness of crystalline materials are illustrated in Figure 18.4.

It is known that high hardness is achieved for maximal concentration of introduced distortions. Grain boundary is an example of a lattice distortion. This mechanism has been observed especially in nanocrystalline ceramics (Szlufarska et al., 2005),

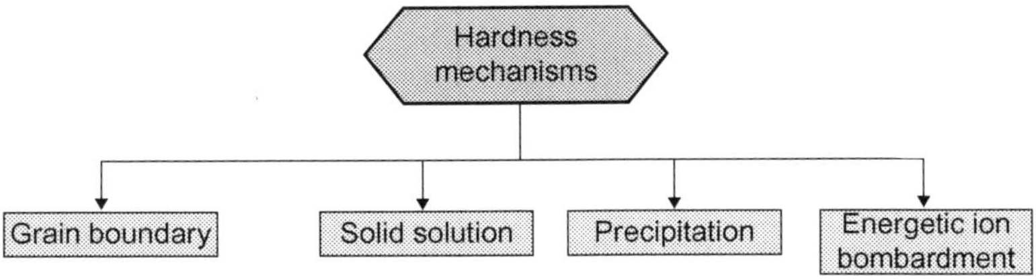

Figure 18.4. Different mechanisms that can influence the hardness of crystalline materials.

nanocrystalline metals (Van Swygenhoven, 2002), and multilayer nanostructures (Mayrhofer et al., 2005).

Solid solution mechanisms are characterized by lattice distortion due to the insertion of atoms of the solute (also called alloy element) either in interstitial or substitutional lattice positions. These mechanisms are limited by the solubility limit of the solute. It has been observed that PVD processes at low substrate temperature increase the solubility limit (Holleck, 1988). Examples of hard coatings that can be obtained from the solid solution mechanism are Ti-C-N, Ti-Al-N, and Cr-Al-N (Zhang et al., 2007).

Precipitation or dispersion strengthening is the process characterized by the precipitation of a second phase from a supersaturated metastable phase. Generally, this mechanism is accelerated and controlled by a post-deposition annealing (Arzt, 1998). The literature reports the precipitation mechanism for nanocrystalline thin films such as Ti-Al-N (Mayrhofer et al., 2003), Ti-B-N (Mayrhofer et al., 2006), and Zr-Al-N (Sanjines et al., 2006).

Energetic ion bombardment is another mechanism used to obtain hard coatings during film growth. It has been employed to modify the morphology of the films. The ion bombardment creates interstitials and vacancies that can lead to the formation of dimensional defects (Mattox, 1989). These defects are associated with the existence of internal stress in the films, which result in hardness increase. It has been reported for different PVD coatings (Mayrhofer et al., 2006).

18.3 Main plasma-based techniques for synthesis of hard nanocoatings

A variety of modern coating techniques have been employed to realize hard nanostructured coatings. Among these technologies are plasma-assisted PVD, CVD, and PECVD processes, which make it possible to grow materials with extremely high indentation hardness (40–80 GPa) (Veprek and Veprek-Heijman, 2007; Veprek et al., 2000; Zou et al., 2010).

Table 18.2 provides a listing of hard thin film materials and compares the microhardness of PECVD and PVD deposited thin films (note that nanocrystalline films have the highest hardness) (Martinu et al., 2009). Several types of films deposited by PECVD have microhardness in the super-hard range while the number is limited for PVD deposited films. Hardness is generally improved by increasing the energy of deposited atoms using high substrate temperatures, substrate bias, plasma bombardment (ion assist), or a combination of all four.

Moreover, as can be seen in Table 18.2 the most widely used hard and superhard coatings are binary and ternary alloys of titanium, boron, chromium, tungsten, silicon, and allotropes of carbon.

The sections that follow describe plasma-assisted techniques for growing hard nanostructured coatings as well as discussing materials synthesis and its resulting properties.

Table 18.2 Microhardness of PECVD and PVD films

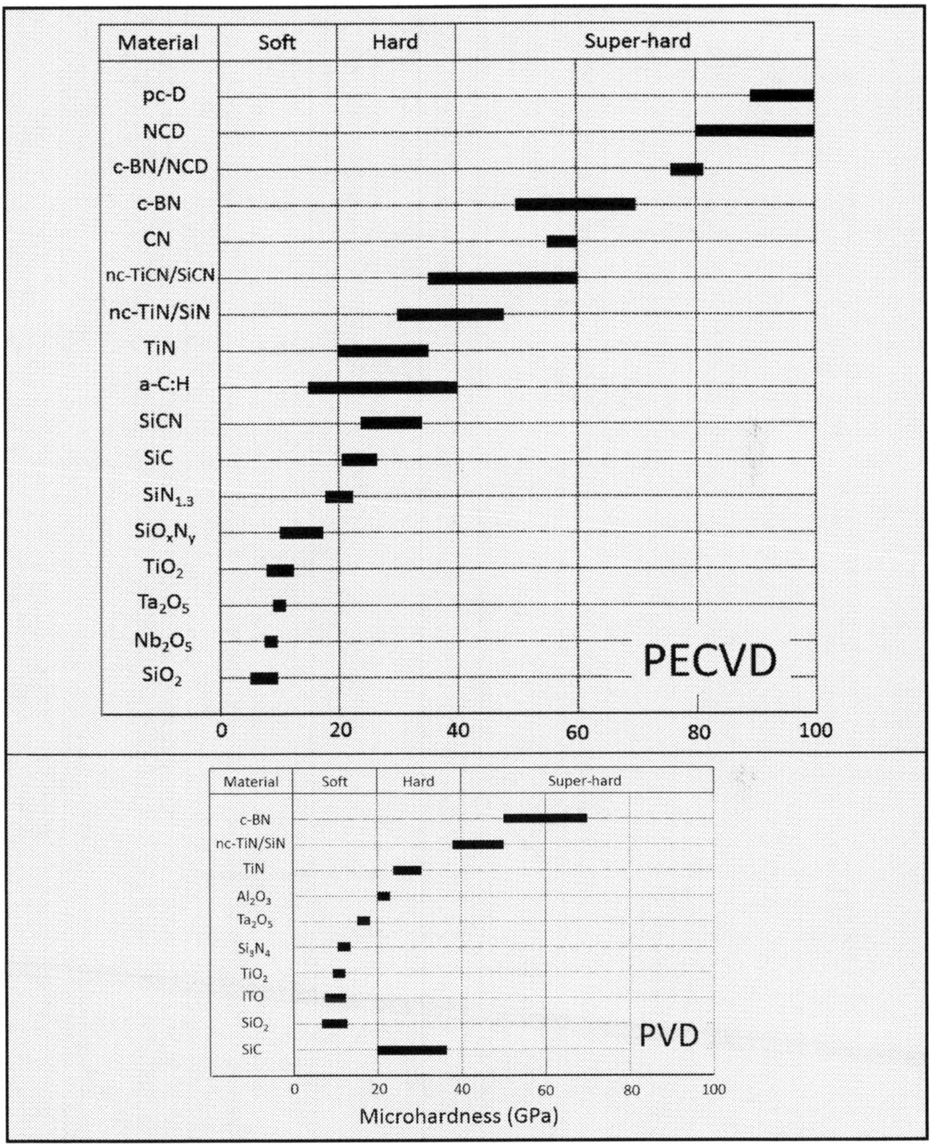

Note: a – amorphous, c – crystalline, pc – polycrystalline, and nc – nanocrystalline.
Source: Adapted from Zhang et al., 2005.

18.3.1 Physical vapor deposition

The PVD process comprises physical and reactive sputtering techniques. Basically, these techniques differ when a neutral gas (physical sputtering) is added together with a reactive gas (reactive sputtering). In physical sputtering, ions impinging on the plasma bombard the target and release atoms (or molecules) of the target material. Argon ions at 500–1000 V are usually used. The sputtered atoms diffuse through the plasma and arrive at the substrate, where they can be deposited (Bogaerts et al., 2003). In reactive sputtering, a molecular gas (for example, N_2 and/or O_2) is used together or without a neutral gas. Besides the positive ions from the plasma that sputter bombard the target, the dissociation products from the reactive gas also react with the target. Hence, the film deposited at the substrate will be a combination of sputtered target material and the reactive gas (Bogaerts et al., 2003; Lieberman and Lichtenberg, 2007; Berg and Nyberg, 2005; Pessoa et al.). The sputter deposition process is schematically presented in Figure 18.5.

The steps of sputtering process are as follows:

1. The neutral gas is ionized by an external power supply (direct current, radio-frequency, etc.), producing a glow discharge or plasma.
2. The source (the cathode, also called the target) is bombarded in a high vacuum by plasma ions due to the potential acceleration drop in the cathode sheath.
3. Atoms from the target are ejected by momentum transfer and diffuse through the vacuum chamber.
4. Atoms are deposited on the substrate to be coated and form a thin film.

Because sputter yields are of the order of unity for almost all target materials, a very wide variety of pure metals, alloys and insulators can be deposited. Physical sputtering, especially of elemental targets, is a well understood process enabling sputtering systems for various applications to be relatively easily designed. Reasonable deposition rates with excellent film uniformity, good surface smoothness and adhesion can be achieved over large areas (Lieberman and Lichtenberg, 2007).

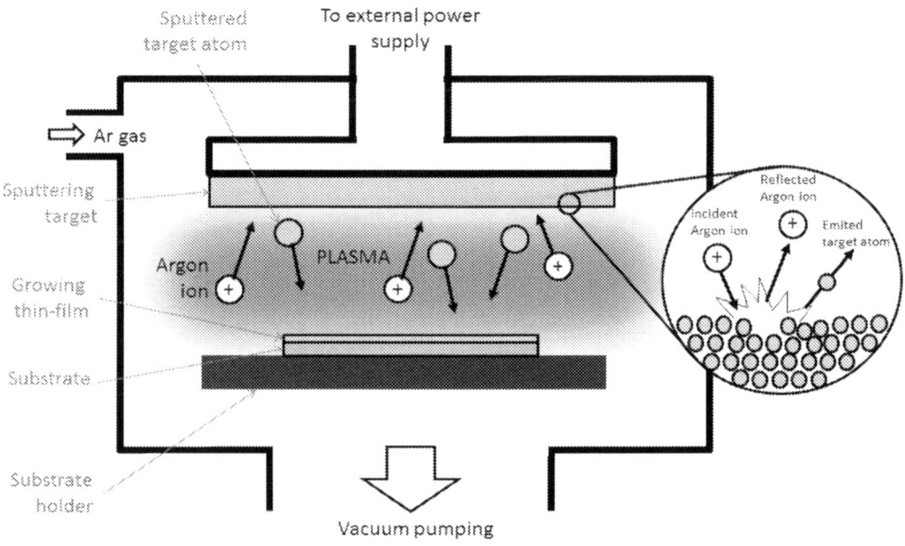

Figure 18.5 Schematics of the sputtering process.

Typically, the sputtering process can be accomplished using a planar configuration of electrodes and a direct current (DC) power supply, where one electrode is biased negatively (cathode) and suffers the sputtering process. However, the sputtering yield is directly dependent on the gas pressure (the best sputtering rates are in the range of mTorr), which compromises the efficiency of planar geometry for this application: it is ideal for pressures above 100 mTorr. To solve this problem, a magnetron discharge where the plasma is magnetically enhanced by placing magnets behind the cathode target was developed, that is, a crossed electric and magnetic field configuration is created. Figure 18.6 shows a schematic drawing of a conventional DC magnetron sputtering discharge. The trapping of the secondary electrons results in higher probability of electron impact ionization and hence higher plasma density, increasing the sputtering flux and allowing operation at lower pressures, below 10 mTorr. Furthermore, the discharge voltage can be lowered into the range of 300–700 V. The main problem with the magnetron sputtering configuration is that the sputtering is confined to a small area of the target cathode governed by the magnetic field. The discharge appears in the form of a high-density annulus of width w and radius R, as seen in Figure 18.6. Sputtering occurs in the corresponding track of the target. This area, known as race track, is created by the uneven ion density.

According to the Hall-Petch equation, the strength properties of the material rise along with the reduction of the grain size. In case of coatings deposited by the PVD processes, the structures obtained, with grain size $\sim$10 nm result in the maximum mechanical properties. Coatings of this structure present very high hardness (>40 GPa), ductility, stability at high temperatures, and so on (Lukaszkowicz, 2011). According to Musil, the addition of one or more selected elements into a base material, for example, the addition of Si in the nitride transition metal (TM), is a very effective way to produce hard nanocomposite coatings with enhanced hardness, here TM = Ti, Zr, Hf, V, Nb, Ta, and so on (Musil, 2012). Moreover, it is highlighted that the energy delivered to the growing film also has a crucial effect on its structure, elemental and phase composition, and physical properties

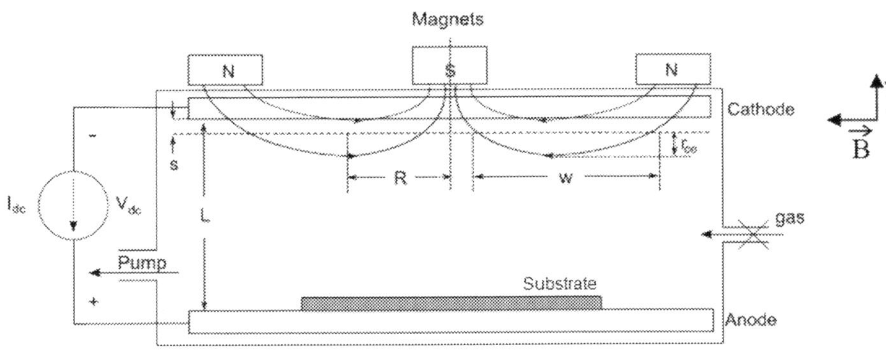

Figure 18.6 Schematic drawing of a conventional dc magnetron sputtering discharge.
Source: Adapted from Bogaerts et al., 2003.

Table 18.3 Summary of hardness values for the B-C-N family (Solozhenko et al., 2009)

Material	Diamond	c-BC_2N	c-BC_5	c-BN	B_4C	ReB_2
Vickers Hardness (GPa)	115	76	71	62	38	22

(Musil, 2012). This energy is delivered from the plasma environment and is composed of (1) heating processes; and (2) conversion of the kinetic energy of bombarding ions and fast neutrals incident on the surface of growing film.

Recently, amorphous coatings with hardness ranging from ~20 – ~40 GPa were reactively sputtered using an unbalanced magnetron deposition technique (Musil, 2012):

1. a-(Si_3N_4/MeN_x) composite coatings with high (≥50 vol.%) content of a-Si_3N_4 phase; here Me = Ta, Zr, Ti, Al, Mo, W, and so on.
2. a-(Si–B–C–N) coatings with strong covalent bonds. Magnetron sputtered films were deposited co-sputtered using high purity graphite and boron targets in 50/50 Ar + N_2 mixtures with formation of hybridized B-C-N bonds (Kim et al., 2004).

Also, super-hard amorphous SiC coatings are deposited by our group using co-sputtering of silicon and carbon targets. Here the SiC coatings were obtained with hardness of up to 50 GPa only modifying the negative bias applied at the substrate holder (Medeiros, 2012).

Relative to the deposition of boron nitride films in cubic form (c-BN), it is reported that the lower adhesion and extreme compressive stresses of these films are the main problems in PVD techniques, which limits film thickness of c-BN up to 1 μm. However, many researchers have reported adherent and thick multilayered films recently (Cesur et al., 2008). Boron carbon nitride (B_xC_yN), also known as "heterodiamond" is essentially c-BN that contains between 5 at% and 30 at% C and is generally deposited by PVD and CVD processes (Yamamoto et al., 2000). Magnetron sputtered films were deposited co-sputtered using high purity graphite and boron targets in 50/50 Ar + N_2 mixtures with formation of hybridized B-C-N bonds. Table 18.3 summarizes hardness values for the B-C-N family (Ulrich et al., 2009).

18.3.2 Chemical vapor deposition

In the field of chemical-based deposition techniques of hard coating thin films liquid phase epitaxy (LPE), CVD techniques and their derivatives, such as PECVD and atomic layer deposition (ALD), can be mentioned, among others. A very desirable feature of CVD-based techniques is that they can be carried out with well controlled atmospheres and at lower temperatures (lower than 1500°C) in comparison with other related techniques.

Currently, the most popular low-pressure thin film deposition techniques are those based on chemical deposition processes such as hot filament chemical vapor

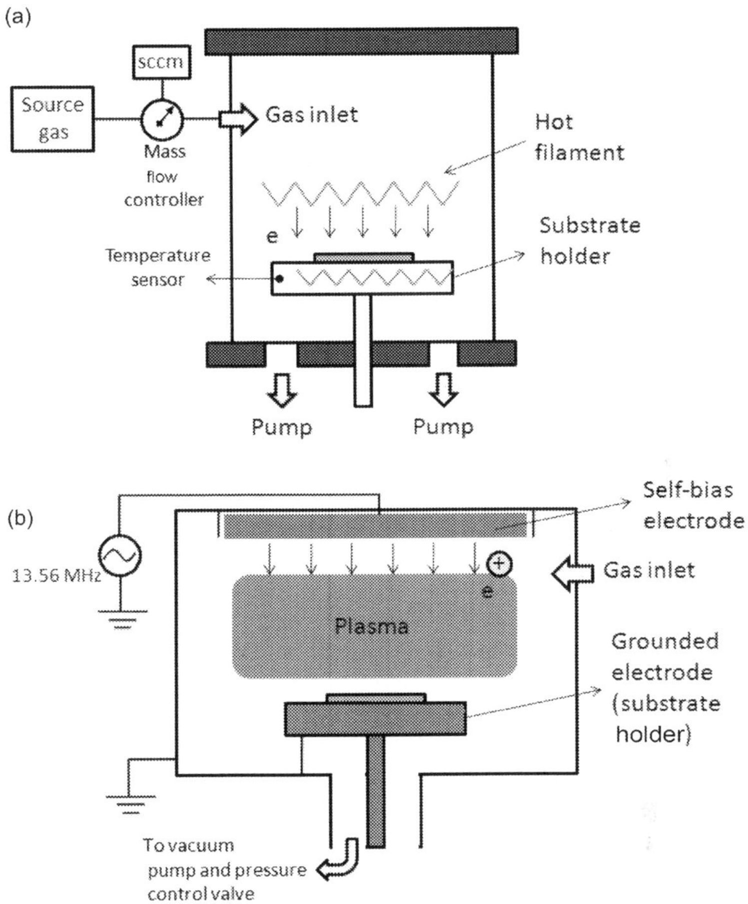

Figure 18.7 Schematic diagram of (a) HFCVD and (b) PECVD systems.

deposition (HFCVD) and PECVD (Veprek et al., 2000; Zou et al., 2010; Bogaerts et al., 2003; Lieberman and Lichtenberg, 2007).

Thermal CVD is the process of gas phase heating (by a hot filament or hot wall, for example (Berg and Nyberg, 2005)) in order to cause the decomposition of the gas, generating radical species that, by diffusion, can reach and be deposited on a suitably placed substrate. It differs from PVD, which relies on material transfer from a condensed-phase evaporant or sputter target sources. A reaction chamber is used for this process, into which the reactant gases are introduced to decompose and react with the substrate to form the film. Figure 18.7(a) illustrates a schematic of the reactor and its main components. Basically, a typical thermal CVD system (HFCVD) consists of the following parts:

1. sources and feed lines of gases;
2. mass flow controllers for metering the gas inlet;

3. a reaction chamber for decomposition of precursor gases;
4. a system for heating up (hot filament or hot wall) the gas phase and wafer on which the film is to be deposited; and
5. temperature sensors.

As a modification to the CVD system, PECVD arose when plasma was used to perform the decomposition of the reactive gas source. By chemical reactions in the plasma (mainly electron impact ionization and dissociation), different kinds of ions and radicals are formed which diffuse toward the substrate where chemical surface reactions are promoted, leading to film growth. The major advantage compared to simple CVD is that PECVD can operate at much lower temperatures. Indeed, the electron temperature of 2–5 eV in PECVD is sufficient for dissociation, whereas in CVD the gas and surface reactions occur by thermal activation. Hence, some coatings, which are difficult to form by CVD due to melting problems, can be deposited more easily with PECVD (Lieberman and Lichtenberg, 2007; Peng et al., 2011). Among the kinds of plasma sources that have been used for this application radiofrequency (RF) discharges (Lieberman and Lichtenberg, 2007), pulsed discharges (Peng et al., 2011), and microwave discharges (Berg and Nyberg, 2005) stand.

Basically, in PECVD the substrate is mounted on one of the electrodes in the same reactor where the species are created, as illustrated in Figure 18.7(b). Here, we focused the RF discharge because it is the configuration more used in research and industry. The RF PECVD reactor essentially consists of two electrodes of different areas, where the substrate is placed on the smaller electrode, to which the power is capacitively coupled. The RF power creates a plasma between the electrodes. Due to the electrons having higher mobility than the ions, a sheath is created next to the electrodes containing an excess of ions. Hence, the sheath has a positive space charge, and the plasma creates a positive voltage with respect to the electrodes. The electrodes therefore acquire a DC self-bias equal to their peak RF voltage (self-bias electrode). The ratio of the DC self-bias voltages is inversely proportional to the ratio of the squared electrode areas, that is, $V_1/V_2 = (A_1/A_2)^2$ (Lieberman and Lichtenberg, 2007).

Therefore, the smaller electrode acquires a larger bias voltage and becomes negative with respect to the larger electrode. The negative sheath voltage accelerates the positive ions towards the substrate which is mounted on this smaller electrode, allowing the substrate to be bombarded by energetic ions facilitating reactions with the substrate surface.

In order to maximize the ion-to-neutral ratio of the plasma, the plasma must be operated at the lowest possible pressure. Nevertheless, the ions are only about 10% of the film-forming flux even at pressures as low as 50 mTorr. Lower pressures cannot be used, as the plasma will no longer strike. A second disadvantage of this source is the energy spread in the ion energy distribution, prohibiting a controlled deposition. This energy spread is due to inelastic collisions as the ions are accelerated towards the substrate. The effect of this energy spread is to lower the mean ion energy to about 0.4 of the sheath voltage. Another disadvantage of the RF PECVD source is that it is not possible to have independent control over the ion energy and the ion current, as they both vary with the RF power. On the other hand, PECVD allows the deposition of uniform films over large areas, and PECVD systems can be easily scaled up (Zhao et al., 2010).

Figure 18.8 illustrates the deposition mechanism of the CVD technique (Zou et al., 2010). The mechanism occurs as a result of the following steps:

1. A predefined mix of reactant gases and diluents inert gases are introduced at a specified flow rate into the reaction chamber.
2. A heat source is applied in order to dissociate the reactant gases.
3. The resulting radical species diffuse to the substrate.
4. The reactants get adsorbed on the surface of the substrate.
5. The reactants undergo chemical reactions with the substrate to form the film.
6. The gaseous by-products of the reactions are desorbed.
7. The gaseous by-products of the reactions are evacuated from the reaction chamber.

In general, SiC PECVD is performed in conventional PECVD reactors with heated substrate holders (Neyts, 2006). Methane and silane are commonly used precursors (Sarro et al., 1998; Pelegrini et al., 2010). It is also possible to use a single precursor such as 1,3-disilabutane DSB or methylsilane. The film properties strongly depend on the process parameters including deposition temperature, pressure, plasma power, and gas phase composition. In general, PECVD SiC films exhibit amorphous form and compressive stress, but process conditions can be optimized to tailor the stress levels to desired values. Typically, low plasma power and high temperature conditions yield low stress films. Sometimes post-deposition annealing at temperatures of order of 1000°C may still be required to crystallize the films. Typical values of hydrogenated SiC films deposited by PECVD ranging from 10 to 30 GPa (El Khakania et al., 1994; Kim et al., 2008).

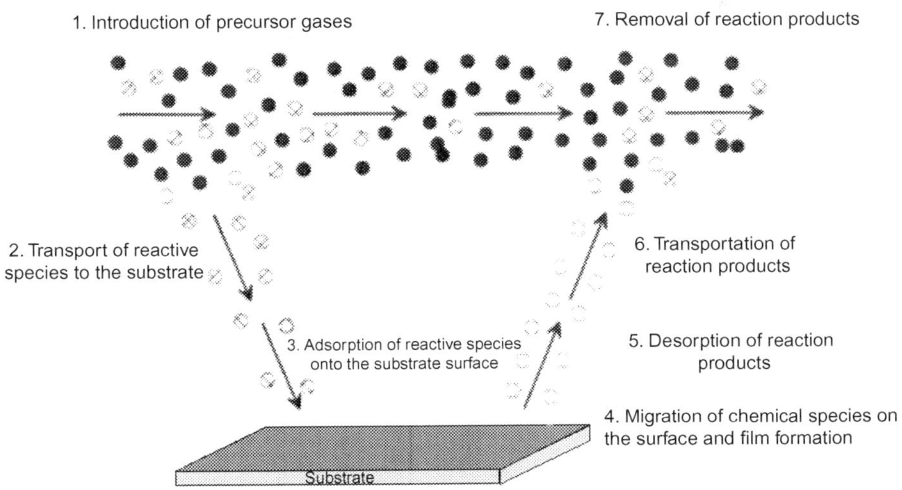

Figure 18.8 Chemical vapor deposition mechanism.
Source: Adapted from Zou et al., 2010.

Other hard materials synthesized by CVD-based techniques are carbon-based materials. The carbon-based family of thin film materials includes diamond-like carbon (DLC), tetrahedral amorphous carbon (ta-C), amorphous carbon (a-C), and hydrogenated amorphous carbon (a-C:H). DLC and ta-C are disputably two of the most widely developed and applied thin film materials used to increase wear resistance and lubricity over the past two decades, and volumes have been published on their properties. DLC is also deposited in microcrystalline, nanocrystalline, and multilayer microstructures. Nanocomposite DLC films are being developed to have lower wear and friction and significantly higher load carrying capability (Erdemir and Voedvodin, 2009). High-quality DLC films have been demonstrated for a number of applications, including low friction and wear, optical and anti-reflection coatings, and barrier coatings. The various forms of DCL can be combined to form a variety of nanoscale structures that are amorphous, flexible, and yet purely sp^3 bonded "diamond." The hardest, strongest, and most lubricious mixture of this type is known as tetrahedral amorphous carbon, or ta-C. For example, a coating of only 2 μm thickness of ta-C increases the resistance of common (i.e. type 304) stainless steel against abrasive wear, changing its lifetime in such service from one week to 85 years (Erdemir and Voedvodin, 2009). It can be considered to be the pure form of DLC, since it consists only of sp^3 bonded carbon atoms.

The properties of DLC and ta-C films depend on the amount of sp^3 and sp^2 bonding present, which can ultimately be related to the deposition process (Pierson, 1993). Figure 18.9 shows the percentage of sp^3 bonding for thin film carbon

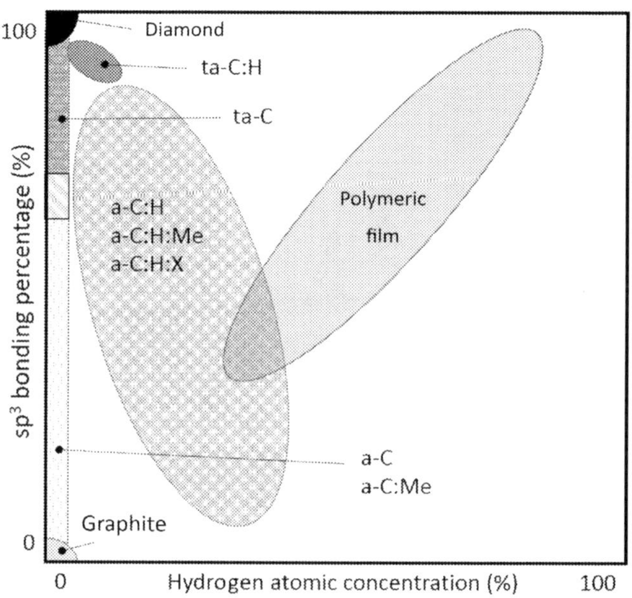

Figure 18.9 Percent sp^3 bonding for thin film carbon allotropes.
Source: Adapted from Sasvvides and Bell, 1992.

allotropes (Sasvvides and Bell, 1992). It shows the two extremes, diamond and graphite, and every combination of bonding in between. Note also that bonding configuration is related to hydrogen content, and eventually back to the deposition process. Microstructures range from purely amorphous to nanocrystalline. Hardness definitely increases with an increase of sp^3 percentage. CVD techniques allow controlling sp^3 percentage only modifying process parameters such as gas flow, substrate bias, discharge power, etc.

Finally, we present some of the main advantages and disadvantages of CVD-based techniques:

18.3.2.1 Advantages

- excellent interface and surface morphology;
- growth of complex heterostructures with many different layers;
- growth on patterned substrates;
- multiple wafer scale-up;
- high purity of the layers.

18.3.2.2 Disadvantages

- large quantities of serious toxic or flammable gases (AsH_3, SiH_4, etc.);
- almost no possibility of in situ monitoring;
- not the best technique for some specific material (lack of adequate precursors or of not suited growth conditions).

18.3.3 Atomic layer deposition

18.3.3.1 Deposition technique

ALD is a nanocoating process based on the self-limiting gas–solid reaction where it is possible to produce thin films of high quality (Violet et al., 2009; Marin et al., 2012; Liang et al., 2008). This process is similar to CVD but in the ALD method the reactions are divided into two separated processes (Marin et al., 2012). Thus, ALD is a variant of the CVD process (Marin et al., 2009).

In the ALD method it is possible to control the film thickness at the atomic level (Marin et al., 2009; Hoivik et al., 2003) over the deposition process. This control is possible because unlike CVD, where the reactants are presented simultaneously during the process, in ALD the precursor gases pulses are separated by purge steps of an inert gas to remove the unreacted precursor and gases by products (Miikkulainen et al., 2013; Wikimedia Commons; Leskela and Ritala, 2002; Puurunen, 2005). In this process, the film thickness increases in each deposition cycle (Leskela and Ritala, 2002). The cycle of the ALD reaction is showed in Figure 18.10.

When the alternating is realized with the ALD steps replaced by a pulsed CVD step, the process is known as partial ALD (Granneman et al., 2007). Figure 18.11 shows a schematic model of ALD and partial ALD processes.

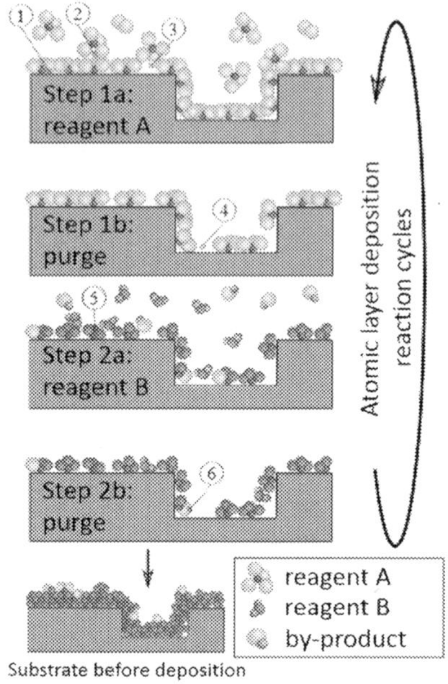

Figure 18.10 Schematic diagram of the ALD reaction cycle (adapted from Miikkulainen et al. and Wikimedia Commons (Miikkulainen et al., 2013; Wikimedia Commons)): (1) adsorbed reagent A molecule; (2) non-adsorbed reagent A molecule; (3) surface without –H or –OH ligant; (4) unreacted site; (5) adsorbed reagent B molecule; (6) reagent A contamination. Depending on temperature of the ALD process the as-deposited film can contain reagent A contaminant.

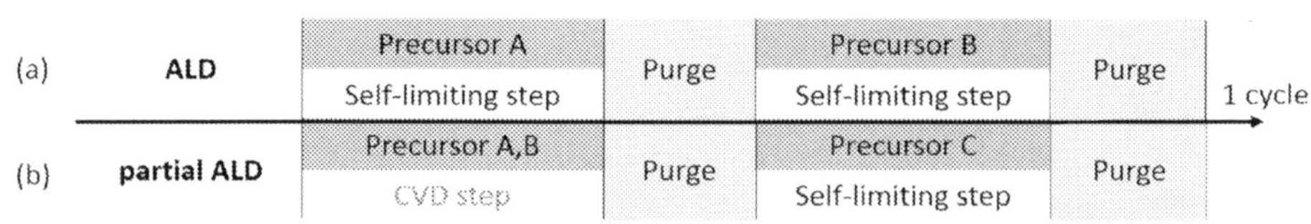

Figure 18.11 Schematic model of: (a) ALD and (b) partial ALD process. (Adapted from Granneman et al., 2007).

Precursor chemistry plays a very important role in the ALD process (Kim, 2011; Kim, 2003). There are many requirements for the precursor used in the ALD process, such as (Miikkulainen et al., 2013):

- it must be gaseous or vaporizable at a temperature lower than the ALD reaction temperature;
- it must react, preferably fast and aggressively, in a saturating, irreversible manner with sites on the surface of the growth substrate;
- the by-products formed should be inert not interfere with the ALD grown;

- its reactants should not dissolve in the film and they must not decompose thermally during storage, vaporization or ALD growth;
- for some applications, high purity is required; and
- it should present safe and easy handling, preferably not toxic or corrosive, thus ensuring the integrity of the reactor and the user. And like any industrial process, low cost will make it attractive to industries.

For reaction between the precursor and substrate surface an activation energy is required. The activation energy is provided by the heating of the substrate and/or the system (reactor chamber). In this case the process is known as "thermal ALD" (Kim, 2011; Kim, 2003). Sometimes, precursors (molecular elements precursors) are not sufficiently reactive, and thus they require more energy than provided in the thermal ALD process. In order to reach the required energy, alternative methods that complement the ALD process are used, for example plasma enhanced ALD (PEALD). The difference between these two processes is that in PEALD the plasma is activated during the step of exposure of the precursor (Kim, 2011; Kim, 2003) and the film growth occurs at a lower temperature than in thermal ALD (Kim, 2011). Moreover, the reactant molecules of the precursor become more reactive and the films deposited by PEALD have better properties, such as improved hardness (Kim, 2011; Profijt et al., 2011).

Finally, the advantages of use the ALD technique include (Violet et al., 2009; Marin et al., 2012; Leskela and Ritala, 2002; Granneman et al., 2007; Kim, 2003):

- deposition of several types of thin films (oxides, nitrides, sulfides and metals);
- possibility of depositing films on various substrates, such as wafers, steel, 3D substrates, micro- and nanoparticles, and others;
- accurate thickness;
- control of film thickness;
- deposition of multilayers of different structures;
- low deposition temperature.

18.3.3.2 Applications of ALD hard coatings

Although the main application of the ALD process is the deposition of ultra-thin films, the hard coatings application is more geared towards the field of microelectronics, more specifically in MEMS. MEMS comprise a variety of functional mechanical devices fabricated on length scales from fractions of a micrometer to millimeters. These devices are finding application in a wide variety of technologies, including motion sensing, optical switching and other photonic applications, RF switches and filters, mechanical actuators, chemical sensors, and others. As the size of these devices shrinks, surface-to-volume ratio increases, and at dimensions typical of MEMS devices, inertial forces become insignificant compared to surface forces acting on these mechanisms. Then interfacial phenomena such as adhesion, friction and wear can dominate the performance and reliability of MEMS devices. The most sophisticated MEMS devices now in development employ sliding contacts in rotary- or linear-motion structures, where friction and wear are a major concern (Sniegowski and de Boer, 2000).

To mitigate these problems, the surfaces of MEMS devices can be modified with coatings to prevent adhesion, and reduce friction and wear. Some of the most commonly used/studied materials are the Al_2O_3, ZnO, WS_2, W, and W/$A1_2O_3$ nanolaminates. WS_2 is a superior solid phase lubricant, W is a hard metallic, wear-resistant material, and W/Al_2O_3 laminate films are potential super-hard coatings with properties determined by the nanometer length scales of the laminate layers.

Al_2O_3 is an advantageous material to use over various forms of silicon nitride, for micro- and nanoelectromechanical systems, due in part to the low deposition temperature that allows for integration with Complementary metal–oxide–semiconductor) CMOS processing. Also, Al_2O_3, unlike silicon nitride, has a high chemical resistance to dry-chemistry Si etchants (Tripp et al., 2006; Bull, 2012). Tripp et al. observed that the hardness of ALD AL_2O_3 (order of 12 GPa) is comparable with the films produced by other deposition techniques (Tripp et al., 2006).

18.4 Conclusion

Engineered nanostructured thin films and low-dimensional structures are present on thin film technology. New advanced deposition processes, such as ALD, are now being developed and used to deposit advanced thin film materials and structures not possible with conventional techniques. As a result, the hardness of thin films has been steadily increasing, due primarily to new materials, improved deposition processes and use of nanostructures and nanocomposites.

Plasma-assisted processes have contributed significantly to this development, especially due to the ease automation of these processes allowing fine control of the properties of matter. An example is the ALD process that has been well suited to deposition of friction and wear-reducing films on MEMS structures. Sequential reactant exposure and self-limiting surface chemistry result in exquisite control of deposition rate and film thickness, and extremely conformal coating of high-aspect ratio structures. However, the challenge remains to examine the mechanical and tribological properties of thin and ultra-thin ALD films, and their suitability for application in functional MEMS devices. Processes for preparation of a wide range of materials, which are potential hard and/or lubricious coatings, have been demonstrated by ALD (Ritala et al., 2002). The ease of fabrication of alloys and nanolaminates of ALD materials adds extra dimensions to the possibilities for tailoring the properties of materials for tribological applications.

Acknowledgments

This work was supported by FAPESP-MCT/CNPq-PRONEX (process no. 2011/50773-0) and PNPD-CAPES (process no. 02765/09-8).

References

Arzt, E., 1998. Size effects in materials due to microstructural and dimensional constraints: a comparative review. Acta Mater. 46, 5611–5626.

Baer, D.R., Burrows, P.E., El-Azab, A.A., 2003. Enhancing coating functionality using nanoscience and nanotechnology. Prog. Org. Coat. 47, 342–356.

Bathe, R., Vispute, R.D., Habersat, D., Takeuchi, I., Sharma, R.P., et al., 2001. BN protective coating for high temperature applications. MRS Proc. 697, P3.3. http://dx.doi.org/10.1557/PROC-697-P3.3.

Berg, S., Nyberg, T., 2005. Fundamental understanding and modeling of reactive sputtering processes. Thin Solid Films. 476, 215–230.

Bogaerts, A., Neyts, E., Gijbels, R., van der Mullen, J., 2003. Gas discharge plasmas and their applications. Spectrochim. Acta Part B. 57, 609–658.

Bull, S.J., 1995. Tribology of carbon coatings: DLC, diamond and beyond. Diamond Related Materials. 4, 827–836.

Bull, S.J., 2012. Mechanical response of atomic layer deposition alumina coatings on stiff and compliant substrates. J. Vac. Sci. Technol. A. 30, 01A160.

Cesur, H., Kaftanoglu, B., Kalkanli, A., Oral, B., 2008. Deposition of boron nitride coatings on steel substrates by rf magnetron sputtering, In: Proceedings of the 7th International Conference Coatings in Manufacturing Engineering, Greece, pp. 409–416.

Drexler, K.E., 1986. Engines of Creation: The Coming Era of Nanotechnology. Doubleday, New York.

Drexler, K.E., 1992. Nanosystems: Molecular Machinery, Manufacturing, and Computation. John Wiley & Sons, New York.

El Khakania, M.A., Chakera, M., Jeana, A., Boilya, S., Kieffera, J.C., et al., 1994. Hardness and Young's modulus of amorphous a-SiC thin films determined by nanoindentation and bulge tests. J. Mater. Res. 9, 96–103.

Erdemir, A., Voedvodin, A.A., 2009. Nanocomposite coatings for severe applications. In: Martin, P.M. (Ed.), Handbook of Deposition Technologies for Films and Coatings, third ed. Elsevier, Amsterdam, pp. 679–715.

Fraga, M.A., Pessoa, R.S., Maciel, H.S., Massi, M., 2011. Recent developments on silicon carbide thin films for piezoresistive sensors applications. In: Mukherjee, M. (Ed.), Silicon Carbide – Materials, Processing and Applications in Electronic Devices. InTech, Croatia, http://dx.doi.org/10.5772/20332. Available from: <http://www.intechopen.com/books/silicon-carbide-materials-processing-and-applications-in-electronic-devices/recent-developments-on-silicon-carbide-thin-films-for-piezoresistive-sensors-applications> (accessed 28.06.13.).

Fraga, M.A., Pessoa, R.S., Massi, M., Maciel, H.S., 2012. Applications of SiC-based thin films in electronic and MEMS devices. In: Hijikata, Y. (Ed.), Physics and Technology of Silicon Carbide Devices. InTech, Croatia, http://dx.doi.org/10.5772/50998. Available from: <http://www.intechopen.com/books/physics-and-technology-of-silicon-carbide-devices/applications-of-sic-based-thin-films-in-lectronic-and-mems-devices> (accessed 05.08.13.).

Gao, F., He, J., Wu, E., Liu, S., Yu, D., et al., 2003. Hardness of covalent crystals. Phys. Rev. Lett. 91, 015502.

Gleiter, H., 2000. Nanostructured materials: basic concepts and microstructure. Acta Mater. 48, 1–29.

Granneman, E., Fischer, P., Pierreux, D., Terhorst, H., Zagwijn, P., 2007. Batch ALD: characteristics, comparison with single wafer ALD, and examples. Surf. Coat. Technol. 201, 8899–8907.

Hernández-Torres, J., García-González, L., Zamora-Peredo, L., Hernández-Quiroz, T., Sauceda-Carvajal, A., et al., 2012. Analysis of hardness of nanocrystalline coatings of aluminum-rich Ti1-xAlxN. Bull. Mater. Sci. 35 (5), 733–738.

Hoivik, N.D., Elam, J.W., Linderman, R.J., Bright, V.M., George, S.M., et al., 2003. Atomic layer deposited protective coatings for micro-electromechanical systems. Sens. Actuators A. 103, 100–108.

Holleck, H., 1988. Metastable coatings – prediction of composition and structure. Surf. Coat. Technol. 36, 151–159.

Holleck, H., 1999. Design of nanostructured thin films for tribological applications. In: Kumar, A., Chung, Y.W., Moore, J.J., Smugeresky, J.E. (Eds.), Surface Engineering: Science and Technology I. The Minerals, Metals and Materials Society, Pennsylvania, pp. 207–218.

Hovsepian, P.E., Munz, W.-D., 2003. Recent progress in large-scale production of nanoscale multilayer/superlattice hard coatings. Vacuum. 69, 27–36.

John, P., Polwart, N., Troupe, C.E., Wilson, J.I.B., 2002. The oxidation of (100) textured diamond. Diamond Related Materials. 11 (3–6), 861–866.

Kasu, M., Kobayashi, N., 2000. Large and stable field-emission current from heavily Si-doped AlN grown by metalorganic vapor phase epitaxy. Appl. Phys. Lett. 76 (20), 2910–2912.

Kim, H., 2003. Atomic layer deposition of metal and nitride thin films: current research efforts and applications for semiconductor device processing. J. Vac. Sci. Technol. B. 21, 2231–2261.

Kim, H., 2011. Characteristics and applications of plasma enhanced-atomic layer deposition. Thin Solid Films. 519, 6639–6644.

Kim, D.H., Byon, E., Lee, S., Kim, J.-K., Ruh, H., 2004. Characterization of ternary boron carbon nitride films synthesized by RF magnetron sputtering. Thin Solid Films. 447–448, 192–196.

Kim, J.H., Lee, H.K., Park, J.-Y., Kim, W.-J., Kim, D.K., 2008. Mechanical properties of chemical-vapor-deposited silicon carbide using a nanoindentation technique. J. Korean Ceram. Soc. 45, 518–523.

Kortright, J.B., Windt, D.L., 1988. Amorphous silicon carbide coatings for extreme ultraviolet optics. Appl. Opt. 27, 2841–2846.

Langdon, T.G., 2010. The impact of bulk nanostructured materials in modern research. Rev. Adv. Mater. Sci. 25, 11–15.

Leskela, M., Ritala, M., 2002. Atomic layer deposition (ALD): from precursors to thin film structures. Thin Solid Films. 409, 138–146.

Li, M.K., Li, C.B., Liu, C.S., Fan, X.J., 2005. Preparation of AlN Films and nc-AlN/a-SiNx nanocomposite films by medium frequency magnetron sputtering. J. Korean Phys. Soc. 46, S83–S87.

Liang, X., Zhan, G.-D., King, D.M., McCormick, J.A., Zhang, J., et al., 2008. Alumina atomic layer deposition nanocoatings on primary diamond particles using a fluidized bed reactor. Diamond Related Materials. 17, 185–189.

Lieberman, M.A., Lichtenberg, A.J., 2007. Principles of Plasma Discharges and Materials Processing. second ed. Wiley, New York.

Lu, C., Mai, Y.-W., Shen, Y.-G., 2006. Recent advances on understanding the origin of superhardness in nanocomposite coatings: a critical review. J. Mater. Sci. 41, 937–950.

Lukaszkowicz, K., 2011. Review of nanocomposite thin films and coatings deposited by PVD and CVD technology. In: Rahman, M. (Ed.), Nanomaterials. InTech, Croatia. Available from: <http://www.intechopen.com/books/nanomaterials/review-of-nanocomposite-thin-films-andcoatings-deposited-by-pvd-and-cvd-technology> (accessed 20.08.13.).

Malshe, A.P., Jiang, W., Dhamdhere, A.R., 2002. Nanostructured coatings for machining and wear-resistant applications. JOM. 54 (9), 28–30.

Marin, E., Guzman, L., Lanzutti, A., Fedrizzi, L., Saikkonen, M., 2009. Chemical and electrochemical characterization of hybrid PVD + ALD hard coatings on tool steel. Electrochem. Commun. 11, 2060–2063.

Marin, E., Lanzutti, A., Lekka, M., Guzman, L., Ensinger, W., et al., 2012. Chemical and mechanical characterization of TiO2/Al2O3 atomic layer depositions on AISI 316 L stainless steel. Surf. Coat. Technol. 211, 84–88.

Martinu, L., Zabeida, O., Klemberg-Sapieha, J.E., 2009. Plasma-enhanced chemical vapor deposition of functional coatings. In: Martin, P.M. (Ed.), Handbook of Deposition Technologies for Films and Coatings, third ed. Elsevier, Amsterdam, pp. 392–465.

Mattox, D.M., 1989. Particle bombardment effects on thin-film deposition: a review. J. Vac. Sci. Technol. A. 7, 1105–1114.

Mayrhofer, P.H., Horling, A., Karlsson, L., Sjolen, J., Larsson, T., et al., 2003. Self-organized nanostructures in the Ti-Al-N system. Appl. Phys. Lett. 83, 2049–2051.

Mayrhofer, P.H., Mitterer, C., Hultman, L., Clemens, H., 2006. Microstructural design of hard coatings. Prog. Mater. Sci. 51, 1032–1114.

Mayrhofer, P.H., Mitterer, C., Wen, J.G., Greene, J.E., Petrov, I., 2005. Self-organized nanocolumnar structure in superhard TiB2 thin films. Appl. Phys. Lett. 86, 131909.

Mayrhofer, P.H., Mitterer, C., Wen, J.G., Petrov, I., Greene, J.E., 2006. Thermally induced self-hardening of nanocrystalline Ti-B-N thin films. J. Appl. Phys. 100, 044301.

Medeiros, H.S., 2012. Obtenção de filmes de SiC, SiCN e AlN por Magnetron Sputtering para aplicação em Microsensores, Master's Thesis, Technological Institute of Aeronautics, Brazil.

Meyers, M.A., Mishra, A., Benson, D.J., 2006. Mechanical properties of nanocrystalline materials. Prog. Mater. Sci. 51, 427–556.

Miao, X.S., Chan, Y.C., Lee, Z.Y., 1997. Optical properties and reactive sputtering conditions of ain and alsin thin films for magneto-optical applications. J. Electron. Mater. 26 (1), 21–24.

Miikkulainen, V., Leskelä, M., Ritala, M., Puurunen, R.L., 2013. Crystallinity of inorganic films grown by atomic layer deposition: overview and general trends. J. Appl. Phys. 113, 021301.

Miskys, C.R., Garrido, J.A., Nebel, C.E., Hermann, M., Ambacher, O., et al., 2003. AlN/diamond heterojunction diodes. Appl. Phys. Lett. 82, 290–292.

Münz, W.-D., Lewis, D.B., Hovsepian, P.E., Schönjahn, C., Ehiasarian, A., et al., 2001. Industrial scale manufactured superlattice hard PVD coatings. Surf. Eng. 17 (1), 15–27.

Musil, J., 2000. Hard and superhard nanocomposite coatings. Surf. Coat. Technol. 125, 322–330.

Musil, J., 2012. Hard nanocomposite coatings: thermal stability, oxidation resistance and toughness. Surf. Coat. Technol. 207, 50–65.

Musil, J., Karvánková, P., Kasl, J., 2001. Hard and superhard Zr–Ni–N nanocomposite films. Surf. Coat. Technol. 139, 101–109.

Nagel, H., Aberle, A.G., Hezel, R., 1999. Optimised antireflection coatings for planar silicon solar cells using remote PECVD silicon nitride and porous silicon dioxide. Prog. Photovoltaics Res. Appl. 7, 245–260.

Neyts, E., 2006. Mathematical Simulation of the Deposition of Diamond-like carbon (DLC) Films, PhD thesis, Universiteit Antwerpen.

Nordmann, A., 2009. Invisible origins of nanotechnology: Herbert Gleiter, materials science, and questions of prestige. Perspect. Sci. 17 (2), 123–143.

Pelegrini, M.V., Rehder, G.P., Pereyra, L., 2010. a-SiC:H films deposited by PECVD for MEMS applications. Physica Status Solidi C. 7, 786–789.

Peng, X., Matthews, A., Xue, S., 2011. Plasma-based processes and thin film equipment for nano-scale device fabrication. J. Mater. Sci. 46, 1–37.

Perooly, S.S., Garg, V., Rahman, M.H., Rosenberger, L.W., Shreve, G.S., et al., 2005. Determination of elastic modulus of piezoelectric aluminum nitride coating. MRS Proc. 888, 0888-V03-12. http://dx.doi.org/10.1557/PROC-0888-V03-12.

Pessoa, R.S., Medeiros, H.S., Fraga, M.A., Galvão, N.K.A.M., Sagas, J.C., et al., 2013. Low pressure deposition techniques of silicon carbide thin films: an overview. In: Withers, M.C. (Ed.), Advances in Materials Science Research, vol. 16. Nova Science Publishers, New York, pp. 1–21.

Pierson, H.O., 1993. Handbook of Carbon, Graphite, Diamond and Fullerines. Noyes Publications, USA, pp. 249–266.

Pogrebnjak, A.D., Beresnev, V.M., 2012. Hard nanocomposite coatings, their structure and properties. In: Ebrahami, F. (Ed.), Nanocomposites – New Trends and Developments. InTech, Croatia, http://dx.doi.org/10.5772/50567. Available from: <http://www.intechopen.com/books/nanocomposites-new-trends-and-developments/hard-nanocomposite-coatings-their-structure-and-properties> (accessed 20.08.13.).

Pogrebnjak, A., Baidak, V., Beresnev, V., Turbin, P., Makhmudov, N., et al., 2013. Physical-mechanical properties of superhard nanocomposite coatings on base Zr-Ti-Si-N. Medziagotyra. 19, 140–143.

Profijt, H.B., Potts, S.E., van de Sanden, M.C.M., Kessels, W.M.M., 2011. Plasma-assisted atomic layer deposition: basics, opportunities and challenges. J. Vac. Sci. Technol. A. 29 (5), 050801.

Puurunen, R.L., 2005. Surface chemistry of atomic layer deposition: a case study for the trimethylaluminum/water process. J. Appl. Phys. 97, 121301.

Ritala, M., Leskela, M., Nalwa, H.S., 2002. Atomic layer deposition. In: Nalwa, H.S. (Ed.), Handbook of Thin Film Materials, vol. 1. Academic Press, San Diego, pp. 103–159.

Rudolph, S., 1993. Composition and application of coatings based on Boron nitride. Interceram. 42, 302–305.

Sandhu, A., 2006. Who invented nano? Nat. Nanotechnol. 1, 87.

Sangkla, T., Bland, S.,. Tuchinda, K., 2011. Feasibility study of using TiCN and CrN thin film coatings to enhance lifetime of grippers used in hard disk drive assembly line, In: Proceedings of the World Congress on Engineering 2011, Vol. 3. WCE 2011, July 6–8, London, UK, pp. 1987–1992.

Sanjines, R., Sandu, C.S., Lamni, R., Levy, F., 2006. Thermal decomposition of Zr1 – xAlxN thin films deposited by magnetron sputtering. Surf. Coat. Technol. 200, 6308–6312.

Sarro, P.M., deBoer, C.R., Korkmaz, E., Laros, J.M.W., 1998. Low-stress PECVD SiC thin films for IC-compatible microstructures. Sens. Actuators A. 67, 175–180.

Sasvvides, N., Bell, T.J., 1992. Microhardness and Young's modulus of diamond and diamondlike carbon films. J. Appl. Phys. 72, 2791–2796.

Schiotz, J., 2001. Simulations of nanocrystalline metals at the atomic scale. What can we do? What can we trust? In: Dinesen, A.R. (Ed.), Science of Metastable and Nanocrystalline Alloys: Structure, Properties and Modelling; Proceedings of the 22nd International Symposium on Materials Science 3–7 September 2001. Risø National Laboratory, Roskilde, Denmark, pp. 127–139.

Sniegowski, J.J., de Boer, M.P., 2000. IC-compatible polysilicon surface micromachining. Annu. Rev. Mater. Sci. 30, 297.

Solozhenko, V.L., Kurakevych, O.O., Andrault, D., Godec, Y.L., Mezouar, M., 2009. Ultimate metastable solubility of boron in diamond: synthesis of superhard diamondlike BC5. Phys. Rev. Lett. 102, 015506.

Steinmann, P.A., Hintermann, H.E., 2009. Adhesion of TiC and Ti(C,N) coatings on steel. J. Vac. Sci. Technol. A. 3, 2394–2400.

Stüber, M., Ulrich, S., Leiste, H., Holleck, H., 2011. Magnetron sputtered nanocrystalline metastable (V,Al)(C,N) hard coatings. Surf. Coat. Technol. 206, 610–616.

Szlufarska, I., Nakano, A., Vashishta, P., 2005. A crossover in the mechanical response of nanocrystalline ceramics. Science. 309, 911–914.

Takeno, T., Shiota, H., Miki, H., Takagi, T., Luo, Y., 2010. Carbon-based nanostructured coatings on NiTi shape memory alloy for biomedical applications. Intell. Rob. Appl. Lect. Notes Comput. Sci. 6425, 742–753.

Tripp, M.K., Stampfer, C., Miller, D.C., Helbling, T., Herrmann, C.F., et al., 2006. The mechanical properties of atomic layer deposited alumina for use in micro- and nano-electromechanical systems. Sens. Actuators A. 130–131, 419–429.

Ulrich, S., Ye, J., Stüber, M., Ziebert, C., 2009. Cubic boron nitride based metastable coatings and nanocomposites. Thin Solid Films. 518, 1443–1450.

Van Swygenhoven, H., 2002. Grain boundaries and dislocations. Science. 296, 66–67.

Veprek, S., Veprek-Heijman, M.G.J., 2007. The formation and role of interfaces in superhard nc-MenN/a-Si3N4 nanocomposites. Surf. Coat. Technol. 201, 6064–6070.

Veprek, S., Niederhofer, A., Moto, K., Bolom, T., Mannling, H.D., et al., 2000. Composition, nanostructure and origin of the ultrahardness in nc-TiN/a-Si3N4/a- and nc-TiSi2 nanocomposites with HV = 80 to $\geq$ 105 GPa. Surf. Coat. Technol. 133–134, 152–159.

Violet, P., Blanquet, E., Monnier, D., Nuta, I., Chatillon, C., 2009. Experimental thermodynamics for the evaluation of ALD growth processes. Surf. Coat. Technol. 204, 882–886.

Vitu, T., Polcar, T., Cvrcek, L., Novak, R., Macak, J., et al., 2008. Structure and tribology of biocompatible Ti–C:H coatings. Surf. Coat. Technol. 202, 5790–5793.

Wikimedia Commons (n.d). Atomic layer deposition-schematic cycle DE.svg. Available from: <http://commons.wikimedia.org/wiki/File:Atomic_layer_deposition_-_schematic_cycle_DE.svg> (accessed 29.01.14.).

Williams, K.R., Gupta, K., Wasilik, M., 2003. Etch rates for micromachining processing – Part II. J. Microelectromech. Syst. 12, 761–778.

Yamamoto, K., Keunecke, M., Bewilogua, K., 2000. Deposition of well adhering cBN films up to 2 μm thickness by B–C–N gradient layer system. Thin Solid Films. 377–378, 331–339.

Zhang, S., Fu, Y., Du, H., Zeng, X.T., Liu, Y.C., 2002. Magnetron sputtering of nanocomposite (Ti,Cr)CN/DLC coatings. Surf. Coat. Technol. 162, 42–48.

Zhang, S., Sun, D., Fu, Y., Du, H., 2003. Recent advances of superhard nanocomposite coatings: a review. Surf. Coat. Technol. 167, 113–119.

Zhang, S., Sun, D., Fu, Y., Du, H., 2005. Toughening of hard nanostructural thin films: a critical review. Surf. Coat. Technol. 198, 2–8.

Zhang, S., Wang, H.L., Ong, S.E., Sun, D., Bui, X.L., 2007. Hard yet tough nanocomposite coatings – Present status and future trends. Plasma Process Polym. 4, 219–228.

Zhao, D., Mourey, D.A., Jackson, T.N., 2010. Low-temperature pulsed-PECVD ZnO thin-film transistors. J. Electron. Mater. 39 (5), 554–558.

Zou, C.W., Wang, H.J., Li, M., Yu, J.F., Liu, C.S., et al., 2010. Characterization and properties of TiN-containing amorphous Ti–Si–N nanocomposite coatings prepared by arc assisted middle frequency magnetron sputtering. Vacuum. 84, 817–822.

Thermal spray nanostructured ceramic and metal-matrix composite coatings

19

George E. Kim and Mathieu Brochu*[†]
*Perpetual Technologies Inc., Canada; [†]McGill University, Canada

19.1 Introduction

Thermal spray nanostructured coatings were first conceived of and developed within a United States Office of Naval Research (ONR) programme entitled, "Thermal Spray Processing of Nanostructured Coatings" (Kabacoff, 2002). The programme has yielded fundamental knowledge critical to the development and advancement of thermal sprayed nanostructured protective coatings used in extreme environments found in both military and industry applications. Although the ONR programme produced useful information concerning thermal and corrosion protective coatings, the programme's flagship results were those related to advancements in protection against wear, which led to maintenance reduction, enhanced performance, and superior reliability of critical components. Since ceramic and metal-matrix composite (MMC) coatings are often used for wear resistance, these materials were selected for the majority of the nanostructure coating studies within the ONR project.

Since the end of the ONR programme in 2002, scientists from many nations have validated and advanced this area of work. To clarify the advantages and limitations of nanostructured coatings, it is important to decipher the vast quantities of information that has been produced in the past decade. This chapter attempts to provide an update on the status of thermal spray nanostructured coatings for protection against wear.

19.1.1 *Wear resistance and nanostructured materials*

There are good sources of information pertaining to the behaviour of nanostructured materials (Koch, 1999), which suggest their attractiveness for protective coating applications. Among nanostructured bulk materials, MMCs and monolithic ceramics are two of the most common materials with properties favourable for wear resistance. A good example is the increased hardness and toughness of WC-Co cermets. These favourable characteristics have led to the use of nanostructured/multimodal (combination of nanostructured and microstructured constituents) bulk WC-Co for industrial applications, for example, cutting tools, dies and drill bits.

Anti-Abrasive Nanocoatings.

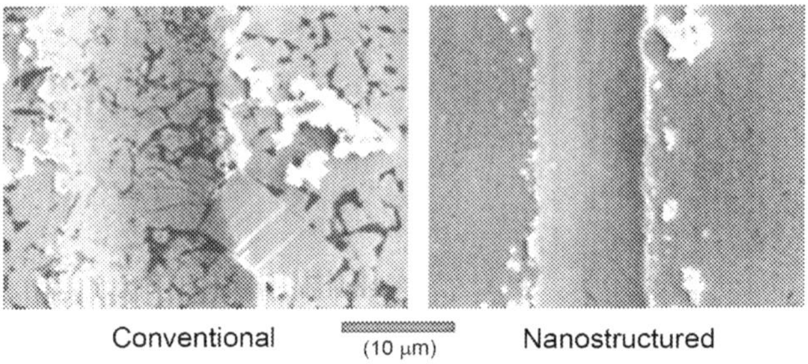

Figure 19.1 Surface scratch scar comparison between sintered WC-Co from nanocomposite powder and coarse grain WC-Co.
Source: Reprinted from Jia and Fischer (1996). Copyright (1996), with permission from Elsevier.

Figure 19.1 (Jia and Fischer, 1996) presents magnified images of the scratch region of microstructured and nanostructured bulk WC-Co. The scar of the nanostructured WC-Co clearly shows a shallower and smoother surface with no evidence of cracking. When compared to the rough surface and cracking observed on the microstructured material, it is evident that the very fine grain structure of the nanostructured sample increases the material's toughness without compromising on its hardness/strength. This enhanced toughness has been observed in numerous studies (Fang and Eason, 1995; Fang et al., 2009; Yao et al., n.d.) and contributes to the notable increases in wear resistance.

Siegel and Kear (2000) have discussed the advantages of nanophase ceramics such as alumina and titania, consolidated from ceramic nanoparticles. These materials possess a significant degree of ductile behaviour in compression. Karch et al. (1987) attributed the enhanced ductility to diffusional flow of atoms along the grain interfaces. They observed how polycrystalline ceramic with very fine (a few nm) crystal size underwent large plastic deformation at low temperature. The challenge for larger components would be in finding a means of fabricating components without excessive grain growth and porosity. Osiko's (2009) work showed how nanostructured partially stabilized zirconium dioxide possesses high fracture toughness and wear resistance conducive for medical and friction coupling applications.

The enhanced wear-resistance observed in bulk nanostructured materials makes transferring its qualities onto coating materials an obvious yet challenging next step.

Table 19.1 summarizes results of nanostructured materials discussed above.

19.1.2 Thermal and cold spray

Thermal spray technology is a very versatile surface engineering method that involves the projection of molten or semi-molten particles of metals, ceramics, or their composites from mostly powder or wire feedstock (Figure 19.2). Any material that has a stable molten phase and can be processed into the appropriate feed specifications can be thermal sprayed. The heating of the in-flight particles originates via oxygen–fuel

Table 19.1 Summary of prior work on nanostructured ceramics and metal-matrix composites

References	Material	Findings
Kabacoff (2002)	Alumina-titania	Superior toughness and wear resistance
Koch (1999)	Carbide cermets, monolithic ceramics	Superior wear resistance and increased hardness
Jia and Fischer (1996)	WC-Co	Superior toughness and wear resistance
Fang and Eason (1995)	WC-Co	Superior toughness and wear resistance
Yao et al. (n.d.)	WC-Co	Superior toughness and wear resistance
Fang et al. (2009)	WC-Co	Superior toughness and wear resistance
Siegel and Kear (2000)	Alumina and titania	Enhanced ductility
Karch et al. (1987)	CaF_2, TiO_2	Enhanced ductility
Osiko (2009)	PSZ	Superior wear resistance

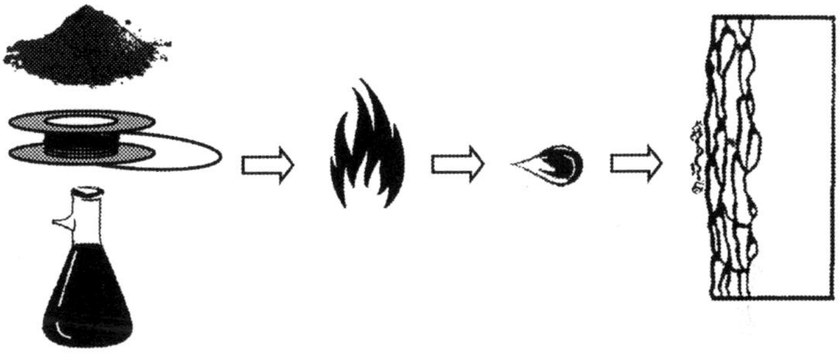

Figure 19.2 Schematic of thermal spray process.

combustion or electric arc. Once heated, the particles are accelerated by the combustion flame or the plasma jet onto a surface, forming a lamellar structure. Multiple passes are frequently required to build up lamellae layers of desired thickness.

Due to the nature of the nanostructured feedstock, spraying nanostructured coatings requires optimization of parameters to attain/maintain a specific structure and/or chemistry within the coating. The very high surface area associated with nanoparticles exacerbates thermal alterations of constituents such as carbides and nitrides. To overcome this thermal alteration, one approach is to replace most, if not all, of the thermal energy (e.g. flame or jet temperature) with kinetic energy (e.g. particle velocity) so as to retain the nanostructure while limiting coating oxidation and/or porosity. For these reasons, spray processes such as high-velocity oxyfuel (HVOF) and cold spray (CSP), which rely on higher kinetic energy in addition to lower or no thermal energy, have played a significant role in the application of nanostructured carbide cermet coatings.

The HVOF process uses a torch design with a nozzle that allows for dramatic gas acceleration. The gas velocities have been measured in the range of 1500 to 2000 m/s while maintaining a gas temperature in the order of 2900°C. The upper temperature limit excludes this process from depositing all monolithic ceramics; however, metals and metal-bearing composites can be readily applied with high density and bond strength. Popular carbide cermets such as WC-Co and Cr_3C_2-NiCr are typically deposited using the HVOF process because the gas temperature is sufficiently low to mitigate decarburization of the carbide phases.

The CSP was introduced to North America in 1994 and remains in its infancy stage. It uses a supersonic gas jet (velocity of 5000 km/h) to accelerate solid fine powders (micron size) of various materials above a critical velocity at which particles impact, deform plastically and bond to the substrate to form the coating. Because of the high kinetic energy with little or no thermal energy imparted onto the particles, CSP holds much promise in the application of metal-based nanostructured coatings and near net-shape forms (Kim, Lee et al., 2005).

For most monolithic ceramic materials, the choice of thermal spray process has been mainly plasma spray (atmospheric or APS, and vacuum or VPS), and sometimes HVOF for lower melting temperature ceramics. Plasma spraying uses a torch where a direct current (DC) arc is struck between a tungsten alloy cathode and an annular copper anode to heat a flowing inert gas stream (usually argon with a small percentage of enthalpy enhancing gas such as hydrogen), forming a plasma jet. The gas temperature just outside the torch exit or the core of the plasma jet is approximately 17,000°C (Unger, 1987). The average particle velocity for plasma spray processes (300–400 m/s) is approximately half that of HVOF particle velocities and can vary depending on the power capacity of the plasma torch. The high gas temperature permits the plasma spray processes to be effective for applying metal, ceramic and composite deposits.

It is important to note that in most of the thermal and cold spray approaches used to deposit nanostructured MMC and ceramic coatings, the objective has been to retain some, if not all, of the nanostructure features of the feedstock material within the coating structure. For thermal spray processes, a certain level of melting is required to provide cohesive integrity in the coating. The key with monolithic nanostructured ceramic material is to partially or fully melt some particles to provide the binding component while retaining some unmelted component of the coating. For nanostructured MMCs, the emphasis is on melting the metal component without melting or decomposing the hard phase component of the coating. With cold spraying of metal-based nanostructured materials, bonding is accomplished by metallurgical bonding associated with the formation of oxide-free surfaces during jetting, and mechanical interlocking of jetted material between the powder particles that form the coating.

19.2 Nanostructured feedstock

As with conventional thermal spray feedstock, there are many features to consider when designing, fabricating or selecting nanostructured powder feedstock. Particle size distribution and particle morphology are two key features. Although there are several creative approaches to processing nanostructured powder feedstock, the focus in this discussion will be on the most readily applied approaches with commercial viability.

Since thermal spray processes require a minimum particle size in the micron range in order for particles to penetrate the jet or flame, direct spray of nanoparticulates is not possible. Hence, nanostructured powders must undergo some type of processing in order to attain a particle size that is conducive to thermal spraying. The processing of nanostructured powders can be top-down, bottom-up, or a combination of the two as illustrated by Eigen et al. (2005) in Figure 19.3. Not all of the approaches are conducive to the processing appropriate for both ceramics and MMC nanostructured powder. Generally, in the top-down approach, nanostructured powders can be processed by refining coarse-grained materials. Severe plastic deformation (SPD), for example ball milling, with or without subsequent agglomeration, is a common top-down approach. In the bottom-up approach, nanostructured particles can be assembled from atoms, molecules and/or nanoparticles. Agglomeration of nanoparticles is an example of a bottom-up approach to forming nanostructured powder.

The milling approach, often used to process nanostructured MMCs, is based on repeated cold welding and fracturing of particles of a powder when impacted between colliding milling media (often balls) or between milling media and the chamber wall. This is a very effective means of attaining grain refinement when a metal is at least one constituent of the powder. In addition, the high energy transfer of milling systems can also be used to activate chemical reactions (i.e. reactive milling, mechanical alloying), to reduce particle size of hard and brittle materials, and to produce MMC powders with uniformly distributed reinforcing particles within a very fine metal matrix. The alteration in composition may be due to various factors, including: contamination from the milling media and chamber walls, reaction between the powder constituents, and reaction with the chamber environment. It is

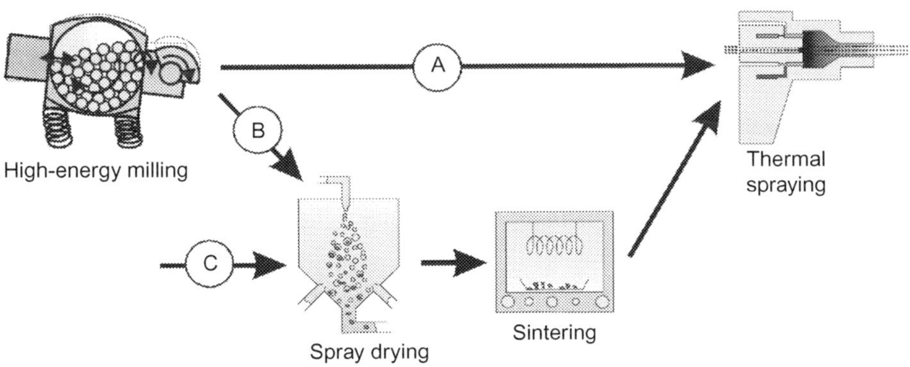

Figure 19.3 Process routes for the production of thermal spray coatings: (A) milling for direct use as feedstock; (B) milling, spray drying and sintering; (C) spray drying, sintering. *Source:* Eigen et al. (2005). Reproduced with permission.

important to note such changes so as to ensure that their presence does not adversely affect the performance of the coating.

The process of grain refinement through milling has been described by Koch (2003) and through work carried out by Fecht et al. (1990) in the following manner:

> *The observed phenomenology of nanocrystallization by mechanical attrition was summarized as occurring in three stages, namely:*
>
> - ***Stage 1****. Deformation localization in shear bands containing a high dislocation density.*
> - ***Stage 2****. Dislocation annihilation/recombination/rearrangement to form cell/ subgrain structures with nanoscale dimensions – further milling extends this structure throughout the sample.*
> - ***Stage 3****. The orientation of the grains becomes random, that is, high angle grain boundaries replace them, by presumably grain boundary rotation/sliding.*

The choice of milling environment (solid, liquid or gas) within the chamber is influenced by considerations such as the ability to control the level of contaminants in the powder, to introduce a reactant to form additional phases, to control the milling efficiency or to control the process temperature. For instance, it is not uncommon when milling metal powder to introduce liquid nitrogen, otherwise known as cryomilling (Witkin and Lavernia, 2006). Cryomilling reduces oxidation, produces nitride/oxynitride nanodispersoids (Luton et al., 1988; Jiang et al., 1999), enhances milling efficiency and controls powder morphology (Shen and Koch, 1995). The drawback to cryomilling is the high cost of liquid nitrogen (Ye and Schoenung, 2004) and the relatively small batch capabilities.

The feasibility of processing metal-based powders into nanostructure form without the use of liquid nitrogen was studied by n-Werkz Inc. (Kim and Addona, 2011). A non-cryogenic milling (NCM) process, using a novel, high-energy milling system, was used to convert conventional metal-based powders into nanostructured powders with equiaxed particles. To compensate for the absence of liquid nitrogen, strain rate exerted onto the particles was increased during milling, thereby increasing the likelihood of fracturing occurring at non-cryogenic temperatures.

The findings of Oleszak and Shingu (1996) may provide further insight into the ability of the NCM system to refine the grain structure of metal-based powder and attain equiaxed particle shape at non-cryogenic temperatures with comparable processing times. Their work suggests that the minimum crystalline grain size may be a function of the total strain energy. Figure 19.4 shows the strain energy introduced to the metal powders during NCM and cryomilling. The NCM values were calculated and compared to those of cryomilled powders using the formula and calculations presented by Lin et al. (2012). It is evident that the NCM process introduces much greater amounts of strain energy to the powder compared to the cryomill process. This, and the very high strain rate, are likely linked to the NCM process's unique ability to produce relatively large quantities of nanostructured metal-based powder within a time comparable to that required for cryomilling.

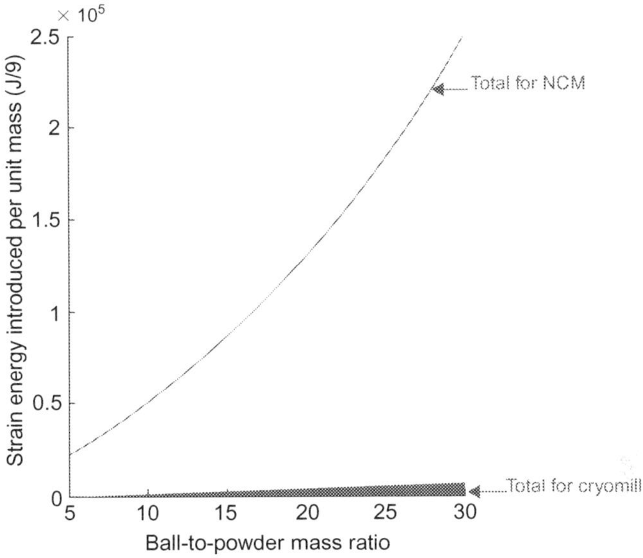

Figure 19.4 Strain energy introduced to the powder.
Source: Eigen et al. (2005). Reproduced with permission.

Carbide-based MMC materials seem to be the preferred feedstock choice for most studies on wear-resistant nanostructured MMC coatings. Various types of carbide and metal matrix feedstock materials and milling environments have been studied and/or applied. The following are some of the milling combinations found in public literature:

- Cunha et al. (2007) and He and Schoenung (2002) used *n*-CrC-NiCr powder processed in hexane and gaseous nitrogen.
- Eigen et al. (2005) used *n*-TiC-Ni-bsaed composite powder processed in vacuum and argon.
- Yandouzi et al. (2012) used *n*-Al5356 + 20% B_4C powder processed in liquid nitrogen.
- He et al. (2000) used *n*-WC-Co powder processed in hexane.
- Liu et al. (2008) used *n*-WC-NiCr powder processed in gaseous nitrogen environment.
- Kim and Addona (2011) used *n*-Al-Mg-B_4C powder processed in gaseous nitrogen using the NCM system.

Agglomerating nanoparticles to form nanostructured powder in micron-scale size is another popular approach that has led to successful thermal spray application of nanostructured coatings. This approach provides better distribution and mixing of the constituents of the composite powders than milling. Spray drying has been the most popular means of agglomerating the nanoparticles of monolithic ceramics and MMCs.

Spray drying is a process by which a fluid is atomized to micron-sized drops and dried into a dry powder by hot gas. To form nanostructured agglomerates, a suspension consisting of nanoparticles, liquid, and organic binder is atomized and dried.

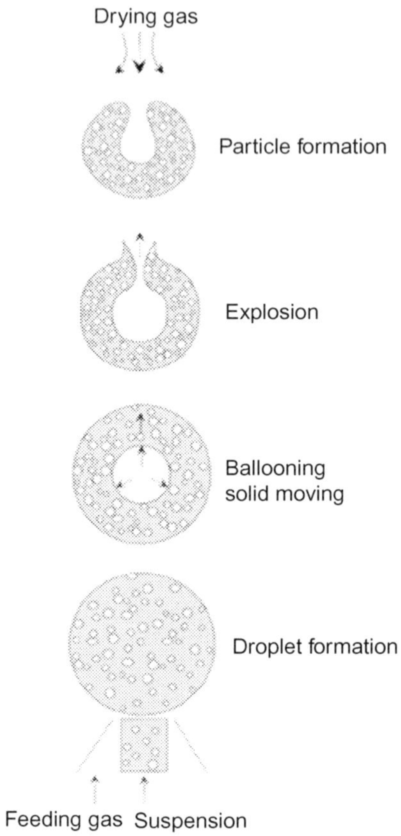

Figure 19.5 Spray drying process of ceramics.

Nanostructured agglomerates of oxides and MMCs have been successfully produced via spray drying. Cao et al. (2000) have proposed a simple mechanism to describe and illustrate (Figure 19.5) the spray drying process of ceramics in four steps:

1. droplet formation;
2. evaporation and ballooning;
3. explosion; and
4. particle formation.

In addition to the suspension preparation, considered to be the most important factor determining the characteristics of spray dried powder, the type of atomizer, feed rate, temperature and atomizing pressure also influence the outcome. Some details on spray dry processing of nanostructured powder have been documented by Lindelov and Wahlberg (2009) and Viswanathan et al. (2008). It is very common following spray drying for the agglomerate particles to undergo a sintering/calcining/heat treatment step to remove the organic binder and any liquid, and to provide additional mechanical integrity. The following list provides examples of nanostructured MMC and monolithic ceramic powder processed for thermal spray via spray drying:

- Cunha et al. (2008) used processed n-CrC-NiCr powder.
- Wang et al. (2009) and Ding et al. (2007) used processed n-WC-Co powder.
- Chwa et al. (2004) used processed n-TiO_2 and n-TiO_2-Al powder.
- Kawakita et al. (2008) used processed n-TiO_2-Fe_2O_3 powder.
- Jiang and Liu (2002) used processed n-Al_2O_3-TiO_2-SiO_2 powder.
- Gell et al. (2001) used processed n-Al_2O_3-TiO_2 powder.
- Lima and Marple (2005) used processed n-TiO_2 powder.
- Turunen et al. (2006) used processed n-Al_2O_3 and n-Al_2O_3-Ni powders.
- Li et al. (2004) used processed n-YPSZ powder.
- Varis et al. (2007) used processed n-Al_2O_3, n-Al_2O_3-TiO_2 and n-Cr_2O_3-TiO_2 powders.
- Wang (2004) used processed n-FeS powder.
- Cellard et al. (2009) used processed Cr_2O_3 powder.

Figure 19.6 shows a scanning electron microscope (SEM) micrograph of a spray dried nanostructured TiO_2 particle from Altair (Reno, Nevada). Spray drying has also been used to agglomerate nanoparticles of WC with Co (Figure 19.7).

There is also a fairly novel means of thermal spraying nanoparticles without prior agglomeration. The approach is called suspension thermal spray and uses a fluid to carry dispersed fine powder into the hot zone within a thermal spray (HVOF or APS) flame or jet. As the suspension enters the flame or jet, it fragments into droplets, liquid vaporizes, and fine particles undergo melting prior to impacting the substrate. Some of the notable advantages with this approach include a wider range of coating morphology, direct use of fine powder, thin and thick coating capability, isotropic coating structure and low coating surface roughness. As with spray drying, suspension preparation plays a key role in the overall process. The fluid is often water, organic solvent (e.g. alcohols) or a mixture of the two. Typical

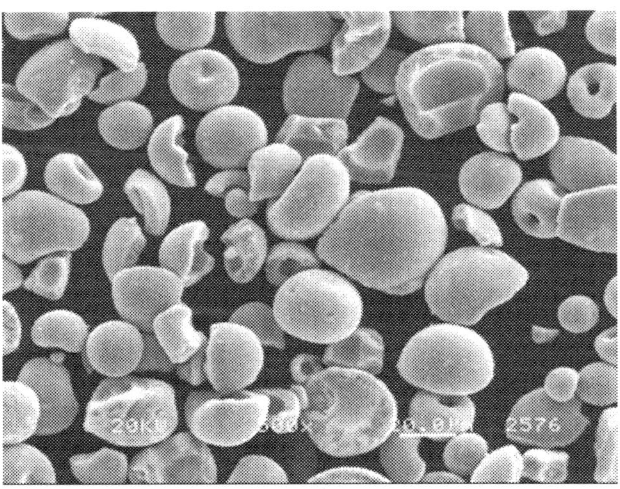

Figure 19.6 Agglomerated nanostructured TiO_2 powder.

Figure 19.7 Agglomerated nanostructured WC-Co powder.

solid load is between 5 and 50 wt% with possible additives such as surfactants and/or acids.

Studies on the use of suspension thermal spray process to deposit nanostructured coatings for wear resistance have been reported:

- Darut, Valette et al. (2010) have plasma sprayed suspensions of Al_2O_3 and Al_2O_3-TiO_2.
- Tingaud et al. (2010) have plasma sprayed suspensions of Al_2O_3 and Al_2O_3-YSZ.
- Darut, Ben-Ettouil et al. (2010) have plasma sprayed suspensions of Al_2O_3, Al_2O_3-SiC and Al_2O_3-ZrO.
- Oliker et al. (2009) have plasma sprayed suspensions of Al_2O_3.

19.3 Nanostructured coatings

As mentioned in the previous section, deposition of nanostructured coatings through thermal spray technology has revolved mostly around using nanostructured, severely deformed or agglomerate powder. It is only in recent years that the ability to spray nanoparticles without agglomeration has been successfully implemented. This section presents the development in nanostructured thermal spray coatings of monolithic ceramic and MMC materials for wear protection.

19.3.1 Ceramic

Some of the first work on thermal spraying nanostructured ceramic coatings took place at the University of Connecticut (UCONN), as part of the ONR programme on 'Thermal Spray Processing of Nanostructured Coatings (Gell et al., 2001; Goberman et al., 2002). UCONN's general approach consisted of agglomerating

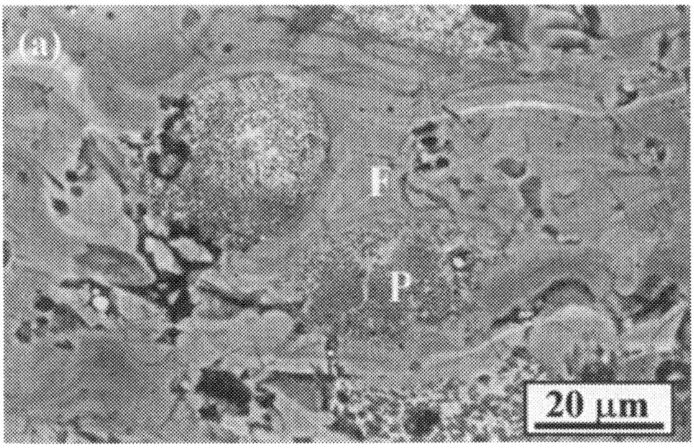

Figure 19.8 SEM micrograph of APS n-AT coating: 'F' is fully melted and 'P' is partially melted regions.
Source: Gell et al. (2001). Reproduced with permission.

nanoparticles of alumina and titania (87:13 wt% ratio) via spray drying and using APS to deposit the nanostructured coatings. The UCONN group observed remarkable and unprecedented enhancements to the following coating attributes:

- adhesion strength;
- spallation resistance; and
- abrasive wear resistance.

These favourable attributes were directly linked to the bimodal microstructure of the nanostructured alumina-titania (n-AT) coating. The bimodal microstructure (Figure 19.8) was identified as regions of fully melted and splat-quenched structure adjacent to regions of partially melted structure, where some microstructural features of the original particles were retained. It was noted that, for both conventional and nanostructured coatings, cracks which formed during indentation tests travelled predominantly along the splat boundaries. However, the cracks formed in the bimodal nanostructured coating would deflect or stop at the partially melted regions. This phenomenon provided the n-AT coating with enhanced toughness, and was the main reason for the increase in wear resistance, even with lower (at least 20% lower) hardness (Jordan et al., 2001). Traditionally, abrasive wear resistance is often superior for harder materials/coatings; however, the enhanced toughness of the n-AT coating seemed to more than compensate for its lower strength or hardness.

Following the discovery of the n-AT coating and of the advantages of having a nanostructured bimodal ceramic coating structure, further advances in the performance of these coatings were made, in particular in the context of severe industrial applications (Kim, 2007). A major advancement was the development of an alternate thermal spray process which optimized the spray process so as to maximize the n-ceramic coating's toughness while increasing the hardness. The resulting n-TiO_2 coatings possessed a more isotropic nanostructure, with considerable

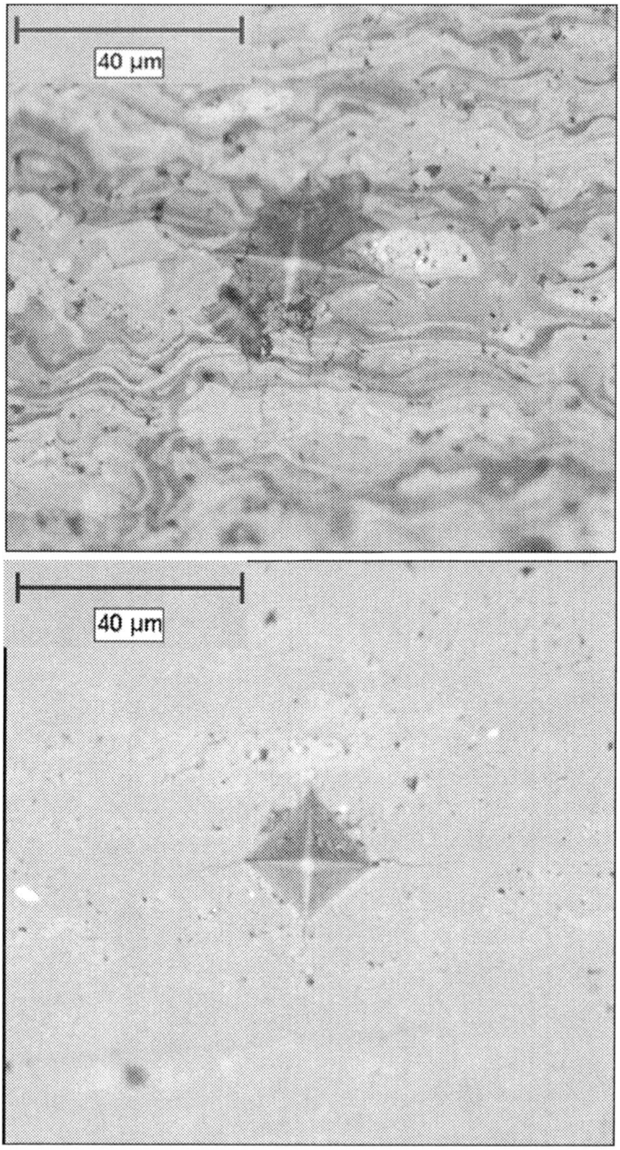

Figure 19.9 SEM micrograph of APS (top) and HVOF (bottom) applied nanostructured TiO_2 coatings following 500 g load Vickers indent.

unmelted nanoporous regions held firmly together by a fully molten matrix (Figure 19.9). The simultaneous and unprecedented increases in both toughness and hardness resulted in substantive improvements in resistance against three-body abrasion, and against low- and high-angle impingement slurry erosion

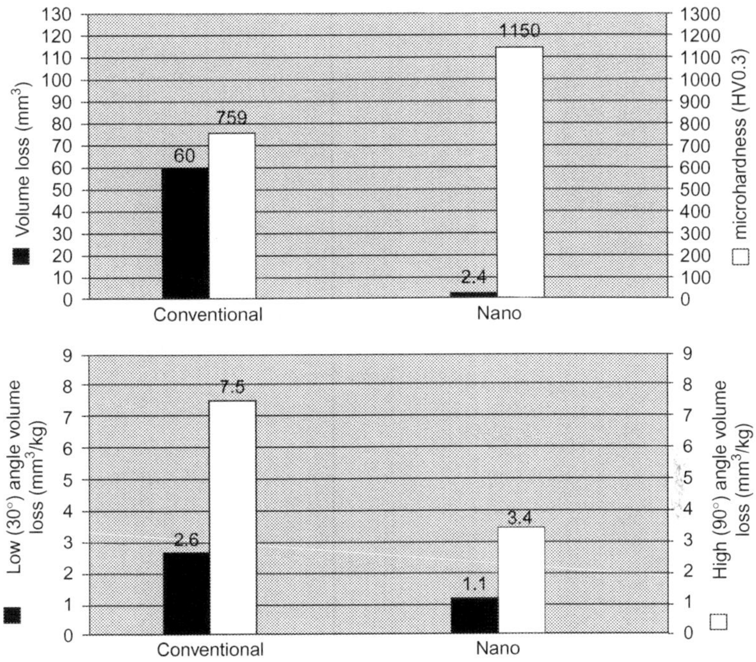

Figure 19.10 Abrasion (top) and slurry erosion (bottom) wear test results on TiO_2 coatings.

(Figure 19.10) (Kim and Walker, 2007). Evidence of the simultaneous enhancements was also evident on the wear scars following abrasion and erosion testing (Figure 19.11). Not only were the scars on the n-TiO_2 coating shallower than on its conventional counterpart (reflective of a harder material), but they also showed ductile directional modes of wear (e.g. micro-ploughing, micro-cutting and plastic deformation) (Kim et al., 2010). The scars on the conventional TiO_2 coating showed more brittle, non-directional modes of wear (e.g. fractures along splat boundaries, random microcracks).

A group at Rutgers University has been using the National Synchrotron Light Source (NSLS) located in Brookhaven National Laboratory (BNL) to characterize nanostructured materials in both bulk and coating form. Using the energy dispersive X-ray diffraction (EDXRD) method with the synchrotron light source, measurement of phase composition and strain can be deduced from its 3D-mapping capabilities (Tsakalakos et al., 2008). Akdogan et al. (2010) were able to identify the response of conventional and nanostructured Al_2O_3-TiO_2 coatings to tensile and compressive stresses. Compared to the conventional Al_2O_3-TiO_2 coating, the n-AT coating had a higher Young's modulus in tension (~300 GPa vs. ~260 GPa) and higher crack initiation load (40 lbs vs. ~0 lbs in tension and 450 lbs vs. 300–400 lbs in compression). These results support prior observations of thermal sprayed nanostructured oxide coatings. Professor Akdogan attributes the results of the nanostructured

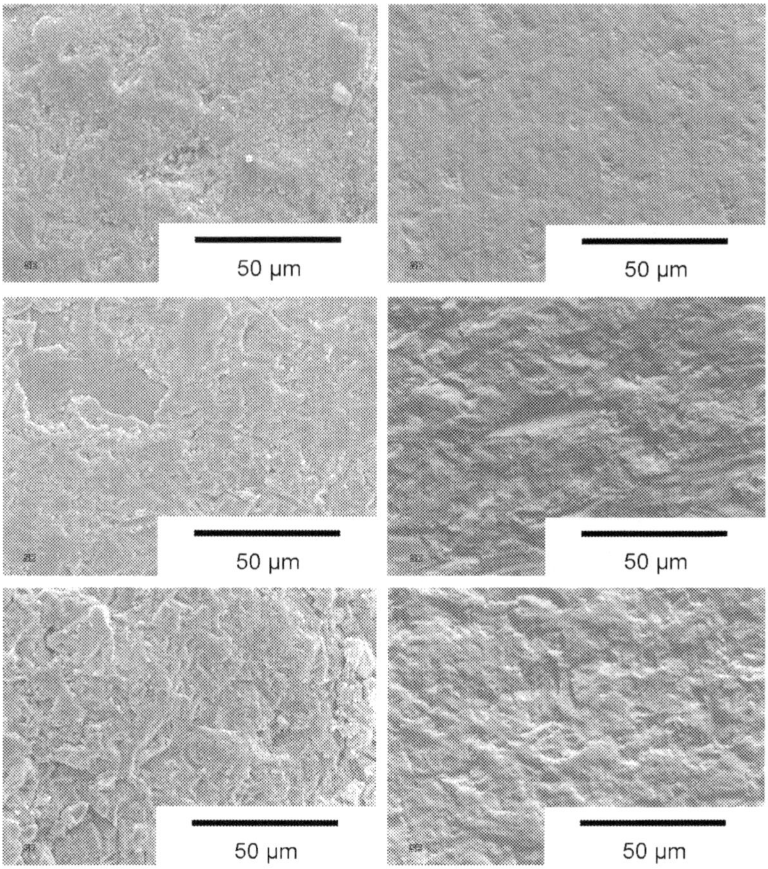

Figure 19.11 Scars on conventional TiO_2 coating following three-body abrasion. Left, conventional coatings, right nanostructured TiO_2 coatings; top, following abrasion test; middle, following low-impingement angle slurry erosion test; bottom following high-impingement angle slurry erosion test.

coating to the presence of nanograins, which cause crack deflection and/or stoppage or slowing down of crack growth, as well as a more uniform structure with finer flaw size.

Overall, the research on thermal spray nanostructured ceramic coatings indicates that the coatings' enhanced toughness is a physical phenomenon that transcends variations in coating composition, at least within oxides. Observations of enhanced toughness and/or superior wear resistance were also found in the following efforts:

- Jiang and Liu (2002) observed enhanced toughness in n-Al_2O_3-TiO_2-SiO_2 coatings.
- Turunen et al. (2006) observed abrasive wear resistance in n-Al_2O_3 coatings.
- Li et al. (2004) observed enhanced sliding wear resistance in n-YPSZ coatings.
- Chwa et al. (2004) observed enhanced sliding wear resistance in n-TiO_2-Al coatings.

- Chen et al. (2003) observed enhanced wear against sliding, reciprocating and vibrating test with water lubricant in their n-YPSZ coatings.
- Cellard et al. (2009) observed enhanced sliding wear resistance in n-Cr_2O_3 coatings.

19.3.2 MMC

The first attempts at depositing nanostructured MMCs were conducted under an ONR programme between 1996 and 2001. The MMC material chosen for preliminary trials was the popular WC-Co. It is accepted that decarburization of the carbide particles can readily occur when thermal spraying conventional carbide-based coatings and, therefore, must be closely monitored. However, for very fine, nano-sized carbides, the issue of decarburization is considerably greater. In fact, until this decarburization issue was addressed through proper selection of spray process and optimization of spray parameters, the resulting coatings' compositions and the properties were inconsistent and unpredictable. Examples of excessive decarburization leading to inconsistent coating properties were found in the following publications:

- Zhu et al. (2001b) sprayed n-WC-Co coatings and found various compositional phases due to decarburization, including α-W_2C, ß-WC_{1-x} and W_3Co_3C.
- Yao et al. (1998) describe the challenges of plasma spraying n-WC-Co powders due to higher levels of decarburization.
- Shipway et al. (2005) sprayed n-WC-Co coatings and found that the excessive decarburization of the fine carbides led to reduced sliding wear resistance.
- Guilemany et al. (2005) sprayed n-WC-Co coatings with higher decarburization which led to similar abrasive wear resistance but higher hardness and lower coefficient of friction (COF), compared to the conventional coating.

Several studies have used approaches which led to reduced decarburization and superior coating properties, such as hardness, wear resistance and toughness. These include:

- Zhu et al. (2001a) used a VPS system to deposit n-WC-Co coating that was superior to sliding, reciprocating and vibrating wear with a lower COF.
- Ibrahim and Berndt (2004) used the relatively low-temperature, high-velocity JP-5000 HVOF system to deposit n-WC-Co coatings with superior fatigue resistance, hardness and abrasive wear resistance.
- Zha et al. (2004) linked an increase in hardness of n-WC-Co coatings to a decrease in decarburization using an HVO-AF spray system.
- Josephson et al. (2005) sprayed n-WC-Co coatings and found that minimizing decomposition, by using HVOF fuel chemistry with a low fuel:oxygen ratio and using grain size inhibitors, resulted in a tougher and harder coating.
- Kim, Lee et al. (2005) cold sprayed n-WC-Co coatings with no decarburization and very high hardness.
- Smirnov et al. (2007) used a D-gun system to spray n-WC-Co coatings with superior sliding wear resistance and found a direct correlation between sliding wear resistance and concentration of WC.
- Liu et al. (2008) used the HVAF process to deposit n-WC-NiCr with very little decarburization and far superior sliding wear resistance.

The following are some interesting characteristics of thermal sprayed nanostructured cermet coatings derived from the literature review (Kim, 2010b):

- Unlike conventional carbide cermet coatings where sliding and abrasive wear resistance improves with carbide size, nanostructured coatings with limited carbide degradation seem to possess very good wear resistance, even with very fine carbides.
- The mechanism of wear is different for nanostructured carbide coatings; the wear scars resemble plastic deformation even though the coating is as hard as or harder than its conventional counterpart.
- Deposition technique and spray parameters play a critical role in the quality, that is, structure, hardness, toughness, of all carbide coatings, but even more so for nanostructured carbide cermets.
- Any means of attaining dense, well-adhered nanostructured coatings with limited decomposition will likely result in a harder and tougher coating.
- Nanostructured carbide powder with a dense structure will likely result in decreased decarburization as opposed to porous agglomerates.
- Although there are mixed results indicating whether nanostructured carbide coatings with a higher degree of decomposition will be harder and/or more wear resistant than the conventional carbide coatings, all results with well-bonded nanostructured coatings with limited or no decarburization indicate superior microhardness and/or superior wear resistance and toughness.
- The preliminary results of cold spray application of n-WC-Co look promising from the perspective of hardness and absence of decarburization.

There is an added benefit with nanostructured carbide coatings that relates to their surface finishing. Zhang and Liu (2001) have shown that nanostructured WC-12Co coatings require larger grinding force due to their higher wear resistance; however, the surface finish is better than their conventional counterpart coatings. In addition, the better surface finish of the nanostructured coating was obtained at a reduced depth of cut. It is important to note that surface finishing can represent a significant cost for the overall coating and finishing process.

As for CrC-based MMC coatings, the information in the public domain points to this coating material consistently providing superior wear performance in nanostructure form. This may be, in part, due to CrC having a higher resistance to decarburization compared to WC (Zorawski et al., 2008). Research on n-CrC-NiCr coatings has led to claims of higher scratch resistance and lower COF (He and Lavernia, 2001), higher hardness and fracture toughness (Cunha et al., 2008) and superior erosion-oxidation resistance at 500°C (Padial et al., 2008).

The data to date on nanostructured carbide MMC coatings indicates the possibility of attaining a notably superior wear-resistant and tougher coating by limiting decarburization within a well-adhered, dense coating. To attain this quality, reliance on greater kinetic energy and lower thermal energy is required through proper spray process selection and optimization of the spray parameters. Thermal spray processes such as HVOF and HVAF have been successfully used to deposit nanostructured metal-based material, including MMCs such as WC cermets. Kim, Lee et al. (2005) have gone so far as to use cold spray to further decrease the process temperature in spraying an n-WC-Co coating. This resulted in a coating with no decarburization and very high (~2050 $HV_{0.5}$) hardness.

19.4 Proven applications

19.4.1 Nanostructured alumina-titania coating for US Navy applications

As mentioned in Section 19.3.1, the n-AT coating was developed within a US Navy research programme. Prior to the programme's end, the coating was put through some rigorous tests that led to its MIL STD 1687A qualification for application onto Navy ship assets. The n-AT coating has been successfully applied onto numerous components, including through-hull submarine ball valves, gear shafts of 80 ton air conditioning units, and titanium submarine doors (Pappalardo, 2004).

One application that stands out in particular is the use of the n-AT coating onto mine countermeasure (MCM) propulsion shafts (Kim and Scandell, 2012). Unlike other ships' propulsion shafts, the MCM uses non-magnetic nickel aluminum bronze cylindrical shafts to minimize magnetic signature that could trigger mines. The use of a soft material such as nickel aluminum bronze meant that abrasive wear from calcareous build-up was quite high. Prior to the application of n-AT coatings, the shafts (two per ship-set) required weld rebuild and machine down to original tolerances every two years. This came at a cost of around $280,000 per ship-set, with a limited number of repair capabilities due to the heat-affected zone.

There were numerous challenges inherent to thermal spraying a ceramic coating onto this component. The most challenging was the high level of torsion strain exerted on the wear regions during start-up.

Currently all MCM propulsion shafts are coated with n-AT at a cost of $75 000 per ship-set. The latest evaluation of the coated shafts was carried out in February of 2009. The non-destructive evaluation (NDE) of the shafts that had been in service for seven years showed no signs of coating wear or spallation. Over 80% of the coating showed no evidence of change from the as-sprayed and machined state (Figure 19.12).

Figure 19.12 Nanostructured alumina-titania coating after seven years of service.

19.4.2 Nanostructured titania coating for hydrometallurgical applications

Nanostructured coatings have also been used in the mining industry to improve the lifespan of ball valves, which are critical process components. Nickel, cobalt and gold are extracted from low-grade ore using hydrometallurgical approaches called high-pressure acid leach (HPAL) or pressure oxidation (POX). HPAL and POX use a very severe process environment to dissolve as much of the ore constituents as possible for the separation and retrieval of the precious metals. Some of the extreme process environments can consist of high pressures (~5500 kPa), high temperatures (~260°C) and high corrosivity (up to 95% sulfuric acid) – all within an abrasive/erosive slurry (~20 wt% solids).

Ball valves play a critical role in containing and directing the flow of the very hot acidic slurry within the autoclaves. They have been widely used in these severe environments due to their angular rather than volumetric displacement capability, their large, low stress, precision-lapped seal and their ability to operate in an environment with crushed, abrasive solids. Thousands of ball valves in HPAL and POX facilities around the world are coated with nanostructured ceramic coating (Williams et al., 2004). An independent study conducted in 2001 of HPAL plants indicated that autoclave valve costs were six times greater than expected and represented 30%–40% of the operating budget. This was the underlying reason for a major thrust towards developing a novel coating for severe-service ball valves.

Even though a tough material such as titanium is used to make the valve components, the seal surfaces (e.g. balls and seats) require additional protection against abrasion, erosion and corrosion to extend their lives. Among the surface engineering options available, thermal spray application of a single- or multi-layer coating is commonly used. Traditionally, a chromia-blend or titania ceramic coating on titanium balls and seats with or without a metallic bond coat was widely used to provide such protection. However, failure analysis carried out on actual thermal sprayed components has revealed chemical attack of some ceramic top coats and all metallic bond coats (Kim, 2007). Since the bond coat does not provide a long-term barrier from the liquid component of the slurry, a galvanic cell is eventually formed between the bond coat and substrate. Hydrogen is formed at the cathodic region of the galvanic cell which in turn accelerates the degradation of both the tantalum or titanium bond coat layer and the titanium substrate.

To address this problem, development work has focussed on identifying new, nanostructured coatings to provide more effective protection for ball valves in HPAL and POX environments. As discussed in Section 19.3.1, a nanostructured titania (n-TiO_2) coating was developed that showed exceptional enhancements against abrasion, low- and high-impingement angle erosion and improved toughness (Kim, 2007). Performance results appear to indicate that this coating has produced remarkable improvements in the lifespan of these critical autoclave components.

Due to constant changes and differences in service conditions within the same autoclave and between different mines respectively, it is challenging to attain reliable and comparable performance data on the coated valves. However, in 2003, Lihir Gold (a mine in Papua New Guinea) installed two 10-inch ID Ti Gr5 ball valves (one with n-TiO_2 and another with Cr_2O_3-blend coatings) in parallel in the same process conditions for the same duration. From the visual inspection of the two valves after 10 months of service, it was clear that the n-TiO_2 coating was in a far superior state when compared to the non-nanostructured Cr_2O_3-blend coating (Figure 19.13). The surface of the Cr_2O_3-blend coated ball had large regions without coating. In contrast, the n-TiO_2 coated ball showed only a few isolated regions without coating.

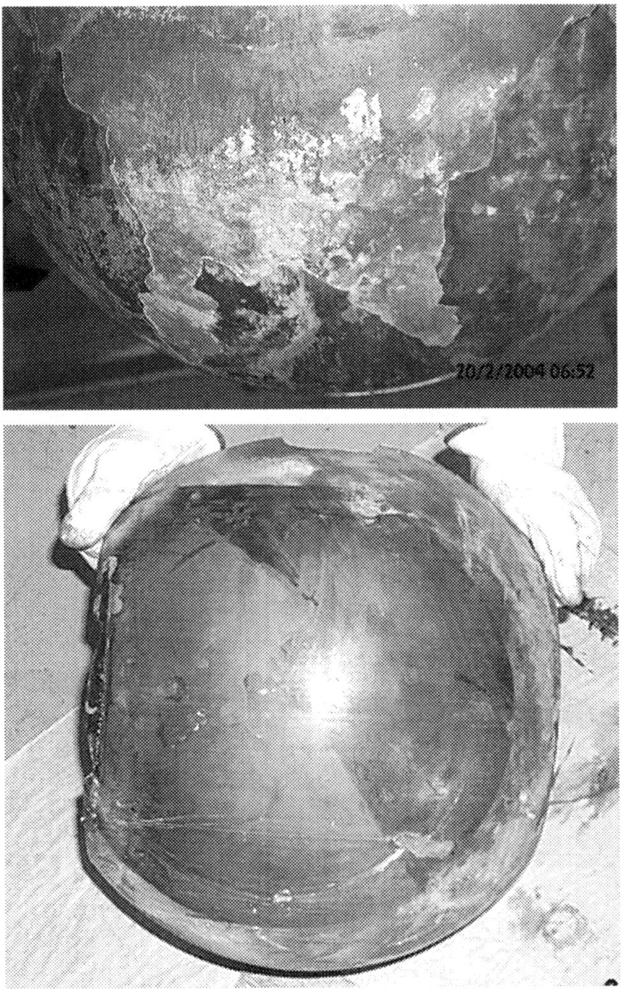

Figure 19.13 Ball valves after 10 month's service in gold autoclave with Cr_2O_3-blend (top) and n-TiO_2 (bottom) coatings.

Two additional valves have recently been removed for viewing during a scheduled shutdown at Lihir Gold. The n-TiO_2 coating on both the autoclave feed (8-inch ID) and discharge (10-inch ID) valves were in very good shape following a 12-month service period, other than some minor scratches and localized chips. The coated component was returned into service without any alterations. There are currently 10 installations around the world using thousands of n-TiO_2 coating ball valve components.

19.5 Possible future applications

As knowledge and understanding of nanostructured coatings has increased over the past decade, so has the scope of potential applications. The most obvious opportunities consist of enhancing existing coatings for applications that already make use of coatings. However, the superior properties of these novel coatings may also make it possible to consider their use for more innovative applications. In the authors' view, note that the possible future applications discussed below are examples of contexts in which the use of nanostructured coatings could offer the opportunity for important improvements in efficiency, productivity, performance and/or product quality.

19.5.1 Ceramics

There are a number of areas in which the use of nanostructured ceramic coatings could offer important improvements or enhancements to current applications.

In the field of gas turbine engines, manufacturers currently use conventional thermal sprayed coatings to extend the life and enhance the performance of their components. Oxide coatings such as alumina are used to protect components such as the knife edge tips and face of labyrinth seal fins and seal teeth, as well as the blade tips or outer airfoil section of compressor blades. Although the main role of these coatings is to protect against wear, they can also mitigate overheating of a component. Because nanostructured oxide coatings offer superior adhesion strength, wear resistance, and fracture toughness compared to conventional coatings, there are good reasons to believe that they may provide superior performance for these components.

In the field of bio-implants, the National Research Council of Canada (NRCC) has developed and evaluated ceramic-based nanostructured coatings such as nanostructured titanium oxide (n-TiO_2) (Lima et al., 2006) and hydroxyapatite (HA) (Lima et al., 2005) coatings for biocompatible implant application. NRCC's n-TiO_2 coating possessed higher hardness, greater bond strength, equivalent or higher osteoblast cell proliferation and higher contact surface between bone and coating, as compared to current commercial HA coating. HVOF-applied nanostructured HA coatings resulted in higher crystallinity, bioactivity and bond strength compared to its conventional counterpart. In addition, the nanostructured ceramic coatings are expected to provide superior wear resistance (e.g. fretting, abrasive, sliding).

In the field of nuclear fusion, like many other industrial fields, materials' survivability in harsh operating environments is a major obstacle towards full commercialization (Kim, 2010a). One of the most challenging applications from a materials point of view involves protection for the plasma-facing first wall (Baharvandi et al., 2009; Begrambekov and Buzhinskij, 2007; Mattox et al., 1979) of a magnetically confined plasma fusion reactor.

The first wall must possess the following qualities:

- resistance to electric arcs;
- resistance to high temperatures;
- resistance to chemical attacks (i.e. chemical sputtering);
- resistance to high temperatures;
- high hardness;
- low specific weight;
- high thermal conductivity;
- high cross section for neutron absorption;
- resistance to physical wear;
- resistance to irradiation.

B_4C provides high hardness, low specific weight, high cross-section for neutron absorption, as well as, resistance to heat, wear, and corrosion. Although B_4C has a relatively low thermal conductivity, this can be compensated for by applying the coating onto a conductive substrate such as graphite. Should a nanostructured coating of B_4C be successfully developed and applied, further enhancements in wear resistance and hardness may be expected. In addition, enhancements such as increases in both electric arc resistance and neutron absorption cross-section may also be possible and relevant to this specific application. Research studies on arc erosion/work function of nanostructured metal and MMC in bulk and coating forms have shown that a finer grain structure leads to reduced work function and, hence, increased arc erosion resistance (Kim, Kwon et al., 2005; Zhang et al., 2006; Rao et al., 2007). Artem'ev (2003) has stated that since the structural units of dispersed nanoparticles are comparable to the wavelength of a slow neutron, the neutron diffusion coefficient may decrease, resulting in an increase in neutron absorption compared to its conventional counterpart powder. To transfer this characteristic to coatings, the key would be to try to attain a structure with uniformly dispersed nanopores to simulate dispersed nanoparticles. It is important to note that depositing B_4C is quite challenging due to issues of decomposition during spraying.

Finally, a weight-saving automotive application that could benefit from the incorporation of a nanostructured coating is Al-MMC brake discs. By using a nanostructured aluminum oxide coating instead of its conventional counterpart, the weight-saving benefits can be retained (e.g. improved fuel economy, better acceleration, exceptional wear resistance, improved braking performance). In addition, however, the nanostructured coating could offer additional benefits such as wear and impact resistance on their discs, thereby extending the life and performance of these components.

Table 19.2 summarizes possible future applications for nanostructured ceramic coatings.

Table 19.2 Summary of possible future applications for nanostructured ceramic coatings

Material	Application	Favourable properties
Oxides, e.g. Al_2O_3	Gas turbine engine components, e.g. knife edge tips and the face of labyrinth seal fins and seal teeth, as well as the blade tips or outer airfoil section of compressor blades	Enhanced adhesion strength, wear resistance and fracture toughness
Ceramics, e.g. TiO_2 and HA	Bio-implants	Higher hardness, greater bond strength, equivalent or higher osteoblast cell proliferation, higher wear resistance
B_4C	Nuclear reactor, e.g. plasma-facing first wall	Superior wear resistance, hardness electric arc resistance and neutron absorption
Al_2O_3	Al-MMC brake discs	Enhanced wear and impact resistance

19.5.2 MMCs

As with the n-ceramic coatings, favourable characteristics of n-MMC coatings include enhanced hardness, wear resistance and toughness. WC-based MMC conventional coatings are widely used for wear resistance; hence, numerous applications can benefit from these coatings in nanostructured form. There are other MMC coating compositions that could also do the same.

Offshore, sub-sea valves are exposed to severe conditions that include erosion, corrosion and high pressures. In extreme cases, valves can be severely eroded within hours of service. To maintain safety and reduce shut-downs, it is necessary to provide protection for such valves. A WC-CrNi coating, found to be corrosion resistant in seawater (Souza and Neville, 2003), may be an interesting coating alternative towards also providing superior abrasion, erosion and/or sliding wear protection in nanostructure form.

The pumps used in the oil and gas industries are often used to transport abrasive solids entrained in fluids. HVOF-applied WC-CoCr coatings have been shown to provide improved service life and reliability on a vertical axis double entry pump with Francis type first stage impeller blades (Cropper et al., 2000). In fact, the authors of that study state that if the coating was applied to the entire runner, the time between overhaul would be increased by at least fourfold. Extraction of oil from the bituminous sands (oil sands) rely on bitumen feed pumps to transport the viscous and erosive fluid. Syncrude introduced a conventional HVOF-applied WC-Co coating on the pump impellers which resulted in a six-fold increase in the life of the pumps, from 3500 service hours to more than 22,000 hours (Lima, 2006). This

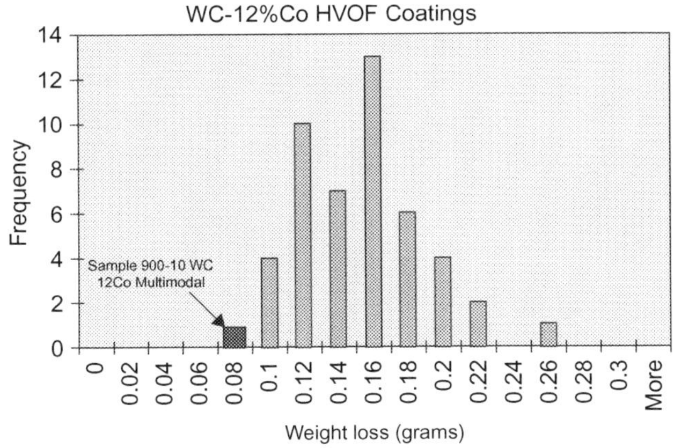

Figure 19.14 Comparative abrasion resistance of WC-12Co coatings from Syncrude Canada. *Source:* Syncrude. Reproduced with permission.

translated into an annual savings of more than $280 000 per pump for Syncrude. Studies and development work were subsequently carried out by third-parties to evaluate n-WC-Co coatings (in multimodal form) (Kim, 2011). Syncrude, with their extensive in-house experience in evaluating surface engineered samples, agreed to perform a three-body abrasion wear test on the n-WC-Co coatings. Their results showed that the n-WC-Co coating out-performed all other thermal sprayed WC-Co coatings in Syncrude's database and even approached the values observed for welded (plasma transferred arc) deposits with similar chemistries (Figure 19.14).

Both WC- and CrC-based MMC coatings are extensively used on gas turbine engine (GTE) components. Their high wear resistance and durability contribute to these coatings' use in all sections of gas turbine engines, including the compressor, combustion, turbine, and afterburner sections. The applications vary from blade tips of the outer airfoil section in the compressor section to bearing surfaces of components found in the turbine and combustion sections. The type of wear exerted on carbide cermet-coated components is mainly fretting (Koiprasert et al., 2004); however, sliding, rubbing, erosion and brush wear are also encountered. For this reason, the enhanced wear properties of n-carbide-based MMC coatings may be attractive for gas turbine engine components.

In addition to the use of n-carbide-based MMC coatings on aero GTEs, other aircraft components benefit from the same coatings. Nanostructured MMC coatings of the same composition as currently applied on each component in conventional form may further extend the life and/or enhance the performance of components such as flap and slat tracks, landing gear and hydraulics, actuators, helicopter drive trains, and propeller assemblies. In fact, there is a good evidence that a n-WC-based MMC coating may further increase the service life of the flap tracks on DHC-8 aircraft beyond the 40,000 cycles (one cycle representing one take-off and landing) provided by conventional WC-based MMC coatings (Ruggiero, 2005). This is based on the improvement in fatigue resistance observed by Ibrahim and Berndt (2004) of n-WC-Co coatings, and the fact that rolling-contact fatigue is a major contributor to the degradation of slat and flap track surfaces (Ahmed and Hadfield, 1997).

Over the past few decades, the replacement of hard chrome plating (HCP) has been a very important environmental concern. Several studies have shown that thermal sprayed (e.g. HVOF) conventional WC-based MMC coatings can provide comparable and in many cases superior properties to HCP without the negative health or environmental implications associated with HCP. Given the superior resistance of n-WC-based MMC coating to impact and sliding/abrasion wear, there is good reason to think that the use of nanostructured coatings could further improve on the results shown by their conventional counterparts, and translate to reduced maintenance and improved reliability.

The automotive industry has also benefited from the use of thermal spray coatings. One very interesting application which has the potential for reducing energy costs and material wear in powertrain components has been studied by several researchers (Ernst and Barbezat, 2008; Gerard, 2006). These studies have demonstrated the advantages of APS-deposited conventional MMC coatings on cylinder liner surfaces of heavy duty diesel truck engines. The advantages include reductions in wear, blow-by and oil consumption. Based on the results observed with n-MMC (i.e. WC-Co) coatings, there may be an opportunity to further improve on the positive attributes of the conventional MMC coating presented by Ernst and Barbezat (2008). In addition to further reducing wear, blow-by and oil consumption, there may also be an additional advantage in attaining a superior finish with reduced depth of cut with the nanostructured MMC coating (Zhang and Liu, 2001). Other automotive components that could benefit from a similar nanostructured coating include cylinder rods, plungers and shafts; diesel valves and bearing areas.

As in the aerospace industry, the automotive sector is interested in weight reduction using light materials and, for this reason, aluminium alloys are gradually being introduced. In many instances the aluminium alloys can replace steel for purposes of matching its strength. However, there is a compromise with respect to wear resistance. To address this issue, a thermal sprayed hard coating such as CrC-NiCr has been applied and tested against wear and corrosion (Wank et al., 2005). According to the authors of that work, the coated prototypes passed both the dynamic and salt fog corrosion tests. Once again, based on the results of nanostructured cermet MMCs, one may expect to improve resistance against wear, impact and localized corrosion by using a nanostructured coating of the same material.

Finally, the refurbishment of worn components is another area where nanostructured coatings could be beneficial. Refurbishment can lead to substantial cost savings. For automotive applications, thermal spray repair of worn shifter forks, crankshafts, bearing regions and transmission shafts have been successfully implemented. By using nanostructured metal coatings, one may be able to, in some cases, surpass the original performance characteristics of the repaired regions with respect to strength, wear and corrosion.

Table 19.3 summarizes possible future applications for nanostructured MMC coatings.

Table 19.3 Summary of possible future applications for nanostructured MMC coatings

Material	Application	Favourable properties
WCCrNi	Offshore, subsea valve	Superior abrasion, erosion, and/or sliding wear protection
WC-CoCr	Oil and gas industry pumps	Superior slurry erosion resistance
WC- and CrC-MMCs	Gas turbine engine components, e.g. blade tips of the outer airfoil section in the compressor section, bearing surfaces of components found in the turbine and combustion sections	Superior wear resistance
WC-based MMCs	Aircraft components, e.g. flap and slat tracks, landing gear, hydraulics, actuators, helicopter drive trains and propeller assemblies	Superior wear and fatigue resistance
WC-based MMCs	Hard chrome replacement	Superior wear resistance, e.g. sliding and abrasion
MMCs	Automotive components, e.g. cylinder liner surfaces of heavy duty diesel truck engines; cylinder rods, plungers, and shafts; diesel valves; and bearing areas	Reductions in wear, blow-by, oil consumption, and superior surface finish
CrC-NiCr	Aluminum alloy automotive components	Improved resistance against wear, impact, and localized corrosion
Metals	Refurbishment of worn components	Superior wear and corrosion resistance

19.6 Summary

Thermal spray nanostructured ceramic coatings were initially evaluated and developed through the US Navy research programmes, and were subsequently successfully applied onto both military and industrial components. The behaviour and attributes of nanostructured MMC coatings, especially the WC-based ones, were more complex than ceramic coatings, and therefore have taken longer to understand. However, research carried out by numerous groups around the world has led to repeatable advancements in this area. As with ceramic coatings, the nanostructured form of MMC coatings offers enhancements in toughness without compromising the coatings' hardness. The ability to attain such new and unique combinations of coating characteristics opens opportunities for enhanced performance and longevity of critical components exposed to harsh environments.

Acknowledgements

The authors thank Jimmy Walker and FW Gartner-Curtiss Wright Surface Technologies for their technical contribution to numerous portions of this chapter. A special thank you goes to Linda Jensen (Juridica Consulting Inc.) for her constructive suggestions in the preparation and review of this manuscript.

References

Ahmed, R., Hadfield, M., 1997. Wear of high-velocity oxy-duel (HVOF)-coated cones in rolling contact. Wear. 203–204, 98–106.

Akdogan, E.K., Balarinni, L., Zhong, Z., Ignatov, A., Croft, M., et al., 2010. 3D quantitative phase analysis and strain mapping of nanostructured coatings by synchrotron energy X-ray diffraction, paper presented at the Twentieth International Offshore and Polar Engineering Conference, Beijing, China.

Artem'ev, V.A., 2003. Estimate of neutron attenuatiuon and moderation by nanostructural materials. At. Energ. 94 (4), 282–285.

Baharvandi, H.R., Talebzadeh, N., Ehsani, N., Aghand, F., 2009. Synthesis of B4C-nano TiB2 composite powder by sol-gel method. J. Mater. Eng. Perform. 18 (3), 273–277.

Begrambekov, L.B., Buzhinskij, O.I., 2007. Features and advantages of boron carbide as a protective coating of the tokamak first wall. Plasma Devices Oper. 15 (3), 193–199.

Cao, X.Q., Vassen, R., Schwartz, S., Jungen, W., Tietz, F., et al., 2000. Spray-drying of ceramics for plasma-spray coating. J. Eur. Ceram. Soc. 20, 2433–2439.

Cellard, A., Garnier, V., Fantozzi, G., Baret, G., Fort, P., 2009. Wear resistance of chromium oxide nanostructured coatings. Ceram. Int. 35, 913–916.

Chen, H., Ding, C., Zhang, P., La, P., Lee, S.W., 2003. Wear of plasma-sprayed nanostructured zirconia coatings against stainless steel under distilled-water conditions. Surf. Coat. Technol. 173, 144–149.

Chwa, S.O., Klein, D., Coddet, C., Bertrand, G., Liao, H., et al., 2004. Mechanical properties of plasma sprayed nanostructured titania coatings with Al addition. Proceedings of Thermal Spray 2004: Advances in Technology and Applications. ASM International, Materials Park, OH, pp. 837–842.

Cropper, M.S., Schmid, R.K., Mohanty, M., 2000. Increasing pump life in abrasive service through state-of-the-art surface protection, In: Proceedings of the Seventeenth International Pump Users Symposium, 87–101. Available from: <http://turbolab.tamu.edu/proc/pumpproc/P17/P1787-101.pdf> (accessed 31.05.13.).

Cunha, C.A., Padial, A.G.F., de Lima, N.B., Martinelli, J.R., Correa, O.V., et al., 2007. Effect of high energy milling parameters on nanostructured Cr_3C_2-Ni20Cr powder characteristics. Proceedings of the Sixth International Latin-American Conference on Powder Technology. SciELO, Sao Carlos, Brazil, pp. 523–529.

Cunha, C.A., de Lima, N.B., Martinelli, J.R., Bressiani, A.H.de.A., Padial, A.G.F., et al., 2008. Microstructure and mechanical properties of thermal sprayed nanostructured Cr_3C_2-Ni20Cr coatings. Mater. Res. 11 (2), 137–143.

Darut, G., Ben-Ettouil, F., Denoirjean, A., Montavon, G., Ageorges, H., et al., 2010. Dry sliding behaviour of sub-micrometer-sized suspension plasma sprayed ceramic oxide coatings. J. Therm. Spray Technol. 19 (1–2), 275–285.

Darut, G., Valette, S., Montavon, G., Ageorges, H., Denoirjean, A., et al., 2010. Comparison of Al2O3 and Al_2O_3-TiO_2 coatings manufactured by aqueous and alcoholic suspension plasma spraying, Proceedings of the International Thermal Spray Conference & Exposition, 197–202.

Ding, Z., Zhang, Y., Zhao, H., 2007. Resistance of HVOF nanostructured WC-12Co coatings to cavitation erosion. Proceedings of Thermal Spray 2007: Global Coating Solutions. ASM International, Materials Park, OH, pp. 633–637.

Eigen, N., Gartner, F., Klassen, T., Aust, E., Bormann, R., et al., 2005. Microstructures and properties of nanostructured thermal sprayed coatings using high-energy milled cermet powders. Surf. Coat. Technol. 195, 344–357.

Ernst, P., Barbezat, G., 2008. Thermal spray applications in powertrain contribute to the saving of energy and material resources. Surf. Coat. Technol. 202, 4428–4431.

Fang, Z., Eason, J.W., 1995. Study of nanostructured WC-Co composites. Int. J. Refract. Met. Hard Mater. 13, 297–303.

Fang, Z.Z., Wang, X., Ryu, T., Hwang, K.S., Sohn, H.Y., 2009. Synthesis, sintering, and mechanical properties of nanocrystalline cemented tungsten carbide – a review. Int. J. Refract. Met. Hard Mater. 27, 288–299.

Fecht, H.J., Hellstern, E., Fu, Z., Johnson, W.L., 1990. Nanocrystalline metals prepared by high-energy ball milling. Metall. Trans. A. 21A, 2333–2337.

Gell, M., Jordan, E.H., Sohn, Y.H., Goberman, D., Shaw, L., et al., 2001. Development and implementation of plasma sprayed nanostructured ceramic coatings. Surf. Coat. Technol. 146–7, 48–54.

Gerard, B., 2006. Application of thermal spraying in the automobile industry. Surf. Coat. Technol. 201, 2028–2031.

Goberman, D., Sohn, Y., Shaw, L., Jordan, E., Gell, M., 2002. Microstructure development of Al_2O_3–13wt.%TiO_2 plasma sprayed coatings derived from nanocrystalline powders. Acta Mater. 50, 1141–1152.

Guilemany, J.M., Dosta, S., Nin, J., Miguel, J.R., 2005. Study of the properties of WC-Co nanostructured coatings sprayed by high-velocity oxyfuel. J. Therm. Spray Technol. 14 (3), 405–413.

He, J., Lavernia, E.J., 2001. Precipitation phenomenon in nanostructured Cr_3C_2-NiCr coatings. Mater. Sci. Eng., A. 301, 69–79.

He, J., Schoenung, J.M., 2002. Nanostructured coatings. Mater. Sci. Eng. A. 336, 274–319.

He, J., Ice, M., Dallek, S., Lavernia, E.J., 2000. Synthesis of nanostructured WC-12 pct Co coating using mechanical milling and high velocity oxygen fuel thermal spraying. Metall. Mater. Trans. A. 31A, 541–553.

Ibrahim, A., Berndt, C.C., 2004. Fatigue and mechanical properties of nanostructured WC-Co coatings. Proceedings of Thermal Spray 2004: Advances in Technology and Applications. ASM International, Materials Park, OH, pp. 878–880.

Jia, K., Fischer, T.E., 1996. Abrasion resistance of nanostructured and conventional cemented carbides. Wear. 200, 206–214.

Jiang, H.G., Ruhle, M., Lavernia, E.J., 1999. On the applicability of the X-ray diffraction line profile analysis in extracting grain size and microstrain in nanocrystalline materials. J. Mater. Res. 14, 549–559.

Jiang, X.-L., Liu, M., 2002. D.C. Plasma-sprayed coatings of nanostructured alumina-titania-silica. Plasma Sci. Technol. 4 (5), 1481–1484.

Jordan, E.H., Gell, M., Sohn, Y.H., Goberman, D., Shaw, L., et al., 2001. Fabrication and evaluation of plasma sprayed nanostructured alumina–titania coatings with superior properties. Mater. Sci. Eng. A. 201, 80–89.

Josephson, M., Larsen, A.H., Larsson, F., 2005. Nanostructured WC-Co coating, Project for the course 'Introduction to Nanotechnology' at Malmo University Sweden, Available from: <http://mixtra.se/nano/PRES-NANO.pdf> (accessed 02.06.05.).

Kabacoff, L.T., 2002. Nanoceramic coatings exhibit much higher toughness and wear resistance than conventional coatings. AMPTIAC Newslett. 6 (1), 37–42.

Karch, J., Birringer, R., Gleiter, H., 1987. Ceramics ductile at low temperature. Nature. 330, 556–558.

Kawakita, J., Shinohara, T., Kuroda, S., Suzuki, M., Sodeoka, S., 2008. Fabrication of nanosized oxide composite coatings and photo-electric conversion/electron storage characteristics. Surf. Coat. Technol. 202, 4028–4035.

Kim, G.E., 2007. Thermal sprayed nanostructured coatings: applications and developments. In: Koch, C.C. (Ed.), Nanostructured Materials Processing, Properties, and Applications, second ed. William Andrew Publishing, Norwich, NY, pp. 91–117.

Kim, G.E., 2010a. Potential for nanostructured coatings for a magnetically confined plasma fusion reactor-first wall application, blog, F.W. Gartner, 8 April. Available from: <http://blog.fwgartner.com/blog/2010/4/8/potential-for-nanostructured-coatings-for-magnetically-confi.html> (accessed 09.04.10.).

Kim, G.E., 2010b. Nanostructured carbide coatings – why and how? blog, F.W. Gartner, 21 April. Available from: <http://blog.fwgartner.com/blog/2010/4/21/nanostructured-carbide-coatings-why-and-how.html> (accessed 22.04.10.).

Kim, G.E., 2011. Effects of carbide size on cermet materials and coatings, Part 2: from bulk to coating, blog, F.W. Gartner, 20 October. Available from: <http://blog.fwgartner.com/blog/2011/10/20/effects-of-carbide-size-on-cermet-materials-and-coatings-par.html> (accessed 21.10.11.).

Kim, G.E., Addona, T., 2011. Synthesis of nanostructured Al alloys via non-cryogenic milling, paper presented at the ONR Program Review on Nanostructured Materials.

Kim, G.E., Scandell, K., 2012. Nanostructured coatings for US navy applications. Adv. Mater. Processes. 170 (2), 34–35.

Kim, G.E., Walker, J., 2007. Successful application of nanostructured titanium dioxide coating for high-pressure acid-leach application. J. Therm. Spray Technol. 16 (1), 34–39.

Kim, G.E., Falzon, J., Walker, J., Franceschini, M., Punga, N., 2010. Use of nano-particle titanium dioxide (n-TiO2) thermal spray coatings for abrasion resistance in severe service applications. Proceedings of the International Titanium Association Conference (DVD). ITA, Northglenn, CO.

Kim, H.-J., Lee, C.-H., Hwang, S.-Y., 2005. Superhard nano WC-12%Co coating by cold spray deposition. Mater. Sci. Eng. A. 391, 243–248.

Kim, J.-S., Kwon, Y.-S., Dudina, D.V., Lomovsky, O.I., Korchagin, M.A., Mali, V.I., 2005. Nanocomposites TiB2-Cu: consolidation and erosion behaviour. J. Mater. Sci. 40 (13), 3491–3495.

Koch, C.C., 1999. Bulk behavior of nanostructured materials. In: Siegel, R.W., Hu, E., Roco, M.C. (Eds.), Nanostructure Science and Technology – R&D Status and Trends in Nanoparticles, Nanostructured Materials, and Nanodevices. Kluwer Academic Publishers, The Netherlands, pp. 93–112.

Koch, C.C., 2003. Top-down synthesis of nanostructured materials: mechanical and thermal processing methods. Rev. Adv. Mater. Sci. 5, 91–99.

Koiprasert, H., Niranatlumpong, P., Dumrongrattana, S., 2004. Thermally sprayed coating for protection of fretting wear. J. Sci. Res. Chula. Univ. 29 (2), 105–118.

Li, J.F., Liao, H., Wang, X.Y., Normand, B., Ji, V., et al., 2004. Improvement in wear resistance of plasma sprayed yttria stabilized zirconia coating using nanostructured powder. Tribol. Int. 37, 77–84.

Lima, R.S., 2006. Success story – HVOF Application of WC-Co on bitumen feed pumps. Available from: <http://archive.nrc-cnrc.gc.ca/eng/ibp/imi/research/surface-engineering.html> (accessed 21.01.07.).

Lima, R.S., Marple, B.R., 2005. Enhanced ductility in thermally sprayed titania coating synthesized using a nanostructured feedstock. Mater. Sci. Eng. A. 395, 269–280.

Lima, R.S., Khor, K.A., Li, H., Cheang, P., Marple, B.R., 2005. HVOF spraying of nanostructured hydroxyapatite for biomedical application. Mater. Sci. Eng. A. 396, 181–187.

Lima, R.S., Marple, B.R., Li, H., Khor, K.A., 2006. Biocompatible nanostructured high-velocity oxyfuel sprayed titania coating: deposition, characterization, and mechanical properties. J. Therm. Spray Technol. 15 (4), 623–627.

Lin, Y., Yao, B., Zhang, Z., Li, Y., Sohn, Y., et al., 2012. Strain energy during mechanical milling: part I. mathematical modeling. Metall. Mater. Trans. A. 43A, 4247–4257.

Lindelov, J.S., Wahlberg, M., 2009. Spray drying for processing of nanomaterials. In: Proceedings of Nanosafe 2008: International Conference on Safe Production and Use of Nanomaterials, Journal of Physics: Conference Series 170: 012027.

Liu, S., Sun, D., Fan, Z., Yu, H.-Y., Meng, H.-M., 2008. The influence of HVAF powder feedstock characteristics on the sliding wear behaviour of WC-NiCr coatings. Surf. Coat. Technol. 202, 4893–4900.

Luton, M.J., Jayanth, C.S., Disko, M.M., Matras, S., Vallone, J., 1988. Cryomilling of nano-phase dispersion strengthened aluminum. MRS Proc. 132, 79–86.

Mattox, D.M., Mullendore, A.W., Pierson, H.O., Sharp, D.J., 1979. Low Z coatings for fusion reactor applications. J. Nucl. Mater. 85 & 86, 1127–1131.

Oleszak, D., Shingu, P.H., 1996. Nanocrystalline metals prepared by low energy ball milling. J. Appl. Phys. 79, 2975–2980.

Oliker, V.E., Terent'ev, A.E., Shvedova, L.K., Martsenyuk, I.S., 2009. Use of aqueous suspensions in plasma spraying of alumina coatings. Powder Metall. Met. Ceram. 48 (1–2), 115–120.

Osiko, V.V., 2009. Extra-strong wear-resistant materials based on nanostructured crystals of partially stabilized zirconium dioxide. Mendeleev Commun. 19, 117–122.

Padial, A., Cunha, C.A., de Lima, N.B., Martinelli, J.R., Correa, O.V., et al., 2008. Structure and properties of thermal sprayed nanostructured Cr_3C_2–25(NiCr) coatings. 18° CBECiMat - Congresso Brasileiro de Engenharia e Ciência dos Materiais, Porto de Galinhas, PE, Brasil. SciELO, Sao Carlos, Brazil.

Pappalardo, J., 2004. US Navy finding new applications for advances in nanotechnology, National Defense Magazine, October. Available from: <http://www.nationaldefensemagazine.org/archive/2004/October/Pages/US_Navy_Finding3402.aspx?PF=1> (accessed 05.12.06.).

Rao, L., Munz, R.J., Meunier, J.-L., 2007. Vacuum arc velocity and erosion rate measurements on nanostructured plasma and HVOF spray coatings. J. Phys. D Appl. Phys. 40, 4192–4201.

Ruggiero, P.F., 2005. High velocity oxygen fuel coatings: a way to extend component life. Available from: <http://www.aviationpros.com/article/10385811/high-velocity-oxygen-fuel-coatings-a-way-to-extend-component-life?print = true> (accessed 04.06.13.).

Shen, T.D., Koch, C.C., 1995. Influence of dislocation structure on formation of nanocrystals by mechanical attrition. Mater. Sci. Forum. 179–181, 19–24.

Shipway, P.H., McCartney, D.G., Sudaprasert, T., 2005. Sliding wear behaviour of conventional and nanostructured HVOF sprayed WC-Co coatings. Wear. 259, 820–827.

Siegel, R.W., Kear, B., 2000. Consolidated nanostructures. In: Roco, M.C., Williams, R.S., Alivisatos, P. (Eds.), Nanotechnology Research Directions – Vision for Nanotechnology in the Next Decade. Kluwer Academic Publishers, The Netherlands, pp. 97–106.

Smirnov, N.I., Prozhega, M.V., Smirnov, N.N., 2007. Study of tribological properties of detonation nanostructured WC-Co-based coatings. J. Frict. Wear. 28 (2), 200–204.

Souza, V.A.D., Neville, A., 2003. Linking electrochemical corrosion behaviour and corrosion mechanisms of thermal spray cermet coatings (WC-CrNi and WC/CrC-CoCr). Mater. Sci. Eng. A. 352, 202–211.

Tingaud, O., Bertrand, P., Bertrand, G., 2010. Microstructure and tribological behaviour of suspension plasma sprayed Al_2O_3 and Al_2O_3-YSZ composite coatings. Surf. Coat. Technol. 205, 1004–1008.

Tsakalakos, T., Croft, M., Zhong, Z., Kao, C.-C., Jisrawi, N., et al., 2008. Strain mapping in nanostructured coatings by synchrotron probe, paper presented at the International Offshore and Polar Engineering Conference 2008, Vancouver, Canada.

Turunen, E., Varis, T., Gustafsson, T.E., Keskinen, J., Falt, T., et al., 2006. Parameter optimization of HVOF sprayed nanostructured alumina and alumina–nickel composite coatings. Surf. Coat. Technol. 200 (16–17), 4987–4994.

Unger, R., 1987. Comparison of thermal spray bond coats. Proceedings of the National Thermal Spray Conference. ASM International, Materials Park, OH, pp. 365–370.

Varis, T., Knuuttila, J., Turunen, E., Leivo, J., Silvonen, J., et al., 2007. Improved protection properties by using nanostructured ceramic powders for HVOF coatings. J. Therm. Spray Technol. 16 (4), 524–532.

Viswanathan, V., Rea, K.E., Vaidya, A., Seal, S., 2008. Role of spray drying of nanoagglomerates in morphology evolution in nanostructured APS coatings. J. Am. Ceram. Soc. 91 (2), 379–386.

Wang, Q., Chen, Z.H., Ding, Z.X., 2009. Performance of abrasive wear of WC-12Co coatings sprayed by HVOF. Tribol. Int. 42, 1046–1051.

Wang, Y., 2004. Nano- and submicron-structured sulfide self-lubricating coatings produced by thermal spraying. Tribol. Lett. 17 (2), 165–168.

Wank, A., Wielage, B., Grund, T., Rupprecht, C., Angermann, K., et al., 2005. Local wear protection for demountable automotive aluminum draw bars by thermal spray coatings. ITSC 2005: Thermal Spray Connects: Explore its Surfacing Potential! ASM International, Metals Park, OH, pp. 118–121.

Williams, J., Kim, G.E., Walker, J., 2004. Ball valves with nanostructured titanium oxide coatings for high-pressure acid-leach service: development to application. Proceedings of Pressure Hydrometallury 2004 (DVD). Canadian Institute of Mining, Metallurgy and Petroleum, Canada.

Witkin, D.B., Lavernia, E.J., 2006. Synthesis and mechanical behavior of nanostructured materials via cryomilling. Prog. Mater Sci. 51, 1–60.

Yao, Z., Stiglich, J.J., Sudarshan, T.S., n.d. Nano-grained tungsten carbide-cobalt (WC/Co). Available from: <http://www.matmod.com/publications/armor_1.pdf> (accessed 23.04.02.).

Yao, Z., Stiglich, J.J., Sudarshan, T.S., 1998. Nanosized WC-Co holds promise for the future. Met. Powder Rep. 53 (3), 26–33.

Yandouzi, M., Bu, H., Brochu, M., Jodoin, B., 2012. Nanostructured Al-based metal matrix composite coating production by pulsed gas dynamic spraying process. J. Therm. Spray Technol. 21 (3–4), 609–619.

Ye, J., Schoenung, J.M., 2004. Technical cost modeling for the mechanical milling at cryogenic temperature (cryomilling). Adv. Eng. Mater. 6 (8), 656–664.

Zha, B., Wang, H., Su, X., 2004. Nano structured WC-12Co coatings sprayed by HVO/AF. Proceedings of Thermal Spray 2004: Advances in Technology and Applications. ASM International, Materials Park, OH, pp. 881–883.

Zhang, B., Liu, X.B., 2001. An investigation of microgrinding of nanostructured material coatings, paper presented at the UEF Conference on Novel Synthesis and Processing of Nanostructured Coatings for Protection Against Degradation, Davos, Switzerland.

Zhang, C., Yang, Z., Ding, B., 2006. Low electrode erosion rate of nanocrystalline CuCr-50 alloy in vacuum. Mod. Phys. Lett. B. 20 (21), 1329–1334.

Zhu, Y.-C., Yukimura, K., Ding, C.-X., Zhang, P.-Y., 2001a. Tribological properties of nanostructured and conventional WC-Co coatings deposited by plasma spraying. Thin Solid Films. 388, 277–282.

Zhu, Y.C., Ding, C.X., Yukimura, K., Xiao, T.D., Strutt, P.R., 2001b. Deposition and characterization of nanostructured WC-Co coating. Ceram. Int. 27, 669–674.

Zorawski, W., Skrzypek, S., Trpcevska, J., 2008. Tribological properties of hypersonically sprayed carbide coatings. FME Trans. 36, 81–86.

Thermally sprayed nanostructured coatings for anti-wear and TBC applications: state-of-the-art and future perspectives

20

Giovanni Di Girolamo and Emanuele Serra
ENEA, UTTMAT, Casaccia Research Centre, Italy

20.1 Introduction

Many metal components of aircraft engines, and land-based and marine turbines are exposed to severe operating conditions, including high temperatures, thermal fatigue, wear, erosion and hot-corrosion which, through different damaging mechanisms, reduce their service performance and life. Replacement and repair of damaged components also involve significant downtimes and costs. To this end, it is well accepted that the application of protective surface coatings makes it possible to increase the durability and, at the same time, provides new perspectives for materials and components with better capability and performance, without compromising the original strength and the integrity of the same structural components (National Research Council, 1996). Several advantages are then expected in terms of higher efficiency and lower environmental impact.

Thermal spraying is a cost-effective industrial technology for depositing thick coatings on large areas and complex components starting from ceramic, metal and composite powder feedstocks, because of its high deposition rates and low manufacturing times (Davis, 2004). However, thermal spraying does not always allow to deposit and thin coatings as well as coatings on internal surfaces, due to the 'line-of-sight' uniform problem.

The use of nanostructured raw materials in conjunction with the search for new compositions allows the development of coatings with better strength, hardness and functional properties than those of their conventional counterparts (He and Schoenung, 2002). Starting from this point of view, much research has recently focused on nanostructured composite coatings for anti-wear applications and ceramic thermal barrier coatings (TBC) for hot section components of aircraft and land-based turbine engines. A summary of previous works focused on these subjects is reported in Table 20.1.

Thermal spraying is a reliable technique for processing nanostructured materials, due to the short residence time of the sprayed particles at high temperature. However, when nanostructured powders are employed, they have to be only

Anti-Abrasive Nanocoatings.

Table 20.1 Some examples of applications of thermally sprayed nanostructured coatings studied in previous research

Application	Coating materials
Wear resistance	Composites: WC-Co (He and Schoenung, 2002; Stewart et al., 1999; Shipway et al., 2005; Zhao et al., 2006; Sanchez et al., 2010; Qiao et al., 2001), WC-Co-Cr (Ghabchi et al., 2010), Cr_3C_2-NiCr (He and Schoenung, 2002; Sasaki et al., 1992); ceramics: Al_2O_3-TiO_2 (Shaw et al., 2000; Sanchez et al., 2011; Gell et al., 2001; Ahn et al., 2006; Wang et al., 2000; Lin et al., 2004), TiO_2 (Lima and Marple, 2007)
Thermal insulation	Ceramics: ZrO_2-Y_2O_3 (Di Girolamo et al., 2011; Lima and Marple, 2008; Liang and Ding, 2005; Kassner et al., 2009; Guignard et al., 2012; Gell et al., 2004), ZrO_2-Y_2O_3-CeO_2 (Jin et al., 2012; Latka et al., 2013), $La_2Zr_2O_7$ (Chen et al., 2009)

partially melted to retain part of the starting nanostructure embedded in the as-deposited coating (Shaw et al., 2000).

Therefore, thermal spraying parameters have to be properly adjusted in order to minimize grain growth and some detrimental phase changes associated with the high temperature experienced by the sprayed particles. Indeed, when the latter are fully melted the coatings tend to behave like their conventional counterparts (Lima and Marple, 2007). Therefore, the particles should experience a temperature slightly above their melting point, in order to guarantee particle adhesion and cohesion as well as coating mechanical integrity.

When these recommendations are followed, nanostructured coatings with unique microstructure and enhanced functional characteristics can be properly manufactured.

The recent findings and trends in the development of advanced thermally sprayed nanostructured coatings for anti-wear and thermal barrier coating applications are reviewed in the sections that follow and, for completeness, their functional properties are presented and compared to those of their conventional counterparts.

20.2 Thermal spraying processes

In thermal spraying processes a heat source is employed to melt the raw material, generally in the form of powder particles. These particles are fed by a carrier gas to the torch, injected in the flame/plasma jet and heated above their melting point. The molten droplets are then accelerated by a gas stream toward a substrate where they impact, and are flattened and quenched to room temperature, thus producing the deposition of overlapped splats and, finally, the build-up of a coating. Both the morphology and the size of the powder particles affect their melting state and then the microstructural, physical and mechanical properties of the as-deposited coatings.

In order to deposit nanostructured coatings, it should be noted that single nanoscale particles cannot be processed by thermal spraying, because they are unable to be carried in a moving gas stream and deposited on a substrate, due to their low mass and flowability. Indeed, they can stick to the sidewall of feed lines and spray nozzle tip and then they are easily dispersed and ablated by flame or plasma stream. In addition, their transport would require very high carrier gas flow, thus destabilizing the flame or the plasma jet.

Therefore, these nanoparticles are usually pre-agglomerated in spherical micron-sized granules by spray drying process, in order to obtain particles with a diameter between 5 and 100 μm, depending on the atomization parameters and on the slurry characteristics. A slurry (water-based or organic solvent-based suspension) is atomized into droplets by using a centrifugal atomizer in a chamber through which heated gases are passed. The liquid is evaporated and the solid is collected continuously from the chamber (Houck, 1985). Therefore, particles with well-determined shape, size and density can be produced at low cost.

Much thermal spraying equipments are commercially available for coating deposition, depending on the heat source, input energy and spraying environment employed. The most suitable deposition method for a given application is also related to the intrinsic characteristics of the raw materials employed, as well as to the functional requirements and to the costs for the equipment. High-velocity oxy-fuel (HVOF) and plasma spraying (PS) are mainly used for industrial production, due to their relatively high performance-to-cost ratio as well as to the high quality of the as-produced coatings.

20.2.1 HVOF spraying

In an HVOF spraying system a combustion jet is produced into a high-pressure chamber by the reaction between the oxygen and the gas (propylene, propane or hydrogen) or liquid (kerosene) fuel at temperatures between 2500 and 3000°C. A supersonic gas stream is generated through a confined cooled convergent-divergent nozzle. The powders are injected into the gas jet, heated and propelled toward the substrate at speed of about 1800 ms^{-1}.

HVOF allows coatings with high microstructural density and bonding strength to be deposited, so it is particularly suitable for fabricating dense ceramic and composite coatings for anti-wear applications as well as metal coatings for high-temperature environmental protection against oxygen and highly corrosive media.

20.2.2 Plasma spraying

In PS a high-temperature plasma is generated by transferring energy into an inert gas until the energy level is sufficient to ionize the gas, allowing the electrons and ions to act independently of one another. The plasma-forming gas mixture is introduced through the gun and enters the arc chamber through a gas injector that imparts a spin or vortex flow. Then it is accelerated as it expands and comes out of the nozzle.

The powder particles are fed by an inert gas, injected in the plasma jet, melted and accelerated in a gas stream towards a substrate to be coated. When they

Figure 20.1 Air plasma spraying equipment during operation.

impinge on the surface of the same substrate, they are flattened and form splats or lamellae that solidify at high quenching rate, thus producing the build-up of a coating with a lamellar microstructure.

Figure 20.1 shows plasma jet and spray particle stream during operation. Plasma spraying is particularly suitable for depositing high melting-point materials, such as ceramic oxides for porous thermal barrier and abradable coatings.

More recently, atmospheric suspension and solution precursor plasma spraying (SPS and SPPS) systems have been proposed for the fabrication of ceramic TBC with a very fine microstructure and enhanced service lifetime compared to that of the same coating produced by conventional atmospheric plasma spray (APS).

In turn, in vacuum plasma spraying (VPS) a low-pressure spraying chamber is employed to reduce the interaction between the sprayed particles and the surrounding environment, which can promote common detrimental effects such as oxidation in the case of metal particles. However the in-vacuum deposition is very time consuming, because for each chamber load the vacuum must be established and refilled with partial pressure inert gas, then coating deposition has to be carried out and followed by cooling down in the vacuum atmosphere.

20.3 Typical nanostructured coatings for technological applications

As previously highlighted, the investigations into thermally sprayed nanostructured coatings are mainly focused on two industrial applications: composite-based anti-wear systems and ceramic thermal barriers, due to their enhanced microstructural, mechanical, thermal and functional properties.

Table 20.2 Typical thermally sprayed nanostructured coatings for technological applications

Type of coatings	Material	Processing
Cermet coatings for anti-wear applications	WC-Co, WC-Co-Cr, Cr_3C_2-NiCr	APS, VPS, HVOF
Composite coatings for anti-wear applications	Al_2O_3-TiO_2	APS, VPS, HVOF
Ceramic TBC	ZrO_2-Y_2O_3 (YSZ), ZrO_2-Y_2O_3-CeO_2 (CYSZ), rare earth zirconates ($La_2Zr_2O_7$, $Sm_2Zr_2O_7$, etc.)	APS, SPS, SPPS

As schematized in Table 20.2, cermet and ceramic coatings, typically characterized by high hardness and toughness, are mainly addressed to protect machine mechanical parts from adhesive and abrasive wear, while ceramic materials with low thermal conductivity are devoted to the development of next-generation TBC for protection of stator blades, vanes and combustion chamber tiles of turbine engines against high temperatures, oxidation, hot corrosion and erosion.

20.3.1 Cermet coatings for anti-wear applications

Cermet coatings are advanced materials composed of two phases: ceramic and metal. They are usually employed for anti-wear applications, because the hardness of the ceramic phase (WC, Cr_3C_2, TiC) and the toughness of the metallic binder (Co, Ni, NiCr, CoCr) are properly matched, thus providing high wear resistance.

WC-Co coatings are commonly adopted to protect fan blades and flanges of jet engine parts, rocker arms, piston rings, cylinder lines, compressor air seals and valve seats, dovetail interlocks, discs, shafts, casings and nozzle wear pads of gas turbines. They are generally used in cold sections because their tribological properties dramatically drop as the temperature increases due to the formation of ternary carbides. Indeed, during the vacuum heat treatment, part of the primary WC and amorphous mixed phases, commonly detected in as-deposited coatings, transforms to η-carbides ($Co_xW_yC_z$), thus increasing hardness and brittleness, and reducing the fracture toughness and the wear resistance (Khameneh Asl et al., 2006). The effects of phase changes become more significant with increasing the temperature of the heat treatment from 800 to 1100°C. It has been reported that after thermal exposure at 1100°C in a vacuum the hardness of as-deposited coatings increased by a factor of about 20%, while the decrease of fracture toughness, measured using the indentation method, was of about 33%. The fracture toughness significantly affects the wear resistance, and the hardness to a lesser extent. The thermal exposure in air environment further restricts the service temperature of WC-Co coatings due to the high-temperature oxidation and to the subsequent formation of brittle ternary oxides. Figure 20.2 shows agglomerated

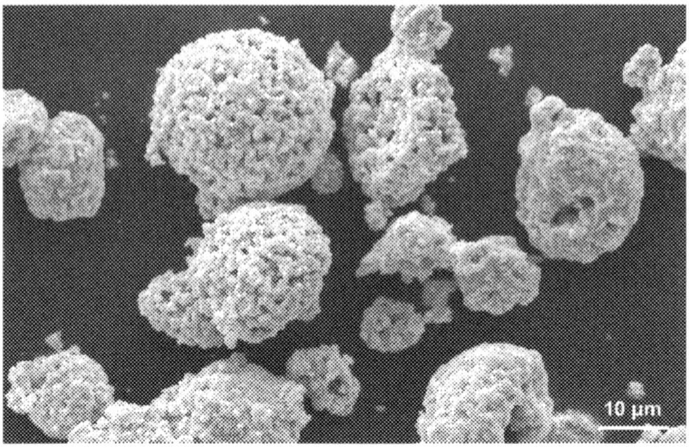

Figure 20.2 Agglomerates of WC-Co particles.

Figure 20.3 A detail of WC-Co powder feedstock showing nanostructured carbides embedded in a Co matrix.

nanostructured WC-Co powder particles, while Figure 20.3 shows carbides with submicrometre size and different morphologies.

It is worth noting that when carbide particles are thermally sprayed to produce coatings they are commonly subjected to some detrimental phase changes which affect the microstructure as well as the mechanical and tribological properties of the same coatings. During spraying, decarburization typically occurs due to the high temperature experienced by the molten particles and the oxidizing environment, which assist the dissolution of carbides in the metal matrix and the carbon loss, promoting the partial depletion of the WC phase and the appearance of new crystalline and amorphous phases (Chen and Hutchings, 1998). Looking at PS, some chemical reactions occur in the plasma core and take place during the flight of the molten droplets and their quenching at the substrate surface. It is well known that the wear resistance decreases with increasing the degree of decarburization,

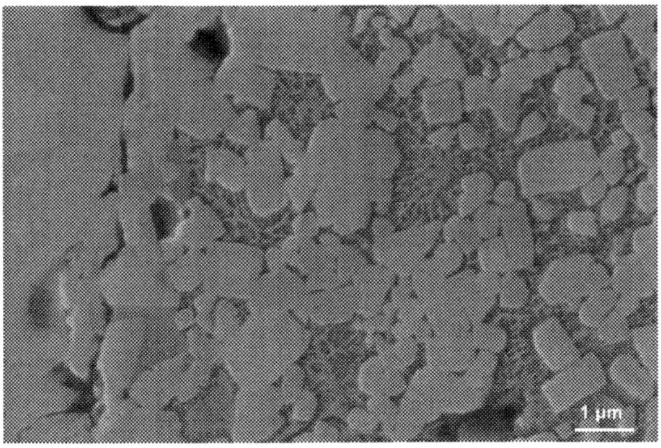

Figure 20.4 Cross-sectional scanning electron microscope (SEM) microstructure of plasma sprayed nanostructured WC-Co coating showing carbides with different sizes embedded in a Co-rich matrix with dendritic nanostructures.

which is proportional to the heat transferred to the powder particles in the plasma core (Stewart et al., 1999; Shipway et al., 2005).

The micrograph in Figure 20.4 presents the cross-sectional microstructure of nanostructured WC-Co coating deposited by air plasma spraying using an Ar-He process gas mixture. The effects of grain growth and carbide decarburization are partially restrained, even if carbide grains with a different stage of coalescence can be observed in the picture. Some strip- and square-shaped structures, approximately in the range between 500 nm and 1 μm probably derive from a second recrystallization during successive torch passes. The carbide grains which are partially dissolved and mixed with the Co-matrix produce ultrafine Co-rich dendritic structures.

Table 20.3 shows the average hardness data for conventional and nanostructured WC-Co coatings produced by APS and HVOF, and measured by indentation.

Sanchez et al. (2011) have obtained similar hardness values for nanostructured and conventional coatings deposited by APS, while in our previous experiments (Di Girolamo et al., 2013) we have found that nanostructured coatings are less hard than microstructured ones due to their higher degree of decarburization and lower retention of primary carbides, associated with higher surface-to-volume ratio of nanostructured agglomerates with respect to fused and crushed micronsized particles. The mechanical properties are meaningfully related to the process parameters and to the final microstructure of the coating, that is, the presence of pores, splat boundaries, microcracks and different phases.

The best results have been obtained when Ar-He plasma gas forming mixtures have been employed, because these involve lower plasma energy and temperature in comparison with conventional $Ar\text{-}H_2$ mixtures as well as higher plasma density, thus reducing the air entrainment in the plasma core and the carbide dissolution. To this purpose, it should be remarked that the dissolution of carbides in the metal matrix starts in the plasma jet core and it keeps going in the first tens of millimetres of the plasma plume. Some reactions with the surrounding oxygen can occur during particle

Table 20.3 Microhardness data for conventional (conv.) and nanostructured (nano) WC-Co coatings produced by APS and HVOF

Coating [Ref]	Hardness$_{300g}$ [GPa]	Processing
Conv.† (Sanchez et al., 2011)	7.47	APS
Nano† (Sanchez et al., 2011)	7.67	APS
Conv. (Di Girolamo et al., 2013)	11.43	APS
Nano (Di Girolamo et al., 2013)	8.33	APS
Conv. (Stewart et al., 1999)	11.10	HVOF
Nano (Stewart et al., 1999)	11.90	HVOF
Conv. (Guilemany et al., 2005)	12.12	HVOF
Nano (Guilemany et al., 2005)	15.38	HVOF
Nano* (Baik et al., 2007)	12.07	HVOF
Nano* (Baik et al., 2007)	12.73	HVOF
Conv.** (Ghabchi et al., 2010)	10.67	HVOF
Nano** (Ghabchi et al., 2010)	11.25	HVOF
Conv.** (Ghabchi et al., 2010)	12.41	HVOF
Nano** (Ghabchi et al., 2010)	12.80	HVOF

†Measured at 200 g.
*Co-Co matrix.
**Co-Cr matrix.

flight and solidification at substrate surface. Increased decarburization is associated to the significant reduction in the content of primary carbides (WC) and to the formation of secondary carbides, such as W_2C and WC_{1-x}, ternary carbides like $Co_xW_yC_z$ and amorphous phases. All these phases are usually characterized by high brittleness.

HVOF can be successfully used to deposit coatings with improved bonding strength and high density, which result in both enhanced hardness and anti-wear performance. Indeed, HVOF-sprayed coatings exhibit a lower degree of carbide decomposition in comparison with air plasma sprayed ones, because the sprayed powder particles experience lower temperature and lower dwell time in the flame. HVOF spraying allows fabrication of coatings with reduced porosity (1–2% or lower) and exhibiting more homogeneous distribution of carbides (Eidelman and Yang, 1997). Other investigators found that HVOF-sprayed nanostructured coatings are harder than conventional ones (Stewart et al., 1999; Guilemany et al., 2005).

These different results can be explained in terms of the different powder feedstock and processing parameters employed. To this purpose, the addition of a second metallic binder phase, that is, the application of a thin protective metal surface layer on the spray dried porous WC-Co particles by immersion in cobalt hydrate sol and subsequent hydrogen reduction, is able to reduce the carbon loss during spraying, thus leading to less decarburized coatings. The coatings obtained using this modified powder feedstock exhibited higher hardness than those obtained from as-received ones (12.73 GPa against 12.07 GPa) (Baik et al., 2007). In turn, the data published by Ghabchi et al. (2010) referring to WC-CoCr coatings, report that the nanostructured coatings are denser (their porosity is about 0.7–0.8% against 2% of the microstructured ones) and harder than the microstructured ones produced using the same processing parameters. The addition of Cr to the metallic binder is addressed to increase the corrosion resistance of these coatings, because this technical requirement is particularly appreciated in those applications where both the corrosion and the wear resistance are critical. At the end of their work the authors deduced that the HVOF processing parameters and the fuel-to-oxygen ratio at a higher extent significantly affected the final properties of the coatings.

However, it is worth noting that the wear resistance of thermally sprayed coatings is not only affected by their hardness, but deeply depends on their fracture toughness. Nanostructured coatings showing a homogeneous dispersion of carbides in the metal binder phase generally have exhibited enhanced wear resistance with respect to that of microstructured analogous coatings because of their better toughness and higher resistance to form cracks under thermo-mechanical stresses (Zhao et al., 2006; Zhu et al., 2001; Kim et al., 2006). The performances of several conventional and nanostructured WC-Co coatings produced by APS and HVOF systems, and tested by sliding and abrasive wear tests are presented in Table 20.4. The anti-wear performance is significantly influenced by the powder feedstock, the processing parameters and the testing conditions (mainly applied load and counterpart) employed.

In previous experiments we studied the effect of plasma gas mixture on the wear resistance of air plasma sprayed conventional and nanostructured WC-Co coatings. The coatings sprayed using traditional Ar-H_2 gas mixtures were highly decarburized, less hard and tougher, and thus their wear resistance was lower than that of the same coatings deposited using Ar-He gas mixtures (Di Girolamo et al., 2013). A similar wear rate was found for conventional and nanostructured coatings sprayed using Ar-He gas mixture ($4.4 \times 10^{-6}\,mm^3\,N^{-1}\,m^{-1}$ against $4.5 \times 10^{-6}\,mm^3\,N^{-1}\,m^{-1}$) and tested during sliding wear tests at 91 N load, using an Al_2O_3-coated ring as counterpart in a block-on-ring configuration. As discussed above, the nanostructured coatings were softer than the conventional ones, due to their higher degree of decarburization associated with their higher surface-to-volume ratio, but exhibited higher toughness and bonding strength.

Sanchez et al. (2010) have also studied the wear properties of conventional and nanostructured WC-Co coatings produced by APS using different process gas mixtures. The wear rate increased with load for the conventional coatings and decreased for nanostructured ones. By considering the coatings sprayed using Ar-He gas mixtures, the nanostructured coatings were harder and tougher than the conventional ones, so that they exhibited higher wear resistance when tested during sliding wear tests at different loads (20 and 50 N) using a Si_3N_4 ball as counterpart. In turn, when an Al_2O_3 ball was employed as counterpart, their wear resistances were comparable at 20 N, while at 50 N the nanostructured coatings exhibited better anti-wear performance.

Looking at other investigations focused on HVOF sprayed coatings, Shipway et al. (2005) found that the nanostructured coatings were less wear resistant than the conventional ones tested during ball-on-disk wear tests using an Al_2O_3 ball as counterpart. Otherwise, according to the studies of Zhu et al. (2001) the nanostructured coatings possessed higher wear resistance. They performed sliding wear tests

Table 20.4 **Sliding and abrasive wear rates of conventional (conv.) and nanostructured (nano) WC-Co coatings produced by APS and HVOF and tested at various conditions**

Coating[Ref]	Sliding wear test	Load [N]	Counterpart	Sliding wear rate [10^{-6} mm^3 N^{-1} m^{-1}]	Abrasive wear test	Load [N]	Counterpart	Abrasive wear rate [10^{-2} mm^3 N^{-1} m^{-1}]	Processing
Conv. (Di Girolamo et al., 2013)	Block on ring	91	Al_2O_3	4.4	–	–	–	–	APS
Nano (Di Girolamo et al., 2013)	Block on ring	91	Al_2O_3	4.5	–	–	–	–	APS
Conv. (Shipway et al., 2005)	Ball on disk	19–35	Al_2O_3	0.1–0.3	–	–	–	–	HVOF
Nano (Shipway et al., 2005)	Ball on disk	19–35	Al_2O_3	1.9–1.1	–	–	–	–	HVOF
Conv. (Zhu et al., 2001)	Ball on disk	20–80	Al_2O_3	0.7–1.6	–	–	–	–	HVOF
Nano (Zhu et al., 2001)	Ball on disk	20–80	Al_2O_3	0.3–0.8	–	–	–	–	HVOF
Nano (Qiao et al., 2001)	Ball on disk	9.8	Si_3N_4	0.2–0.6	Pin on disk	2–17	SiC	0.5–1.1	HVOF
Nano (Liu et al., 2002)	Ball on disk	9.8	Si_3N_4	0.19–0.38	Pin on disk	7	SiC	0.78–1.50	HVOF

at different loads (from 20 to 80 N), but no linear dependence of wear rate with load was detected for both the conventional and the nanostructured coatings. The resistance of nanostructured coatings remained rather constant up to 60 N and then decreased, whereas the wear resistance of the conventional coatings decreased between 20 and 40 N and then increased. Moreover, the wear rate of the nanostructured coatings between 40 and 60 N was of about one-sixth of that exhibited by the conventional coatings, and half of that of the conventional coatings at 20 and 80 N.

A Si_3N_4 counterpart was employed in the experiments published by Qiao et al. (2003). The results discussed by these investigators are affected by both the fuel chemistry and the grain size of the starting powder feedstock. The wear resistance was strongly affected by the porosity and the hardness of the as-deposited coatings. The best anti-wear performance was obtained when a neutral flame was used (fuel-to-oxygen ratio = 0.23), while lower wear resistance was achieved using both oxidizing and reducing environments.

The effect of fuel and the fuel-oxygen ratio has been also investigated by Liu et al. (2002). They found that the sliding and abrasive wear resistance of nanostructured coatings could be improved using C_3H_6 as fuel rather than H_2. In this case a good correlation between hardness and wear resistance was detected.

It has been shown that in the conventional coatings the mechanism of wear occurs as follows. The application of a load on the coating surface promotes the removal of the soft metallic binder phase, followed by the fragmentation or uprooting of carbide grains. The absence of grooves in the wear scar suggests that the fracture is typically brittle without any plastic deformation. It has been demonstrated that with increasing the surface load the carbide fragmentation becomes more severe and the detached carbides act as abrasives, promoting a three-body wear mechanism which involves remarkable subsurface cracking and detachment of entire splats from the coating surface (Zhu et al., 2001).

In contrast, the damaging of nanostructured coatings under wear is different: under mild testing conditions plastic deformation typically occurs on the coating surface and the wear on a microscale produces a smooth surface. The coating works as a hard ductile material. The propagation of cracks assists delamination or spallation, which are more significant at higher loads due to higher shear stresses suffered by the subsurface of the same coating. Grooving is more important in nanostructured coatings, where WC grain scratching and fragmentation as well as subsurface cracking are generally restricted. Therefore, according to Zhu et al. (2001) severe adhesion dominates the wear of conventional coatings, while mild adhesion and frictional polishing generally govern the wear damage of the nanostructured coatings.

Otherwise, according to the studies published by Sanchez et al. (2010) subsurface cracking is the main wear mechanism for both the conventional and the nanostructured coatings, while the removal of the Co binder as well as the fragmentation and the pull-out of WC grains are only secondary effects. The contact between the asperities produces high local stresses, so that wear particles are generated by fatigue propagation of cracks. The formation and propagation of subsurface cracks lead to the removal of large coating areas. A primary crack generally appears at the surface and propagates through the tungsten-rich binder phase or along the splat boundaries (Shipway et al., 2005). Then a secondary crack grows from the primary one or the primary crack can connect with an existing subsurface crack. When the resulting crack reaches the surface again a wear particle is released. This mechanism occurs more frequently in highly decarburized coatings because of the highest content of W-rich binder phase, while the removal of Co-rich binder phase is predominant in the coatings showing higher content of primary WC. In substance wear damaging generally follows a three-body abrasive mechanism. Moreover, there were no significant differences in the wear mechanism when different ball materials were used, even if groove formation was more significant when a Si_3N_4 ball was employed, while WC fragmentation was more important when an Al_2O_3 ball was used, probably due to the highest hardness of the silicon nitride ball compared to the alumina ball.

Similar remarks can be made about conventional abrasive wear tests where large particles of abradant are employed to produce severe wear. Abrasive wear is generally 50,000 higher than sliding wear (Qiao et al., 2001) due to the size and the quantity of particles rolling in between the surfaces in motion. Abrasive particles are much larger than the debris produced by sliding wear. It has been shown that the wear rate increases with the abrasive size. During conventional abrasive wear tests the abradant used (Al_2O_3) is usually harder than the coating and promotes removal of material and groove formation on the flat surface (Stewart et al., 1999).

Figures 20.5 and 20.6 show the cross-sectional view of wear scars for nanocomposite and conventional coatings. For both the coatings subsurface cracks propagate parallel to the surface through W-rich phase or splat boundaries. In addition, as shown in Figure 20.6, vertical cracks can be clearly detected in the conventional coating due to the indentation effect. This is the initial stage of the material loss followed by crack propagation through the W-rich phase and horizontal cracking. If a softer abradant (SiO_2) is employed the wear rate is an order of magnitude lower, and less coarse grooving takes place on the wear surface, without significant subsurface cracking. The wear behaviour is affected by coating microstructure, hardness and toughness as well as by the testing conditions employed.

It should be noted that some industrial applications require both high-temperature and corrosion resistance in conjunction with wear resistance. To this purpose, Cr_3C_2 coatings were found to exhibit better phase stability than WC-Co ones during thermal spraying. The hardness of HVOF-sprayed nanostructured Cr_3C_2-25 (Ni20Cr) coatings was higher than that of their conventional counterparts (10.0 against 8.3 GPa) due to better homogeneous dispersion of carbides and lower degree of decarburization, so that they also exhibited higher fracture toughness (He and Schoenung, 2002). In other works focused on conventional HVOF coatings the hardness values varied between 6.9 and 9.3 GPa, depending on the spraying conditions as well as on the characteristics of the powder feedstock, that is, particle morphology and size (He and Schoenung, 2002; Sasaki et al., 1992) while the microhardness of APS coatings was found to be in the range 8.1–9.2 GPa. As also observed for WC-Co coatings, the best mechanical properties were found when Ar-He gas mixtures were employed during deposition that typically involves lower decarburization and formation of brittle sub-carbides (Vuoristo et al., 1994; Niemi et al., 1992).

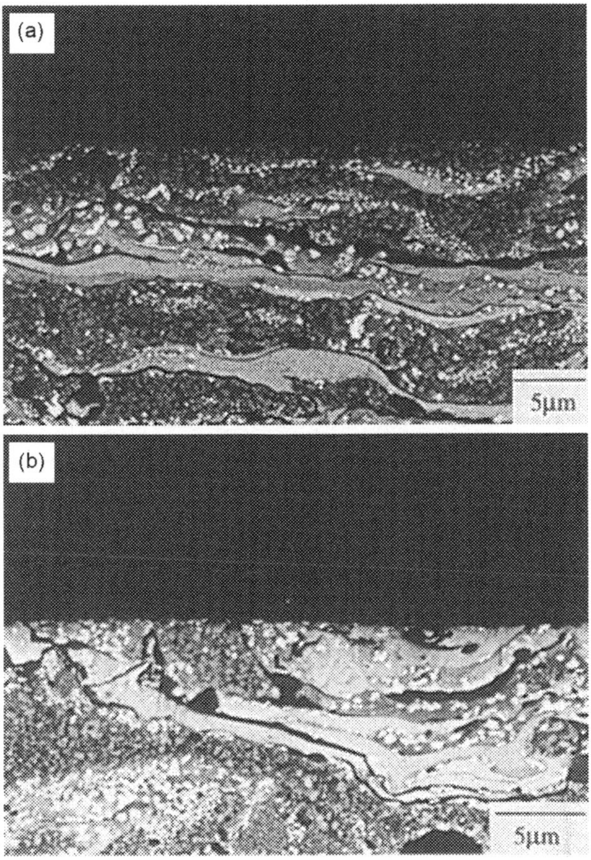

Figure 20.5 SEM backscattered electron images (BEI) cross-sectional view of wear scars in nanocomposite coatings with 125–150 μm alumina abrasive under (a) 23 N and (b) 75 N load showing crack propagation through tungsten-rich binder phase.
Source: Reprinted with permission from Stewart et al. (1999). © 1999, Elsevier.

20.3.2 Alumina-titania composite coatings for anti-wear applications

Plasma-sprayed nanostructured alumina-titania (Al_2O_3-$13TiO_2$) coatings are commonly used to enhance the wear resistance and the durability of turbine and marine engine components and represent a short-term successful application of nanostructured thermally sprayed coatings. They are obtained from agglomerated nanostructured particles which are partially melted to reduce the thermal effects of nucleation and grain growth promoted by high-temperature processing, thus preserving their starting nanostructure.

The agglomerated particles are pressurized and eventually destroyed, blowing up into several smaller ones, while their core remains unmelted. The partial melting of

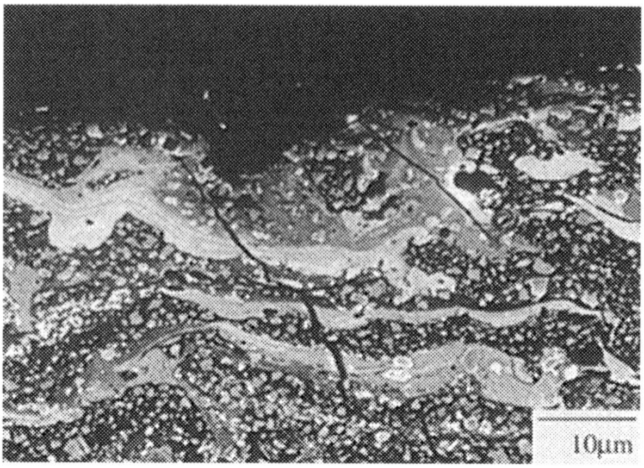

Figure 20.6 SEM BEI cross-sectional view of wear scar in conventional coating with 125−150 μm alumina abrasive under 23 N load showing vertical cracks within the coating. *Source:* Reprinted with permission from Stewart et al. (1999). © 1999, Elsevier.

these agglomerates makes it possible to control both the percentage and the morphology of the retained nanostructured zones in the as-deposited coatings and to engineer coatings with multifunctional properties. Obviously these powder feedstocks are particularly sensitive to the process parameters.

It has been clearly demonstrated that nanostructured alumina-titania coatings exhibit a bimodal microstructure composed of fully melted areas and partially melted regions (Gell et al., 2001; Lima and Marple, 2007). Figures 20.7 and 20.8 show the cross-sectional microstructure of nanostructured and conventional Al_2O_3-13TiO_2 coatings deposited by APS: the nanostructured coating is formed by a melted matrix of γ-Al_2O_3 with titanium rich areas (brighter in the micrograph) and partially melted agglomerates, containing α-Al_2O_3 rich zones embedded in the γ-Al_2O_3 matrix. The nanostructured coatings exhibit a higher α- to γ-Al_2O_3 ratio, because the partial melting of the porous agglomerates avoids the total transformation of α to γ-alumina. In the nanostructured coatings TiO_2 is uniformly distributed along the microstructure, while in the conventional coatings the white areas contain a lower content of aluminum and higher content of Ti in comparison with the matrix, as well observable in the micrograph reported in Figure 20.8. This segregation commonly promotes delamination and reduces wear resistance in comparison with nanostructured coatings.

Previous investigations demonstrated that nanostructured alumina-titania coatings possess high sliding and abrasive wear resistance in comparison with conventional coatings (Shaw et al., 2000). Their better performance can be associated to their enhanced fracture toughness, as it has been generally reported that the nanostructured coatings are less hard than conventional ones (Ahn et al., 2006; Wang et al., 2000; Luo et al., 2003). Sanchez et al. (2011) reported hardness of 7.6 GPa for conventional coatings and 8.4 GPa for nanostructured coatings produced by APS and tested at 200 g. The reduction in the wear rate levels was 39% for Al_2O_3-3TiO_2 coatings and

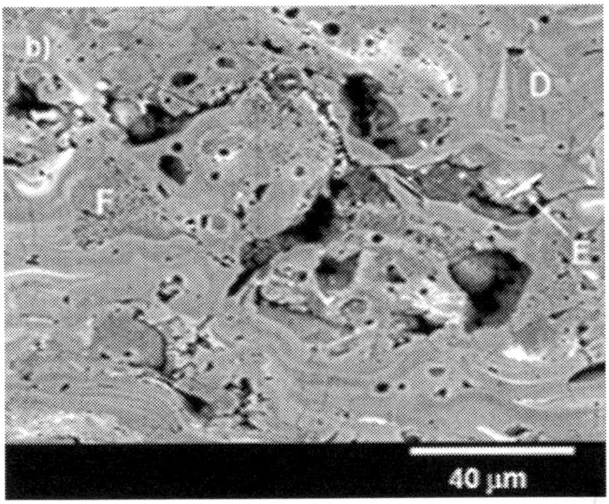

Figure 20.7 SEM micrographs of nanostructured Al_2O_3-TiO_2 coatings sprayed on AISI 304 steel. The coatings are formed by a melted matrix made of γ-Al_2O_3 with Ti + 4 ((a) and (d)), titanium-rich areas ((b) and (e)) and, in the nanostructured case, by partially melted agglomerates ((f)).
Source: Reprinted with permission from Sanchez et al. (2011). © 2011, Wiley.

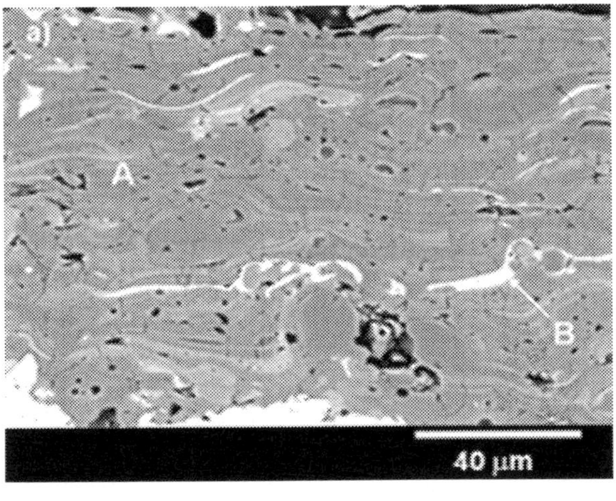

Figure 20.8 SEM micrographs of conventional Al_2O_3-TiO_2 coatings sprayed on AISI 304 steel. The coatings are formed by a melted matrix made of γ-Al_2O_3 with Ti + 4 (A) and titanium-rich areas (B).
Source: Reprinted with permission from Sanchez et al. (2011). © 2011, Wiley.

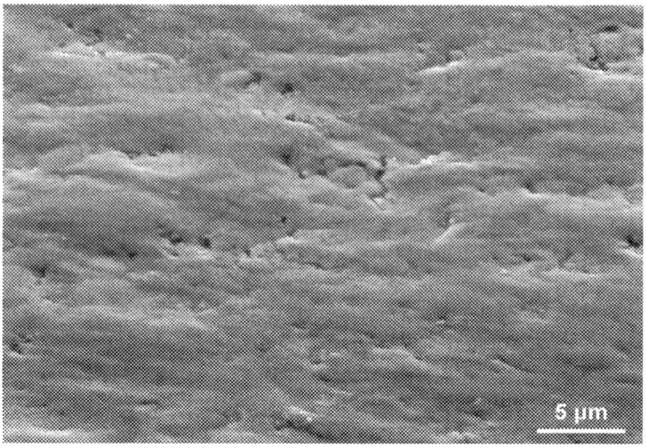

Figure 20.9 SEM picture (taken at a 50° angle) of the wear scar of an HVOF-sprayed nanostructured titania coating.
Source: Reprinted with permission from Lima and Marple (2007). © 2007, Springer.

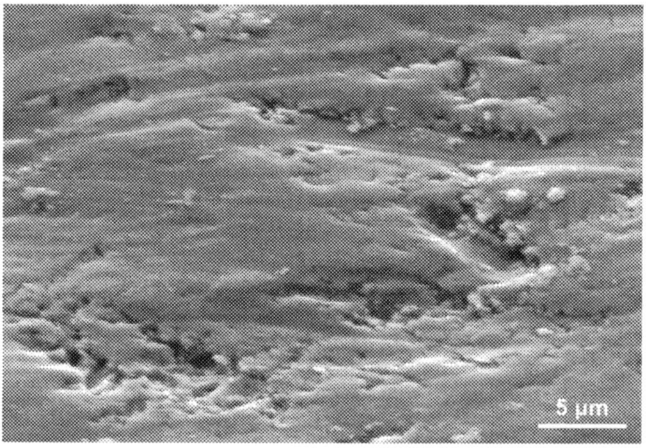

Figure 20.10 SEM picture (taken at a 50° angle) of the wear scar of an HVOF-sprayed conventional titania coating.
Source: Reprinted with permission from Lima and Marple (2007). © 2007, Springer.

between 71% and 75% for Al_2O_3-13TiO_2 coatings (Gell et al., 2001; Lin et al., 2004; Lima and Marple, 2007; Jordan et al., 2001).

The presence of retained nanostructured areas makes the microstructure more ductile and enhances the resistance to crack propagation, as it has been studied by various authors using Vickers indentation technique (Gell et al., 2001; Luo et al., 2003; Bansal et al., 2003).

Figures 20.9 and 20.10 show the wear track of HVOF nanostructured and conventional coatings after abrasion tests. The wear scar of the nanostructured coatings is smoother due to their plastic-like behaviour. Plastic deformation is the main wear mechanism where cracking, delamination and microfracture typically characterize the

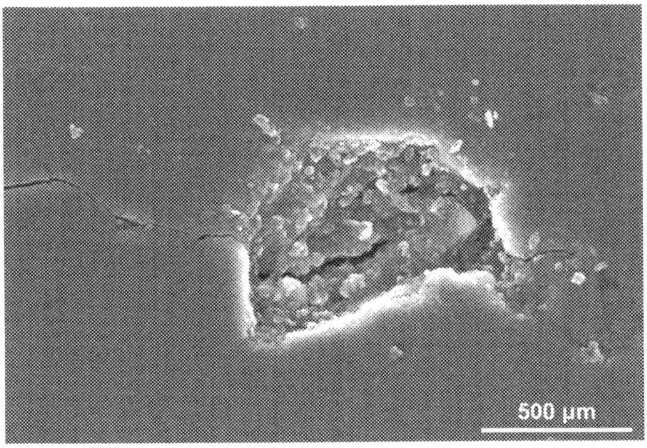

Figure 20.11 Vickers indentation crack tip arresting after passing through a nanozone in the nanostructured coating.
Source: Reprinted with permission from Lima and Marple (2007). © 2007, Springer.

wear behaviour of conventional coatings (Gell et al., 2001). Large-scale cracking and material removal can be observed in the conventional coatings, while no cracking occurs in the nanostructured coatings. Indeed, after indentation tests performed on HVOF-sprayed titania and alumina-titania coatings it has been found that when cracks generated at the corners of the imprint propagate along the microstructure and pass through partially melted nanosized areas they are arrested or deflected (see Figure 20.11). These nanozones are very dense, because the infiltration of the molten material of a semi-melted particle into the small capillaries of its unmelted core during processing is able to reduce the pore volume (Lima and Marple, 2007).

The control of the distribution and the percentage of nanostructured areas embedded in coating microstructure represents a significant key point for tailoring the wear properties of the same coating for well-determined industrial applications. To this purpose, there is no consensus about the percentage of retained nanozones in order to obtain the best wear resistance. Some studies that focused on APS alumina-titania coatings suggested a percentage close to 15% or 20% (Gell et al., 2001; Luo et al., 2003; Guilemany et al., 2005; Goberman et al., 2002).

Other studies suggested that the improved crack resistance of nanocomposite coatings can be related to better splat-to-splat cohesion, associated with the presence of a higher amount of TiO_2 (Khameneh Asl et al., 2006) or enhanced interfacial toughness and bonding strength, as well as to better plasticity. The bonding strength of nanostructured Al_2O_3-13TiO_2 APS coatings doubles with respect to that of the conventional ones (Jordan et al., 2001).

VPS and HVOF spraying systems have been also employed to fabricate coatings with anti-wear performance better or equal to that of APS coatings (Leblanc, 2003). For HVOF coatings the reduction of abrasive wear rate was of about 90%, even if the percentage of semi-molten zones was close to 52%, due to the lowest temperature of the HVOF process (Lima and Marple, 2007).

It has been also reported that the nanostructured coatings have much higher thermal shock resistance than that of the conventional coatings at temperatures of 650°C and 850°C (Wang et al., 2007).

Air plasma-sprayed nanostructured alumina titania coatings have been developed and approved by the US Navy for shipboard and submarine applications in propulsion shafts, bullhead pivot arms, bearing sleds, magnet arms, sockets and arm pivot pins, hydraulic pistons and reduction gear sets (Lima and Marple, 2007). For example, shafts are commonly subjected to severe abrasion on the bearing surfaces. To this end, it has been found that conventional coatings were not able to tolerate the high level of bending and fatigue experienced by these components. Otherwise, these components, when coated with nanostructured coatings, did not suffer significant wear or delamination after four years of service.

20.3.3 Ceramic thermal barrier coatings

Nanostructured ceramic coatings are also emerging as potential candidates for thermal barrier applications. TBCs are commonly applied to protect the surface of Ni-based hot-section components (vanes, blades, cylinder liners, transition pieces and exhaust ports) of aircraft engines and gas turbines against severe temperature and environmental attack promoted by oxygen and other corrosive media, such as sulphates, phosphates, vanadates and calcium-magnesium-aluminum-silicates, with the purpose of increasing their durability and improving efficiency.

Up to now partially yttria-stabilized zirconia (YSZ) has been recognized as the best ceramic material for TBC development because of its good mechanical and thermal properties. However, further improvements are expected based on the constant demand for highly performant and green engines, so that research is currently focused on the development of further TBC materials as well as on the design of systems with enhanced thermal cycling lifetime. This last is strongly related to the microstructure of the TBC.

To this purpose, the refinement of the microstructure and the fabrication of TBC with better strain tolerance under thermal fatigue are expected to provide significant enhancements at short term. The micrograph in Figure 20.12 shows nanostructured YSZ grains agglomerated in micronsized particles. As shown in Figures 20.13, 20.14 and 20.15, plasma-sprayed nanostructured YSZ coating exhibits a typical bimodal microstructure composed of well-melted thin splats and retained porous and semi-molten nanozones resembling the morphology of the starting powder feedstock (Figure 20.14). Figure 20.16 shows the columnar grains embedded in the melted splats and some equiaxed grains generally located at the splat boundary. This microstructure can also imply a bimodal distribution of the mechanical properties (Di Girolamo et al., 2011; Lima et al., 2002).

It has been demonstrated that nanostructured YSZ coatings possess higher phase stability, lower thermal conductivity (0.7–0.8 against 0.8–1.2 $Wm^{-1} K^{-1}$) and higher thermal expansion coefficient (11 against $10.5-11 \times 10^{-6} K^{-1}$) than their conventional counterparts (Lima and Marple, 2007; Chen et al., 2002; Trice et al., 2002; Di Girolamo et al., 2010a). Table 20.5 reports the thermal conductivity and

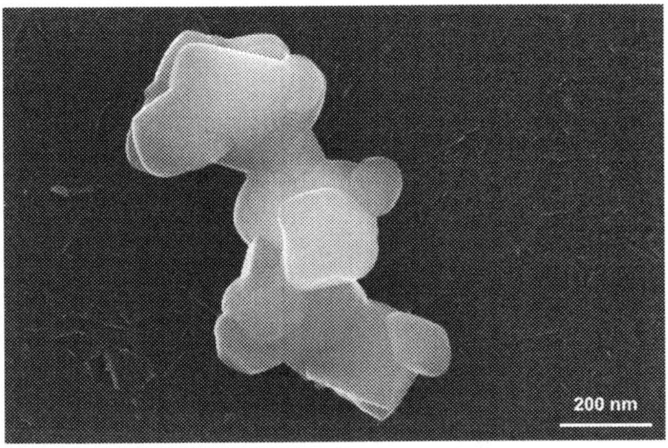

Figure 20.12 Nanostructured YSZ particles.

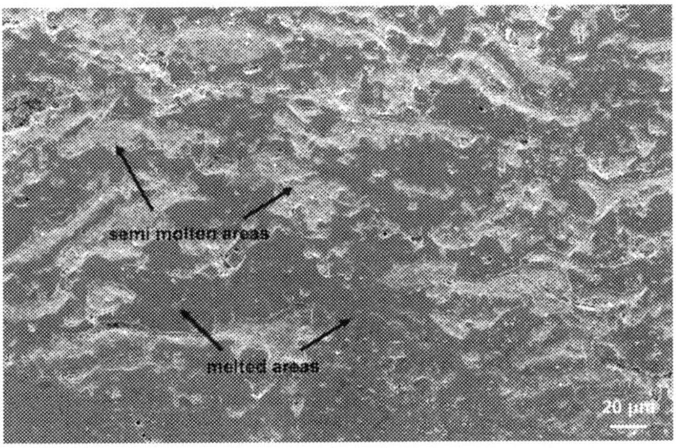

Figure 20.13 Cross-sectional microstructure of plasma-sprayed nanostructured YSZ coating composed of well-melted and semi-molten areas.

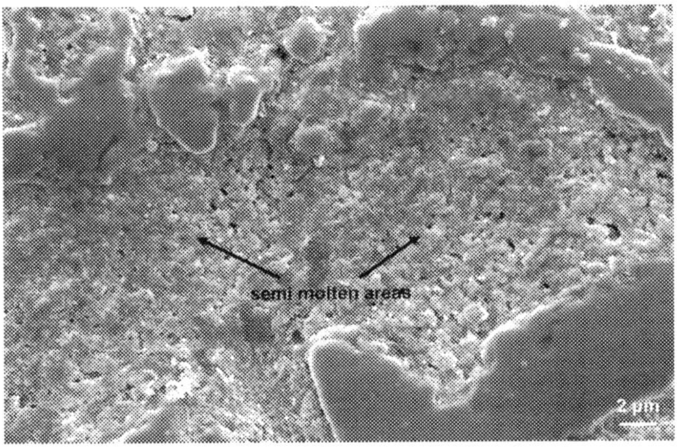

Figure 20.14 Nanostructured unmelted particles embedded in a YSZ coating microstructure.

Figure 20.15 Lamellar microstructure of plasma-sprayed nanostructured YSZ coating.

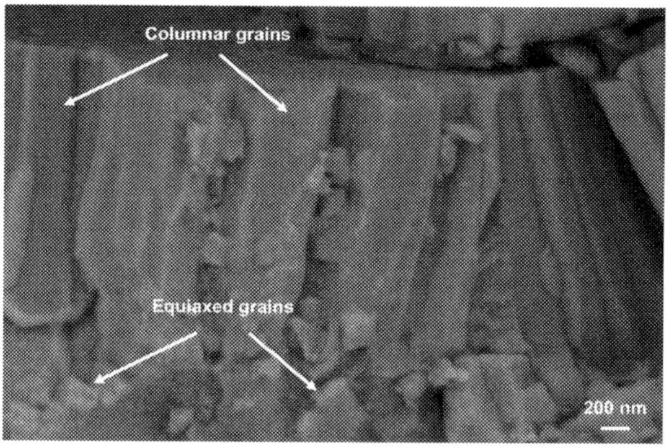

Figure 20.16 Columnar and equiaxed grains embedded in a YSZ coating microstructure.

Table 20.5 Thermal conductivity and thermal expansion coefficient of plasma-sprayed YSZ thermal barrier coatings

	Thermal conductivity ($Wm^{-1} K^{-1}$) [Ref]	CTE ($10^{-6} K^{-1}$) [Ref]	Processing
Conv.	0.8–1.2 (Trice et al., 2002; Zhu et al., 2004)	10.7 (Di Girolamo et al., 2010a; Zhu et al., 2004)	APS
Nano	0.7–0.8 (Chen et al., 2002)	11 (Chen et al., 2002)	APS
Nano	0.55–0.65 (Ma et al., 2006)	–	SPPS

Table 20.6 Thermal shock resistance of APS conventional and nanostructured YSZ coatings

Coating [Ref]	Testing temperature (°C)	Number of cycles
Conv. (Di Girolamo et al., 2011)	1020	50
Nano (Di Girolamo et al., 2011)	1020	80–90
Conv. (Liang and Ding, 2005)	1000	48
Nano (Liang and Ding, 2005)	1000	120
Conv. (Wang et al., 2006)	1200	2
Nano (Wang et al., 2006)	1200	11

the thermal expansion coefficient (CTE) of conventional and nanostructured YSZ coatings deposited by APS and SPPS processes.

In previous work we also found that the hardness of nanostructured YSZ coatings having a percentage of retained nanozones close to 22% is similar to that of conventional coatings tested at 50 g indentation load, while their thermal shock resistance is better because of their better toughness and higher resistance to crack propagation (the durability increases from 60% to 80%). In the case of nanostructured coatings the formation of crack networks and the beginning of surface delamination require more cycles than conventional coatings. The nanostructured ceramic coating peels off and delaminates splat by splat, while the formation of the thermally grown oxide (TGO) and the spallation at the top coat/bond coat interface play a minor role in the mechanism of damaging. Table 20.6 summarizes the results of thermal shock tests carried out on conventional and nanostructured YSZ TBCs produced by APS and tested at different temperatures (Di Girolamo et al., 2011; Liang and Ding, 2005; Wang et al., 2006).

The retention of embedded porous nanozones increases the compliance and can also counteract the sintering effects (Lima and Marple, 2008). It has been reported that the retention of partially melted particles and the presence of areas with different sintering rates are able to stop crack propagation and to prevent and retard the sintering effects occurring at high temperature (>1200°C) (Lima and Marple, 2008). The partially melted areas act as crack arresters (Lima and Marple, 2005).

More recently, the research on TBCs has been focused on alternative compositions such as ceria and yttria co-stabilized zirconia. Indeed, in many works it has been found that conventional CYSZ coatings exhibit higher CTE, better hot-corrosion resistance, higher hardness and phase stability than those of YSZ coatings (Di Girolamo et al., 2010a,b; Alfano et al., 2010; Park et al., 2005).

As shown for YSZ coatings, nanostructured CYSZ coatings also exhibited a microstructure composed of well-melted splats with columnar grains and partially melted areas which resemble the morphology of the starting powder: this microstructure is beneficial for enhancing phonon scattering and reducing thermal diffusivity. Indeed, the thermal diffusivity of nanostructured CYSZ coatings was lower than that of YSZ coatings ($0.33–0.55 \times 10^{-6}\ m^2\ s^{-1}$against

$0.51-0.75 \times 10^{-6}\ m^2\ s^{-1}$) (Jin et al., 2012). In turn, the thermal cycling lifetime of CYSZ nanostructured coatings was of about 860 cycles at 1050°C, higher than that of YSZ coatings (400–600 cycles). This result can be explained in terms of higher CTE and lower thermal conductivity.

Nanostructured lanthanum zirconate TBCs have been also fabricated by air plasma spraying. Their thermal conductivity was between 0.73 and 0.79 $Wm^{-1}\ K^{-1}$ in the range from 200°C to 1200°C (0.8–1.2 $Wm^{-1}\ K^{-1}$ for YSZ coatings) (Trice et al., 2002) while the CTE was slightly lower than that of YSZ TBCs ($9.45 \times 10^{-6}\ K^{-1}$ vs. $10.7 \times 10^{-6}\ K^{-1}$) (Chen et al., 2009). The thermal aging at 1300°C and 1400°C for 50 hours caused an increase in their thermal conductivity of 0.3 $Wm^{-1}\ K^{-1}$ and 0.6 $Wm^{-1}\ K^{-1}$, respectively (Chen et al., 2009).

The use of nanostructured materials in conjunction with the search for alternative compositions, for example based on rare-earth zirconates, represents an interesting perspective in order to enhance the functional properties of these thermal protection systems and the performance of the related coated components. For example, conventional samarium zirconate TBCs have exhibited thermal conductivity of about 0.35–0.45 $Wm^{-1}\ K^{-1}$ and a thermal expansion coefficient equal to $10.8 \times 10^{-6}\ K^{-1}$ (Hu et al., 2010; Qu et al., 2007).

It should be noted that the first nanostructured powders were introduced commercially in the early 2000s. The processing routes used to produce them are somewhat sophisticated and time consuming, so they are often produced at laboratory scale for research purposes.

However, more interestingly, in recent years SPS and SPPS systems have been proposed for deposition of ceramic TBC with finer microstructure, and the thermal spray community has a great interest and confidence in the development of these processes. The suspensions are composed of fine solid particulates dispersed in liquid (water or organic solvent), whereas the solutions are inorganic salts and metallorganic compounds dissolved in water or organic solvent.

The suspensions are injected by a pressurized feed system in the plasma jet and atomized in a stream of fine droplets, which impinge on the substrate after a sequence of physical interactions (aerodynamic breaking up, evaporation of liquid, sintering and melting of fine solids and agglomerates). The suspension liquid evaporates after injection in the plasma jet and the solid may agglomerate, melt and impact the substrate. The suspension can be introduced to the plasma torch with a radial feedstock injector through an atomizer or a nozzle (Marchand et al., 2011).

In turn, in SPS the droplets go through several chemical and physical changes (breaking up, solvent evaporation, solute decomposition and melting), so that they can impact on the substrate as fully melted or unpyrolized particles. The fine microstructure of SPS coatings is associated with the small size of the molten droplets and with their in-flight fragmentation in the plasma jet (Padture et al., 2001; Bhatia et al., 2002; Pawlowski, 2008).

The thermal conductivity of YSZ coatings produced by the SPS method was of about 0.5–1 $Wm^{-1}\ K^{-1}$, close to that of APS coatings having the same porosity level. However, the SPS method makes it easy to control the final microstructure and to fabricate coatings with very high porosity. Their hardness was found to be higher than that of APS coatings, while the Young's modulus, measured by three-point bending test, was equal to 3.3 GPa (for APS coatings it was 15–20 GPa at room temperature and decreased to 3 GPa when measured at 1400°C) (Kassner et al., 2009; Guignard et al., 2012; Ahrens et al., 2012).

SPS TBCs survived 660 cycles at 1400°C (the temperature on the metal substrate surface was 1050°C), then the partial spallation of the top coat, clearly detectable in Figure 20.17(a), occurred because of TGO formation (see Figure 20.17(b)) and the propagation of delamination cracks. As shown in Figure 20.17(c), the typical segmented cracks of SPS coatings survived, while the thermal cycling promoted sintering, pore coarsening and stiffening (Figure 20.17(d)). The thermal cycling resistance was 50% lower than that of APS coatings because of the highest sintering rate of overspray particles with very small size, assisting neck formation and sintering (Guignard et al., 2012).

According to other investigations, SPPS offers the prospect of depositing highly durable nanostructured YSZ TBCs characterized by low thermal conductivity (0.55–0.65 $W m^{-1} K^{-1}$). The presence of fine porosity and extended vertical segmentation cracks increases the strain tolerance and reduces the Young's modulus, thus improving the thermal cycling resistance by a factor of 2.5 with respect to APS TBCs (Gell et al., 2004; Ma et al., 2006; Jadhav et al., 2006; Jadhav and Padture, 2008).

Actually there are a few results for these coating systems and further efforts are needed to understand the potential of SPPS/SPS coatings with very fine microstructure. It would be very useful to combine the advantages of the SPS or SPPS process with the synthesis of new TBC materials characterized by low thermal conductivity and better sintering resistance than YSZ. Indeed, at temperatures higher than 1200°C the performance and durability of YSZ TBCs drop because of the sintering of the porous microstructure and the tetragonal-to-monoclinic transition during the subsequent cooling to ambient temperature. To this end, the processing and the mechanical properties of SPS YSZ and CYSZ coatings have been studied in a recent work (Latka et al., 2013).

Finally, nanostructured zirconia-based coatings can be also engineered as abradable coatings and employed in gas turbine engines to minimize the bypass flow of hot combustion or cold compressor gases through the spaces between the rotating blade tips and the walls of the shroud, to provide a seal and thus improve the efficiency. These coatings are typically characterized by low structural integrity so that they can be readily abraded when they come into contact with the rotating blade tips during start-up and service, but at the same time they should exhibit good mechanical and thermal stability. By starting from agglomerated nanostructured particles and controlling their degree of melting during plasma spray processing, the original nanostructure can be largely retained in the final coating. Obviously the percentage of embedded nanosized areas and the porosity level are much higher in comparison with similar coatings designed for TBC applications (Lima and Marple, 2007).

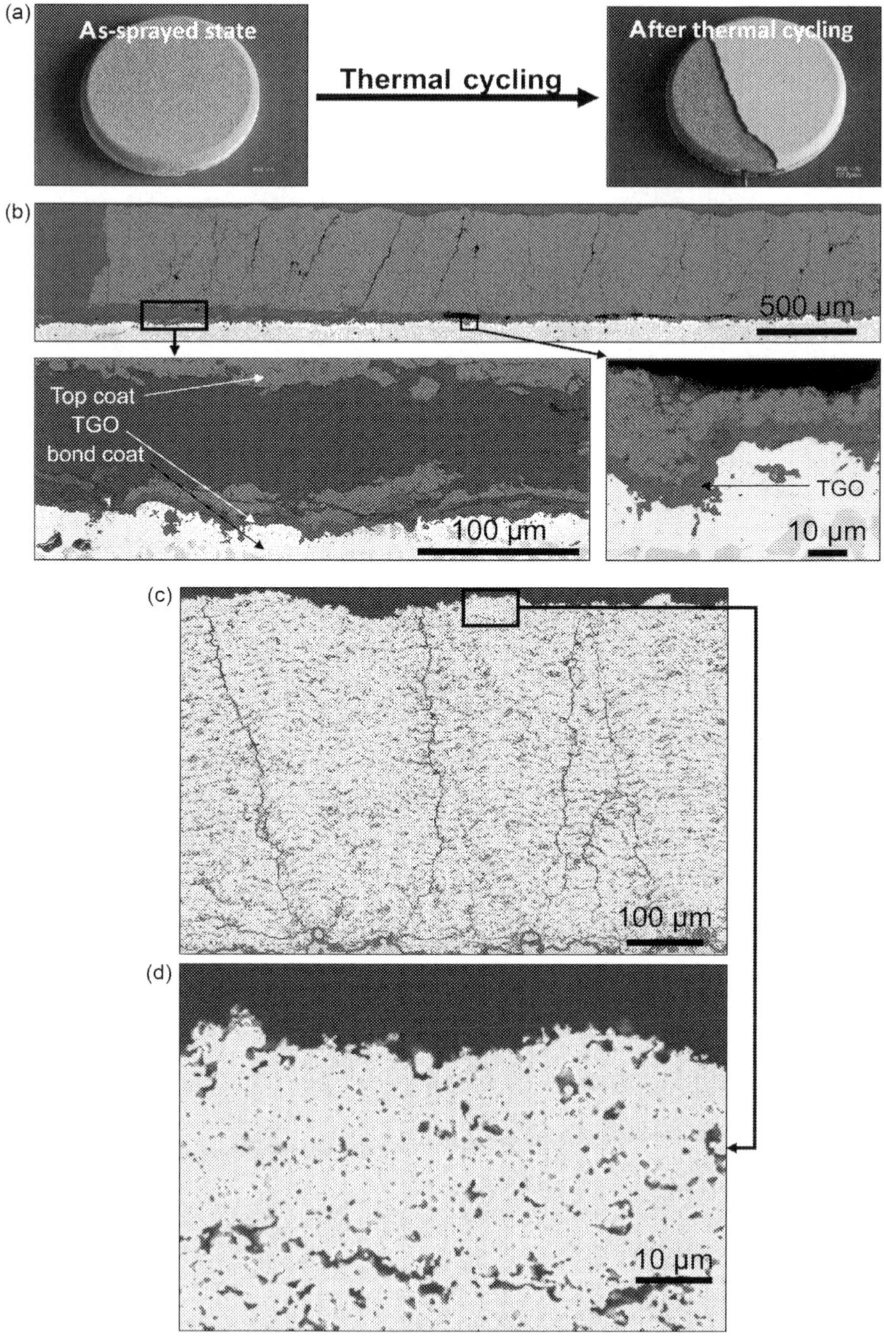

Figure 20.17 Images of SPS TBC subjected to thermal cycling test: (a) photographs of the specimen in the as-sprayed state and after thermal cycling, (b) confocal laser microscope images of polished cross section after thermal cycling showing coating spallation and TGO layer, (c) SEM image (back scattered electron (BSE) mode) after thermal cycling showing segmentation cracks that survived and (d) SEM image (BSE mode) at higher magnification after thermal cycling pointing out beginning of sintering.
Source: Reprinted with permission from Guignard et al. (2012). © 2012, Springer.

20.4 Conclusion

In this chapter an overview of thermally sprayed nanostructured coatings currently employed for anti-wear and thermal barrier applications has been presented. Nanostructured coatings are generally obtained starting from agglomerates of nano-particles, because single nanoparticles are not suitable to be carried by a moving gas stream and deposited on a substrate. The partial melting of these agglomerates allows the fabrication of coatings with unique microstructure and multi-functional properties.

Cermet coatings are commonly used at relatively high temperature for wear protection of machine and engine parts, because at higher temperatures (700–800°C) their mechanical and tribological properties drop dramatically, so ceramic coatings are better candidates for severe operating conditions when high-temperature capability and wear resistance have to be properly matched. Nanostructured alumina-titania coatings represent a recent success case because they have been approved by the US Navy for marine and submarine applications and are currently employed in this way.

In turn, nanostructured ceramic coatings are very promising for thermal barrier applications because of their finer microstructure in comparison to conventional microstructured coatings, involving better toughness, improved high-temperature sintering resistance and enhanced thermal cycling lifetime. These coatings can be obtained using APS, SPS and SPPS systems and using different zirconia-based raw materials.

It should be pointed out that many aspects about the processing of nanostructured coatings and the fabrication of well-determined microstructures are not clear yet. Therefore, further efforts are needed to validate the improved multifunctional properties of nanostructured coatings, to make their manufacturing methods easier and to reduce their production costs, thus allowing high-volume production and creating an interesting niche market.

References

Ahn, J., Hwang, B., Song, E.P., Lee, S., Kim, N.J., 2006. Correlation of microstructure and wear resistance of Al_2O_3-TiO_2 coatings plasma sprayed with nanopowders. Metall. Mater. Trans. A. 37, 1851–1861.

Ahrens, M., Lampensherf, S., Vassen, R., Stöver, D., 2012. Deposition and characteristics of sumicrometer-structured thermal barrier coatings by suspension plasma spraying. J. Therm. Spray Technol. 21 (3–4), 416–424.

Alfano, M., Di Girolamo, G., Pagnotta, L., Sun, D., Zekonyte, J., et al., 2010. The influence of high-temperature sintering on microstructure and mechanical properties of free-standing APS CeO_2-Y_2O_3-ZrO_2 coatings. J. Mater. Sci. 45, 2662–2669.

Baik, K.H., Kim, J.H., Seong, B.G., 2007. Improvements in hardness and wear resistance of thermally sprayed WC-Co nanocomposite coatings. Mater. Sci. Eng. A. 449–451, 846–849.

Bansal, P., Padture, N.P., Vasiliev, A., 2003. Improved interfacial mechanical properties of Al_2O_3−13wt% TiO_2 plasma-sprayed coatings derived from nanocrystalline powders. Acta Mater. 51, 2959−2970.

Bhatia, T., Ozturk, A., Xie, L., Jordan, E.H., Cetegen, B., et al., 2002. Mechanisms of ceramic coating deposition in solution precursor plasma spray. J. Mater. Res. 17, 2363−2372.

Chen, H., Hutchings, I.M., 1998. Abrasive wear resistance of plasma-sprayed tungsten carbide-cobalt coatings. Surf. Coat. Technol. 107, 106−114.

Chen, H., Zhou, X., Ding, C., 2002. Investigation of the thermomechanical properties of a plasma sprayed nanostructured zirconia coating. J. Eur. Ceram. Soc. 23, 1449−1455.

Chen, H., Gao, Y., Tao, S., Liu, Y., Luo, H., 2009. Thermophysical properties of lanthanum zirconate coating prepared by plasma spraying and the influence of post-annealing. J. Alloys. Compd. 486, 391−399.

Davis, J.R., 2004. In: Coutsouranis, D. (Ed.), Handbook of Thermal Spray Technology. ASM International, Materials Park, OH.

Di Girolamo, G., Blasi, C., Pagnotta, L., Schioppa, M., 2010a. Phase evolution and thermophysical properties of plasma sprayed thick zirconia coatings after annealing. Ceram. Int. 36, 2273−2280.

Di Girolamo, G., Blasi, C., Schioppa, M., Tapfer, L., 2010b. Structure and thermal properties of heat treated plasma sprayed ceria-yttria co-stabilized zirconia coatings. Ceram. Int. 36, 961−968.

Di Girolamo, G., Marra, F., Blasi, C., Serra, E., Valente, T., 2011. Microstructure, mechanical properties and thermal shock resistance of plasma sprayed nanostructured zirconia coatings. Ceram. Int. 37, 2711−2717.

Di Girolamo, G., Marra, F., Pilloni, L., Pulci, G., Tirillò, J., et al., 2013. Microstructure and wear behavior of plasma-sprayed nanostructured WC-Co coatings. Int. J. Appl. Ceram. Technol. 10 (1), 60−71.

Eidelman, S., Yang, X., 1997. Three dimensional simulation of HVOF spray deposition of nanoscale materials. Nanostruct. Mater. 9, 79−84.

Gell, M., Jordan, E.H., Sohn, Y.H., Goberman, D., Shaw, L., et al., 2001. Development and implementation of plasma sprayed nanostructured ceramic coatings. Surf. Coat. Technol. 146−147, 48−54.

Gell, M., Xie, L., Ma, X., Jordan, E.H., Padture, N.P., 2004. Highly durable thermal barrier coatings made by the solution precursor plasma spray process. Surf. Coat. Technol. 177−178, 97−102.

Ghabchi, A., Varis, T., Turunen, E., Suhonen, T., Liu, X., et al., 2010. Behavior of HVOF WC-10Co4Cr coatings, with different carbide size in fine and coarse particle abrasion. J. Therm. Spray Technol. 19, 368−377.

Goberman, D., Sohn, Y.H., Shaw, L., Jordan, E., Gell, M., 2002. Microstructure development of Al_2O_3−13wt.% TiO_2 plasma sprayed coatings derived from nanocrystalline powders. Acta Mater. 50, 1141−1152.

Guignard, A., Mauer, G., Vassen, R., Stöver, D., 2012. Deposition and characteristics of submicrometer-structured thermal barrier coatings by suspension plasma spraying. J. Therm. Spray Technol. 21 (3−4), 416−424.

Guilemany, J.M., Dosta, S., Nin, J., Miguel, J.R., 2005. Study of the properties of WC-Co nanostructured coatings sprayed by high-velocity oxy-fuel. J. Therm. Spray Technol. 14, 405−413.

He, J., Schoenung, J.M., 2002. Nanostructured coatings. Mater. Sci. Eng. A. 336, 274−319.

Houck, D.L., 1985. Spray drying in the metals industries. Met. Powder Rep. 40, 707−709.

Hu, J., Zhao, H., Tao, S., Zhou, X., Ding, C., 2010. Thermal conductivity of plasma sprayed $Sm_2Zr_2O_7$ coating. J. Eur. Ceram. Soc. 30, 799–804.

Jadhav, A.D., Padture, N.P., 2008. Mechanical properties of solution-precursor plasma sprayed thermal barrier coatings. Surf. Coat. Technol. 202, 4976–4979.

Jadhav, A.D., Padture, N.P., Jordan, E.H., Gell, M., Miranzo, P., et al., 2006. Low-thermal-conductivity plasma-sprayed thermal barrier coatings with engineered microstructures. Acta Mater. 54, 3343–3349.

Jin, L., Yu, Q., Ni, L., Zhou, C., 2012. Microstructure and thermal properties of nanostructured 8 wt.% CeO_2 doped YSZ coatings prepared by atmospheric plasma spraying. J. Therm. Spray Technol. 21, 928–934.

Jordan, E.H., Gell, M., Sohn, Y.H., Goberman, D., Shaw, L., et al., 2001. Fabrication and evaluation of plasma sprayed nanostructured alumina-titania coatings with superior properties. Mater. Sci. Eng. A. 301, 80–89.

Kassner, H., Stuke, A., Rödig, M., Vassen, R., Stöver, D., 2009. Influence of porosity on thermal conductivity and sintering in suspension plasma sprayed thermal barrier coatings. Ceram. Eng. Sci. Proc. 29 (4), 147–158.

Khameneh Asl, S., Heydarzadeh Sohi, M., Hokamoto, K., Uemura, M., 2006. Effect of heat treatment on wear behavior of HVOF thermally sprayed WC-Co coatings. Wear. 260, 1203–1208.

Kim, J.H., Yang, H.S., Baik, K.H., Seong, B.G., Lee, C.H., et al., 2006. Development and properties of nanostructured thermal spray coatings. Curr. Appl. Phys. 6, 1002–1006.

Latka, L., Cattini, A., Chicot, D., Pawlowski, L., Kozerski, et al., 2013. Mechanical properties of yttria-and ceria-stabilized zirconia coatings obtained by suspension plasma spraying. J. Therm. Spray Technol. 22 (2–3), 125–130.

Leblanc, L., 2003. Abrasion and sliding wear of nanostructured ceramic coatings. In: Moreau, C., Marple, B. (Eds.), Advancing the Science & Applying the Technology. ASM International, Materials Park, OH, pp. 291–299.

Liang, B., Ding, C., 2005. Thermal shock resistance of nanostructured and conventional zirconia coatings deposited by atmospheric plasma spraying. Surf. Coat. Technol. 197, 185–192.

Lima, R.S., Marple, B.R., 2005. Enhanced ductility in thermally sprayed titania coating synthesized using a nanostructured feedstock. Mater. Sci. Eng. A. 395, 269–280.

Lima, R.S., Marple, B.R., 2007. Thermal spray coatings engineered from nanostructured ceramic agglomerated powders for structural, thermal barrier and biomedical applications: a review. J. Therm. Spray Technol. 16, 40–63.

Lima, R.S., Marple, B.R., 2008. Nanostructured YSZ thermal barrier coatings engineered to counteract sintering effects. Mater. Sci. Eng. A. 485, 182–193.

Lima, R.S., Kucuk, A., Berndt, C.C., 2002. Bimodal microstructure of mechanical properties on plasma sprayed nanostructured partially stabilized zirconia. Mater. Sci. Eng. A. 327, 224–232.

Lin, X., Zeng, Y., Lee, S.W., Ding, C., 2004. Characterization of alumina-3wt.% titania coating prepared by plasma spraying of nanostructured powders. J. Eur. Ceram. Soc. 24, 627–634.

Liu, Y., Qiao, Y., He, J., Lavernia, E.J., Fischer, T.E., 2002. Near-nanostructured WC-18 pct Co coatings with low amounts of non-WC carbide phase: part II. Hardness and resistance to sliding and abrasive wear. Metall. Mater. Trans. A. 33, 159–164.

Luo, H., Goberman, D., Shaw, L., Gell, M., 2003. Indentation fracture behaviour of plasma sprayed nanostructured Al_2O_3–13wt% TiO_2 coatings. Mater. Sci. Eng. A. 346, 237–245.

Ma, X., Wu, F., Roth, J., Gell, M., 2006. Low thermal conductivity thermal barrier coating deposited by the solution plasma spray process. Surf. Coat. Technol. 201, 4447–4452.

Marchand, O., Girardot, L., Planche, M.P., Bertrand, P., Bailly, Y., et al., 2011. An insight into suspension plasma spray: injection of the suspension and its interaction with the plasma flow. J. Therm. Spray Technol. 20, 1310–1320.

National Research Council, 1996. Coatings for High-temperature Applications: Trends and Opportunities. National Academy of Sciences, Washington, DC.

Niemi, K., Vuoristo, P., Mantyla, T., Barbezat, G., Nicoll, A.R., 1992. Abrasion wear resistance of carbide coatings deposited by plasma and high velocity combustion process. In: Berndt, C.C. (Ed.), Thermal Spray: International Advances in Coatings Technology. ASM International, Materials Park, OH, pp. 685–689.

Padture, N.P., Schlichting, K.W., Bhatia, T., Ozturk, A., Cetegen, B., et al., 2001. Toward durable thermal barrier coatings with novel microstructure deposited by solution precursors plasma spray. Acta Mater. 49, 2251–2257.

Park, S.Y., Kim, J.H., Kim, M.C., Song, H.S., Park, C.G., 2005. Microscopic observation of degradation behaviour in yttria and ceria stabilized zirconia thermal barrier coatings under hot corrosion. Surf. Coat. Technol. 190, 357–365.

Pawlowski, L., 2008. Finely grained nanometric and submicrometric coatings by thermal spraying: a review. Surf. Coat. Technol. 202, 4318–4328.

Qiao, Y., Liu, Y., Fischer, T.E., 2001. Sliding and abrasive wear resistance of thermal-sprayed WC-Co coatings. J. Therm. Spray Technol. 10, 118–125.

Qiao, Y., Fischer, T.E., Dent, A., 2003. The effect of fuel chemistry and feedstock powder on the mechanical and tribological properties of HVOF thermal-sprayed WC-Co coatings with very fine structures. Surf. Coat. Technol. 172, 24–41.

Qu, Z., Wan, C., Pan, W., 2007. Thermal expansion and defect chemistry of MgO doped $Sm_2Zr_2O_7$. Chem. Mater. 19, 4913–4918.

Sanchez, E., Bannier, E., Salvador, M.D., Bonache, V., Garcia, J.C., et al., 2010. Microstructure and wear behavior of conventional and nanostructured plasma-sprayed WC-Co coatings. J. Therm. Spray Technol. 19, 964–974.

Sanchez, E., Bannier, E., Vicent, M., Moreno, A., Salvador, M.D., et al., 2011. Characterization of nanostructured ceramic and cermet coatings deposited by plasma spraying. Int. J. Appl. Ceram. Technol. 8, 1136–1146.

Sasaki, M., Kawakami, F., Komaki, C., Ishida, M., 1992. Characterization of HVOF sprayed Cr_3C_2 coating. In: Berndt, C.C. (Ed.), Thermal Spray: International Advances in Coatings Technology. ASM International, Materials Park, OH, pp. 165–170.

Shaw, L.L., Goberman, D., Ren, R., Gell, M., Jiang, S., et al., 2000. The dependency of microstructure and properties of nanostructured coatings on plasma spray conditions. Surf. Coat. Technol. 130, 1–8.

Shipway, P.H., McChartney, D.G., Sudaprasert, T., 2005. Sliding wear behaviour of conventional and nanostructured HVOF sprayed WC-Co coatings. Wear. 259, 820–827.

Stewart, D.A., Shipway, P.H., McChartney, D.G., 1999. Abrasive wear behaviour of conventional and nanocomposite HVOF-sprayed WC-Co coatings. Wear. 225–229, 789–798.

Trice, R.W., Su, Y.J., Mawdsley, J.R., Faber, K.T., Arellano-Lopez, A.R., et al., 2002. Effect of heat treatment on phase stability, microstructure, and thermal conductivity of plasma-sprayed YSZ. J. Mater. Sci. 37, 2359–2365.

Vuoristo, P., Niemi, K., Makela, A., Mantyla, T., 1994. Abrasion and erosion wear resistance of Cr_3C_2-NiCr coatings prepared by plasma, detonation and HVOF spraying. In: Berndt, CC., Sampath, S. (Eds.), Thermal spray industrial applications. ASM International, Materials Park, OH, pp. 121–126.

Wang, W.Q., Sha, C.K., Sun, D.Q., Gu, X.Y., 2006. Microstructural feature, thermal shock resistance and isothermal oxidation resistance of nanostructured zirconia coating. Mater. Sci. Eng. A. 424, 1–5.

Wang, Y., Jiang, S., Wang, M., Wang Xiao, S.T.D., Strutt, P.R., 2000. Abrasive wear characteristics of plasma sprayed nanostructured alumina/titania coatings. Wear. 237, 176–185.

Wang, Y., Tian, W., Yang, Y., 2007. Thermal shock behavior of nanostructured and conventional Al_2O_3/13 wt%TiO_2 coatings fabricated by plasma spraying. Surf. Coat. Technol. 201, 7746–7754.

Zhao, X.Q., Zhou, H.D., Chen, J.M., 2006. Comparative study of the friction and wear behaviour of plasma sprayed conventional and nanostructured WC-12%Co coatings on stainless steel. Mater. Sci. Eng. A. 431, 290–297.

Zhu, D., Nesbitt, J.A., Barrett, C.A., McCue, T.R., Miller, R.A., 2004. Furnace cyclic oxidation behavior of multi-component low conductivity thermal barrier coatings, NASA/TM-2004–212962.

Zhu, Y.C., Yukimura, K., Ding, C.X., Zhang, P.Y., 2001. Tribological properties of nanostructured and conventional WC-Co coatings deposited by plasma spraying. Thin Solid Films. 388, 277–282.

Hard thin films: applications and challenges

21

Kausala Mylvaganam, Yiqing Chen, Weidong Liu, Mei Liu, and Liangchi Zhang
The University of New South Wales, Australia

21.1 Introduction

Materials with thicknesses ranging from nanometres to micrometres and high surface-to-volume ratios are called thin films; they behave differently from bulk materials of the same chemical composition in several different ways. Thin films are becoming increasingly important as means of altering the properties of surfaces and interfaces. The first use of thin films for optical purposes can be dated back to 1912. Pohl and Pringsheim published the well-known work about the production of mirrors in which they vaporize metals like Ag and Al in high vacuum (Pohl and Pringsheim, 1912). Traditionally *hard* thin films are characterized with extreme hardness (>17 GPa), high wear resistance, high lubricity, high corrosion resistance and so forth. Thin films of diamond, TiC, TiN, WC and TaC are some examples.

The concept of *hard* thin films can be extended from its traditional tribological connotation to include the capabilities of resisting harsh operating conditions such as high temperature, high pressure or excessive radiation exposures. Properties of thin films are very sensitive to the techniques used in their preparation. Depending on the specific film/substrate pair and on processing and service conditions, thin films may be subjected to very large compressive/tensile stresses. Although a certain level of stress on a thin film is desirable for certain applications (e.g. compressive stresses in the order of 0.01–0.05 GPa are considered beneficial in the integrated circuit manufacturing industry (Granneman, 1993)), large stresses may lead to its failure.

In general, thin films are produced by vapour deposition processes which can principally be divided into physical vapour deposition (PVD) and chemical vapour deposition (CVD). PVD involves purely physical processes such as high-temperature vacuum evaporation with subsequent condensation. The overall process involves four different steps: evaporation, transportation, reaction and deposition. First, the material to be deposited is bombarded by a high energy source such as a beam of electrons or ions. This dislodges atoms from the surface and vaporizes them. Then these atoms are directly transported to the substrate to be coated. If metal oxide, metal nitride or metal carbide films are required, the corresponding metal is bombarded by a high energy source in the presence of appropriate gas (for example O_2 to get metal oxide, N_2 to get metal nitride and methane to get metal carbide) so that they react to form the required material during transportation.

Anti-Abrasive Nanocoatings.

Finally, during the deposition stage the film is built up on the substrate surface. PVD films are harder and more corrosion resistant compared to films produced by other methods, have a good impact strength, and possess excellent abrasion resistance and durability; the process is environmentally friendly. However, very high operating temperatures and vacuums require appropriate cooling systems and special attention, and as a result the process involves high capital costs. The diamond-like thin film and metallic glass thin film discussed later in this chapter can be prepared by the PVD method.

In the CVD process, the substrate is exposed to one or more volatile 'precursor' substances, which react and/or decompose on the substrate, to produce the desired deposit. Depending on the conditions and the precursors used during the deposition process, several acronyms are used. The CVD process that utilizes metal-organic precursors is known as MOCVD. If electrical energy rather than thermal energy is used to initiate reactions for the production of ions and radicals, the technique is known as plasma-enhanced CVD (PECVD) and so forth. In a basic CVD process, a predefined mix of reactant gases is first introduced into a reaction chamber together with a diluent inert gas. Then the reactants get adsorbed on the substrate and undergo chemical reactions with the substrate to form the film. Finally, the volatile by-products produced are desorbed and evacuated from the reaction chamber. During the CVD process the reactant gases not only react with the substrate on its surface, but also in the gas phase in the chamber. The reactions at the substrate surface occur selectively and create good-quality films. However, the reactions that occur in the gas phase form gas phase aggregates of the depositing material, and may not adhere to the surface well, leading to the formation of low-density films with defects. Hence, reactions occurring only at the surface are more desirable. The substrate temperature is critical and can influence the types of reactions that will take place. The CVD process used to produce high-purity and high-performance solid materials is often used in the semiconductor industry to produce thin films. The diamond film, AlN film and GaAs film discussed in this chapter can be produced by the CVD method.

Thin films with super-high hardness alone are not very useful. Appropriate combinations of high hardness with other properties such as high toughness, low residual stress, good adhesion with the substrate and oxidation resistance is of vital importance (Zhang et al., 2005). Toughness is the ability of a film to absorb energy during deformation till its failure; this energy encompasses the energy required to both create the crack and to propagate it until fracture. Various toughening designs and methodologies are used for thin films, of which the common basis is to increase the critical failure stress and to hinder the crack propagation once a crack forms. Toughening can also be achieved if the accumulated strain is properly released via dislocation propagation or grain boundary sliding. Residual stresses (composed of a thermal stress, a lattice mismatch stress and an intrinsic stress) usually exist within a thin film system even in the absence of external loading or thermal gradients (Gioia and Ortiz, 1997).

This chapter will discuss the properties and applications of three types of thin films: those that are hard in terms of traditional tribology (e.g. diamond, diamond-like and

Table 21.1 Hard thin films – an overview

Classification	Hard in terms of	Method of production
Diamond and diamond-related thin films (diamond, diamond-like carbon (DLC) and NCD)	Traditional tribology (high hardness, wear resistance, lubricity, etc.)	CVD/PVD
Semiconductor thin films (AlN, GaAs)	Resisting high T, P and excessive radiation	CVD
Metallic glass thin film (e.g. $Zr_{47}Cu_{31}Al_{13}Ni_{9}$)	Ultra-high strength, excellent elasticity, and super-thermal formability	PVD

nanocrystalline diamond (NCD) films); those that are not hard in general terms of being anti-abrasive under contact sliding but are capable of resisting high pressure (e.g. AlN film), excessive radiation (e.g. GaAs film) and so forth; and those that have ultra-high strength and excellent elasticity while maintaining the structural features of the liquid state (e.g. metallic glass thin film). For convenience, these are briefly summarized in Table 21.1 above.

Following this, the challenges due to residual stresses, surface roughness, defects and delamination will be discussed. Finally, the present state of knowledge and the future prospects in practical applications in the near future will be described.

21.2 Characterization of thin films

21.2.1 Diamond and diamond-related films

Since the last century, there have been very important advances in the scientific understanding of carbon, including the developments of CVD of diamond, and in the field of disordered carbon structures with carbon hybridizations sp^3, sp^2 and sp^1. This section will focus on the material properties, synthesis and applications of diamond, DLC and NCD films, which are traditional extremely hard thin film materials. In diamond structure, each of a carbon atom's four valence electrons forms a strong bond to an adjacent atom (each C is sp^3 hybridized), and in DLC and NCD, a significant fraction of C atoms are sp^3 hybridized.

21.2.1.1 Properties and productions

Diamond possesses a unique combination of excellent optical, thermal, mechanical, chemical and semiconducting properties (Prelas et al., 1998), as highlighted in Table 21.2, of which the hardness, thermal conductivity, bulk modulus, atomic density and sound propagation velocity are among the highest of any known materials. Diamond also has other outstanding properties, including its high electrical resistance, large optical band gap, high transmission in the ultraviolet to infrared regions

Table 21.2 Some outstanding properties of diamond (Prelas et al., 1998; Field, 1992; May, 2000; Chen and Zhang, 2013)

Highest hardness (60–120 GPa, depends on crystallographic plane)
Highest room temperature thermal conductivity (2×10^3 W m^{-1} K^{-1})
Highest bulk modulus (1.2×10^{12} N m^{-2})
Highest sound propagation velocity (17.5 km s^{-1})
High room temperature electrical resistivity (c. 10^{13} Ω cm)
Broad optical transparency from the ultraviolet to the far infrared
Lowest compressibility (8.3×10^{-13} m^2 N^{-1})
Extremely low thermal expansion coefficient at room temperature (1×10^{-6} K)
Low electron work function, can be doped to become a semiconductor with a wide band gap of 5.4 eV

of the spectrum, chemical inertness to most corrosive environments, low adhesion and friction, extremely low thermal expansion coefficient and compressibility and excellent biological compatibility. The full industrial exploitation of the exceptional properties of diamond has taken a new leap forward since the successful synthesis of diamond, especially the rapid development of CVD technology to produce diamond thin films.

Diamond films can be produced by the CVD method, which requires the activation of the carbon species and the action of atomic hydrogen (Granneman, 1993). The activation can involve thermal methods (e.g. a hot filament), electric discharge (e.g. direct current (DC), radio frequency or microwave) or a combustion flame (e.g. an oxyacetylene torch). The growth of diamond requires that the carbon precursor gas (e.g. CH_4) is diluted in an excess of hydrogen (usually in a mixing ratio of 1 to 5 vol.% of CH_4) and the temperature of the substrate is normally greater than 700°C to ensure the formation of diamond rather than amorphous carbon. CVD diamond films are polycrystalline diamond (PCD) with many grain boundaries and defects, and exhibits a columnar structure extending upward from the substrate. As the film becomes thicker, the crystal size increases while the number of defects and grain boundaries decreases. The CVD diamonds are available in the form of extended thin films and free-standing plates or windows.

The diamond coatings prepared by CVD techniques are generally rough and non-uniform over large areas, which cannot be used directly in many applications. Potential solutions are post-processing in the form of polishing, or of the deposition of diamond-related films, for example DLC, tetrahedral amorphous carbon (ta-C) or NCD. These films are much smoother, and to some extent, contain similar properties of PCD coatings, such as their hardness, elastic modulus, chemical inertness and biocompatibility (Popov et al., 2004).

DLC or ta-C films are a metastable form of amorphous carbon containing a significant fraction of sp^3 type carbon bonds and up to 50 percentage of hydrogen (Robertson, 2002). DLC has attractive properties that are similar to diamond, such as its high mechanical hardness, chemical inertness, optical transparency, and is a wide band gap semiconductor. Compared to CVD diamond films, DLC films are extremely smooth because DLC is amorphous with no grain boundaries. DLC films can be made by PECVD or PVD techniques, including ion deposition, ion assisted sputtering, sputtering, catholic vacuum arc and pulsed laser deposition. By using PVD methods, DLC films can be deposited at room temperature, which can be applied on temperature-sensitive substrates such as plastics. It is much cheaper to produce DLC than diamond films, and thus has great advantages for many applications. The disadvantages of DLC films are their intrinsic stress and thermal instability, as the films are prepared by ion-assisted methods leading to high compressive stresses. The stress in the films directly affects their adhesion to the substrate and causes delamination.

NCD films are composed of diamond crystallites of nano size, for example 3–5 nm separated by grain boundaries of about 1–1.5 nm; thus they are very smooth and retain to a great extent the outstanding properties of PCD coatings (Popov et al., 2004). The optical properties of the NCD/a-C films are far from those of diamond film because they are determined by the properties of the amorphous matrix rather than by those of diamond nanocrystals (Popov et al., 2006).

In general, there are two main approaches towards the deposition of NCD films, either in a 'pure' form or as nanocrystallites embedded in an amorphous matrix. The first is 'the polycrystalline diamond route' (Popov et al., 2004) which is based on the standard CVD diamond techniques, but carried out at deviated process parameters to interrupt the crystal growth and to enhance the nucleation leading to nanocrystallites. The variant conditions include lower working pressure, higher methane concentrations, partial or complete substitution of H_2 by N_2 or Ar and application of bias voltage during the entire process. By using CVD techniques without ion impact on the growing films, the stress can be minimized. The second approach is 'the DLC and ta-C route', based on ion-assisted techniques such as sputtering or PECVD. Diamond nanocrystals in an amorphous matrix can be achieved by using high pressures, increased concentrations of Ar in the plasma as well as the addition of H_2 and provision of sp^3-rich under-layers.

21.2.1.2 *Applications*

CVD diamond, DLC and NCD films have widespread mechanical, optical, biomedical, thermal and electronic applications as protective coatings in many areas, such as cutting tools, wear components, optical windows, thermal management, detectors and sensors and bio-related components.

21.2.1.2.1 Cutting tools and wear components

The comprehensive properties of extreme hardness, high abrasion resistance, high thermal conductivity and low coefficient of friction make CVD diamonds versatile industrial materials for cutting, drilling and milling tools. The CVD diamond films can be directly coated onto a cutting tool material, such as on tungsten carbide (WC) alloys. Commercial diamond coatings generally fall in the thickness range of 5 to 35 μm. Such CVD diamond-coated tools can offer not only the promise of longer tool life under standard operating conditions, but also opportunities for cost-effective high-speed and reduced coolant operations for machining non-ferrous metals, plastics, chip-board and composite materials which are difficult-to-machine by conventional materials. In addition, both the unique characteristics of diamond and the flexibility of CVD diamond processes allow CVD diamond films for unique wear applications, such as in bearings, seals, wire dies, engraving tools and dressing tools for grinding wheels.

DLC coatings are also used in higher temperature cutting applications and have an advantage over diamond coatings in edge sharpness, coating surface roughness and much lower sensitivity to substrate selection due to low-temperature coating process. Although the costs of DLC tooling are markedly lower than those of CVD diamond coatings, the cost-performance advantage of crystalline CVD diamond is obvious in many applications. DLC films also have extremely good coverage; so they could act as good corrosion barriers. The most widespread use of DLC films is in wear and corrosion protection of magnetic storage media. Nano-smooth and ultra-thin (<50 nm, and even <10 nm) DLC films are used as corrosion and wear-protective coatings for both the magnetic disks and the magnetic heads to reduce wear and friction and to extend their life and reliability (Robertson, 2002; Grill, 1999).

21.2.1.2.2 Optical windows

Diamond is a desirable material for optical application because of its widest spectral optical transmission range of all known solid materials, coupled with its remarkable thermal properties, scratch resistance and chemical inertness (Gicquel et al., 2001; Asmussen and Reinhard, 2002). Due to the high cost of high-quality diamond, most of the optical applications are related to operations in severe environments, such as missile radomes, output gyratron windows and so on (Gicquel, et al., 2001). In addition, diamond thin films can be coated on optical substrates. For example, the direct CVD of diamond on a common optical material, silicate-based glass, which lacks the resistance to erosion and abrasion, is of interest in enhancing its resilience in hostile environments. Diamond films with a thickness of less than 10 μm can be used for other optical applications, such as polarizers and membranes for X-ray photolithography masks (Gicquel et al., 2001). For its applications in the visible to infrared range, the transparency can be optimized during growth by finding a compromise between intrinsic quality and roughness, such as producing 1–2 μm thick membranes for applications in the visible range.

A DLC film can be used as an anti-reflective and scratch-resistant wear-protective coating for infrared optics made of Ge, ZnS and ZnSe (Grill, 1999). The low deposition temperature of DLC allows its use as a wear-protective layer on plastics, for example, for the protection against abrasion of sunglass lenses made of polycarbonate.

21.2.1.2.3 Biomedical components

Diamond is known to be inherently biocompatible, together with its other exceptional properties mentioned above. Thus diamond thin films are ideal for medical applications, and, in recent years, have attracted significant research efforts into exploring their applications in biomedicine, biotechnology and biosensorics (Popov et al., 2006, 2008; Hamers et al., 2005). For example, diamond films have been proposed as a material for glucose sensors, for catheter ablation in the heart (Muller et al., 2004), and as encapsulation for retinal implant microchips. In addition, diamond films are being explored as a template for the immobilization of biomolecules in biological investigations or in biosensoric applications (Popov et al., 2007).

However, the rough and non-uniform surfaces of diamond coatings prepared by CVD techniques prevent their applications in many bio-related fields. For example, the rough diamond coatings on non-plane surface implants are extremely difficult to achieve through the post-deposition process. NCD films are the best candidates for these applications, as they (1) retain the diamond properties to a great extent; (2) the surfaces are smooth and can be adjusted to optimize friction; and (3) can be made conductive by boron doping if properly designed. Recently, NCD films have been investigated for application in biosensorics, as coatings for implants and templates for the immobilization of biomolecules (Popov et al., 2006, 2008; Carlisle and Auciello, 2003; Yang et al., 2002; Kulisch et al., 2011).

DLC films are also being developed for their applications in biological environments due to their biocompatibility, chemical inertness and impermeability to liquids (Robertson, 2002; Grill, 1999). Various forms of DLC have been investigated as biocompatible protective coatings, deposited on stainless steel and titanium alloys used for the components of artificial heart valves, replacement hip joints and stents. It has been found that DLC coatings can satisfy both mechanical and biological requirements and can be capable of improving the performance of these components (Grill, 1999; Roy and Lee, 2007). In addition, DLC films are being studied as coatings of metallic and polymeric bio-components to improve their compatibility with body tissues. These coatings show improved anti-wear and anti-corrosion performance and excellent in-vitro cyto-compatibility, when compared with the currently used uncoated Co-Cr-Mo and Ti-Al-V implant materials (Hinuber et al., 2010).

21.2.1.2.4 Thermal management and electronic devices

The high thermal conductivity of diamond, coupled with a low thermal expansion coefficient, chemical inertness and high electrical resistivity, makes diamond the best material for a variety of thermal management applications, especially in electronic devices (Gicquel et al., 2001; Malshe and Brown, 2002). The CVD polycrystalline diamond films can be used as sub-mounts for integrated circuits, heat spreaders for high-power laser diodes, or as substrate materials for multi-chip modules. In addition, diamond films are of interest in thermal management of electrical packaging applications, as they provide efficient paths for heat flow without compromising the electrical isolation of individual components.

For electronic applications, diamond has a set of characteristics that can improve performances of the devices and push further their limits of use (Gicquel et al., 2001; Asmussen and Reinhard, 2002): (1) the high resistivity of diamond allows a substantial decrease in current leakage; (2) the excellent thermal conductivity is crucial to high power and high integration devices; (3) the high mobility of carriers, low dielectric constant and high electronic gap (5.5 eV) enable operations at very high frequencies with low parasite capacitance and high protection to noise, as well as at high temperature (up to 500°C). In addition, the possibility of doping diamond and changing it from an electrical insulator to a semiconductor opens up a whole range of potential electrical applications. Diamond films can be used not only for applications involving semi-conductors, such as field effect transistors, Schottky diodes which can operate at up to 1000°C, photo-conductor switch-off, high voltage–high current switches, but also for realizing specific devices such as electrochemical sensors and detectors. For most of the electronic applications, the crystallographic perfection of the films, purity and the level of doping govern the performances of the devices. Non-diamond phases and any defect will influence the electronic behaviour.

21.2.2 Semiconductor hard thin films

In the last two decades, the group III–V compound semiconductor thin films have attracted tremendous attention, driven by the rapid development of electromechanical and optoelectronic systems. The development of these advanced materials with superior environmental durability (e.g. high pressure and temperature resistance) is the subject of considerable interest. Semiconductor thin films applied in electronic or photovoltaic devices have been extended to outer space and high-radiation environments. In this context, the high-energy electron or proton irradiation will produce atomic displacements in semiconductor materials, resulting in lattice defects such as vacancies, interstitials and so on. These defect sites will bring about an increase of recombination or trapping centres of carriers, which degrade the electronic properties of these devices. Therefore, a material with high resistance to radiation damage is an essential requirement in space applications. The group III–V compound semiconductor thin film materials, AlN and GaAs, have been widely exploited as the components of electromechanical, optoelectronic and photovoltaic devices.

21.2.2.1 AlN film

21.2.2.1.1 Properties and productions

AlN is the group III–V compound semiconductor with a stable wurtzite (WZ) structure, and the wurtzite to rock salt phase transformation occurs at a pressure as high as 22.9 GPa (Ueno et al., 1992). The large band gap (~6.2 eV) makes AlN an ideal material for optoelectronic applications in the deep ultraviolet (UV) spectral region. AlN has a high piezoelectric coefficient ($d_{33} = 5.3$ in the longitudinal direction), high electrical resistivity ($10^{11}-10^{14}$ Ω cm) and a high sound acoustic wave velocity (>5500 m/s), making it an excellent piezoelectric material for microelectromechanical systems (MEMS) applications. The inherent physical properties of AlN such as its extremely high microhardness (~20 GPa), high thermal conductivity (320 W/m K) and good chemical stability allow for its extensive applications in aggressive conditions.

The common growth techniques for AlN films are MOCVD (Dovidenko et al., 1996), plasma-assisted molecular beam epitaxy (PAMBE) (Rowland et al., 1993), pulsed laser deposition (PLD) (Vispute et al., 1995) and reactive magnetron sputtering (RMS) (Ishihara et al., 1998), on the substrates of sapphire, silicon and 6H-SiC. PLD and RMS are the most frequently utilized techniques in AlN growth because of their relatively lower deposition temperatures and greater controllability compared to PAMBE and MOCVD. PLD has the advantage of retaining the electronic properties of the original target material; however, this technique suffers from a relatively smaller deposition area. In contrast, the RMS is more effective for large-scale depositions with a higher reliability and a good control of film properties. A comparative study (Jagannadham et al., 1998) showed that under similar deposition conditions, the PLD technique produced a larger compressive stress than the RMS in an AlN film, resulting in a higher value of hardness but a lower interface adhesion strength. The surface tribological properties of sputtered AlN films could be modified by ion (e.g. B^+ and C^+) bombardment at a low-dose density. The improvement is due to a chemical effect rather than radiation damage (Alkjaersig et al., 1992).

21.2.2.1.2 Applications

- **MEMS:** Because of the excellent piezoelectric property and large surface acoustic wave velocity, AlN thin films show their promise in the applications of MEMS devices such as resonators, microphones and switches, and so on. To optimize the electromechanical coupling coefficient, the growth of high-quality c-axis textured polycrystalline (column structure) AlN films is essential, such that the piezoelectric coefficient along the film growth direction (d_{33}) is maximized. AlN films have demonstrated various advantages over other traditional piezoelectric materials with respect to their exceptional environmental resistance. For example, an AlN thin film is regarded as an ideal futuristic piezoelectric material in high-altitude aerospace applications, because it can withstand extremely high pressure, is insensitive to temperature variation, and possesses high stiffness and low dielectric loss. These features make an AlN thin film a potentially promising material for acoustic microphone components employed in the aircraft fuselage arrays to test and reduce cabin noises (Williams et al., 2012). Experimental results have testified that the sensitivity of microphones could be improved by a factor of 50 with an AlN thin film compared with that of the traditional PZT (lead zirconate titanate) based microphone (Horowitz et al., 2007). Besides, attributed to its excellent biocompatibility, AlN has shown more prospects than for bioMEMS device applications. For example, the integration of a piezoelectric AlN layer with an ultrananocrystalline diamond (UNCD) membrane shows excellent potential for applications in implantable drug delivery systems (Luong and Nguyen, 2010), because a very thin AlN layer of 200 nm with a very good crystalline quality could be obtained by chemical mechanical polishing on the UNCD surface prior to deposition. The high crystal quality guarantees a high piezoelectric coefficient ($\sim$5.3 pm/V) in MEMS devices, which promises a significant spectrum of applications in biocompatible piezoelectric devices implantable in the human body.

- **Deep ultraviolet (DUV) light emitting diode (LED):** Recently, the group III−V nitride LEDs have generated much interest, stemming from their ability to emit DUV light, which can be of benefit in environmental protection and biochemical engineering (Kneissl et al., 2011). The application of DUV emitters in water purification (Shur et al., 2005) relies on its high-energy photons, which can kill bacteria and viruses or decompose them into very stable substances. The semiconductor LEDs have more advantages compared to traditional gas sources because they are free of toxic matter and are relatively more energy efficient. Being the largest band gap nitride, thin film AlN appears exceedingly attractive for DUV applications. Recently, the shortest wavelength (210 nm) in the DUV region was obtained from an AlN thin film (Taniyasu et al., 2006), which was achieved by controlling the n-type and p-type doping concentrations in the AlN PIN (p-type/intrinsic/n-type) junctions. The high density of intrinsic defects and impurities from the AlN growth are the main obstacles against achieving higher output power and narrower wavelengths. To suppress the movement of atoms on the crystal surface, AlN is normally deposited by the MOVPE technique at a temperature lower than 1000°C. Other attempts have been made, for example, by increasing film thickness (Pantha et al., 2007) or by applying pre-growth 'nitridation' and 'alumination' to the substrate surface (Reentilä et al., 2008) and so on. These approaches have demonstrated significant enhancements of the film quality, which, in turn, has provided more potential in upgrading the AlN-based DUV-LEDs with respect to emission wavelength and power efficiency. In addition, AlN films have pronounced thermal stability and high radiation hardness, resulting in superior electronic properties such as low dark current and high breakdown-voltage for the photodetector applications. These merits have broadened the scope of AlN-based DUV-LEDs for use under extreme conditions such as in space communications and for solar vacuum UV photon detection (Li et al., 2006).
- **Protective coatings:** Multilayer coatings with nanoscaled periodicity (i.e. superlattice) for improving wear resistance has been a subject of great interest in surface engineering. The integration of different materials could result in synthesized protective properties (Stueber et al., 2009). Pioneering work has shown that introducing nano-layer AlN between traditional protective layers such as TiN (Thobor et al., 2000) or CrN (Park and Baik, 2005) to create multilayered coatings allows a remarkable level of optimization of the chemical inertness and tribological properties of these single-layer coatings to be achieved. The enhanced wear resistance is attributed to dislocations being prevented from propagating or increased adhesion and stress relaxation with the increased number of interfaces (Chu and Barnett, 1995). Besides, the toughness of the multilayered coatings are also improved by crack deflection or dissipation of crack energy at the interfaces (Was and Foecke, 1996). It was found that the bilayer period, Λ , is essential for the tribological properties of the multilayered film. For example, as shown in Figure 21.1(a), the hardness of the CrN/AlN multilayered film increases with the decrease of Λ (Park and Baik, 2005). The maximum hardness of 40 GPa was observed when $\Lambda_c = 3.8$ nm, which is regarded as the critical bilayer period in this multilayer system. The hardness maxima at Λ_c is driven by the wurtzite-rocksalt phase transformation due to pseudomorphic stabilization of the non-equilibrium rocksalt-type (B1) AlN phase. When $\Lambda < \Lambda_c$, the breakage of the nitrogen bond will eventually result in a drop in hardness. The results of chemical compositional auger electron spectroscopy (AES) (Figure 21.1(b)) show that the oxidation resistance was largely improved by the multilayered CrN/AlN film, evidenced by the reduction of the oxidation layer compared to the AlN and CrN monolithic layers. The chemical stability of multilayers integrated with AlN has brought about a wide range of uses for high-temperature applications.

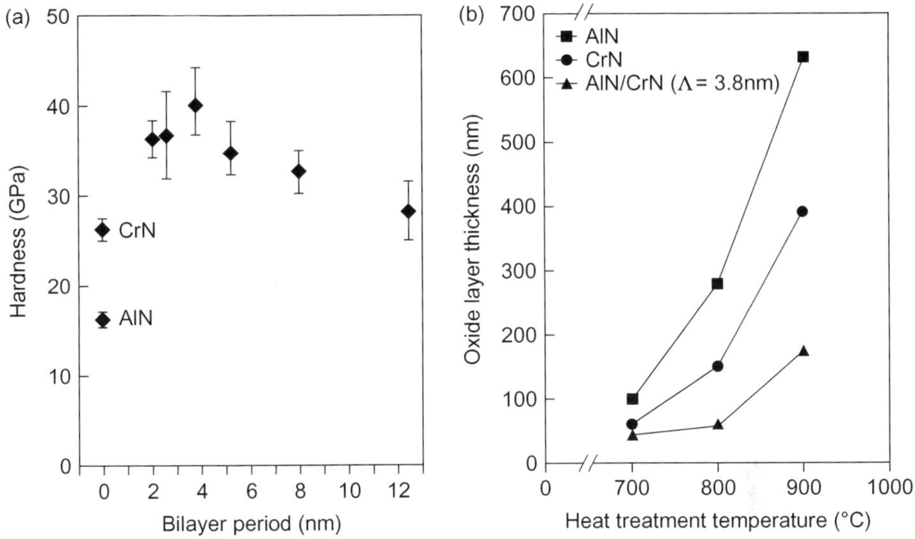

Figure 21.1 Enhanced properties of AlN/CrN coatings (a) hardness as a function of bilayer period, and (b) surface oxide layer thicknesses of AlN, CrN single layers and AlN/CrN superlattice coatings (Λ = 3.8 nm) in air at various temperatures for 1 hour (after Park and Baik, 2005).

21.2.2.2 GaAs film

21.2.2.2.1 Properties and production

GaAs is a group III–V compound semiconductor with a zinc blende structure. The thin film GaAs has tremendous potential as a solar cell material owing to its outstanding photovoltaic properties. For example, GaAs exhibit high irradiation resistance mainly attributed to its immunity to accumulated radiation doses. The total-dose tolerance is up to 1 Gad, which is about two orders of magnitude higher than that of Si-based devices. More promisingly, the radiation damage could be readily recovered by annealing at the low temperature of 200°C. This special feature has facilitated GaAs the best fit for space applications (Loo et al., 1990; Iles, 2001; Law et al., 2010). Besides, the direct band gap (1.42 eV) has provided GaAs a wide range of applications in optoelectronic devices such as LEDs, lasers and detectors. The GaAs films grown at a relatively low temperature (LT-GaAs) demonstrate exceptional characteristics, e.g. high resistivity and sub-picosecond (<0.4 ps) carrier lifetime (Gupta et al., 1991), high breakdown field (Yin et al., 1990) and high carrier mobility (~200 $cm^2 V^{-1} S^{-1}$). These fascinating optoelectronic properties are the primary driving force for the growing interest in GaAs thin films for generating terahertz-frequency radiation. Moreover, as a transistor, GaAs has demonstrated overwhelming advantages over Si in terms of a higher electron mobility and a higher breakdown voltage, which allows the GaAs-based transistors to tolerate high-power and high-frequency (250 GHz) conditions but with less noise.

The development of GaAs thin-film technology was sluggish before the 1980s, primarily due to the lack of low-cost structurally compatible substrate materials which are needed for producing high-quality GaAs films. A wealth of applications emerged after the success of hetero-epitaxially grown high-quality GaAs films on Si, Ge or GaAs substrates by MOCVD (Manasevit, 1981) and MBE (Kawabe and Ueda, 1987). GaAs grown by MBE at low temperatures of about 200°C (LT-GaAs) with excess arsenic (Kaminska et al., 1989). possesses unique electrical properties such as ultra-short carrier lifetime and high electric breakdown field. By subsequent annealing, arsenic clusters are formed and act as efficient traps for photo-induced carriers. The superlattice intermixing and impurity diffusion-induced gallium vacancies in LT-GaAs are the main technical barrier, resulting in thermal instability and reduced operating times. Annealing at 600°C can reduce the vacancies by controlling the precipitates (Melloch et al., 1990) and produce good quality GaAs film for various applications.

- **GaAs solar cell:** The higher absorption coefficient and tuneable direct band gap by alloying the multi-junctions provide GaAs solar cells with a higher conversion efficiency (29.1% for single-junction GaAs and over 30% in the multi-junction cell), compared with the traditional silicon solar cells (27.6%) (Green et al., 2012). When subjected to radiation exposure, a GaAs cell is more resistant than a silicon cell due to its direct band gap and resultant high photon absorption coefficient. As such, less film thickness is required in the GaAs cell to absorb solar radiation. In this context, most high-energy radiation particles such as protons and electrons will pass through the active region of the GaAs solar cell and produce the least radiation damage. However, high material cost is the primary challenge for GaAs-based solar cells. The innovative epitaxial lift off (ELO) technique (Yablonovitch et al., 1987; Schermer et al., 2006; Cheng et al., 2013) has largely reduced the material cost of GaAs by enabling the reuse of the expensive substrate. More importantly, the reduction of the film thickness resulting from ELO could considerably improve the radiation hardness (Schermer et al., 2006), allowing for more perspectives of GaAs in space applications (Schermer et al., 2005). The recent concept of 'releasable multilayer epitaxial assemblies' (Yoon et al., 2010) (Figure 21.2) could prevent the problems (e.g. film handling and low etch rate) in the ELO technique. In this approach, the multilayered GaAs are stacked with AlAs films, and could be easily separated by selective etching of

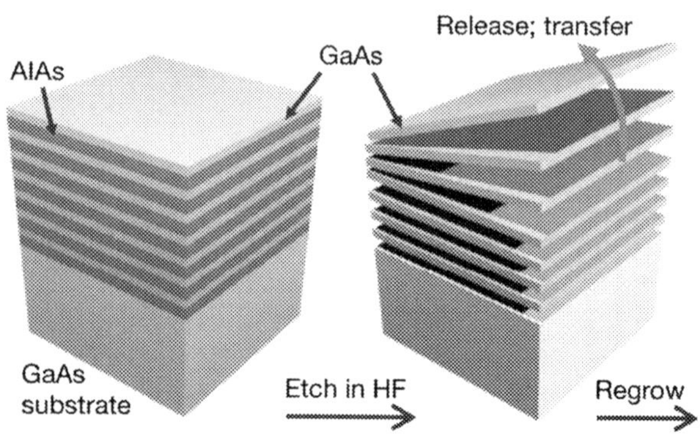

Figure 21.2 Schematics of fabrication of releasable GaAs solar cells (after Yoon et al., 2010).

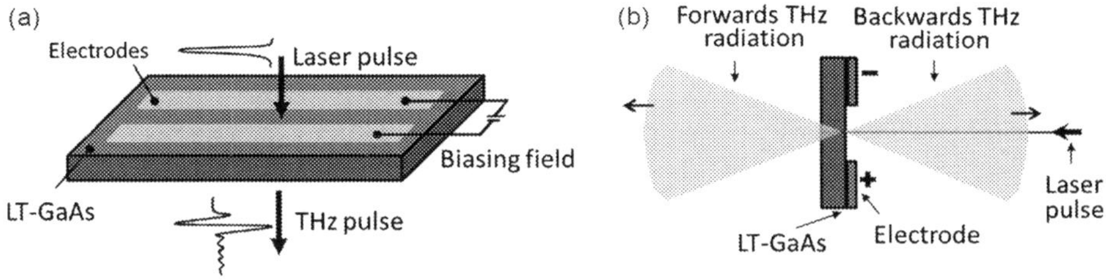

Figure 21.3 (a) Schematic of an LT-GaAs photoconductive switch THz emitter, (b) the forwards and backwards THz radiation scheme (after Shen et al., 2003).

the AlAs layers. In addition, the approaches based on plasmonic nanostructures (Pillai et al., 2007) for light trapping provide a further reduction of the thickness of the absorber layer, leading to an enhanced absorption enhancement of ~31% and an improved radiation resistance (Hong et al., 2012).

- **LT-GaAs Terahertz (THz) wave emitter:** The THz radiation (from 0.1 THz to 10 THz) is attractive in the scientific and technical applications of time-domain imaging and detecting (Jepsen et al., 2011; Davies et al., 2008). To date, LT-GaAs is regarded as the best material for broad band THz generation and detection because of its short carrier lifetime and high resistivity. As shown in Figure 21.3(a), the THz radiation is generated from the ultrafast impulses, induced by a femtosecond laser pulse impinging onto a biasing field on the semiconductor (LT-GaAs) surface. The femtosecond laser pulse excitation produces a physical separation of electrons and holes and consequently a sudden electrical current opposite to the applied biased field, and fast temporal changes of the electric field produces electromagnetic waves (0.3 to 3 THz for LT-GaAs).

Enormous efforts have been made to generate tuneable THz radiation, which was achieved by the 'photomixing' of multiple lasers (Brown et al., 1993; Tani et al., 2005). One of the attempts was to alter the collection method from the conventional forwards to the backwards mode (Shen et al., 2003) (Figure 21.3(b)), resulting in the maxima of THz frequency up to 20 THz. Other approaches were to enhance the efficiency and output power of THz radiation through the innovative structural design of LT-GaAs, for example, by applying multilayered GaAs (Rihani et al., 2010) and GaAs/n-GaSb heterostructure (Sadia et al., 2012). These technological improvements (higher efficiency and broader bandwidth) have enabled the generation of coherent radiation for a wide spectrum of technical applications in scientific analysis (Jepsen et al., 2011), hazards detection (Davies et al., 2008), and skin cancer identification and so on.

21.2.3 Metallic glass thin film

Compared to other thin films discussed in this chapter, metallic glass thin film (TFMG) is fairly new, and has been the subject of a growing number of research studies in both academia and industry ever since the first $Au_{75}Si_{25}$ sample was produced by Duwez at Caltech, USA, in 1960 (Klement et al., 1960). TFMGs are generally fabricated by fast cooling from a melt, thus maintaining the structural features at the material's liquid state. Different from traditional crystalline metals, the microstructure of TFMGs is amorphous, i.e. the material does not have a crystal grain or any long-range atomic periodicity (Greer et al., 2002; Johnson, 2002; Wang et al., 2004). Because the metallic bond is much freer than a covalent or an ionic bond, TFMGs have the advantages of both glass and metal, demonstrated by their excellent electrical and thermal conductivity, which are far superior to that of common glass, and the ultrahigh strength of TFMGs which is far higher than their crystalline counterparts. Also, their super-plasticity allows MEMS components to be easily formed in a supercooled liquid regime (Chen, 2008; Inoue, 2000).

The following sections will briefly introduce the production, properties and potential applications of TFMGs. The unique properties of TFMGs discussed below include elasticity, strength, and superthermal formability. These properties lead directly to useful applications in MEMS/nano-electro-mechanical systems (NEMS), thermal replication, biomedical tools, and protective coating.

21.2.3.1 Properties and production

TFMGs were developed in the 1960s and 1970s were based on binary Au\Si, ternary Pd\X\Si (X could be Ni, Co or Fe), and Pd\Si\Y (Y could be Ag, Cu or Au) systems (Daveis, 1983; Johnson, 1986; Luborsky, 1983), fabricated by some rapid solidification methods such as 'splat quenching'. In order to inhibit the formation of a crystalline structure, a fast quenching rate should be applied, which will in turn limit the thickness of the TFMG produced, typically in the order of micrometres. Although the breakthrough in size has expanded their potential applications since the late 1980s, metallic glass does not seem to have overcome the classical property evolution barrier of its ductility being traded off by its increase in strength. Recently, some studies (Greer et al., 2005; Jang and Greer, 2010) have revealed that metallic glass can have a very high toughness at the micro-scale, indicating that TFMGs can be promising as protective coating materials (Chu et al., 2012). In this context, TFMGs are usually deposited onto a substrate directly, to take full advantage of their unique properties such as wear resistance, high-level hardness and resistance to bacterial action (Chiang et al., 2006).

The investigations of the mechanical properties of TFMGs are one of the most interesting research topics because of its ultrahigh strength and excellent elasticity. As summarized in Figure 21.4, the phenomenological studies of its deformation, by both theoretical and experimental means, have led to the conclusion that the plasticity of TFMGs is inhomogeneous shear-banding mediated at a low temperature and becomes rheological homogenous flow at a high temperature (Schuh et al., 2007; Spaepen, 1977). TFMG is very brittle at room temperature, especially under tension (Lu et al., 2003). Besides, its strength is insensitive to a wide range of strain rates, but will become weaker as the strain rate is beyond $10^3\ s^{-1}$ (Liu and Liu, 2010). In the supercooled liquid regime, however, MG demonstrates super-plasticity and rheological flow at high temperature and is dependent on strain rate (Schuh et al., 2007; Lu et al., 2003). Moreover, in the transition regime, the viscosity of MG changes across about a dozen of orders of magnitude. These unique properties and

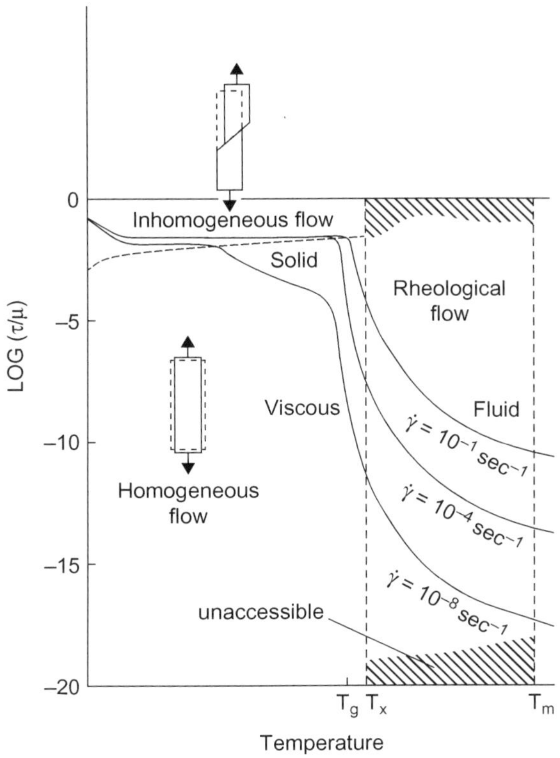

Figure 21.4 Mechanical deformation map of metallic glass (after Spaepen, 1977).

dramatic transition characteristics have stimulated a vast wave of MG applications in making MEMS, and of developing new experimental tests for revealing the physics and mechanisms behind the extraordinary deformation behaviours of MG.

The high strength of MG arises from its particular defect-free atomic structure. In contrast with crystalline materials, MGs do not have common plastic-based intrinsic defects such as dislocation and grain boundaries. This makes their actual strength much closer to their theoretical strength. In some metal-metalloid systems, the yield strength of MGs almost exceeds 4 GPa. The yield strength of metallic glass is determined by the cooperative shear motion of shear transformation zone (STZ), which is analogous to the movement of dislocations in crystalline materials. Recently Johnson and Samwer (Johnson and Samwer, 2005) have analysed the room temperature elastic constants and compressive yield stresses of about 30 types of metallic glass, based on which they reported that the plastic yielding in metallic glass is related to the glass transition temperature.

Although TFMGs are less ductile at room temperature, they demonstrate superplasticity in their glass transition region, making them excellent materials for the thermal forming process. Homogeneous deformation occurs in the glass transition region. The transition temperature between the inhomogeneous and homogeneous deformation is considered a brittle-to-ductile transition, and is strongly dependent on the applied strain rate, suggesting that homogeneous deformation is rate-dependent (Schuh et al., 2007). By conducting steady-state flow tests over a large range of stresses and strain rates, and in a number of different alloy systems at different temperatures in the glass transition region, it was observed that at high temperatures and low applied stresses, the fluid is Newtonian. With increasing stress, however, the stress sensitivity of deformation drops rapidly and the fluid becomes non-Newtonian (Lu et al., 2003).

21.2.3.2 Applications

21.2.3.2.1 MEMS/NEMS

TFGMs have many attractive properties, for example, very high strength of 1 to ~4 GPa, defect-free amorphous structure, high formability in the supercooled liquid range and high corrosion resistance. These features make the TFMGs suitable for the fabrication of robust MEMS/NEMS components (Schroers, 2005). The polymer-like formability of metallic glasses in their supercooled liquid region is already utilized for the fabrication of 3D microstructures (Hata et al., 1999; Jeong et al., 2003). For such applications, an amorphous alloy synthesized by a sputtering process can provide a wide compositional window for the glass forming ability without micro segregation (Schroers, 2005). Hata et al., 2003, utilized the super-plastic forming ability of Pd-based TFMG to produce micro-actuators as shown in Figure 21.5. The tensile strength of this material is about 1.14 GPa, nearly the same as that of polysilicon, and is two times larger than that of stainless steel (~500 MPa). Its Young's modulus is 63.4 GPa, which is lower than that of silicon (~170 GPa) and stainless steel (~210 GPa). Furthermore, the elastic limit is about 2%, which is much higher than that of other conventional alloys (~0.2%). In addition, since the TFMG has an amorphous structure, the effect of anisotropy and

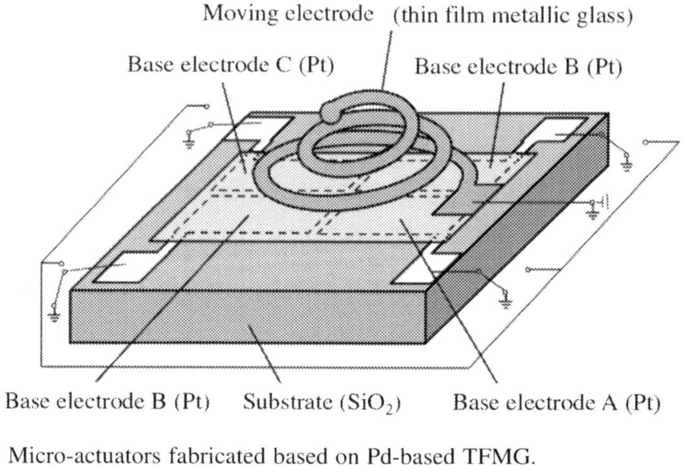

Figure 21.5 Micro-actuators fabricated based on Pd-based TFMG.
Source: Modified from Hata et al., 2003.

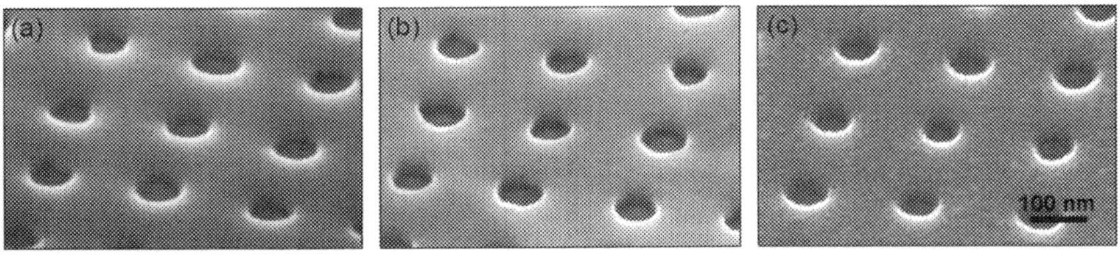

Figure 21.6 SEM surface images of (a) Zr-based, (b) Pb-based and (c) Cu-based TFGMs in the as-imprinted states (after Takenaka et al., 2011).

inhomogeneity are insignificant, making it more advantageous than polysilicon and other commonly used metals for application in springs for the moving electrodes in MEMS.

21.2.3.2.2 Thermal replication

Replication of devices, parts and patterns with features can be done by imprinting features from a hard mould. The imprint process promises low material costs and allows mass production. TFMGs exhibit viscous flow in the supercooled liquid region. Thus, they are sufficiently malleable for forming micro- and nanoscaled components with complex surface features, and are therefore of commercial promise (Schroers, 2010). Figure 21.6 presents an imprinted hole array on 100 nm-thick Zr, Cu, and Pd-based TFMGs (Takenaka et al., 2011). The main advantages of using TFMGs over conventional moulding materials are the high strength, excellent wear and corrosion resistance, low process force, ease of demoulding, good surface quality and intrinsic homogeneity at the nanoscale due to the absence of grain boundaries (Chu et al., 2012). In addition, replicated surface features of TFMG components are being used as moulding dies to further create second-generation replicas in conventional polymethylmethacrylate (PMMA) (Chu et al., 2007; Henann et al., 2009).

21.2.3.2.3 Biomedical tools

Potential applications of Zr-based TFMGs can be extended to medical tools such as surgical blades and micro-surgery scissors. The sharpness of the blades and scissors is a key parameter in cutting biological tissues, and it is related to the quality of the cut surface, and the life of the cutting instrument. Commercial blades made of martensitic stainless steel always present micron-scale roughness on the surface and tip of the cutting edge. This rough blade-edge directly influences the blade sharpness, surgical quality, and eventually the performance and life of the tool. Due to the defect-free structure and self-smoothing properties, TFMGs coatings have been used to decrease the blade sharpness index value from $\sim$0.36 to $\sim$0.25 and a 26.5% improvement in sharpness (Tsai et al., 2012). Additionally, the films have extremely smooth surface profiles (roughness $\sim$1 nm) and a homogenous composition distribution, and exhibit strong antimicrobial effects for different microbes including *E. coli, S. aureus, Pseudomonas aeruginosa, A. baumannii,* and *Candida albicans* for at least 24 hours (Tsai et al., 2012).

21.2.3.2.4 Protective coatings

Beneficial effects of TFMGs on fatigue properties of crystalline metallic substrates have been reported, including sputter-deposited Zr, Cu and Fe-based materials on substrates such as stainless steels, Ni and Al-based alloys. Chiang et al. have reported that the fatigue life of a 316L stainless steel coated with a 200 nm thick $Zr_{47}Cu_{31}Al_{13}Ni_9$ film can be considerably improved by at least 30 times (Chiang et al., 2006). The application of TFMGs yields an increase of the fatigue-endurance limit by 30%. Coupled with a 50 nm titanium thin film buffer layer, Chang et al. coated a 200 nm Zr-based TFMG on to the 7075-T6 aluminum alloy by DC sputtering (Chu et al., 2012). They found that the fatigue limit of the coated samples was improved by 235 MPa (56.7% increase), in comparison with 150 MPa for uncoated samples. The fatigue improvements are attributed to several factors such as higher strength and ductility of the TFMG. The good adhesion and compressive residual stress are found to play beneficial roles in achieving superior fatigue properties. In addition, a smooth surface is obtained with the TFMG coating, which is beneficial for preventing crack formation.

21.3 Challenges

While thin films have major applications in a variety of fields as discussed above, there are always challenges during production and usage. For example, virtually all vapour-deposited coatings are in a state of stress, independent of the method of production and, as a result, films crack, buckle and/or delaminate. Moreover, different conditions and growth methods lead to a wide variety of different surface/interface morphologies and film microstructures. These, in turn, have a significant influence on surface roughness. Thus the effect of residual stress and surface roughness and the problems encountered are discussed below in detail with examples.

21.3.1 Residual stress

Thin films deposited on substrates are often in a state of residual stress that may affect adhesion, generation of crystallographic defects, and formation of surface growth such as hillocks and whiskers. The total residual stress is composed of a thermal stress, a lattice mismatch stress and an intrinsic stress. Thermal stress arises as a result of mismatches in the thermal expansion coefficient of film and substrate. For example, in the diamond/silicon system, the thermal stress acted as a compressive stress around -0.25 GPa. The intrinsic stress of films is very much dependent on growth conditions. For CVD diamond film deposited on silicon, intrinsic stress changed from compressive to tensile with increased film thickness (Kim and Yu, 1998). On the other hand, when depositing AlN films by reactive magnetron sputtering, a steep transition of the stress from tensile to compressive was observed by varying the sputtering parameters (Dubois and Muralt, 2001). For a Zr based metallic glass film with 200 nm thickness the residual stress was measured to be −88.2 MPa (compressive) (Chu et al., 2012). In addition, it has been demonstrated that the internal stress can be controlled by adjusting the bombardment energy of the plasma species and by the sensitivity of the piezoelectric coefficient to the bombardment.

The films with large compressive stresses may lead to failure by a variety of stress-release mechanisms such as hillock formation, peeling at the film/substrate interface, creep, and plastic flow and buckling driven delamination. Films partially or totally delaminate at the interface or within the substrate. For example, when the compressive stress in a film exceeds the ultimate tensile strength of the substrate, failure occurs in the substrate itself (Freller et al., 1992). Zehnder et al. reported that amorphous carbon film deposited on silicon at 20°C showed no evidence of failure in the bulk, but delaminated at the interface, all over the sample (Zehnder et al., 1995). On the other hand, film grown at 500°C would adhere very well to the silicon substrate (nevertheless it should be noted that the film would delaminate when its thickness exceeds 600 nm).

One of the challenges in working with thin films is to prevent their buckling/delamination. Various experimental and theoretical methods have been used to understand the buckling modes and the stress levels at which buckling occurs. Mylvaganam et al. used atomistic simulations to investigate the buckling modes of diamond-like carbon films deposited on silicon (Mylvaganam and Zhang, 2003). They found that the atomic scale buckling pattern varies with the ratio of the residual stresses in the two directions, which indicates that the stress ratio is an important factor. Figure 21.7(b) shows the telephone-cord-like buckling pattern of the film on releasing the equal biaxial stresses and is similar to that observed in microscale experiments, Figure 21.7(a).

Som et al. studied the delamination of CVD diamond films deposited on silicon by scanning electron microscopy. Micro-Raman measurements showed that the biaxial compressive stress to initiate the film buckling was 1.19 GPa (Som et al., 1998).

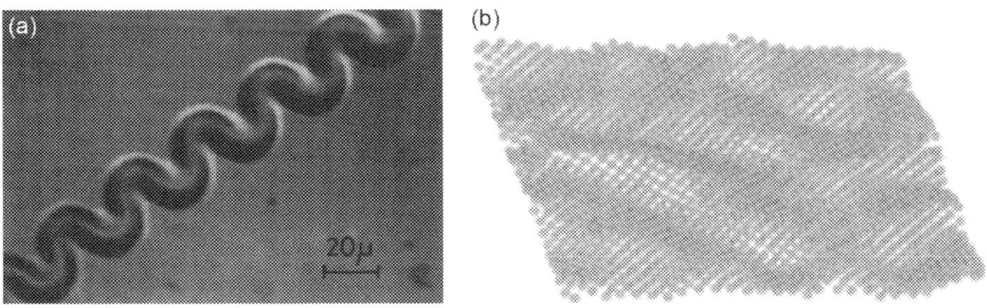

Figure 21.7 (a) Telephone-cord blister (after Iyer et al., 1995) and (b) buckling mode of diamond-like carbon films on silicon substrate (after Mylvaganam and Zhang, 2003).

21.3.2 Surface roughness

Surface roughness has a direct effect on the optical, electrical and mechanical quality of solid-thin film devices and becomes a significant issue for thin films that are prepared for optical wave-guides, multilayer interference filters and optical resonant filters. Different preparation conditions and different growth methods would lead to a wide variety of surface/interface morphologies and film microstructures that are inherently related to different growth mechanisms; the film growth may not occur in a layer-by-layer manner and instead the growth-front can be rough. For example, when using the magnetron sputtering method for film deposition, alternating bias can modulate the plasma flow of the sputtered material by modulating the sputtering energy. Controlling the plasma flow at a frequency that is consistent with the rate of deposition of a single atomic layer can produce a more even deposition, and thus less surface roughness in the deposited film. It has been found that the Zr-based TFMG deposited by the radio frequency magnetron sputtering system has a low surface toughness of 0.2 nm when the nitrogen content is zero, but it increased with an increase of the nitrogen content. The 13.8% N containing TFMG provided a good wear resistance, high hardness and excellent adhesion properties Chuang et al. (2013). In the preparation of NCD films surface roughness becomes larger with the increase of temperature and pressure.

21.4 Summary

This chapter has discussed on the production, properties and applications of three types of thin films: (1) diamond and diamond-related thin films that are anti-abrasive under contact sliding; (2) semiconductor thin films (AlN and GaAs) that are anti-abrasive to operating conditions; and (3) metallic glass thin films, though fairly new, that have unique properties and characteristics that are not readily achievable with other thin films. All thin films function as protective coatings on different components with widespread mechanical, optical, electronic, and biomedical applications. Yet, the residual stress causes thin films to buckle/delaminate/crack under various conditions. For CVD diamond thin films, the residual stress changes from compressive to tensile with increased film thickness; for semiconductor AlN thin films, the transition was obtained by varying the sputtering parameters; for metallic glass thin films with 200 nm thickness the residual stress was compressive.

Being an excellent piezoelectric material and also having exceptional resistance to pressure, temperature and radiation, AlN film shows great promise in aerospace applications. In addition, new efforts have recently been made to develop implantable drug delivery devices by integrating the piezoelectric AlN film with biocompatible UNCD thin film. These efforts are soon expected to enter the stage of practical application.

References

Alkjaersig, K., Christensen, H., Gupta, B., Jensen, H., Jensen, U., et al., 1992. Ion bombardment of AlN films deposited in a reactive sputtering process with accurate control of the mass flow of the reactive gas. Surf. Coat. Technol. 51, 500–508.

Asmussen, J., Reinhard, D.K. (Eds.), 2002. Diamond Films Handbook. Marcel Dekker, New York, pp. 56.

Brown, E., Smith, F., McIntosh, K., 1993. Coherent millimeter-wave generation by heterodyne conversion in low-temperature-grown GaAs photoconductors. J. Appl. Phys. 73, 1480–1484.

Carlisle, J.A., Auciello, O., 2003. Ultrananocrystalline diamond: properties and applications in biomedical devices. Electrochem. Soc. Interface. 12, 28–31.

Chen, M.W., 2008. Mechanical behavior of metallic glasses: microscopic understanding of strength and ductility. Annu. Rev. Mater. Res. 38, 445–469.

Chen, Y., Zhang, L.C., 2013. Polishing of Diamond Materials: Mechanisms, Modelling and Implementation. Springer-Verlag, London.

Cheng, C.-W., Shiu, K.-T., Li, N., Han, S.-J., Shi, L., et al., 2013. Epitaxial lift-off process for gallium arsenide substrate reuse and flexible electronics. Nat. Commun. 4, 1577.

Chiang, C.L., Chu, J.P., Liu, F.X., Liaw, P.K., Buchanan, R.A., 2006. A 200 nm thick glass-forming metallic film for fatigue-property enhancements. Appl. Phys. Lett. 88.

Chu, J.P., Wijaya, H., Wu, C.W., Tsai, T.R., Wei, C.S., et al., 2007. Nanoimprint of gratings on a bulk metallic glass. Appl. Phys. Lett. 90.

Chu, J.P., Jang, J.S.C., Huang, J.C., Chou, H.S., Yang, Y., et al., 2012. Thin film metallic glasses: unique properties and potential applications. Thin Solid Films. 520, 5097–5122.

Chu, X., Barnett, S.A., 1995. Model of superlattice yield stress and hardness enhancements. J. Appl. Phys. 77, 4403–4411.

Chuang, C.-Y., Lee, J.-W., Li, C.-L., Chu, J.P., 2013. Mechanical properties study of a magnetron-sputtered Zr-based thin film metallic glass. Surf. Coat. Technol. 215, 312–321.

Daveis, H.A., 1983. Amorphous Metallic Alloys. Butterworth, London.

Davies, A.G., Burnett, A.D., Fan, W., Linfield, E.H., Cunningham, J.E., 2008. Terahertz spectroscopy of explosives and drugs. Mater. Today. 11, 18–26.

Dovidenko, K., Oktyabrsky, S., Narayan, J., Razeghi, M., 1996. Aluminum nitride films on different orientations of sapphire and silicon. J. Appl. Phys. 79, 2439–2445.

Dubois, M.-A., Muralt, P., 2001. Stress and piezoelectric properties of aluminum nitride thin films deposited onto metal electrodes by pulsed direct current reactive sputtering. J. Appl. Phys. 89, 6389–6395.

Field, J.E. (Ed.), 1992. The Properties of Natural and Synthetic Diamond. Academic Press, Waltham, MA.

Freller, H., Hempel, A., Lilge, J., Lorenz, H.P., 1992. Influence of intermediate layers and base materials on adhesion of amorphous-carbon and metal-carbon coatings. Diam. Relat. Mater. 1, 563–569.

Gicquel, A., Hassouni, K., Silva, F., Achard, J., 2001. CVD diamond films: from growth to applications. Curr. Appl. Phys. 1, 479–496.

Gioia, G., Ortiz, M., 1997. Delamination of compressed thin films. In: Hutchinson, J.W., Wu, T.Y. (Eds.), Advances in Applied Mechanics, vol. 33. Elsevier, Amsterdam, pp. 119–192.

Granneman, E.H.A., 1993. Thin-films in the integrated-circuit industry - requirements and deposition methods. Thin Solid Films. 228, 1–11.

Green, M.A., Emery, K., Hishikawa, Y., Warta, W., Dunlop, E.D., 2012. Solar cell efficiency tables (version 39). Prog. Photovolt. Res. Appl. 20, 12–20.

Greer, A.L., Rutherford, K.L., Hutchings, M., 2002. Wear resistance of amorphous alloys and related materials. Int. Mater. Rev. 47, 87–112.

Greer, J.R., Oliver, W.C., Nix, W.D., 2005. Size dependence of mechanical properties of gold at the micron scale in the absence of strain gradients. Acta Mater. 53, 1821–1830.

Grill, A., 1999. Diamond-like carbon: state of the art. Diam. Relat. Mater. 8, 428–434.

Gupta, S., Frankel, M., Valdmanis, J., Whitaker, J., Mourou, G., et al., 1991. Subpicosecond carrier lifetime in GaAs grown by molecular beam epitaxy at low temperatures. Appl. Phys. Lett. 59, 3276–3278.

Hamers, R.J., Butler, J.E., Lasseter, T., Nichols, B.M., Russell Jr, J.N., et al., 2005. Molecular and biomolecular monolayers on diamond as an interface to biology. Diam. Relat. Mater. 14, 661–668.

Hata, S., Sato, K., Shimokohbe, A., 1999. Fabrication of thin film metallic glass and its application to microactuator. Device Process Technol. MEMS Microelectron. 3892, 97–108.

Hata, S., Kato, T., Fukushige, T., Shimokohbe, A., 2003. Integrated conical spring linear actuator. Microelectron. Eng. 67–8, 574–581.

Henann, D.L., Srivastava, V., Taylor, H.K., Hale, M.R., Hardt, D.E., et al., 2009. Metallic glasses: viable tool materials for the production of surface microstructures in amorphous polymers by micro-hot-embossing. J. Micromech. Microeng. 19.

Hinuber, C., Kleemann, C., Friederichs, R.J., Haubold, L., Scheibe, H.J., et al., 2010. Biocompatibility and mechanical properties of diamond-like coatings on cobalt-chromium-molybdenum steel and titanium-aluminum-vanadium biomedical alloys. J. Biomed. Mater. Res. A. 95, 388–400.

Hong, L., Wang, X., Zheng, H., He, L., Xu, X., et al., 2012. Design principles for plasmonic thin film GaAs solar cells with high absorption enhancement. J. Appl. Phys. 112, 054326.

Horowitz, S., Nishida, T., Cattafesta, L., Sheplak, M., 2007. Development of a micromachined piezoelectric microphone for aeroacoustics applications. J. Acoust. Soc. Am. 122, 3428.

Iles, P.A., 2001. Evolution of space solar cells. Sol. Energy Mater. Sol. Cells. 68, 1–13.

Inoue, A., 2000. Stabilization of metallic supercooled liquid and bulk amorphous alloys. Acta Mater. 48, 279–306.

Ishihara, M., Li, S., Yumoto, H., Akashi, K., Ide, Y., 1998. Control of preferential orientation of AlN films prepared by the reactive sputtering method. Thin Solid Films. 316, 152–157.

Iyer, S.B., Harshavardhan, K.S., Kumar, V., 1995. Buckling patterns in diamond-like carbon-films. Thin Solid Films. 256, 94–100.

Jagannadham, K., Sharma, A., Wei, Q., Kalyanraman, R., Narayan, J., 1998. Structural characteristics of AlN films deposited by pulsed laser deposition and reactive magnetron sputtering: a comparative study. J. Vac. Sci. Technol. A. 16, 2804–2815.

Jang, D.C., Greer, J.R., 2010. Transition from a strong-yet-brittle to a stronger-and-ductile state by size reduction of metallic glasses. Nat. Mater. 9, 215–219.

Jeong, H.W., Hata, S., Shimokohbe, A., 2003. Microforming of three-dimensional microstructures from thin-film metallic glass. J. Microelectromech. Syst. 12, 42–52.

Jepsen, P.U., Cooke, D.G., Koch, M., 2011. Terahertz spectroscopy and imaging–modern techniques and applications. Laser Photonics Rev. 5, 124–166.

Johnson, W.L., 1986. Thermodynamic and kinetic aspects of the crystal to glass transformation in metallic materials. Prog. Mater. Sci. 30, 81–134.

Johnson, W.L., 2002. Bulk amorphous metal – an emerging engineering material. JOM. 54, 40–43.

Johnson, W.L., Samwer, K., 2005. A universal criterion for plastic yielding of metallic glasses with a (T/T-g)(2/3) temperature dependence. Phys. Rev. Lett. 95.

Kaminska, M., Liliental-Weber, Z., Weber, E., George, T., Kortright, J., et al., 1989. Structural properties of As-rich GaAs grown by molecular beam epitaxy at low temperatures. Appl. Phys. Lett. 54, 1881–1883.

Kawabe, M., Ueda, T., 1987. Self-annihilation of antiphase boundary in GaAs on Si (100) grown by molecular beam epitaxy. Jpn. J. Appl. Phys. 26, L944–L946.

Kim, J.G., Yu, J., 1998. Behavior of residual stress on CVD diamond films. Mater. Sci. Eng. B Solid State Mater. Adv. Technol. 57, 24–27.

Klement, W., Willens, R.H., Duwez, P., 1960. Non-crystalline structure in solidified gold-silicon alloys. Nature. 187, 869–870.

Kneissl, M., Kolbe, T., Chua, C., Kueller, V., Lobo, N., et al., 2011. Advances in group III-nitride-based deep UV light-emitting diode technology. Semicond. Sci. Technol. 26, 014036.

Kulisch, W., Popov, C., Gilliland, D., Ceccone, G., Reithmaier, J.P., et al., 2011. UNCD/a-C nanocomposite films for biotechnological applications. Surf. Coat. Technol. 206, 667–675.

Law, D.C., King, R., Yoon, H., Archer, M., Boca, A., et al., 2010. Future technology pathways of terrestrial III–V multijunction solar cells for concentrator photovoltaic systems. Sol. Energy Mater. Sol. Cells. 94, 1314–1318.

Li, J., Fan, Z., Dahal, R., Nakarmi, M., Lin, J., et al., 2006. 200 nm deep ultraviolet photodetectors based on AlN. Appl. Phys. Lett. 89, 213510.

Liu, W.D., Liu, K.X., 2010. Mechanical behavior of a Zr-based metallic glass at elevated temperature under high strain rate. J. Appl. Phys. 108.

Loo, R.Y., Kamath, G.S., Li, S.S., 1990. Radiation damage and annealing in GaAs solar cells. IEEE Trans. Electron Devices. 37, 485–497.

Lu, J., Ravichandran, G., Johnson, W.L., 2003. Deformation behavior of the Zr41.2Ti13.8CU12.5Ni10Be22.5 bulk metallic glass over a wide range of strain-rates and temperatures. Acta Mater. 51, 3429–3443.

Luborsky, F.E., 1983. Amorphous Metallic Alloys. Butterworth, London.

Luong, T.-D., Nguyen, N.-T., 2010. Surface acoustic wave driven microfluidics – a review. Micro Nanosyst. 2, 217–225.

Malshe, A., Brown, W.D., 2002. Diamond heat spreaders and thermal management. In: Asmussen, J., Reihard, D.K. (Eds.), Diamond Films Handbook. Marcel Dekker, New York, pp. 511–578.

Manasevit, H.M., 1981. Recollections and reflections of MO-CVD. J. Cryst. Growth. 55, 1–9.

May, P.W., 2000. Diamond thin films: a 21st-century material. Philos. Trans. R. Soc. A. 358, 473–495.

Melloch, M., Otsuka, N., Woodall, J., Warren, A., Freeouf, J., 1990. Formation of arsenic precipitates in GaAs buffer layers grown by molecular beam epitaxy at low substrate temperatures. Appl. Phys. Lett. 57, 1531–1533.

Muller, R., Adamschik, M., Steidl, D., Kohn, E., Thamasett, S., et al., 2004. Application of CVD-diamond for catheter ablation in the heart. Diam. Relat. Mater. 13, 1080–1083.

Mylvaganam, K., Zhang, L.C., 2003. Residual stress induced atomic scale buckling of diamond carbon coatings on silicon substrate. Thin Solid Films. 425, 145–149.

Pantha, B., Dahal, R., Nakarmi, M., Nepal, N., Li, J., et al., 2007. Correlation between optoelectronic and structural properties and epilayer thickness of AlN. Appl. Phys. Lett. 90, 241101–241101-3.

Park, J.-K., Baik, Y.-J., 2005. The crystalline structure, hardness and thermal stability of AlN/CrN superlattice coating prepared by DC magnetron sputtering. Surf. Coat. Technol. 200, 1519–1523.

Pillai, S., Catchpole, K., Trupke, T., Green, M., 2007. Surface plasmon enhanced silicon solar cells. J. Appl. Phys. 101, 093105.

Pohl, R., Pringsheim, P., 1912. Uber die Herstellung von Metallspieglen durch distillation in Vakuum. Verhandle Deut Physik Ges. 14, 506.

Popov, C., Kulisch, W., Boycheva, S., Yamamoto, K., Ceccone, G., et al., 2004. Structural investigation of nanocrystalline diamond/amorphous carbon composite films. Diam. Relat. Mater. 13 (2004), 2071–2075.

Popov, C., Kulisch, W., Jelinek, M., Bock, A., Strnad, J., 2006. Nanocrystalline diamond/ amorphous carbon composite films for applications in tribology, optics and biomedicine. Thin Solid Films. 494, 92–97.

Popov, C., Kulisch, W., Reithmaier, J.P., Dostalova, T., Jelinek, M., et al., 2007. Diam. Relat. Mater. 16, 735–739.

Popov, C., Bliznakov, S., Boycheva, S., Milinovik, N., Apostolova, M.D., et al., 2008. Nanocrystalline diamond/amorphous carbon composite coatings for biomedical applications. Diam. Relat. Mater. 17, 882–887.

Prelas, M.A., Popovicci, G., Bigelow, L.K., 1998. Handbook of Industrial Diamonds and Diamond Films. Marcel Dekker, New York, pp. 1232.

Reentilä, O., Brunner, F., Knauer, A., Mogilatenko, A., Neumann, W., et al., 2008. Effect of the AlN nucleation layer growth on AlN material quality. J. Cryst. Growth. 310, 4932–4934.

Rihani, S., Faulks, R., Beere, H.E., Farrer, I., Evans, M., Ritchie, D.A., et al., 2010. Enhanced terahertz emission from a multilayered low temperature grown GaAs structure. Appl. Phys. Lett. 96 (2010), 091101–091101–3.

Robertson, J., 2002. Diamond-like amorphous carbon. Mater. Sci. Eng. R. 37, 129–281.

Rowland, L., Kern, R., Tanak, S., Davis, R.F., 1993. Epitaxial growth of AlN by plasma-assisted, gas-source molecular beam epitaxy. J. Mater. Res. 8 (1993), 2310–2314.

Roy, R.K., Lee, K.R., 2007. Biomedical applications of diamond-like carbon coatings: a review. J. Biomed. Mater. Res. B. 83, 72–84.

Sadia, C.P., Maria Laganapan, A., Agatha Tumanguil, M., Estacio, E., Somintac, A., et al., 2012. Intense terahertz emission from molecular beam epitaxy-grown GaAs/GaSb (001). J. Appl. Phys. 112, 123514–123515.

Schermer, J., Mulder, P., Bauhuis, G., Larsen, P., Oomen, G., et al., 2005. Thin-film GaAs epitaxial lift-off solar cells for space applications. Prog. Photovolt. Res. Appl. 13, 587–596.

Schermer, J., Bauhuis, G., Mulder, P., Haverkamp, E., Van Deelen, J., et al., 2006. Photon confinement in high-efficiency, thin-film III–V solar cells obtained by epitaxial lift-off. Thin Solid Films. 511, 645–653.

Schroers, J., 2005. The superplastic forming of bulk metallic glasses. JOM. 57, 35–39.

Schroers, J., 2010. Processing of bulk metallic glass. Adv. Mater. 22, 1566–1597.

Schuh, C.A., Hufnagel, T.C., Ramamurty, U., 2007. Overview no. 144 – mechanical behavior of amorphous alloys. Acta Mater. 55, 4067–4109.

Shen, Y., Upadhya, P., Linfield, E., Beere, H., Davies, A., 2003. Ultrabroadband terahertz radiation from low-temperature-grown GaAs photoconductive emitters. Appl. Phys. Lett. 83, 3117–3119.

Shur, M., Gaska, R., Bilenko, Y., 2005. Ultraviolet radiation-based media purification. Google Patents.

Som, T., Bhargava, S., Malhotra, M., Bist, H.D., Kulkarni, V.N., et al., 1998. MeV He + ion induced delamination of diamond films. Appl. Phys. Lett. 72, 3014–3016.
Spaepen, F., 1977. A microscopic mechanism for steady state inhomogeneous flow in metallic glasses. Acta Metall. 25, 407–415.
Stueber, M., Holleck, H., Leiste, H., Seemann, K., Ulrich, S., et al., 2009. Concepts for the design of advanced nanoscale PVD multilayer protective thin films. J. Alloys Compd. 483, 321–333.
Takenaka, K., Saidoh, N., Nishiyama, N., Inoue, A., 2011. Fabrication and nano-imprintabilities of Zr-, Pd- and Cu-based glassy alloy thin films. Nanotechnology. 22, 105302.
Tani, M., Morikawa, O., Matsuura, S., Hangyo, M., 2005. Generation of terahertz radiation by photomixing with dual-and multiple-mode lasers. Semicond. Sci. Technol. 20, S151.
Taniyasu, Y., Kasu, M., Makimoto, T., 2006. An aluminium nitride light-emitting diode with a wavelength of 210 nanometres. Nature. 441, 325–328.
Thobor, A., Rousselot, C., Clement, C., Takadoum, J., Martin, N., et al., 2000. Enhancement of mechanical properties of TiN/AlN multilayers by modifying the number and the quality of interfaces. Surf. Coat. Technol. 124, 210–221.
Tsai, P.H., Lin, Y.Z., Li, J.B., Jian, S.R., Jang, J.S.C., et al., 2012. Sharpness improvement of surgical blade by means of ZrCuAlAgSi metallic glass and metallic glass thin film coating. Intermetallics. 31, 127–131.
Ueno, M., Onodera, A., Shimomura, O., Takemura, K., 1992. X-ray observation of the structural phase transition of aluminum nitride under high pressure. Phys. Rev. B. 45, 10123.
Vispute, R., Narayan, J., Wu, H., Jagannadham, K., 1995. Epitaxial growth of AlN thin films on silicon (111) substrates by pulsed laser deposition. J. Appl. Phys. 77, 4724–4728.
Wang, W.H., Dong, C., Shek, C.H., 2004. Bulk metallic glasses. Mater. Sci. Eng. R. 44, 45–89.
Was, G., Foecke, T., 1996. Deformation and fracture in microlaminates. Thin Solid Films. 286, 1–31.
Williams, M.D., Griffin, B.A., Reagan, T.N., Underbrink, J.R., Sheplak, M., 2012. An AlN MEMS piezoelectric microphone for aeroacoustic applications. J. Microelectromech. Syst. 21, 270–283.
Yablonovitch, E., Gmitter, T., Harbison, J., Bhat, R., 1987. Extreme selectivity in the lift-off of epitaxial GaAs films. Appl. Phys. Lett. 51, 2222.
Yang, W., Auciello, O., Butler, J.E., Cai, W., Carlisle, J.A., et al., 2002. DNA-modified nanocrystalline diamond thin-films as stable, biologically active substrates. Nat. Mater. 1, 253–257.
Yin, L.-W., Hwang, Y., Lee, J., Kolbas, R.M., Trew, R.J., et al., 1990. Improved breakdown voltage in GaAs MESFETs utilizing surface layers of GaAs grown at a low temperature by MBE. IEEE Electron Device Lett. 11, 561–563.
Yoon, J., Jo, S., Chun, I.S., Jung, I., Kim, H.-S., et al., 2010. GaAs photovoltaics and optoelectronics using releasable multilayer epitaxial assemblies. Nature. 465, 329–333.
Zehnder, T., Balmer, J., Luthy, W., Weber, H.P., 1995. Determination of limits in deposition of adhering a-c films on silicon produced by pulsed-laser deposition. Thin Solid Films. 263, 198–202.
Zhang, S., Sun, D., Fu, Y.Q., Du, H.J., 2005. Toughening of hard nanostructural thin films: a critical review. Surf. Coat. Technol. 198, 2–8.

Index

Note: Page numbers followed by '*f*' and '*t*' refer to figures and tables, respectively.

Q

R

Printed in the United States
By Bookmasters